Fishes of the Dakotas

Kathryn E. Schlafke • Matthew D. Wagner • Chelsey A. Pasbrig
Authors

Michael Barnes
Editors

Fishes of the Dakotas

 Springer

Authors
Kathryn E. Schlafke
Montana Fish, Wildlife and Parks, Bozeman, MT, USA

Matthew D. Wagner
U.S. Fish and Wildlife Service, Jackson, MS, USA

Chelsey A. Pasbrig
South Dakota Game, Fish and Parks, Pierre, SD, USA

Editor
Michael Barnes
South Dakota Department of Game, Fish and Parks
Pierre, SD, USA

ISBN 978-3-031-38039-6   ISBN 978-3-031-38040-2   (eBook)
https://doi.org/10.1007/978-3-031-38040-2

This Springer imprint is published by the registered company Springer Nature Switzerland AG
The registered company address is: Gewerbestrasse 11, 6330 Cham, Switzerland

Paper in this product is recyclable.

Acknowledgments

The following people were instrumental in capturing and aiding in photographing fish used in this book: Savannah Allard, Ethan Coombes, Jake Davis, Eli Felts, Carlee Fletcher, Seth Fopma, Josh Freeseman, Gene Gallinat, Dylan Gravenhof, Chris Greene, Logan Gutzmer, Jenna Haag, Zachary Jessee, Stephen Jones, Chad Kaiser, Morgan Kauth, Jeremy Keintz, Nicholas Kludt, Rebecca Kolstrom, Jason Kral, Jacob Krause, Megan Leonard, John Lorenzen, Dave Lucchesi, Chuck Mordhorst, Matthew Neff, Dan Nelson, Jordan Redmond, Travis Rehm, Calvin Rezac, Scott T. Sabo, Benjamin Schall, David Schumann, Greg Simpson, Brad Smith, Sam Stukel, Brandon Vanderbush, Chris Hull, Allie Ellingson, Nick Harrington, Jason Jungwirth, and Haley Wagner.

Some species of fish were not able to be photographed for this project and some photos were used with permission. Credit and acknowledgment for these borrowed photos goes to Uland Thomas for mooneye, skipjack herring, threadfin shad, mimic shiner, suckermouth minnow, highfin carpsucker, black buffalo, river redhorse, greater redhorse, brown bullhead, and striped bass; Brian Zimmerman for silver lamprey, emerald bowfin, alewife, and river shiner; Robert Hrabik for channel shiner; Konrad Schmidt for lake whitefish, and the mouths of silver lamprey and chestnut lamprey; and Matt Thomas with Kentucky Department of Fish and Wildlife Resources for yellow bass.

Brinley Adams, Geno Adams, Brian Blackwell, Michelle Bucholz, Jake Davis, Mark Ermer, Mark Fincel, Gene Galinat, Amy Gebhard, Cameron Goble, Todd Kaufmann, Jeremy Kientz, Nathan Loecker, Dave Lucchesi, Tyrel Moos, Benjamin Schall, Greg Simpson, Mike Smith, and Todd St. Sauver contributed to reviewing the species accounts.

Geno Adams, Jake Davis, Scott Gangl, Tyrel Moos, and Casey Williams assisted with quality control checking the databases used to produce the species maps. Brian and Katie Graeb helped at the start of this project.

Preliminary species profile maps were made by Nicholas Kludt. Final species profile maps could not have been completed without the assistance of Heather Berg, South Dakota Game, Fish and Parks GIS Program Specialist.

Jill Voorhees, Eric Krebs, Nicole McCarty, Maggie Erlenbusch, and Michael Robidoux assisted greatly with formatting.

Finally, the following individuals provided review of this book prior to publication: Geno Adams, Brian Blackwell, Alison Coulter, Scott Gangl, Chris Hoagstrom, Mark Kaemingk, Dave Lucchesi, David Schumann, Jill Voorhees, and Casey Williams.

Acknowledgements

Disclaimer

The findings and conclusions in the book are those of the authors and do not necessarily represent the views of the South Dakota Game, Fish, and Parks or the U.S. Fish and Wildlife Service, Montana Fish, Wildlife and Parks.

Contents

Chapter 1
Waters and Geology of the Dakotas

Kathryn E. Schlafke, Matthew D. Wagner, and Chelsey A. Pasbrig

1.1 The Dakotas

North Dakota (ND) and South Dakota (SD), often referred to together as the Dakotas, lie entirely within the Great Plains region of North America. Historically, they were part of the Louisiana Territory purchased by the United States (US) from France in 1803. The Missouri River, which meanders from northwest ND to southeast SD, was initially a territory boundary. From 1812 to 1861, the land west of the Missouri River was included in the Nebraska Territory. In contrast, administration of the land east of the Missouri frequently changed, belonging to the territories of Missouri (MO) from 1812 to 1834, Michigan (MI) from 1834 to 1836, Wisconsin (WI) from 1836 to 1838, Iowa (IA) from 1838 to 1849, and Minnesota (MN) from 1849 to 1858 (Visher 1918). Between 1858 and 1861, this land was classified as unorganized and became known as the "Land of the Dakotas" (Visher 1918). In 1861, the Dakota Territory was established, including what is now ND, SD, most of Montana (MT), and sections of Nebraska (NE) and Wyoming (WY) (Visher 1918). By 1868, the land included in the Dakota Territory had been reduced to only include the current states of ND and SD, and, in 1889, these two states were admitted into the Union.

The geological history of the Dakotas began nearly 2.5 billion years ago during the Precambrian Era with the formation of layers of granite and indigenous rock. Glaciers once covered most of eastern and central ND and extended into eastern SD. After retreating, the glaciers left thick deposits of glacial till, a mixture of ground-up gravel, grit, and bedrock, where grass could thrive. Glacial water filled shallow depressions to form wetlands and smaller bodies of water called prairie potholes or kettle lakes. Glaciers did not cover the western part of the Dakotas, most of which was covered by ancient, shallow inland seas. The SD badlands were formed by water and wind erosion. The Black Hills in western SD are ancient rocks formed by an uplift of volcanic activity. This dome, or island within the prairie, is also known as the Middle Rocky Mountains.

A frequently unnoticed geological feature of the Dakotas is the Northern Divide, which is also known as the Laurentian Divide in northern MN (Gonzalez 2003). The Northern Divide is one of four continental divides in North America, along with the Great Divide (often incorrectly referred to as the Continental Divide), the Eastern Divide, and the St. Lawrence Seaway Divide. The Northern Divide originates deep within the interior of the North American continent. It shares the same path as the Great Divide from the Seward Peninsula to Triple Divide Peak, MT. From Triple Divide Peak, the Northern Divide extends east through MT, Alberta, Saskatchewan, ND, SD, and northern MN until it continues north and east across Ontario and Quebec (Gonzalez 2003). The Northern Divide separates a large portion of North America, including the Dakotas, into two gulfs: the Hudson Bay and the Gulf of Mexico (Fig. 1.1). The Hudson Bay drains to the north and empties into the Labrador Sea and Arctic Ocean, whereas the Gulf of Mexico drains to the south emptying into the Gulf of Mexico and Atlantic Ocean.

K. E. Schlafke
Montana Fish, Wildlife & Parks, Bozeman, MT, USA

M. D. Wagner
U.S. Fish and Wildlife Service, Jackson, MS, USA

C. A. Pasbrig (✉)
South Dakota Game, Fish and Parks, Pierre, SD, USA
e-mail: Chelsey.Pasbrig@state.sd.us

Fig. 1.1 The Northern Divide separates a large portion of the Dakotas into two gulfs: the Hudson Bay gulf and the Gulf of Mexico gulf. Two river basins span across ND and SD: (1) the Nelson River basin, which includes the Red River of the North drainage, and (2) the Mississippi River basin, which encompasses the Missouri and Upper Mississippi river drainages. Devils Lake is an endorheic (closed) watershed that occasionally overflows, eventually flowing into the Sheyenne River system and Red River of the North drainage

The Hudson Bay gulf primarily encompasses the central territories of Canada but also extends south into MT, ND, SD, and northern MN. In ND, the Assiniboine, Pembina, Park, Forest, Turtle, Goose, Elm, Sheyenne, Wild Rice, and Bois de Sioux River systems are all part of the Red River of the North drainage, which, subsequently, flows into the Nelson River basin before emptying into the Hudson Bay. In SD, the Hudson Bay gulf only extends into the extreme northeast corner of the state, encompassing the Bois de Sioux River system, which flows into the Red River of the North drainage.

The Gulf of Mexico contains the Mississippi River basin and most of the United States. The Mississippi River basin includes southwestern ND and nearly all of SD (Hoagstrom et al. 2007). The Mississippi River basin in the Dakotas is divided into the Upper Mississippi River drainage and the Missouri River drainage. The Upper Mississippi River basin covers less than 3% of SD and is drained by the Minnesota River system (Hoagstrom et al. 2007). The remainder of SD is part of the Missouri River drainage (Goolsby et al. 1999; Hoagstrom et al. 2007).

In ND, the James, Cannonball, Heart, Knife, and Little Missouri River systems, and other waters to the west and south of the Northern Divide, flow into the Missouri River and eventually into the Gulf of Mexico. All the rivers and streams within SD, except for the few streams in the Red River of the North drainage and Minnesota River system, are a part of the Missouri River drainage. The Devils Lake watershed in ND is recognized as a closed basin surrounded by the Hudson Bay gulf because there is no natural external drainage leading to the Hudson Bay gulf (Gonzalez 2003). However, because of flooding issues, pumps and an outlet have been installed on the lower end of Devils Lake to divert water into the Sheyenne River. This anthropogenic addition connects the Devils Lake watershed to the Hudson Bay gulf via the Sheyenne River system.

The fish assemblages of the Hudson Bay and Gulf of Mexico gulfs are remarkably similar, despite flowing in different directions and emptying into different oceans. Historical geology explains this similarity. The history of the earth is divided into four main time periods or eras: the Precambrian, Paleozoic, Mesozoic, and Cenozoic. The Cenozoic is the most recent era and is divided into the Tertiary and the Quaternary periods. The Tertiary period began 65 million years ago and ended 1.8 million years ago with the start of the current Quaternary period. The Pleistocene Epoch, or Ice Age, encompassed most of the Quaternary period and ended approximately 11,700 years ago until the start of the current Holocene Epoch (Hendon and Matthews 2014). During the Pleistocene Epoch, glaciers and large sheets of ice formed. The glaciers spanned from northern Canada to northern and central United States. The Wisconsin glacier covered the majority of ND and the eastern portion of SD. It formed approximately 40,000 years ago, advancing and retreating several times before melting entirely about 12,000 years ago (Herman and Johnson 2007). As the glacier advanced, fish species were driven into the southern parts of the Mississippi River basin, and, as the glacier retreated, fish were able to recolonize areas using north-south rivers and streams (Oberdorff et al. 1995). More specifically, fish species indigenous to the ancient Missouri River and its tributaries, which at one time were part of the Hudson Bay gulf, were free to roam within the Mississippi River basin, and vice versa (Owen et al. 1981).

The historical back-and-forth shifting patterns of the Pleistocene Epoch glaciers contributed to the current occurrence of lower fish species diversity and richness in northern and western states compared to southern and eastern states. For example, native fish species richness in the Missouri River is higher in downstream states, with 110 species in MO compared to 64 species in MT (Galat et al. 2005). This pattern is also evident in the Dakotas, with species diversity increasing from the west to the east (Hoagstrom et al. 2007). These changes in species composition and diversity are likely caused by interactions between natural factors, such as climate, physiography, hydrology, and zoology, and anthropogenic factors, such as impoundment, flow changes, temperature alterations, and introduced species (Galat et al. 2005). The distribution and abundance of fishes within the Dakotas continues to change with range expansions, species declines, and fish stocking (Hoagstrom et al. 2007).

1.2 Level III Ecoregions

Ecoregions classify areas with similar quantities and qualities of aquatic, terrestrial, biotic, and abiotic factors (such as hydrology, geology, wildlife, vegetation, soils, climate, and land use) that assist state and federal agencies and nongovernmental organizations in managing natural resources (Omernik 1987). Ecoregions are used to define waterways and often help explain fish distributions. The United States Environmental Protection Agency has divided the Dakotas into eight level III ecoregions (Bryce et al. 1996). From east to west, these ecoregions are the Lake Agassiz Plain, Western Corn Belt Plains, Northern Glaciated Plains, Northwestern Glaciated Plains, Northwestern Great Plains, Nebraska Sand Hills, High Plains, and the Middle Rockies (Fig. 1.2).

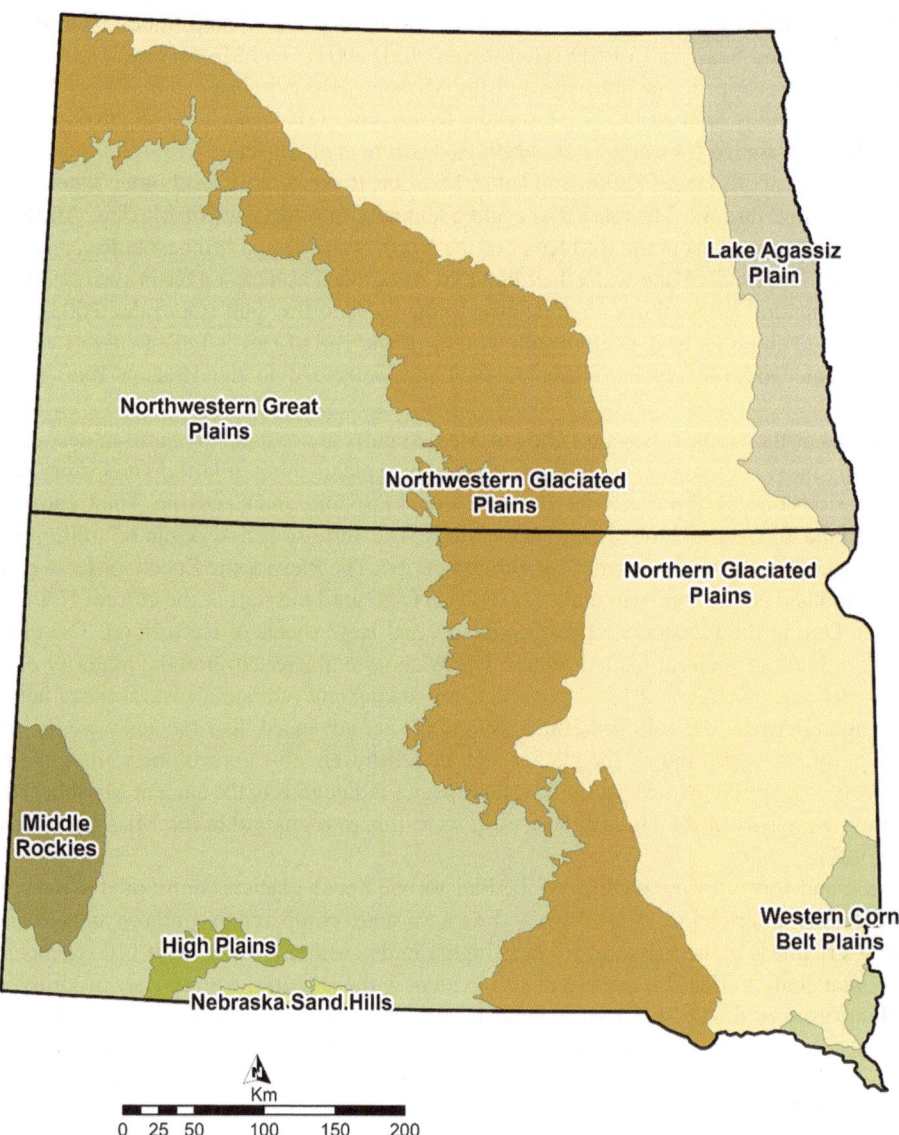

Fig. 1.2 The eight level III ecoregions in the Dakotas as defined by the Environmental Protection Agency

1.2.1 Lake Agassiz Plain

As the Wisconsin glacier melted, Canadian glaciers acted as a dam, preventing northernly water flow (Herman and Johnson 2007). The melt water accumulated and eventually formed the massive Lake Agassiz, named after the Swiss geologist Louis Agassiz. Lake Agassiz was the largest of the many Pleistocene lakes in North America at approximately 1,126 km long, 322 miles wide, and 91 m deep (Herman and Johnson 2007). It was mostly located in the current Canadian provinces of Saskatchewan, Manitoba, and Ontario. Only about one-fifth of the lake occurred in what is now ND, SD, and MN (Upham 1895). Large waves created the well-defined shoreline of Lake Agassiz, making it easily recognized and mapped today. The very southwestern shoreline of the lake is the current western border of the Red River Valley in the Dakotas (Upham 1895). As the lake slowly retreated northward, extremely flat land with highly fertile soil was left behind. The remnants of Lake Agassiz are Lakes Winnipeg and Winnipegosis in Manitoba, Canada (Upham 1895).

Level III ecoregion 48 is the Lake Agassiz Plain, located on the eastern edge of ND and extreme northeastern SD. It is a low-relief, 7101 square mile area comprised of flat-to-low-rolling plains covered by moraine and lacustrine deposits (Griffith 2010). Calcareous glacial till is found in the northern region (Griffith 2010). Thick beds of lacustrine sediments over glacial till have created a flat landscape with few lakes and wetlands. Once mostly tallgrass prairie and riparian areas, the land has now been extensively converted to row crop agriculture (Bryce et al. 1996; Griffith 2010). Regional hydrology consists of low densities of low-gradient stream and river networks that drain into the Red River of the North drainage (Griffith 2010).

1.2.2   Western Corn Belt Plains

Level III ecoregion 47 is the Western Corn Belt Plains. It is 1453 square miles of mostly grass-dominated land across SD, MN, Iowa (IA), NE, and Kansas (KS) (Bryce et al. 1996). Located in the southeastern corner of SD, this ecoregion is characterized by a temperate climate and fertile soils, with nearly level to gently rolling glaciated till plains over Mesozoic and Paleozoic deposits of shale, sandstone, and limestone (Bryce et al. 1996; Griffith 2010). Once filled with tall grass prairie and small oak woodlands, approximately three-fourths of this ecoregion has now been converted to row crop agriculture (Omernik and Gallant 1988; Griffith 2010; Wright and Wimberly 2013). Some pastures with small parcels of native prairie and anthropogenically modified grass and legumes also occur. The hydrology is characterized by a few natural lakes and a system of intermittent and perennial streams frequently channelized within shallow, narrow valleys (Griffith 2010). Patches of remnant deciduous woodlands sometimes border these prairie streams (Omernik and Gallant 1988). In SD, this ecoregion contains the lower sections of the Big Sioux, Vermillion, James, and Missouri rivers.

1.2.3   Northern Glaciated Plains

Level III ecoregion 46 is the Northern Glaciated Plains, an area extending southward from Saskatchewan and Manitoba, Canada, to north-central ND, eastern SD, and southwest MN (Omernik and Gallant 1988). The terrain of this subhumid, 48,511 square mile ecoregion is characterized by nearly level to gently rolling glacial till plains, with glacial lacustrine and ridged fluvial deposits over Tertiary and Cretaceous sandstone and shale (Bryce et al. 1996; Griffith 2010). Historically covered with prairie and groves of aspen and oak, this ecoregion has mostly been converted to agricultural use because of its highly fertile soil (Omernik and Gallant 1988; Griffith 2010). Areas of native woodlands, pastures, and seasonal wetlands also occur (Bryce et al. 1996; Griffith 2010). Its hydrology is characterized by low densities of intermittent streams, high concentrations of temporary and seasonal prairie pothole wetlands, and many small glacial or kettle lakes (Omernik and Gallant 1988; Griffith 2010). The intermittent waterways and wetlands that occur throughout the grassland landscape drain into prairie pothole basins (Bryce et al. 1996; Krause et al. 2013). In the Dakotas, this ecoregion contains the headwaters of numerous tributaries of the Red River of the North drainage as well as large portions of the Souris, Sheyenne, James, Vermillion, Big Sioux, and Minnesota rivers.

1.2.4   Northwestern Glaciated Plains

Level III ecoregion 42, the Northwestern Glaciated Plains, covers 28,378 square miles from southwestern Saskatchewan and southeastern Alberta, Canada, to northern MT, the central Dakotas, and northern NE (Griffith 2010). In the Dakotas, the Northwestern Glaciated Plains lie entirely east of the Missouri River and is the westernmost continental glaciation during the Pleistocene Epoch. This ecoregion is essentially a transition between the more level and moist Northern Glaciated Plains to the east and the irregular and dryer Northwestern Great Plains to the west (Bryce et al. 1996; Chapman et al. 2001; Griffith 2010). It is characterized by nearly level plains and gently rolling hills that were once covered with tall grass and shrubs growing in glacial till soils of shale, sandstone, and limestone (Bryce et al. 1996; Griffith 2010). Its hydrology is a high concentration of intermittent streams, with some perennial streams and larger rivers present on the western edge (Griffith 2010). These waterways are drained by the Missouri River to the south and the Saskatchewan River to the north (Griffith 2010). In parts of this ecoregion, the landscape is dotted with abundant seasonal and semipermanent wetlands commonly called prairie potholes. While the prairie potholes are used by waterfowl for nesting and migration, most of the land is either used for grazing or row crop agriculture (Bryce et al. 1996; Griffith 2010).

1.2.5   Northwestern Great Plains

The Northwest Great Plains, level III ecoregion 43, is entirely unglaciated and includes the base of the Rocky Mountains in MT, the southwestern corner of ND, the northeast corner of WY, and most of western SD. In the Dakotas, this ecoregion is entirely west of the Missouri River and surrounds the ecoregions of Nebraska Sand Hills, High Plains, and Middle Rockies in southwestern SD. It is largely characterized by gently rolling plains, occasional buttes, separated badland terrain, and steep river breaks over Cretaceous shale and Tertiary sandstone deposits (Bryce et al. 1996; Griffith 2010). Mixed-grass prairie and shrub land are the primary vegetation types within MT and the Dakotas, with more shortgrass prairie occurring in WY (Biondini et al. 1998; Tan et al. 2005). The native shortgrass and mixed-grass prairies are now either used for grazing or have been converted to row crop agriculture (Bryce et al. 1996; Griffith 2010). The hydrology of this ecoregion is characterized by intermittent and ephemeral streams, along with a few perennial rivers that are the western tributaries of the Missouri River (Griffith 2010). In the Dakotas, these perennial rivers include the lower Yellowstone, Little Missouri, Knife, Heart, Cannonball, Grand, Moreau, Cheyenne, Bad, White, Upper Niobrara, and Keya Paha rivers. Most of these rivers have turbid, opaque water caused by the sediments and clay of eroding soils and weathering rock upstream near the badlands (Bryce et al. 1996). Small impoundments also occur semi-frequently over the landscape, and there are some large reservoirs along the Missouri River (Griffith 2010).

1.2.6   Nebraska Sand Hills

Level III ecoregion 44 is the Nebraska Sand Hills, one of the most distinct and homogeneous ecoregions in North America. It spans over north-central and northwestern NE as well as a small strip along the south-central edge of SD, west of the Missouri River (Griffith 2010). The Nebraska Sand Hills were formed when climate changes killed native grass, exposing the underlying sand (Billesbach and Arkebauer 2012). This ecoregion is one of the largest areas of grass-stabilized sand dunes in the world. Beneath this ecoregion are some of the deepest sand and gravel deposits of the High Plains (or Ogallala) aquifer, which supplies approximately 30% of the commercial agricultural irrigation water consumed annually in the United States (Bryce et al. 1996; USGS 1997; Bleed and Flowerday 1998; Chapman et al. 2001). The rolling-to-steep, irregular, wind-sculpted sand dunes of Quaternary aeolian sand cover Tertiary sandstones and conglomerates (Griffith 2010). The mixed vegetation of short, mid, and tall grass in this mostly treeless ecoregion primarily exists in the dune valleys. Riparian areas in the north and east contain some trees (Billesbach and Arkebauer 2012). The Nebraska Sand Hills' ecoregion hydrology is comprised of numerous lakes, alkaline wetlands, and a few streams (Griffith 2010). Several larger perennial streams and small rivers, such the Niobrara, also flow through this region. Due to the soil composition, row crop agriculture is limited and most of the land is used for grazing livestock (Griffith 2010).

1.2.7   High Plains

The High Plains, level III ecoregion 25, covers only a small area of south-central SD. It spans a large latitudinal area, including southeastern WY, western NE, eastern Colorado (CO), western KS, western Oklahoma (OK), western Texas (TX), and eastern New Mexico (NM) (Griffith 2010). In SD, this ecoregion is bordered by the Northwestern Great Plains and the Missouri River. The High Plains were originally formed by erosion of the preglacial ancestral Rocky Mountains leaving behind nearly smooth to slightly irregular rolling plains and tablelands above Tertiary and Cretaceous sandstones, siltstones, clay stones, and caliche (Bryce et al. 1996; Griffith 2010). The vegetation in this ecoregion is primarily drought-resistant short and mixed grass and a small area of ponderosa pine in SD (Bryce et al. 1996). Most of the hydrology in the northern and central portions of this ecoregion is characterized by intermittent and ephemeral streams, whereas the southern portion has few, if any, streams (Griffith 2010). Some larger rivers, such as the Platte, also occur within this ecoregion. Today, the landscape is used for both grazing and row crop agriculture, with irrigation from the High Plains aquifer due to limited precipitation (Bryce et al. 1996).

1.2.8 Middle Rockies

Level III ecoregion 17, the Middle Rockies, spans most of southwestern MT, eastern Idaho (ID), and northern WY. However, this ecoregion also includes an outlier that shares the same characteristics, the Black Hills of western SD and northeastern WY (Bryce et al. 1996; Griffith 2010). Geologically, the Middle Rockies ecoregion is a dome-shaped uplift composed of high-elevation alpine mountain ranges, grassy glacial basins, and plateaus (DeWitt et al. 1989; Bryce et al. 1996). These geological features lie over a variety of rock types created during numerous geological periods.

The Middle Rockies were created 60–65 million years ago when a dome of Precambrian metamorphic rock was uplifted through Cretaceous sedimentary formations that now completely encircle the core (DeWitt et al. 1989; Williamson and Carter 2001). Within the SD portion of this ecoregion, vegetation consists of ponderosa pine forests, partly wooded and shrub-covered foothills, alpine grasslands, and meadows (Griffith 2010). The varying hydrological features of the Middle Rockies include a mixture of small alpine glacial lakes, larger lakes, and high-gradient perennial streams (Griffith 2010). The sedimentary formations and underlying geology of this ecoregion have a large impact on this area's hydrology (Williamson and Carter 2001). In the Black Hills, limestone formations that lie west of the center act as groundwater recharge zones with only small amounts of discharge (Williamson and Carter 2001). The surface discharge of streams that originate in the center of the Black Hills is primarily lost as they flow north or east away from the core (Williamson and Carter 2001). Streams within the region have variable morphological and biological characteristics but generally have cobble and rubble substrates, along with slack water sections that accumulate fine sediments and organic debris (Schultz 2011). Over the last century, water quality in many Black Hills streams has been impacted by mining activities (Rahn et al. 1996). Land use in this ecoregion, in addition to mining, includes forestry, ranching, grazing, and recreation.

1.3 Major Rivers of the Dakotas

Currently, the level III ecoregions within ND and SD contain 27 different river systems (Fig. 1.3). Each system belongs to a larger drainage (Missouri, Upper Mississippi, and Red River of the North drainage), a basin (Mississippi River or Nelson River), and a gulf (Hudson Bay and Gulf of Mexico), which eventually drain into different oceans. Table 1.1 details the classification of rivers within ND and SD.

1.3.1 Mississippi River Basin

The Mississippi River basin is the largest in the United States and the third largest in the world, draining an area of more than 3,270,000 km². This basin drains part or all of 2 Canadian provinces and 31 states. The Mississippi River originates in Lake Itasca in northern MN and flows 3,730 km south before emptying into the Gulf of Mexico. It has hundreds of tributaries, including the Missouri River in ND and SD. In the Dakotas, the Mississippi River basin includes the Missouri River drainage and the Upper Mississippi River drainage.

Missouri River Drainage (Mississippi River Basin)

The Missouri River drainage is the second largest in the United States, draining about one-sixth of the continental United States (Pegg et al. 2003). In the United States, the Missouri River drainage is 1,371,017 km², and, in Canada, it is 25,122 km² (USACE 1998; Galat et al. 2005). Spanning 3,768 km, the Missouri River is the longest river in North America, 224 km longer than the Mississippi River (Benke and Cushing 2005; Galat et al. 2005). Nicknamed the "Big Muddy" because of its high turbidity from frequent flooding and shifting channels, the Missouri originates near Three Forks, MT, at the confluence of the Madison, Jefferson, and Gallatin rivers (Galat et al. 2005). It then flows southeast across the continent until merging with the Mississippi River near St. Louis, MO.

Fig. 1.3 Major river systems in ND and SD. *Souris River is a smaller watershed of the Assiniboine River system located in Manitoba, Canada. **Devils Lake is an endorheic, or closed, watershed that occasionally overflows into the Sheyenne River system in the Red River of the North drainage

Missouri River System (Missouri River Drainage)

The Missouri River in the Dakotas runs from the northwest corner of ND downstream to the near center of the state (Fig. 1.3). It then flows through SD, dividing the state into nearly equal eastern and western halves known colloquially as East River and West River, respectively. Within the Missouri River drainage, the Missouri River system includes the mainstem Missouri River and its associated tributaries. It drains approximately 87,612 km^2 of land on both sides of the Missouri River in the Dakotas.

The Missouri River is no longer free-flowing. Between the late 1930s and the early 1960s, six mainstem dams were constructed for flood control, commercial navigation, power generation, irrigation, water quality management, and recreation (Galat et al. 1996; Fig. 1.4). Fort Peck Dam occurs in northeast MT and Garrison Dam is in ND, with the four remaining dams in SD. The reservoirs created by the six dams impound approximately half of the upper 2,500 km of the Missouri River (Morris et al. 1968). The natural flow regime of the Missouri River, particularly within the Dakotas, has been significantly altered by these dams and channelization (Pegg et al. 2003).

Table 1.1 Classification of rivers within ND and SD

Gulf	Basin	Drainage	System	Watershed
Gulf of Mexico	Mississippi River	Missouri River	Little Missouri River	
			Knife River	
			Heart River	
			Cannonball River	
			Grand River	
			Moreau River	
			Cheyenne River	Belle Fourche River
				Cherry Creek
				Fall River
			Bad River	
			White River	
			Niobrara River	Keya Paha River
			Ponca Creek	
			James River	
			Vermillion River	
			Big Sioux River	
		Upper Mississippi River	Minnesota River	
Hudson Bay	Nelson River	Red River of the North drainage	Assiniboine River	Souris River
			Pembina River	Rock Lake
			Park River	
			Forest River	
			Turtle River	
			Goose River	
			Elm River	
			Sheyenne River	Maple River
				Devils Lake[a]
			Wild Rice River	
			Bois de Sioux River	

[a]Devils Lake is an endorheic, or closed, watershed that occasionally overflows into the Sheyenne River system in the Red River of the North drainage

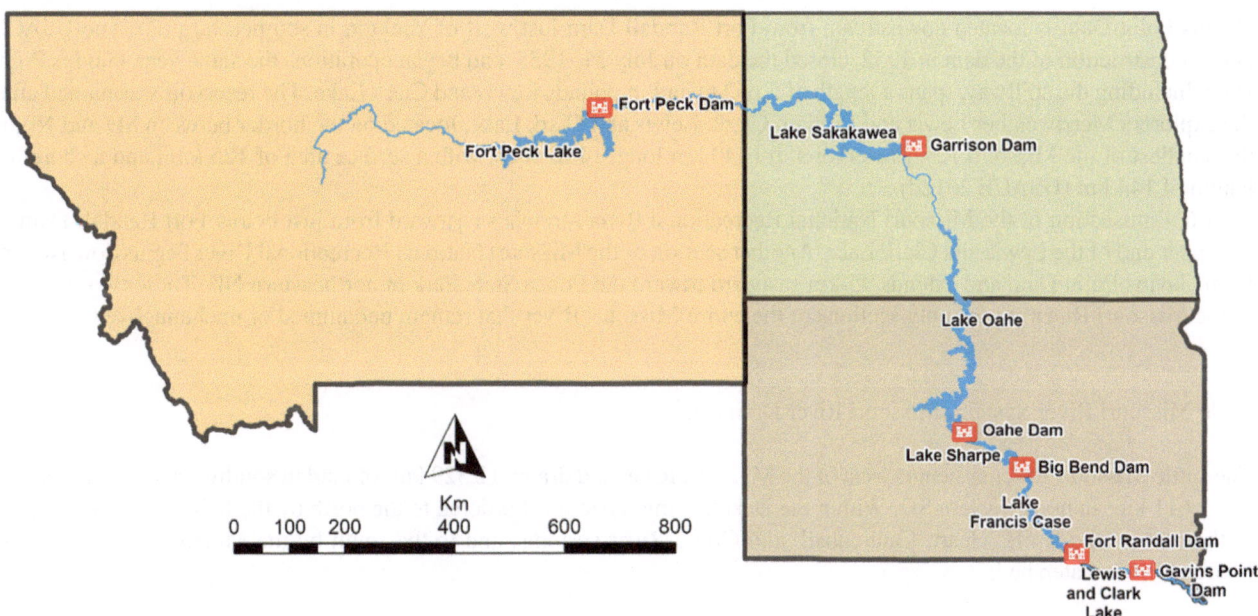

Fig. 1.4 Locations of the Missouri River's six mainstem dams

Garrison Dam (Lake Sakakawea)

Garrison Dam, named after the nearby community of Garrison, ND, is located between Park City and Riverdale in the central part of the state. The US Army Corps of Engineers (USACE) began construction of the dam in 1946, and the dam was closed on April 15, 1953. The USACE has officially operated the dam since 1955. Excluding the spillway, Garrison Dam spans 3,444 m and impounds Lake Sakakawea, the third largest reservoir in the United States. The reservoir is named after Sakakawea, the Shoshone-Hidatsa Native American woman who served as a translator during the Lewis and Clark expedition. Lake Sakakawea is 286 km long, 55 m deep, with a surface area of 1,546 km^2, and 2,156 km of shoreline (USACE 2012c).

Oahe Dam (Lake Oahe)

Oahe Dam is located just north of Pierre in central SD. Construction by the USACE began in 1948, the dam was closed on August 3, 1958, and official operations began in 1962. Excluding the spillway, Oahe Dam spans a length of 2,835 m and impounds Lake Oahe, the fourth largest reservoir in the United States. Lake Oahe, named after the 1874 Oahe Indian Mission, is 372 km long, 62 m deep, with a surface area of 834 km^2, and a shoreline length of 3,621 km (USACE 2012e).

Big Bend Dam (Lake Sharpe)

Big Bend Dam is located near Fort Thompson in south-central SD. The USACE began construction in 1959, closed the dam on July 24, 1963, and began official operation in 1964. Excluding the spillway, Big Bend Dam spans a length of 3,222 m and impounds Lake Sharpe, the 54th largest reservoir in the United States. Lake Sharpe is named after the 17th South Dakota governor, Merrill Q. Sharpe. Lake Sharpe is the first reservoir below Lake Oahe and is 129 km long, 24 m deep, with a surface area of 230 km^2, and a shoreline length of 322 km (USACE 2012a).

Fort Randall Dam (Lake Francis Case)

Fort Randall Dam is located downstream from Big Bend Dam near Pickstown in southeastern SD. The USACE began construction of the dam in 1946, closed the dam on July 20, 1952, and began operations in 1953. Including the spillway, Fort Randall Dam spans a length of 3,261 m and impounds Lake Francis Case, the 11th largest reservoir in the United States. Lake Francis Case is named after Francis Higbee Case, a former US senator from SD. The reservoir is 172 km long, 43 m deep, with a surface area of 413 km^2, and a shoreline length of 869 km (USACE 2012b).

Gavins Point Dam (Lewis and Clark Lake)

Gavins Point Dam is located downstream from Fort Randall Dam just west of Yankton in southeastern SD. The USACE began construction of the dam in 1952, closed the dam on July 31, 1955, and began operations the same year. Gavins Point Dam, including the spillway, spans a length of 2,652 m and impounds Lewis and Clark Lake. The reservoir was named after the explorers Merriweather Lewis and William Clark. Lewis and Clark Lake, located on the border between SD and NE, is the smallest of the Missouri River reservoirs. It is 40 km long, 14 m deep, with a surface area of 125 km^2, and a shoreline length of 144 km (USACE 2012d).

A 63-km section of the Missouri National Recreational River stretches eastward from just below Fort Randall Dam to the upper end of the Lewis and Clark Lake. Another section of the Missouri National Recreational River begins downstream from Gavins Point Dam and extends 95-km eastward toward the Ponca State Park in northeastern NE. These two stretches of the Missouri River are the only sections of the entire Missouri River that remain undammed or unchannelized.

Little Missouri River System (Missouri River Drainage)

The Little Missouri River system is west of the Missouri River and drains 12,329 km^2 of land in southwestern North Dakota and 1,611 km^2 in northwestern SD. Within the Dakotas, this system is bordered to the north by the Missouri River system, to the east by the Knife, Heart, Cannonball, and Grand River systems, and to the south by the Moreau River and Belle Fourche River watershed.

Knife River System (Missouri River Drainage)

The Knife River system drains 6,488 km^2 of land west of the Missouri River in west-central ND. It is bordered to the east and north by the Missouri River system, to the west by the Little Missouri River system, and to the south by the Heart River system. The Knife River, the primary river in the basin, begins at the convergence of the Knife River and the Little Knife River just west of Manning, ND. The Knife then flows east until it converges with the Missouri River just south of Lake Sakakawea near Stanton, ND. The Branch Knife River, a main tributary of the Knife, begins near Hebron, ND, and flows northward before joining with the Knife. Another major tributary, Spring Creek, extends from near the Killdeer Mountains, ND, and flows east until it merges with the Knife River just west of Beulah, ND.

Heart River System (Missouri River Drainage)

The Heart River system drains 8,676 km^2 of land west of the Missouri River in the central-southwest region of ND. This system is bordered to the east by the Missouri River system, to the north by the Knife River system, to the west by the Little Missouri River system, and to the south by the Cannonball River system. The eastern half of the basin may have once been a glaciated plateau, based on the abundant hills with moraine remnants, whereas the western half of the basin was likely an unglaciated plateau with rolling plains interrupted by buttes (Maderak 1966).

Cannonball River System (Missouri River Drainage)

The Cannonball River system drains 11,184 km^2 of land west of the Missouri River in southwest-central ND and northwest-central SD. This system is bordered to the east by the Missouri River system, to the north by the Heart River system, to the west by the Little Missouri River system, and to the south by the Grand River system. The Cannonball is the primary river in the system, originating at White Lake just east of Amidon, ND. From White Lake, the Cannonball first flows northeast and then southeast until it drains into the Missouri River near Cannonball, ND. Its major tributary, Cedar Creek, begins south of Amidon, ND, and mostly flows parallel to the Cannonball River before converging north of McIntosh, SD.

Grand River System (Missouri River Drainage)

The Grand River system drains 13,967 km^2 of land west of the Missouri River in southwest ND and central-northwest SD. This system is bordered to the east by the Missouri River system, to the north by the Cannonball River system, to the west by the Little Missouri River system, and to the south by the Moreau River system. This system drains 2,395 km^2 in ND and 11,572 km^2 in SD. The North Fork of the Grand River begins south of the town of Rhame in southwestern ND. It flows east near the ND–SD border, through the Bowman-Haley Lake, just west of Haley, ND, and further southeast into the Shadehill Reservoir near Shadehill, SD. The South Fork of the Grand River and the Little Grand River converge southwest of Buffalo, SD, and the South Fork continues to flow northeast into the Shadehill Reservoir. The North and South Forks of the Grand River exit on the east side of Shadehill, converging to create the mainstem of the Grand River. The Grand then flows east until it drains into Lake Oahe near Little Eagle, SD.

Moreau River System (Missouri River Drainage)

The Moreau River system drains 8,105 km^2 west of the Missouri River in northwestern SD (Hoagstrom et al. 2007). This system is bordered to the north by the Grand River system, to the west by the Little Missouri River system and the Belle Fourche River watershed, and to the south by the Cheyenne River system. The North and South Forks of the Moreau begin in the Short Pine Hills near the SD–MT border. The forks converge near Zeona, SD, to form the Moreau River, which flows eastward until converging with the Missouri River, south of Mobridge, SD.

Cheyenne River System (Missouri River Drainage)

The Cheyenne River system drains approximately 38,455 km^2 west of the Missouri River in southwestern SD (Hoagstrom et al. 2007). It spans 56,047 km^2 across WY, MT, SD, and NE (Duehr 2004; Galat et al. 2005). Within SD, this system extends through diverse landscapes, including the Black Hills and the badlands, and is bordered by the Moreau, Bad, and White River systems. Water quality is generally poor because of highly erodible soils that contribute to high levels of sedimentation during heavy rainfall and because of agricultural practices (Pirner 2018). The Cheyenne is the main river within this system and has two main tributaries, the Upper Cheyenne and Belle Fourche rivers. These flow to the north and south of the Black Hills, converging on the northeastern side of the Black Hills near Pedro, SD. The lower portion of the Cheyenne River then flows east 386 km before draining into Lake Oahe (Duehr 2004). The Upper Cheyenne River also flows through the Angostura Reservoir near Cascade Springs, SD. The major watersheds within the Cheyenne River system include the Belle Fourche, Cherry Creek, and Fall River watersheds. The Belle Fourche watershed includes the Belle Fourche River ("beautiful fork" in French), the largest tributary of the Cheyenne River. To the east of the Belle Fourche watershed is the Cherry Creek watershed. It includes Cherry Creek, another tributary of the Cheyenne River, which begins in northern Meade County, SD, and flows southeast until it converges with the Cheyenne River in the town of Cherry Creek, SD. The Fall River watershed includes the Fall River, another main tributary of the Cheyenne River. Fed by hot springs that maintain water temperatures above 21 °C throughout the year, the Fall River is a unique component of the Cheyenne River system.

Bad River System (Missouri River Drainage)

The Bad River system drains 8,622 km^2 in west-central SD. This system is bordered to the northeast by the Missouri River system, to the northwest by the Cheyenne River system, and to the south by the White River system. The North Fork of the Bad River receives water from badlands surface runoff and artesian wells near Philip, SD, just prior to converging with the South Fork. The South Fork of the Bad River originates from Big Buffalo Creek and Whitewater Creek in the White River Badlands (Hoagstrom et al. 2007). The forks contribute minimal inflow to the mainstem Bad River, contributing to its primary characteristic of having little to no continuous flow (Pirner 2018). After convergence of the forks, the Bad River flows east into the Missouri River near Fort Pierre, SD (Hoagstrom et al. 2007). The major tributaries of the Bad River system include Dry Creek, Plumb Creek, and White Willow Creek (Hoagstrom et al. 2007).

White River System (Missouri River Drainage)

The White River system drains 21,357 km^2 west of the Missouri River in south-central SD (Hoagstrom et al. 2007). It is bordered to the north by the Bad River system, to the west by the Cheyenne River system, to the south by the Niobrara River system, and to the east by the Ponca Creek and Missouri River systems. Much of the White River system receives runoff from the soils of the western badlands, creating extremely high turbidity. The White River originates in the very northwest corner of NE, just east of Harrison. It flows northeast, crossing the NE–SD border just west of Pine Ridge, SD, and continues to flow northeast along the badlands before eventually emptying into Lake Francis Case south of Oacoma, SD.

Niobrara River System (Missouri River Drainage)

The Niobrara River system originates in eastern WY and runs along the unglaciated terrain of the northern border of NE. A small portion extends into south-central SD. In its entirety, the Niobrara River system drains 4,464 km^2 but only 3,589 km^2 in SD. It is bordered to the west by the Missouri River system, to the east by the Ponca Creek system, to the north and east by the White River system, and to the south by the Loup and Elkhorn River systems in NE. The Keya Paha River begins near Hidden Timber, SD, flows southeast, and eventually merges with the Niobrara River west of Butte, NE.

Ponca Creek System (Missouri River Drainage)

The Ponca Creek system drains 2,137 km^2 west of the Missouri River in south-central SD and northern NE. Within SD, the basin drains 1,184 km^2. This system is bordered to the west and south by the Niobrara River system, to the northwest by the White River system, and to the west by the Missouri River system. Ponca Creek begins near Colome, SD, and flows southeast near the town of Naper, NE. It then flows into Lewis and Clark Lake approximately 6 miles upstream from the mouth of the Niobrara River near Verdel, NE (Newport and Krieger 1959; Hoagstrom et al. 2007). Within SD, Ponca Creek does not have any major tributaries. Whiskey Creek is its largest tributary in NE (Hoagstrom et al. 2007).

James River System (Missouri River Drainage)

The James River system drains 54,390 km^2 of land east of the Missouri River, with 21,756 km^2 drained in east-central ND and 32,634 km^2 in east-central SD (Owen et al. 1981; Shearer and Berry Jr 2002). This system is bordered to the west by the Missouri River system and to the east by the Vermillion, Big Sioux, Minnesota, Wild Rice, and Sheyenne River systems. It was formed from the Lake Dakota Plain during the Pleistocene glaciation period and was completely glaciated. The James River, the primary river within this system, is one of the lowest-gradient rivers in the United States, making it particularly prone to flooding (Benson 1983; Shearer and Berry Jr 2002). The James River begins near Fessenden in central ND and travels 1,202 km south to the Missouri River near Yankton, SD (Benson 1983).

Vermillion River System (Missouri River Drainage)

The Vermillion River system drains 4,267 km^2 east of the Missouri River in southeastern SD. This system is positioned between the James River system to the west and the Big Sioux River system to the east. Once entirely glaciated, the river valley was formed around 12,000 years ago by melting ice (Christensen and Stephens 1967; Hoagstrom et al. 2007). The East and West Forks of the Vermillion River begin at the headwaters of the system in Kingsbury County, SD, near Cherry Lake, Lake Preston, Lake Thompson, Lake Whitewood, and Spirit Lake. The headwaters of the Little Vermillion River are between Howard and Madison, SD. The Little Vermillion merges with the East Fork near Montrose, SD, and the East and West Forks converge to form the Vermillion River just east of Parker, SD. The Vermillion River then flows south, merging with the Missouri River along the border between SD and NE near Vermillion, SD.

Big Sioux River System (Missouri River Drainage)

The Big Sioux River system drains approximately 10,056 km^2 in eastern SD and is the most populated river basin in the state (Hoagstrom et al. 2007; Pirner 2018). It is bordered to the northeast by the Minnesota River system, to the west by the James River system, and to the south by the Vermillion River system. The lower stretch of the Big Sioux River is the border between IA and SD. The river eventually converges with the Missouri River near Sioux City, IA.

Upper Mississippi River Drainage (Mississippi River Basin)

The Upper Mississippi River drainage drains 491,894 km^2, which includes the floodplain emerging from Lake Itasca and all the Mississippi River tributaries north of the confluence of the Mississippi and Ohio rivers near Cairo, Illinois (IL). Although most of the Upper Mississippi River drainage is in MN, WI, IA, IL, and MO, it also includes a small area in northeastern SD. As part of the Minnesota River system, this is the only river system belonging to the Upper Mississippi River drainage in the Dakotas.

Minnesota River System (Upper Mississippi River Drainage)

The Minnesota River system drains 2,529 km^2 east of the Missouri River in northeastern SD (Hoagstrom et al. 2007). In the Dakotas, this system is bordered to the northeast by the Bois de Sioux River system, to the north and northwest by the Wild Rice River system, to the west by the James River system, and to the southwest by the Big Sioux River system. The Minnesota River originates from Big Stone Lake on the MN–SD border and flows southeast through MN, eventually converging with the Mississippi River at Fort Snelling, MN. The primary tributaries of the Minnesota River system in SD include the Little Minnesota, Whetstone, Yellow Bank, and Lac qui Parle rivers.

1.3.2 Nelson River Basin

The Nelson River basin drains 1,149,954 km^2 in parts of four Canadian provinces (British Columbia, Saskatchewan, Manitoba, and Ontario) and four states (MT, ND, SD, and MN). The Nelson River drains Lake Winnipeg in Manitoba, Canada, and flows north 640 km until emptying into the Hudson Bay gulf. The Saskatchewan River, Assiniboine River, and Red River of the North drainage are its main tributaries that feed into Lake Winnipeg and the Nelson River. Within the Dakotas, the Red River of the North drainage is the only component of the Nelson River basin.

Red River of the North Drainage (Nelson River Basin)

The Red River of the North drainage includes 12,767 km^2 in the Dakotas. The drainage is part of the Nelson River basin, encompassing much of eastern ND and a small portion of northeastern SD. The Red River of the North drainage flows north into Lake Winnipeg near Manitoba, Canada, and the Nelson River drains Lake Winnipeg into Hudson Bay.

Red River of the North System (Red River of the North Drainage)

Within the Red River of the North drainage, the Red River of the North system includes the mainstem of the Red River of the North drainage and its smaller tributaries that do not drain into any other system. In the Dakotas, the Red River of the North system drains 80,519 km^2 and is bordered to the west by the Pembina, Park, Forest, Turtle, Goose, Elm, Sheyenne, and Wild Rice River systems and to the south by the Bois de Sioux River system. The Red River of the North system was entirely glaciated and formed when melting ice created a deep channel within the broad, flat valley of the basin (Horton and Follansbee 1906). The Red River of North originates at the confluence of the Bois de Sioux River and Otter Tail River near Wahpeton, ND and forms the border between MN and ND.

Assiniboine River System (Red River of the North Drainage)

The Assiniboine River system drains 162,000 km^2 within the Red River of the North drainage in parts of northern and eastern ND and south-central Manitoba and Saskatchewan, Canada. The Assiniboine is the main river within this system and is a main tributary of the Red River of the North drainage. The Assiniboine River begins in eastern Saskatchewan, flows southeast until joining the Red River of the North drainage at The Forks in Winnipeg, Manitoba, and then empties into Lake Winnipeg. The Assiniboine River is the only tributary of the Souris River watershed in ND.

Souris River Watershed

The Souris River watershed drains 22,607 km^2 east of the Missouri River in north-central ND. This watershed is bordered to the southwest by the Missouri River system, to the southeast by the Sheyenne River system, and to the east by the Pembina River system and the Devils Lake watershed of the Sheyenne River system. The Souris (or Mouse) is the main river that flows through the watershed. It begins in southeast Saskatchewan, Canada, flows south to Velva, ND, and then turns back north, flowing into Manitoba, Canada, just north of Westhope, ND. The Souris River spans 574 km in ND and joins the Assiniboine River near Wawanesa, Manitoba (Owen et al. 1981). Once entirely glaciated and covered by glacial Lake Souris, the basin is primarily comprised of forested hills in the northeast, vast glacial plains in the east, and hills in the southwest.

Pembina River System (Red River of the North Drainage)

The Pembina River system drains 10,141 km^2 east of the Missouri River in northeastern ND. This system is bordered to the west by the Souris watershed of the Assiniboine River system, to the south by the Devils Lake watershed of the Sheyenne River system, to the southeast by the Park River system, and to the east by the Red River of the North system. The Rock Lake watershed drains north where it joins the upper portion of the Pembina River in Canada.

Park River System (Red River of the North Drainage)

The Park River system drains 2,472 km^2 east of the Missouri River in northeastern ND. This system is bordered to the north by the Pembina River system, to the west by the Devils Lake watershed of the Sheyenne River system, to the south by the Forest River system, and to the east by the Red River of the North system.

Forest River System (Red River of the North Drainage)

The Forest River system drains 2,425 km^2 east of the Missouri River in northeastern ND. This system is bordered to the north by the Park River system, to the west by the Devils Lake watershed of the Sheyenne River system, to the south by the Turtle River system, and to the east by the Red River of the North system.

Turtle River System (Red River of the North Drainage)

The Turtle River system drains 1,770 km^2 east of the Missouri River in northeastern ND. This system is bordered to the north and west by the Forest River system, to the southwest by the Goose River system, and to the east by the Red River of the North system.

Goose River System (Red River of the North Drainage)

The Goose River system drains 3,277 km^2 east of the Missouri River in eastern ND. This system is bordered to the north by the Turtle River system and Devils Lake watershed, to the southwest by the Sheyenne River system, to the southeast by the Elm River system, and to the east by the Red River of the North system.

Elm River System (Red River of the North Drainage)

The Elm River system drains 2,838 km^2 east of the Missouri River in eastern ND. This system is bordered to the north by the Goose River system, to the southwest by the Sheyenne River system, and to the east by the Red River of the North system.

Sheyenne River System (Red River of the North Drainage)

The Sheyenne River system drains 27,712 km^2 east of the Missouri River in east-central ND (Owen et al. 1981). This system is bordered to the north by the Devils Lake watershed and Souris River watershed of the Assiniboine River system, to the west by the Missouri River and James River systems, to the south by the Wild Rice River system, and to the east by the Red River of the North, Elm, and Goose River systems. The Sheyenne is the primary river within this system and is 885 km long. It begins in the drift prairie, north of McClusky, ND, and flows east until converging with the North Fork of the Sheyenne River near East Fork Township, ND (Owen et al. 1981). The Sheyenne River then flows east along the southern border of the Devils Lake watershed and continues south until flowing into Lake Ashtabula, behind Baldhill Dam, just north of Valley City, ND. It then turns northeast near Lisbon, ND, and eventually converges with the Red River of the North drainage, just north of Fargo, ND. The Maple River watershed is a large component to the Sheyenne River system just northwest of Fargo.

Devils Lake Watershed

The Devils Lake watershed drains 9,938 km^2 east of the Missouri River in northeast-central ND. It is bordered to the north by the Pembina River system, to the west by the Souris River watershed of the Assiniboine River system, to the south by the Sheyenne River system, and to the east by the Goose, Turtle, Forest, and Park River systems. This system is named after Devils Lake, the largest natural body of water in ND, second in size to only Lake Sakakawea. Devils Lake is located at the southern end of a glaciated basin and is an endorheic, or closed lake, with natural water fluctuations from dry to overflowing. During high water periods, it naturally connects to Stump Lake and eventually the Sheyenne River. Pumps and a constructed outlet can also divert water into the Sheyenne River, artificially connecting the Devils Lake watershed to the Hudson Bay gulf.

Wild Rice River System (Red River of the North Drainage)

The Wild Rice River system drains 5,793 km^2 east of the Missouri River in southeastern ND and northeastern SD. This system is bordered to the north by the Sheyenne River system, to the west by the James River system, to the southeast by the Minnesota and Bois de Sioux River systems, and to the east by the Red River of the North system.

Bois de Sioux River System (Red River of the North Drainage)

The Bois de Sioux River system drains 774 km^2 in the extreme southeastern corner of ND, northeastern corner of SD, and west-central MN. In the Dakotas, this system is bordered to the northwest by the Wild Rice River system and to the southwest by the Minnesota River system. This system was once glaciated and part of the southern portion of Lake Agassiz. The Bois de Sioux River system is the most upstream system in the Red River of the North drainage. The main river within this system, the Bois de Sioux River (French for "Woods of the Sioux"), begins at the outlet of Lake Traverse, the southernmost body of water in the Hudson Bay gulf located on the SD–MN border. The Bois de Sioux River flows north and joins the Otter Tail River near Wahpeton, ND, forming the Red River of the North. The Mustinka and Rabbit rivers are the main tributaries of the Bois de Sioux River, and both enter from MN.

References

Benke AC, Cushing CE (2005) Rivers of North America. Elsevier Academic Press, Burlington

Benson RD (1983) A preliminary assessment of the hydrologic characteristics of the James River in South Dakota. U.S. Geological Survey Water Resources Investigations Report 83-4077, Huron

Billesbach DP, Arkebauer TJ (2012) First long-term, direct measurements of evapotranspiration and surface water balance in the Nebraska Sand Hills. Agric For Meteorol 156:104–110. https://doi.org/10.1016/j.agrformet.2012.01.001

Biondini ME, Patton BD, Nyren PE (1998) Grazing intensity and ecosystem processes in a northern mixed-grass prairie, USA. Ecol Appl 8:469–479. https://doi.org/10.1890/1051-0761(1998)008[0469:GIAEPI]2.0.CO;2

Bleed AS, Flowerday CA (1998) An atlas of the Sand Hills conservation and survey division. University of Nebraska, Lincoln

Bryce SA, Omernik JM, Pater DA, Ulmer M, Schaar J, Freeouf J, Johnson R, Kuck P, Azevedo SH (1996) Ecoregions of North and South Dakota, (color poster with map, descriptive text, summary tables, and photographs): Reston Virginia, U.S. Geological Survey (map scale 1:1,500,000)

Chapman SS, Omernik JM, Freeouf JA, Huggins DG, McCauley JR, Freeman CC, Steinauer G, Angelo RT, Schlepp RL (2001) Ecoregions of Nebraska and Kansas (color poster with map, descriptive text, summary tables and photographs): Reston, Virginia, U.S. Geological Survey (map scale 1:1,950,000)

Christensen C, Stephens J (1967) Geology and hydrology of Clay County, South Dakota. South Dakota Geological Survey Bulletin 19, Vermillion. http://www.sdgs.usd.edu/pubs/pdf/B-19(part1).pdf. Accessed 24 Apr 2023

DeWitt E, Redden JA, Buscher D, Wilson AB (1989) Geologic map of the Black Hills area, South Dakota and Wyoming: U.S. Geological Survey Miscellaneous Investigations Series Map I-1910, I sheet (map scale 1:250,000)

Duehr JP (2004) Fish and habitat relations at multiple spatial scales in Cheyenne River basin, South Dakota. Dissertation, South Dakota State University

Galat DL, Robinson JW, Hesse LW (1996) Restoring aquatic resources to the Lower Missouri River: issues and initiatives. Pages 49–72 in D. L. Galat and A. G. Frazier (eds.), Overview of river-floodplain ecology in the Upper Mississippi River Basin. V. 3 of J. A. Kelmelis (ed.), Science for floodplain management into the 21st century. U.S. Govt. Printing Off., Washington, DC

Galat DL, Berry CR, Gardner WM, Henderson JC, Mestl GE, Power GJ, Stone C, Winston MR (2005) Spatiotemporal patterns and changes in Missouri River fishes. In: Rinne JN, Hughes RM, Calamusso B (eds) Historical changes in large river fish assemblages of the Americas. Am Fish Soc Symp 45, pp 249–291. https://doi.org/10.47886/9781888569728.ch15

Gonzalez MA (2003) Continental divides in North Dakota and North America. North Dakota Geological Survey Newsletter 30 https://www.dmr.nd.gov/ndgs/documents/newsletter/2003Summer/pdf/Divide.pdf

Goolsby DA, Battaglin WA, Lawrence GB, Artz RS, Aulenbach BT, Hooper RP, Keeny DR, Stensland GJ (1999) Flux and sources of nutrients in the Mississippi-Atchafalaya River Basin: Topic 3 report for the integrated assessment on hypoxia in the Gulf of Mexico. US National Ocean Service, NOAA Coastal Ocean Program Decision Analysis Series No 17, Silver Spring. https://repository.library.noaa.gov/view/noaa/21437/noaa_21437_DS1.pdf

Griffith G (2010) Level III North American terrestrial ecoregions: United States descriptions. Prepared for the North American Commission for Environmental Cooperation, pp 1–64

Hendon D, Matthews JA (2014) Pleistocene. Encyclopedia of environmental change. https://doi.org/10.4135/9781446247501

Herman GS, Johnson LA (2007) Geology, geography, and climate. North Dakota Division of Independent Study, Bismark

Hoagstrom CW, Wall SS, Kral JG, Blackwell BG (2007) Recent zoogeography of South Dakota fishes. In: Berry C, Higgins K, Willis D, Chipps S (eds) History of fisheries and fishing in South Dakota. South Dakota Department of Game, Fish and Parks, Pierre, pp 37–89

Horton AH, Follansbee R (1906) Surface water supply of upper Mississippi river and Hudson Bay drainages. US Dep Int, US Geological Survey, Water Supply Paper: 207

Krause JR, Bertrand KN, Kafle A, Troelstrup NH Jr (2013) A fish index of biotic integrity for South Dakota's Northern Glaciated Plains Ecoregion. Ecol Indic 34:313–322. https://doi.org/10.1016/j.ecolind.2013.05.011

Maderak ML (1966) Sedimentation and chemical quality of surface water in the Heart River drainage basin, North Dakota. Geological Survey Water-Supply Paper 1823, GS 66-242

Morris LA, Langemeier RN, Russell TR, Witt A Jr (1968) Effects of main stem impoundments and channelization upon the limnology of the Missouri River, Nebraska. Trans Am Fish Soc 97:380–388. https://doi.org/10.1577/1548-8659(1968)97[380,EOMSIA]2.0.CO;2

Newport TG, Krieger RA (1959) Ground-water resources of the lower Niobrara River and Ponca Creek basins, Nebraska and South Dakota. US Government Print Office. No. 1460-G

Oberdorff T, Guégan JF, Hugueny B (1995) Global scale patterns of fish species richness in rivers. Ecography 18:345–352. https://doi.org/10.1111/j.1600-0587.1995.tb00137.x

Omernik JM (1987) Ecoregions of the conterminous United States. Ann Assoc Am Geogr 77:118–125. https://doi.org/10.1111/j.1467-8306.1987.tb00149.x

Omernik JM, Gallant AL (1988) Ecoregions of the upper Midwest states. U.S. Environmental Protection Agency, Corvallis

Owen JB, Elsen DS, Russell GW (1981) Distribution of fishes in North and South Dakota basins affected by the Garrison Diversion Unit. University of North Dakota, Grand Forks

Pegg MA, Pierce CL, Roy A (2003) Hydrological alteration along the Missouri River Basin: a time series approach. Aquat Sci 65:63–72. https://doi.org/10.1007/s000270300005

Pirner SM (2018) The 2018 South Dakota integrated report for surface water quality assessment. S D Dep Environ Nat Res. https://danr.sd.gov/Conservation/WatershedProtection/ReportsPublications/DANR_18irfinal.pdf. Accessed 24 Apr 2023

Rahn PH, Davis AD, Webb CJ, Nichols AD (1996) Water quality impacts from mining in the Black Hills, South Dakota, USA. Environ Geol 27:38–53. https://doi.org/10.1007/BF00770601

Schultz LD (2011) Environmental factors associated with long-term trends of mountain sucker populations in the Black Hills, and an assessment of their thermal tolerance. Dissertation, South Dakota State University

Shearer JS, Berry CR Jr (2002) Index of biotic integrity utility for the fishery of the James River of the Dakotas. J Freshw Ecol 17:575–588. https://doi.org/10.1080/02705060.2002.9663935

Tan Z, Liu S, Johnston CA, Loveland TR, Tieszen LL, Liu J, Kurtz R (2005) Soil organic carbon dynamics as related to land use history in the northwestern Great Plains. Global Biogeochem Cycles 19(3). https://doi.org/10.1029/2005GB002536

Upham (1895) The Glacial Lake Agassiz, vol 25. United States Government Printing Office. https://doi.org/10.3133/m25. Accessed 24 Apr 2023

USACE (United States Army Corps of Engineers) (1998) Missouri river main stem reservoirs system description and operation. Northwestern Division Missouri River Region Reservoir Control Center

USACE (United States Army Corps of Engineers) (2012a) Big bend project statistics. US Army Corps of Engineers Omaha District. https://www.nwo.usace.army.mil/Media/Fact-Sheets/Fact-Sheet-Article-View/Article/487637/big-bend-project-statistics/. Accessed 24 Apr 2023

USACE (United States Army Corps of Engineers) (2012b) Fort Randall project statistics. US Army Corps of Engineers Omaha District. https://www.nwo.usace.army.mil/Media/Fact-Sheets/Fact-Sheet-Article-View/Article/487642/fort-randall-project-statistics/. Accessed 24 Apr 2023

USACE (United States Army Corps of Engineers) (2012c) Garrison project statistics. US Army Corps of Engineers Omaha District. https://www.nwo.usace.army.mil/Media/Fact-Sheets/Fact-Sheet-Article-View/Article/487634/garrison-project-statistics/. Accessed 24 Apr 2023

USACE (United States Army Corps of Engineers) (2012d) Gavins point project statistics. US Army Corps of Engineers Omaha District. https://www.nwo.usace.army.mil/Media/Fact-Sheets/Fact-Sheet-Article-View/Article/487639/gavins-point-project-statistics/. Accessed 24 Apr 2023

USACE (United States Army Corps of Engineers) (2012e) Oahe project statistics. US Army Corps of Engineers Omaha District. https://www.nwo.usace.army.mil/Media/Fact-Sheets/Fact-Sheet-Article-View/Article/487631/oahe-project-statistics/. Accessed 24 Apr 2023

USGS (United States Geographical Survey) (1997) Groundwater atlas of the United States, HA-730D. https://pubs.usgs.gov/ha/730d/report.pdf. Accessed 24 Apr 2023

Visher SS (1918) The geography of South Dakota. Dissertation, University of Chicago

Williamson JE, Carter JM (2001) Water-quality characteristics in the Black Hills area, South Dakota. U.S. Geological Survey, Rapid City, South Dakota. Water-Resources Investigation Report 01-4194, Rapid City. https://pubs.usgs.gov/wri/wri014194/pdf/wri014194.pdf. Accessed 24 Apr 2023

Wright CK, Wimberly MC (2013) Recent land use in the western corn belt threatens grasslands and wetlands. Proc Natl Acad Sci 110:4134–4139. https://doi.org/10.1073/pnas.1215404110

Chapter 2
History of Ichthyology in the Dakotas

Kathryn E. Schlafke, Matthew D. Wagner, and Chelsey A. Pasbrig

2.1 Prehistoric Fishes

Fossils scattered across North Dakota (ND) and South Dakota (SD) reveal the prehistoric presence of freshwater and marine fishes. Pleistocene Epoch glaciers covered much of the eastern Dakotas, resulting in few fossil deposits. However, the western half of the Dakotas was mostly unglaciated, and fish fossils are commonly found in exposed rocks. The oldest fish in the world (*Anatolepis*) was found in the Bear Lodge Mountains of the Black Hills in eastern Wyoming (WY), just west of the SD border (Parris et al. 2007). Fossils from bradyodont sharks were also discovered in the more than 300-million-year-old rock formations in the Black Hills (Parris et al. 2007). These fossils are primarily teeth because the cartilaginous skeletons of sharks (class Chondrichthyes) do not fossilize well. Fossils of sharks have also been found in rock quarries in Grant County in northeastern SD (Parris et al. 2007). Many of the fossils in the western part of the Dakotas are from bony fishes (class Osteichthyes) and are found within Paleozoic rocks.

2.2 1800s

The expedition journals of Meriwether Lewis and William Clark in 1804 provide the first descriptions of fishes from the Dakotas. One of their journal entries notes a "white catfish…with its eyes small and a tail much like that of a dolphin," which likely was a channel catfish (*Ictalurus punctatus*) (Johnsgard 2003). Around the same time as the Lewis and Clark expedition, Alexander Henry the Younger journaled about the Red River and his trek to the Missouri, documenting the presence of sturgeon and other fish (Henry 1988).

Between 1853 and 1855, the physician John Evans was commissioned by the US government to conduct fish surveys for the Pacific Railroad. While at Fort Pierre, South Dakota, he described goldeye (*Hiodon alosoides*), shorthead redhorse (*Moxostoma macrolepidotum*), western silvery minnow (*Hybognathus argyritis*), flathead chub (*Platygobio gracilis*), creek chub (*Semotilus atromaculatus*), channel catfish, river carpsucker (*Carpiodes carpio*), and paddlefish (*Polyodon spathula*). Evans sent his fish collections to Charles Girard, who officially published the findings and established the first species list for the Dakotas (Girard 1856, 1858).

During the same period, Robert Kennicott of the Smithsonian Institution recorded new discoveries of the Johnny darter (*Etheostoma nigrum*) from the Goose River and of the blackside darter (*Percina maculata*) from the Maple River in ND. In 1857, 10 fish species were documented in the Dakotas. In 1859, Girard published the first known accounts of the Iowa darter (*Etheostoma exile*) from ND, listing a collection from the Little Muddy River by George Suckley and another from the Cannonball River by F.V. Hayden (Fig. 2.1; Girard 1859).

K. E. Schlafke
Montana Fish, Wildlife & Parks, Bozeman, MT, USA

M. D. Wagner
U.S. Fish and Wildlife Service, Jackson, MS, USA

C. A. Pasbrig (✉)
South Dakota Game, Fish and Parks, Pierre, SD, USA
e-mail: Chelsey.Pasbrig@state.sd.us

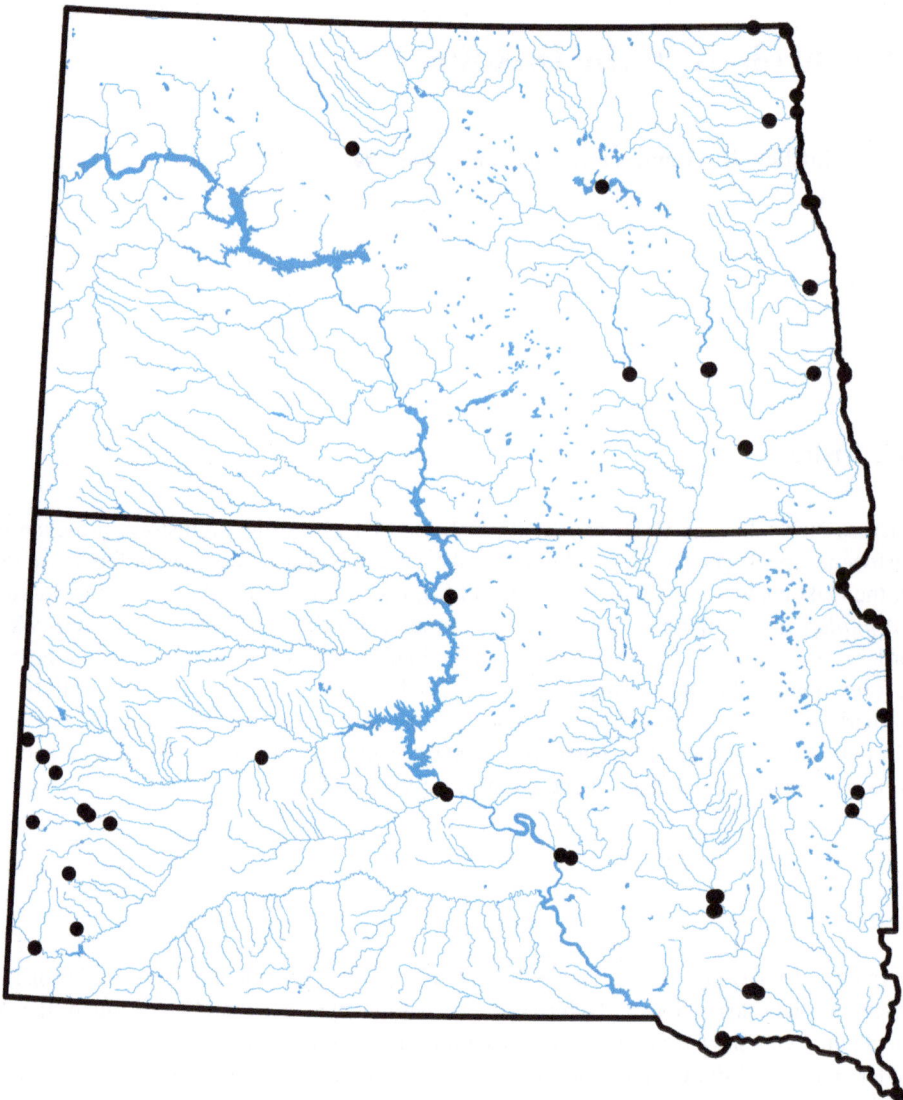

Fig. 2.1 Recorded fish collection sites in the Dakotas prior to 1900

Elliot Coues of the US Geological and Geographical Survey of the Territories added to the fish reports of the Dakotas in 1873 and 1874, sending new specimens to David Starr Jordan, a renowned ichthyologist. Jordan published the *Report on the Collection of Fishes Made by Elliot Coues, U.S.A. in Dakota and Montana during the Seasons of 1873 and 1874* (Jordan 1878), which included the first records in the Dakotas of shovelnose sturgeon (*Scaphirhynchus platorynchus*), northern pike (*Esox lucius*), and lake chub (*Couesius plumbeus*). Jordan named the lake chub genus (*Couesius*) after Coues (Jordan 1878).

In 1876, Edward D. Cope added to the documentation of Dakota fish species while on a trip to collect fossils from the Moreau River. He collected fishes from Battle Creek, a tributary of the Missouri River drainage in SD. This sampling site has also been recorded as Blue Blanket Creek (Bailey and Allum 1962) and LeBeau Creek (Berry et al. 2007). Cope was the first to document burbot (*Lota lota*), longnose dace (*Rhinichthys cataractae*), northern redbelly dace (*Chrosomus eos*), longnose gar (*Lepisosteus osseus*), and shortnose gar (*Lepisosteus platostomus*) in the Dakotas (Cope 1879).

In 1889, the Dakota Territory became the states of ND and SD. In 1891, Seth E. Meek, conducted the first fish surveys after statehood in southeastern SD. Meek collected fishes between Sioux City, IA, and Sioux Falls, SD, in the Missouri River drainage and Big Sioux River system (Meek 1892). This survey was important to SD, reporting 24 new species of fish.

In 1892, Ulysses S. Cox and Albert Jefferson Woolman conducted stream surveys in western Minnesota (MN) and the eastern Dakotas where they reported six new species for ND and SD. The areas surveyed included streams in the Minnesota and James River systems, the Red River of the North drainage, and Lake Traverse and Big Stone Lake (Woolman 1896).

Woolman reported the bleeding shiner (*Notropis zonatus*) from the Sheyenne River in ND. However, this was a misidentification. Carl Hubbs, University of Michigan Museum of Zoology (UMMZ), later correctly identified the specimen as the pugnose shiner (*Miniellus anogenus*) (UMMZ, catalog number: 61945).

Moreover, in 1892, Barton Warren Evermann surveyed Black Hills stream fish during a trip to assess the potential for creating trout fisheries (Evermann 1892). In 1896, Evermann and Ulysses Orange Cox published the *Report Upon the Fishes of the Missouri River Basin*, which compiled official species lists for each of the states within the Missouri river drainage, including ND and SD (Evermann and Cox 1896). This report added 13 species to the list of fishes in the Dakotas and summarized all the previous fish survey data. Trout were stocked into Black Hills streams shortly after the Evermann and Cox assessment.

2.3   Early 1900s

In 1926, William Over and Edward Churchill from the University of South Dakota were paid by the South Dakota Department of Game and Fish (now South Dakota Department of Game, Fish, and Parks (SDGFP)) to survey eastern SD lakes (Churchill and Over 1933). This project was expanded in 1927 and 1928 to include a statewide survey of all lakes and streams in SD. They published their results in the first book dedicated to SD fishes, *Fishes of South Dakota*. This book described 81 species, listed 12 new SD species records, and contained a dichotomous key to the family level. It also provided brief species summaries along with a few illustrations (Churchill and Over 1933).

Thomas Leroy Hankinson published *Fishes of North Dakota* in 1929 (Hankinson 1929). It was the first paper presented at the Michigan Academy annual meeting and included 57 species, along with 13 new species for ND. *Fishes of North Dakota* summarized all ND collections from 1921 to 1928 by R.T. Young, Robert Johnstone, and Hankinson from the University of North Dakota Biological Station and Carl Hubbs, Leonard P. Schultz, and Crystal Thompson from the University of Michigan Museum of Zoology.

The economic depression and prolonged drought of the 1930s greatly impacted fisheries in the Dakotas. Many lakes and ponds dried during this period. Once water levels recovered, fish stocking was prevalent. The National Industrial Recovery Act of 1933 included authorization of the construction of Fort Peck Dam in northeastern Montana (MT). Completed in 1940, Fort Peck was the first dam on the Missouri River. Success with Fort Peck Dam construction, coupled with extreme flooding in the lower Missouri River drainage and the desire for water in the arid west led to passage of the Flood Control Act of 1944, which included the Pick–Sloan Plan. Pick–Sloan authorized construction of five additional dams on the Missouri River from Riverdale, ND, to Yankton, SD. The Missouri River dams and resulting reservoirs were built to provide flood control, water for irrigation and municipal uses, electrical power, downstream navigation, and recreational use such as fishing, hunting, and boating (Berry Jr 2003).

Figure 2.2 illustrates fish collection locations in the Dakotas from 1900 to 1949.

2.4   Post-Reservoir Construction: 1950s to Today

In 1962, Reeve M. Bailey and Marvin O. Allum of the University of Michigan Museum of Zoology published a revised edition of *Fishes of South Dakota* that incorporated more recent data from SD (Bailey and Allum 1962). This edition included 10 new species, more detailed summaries of each of the 110 fish species listed, a dichotomous key, and fisheries information before and after reservoir construction.

In 1981, John B. Owen of the University of North Dakota published *Distribution of Fishes in North and South Dakota Basins Affected by the Garrison Diversion Unit*, the first book of fish species in ND (Owen et al. 1981). This book only focused on the James, Souris, Sheyenne, and Wild Rice River systems but was one of the more complete publications on ND fishes. It listed 10 new species of fish, along with species summaries, distribution maps, and a dichotomous key written by Stephen Kelsch, also of the University of North Dakota.

The North Dakota Game and Fish Department (NDGF) published *Fishes of North Dakota* in 1994. This booklet included 36 species and was written for public outreach and education. It had neither distribution maps nor a dichotomous key. Another public education booklet, *The Guide to the Common Fishes of South Dakota*, written by Robert M. Neuman and David W. Willis, was also published in 1994 (Neumann and Willis 1994). Published in 2011, the *Guide to the Fishes of South Dakota* lists 125 species, but illustrations, life history summaries, and river system-level distribution maps are only included for 53 species, and the maps are not point-specific (Hoagstrom et al. 2011).

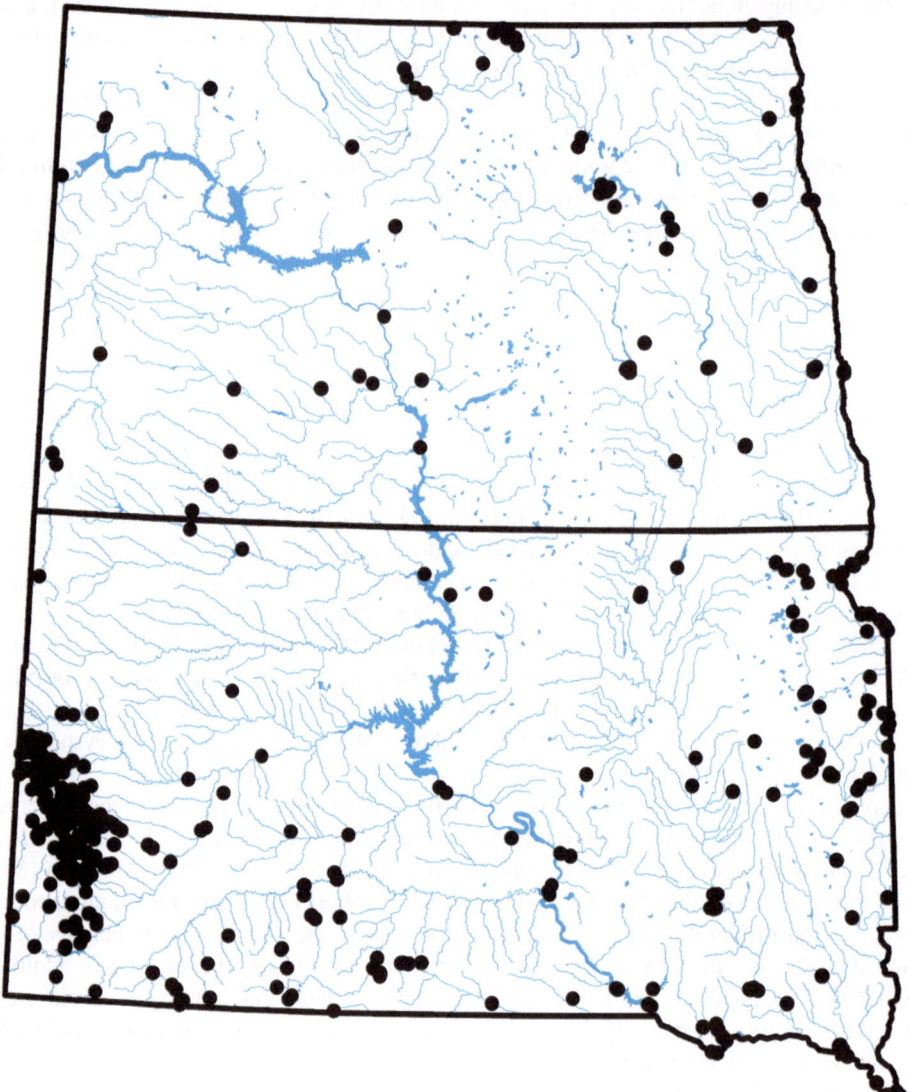

Fig. 2.2 Fish collection sites in the Dakotas prior to 1950

After reservoir construction, fisheries work expanded in the Dakotas. Around 1990, sampling and surveying techniques became more standardized in both states. The following paragraphs describe some of the individuals who have significantly contributed to the advancement of fisheries science in ND and SD.

James C. Schmulbach taught ichthyology from 1958 to 1994 at the University of South Dakota. He directed studies on shovelnose sturgeon's feeding habits (Modde and Schmulbach 1977) and larval drift of fishes (Muth and Schmulbach 1984). Additionally, Schmulbach conducted research on the unchannelized reaches of the Missouri River drainage, downstream of Gavins Point Dam (Schmulbach et al. 1975).

John Peterka taught ichthyology and fisheries biology from 1966 to 1997 at North Dakota State University. He conducted surveys on the Sheyenne (Peterka 1977) and Knife River systems (Peterka and Koel 1997). Peterka also assessed the distribution of fishes in the Red River of the North drainage (Koel 1997).

John Owen directed stream surveys throughout western North Dakota while at the University of North Dakota in the 1970s and 1980s. Owen also maintained an extensive fish collection at the university.

Steven W. Kelsch taught ichthyology and fisheries management from 1991 to 2019 at the University of North Dakota. He led surveys on the Red River of the North drainage (Wendel and Kelsch 1999; Kelsch and Wendel 2004), its western tributaries (Resseguie and Kelsch 2008; Kludt et al. 2017), and the Little Missouri River (Kelsch 1994).

Charles G. Scalet taught ichthyology from 1973 to 2007 at South Dakota State University. As the Chair of the Department of Wildlife and Fisheries Sciences, Scalet had a major influence on fisheries surveys and research in SD. He advanced the knowledge of the Little Missouri River system (Bich and Scalet 1977) and, in conjunction with his student Brian Van Zee, documented the redfin pickerel (*Esox americanus*) for the first time in SD (Van Zee and Scalet 1997).

As leader of the Cooperative Fish and Wildlife Research Unit at South Dakota State University, Charles R. Berry, Jr., made major contributions to the known distribution of fishes in SD. Working together, Berry and Scalet were responsible for surveys on the Moreau (Loomis 1997), Cheyenne (Hampton 1998; Duehr 2004), Belle Fourche (Doorenbos 1998), Vermillion (Braaten 1993), James (Tol 1976; Kubeny 1992; Walsh 1992; Shearer 2001), Big Sioux (Dieterman 1995; Fisher 1996; Arterburn 2001; Kirby 2001; Milewski 2001), Bad (Milewski 2001), and White River systems (Fryda 2001). They also directed a study to assess the status and distribution in eastern SD of the federally endangered Topeka shiner (*Miniellus topeka*) (Blausey 2001). Their collaboration extended to summarizing the fish community across the entire Missouri River drainage that included zoogeography of SD fishes, establishing native or nonnative status for all SD fish species based on both historic and more recent data (Hoagstrom 2006; Hoagstrom et al. 2007). Not confined to just SD, Berry also directed a survey of the Red River of the North drainage and Forest River system in ND (Borgstrum 2010).

David Willis and Brian Graeb from South Dakota State University directed projects on fish distribution in the Missouri River drainage. Willis oversaw one of the few surveys in the backwaters of the Upper Missouri River drainage in ND (Fisher 1999), whereas Graeb was involved in surveying Missouri River deltas formed by inflows from the Niobrara and White River systems (Kaemingk et al. 2007; Schreck 2010; Carlson 2015). Additionally, Graeb directed surveys of the Knife, Heart, and Cannonball River systems in ND (Johnson 2010).

While at South Dakota State University, Katie Bertrand directed surveys in the tributaries of the James River (Krause 2013; Hayer 2014; Schumann 2017; Garcia et al. 2018; Schumann et al. 2021), Big Sioux River (Krause 2013; Hayer 2014), Vermillion River (Krause 2013; Hayer 2014), and the SD portion of the Nebraska Sandhills (Felts 2013). Additionally, Bertrand led surveys on waters throughout the Northwestern Great Plains level III ecoregion in SD, including the Grand, Moreau, Cheyenne, Bad, and White River systems (Schultz 2011; Kaiser 2017; Jones 2018; Fopma 2020).

Mark Pegg of the University of Nebraska-Lincoln led sturgeon chub (*Macrhybopsis gelida*) and sicklefin chub (*Macrhybopsis meeki*) surveys on the Little Missouri, Missouri, Moreau, Grand, Bad, Cheyenne, and White River systems (Magruder and Pegg 2020). Information collected in the first year of this study is included in the distributional data in this book.

Approximately 400 sites throughout ND between 2015 and 2020 were surveyed by Casey S. Williams of Valley City State University. The sampling locations included the mainstem and tributary sites of the Red River of the North drainage (Williams and Anderson 2018), James (Williams and Anderson 2020), Missouri (Williams and Dahl 2015; Williams et al. 2016), Souris (Williams and Coenen 2017), and the Little Missouri River systems (Williams et al. 2016). Voucher specimens of all the ND fish species collected by Williams are preserved, organized by site, and housed at Valley City State University.

The South Dakota State University Fish Collection in McFadden Biostress Laboratory Room 168 contains fish specimens vouchered from past surveys across the Dakotas. The collection holds more than 6000 lots organized and cataloged by phylogenetic order to the familial level and then in alphabetical order to the genus and species level. Each lot is a collection of one or more single species collected at a single location at a single point in time. Lots are labeled with the following information: catalog number, common name, family, genus, species, locality, latitude, longitude, state, county, collectors, field number, date of collection, and count. The collection is a physical record for identification verification and is used for both research and education. Figure 2.3 is a map showing collection locations of the vouchered and cataloged specimens of the South Dakota State University Fish Collection.

Numerous employees from state and federal agencies, along with those from nongovernmental organizations, have also collected considerable information on the fishes of the Dakotas. These data were carefully added to the published literature to establish a fisheries distribution database containing more than 500,000 records of fishes sampled from the 1800s to the present (Fig. 2.4). This database was used to create individual fish species distribution maps and is the basis for this book. Although the database contains comprehensive information for both ND and SD, it may not accurately represent the presence and distribution of each fish species. The Dakotas are undersurveyed, and, some areas, such as the Missouri River reservoirs, lack the georeferencing required for accuracy.

Figure 2.4 illustrates all the fish collection locations in the Dakotas from prior to 2022.

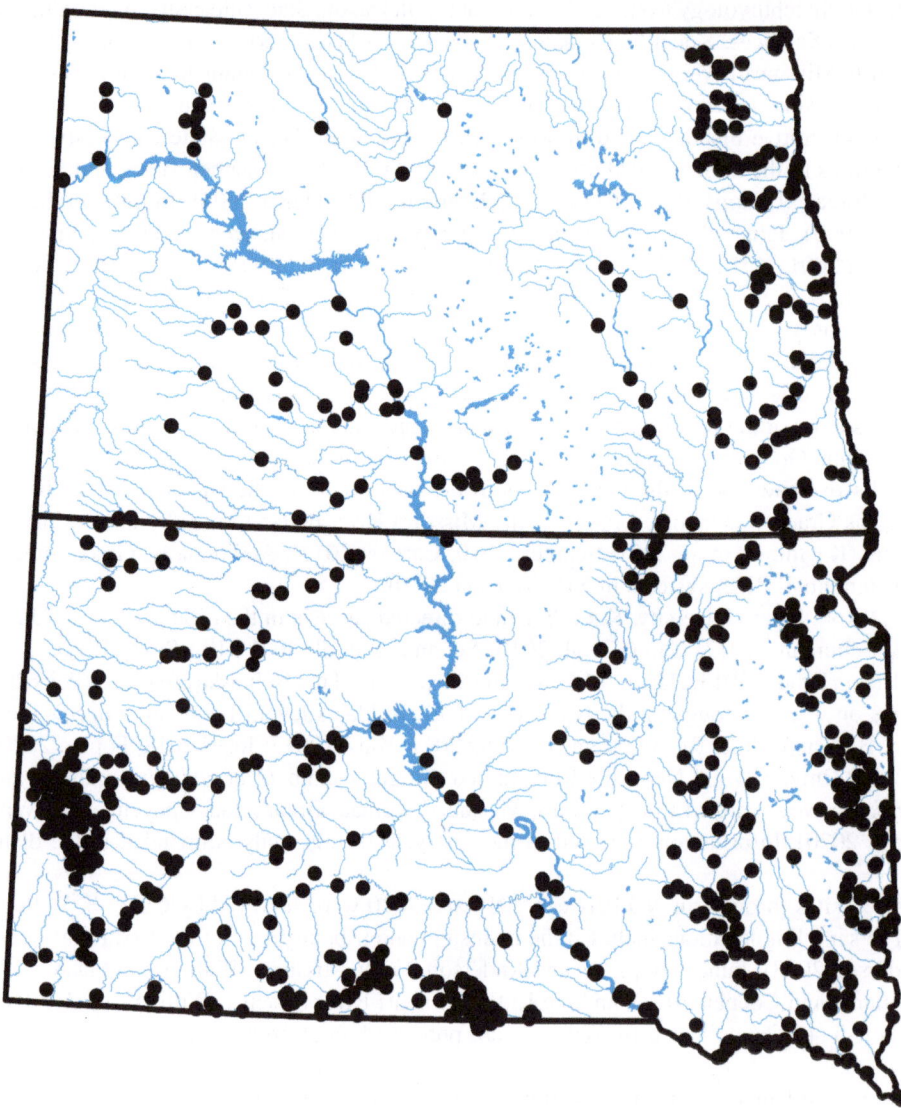

Fig. 2.3 Collection locations of the fish specimens vouchered and cataloged at the South Dakota State University Fish Collection

Fig. 2.4 All recorded fish collection sites in the Dakotas prior to 2022

References

Arterburn JE (2001) Population characteristics and sampling methods of catfish for the James and Big Sioux Rivers. Dissertation, South Dakota State University

Bailey RM, Allum MO (1962) Fishes of South Dakota. University of Michigan, Ann Arbor. https://doi.org/10.3998/mpub.9690435

Berry CR Jr (2003) Committee on Missouri River Ecosystem Science. 2002. The Missouri River ecosystem: exploring the prospects for recovery. Wetlands 23:208–211

Berry CR, Higgins KF, Willis DW, Chipps SR (eds) (2007) History of fisheries and fishing in South Dakota. South Dakota Department of Game, Fish and Parks, Pierre

Bich JP, Scalet CG (1977) Fishes of the Little Missouri River, South Dakota. Proc S D Acad Sci 56:163–177

Blausey CM (2001) The status and distribution of the Topeka shiner *Notropis topeka* in eastern South Dakota. Dissertation, South Dakota State University

Borgstrum LJ (2010) Fish community assembly in the Forest River, North Dakota, and resolution of *Campostoma* species presence. Dissertation, South Dakota State University

Braaten PJ (1993) The influence of habitat structure and environmental variability on habitat use by fish in the Vermillion River, South Dakota. Dissertation, South Dakota State University

Carlson AK (2015) Rapid response to a catastrophic flood: effects on aquatic resources in Missouri River reservoirs. Dissertation, South Dakota State University

Churchill EP, Over WH (1933) Fishes of South Dakota. S.D. Dept. Game and Fish, Pierre. 87 pp.

Cope ED (1879) Eleventh contribution to the herpetology of tropical America. Proc Am Philos Soc 18:261–277

Dieterman DJ (1995) The influence of the clean water act and tributaries on the fish community of the Big Sioux River, South Dakota. Dissertation, South Dakota State University

Doorenbos RD (1998) Fish and habitat of the Belle Fourche River, South Dakota. Dissertation, South Dakota State University

Duehr JP (2004) Fish and habitat relations at multiple spatial scales in Cheyenne River Basin, South Dakota. Dissertation, South Dakota State University Dakota

Evermann BW (1892) The ichthyologic features of the Black Hills. Proc Indiana Acad Sci 1893:73–78

Evermann BW, Cox UO (1896) Report upon the fishes of the Missouri River basin. US Government Printing Office

Felts E (2013) Ecology of glacial relict fishes in South Dakota's Sandhills region. Dissertation, South Dakota State University

Fisher CJ (1996) Population characteristics and habitat selection of walleye in the Big Sioux River, South Dakota. Dissertation, South Dakota State University

Fisher SJ (1999) Seasonal investigation of native fishes and their habitats in Missouri River and Yellowstone River backwaters. Dissertation, South Dakota State University

Fopma SJ (2020) Distribution, density, movement, and support for management of mountain sucker, *Pantosteus jordani*, in the Black Hills of South Dakota. Dissertation, South Dakota State University

Fryda DD (2001) A survey of the fishes and habitat of the White River, South Dakota. Dissertation, South Dakota State University

Garcia C, Schumann DA, Howell J, Graeb BDS, Bertrand KN, Klumb RA (2018) Seasonality, floods and droughts structure larval fish assemblages in prairie rivers. Ecol Freshw Fish 27:389–397. https://doi.org/10.1111/eff.12354

Girard C (1856) Researches upon the cyprinoid fishes inhabitation the fresh waters of the United States, west of the Mississippi Valley, from specimens in the museum of the Smithsonian Institution. Proc Acad Natl Sci Phila 8:165–213

Girard C (1858) Notes upon various new genera and new species of fishes, in the museum of the Smithsonian Institution, and collected in connection with the United States and Mexican boundary survey: Major William Emory, Commissioner

Girard C (1859) Ichthyology of the boundary, vol 2. US Government Printing Office

Hampton DR (1998) A survey of the fishes and habitat of the Cheyenne River in South Dakota. Dissertation, South Dakota State University

Hankinson TL (1929) Fishes of North Dakota. Michigan Academy of Science, Arts, and Letters

Hayer CA (2014) Fish assemblage structure, trophic ecology, and potential effects of invading Asian carps in three Missouri River tributaries, South Dakota. Dissertation, South Dakota State University

Henry A (1988) The journal of Alexander Henry the younger 1799–1814. The Champlain Society, University of Toronto Press

Hoagstrom CW (2006) Fish community assembly in the Missouri River basin. Dissertation, South Dakota State University

Hoagstrom CW, Wall SS, Kral JG, Blackwell BG, Berry CR (2007) Zoogeographic patterns and faunal change of South Dakota fishes. West N Am Nat 67:161–184. https://doi.org/10.3398/1527-0904(2007)67[161:ZPAFCO]2.0.CO;2

Hoagstrom CW, Adams G, Neumann RM, Willis DW (2011) Guide to the fishes of South Dakota. South Dakota Game, Fish and Parks, Pierre

Johnsgard PA (2003) Lewis and Clark on the Great Plains: a natural history. University of Nebraska Press, Lincoln

Johnson MS (2010) Status and distribution of fishes in tributaries of the Garrison reach of the Missouri River, North Dakota. Dissertation, South Dakota State University

Jones S (2018) Western prairie stream fishes: an assessment of past and present fish assemblage structure, biotic homogenization, and population dynamics in western South Dakota streams. Dissertation, South Dakota State University

Jordan DS (1878) A catalogue of the fishes of the fresh waters of North America. U.S. Department of the Interior

Kaemingk MA, Graeb BDS, Hoagstrom CW, Willis DW (2007) Patterns of fish diversity in a mainstem Missouri River reservoir and associated delta in South Dakota and Nebraska, USA. River Res Appl 23:786–791. https://doi.org/10.1002/rra.1002

Kaiser C (2017) Biotic integrity in the northwestern Great Plains and mechanisms regulating stream condition in South Dakota. Dissertation, South Dakota State University

Kelsch SW (1994) Lotic fish-community structure following transition from severe drought to high discharge. J Freshw Ecol 9:331–341. https://doi.org/10.1080/02705060.1994.9664903

Kelsch SW, Wendel JL (2004) Lotic habitat availability and use by channel catfish in the Red River of the North. J Freshw Ecol 19:473–484. https://doi.org/10.1080/02705060.2004.9664922

Kirby DJ (2001) An assessment of the channel catfish population in the Big Sioux River, South Dakota. Dissertation, South Dakota State University

Kludt N, Kelsch S, Newman R, Rundquist B (2017) Riparian and landscape disturbance effects on stream fish community composition in an agriculturally-dominated drainage. Aquat Ecosyst Health Manage 20:445–456

Koel TM (1997) Distribution of fishes in the Red River of the North basin on multivariate environmental gradients. Dissertation, North Dakota State University

Krause JR (2013) Biotic integrity and northern pike ecology in eastern South Dakota. Master's thesis, Dissertation, South Dakota State University

Kubeny SJ (1992) Population characteristics and habitat selection of channel catfish (*Ictalurus punctatus*) in the lower James River, South Dakota. Dissertation, South Dakota State University

Loomis TM (1997) Survey of the fishes and habitat in the upper Moreau River, Perkins County, South Dakota. Dissertation, South Dakota State University

Magruder M, Pegg M (2020) Population structure and habitat use of benthic fishes of the Missouri River and its major tributaries with an emphasis on sicklefin and sturgeon chub in South Dakota. South Dakota Game, Fish and Parks Interim Report. State Wildlife Grant T-89-R-1, Pierre

Meek SE (1892) Report upon the fishes of Iowa, based upon observations and collections made during 1889, 1890, and 1891, vol 10. US Government Printing Office

Milewski CL (2001) Local and systemic controls on fish and fish habitat in South Dakota Rivers and streams: implications for management. Dissertation, South Dakota State University

Modde T, Schmulbach JC (1977) Food and feeding behavior of the shovelnose sturgeon, *Scaphirhynchus platorynchus*, in the unchannelized Missouri River, South Dakota. Trans Am Fish Soc 106:602–608. https://doi.org/10.1577/1548-8659(1977)106<602:FAFBOT>2.0.CO;2

Muth RT, Schmulbach JC (1984) Downstream transport of fish larvae in a shallow prairie river. Trans Am Fish Soc 113:224–230. https://doi.org/1 0.1577/1548-8659(1984)113<224:DTOFLI>2.0.CO;2

Neumann RM, Willis DW (1994) Guide to the common fishes of South Dakota. South Dakota State University Press, Brookings

Owen JB, Elsen DS, Russell GW (1981) Distribution of fishes in North and South Dakota basins affected by the Garrison Diversion Unit. Fisheries Research Unit, University of North Dakota, Grand Forks

Parris DC, Grandstaff BS, Gallagher WB (2007) Fossil fish from the Pierre Shale Group (Late Cretaceous): clarifying the biostratigraphci record. In: Martin JE, Parris DC (eds) The geology and paleontology of the late crustaceous marine deposits of the Dakotas, vol 427. Geological Society of American, pp 99–109. https://doi.org/10.1130/2007.2427(07)

Peterka JJ (1977) An analytical evaluation of the utilization and management of water resources in the Lake Metigoshe watershed, North Dakota. North Dakota Water Resources Research Institute, Fargo

Peterka JJ, Koel TM (1997) Fishes of the Knife River drainage in southwestern North Dakota. Dissertation, North Dakota State University, Fargo

Resseguic T, Kelsch SW (2008) Influence of temperature and discharge on reproductive timing of common carp in a northern Great Plains River. Prairie Nat 41:23–36

Schmulbach JC, Gould G, Groen CL (1975) Relative abundance and distribution of fishes in the Missouri river, Gavins Point Dam to Rulo, Nebraska. Proc SD Acad Sci 54:194–222

Schreck WJ (2010) Seasonal use of Missouri River reservoir deltas by fishes. Dissertation, South Dakota State University

Schultz LD (2011) Environmental factors associated with long-term trends of mountain sucker populations in the Black Hills, and an assessment of their thermal tolerance. Dissertation, South Dakota State University

Schumann DA (2017) Measuring aquatic organism responses to grassland restoration: does the field of dreams really exist? Dissertation, South Dakota State University

Schumann DA, Graeb KNB, Pfimmer J, Stafford JD, Chipps SR (2021) The local responses of aquatic ecosystems to adjacent grassland conservation: can streams of dreams exist in a degraded riverscape? Aquat Conserv: Mar Freshw Ecosyst 31:2481–2495. https://doi.org/10.1002/aqc.3656

Shearer JS (2001) Temporal change in fish communities and modification of the index of biotic integrity for the James River of the Dakotas. Dissertation, South Dakota State University

Tol D (1976) An evaluation of the fishery resource in a portion of the James River, South Dakota scheduled for channel modification. Dissertation, South Dakota State University

Van Zee BE, Scalet CG (1997) Range extension of the grass pickerel into South Dakota. Prairie Nat 29:277–278

Walsh RJ (1992) Differences in fish abundance among habitat types in a warmwater stream; the James River, South Dakota. Dissertation, South Dakota State University

Wendel JL, Kelsch SW (1999) Summer range and movement of channel catfish in the Red River of the North. In: Irwin ER, Hubert WA, Rabeni CF, Schramm HL Jr, Coon T (eds) Catfish 2000: proceedings of the international ictalurid symposium, symp 24. American Fisheries Society, Bethesda, pp 203–214

Williams CS, Anderson EM (2018) Fish survey of select streams from North Dakota. Report to the North Dakota Game and Fish Department, Bismarck

Williams CS, Anderson EM (2020) Fish survey of select streams from North Dakota. Report to the North Dakota Game and Fish Department, Bismarck

Williams CS, Coenen T (2017) Fish survey of select streams from North Dakota. Report to the North Dakota Game and Fish Department, Bismarck

Williams CS, Dahl LD (2015) Fish survey of select streams from Western North Dakota. Report to the North Dakota Game and Fish Department, Bismarck

Williams CS, Coenen T, Folmer C, Marvel C (2016) Fish survey of select streams from Southwestern North Dakota. Report to the North Dakota Game and Fish Department, Bismarck

Woolman AJ (1896) Report upon the ichthyological investigations in western Minnesota and eastern North Dakota. In: Part 19 Report of the commissioner. US Government Printing Office, Washington, DC, pp 343–373

Chapter 3
Distribution and Status of Fishes in the Dakotas

Kathryn E. Schlafke, Matthew D. Wagner, and Chelsey A. Pasbrig

3.1 Distribution

The Dakotas are collectively home to 127 fish species. North Dakota (ND) has 75 native and 20 introduced fish species, whereas South Dakota (SD) has 95 native and 24 introduced species. In the Dakotas, as with the entire Great Plains, species richness increases from west to east. Tables 3.1 and 3.2 detail the distribution across the river drainages and systems in ND and SD for each species from 1990 through 2021 and denote whether the species is native or introduced.

K. E. Schlafke
Montana Fish, Wildlife & Parks, Bozeman, MT, USA

M. D. Wagner
U.S. Fish and Wildlife Service, Jackson, MS, USA

C. A. Pasbrig (✉)
South Dakota Game, Fish and Parks, Pierre, SD, USA
e-mail: Chelsey.Pasbrig@state.sd.us

© The Author(s), under exclusive license to Springer Nature Switzerland AG 2024
M. Barnes (ed.), *Fishes of the Dakotas*, https://doi.org/10.1007/978-3-031-38040-2_3

Table 3.1 ND and SD native and introduced fishes in phylogenetic order, listed by family and both common and scientific names

	SD		ND		River															
	native	status	native	status	Missouri	Little Missouri	Knife	Heart	Cannonball	Grand	Moreau	Cheyenne	Bad	White	Niobrara	Ponca Creek	James	Vermillion	Big Sioux	Minnesota
Petromyzontidae: Lamprey																				
Chestnut lamprey (*Ichthyomyzon castaneus*)	NA		Yes	SCP3																
Silver lamprey (*Ichthyomyzon unicuspis*)	Yes		Yes	SCP3	x													x	x	
Acipenseridae: Sturgeon																				
Lake sturgeon (*Acipenser fulvescens*)	Yes	SGCN	Yes		x												x	x	x	x
Pallid sturgeon (*Scaphirhynchus albus*)	Yes	FE, SE, SGCN	Yes	FE, SE, SCP2	x													x	x	
Shovelnose sturgeon (*Scaphirhynchus platorynchus*)	Yes	FT, SGCN	Yes	FT	x													x	x	
Polyodontidae: Paddlefish																				
Paddlefish (*Polyodon spathula*)	Yes	SGCN	Yes	SCP2	x									x			x			
Lepisosteidae: Gar																				
Longnose gar (*Lepisosteus osseus*)	Yes		NA		x												x			
Shortnose gar (*Lepisosteus platostomus*)	Yes		Yes	SCP2	x			x	x	x	x		x				x	x	x	x
Amiidae: Bowfin																				
Emerald bowfin (*Amia ocellicauda*)	NA		Yes																	
Anguillidae: Eel																				
American eel (*Anguilla rostrata*)	Yes		NA		x															
Hiodontidae: Mooneye																				
Goldeye (*Hiodon alosoides*)	Yes		Yes		x		x	x	x	x	x	x	x	x			x	x	x	
Mooneye (*Hiodon tergisus*)	Yes		Yes															x	x	
Alosidae: Shad																				
Skipjack herring (*Alosa chrysochloris*)	Yes		NA		x															
Alewife (*Alosa pseudoharengus*)	No		NA		x															
Dorosomatidae: Thread herring																				
Gizzard shad (*Dorosoma cepedianum*)	Yes		Yes		x		x	x	x	x	x	x	x	x			x	x		x
Threadfin shad (*Dorosoma petenense*)	No		NA		x															
Catostomidae: Sucker																				
River carpsucker (*Carpiodes carpio*)	Yes		Yes		x	x	x	x	x	x	x	x	x	x	x		x	x	x	x
Quillback (*Carpiodes cyprinus*)	Yes		Yes		x				x	x	x	x	x	x	x		x	x	x	x
Highfin carpsucker (*Carpiodes velifer*)	Yes		NA		x															
Longnose sucker (*Catostomus catostomus*)	Yes	ST, SGCN	Yes		x	x						x								
White sucker (*Catostomus commersonii*)	Yes		Yes		x	x	x	x	x	x	x	x	x	x	x	x	x	x	x	x
Blue sucker (*Cycleptus elongatus*)	Yes	SGCN	Yes	SCP1	x		x													
Smallmouth buffalo (*Ictiobus bubalus*)	Yes		Yes		x			x	x								x	x	x	
Bigmouth buffalo (*Ictiobus cyprinellus*)	Yes		Yes		x		x	x									x	x	x	
Black buffalo (*Ictiobus niger*)	Yes		Yes		x												x			x

Species		Status		SCP														
Silver redhorse (*Moxostoma anisurum*)	Yes		Yes												x	x	x	x
River redhorse (*Moxostoma carinatum*)	Yes		NA			x												
Golden redhorse (*Moxostoma erythrurum*)	Yes		Yes											x	x	x	x	x
Shorthead redhorse (*Moxostoma macrolepidotum*)	Yes		Yes			x	x	x	x	x	x	x	x	x	x	x	x	x
Greater redhorse (*Moxostoma valenciennesi*)	Yes		Yes												x	x		x
Plains sucker (*Pantosteus jordani*)	Yes	SGCN	Yes						x									
Cyprinidae: Barb and carp																		
Goldfish (*Carassius auratus*)	No		NA			x				x		x		x	x			x
Common carp (*Cyprinus carpio*)	No		No			x	x	x	x	x	x	x	x	x	x	x	x	x
Xenocyprididae: Sharpbelly																		
Grass carp (*Ctenopharyngodon idella*)	No		No			x									x			
Silver carp (*Hypophthalmichthys molitrix*)	No		No			x									x	x		
Bighead carp (*Hypophthalmichthys nobilis*)	No		NA			x									x	x		
Leuciscidae: Minnow																		
River shiner (*Alburnops blennius*)	Yes		Yes			x	x				x				x	x	x	x
Central stoneroller (*Campostoma anomalum*)	Yes		NA			x	x			x	x	x	x	x	x	x	x	x
Largescale stoneroller (*Campostoma oligolepis*)	NA	ST, SGCN	Yes	SCP3									x	x				
Northern redbelly dace (*Chrosomus eos*)	Yes	SGCN	Yes	SCP2		x	x	x	x	x	x	x	x	x	x	x	x	x
Southern redbelly dace (*Chrosomus erythrogaster*)	Yes	SE, SGCN	NA												x	x		
Finescale dace (*Chrosomus neogaeus*)	Yes	SGCN	Yes	SCP3		x	x		x		x	x			x			x
Lake chub (*Couesius plumbeus*)	Yes	SGCN	Yes			x		x	x		x				x			
Red shiner (*Cyprinella lutrensis*)	Yes		No			x	x	x	x	x	x	x	x	x	x	x	x	x
Spotfin shiner (*Cyprinella spiloptera*)	No		Yes			x									x	x	x	
Bigmouth shiner (*Ericymba dorsalis*)	Yes		Yes			x				x	x		x	x	x	x	x	x
Spottail shiner (*Hudsonius hudsonius*)	Yes		Yes			x									x	x		x
Western silvery minnow (*Hybognathus argyritis*)	Yes		Yes			x	x	x	x	x	x	x			x	x		x
Brassy minnow (*Hybognathus hankinsoni*)	Yes		Yes			x	x	x	x	x	x	x	x	x	x	x	x	x
Plains minnow (*Hybognathus placitus*)	Yes		Yes			x	x	x	x	x	x	x			x			x
Common shiner (*Luxilus cornutus*)	Yes		Yes			x	x	x	x	x	x	x			x	x	x	x
Sturgeon chub (*Macrhybopsis gelida*)	Yes	C, ST, SGCN	Yes	SCP1		x	x	x	x	x	x	x			x	x	x	x
Shoal chub (*Macrhybopsis hyostoma*)	Yes		NA			x									x			
Sicklefin chub (*Macrhybopsis meeki*)	Yes	C, SE, SGCN	Yes	SCP1		x				x					x			
Silver chub (*Macrhybopsis storeriana*)	Yes		Yes	SCP2		x							x				x	x
Northern pearl dace (*Margariscus nachtriebi*)	Yes	ST, SGCN	Yes	SCP1		x	x	x	x	x	x	x	x	x	x	x	x	x
Pugnose shiner (*Miniellus anogenus*)	NA		Yes	SCP3		x							x		x			
Sand shiner (*Miniellus stramineus*)	Yes		Yes			x	x	x	x	x	x	x	x	x	x	x	x	x
Topeka shiner (*Miniellus topeka*)	Yes	FE, SGCN	NA												x	x		x
Hornyhead chub (*Nocomis biguttatus*)	Yes	SGCN	Yes	SCP3		x												x
Golden shiner (*Notemigonus crysoleucas*)	Yes		Yes			x	x	x	x	x	x	x	x	x	x	x	x	x
Emerald shiner (*Notropis atherinoides*)	Yes		Yes			x	x	x	x	x	x	x	x	x	x	x	x	x

(continued)

Table 3.1 (continued)

	SD native	SD status	ND native	ND status	Missouri	Little Missouri	Knife	Heart	Cannonball	Grand	Moreau	Cheyenne	Bad	White	Niobrara	Ponca Creek	James	Vermillion	Big Sioux	Minnesota
																				(River)
Blacknose shiner (*Notropis heterolepis*)	Yes	SE, SGCN	Yes	SCP3	x										x		x		x	x
Carmine shiner (*Notropis percobromus*)	Yes	SGCN	Yes	SCP3															x	x
Mimic shiner (*Notropis volucellus*)	NA		Yes		x												x			
Silverband shiner (*Paranotropis shumardi*)	Yes		NA		x									x			x		x	
Suckermouth minnow (*Phenacobius mirabilis*)	Yes		NA		x										x				x	x
Bluntnose minnow (*Pimephales notatus*)	Yes		Yes		x			x				x					x	x	x	x
Fathead minnow (*Pimephales promelas*)	Yes		Yes		x	x	x	x	x	x	x	x	x	x	x	x	x	x	x	x
Flathead chub (*Platygobio gracilis*)	Yes	SGCN	Yes	SCP2	x	x	x	x	x	x	x	x	x	x	x		x	x	x	x
Longnose dace (*Rhinichthys cataractae*)	Yes		Yes		x	x	x	x	x	x	x	x	x	x	x		x	x	x	x
Western blacknose dace (*Rhinichthys obtusus*)	Yes		Yes		x							x		x	x		x	x	x	x
Rudd (*Scardinius erythrophthalmus*)	No		NA		x															
Creek chub (*Semotilus atromaculatus*)	Yes		Yes		x	x	x	x	x	x	x	x	x	x	x	x	x	x	x	x
Ictaluridae: Catfish																				
Black bullhead (*Ameiurus melas*)	Yes		Yes		x	x	x	x	x	x	x	x	x	x	x	x	x	x	x	x
Yellow bullhead (*Ameiurus natalis*)	Yes		Yes	SCP3	x				x	x	x	x	x	x	x	x	x	x	x	x
Brown bullhead (*Ameiurus nebulosus*)	Yes		Yes		x						x	x					x	x		x
Blue catfish (*Ictalurus furcatus*)	Yes	SGCN	NA		x												x			
Channel catfish (*Ictalurus punctatus*)	Yes		Yes		x	x	x	x	x	x	x	x	x	x	x		x	x	x	x
Stonecat (*Noturus flavus*)	Yes		Yes		x	x	x	x	x	x	x	x	x	x	x		x	x	x	x
Tadpole madtom (*Noturus gyrinus*)	Yes		Yes		x	x	x	x	x	x							x	x	x	x
Flathead catfish (*Pylodictis olivaris*)	Yes		No		x	x	x	x	x	x							x	x	x	x
Esocidae: Pike and mudminnow																				
Redfin pickerel (*Esox americanus*)	Yes		NA		x															
Northern pike (*Esox lucius*)	Yes		Yes		x	x	x	x	x	x	x	x	x	x	x		x	x	x	x
Muskellunge (*Esox masquinongy*)	No		No		x					x	x	x							x	
Central mudminnow (*Umbra limi*)	Yes	SGCN	Yes		x												x		x	x
Salmonidae: Salmon, trout, and char																				
Cisco (*Coregonus artedi*)	No		No		x															x
Lake whitefish (*Coregonus clupeaformis*)	No		No		x															
Coastal cutthroat trout (*Oncorhynchus clarkii*)	No		No		x	x	x	x	x	x	x	x	x	x			x		x	x
Rainbow trout (*Oncorhynchus mykiss*)	No		No		x	x	x	x	x	x	x	x	x	x	x			x	x	x
Kokanee salmon (*Oncorhynchus nerka*)	No		No		x							x								
Chinook salmon (*Oncorhynchus tshawytscha*)	No		No		x												x			
Atlantic salmon (*Salmo salar*)	No		No		x															
Brown trout (*Salmo trutta*)	No		No		x	x	x	x	x	x	x	x	x	x	x		x	x	x	x

Common name (Scientific name)	Status	Native	SCP														
Brook trout (*Salvelinus fontinalis*)		No				x					x	x	x	x	x	x	x
Lake trout (*Salvelinus namaycush*)		No				x					x						
Osmeridae: Smelt																	
Rainbow smelt (*Osmerus mordax*)		No			x	x			x								
Percopsidae: Trout-perch																	
Trout-perch (*Percopsis omiscomaycus*)	SGCN	Yes	SCP2									x	x	x	x	x	x
Gadidae: Cod																	
Burbot (*Lota lota*)	SGCN	Yes	SCP2		x	x			x		x	x	x	x	x		
Cichlidae: Cichlid																	
Jack dempsey (*Rocio octofasciata*)		No						x									
Fundulidae: Topminnow and killifish																	
Banded killifish (*Fundulus diaphanus*)	SE, SGCN	Yes			x											x	x
Northern plains killifish (*Fundulus kansae*)		NA			x			x									
Plains topminnow (*Fundulus sciadicus*)	SGCN	Yes			x	x		x	x	x	x	x	x	x	x		
Centrarchidae: Sunfish																	
Rock bass (*Ambloplites rupestris*)		Yes			x		x		x	x	x	x	x	x	x	x	x
Green sunfish (*Lepomis cyanellus*)		No			x	x	x	x	x	x	x	x	x	x	x	x	x
Pumpkinseed (*Lepomis gibbosus*)		Yes			x	x	x	x	x	x	x	x	x	x	x		
Orangespotted sunfish (*Lepomis humilis*)		Yes			x		x	x	x	x	x	x	x	x	x	x	x
Bluegill (*Lepomis macrochirus*)		Yes			x	x	x	x	x	x	x	x	x	x	x	x	x
Redear sunfish (*Lepomis microlophus*)		No			x			x				x				x	
Smallmouth bass (*Micropterus dolomieu*)		No			x	x	x		x	x	x	x	x	x	x	x	x
Largemouth bass (*Micropterus nigricans*)		Yes			x	x	x	x	x	x	x	x	x	x	x	x	x
White crappie (*Pomoxis annularis*)		Yes			x	x	x	x	x	x	x	x	x	x	x	x	x
Black crappie (*Pomoxis nigromaculatus*)		Yes			x	x	x	x	x	x	x	x	x	x	x	x	x
Moronidae: Temperate bass																	
White bass (*Morone chrysops*)		No			x	x	x	x	x	x	x	x	x	x	x	x	x
Yellow bass (*Morone mississippiensis*)		NA			x												x
Striped bass (*Morone saxatilis*)		NA			x												
Percidae: Perch and darter																	
Iowa darter (*Etheostoma exile*)		Yes			x	x	x		x	x	x	x	x	x	x	x	x
Johnny darter (*Etheostoma nigrum*)		Yes			x	x	x		x		x	x	x	x	x	x	x
Yellow perch (*Perca flavescens*)		Yes			x	x	x		x	x	x	x	x	x	x	x	x
Logperch (*Percina caprodes*)	SGCN	Yes	SCP3									x			x		
Blackside darter (*Percina maculata*)	SGCN	Yes				x	x								x	x	x
Slenderhead darter (*Percina phoxocephala*)		Yes															x
River darter (*Percina shumardi*)		NA	SCP3														
Sauger (*Sander canadensis*)	SGCN	Yes			x	x	x		x	x	x	x	x	x	x	x	x
Zander (*Sander lucioperca*)		NA											x				
Walleye (*Sander vitreus*)		Yes			x	x	x		x	x	x	x	x	x	x	x	x

(continued)

Table 3.1 (continued)

	SD		ND		River															
	native	status	native	status	Missouri	Little Missouri	Knife	Heart	Cannonball	Grand	Moreau	Cheyenne	Bad	White	Niobrara	Ponca Creek	James	Vermillion	Big Sioux	Minnesota
Sciaenidae: Drum																				
Freshwater drum (*Aplodinotus grunniens*)	Yes		Yes		×	×	×	×	×	×	×	×	×	×	×		×	×	×	×
Gasterosteidae: Stickleback																				
Brook stickleback (*Culaea inconstans*)	Yes		Yes		×	×	×	×	×	×	×	×	×	×	×		×	×	×	×

Status as a native species is denoted by Yes, introduced denoted by No, and NA for not applicable (meaning that the species distribution does not occur within the state). Distribution across the 16 river systems is denoted by X within the Missouri and Upper Mississippi River basins. Native species status by state is denoted as endangered, threatened, or rare as follows:

Key

FE = Federal endangered species: species in danger of extinction throughout all or a significant part of its range

FT = Federal threatened species: species likely to become an endangered species throughout all or a significant part of its range

C = Federal candidate species: species for which the US Fish and Wildlife Service has sufficient information to propose them as endangered or threatened but for which listed is precluded by other higher-priority listing activities

SE = State endangered species: any species of wildlife that is in danger of extinction throughout all or a significant part of its range

ST = State threatened species: any species that is likely to become an endangered species within the foreseeable future throughout all or a significant portion of its range

SGCN = Species of Greatest Conservation Need are animal species identified as the name indicates as the successful implementation of the SD Wildlife Action Plan. The list includes species that are state- or federal-listed or under consideration for federal listing as threated or endangered species; species for which SD represents an important part of the remaining species' range; and/or a species with characteristics that may make it vulnerable

SCP = Species of Conservation Priority are animal species identified as the name indicates in ND to help assess the successful implementation of the ND Wildlife Action Plan. Species of conservation priority were categorized into three levels according to their conservation need (e.g., SCP1, SCP2, and SCP3). Conservation priority levels are as follows:

*Level 1: These are species that are in decline and receive little or no monetary support or conservation efforts. The NDGF has a clear obligation to use State Wildlife Grant (SWG) funding to implement conservation actions that directly benefit these species. Level I species are those having a high level of conservation priority because of their decline status either here or across their range or a high rate of occurrence in ND constituting the core of the species' breeding range but are at-risk range-wide

*Level 2: The NDGF uses SWG funding to implement conservation actions to benefit these species if SWG funding for level I species is sufficient or conservation needs have been met. Level II species are those having a moderate level of conservation priority or a high level of conservation priority but a substantial level of non-SWG funding is available to them. Federally threatened and endangered species are assigned the level II category because other non-SWG funding is available, such as the Cooperative Endangered Species Conservation Fund

*Level 3: These are ND species having a moderate level of conservation priority but are believed to be peripheral or nonbreeding in ND

Table 3.2 ND and SD native and introduced fishes in phylogenetic order and listed by family and both common and scientific names

	SD		ND		River											
	native	status	native	status	Red River of the North	Souris	Pembina	Park	Forest	Turtle	Goose	Elm	Sheyenne	Devils Lake	Wild Rice	Bois de Sioux
Petromyzontidae: Lamprey																
Chestnut lamprey (*Ichthyomyzon castaneus*)	NA		Yes	SCP3	x						x	x				x
Silver lamprey (*Ichthyomyzon unicuspis*)	Yes		Yes	SCP3	x							x				x
Acipenseridae: Sturgeon																
Lake sturgeon (*Acipenser fulvescens*)	Yes	SGCN	Yes		x											
Pallid sturgeon (*Scaphirhynchus albus*)	Yes	FE, SE, SGCN	Yes	FE, SE, SCP2												
Shovelnose sturgeon (*Scaphirhynchus platorynchus*)	Yes	FT, SGCN	Yes	FT												
Polyodontidae: Paddlefish																
Paddlefish (*Polyodon spathula*)	Yes	SGCN	Yes	SCP2												
Lepisosteidae: Gar																
Longnose gar (*Lepisosteus osseus*)	Yes		NA													
Shortnose gar (*Lepisosteus platostomus*)	Yes		Yes													
Amiidae: Bowfin																
Emerald bowfin (*Amia ocellicauda*)	NA		Yes												x	
Anguillidae: Eel																
American eel (*Anguilla rostrata*)	Yes		NA													
Hiodontidae: Mooneye																
Goldeye (*Hiodon alosoides*)	Yes		Yes		x		x	x	x	x	x	x	x		x	x
Mooneye (*Hiodon tergisus*)	Yes		Yes		x				x			x	x		x	x
Alosidae: Shad																
Skipjack herring (*Alosa chrysochloris*)	Yes		NA													
Alewife (*Alosa pseudoharengus*)	No		NA													
Dorosomatidae: Thread herring																
Gizzard shad (*Dorosoma cepedianum*)	Yes		Yes													
Threadfin shad (*Dorosoma petenense*)	No		NA													
Catostomidae: Sucker																
River carpsucker (*Carpiodes carpio*)	Yes		Yes					x	x	x	x	x	x		x	x
Quillback (*Carpiodes cyprinus*)	Yes		Yes		x		x	x	x	x	x	x	x		x	x
Highfin carpsucker (*Carpiodes velifer*)	Yes		NA													
Longnose sucker (*Catostomus catostomus*)	Yes	ST, SGCN	Yes													
White sucker (*Catostomus commersonii*)	Yes		Yes		x		x	x	x	x	x	x	x	x	x	x
Blue sucker (*Cycleptus elongatus*)	Yes	SGCN	Yes	SCP1			x	x	x	x			x			
Smallmouth buffalo (*Ictiobus bubalus*)	Yes		Yes		x	x	x	x	x				x		x	

(continued)

Table 3.2 (continued)

	SD native	SD status	ND native	ND status	Red River of the North	Souris	Pembina	Park	Forest	Turtle	Goose	Elm	Sheyenne	Devils Lake	Wild Rice	Bois de Sioux
Bigmouth buffalo (*Ictiobus cyprinellus*)	Yes		Yes		x		x	x	x	x		x	x		x	x
Black buffalo (*Ictiobus niger*)	Yes		Yes					x	x			x	x			x
Silver redhorse (*Moxostoma anisurum*)	Yes		Yes		x					x	x	x	x			x
River redhorse (*Moxostoma carinatum*)	Yes		NA													
Golden redhorse (*Moxostoma erythrurum*)	Yes		Yes		x		x	x	x	x	x	x	x			x
Shorthead redhorse (*Moxostoma macrolepidotum*)	Yes		Yes		x		x	x	x	x	x	x	x		x	x
Greater redhorse (*Moxostoma valenciennesi*)	Yes		Yes		x				x			x	x			
Plains sucker (*Pantosteus jordani*)	Yes	SGCN	Yes													
Cyprinidae: Barb and minnow																
Goldfish (*Carassius auratus*)	No		NA													
Common carp (*Cyprinus carpio*)	No		No		x		x	x	x	x	x	x	x		x	x
Xenocyprididae: Sharpbelly																
Grass carp (*Ctenopharyngodon idella*)	No		No		x											
Silver carp (*Hypophthalmichthys molitrix*)	No		No													
Bighead carp (*Hypophthalmichthys nobilis*)	No		NA													
Leuciscidae: Minnow																
River shiner (*Alburnops blennius*)	Yes		Yes		x		x	x			x	x	x			
Central stoneroller (*Campostoma anomalum*)	Yes		NA													
Largescale stoneroller (*Campostoma oligolepis*)	NA	ST, SGCN	Yes	SCP3				x	x				x			
Northern redbelly dace (*Chrosomus eos*)	Yes	SGCN	Yes	SCP2		x	x	x		x	x		x		x	x
Southern redbelly dace (*Chrosomus erythrogaster*)	NA	SE, SGCN	NA													
Finescale dace (*Chrosomus neogaeus*)	Yes	SGCN	Yes	SCP3		x						x				
Lake chub (*Couesius plumbeus*)	Yes	SGCN	Yes				x									
Red shiner (*Cyprinella lutrensis*)	Yes		No													
Spotfin shiner (*Cyprinella spiloptera*)	No		Yes		x	x	x	x	x	x	x	x	x		x	x
Bigmouth shiner (*Ericymba dorsalis*)	Yes		Yes		x	x	x	x	x	x	x	x	x		x	x
Spottail shiner (*Hudsonius hudsonius*)	Yes		Yes		x	x		x	x	x		x	x		x	
Western silvery minnow (*Hybognathus argyritis*)	Yes		Yes													
Brassy minnow (*Hybognathus hankinsoni*)	Yes		Yes		x	x	x					x	x		x	x
Plains minnow (*Hybognathus placitus*)	Yes		Yes													
Common shiner (*Luxilus cornutus*)	Yes		Yes		x	x	x	x	x	x	x	x	x		x	x
Sturgeon chub (*Macrhybopsis gelida*)	Yes	C, ST, SGCN	Yes	SCP1												
Shoal chub (*Macrhybopsis hyostoma*)	Yes		NA													
Sicklefin chub (*Macrhybopsis meeki*)	Yes	C, SE, SGCN	Yes	SCP1												

Species				
Silver chub (*Macrhybopsis storeriana*)	Yes		Yes	SCP2
Northern pearl dace (*Margariscus nachtriebi*)	Yes	ST, SGCN	Yes	SCP1
Pugnose shiner (*Miniellus anogenus*)	NA		Yes	SCP3
Sand shiner (*Miniellus stramineus*)	Yes		Yes	
Topeka shiner (*Miniellus topeka*)	Yes	FE, SGCN	NA	
Hornyhead chub (*Nocomis biguttatus*)	Yes	SGCN	Yes	SCP3
Golden shiner (*Notemigonus crysoleucas*)	Yes		Yes	
Emerald shiner (*Notropis atherinoides*)	Yes		Yes	
Blacknose shiner (*Notropis heterolepis*)	Yes	SE, SGCN	Yes	SCP3
Carmine shiner (*Notropis percobromus*)	Yes	SGCN	Yes	SCP3
Mimic shiner (*Notropis volucellus*)	NA		Yes	
Silverband shiner (*Paranotropis shumardi*)	Yes		NA	
Suckermouth minnow (*Phenacobius mirabilis*)	Yes		NA	
Bluntnose minnow (*Pimephales notatus*)	Yes		Yes	
Fathead minnow (*Pimephales promelas*)	Yes		Yes	
Flathead chub (*Platygobio gracilis*)	Yes	SGCN	Yes	SCP2
Longnose dace (*Rhinichthys cataractae*)	Yes		Yes	
Western blacknose dace (*Rhinichthys obtusus*)	Yes		Yes	
Rudd (*Scardinius erythrophthalmus*)	No		NA	
Creek chub (*Semotilus atromaculatus*)	Yes		Yes	
Ictaluridae: Catfish				
Black bullhead (*Ameiurus melas*)	Yes		Yes	
Yellow bullhead (*Ameiurus natalis*)	Yes		Yes	SCP3
Brown bullhead (*Ameiurus nebulosus*)	Yes		Yes	
Blue catfish (*Ictalurus furcatus*)	Yes	SGCN	NA	
Channel catfish (*Ictalurus punctatus*)	Yes		Yes	
Stonecat (*Noturus flavus*)	Yes		Yes	
Tadpole madtom (*Noturus gyrinus*)	Yes		Yes	
Flathead catfish (*Pylodictis olivaris*)	No		Yes	
Esocidae: Pike and mudminnow				
Redfin pickerel (*Esox americanus*)	Yes		NA	
Northern pike (*Esox lucius*)	Yes		Yes	
Muskellunge (*Esox masquinongy*)	No		No	
Central mudminnow (*Umbra limi*)	Yes	SGCN	Yes	
Salmonidae: Salmon, trout, and char				
Cisco (*Coregonus artedi*)	No		No	
Lake whitefish (*Coregonus clupeaformis*)	No		No	
Coastal cutthroat trout (*Oncorhynchus clarkii*)	No		No	
Rainbow trout (*Oncorhynchus mykiss*)	No		No	
Kokanee salmon (*Oncorhynchus nerka*)	No		No	

(continued)

Table 3.2 (continued)

	SD		ND		River											
	native	status	native	status	Red River of the North	Souris	Pembina	Park	Forest	Turtle	Goose	Elm	Sheyenne	Devils Lake	Wild Rice	Bois de Sioux
Chinook salmon (*Oncorhynchus tshawytscha*)	No		No													
Atlantic salmon (*Salmo salar*)	No		No													
Brown trout (*Salmo trutta*)	No		No		x		x					x	x		x	
Brook trout (*Salvelinus fontinalis*)	No		No										x			
Lake trout (*Salvelinus namaycush*)	No		No													
Osmeridae: Smelt																
Rainbow smelt (*Osmerus mordax*)	No		No													
Percopsidae: Trout-perch																
Trout-perch (*Percopsis omiscomaycus*)	Yes	SGCN	Yes	SCP2	x	x	x	x			x	x	x		x	
Gadidae: Cod																
Burbot (*Lota lota*)	Yes	SGCN	Yes	SCP2	x		x		x			x	x			
Cichildae: Cichlid																
Jack dempsey (*Rocio octofasciata*)	No		NA													
Fundulidae: Topminnow and killifish																
Banded killifish (*Fundulus diaphanus*)	Yes	SE, SGCN	Yes					x		x		x	x			
Northern plains killifish (*Fundulus kansae*)	Yes		NA													
Plains topminnow (*Fundulus sciadicus*)	Yes	SGCN	NA													
Centrarchidae: Sunfish																
Rock bass (*Ambloplites rupestris*)	Yes		Yes		x		x	x	x	x	x	x	x		x	x
Green sunfish (*Lepomis cyanellus*)	Yes		No		x	x				x			x		x	x
Pumpkinseed (*Lepomis gibbosus*)	Yes		Yes		x	x						x	x		x	x
Orangespotted sunfish (*Lepomis humilis*)	Yes		Yes		x				x			x	x		x	x
Bluegill (*Lepomis macrochirus*)	Yes		No		x	x	x	x	x	x	x	x	x		x	x
Redear sunfish (*Lepomis microlophus*)	No		NA													
Smallmouth bass (*Micropterus dolomieu*)	Yes		No		x	x	x	x	x	x	x	x	x	x	x	x
Largemouth bass (*Micropterus nigricans*)	Yes		Yes		x	x	x	x	x	x	x	x	x	x	x	x
White crappie (*Pomoxis annularis*)	Yes		Yes		x		x	x			x	x	x		x	x
Black crappie (*Pomoxis nigromaculatus*)	Yes		Yes		x	x	x	x	x	x	x		x	x	x	x
Moronidae: Temperate bass																
White bass (*Morone chrysops*)	Yes		No		x		x	x		x	x		x	x	x	x
Yellow bass (*Morone mississippiensis*)	No		NA										x			
Striped bass (*Morone saxatilis*)	NA		No													

Species	Native	Introduced	SGCN	SCP	1	2	3	4	5	6	7	8	9	10	11	12
Percidae: Perch and darter																
Iowa darter (*Etheostoma exile*)	Yes				X	X	X	X	X	X	X	X		X	X	X
Johnny darter (*Etheostoma nigrum*)	Yes				X	X	X	X	X	X	X	X		X	X	X
Yellow perch (*Perca flavescens*)	Yes				X	X	X	X	X	X	X	X	X	X	X	X
Logperch (*Percina caprodes*)	Yes		SGCN	SCP3	X	X	X	X	X	X	X	X				
Blackside darter (*Percina maculata*)	Yes		SGCN		X	X	X	X	X	X	X	X		X	X	X
Slenderhead darter (*Percina phoxocephala*)	Yes				X											
River darter (*Percina shumardi*)	NA			SCP3												
Sauger (*Sander canadensis*)	Yes		SGCN		X	X	X	X	X	X	X	X		X	X	X
Zander (*Sander lucioperca*)	NA	No														
Walleye (*Sander vitreus*)	Yes				X	X	X	X	X	X	X	X	X	X	X	X
Sciaenidae: Drum																
Freshwater drum (*Aplodinotus grunniens*)	Yes				X	X	X	X	X	X	X	X		X	X	X
Gasterosteidae: Stickleback																
Brook stickleback (*Culaea inconstans*)	Yes				X	X	X	X	X	X	X	X	X	X	X	X

Status as a native species is denoted by Yes, introduced denoted by No, and NA for not applicable (meaning that the species distribution does not occur within the state). Distribution across the 12 river systems is denoted by X within the Red River of the North basin. Native species status by state is denoted as endangered, threatened, or rare as follows:

Key

FE = Federal endangered species: species in danger of extinction throughout all or a significant part of its range

FT = Federal threatened species: species likely to become an endangered species throughout all or a significant part of its range

C = Federal candidate species: species for which the US Fish and Wildlife Service has sufficient information to propose them as endangered or threatened, but for which listed is precluded by other higher-priority listing activities

SE = State endangered species: any species of wildlife that is in danger of extinction throughout all or a significant part of its range

ST = State threatened species: any species that is likely to become an endangered species within the foreseeable future throughout all or a significant portion of its range

SGCN = Species of Greatest Conservation Need are animal species identified as the name indicates to help assess the successful implementation of the SD Wildlife Action Plan. The list includes species that are state- or federal-listed or under consideration for federal listing as threated or endangered species; species for which SD represents an important part of the remaining species' range; and/or a species with characteristics that may make it vulnerable

SCP = Species of Conservation Priority are animal species identified as the name indicates in ND to help assess the successful implementation of the ND Wildlife Action Plan. Species of conservation priority were categorized into three levels according to their conservation need (e.g., SCP1, SCP2, and SCP3). Conservation priority levels are as follows:

*Level I: These are species that are in decline and receive little or no monetary support or conservation efforts. The NDGF has a clear obligation to use State Wildlife Grant (SWG) funding to implement conservation actions that directly benefit these species. Level I species are those having a high level of conservation priority because of their decline status either here or across their range or a high rate of occurrence in ND constituting the core of the species breeding range but are at-risk range-wide

*Level II: The NDGF will use SWG funding to implement conservation actions to benefit these species if SWG funding for level I species is sufficient or conservation needs have been met. Level II species are those having a moderate level of conservation priority or a high level of conservation priority but a substantial level of non-SWG funding is available to them. Federally threatened and endangered species are assigned the level II category because other non-SWG funding is available, such as the Cooperative Endangered Species Conservation Fund

*Level III: These are ND species having a moderate level of conservation priority but are believed to be peripheral or nonbreeding in ND

3.2 Unsubstantiated Species

Some species have been previously reported as present in ND or SD but lack definitive records of ever having been collected, sampled, or occurring in the Dakotas. These unsubstantiated species are not included in the dichotomous key, species accounts, or distribution tables, except for the mimic shiner. These species include, in chronological order:

1. Banded darter (*Etheostoma zonale*) (Meek 1892; Hoagstrom et al. 2007)
2. Northern hogsucker (*Hypentelium nigricans*) (Churchill and Over 1933; Bailey and Allum 1962; NDGF 1994; Hoagstrom et al. 2007)
3. Slender madtom (*Noturus exilis*) (NDGF 1994)
4. Mississippi silvery minnow (*Hybognathus nuchalis*) (NDGF 1994)
5. Blackchin shiner (*Miniellus heterodon*) (NDGF 1994; Hoagstrom et al. 2007)
6. Mottled sculpin (*Cottus bairdii*) (Smith and Brown 2002; Hoagstrom et al. 2007)
7. Rainbow darter (*Etheostoma caeruleum*) (Hoagstrom et al. 2007)
8. Striped fantail darter (*Etheostoma flabellare lineolatum*) (Hoagstrom et al. 2007)
9. Least darter (*Etheostoma microperca*) (Hoagstrom et al. 2007)
10. Mimic shiner (*Paranotropis volucellus*) (Hoagstrom et al. 2007)

Seth E. Meek collected fishes from the Big Sioux River at Sioux Falls, SD, and listed the banded darter in the *Report Upon the Fishes of Iowa, Based Upon Observations and Collections Made During 1889, 1890, and 1891*. However, no vouchers exist for this species and no records of the species exist within the Missouri River drainage. This is considered a misidentification.

In 1933, Edward P. Churchill and William H. Over listed the northern hogsucker (referred to as the black sucker) in SD in their book *Fishes of South Dakota* (Churchill and Over 1938). Reeve M. Bailey and Marvin O. Allum stated in their 1962 publication *Fishes of South Dakota* that the northern hogsucker only occurs in SD in the Minnesota River system. However, collection records of this species from the Minnesota River system in SD are nonexistent. Bailey and Allum stated that James C. Underhill said he collected the northern hogsucker in the Yellow Bank River near Milbank, SD. He also loaned Bailey and Allum a specimen collected from the same stream across the border in Minnesota (MN). Finally, Bailey and Allum (1962) also state that the northern hogsucker is not present in the Red River of the North drainage and likely enters the Missouri River drainage only in the state of Missouri (MO).

In 1994, a pamphlet created by the The North Dakota Game and Fish Department (NDGF) and the Dakota Chapter of the American Fisheries Society indicated that the Mississippi silvery minnow was found in both ND and SD and that the slender madtom was found in SD. The blackchin shiner was also listed in SD by the pamphlet and by Hoagstrom et al. (2007). However, no vouchers or records exist for any of these species in either state.

Smith and Brown (2002) produced a single record for the mottled sculpin during their study on ichthyoplankton entrainment through Big Bend Dam on the Missouri River. Because the closest known population of the mottled sculpin in the Missouri River drainage is more than 600 miles upstream in western Montana, and because no voucher exists, this single report was likely a misidentification.

No records of the rainbow darter, striped fantail darter, and least darter are known from either ND or SD. However, Hoagstrom et al. (2007) considered these species to be "species or subspecies native to specific river drainages but only known outside the borders of SD and thus representing potential former or future inhabitants." These species have still not been documented within the Dakotas.

Hoagstrom et al. (2007) lists the mimic shiner as a species found in SD. However, no vouchers exist for this species from the state. In 2007 and 2008, the South Dakota Game, Fish, and Parks (SDGFP) reported 10 records of the mimic shiner in the Missouri River below Gavins Point Dam, but these were only field identifications, and no voucher specimens were collected. It is notoriously difficult to identify mimic shiners from other minnow Leuciscidae, as the name implies. Without a voucher specimen, it cannot be recognized as occurring in SD. However, both the mimic shiner and its congener, the channel shiner (*Paranotropis wickliffi*), are found in the mainstem Missouri River downstream in MO. Thus, both species are part of the dichotomous key to aid in identification if vouchers are collected in the future. Regardless of the unvouchered reports from SD, a single mimic shiner specimen was collected and vouchered (UT 44.14073) from the Red River of the North drainage near Fargo, ND, in 2018. Identification of this specimen was confirmed, establishing the presence of the mimic shiner in North Dakota. It is included in the distribution tables and given a full species account.

It should also be noted that state agencies have intentionally stocked numerous fish species in the Dakotas, but populations were not established. These species have vouchers and are included in the dichotomous key but do not have species accounts. These introduced but not established species include the Bonneville cisco (*Prosopium gemmifer*) (Sigler and

Sigler 1987; NDGF 1994; Hoagstrom et al. 2007; Fuller and Neilson 2021; South Dakota State University Fish Collection, SDSTATE 3230) and the coho salmon (*Oncorhynchus kisutch*) (NDGF 1994; Hoagstrom et al. 2007; South Dakota State University Fish Collection, SDSTATE 3215). In addition, a single record exists for the redhead cichlid (*Vieja melanura*), from Gateway Lake, Ellsworth Air Force Base, Pennington County, SD (Nico and Neilson 2021; Florida Museum Fish Collection, UF96525). This specimen was likely an illegal aquarium fish stocking by an individual. It was not stocked by the SDGFP, and no population was established.

Finally, state agencies have intentionally stocked fish species that did not establish populations and for which no specimens exist. The Sacramento perch (*Archoplites interruptus*), stocked in ND in the 1960s, are an example of a failed stocking and therefore are not included in the dichotomous key (McCarraher and Gregory 1970; Fuller 2021).

References

Bailey RM, Allum MO (1962) Fishes of South Dakota, University of Michigan, Ann Arbor. https://doi.org/10.3998/mpub.9690435

Churchill EP, Over WH (1933) Fishes of South Dakota. Brown & Saenger Printers, Sioux Falls

Fuller P (2021) *Archoplites interruptus* (Girard, 1854). US Geological Survey, Nonindigenous Aquatic Species Database, Gainesville. https://nas.er.usgs.gov/queries/FactSheet.aspx?speciesID=374. Accessed 24 Apr 2023

Fuller P, Neilson M (2021) *Prosopium gemmifer* (Snyder, 1919). US Geological Survey, Nonindigenous Aquatic Species Database, Gainesville. https://nas.er.usgs.gov/queries/FactSheet.aspx?SpeciesID=922. Accessed 24 Apr 2023

Hoagstrom CW, Wall SS, Kral JG, Blackwell BG, Berry CR (2007) Zoogeographic patterns and faunal change of South Dakota fishes. West N Am Nat 67:161–184. https://doi.org/10.3398/1527-0904(2007)67[161,ZPAFCO]2.0.CO;2

McCarraher DB, Gregory RW (1970) Adaptability and status of introductions of Sacramento perch, *Archoplites interruptus*, in North America. Trans Am Fish Soc 99:700–707. https://doi.org/10.1577/1548-8659(1970)99<700:AACSOI>2.0.CO;2

Meek SE (1892) Report upon the fishes of Iowa, based upon observations and collections made during 1889, 1890, and 1891, vol 10. US Government Printing Office, Washington, DC

Nico LG, Neilson ME (2021) *Vieja melanura* (Günther, 1862). US Geological Survey, Nonindigenous Aquatic Species Database, Gainesville. https://nas.er.usgs.gov/queries/factsheet.aspx?SpeciesID=451. Accessed 24 Apr 2023

North Dakota Game and Fish (1994) Fishes of the Dakotas. Brochure. North Dakota Game and Fish Department, Bismark

Sigler WF, Sigler JW (1987) Fishes of the Great Basin: a natural history. University of Nevada Press, Reno

Smith KA, Brown ML (2002) Seasonal composition and abundance of ichthyoplankton entrained through Big Bend Dam, South Dakota. J Freshw Ecol 17:199–207. https://doi.org/10.1080/02705060.2002.9663888

Chapter 4
Collection, Preservation, and Identification

Kathryn E. Schlafke, Matthew D. Wagner, and Chelsey A. Pasbrig

4.1 Collection

Individuals should learn what state and federal licenses and permits are required before collecting any fish. Scientific collecting permits for non-federally listed fish species can be obtained in South Dakota (SD) from the South Dakota Game, Fish, and Parks (SDGFP) and in North Dakota (ND) from the North Dakota Game and Fish (NDGF). The US Fish and Wildlife Service issues additional permits for work involving federally protected species. To avoid legal consequences from incidentally taking protected species, fish of unknown identity must not be collected. The practice of capture, photograph, and release (CPR) is commonly practiced as an alternative to collection. When using CPR, it is important to obtain a lateral photograph showing any identifying features.

Fish can be collected using four main gear types: (1) nets (seines, fyke/trap nets, gill nets, minnow traps), (2) electrofishing (backpack, barge, or boat), (3) angling, and (4) ichthyocide. The law concerning gear use varies among states and must be consulted prior to collection.

4.2 Preservation

Collected fish should be preserved in a buffered solution containing a minimum of 10% formalin in a sealed container. Formalin is a potential carcinogen and should be handled carefully. Fish should ideally be preserved while alive but can be anesthetized or euthanized using tricaine methanesulfonate (MS-222) prior to preservation. Fish should not be overcrowded in collection containers and should be kept in buffered formalin for at least 1 week. After formalin preservation, specimens should be thoroughly rinsed and soaked in water for at least 24 hours. Rinsing and soaking should be repeated multiple times until the smell of formalin is reduced, but specimens should not be stored in water for more than week to avoid mold. After soaking, specimens should be transferred to 70% ethanol for long-term storage. On freshly preserved specimens, color (reds, greens, blues, yellows) may still be present. However, these colors will fade with storage in ethanol, leaving only brown and black. Guidelines for the use of fishes in research can be found at https://fisheries.org/docs/wp/Guidelines-for-Use-of-Fishes.pdf.

K. E. Schlafke
Montana Fish, Wildlife & Parks, Bozeman, MT, USA

M. D. Wagner
U.S. Fish and Wildlife Service, Jackson, MS, USA

C. A. Pasbrig (✉)
South Dakota Game, Fish and Parks, Pierre, SD, USA
e-mail: Chelsey.Pasbrig@state.sd.us

© The Author(s), under exclusive license to Springer Nature Switzerland AG 2024
M. Barnes (ed.), *Fishes of the Dakotas*, https://doi.org/10.1007/978-3-031-38040-2_4

4.3 Identification

A dichotomous key is best for fish identification. It uses a binary choice system based on the physical characters of the fish. Each numbered step in a key is called a couplet because it allows the choice between two different characters. The dichotomous key in this book functions by first identifying the fish at the family level. The terminal couplet of the family-level key has the page number of the key to the specific family based on which the fish species can be identified. For example, if a fish has no dorsal spines, no barbels, and no dark stripes, then it will be keyed under *Species D* as in the following example.

Example Key

1. Dorsal fin with a spine..*Species A*
Dorsal fin without a spine..2

2. Barbel present...*Species B*
Barbel absent...3

3. Side with dark stripes..*Species C*
Side without dark stripes...*Species D*

On some specimens, there may be characters that do not perfectly fit into the key. This may be because of the specimen's condition, preservation issues, or user unfamiliarity with the specimen's physical features. For instance, it can be difficult to see the small flap-like barbel in the maxillary groove of a creek chub (*Semotilus atromaculatus*). To alleviate this issue, the key uses multiple characters in a couplet whenever possible and provides illustrations. If there is uncertainty about a character in a couplet, the specimen should be keyed along both potential paths, comparing the final choices to the species accounts for species confirmation.

Knowledge of the locality, river system, or ecoregion of the specimen, along with the range maps in the species accounts, may aid in species identification. However, ranges change for a variety of reasons, and identification should not be based on just the location. Specimens should always be completely keyed to species. Range expansion is always a possibility, as is the discovery of a new species in either state.

4.4 Equipment

Fine forceps, a fine point probe, dissection scissors, and a dissecting microscope are needed to properly use dichotomous keys. The dissection microscope must be able to provide underlighting to aid in the counts of fin ray elements and overlighting to examine all other external characters. Maneuverable gooseneck lights are extremely useful for overlighting different features and fish sizes because they can be positioned at almost any angle.

4.5 Body Plan

Fishes are a diverse group of organisms with a variety of body shapes and fin arrangements. Specimen orientation definitions include anterior (toward the head), posterior (toward the tail), dorsal (toward the back/top), and ventral (toward the belly) (Fig. 4.1).

The fish body generally consists of a head, a trunk, and a tail. The head will have eyes (unless it is a larval lamprey) and a mouth. Both eye location and mouth position and shape vary greatly and are associated with feeding strategy. The mouth in adult lampreys (family Petromyzontidae) is composed of a buccal funnel (disc-shaped sucking mouthpart usually containing keratinized teeth for gripping), and the tongue is used for rasping. The rest of the fishes found in ND and SD have a more typical mouth design with both an upper and lower jaw, but mouth position and shape vary greatly among species. Mouth positions are superior (directed upward), terminal (directed forward), subterminal (snout projecting beyond the upper jaw), or inferior (on the underside of the head) (Fig. 4.2).

The head also includes nares (nostrils used in smell/taste but not respiration), which usually have both an anterior and posterior opening. A respiratory opening exists posterior to the mouth to allow for water to go over the gills to oxygenate the blood. This opening can vary from a gill cleft/pores in lampreys (family Petromyzontidae), a restricted gill opening in the American eel (family Anguillidae), to a large operculum found in most of the other ND and SD fishes. The trunk of a

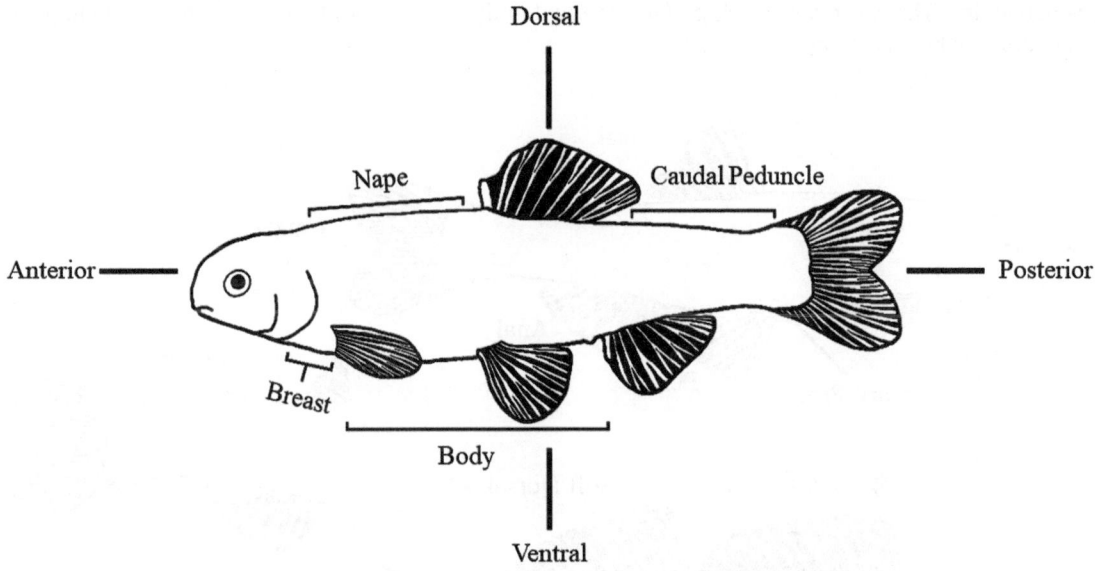

Fig. 4.1 Generalized fish body plan with directions for orientation

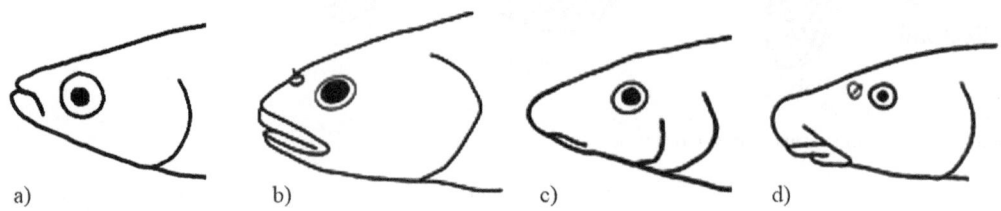

Fig. 4.2 Mouth positions: (**a**) superior, (**b**) terminal, (**c**) subterminal, and (**d**) inferior

fish is the body cavity with the internal organs and muscles and is posterior to the head and anterior to the anus. The tail of the fish is the fleshy part of the body posterior to the anus that includes all the posterior vertebrae. The tail is not to be confused with the caudal fin. Tails typically follow either a symmetrical (homocercal) or asymmetrical (heterocercal) configuration of the fleshy part including the vertebrae. In homocercal tails, the fleshy portion will be rounded or squared with the terminus being at a central point. In heterocercal tails, the fleshy portion will typically taper and terminate dorsally.

4.6 Fins

The structural support of fins is comprised of either spines or rays with membranes. Rays are flexible, segmented, and sometimes branched toward the edge of the fin. They can be found in all fishes in the Dakotas except for lampreys (family Petromyzontidae), which are supported by cartilaginous unsegmented rods. True spines are unbranched, unsegmented, and rigid. In the Dakotas, true spines are found in representatives from the families of Gasterosteidae, Centrarchidae, Cichlidae, Moronidae, Percidae, and Sciaenidae. Although not true spines, a spinous ray derived from fused soft ray elements can be found in the dorsal and pectoral fins of North American catfishes (family Ictaluridae) and in the dorsal fin of species of barbs and carps (family Cyprinidae).

Fish have paired fins and unpaired medial fins. Paired fins, when they exist, are at both the pectoral and pelvic girdles. Depending on the species, the pelvic fins are either in an abdominal (e.g., family Leuciscidae: minnows) or in a thoracic position (e.g., family Percidae). The brook stickleback (*Culaea inconstans*; family Gasterosteidae) has only a rudimentary structure as pelvic fins. The American eel (*Anguilla rostrata*; family Anguillidae) lacks pelvic fins, and lampreys (family Petromyzontidae) lack both pectoral and pelvic fins.

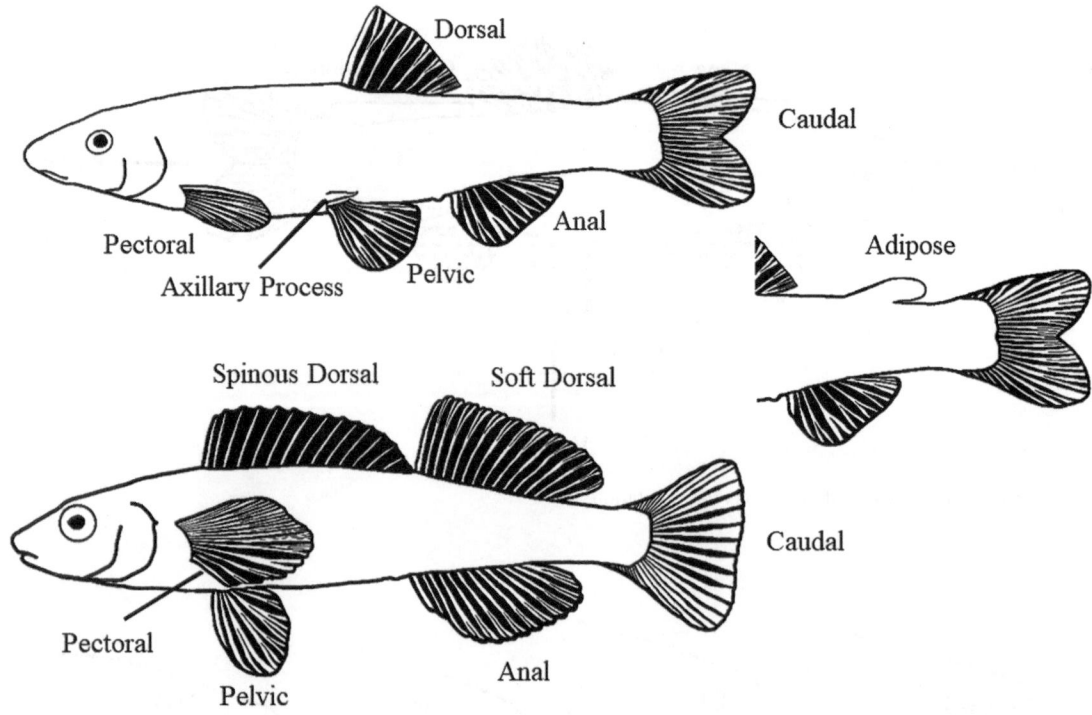

Fig. 4.3 Examples of fins of spinous and non-spinous fishes

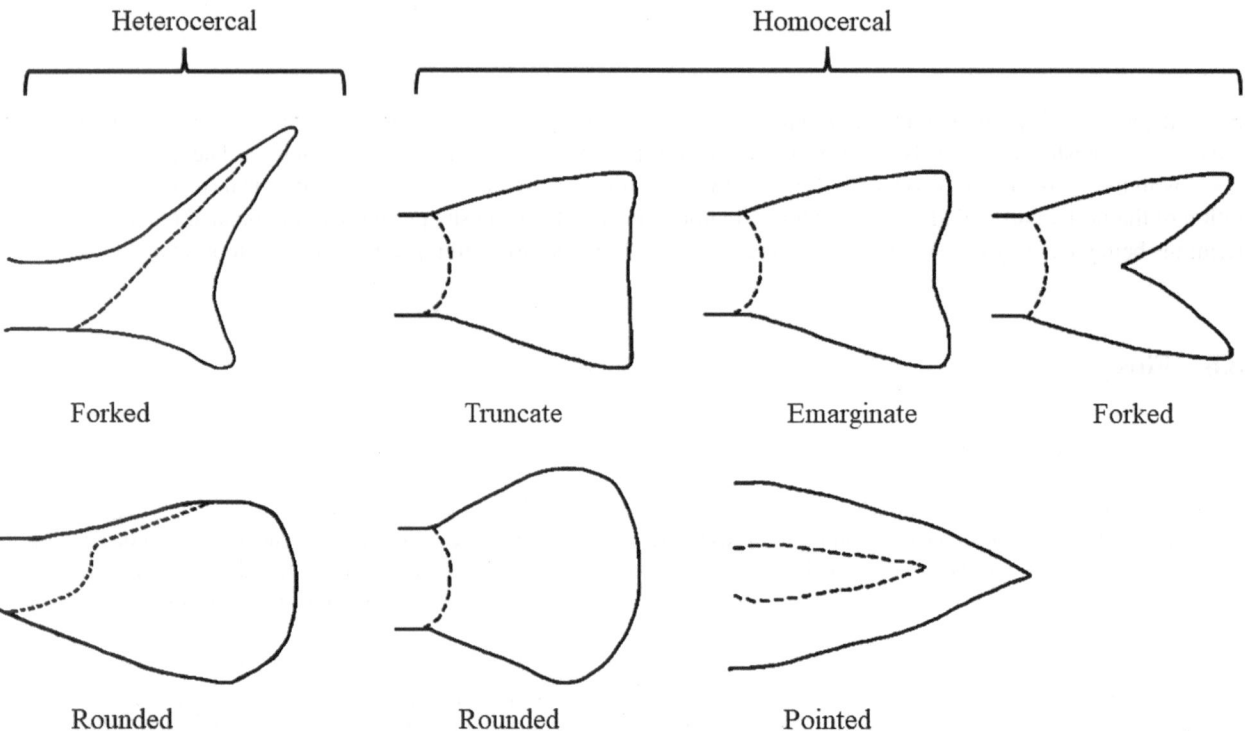

Fig. 4.4 Shapes of heterocercal and homocercal tails

Unpaired median fins include the dorsal fin (on the back), anal fin (on the ventral side posterior to the vent), caudal fin (most posterior fin creating a tail), and, sometimes, an adipose fin (between the dorsal fin and the caudal fin) (Fig. 4.3). Dorsal fins can be composed of combinations of spines and rays and can be one or more fins. In the Dakotas, some species in the order Perciformes have two dorsal fins present, with an anterior spinous dorsal fin and a posterior soft dorsal fin made of rays. A two-dorsal-fin body plan can also be found in lampreys (family Petromyzontidae) and burbot *Lota lota* (family Gadidae), although they lack the true spines found in Perciformes. The brook stickleback (family Gasterosteidae) has a unique dorsal fin plan, with five small dorsal finlets (isolated fins with a spine with its own membrane) followed posteriorly by a soft dorsal fin. Adipose fins hold a similar position to posterior dorsal fins and are fleshy, lacking spines or rays. They can be found in North American catfishes (family Ictaluridae), salmonids (family Salmonidae), rainbow smelt (*Osmerus mordax*; family Osmeridae), and trout-perch (*Percopsis omiscomaycus*; family Percopsidae). The caudal fins of fishes in ND and SD are typically truncate, emarginate, forked, rounded, or pointed (Fig. 4.4).

4.7 Scales

Apart from lampreys (family Petromyzontidae) and North American catfishes (family Ictaluridae), most fishes in the Dakotas have scales.

Ganoid scales are typically extremely thick and rigid, with the exposed portion rhomboidal in shape (Fig. 4.5). They are found on gars (family Lepisosteidae), sturgeons (family Acipenseridae), and paddlefish (*Polyodon spathula*; family Polyodontidae). Ganoid scales on gars cover most of the body as interlocking rhomboidal plates. They occur in small patches on the tail of sturgeons and paddlefish and on the body near the pectoral fin and throat of paddlefish. Sturgeons also have five rows of large sharp bony scutes that run longitudinally along the body.

Bony ridge scales can be found in most bony fishes and are either cycloid or ctenoid. These scales tend to be thin, flexible, and somewhat rounded in shape. Ctenoid scales have small comblike protrusions (ctenii) at the exposed margin of the scale, whereas cycloid scales have a smooth rounded margin. The modified elongated bony ridge scale just above the base of the pelvic fin in mooneyes (family Hiodontidae), shads (family Alosidae), thread herrings (family Dorosomatidae), salmonids (family Salmonidae), and minnows (family Leuciscidae) is referred to as the axillary process (Fig. 4.5).

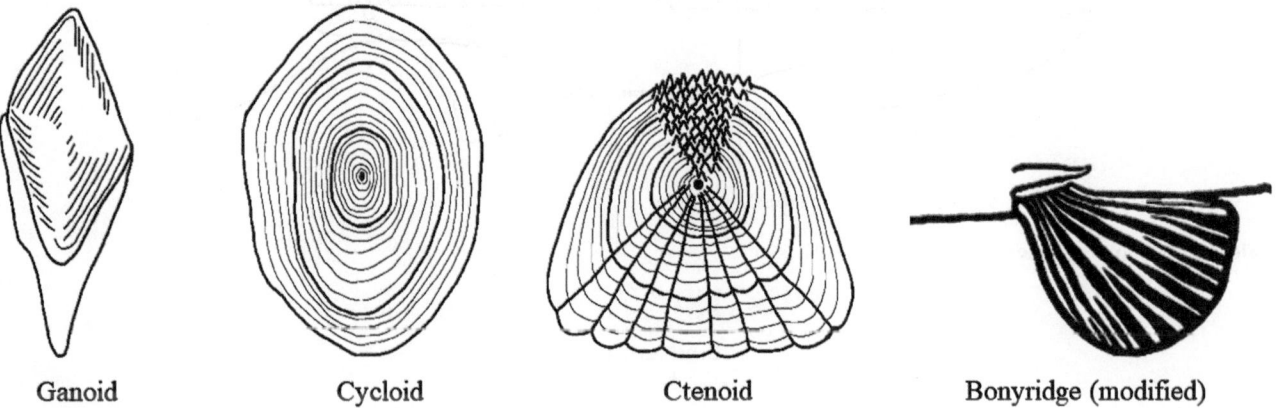

Ganoid   Cycloid   Ctenoid   Bonyridge (modified)

Fig. 4.5 Types of scales

4.8 Counts and Measurements

Counts and measurements of features on specimens are commonly used to identify fishes. All counts and measurements are conventionally made on the left side of the body, when possible. Further details on the counts and measurements can be found in Hubbs and Lagler (1958) and Strauss and Bond (1990).

4.8.1 Counts

The counting of quantitative features of the fish is termed "meristics." In this book, meristics includes scale rows, fin rays, fin spines, gill rakers, pharyngeal teeth, branchiostegal rays, and submandibular pores. Scale counts of circumpenduncular scale rows, lateral scale rows, predorsal scale rows, and scale rows above the lateral line are used in the dichotomous key (Fig. 4.6). Counts of the dorsal, pelvic, and anal fin rays and those of the dorsal and anal spines are also used. The use of underlighting to count these characters makes the fin elements easier to see. All counts of these fin elements are made at the base of the fin close to the body, thereby preventing the double counting of branched rays. For counts of fin ray elements in unpaired fins, such as the dorsal and anal fins, all branched rays and the well-developed leading unbranched ray are counted. The small rays anterior to the well-developed unbranched ray are considered rudimentary or procurrent rays and are not counted. The last two posterior rays are counted as one if they share an origin. For counts of fin ray elements in paired fins, such as the pelvic and pectoral fins, all rays are counted. Unless noted, all spines are counted (Fig. 4.7).

For some species, the key uses the count or shape of the gill rakers on the most anterior arch. The opercle can be bent forward and/or upward in some species to best see the gill rakers. When the rakers are hard to see or a count must be completed, the entire gill arch can be removed. In small specimens, the arch can be excised using forceps to pull the arch where it connects at the base and top. In larger specimens, the arch can be removed by cutting where it connects ventrally and dorsally. In some cases, a ventral and dorsal cut to the gill covering will be necessary to allow access to the arches. This is typically done on the right side of the body to preserve the integrity of the left side for other counts and measurements. All gill rakers are counted, no matter the size (Fig. 4.8).

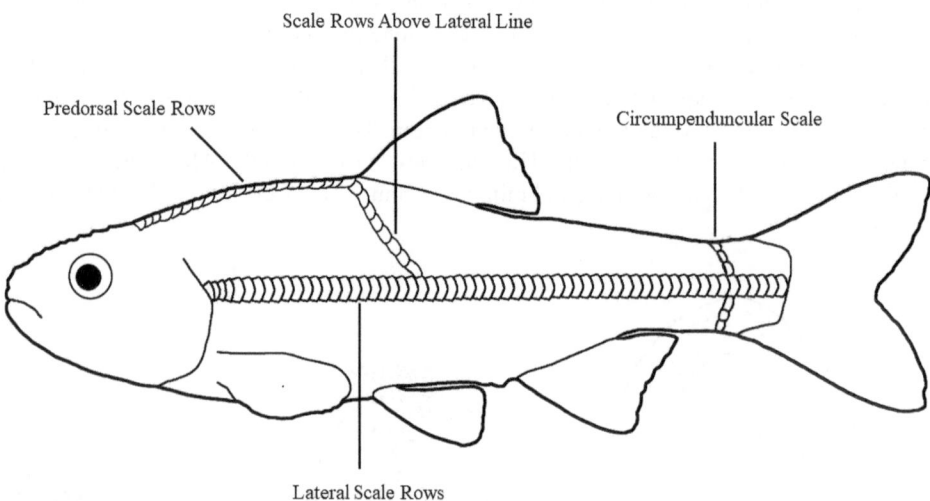

Fig. 4.6 Examples of scale counts

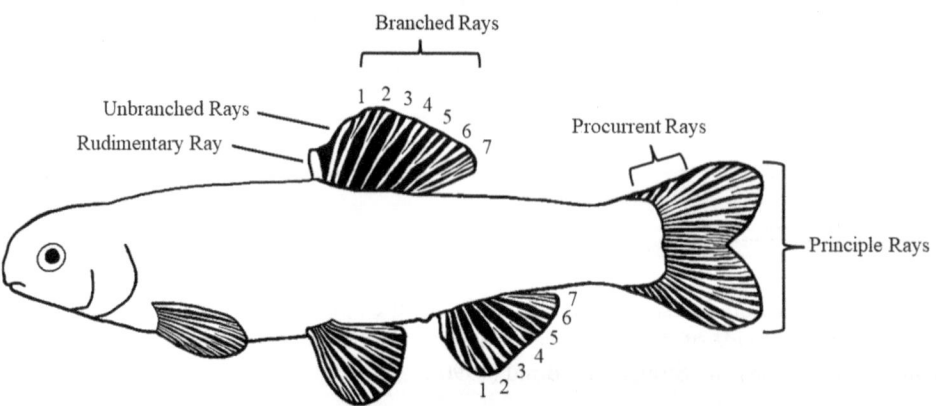

Fig. 4.7 Examples of fin ray counts. *Remember to count branched rays and the well-developed leading unbranched ray in dorsal and anal fins.

Fig. 4.8 Examples of gill raker counts. Both illustrated examples have 14 gill rakers

In barbs and carps (family Cyprinidae), sharpbellies (family Xenocyprididae), minnows (family Leuciscidae), and suckers (family Catostomidae), the tooth shape and formula of the modified fifth gill arch, known as the pharyngeal arch, can be used to distinguish hard-to-identify species. The pharyngeal arch can be removed behind the gills on either side of the fish. It will appear as a tissue-covered swelling behind the last gill arch. Pharyngeal arches are extremely delicate structures and must be extracted carefully by gently freeing the top and bottom of the arch with forceps. Extracted arches are commonly covered in tissue. Removing the tissue with forceps will typically result in broken teeth. Alternatively, soaking in hydrogen peroxide for a few hours will remove the tissue without destroying the arch. Pharyngeal teeth on minnows are counted for both arches going from the left to right outermost row. Figure 4.9a illustrates arches with two teeth on the outer row and four teeth on the inner row of each arch, for which the count would be listed as 2,4-4,2. The shape of the pharyngeal arch and of the pharyngeal teeth (molariform vs. comblike) can also be diagnostic. After being counted, the arches can be carefully placed back into the specimen or put in a small vial added to the larger container with the specimen.

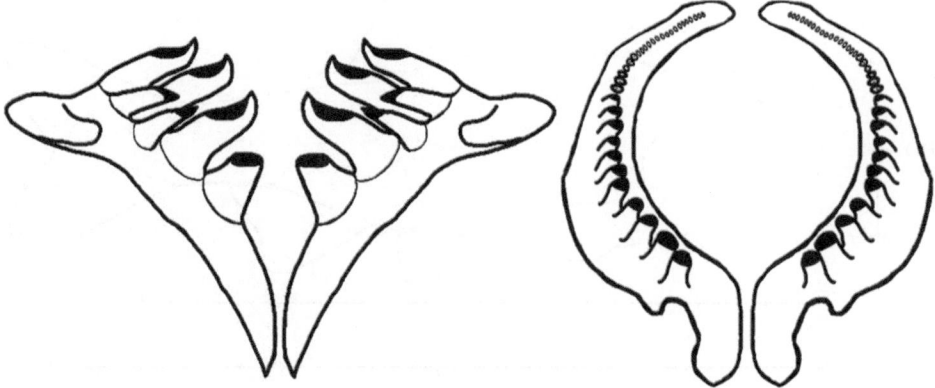

Fig. 4.9 Examples of pharyngeal teeth

Counts on the bottom of the head include branchiostegal rays and submandibular pores. The branchiostegal rays are typically counted on the left side of the body. The most anterior and posterior branchiostegal rays are easily missed and must be counted carefully. To aid in counting, the opercle can be gently pulled away from the head, spreading the branchiostegal rays and their connective membranes. Submandibular pores run along the bottom of the jaw of the fish and are particularly useful for identifying pikes (family Esocidae; Fig. 4.10).

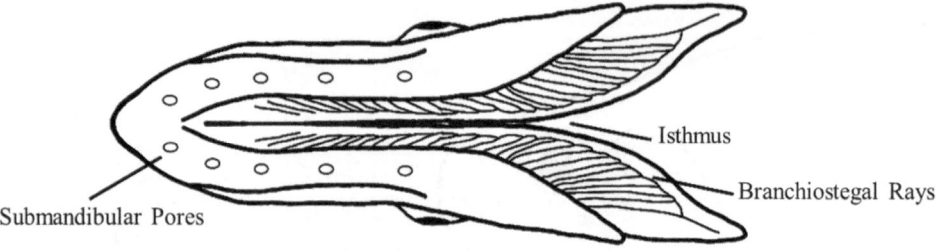

Fig. 4.10 Examples of branchiostegal rays and submandibular pores

4.8.2 Measurements

Measurements are typically performed with calipers and measured in millimeters (Fig. 4.11). Ratios of two measurements are used in the dichotomous key, along with the following seven measurements:

Head length (HL): Distance from the tip of the snout to the posterior edge of the opercle

Snout length (SN): Distance from the front of the eye to the tip of the snout

Eye diameter (ED): Distance from the anterior edge of the eye to the posterior edge of the eye

Body depth (BD): Distance at the maximum vertical depth of the body

Standard length (SL): Distance from the most anterior part of the body (snout or lips) to the end of the vertebral column (at the fleshy base of the caudal fin)

Fork length (FL): Distance from the most anterior part of the body (snout or lips) to the fork of the caudal fin

Total length (TL): Distance from the most anterior part of the body (snout or lips) to the posterior tip of the caudal fin. The upper and lower lobes of the caudal fin are squeezed together for measurement.

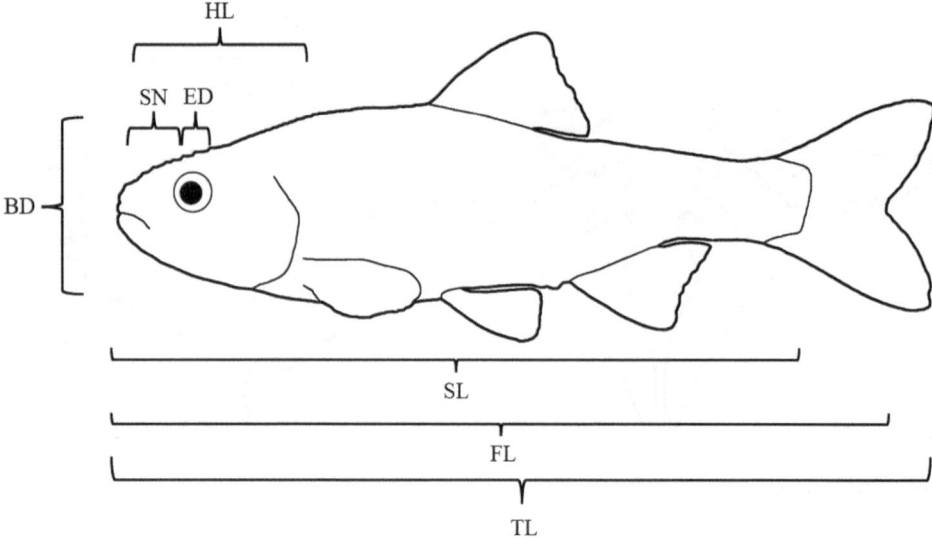

Fig. 4.11 Examples of measurements

4.9 Dissection

Some species require an examination of the intestinal tract or the peritoneum, the membrane lining the abdominal cavity. The peritoneum is usually black, silvery, or silvery with black flecks. A black peritoneum can sometimes be seen through the belly, eliminating the need for dissection. The intestinal tract will typically be an S-shaped loop in carnivorous fishes and long and coiled in herbivorous fishes. To examine these features, an incision should be made from the anus toward the head of the fish as far ventrally as possible, making a cut shallow enough to open the abdominal cavity and not cut the organs. The cut should be made anterior to the pectoral fin base. If the longitudinal cut does not allow for examination of the needed structures, two additional cuts to create a flap can be made on the right side of the fish (Fig. 4.12). The left side should remain intact for counts.

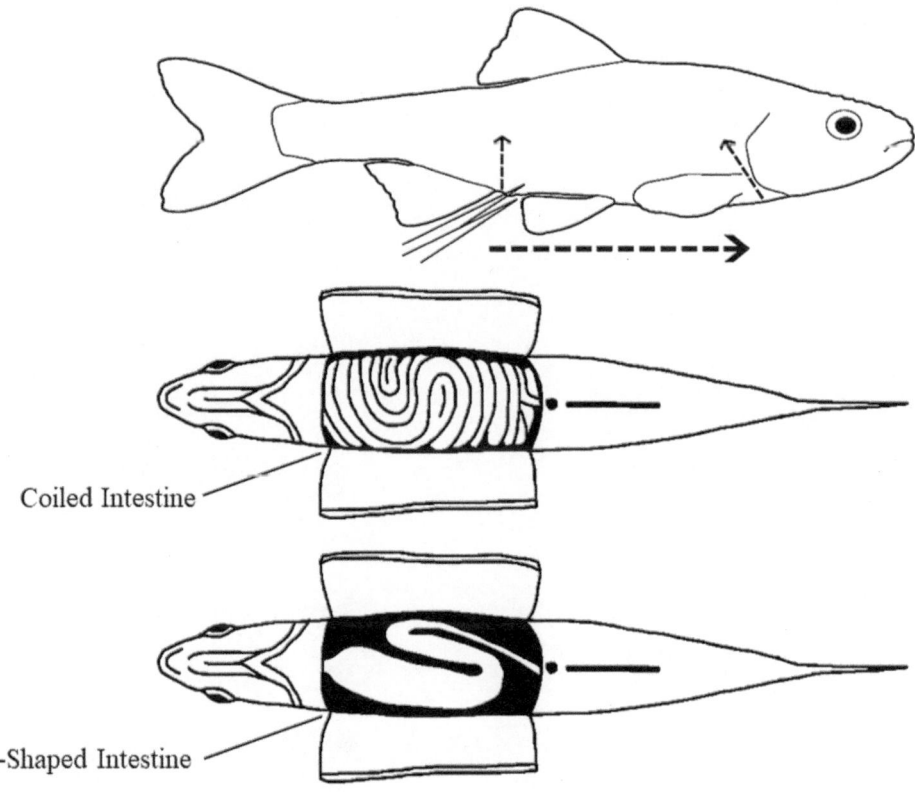

Coiled Intestine

S-Shaped Intestine

Fig. 4.12 Examples of dissection with intestine types

References

Hubbs CL, Lagler KF (1958) Fishes of the Great Lakes region. University of Michigan Press, Ann Arbor

Strauss RE, Bond CE (1990) Taxonomic methods: morphology. In: Schreck CS, Moyle PB (eds) Methods for fish biology. American Fisheries Society, Bethesda, pp 109–140

Chapter 5
Dichotomous Keys

Kathryn E. Schlafke, Matthew D. Wagner, and Chelsey A. Pasbrig

5.1 Key to the Families of Fishes in the Dakotas

1. Pectoral and pelvic fins absent; lacking jaws; seven gill clefts (openings) on each side; single median naris
....Petromyzontidae, p. 86

Fig. 5.1 Petromyzontidae

  Paired fins (pectoral and/or pelvic) present; jaws present; single external gill opening present on each side; one nare present on each side 2

2. Tail heterocercal; fleshy base of the caudal fin asymmetrical in lateral view (the dorsal portion extends farther posteriad than the ventral portion) 3
  Tail homocercal; fleshy base of the caudal fin symmetrical..6

3. Tail slightly heterocercal; body completely covered by scales; mouth terminal; teeth large and numerous 4
  Tail strongly heterocercal; body not completely covered by scales; teeth minute or absent; mouth inferior 5

4. Dorsal fin over half of the body length; body covered with cycloid scales; jaws not long and beak-like; gular plate (bony plate in the throat region) present ....Amiidae, p. 114

K. E. Schlafke
Montana Fish, Wildlife & Parks, Bozeman, MT, USA

M. D. Wagner
U.S. Fish and Wildlife Service, Jackson, MS, USA

C. A. Pasbrig (✉)
South Dakota Game, Fish and Parks, Pierre, SD, USA
e-mail: Chelsey.Pasbrig@state.sd.us

© The Author(s), under exclusive license to Springer Nature Switzerland AG 2024
M. Barnes (ed.), *Fishes of the Dakotas*, https://doi.org/10.1007/978-3-031-38040-2_5

Fig. 5.2 Amiidae

Dorsal fin short and close to the caudal fin; scales diamond shaped, not overlapping; jaws long and beak-like; gular plate absent..Lepisosteidae, p. 107

Fig. 5.3 Lepisosteidae

5. Body with several rows of large bony plates; snout moderately elongated..Acipenseridae, p. 93

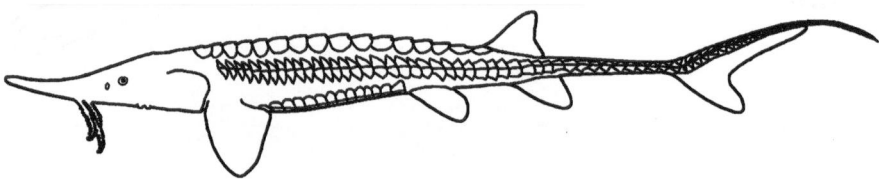

Fig. 5.4 Acipenseridae

Body naked, except for small scales on the dorsal lobe of the caudal fin; snout long and paddlelike.................
....Polyodontidae, p. 103

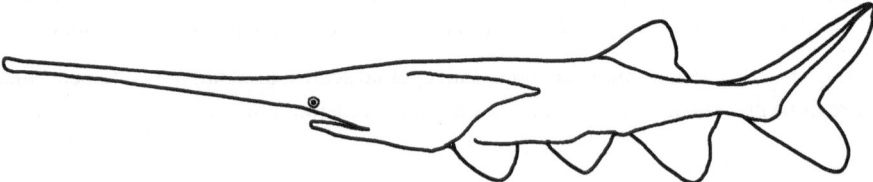

Fig. 5.5 Polyodontidae

6. Pelvic fins absent or only present as reduced spines (sticklebacks)..7
   Pelvic fins present...8

7. Body greatly elongated and snakelike; dorsal, caudal, and anal fins continuous; pelvic fins absent........................
   ....Anguillidae, p. 118

Fig. 5.6 Anguillidae

Body laterally compressed; dorsal spines free, stiff, and not connected by membranes; body naked...............................
....Gasterosteidae, p. 496

Fig. 5.7 Gasterosteidae

8. Single barbel at the tip of the chin; the second dorsal fin much longer than the first dorsal fin and well separated..Gadidae, p. 400

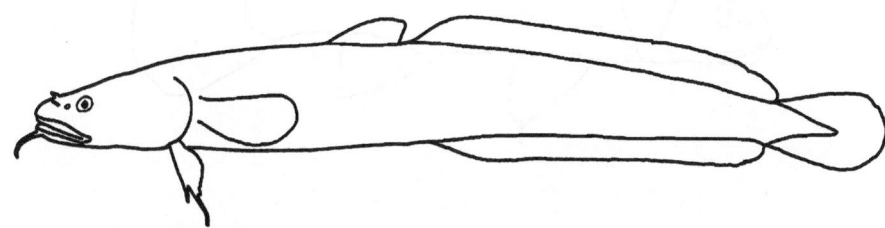

Fig. 5.8 Gadidae

No single median barbel at the tip of the chin...9

9. Body without scales, pectoral fins with strong spines; four pairs of barbels present around the mouth..........................
   ....Ictaluridae, p. 322

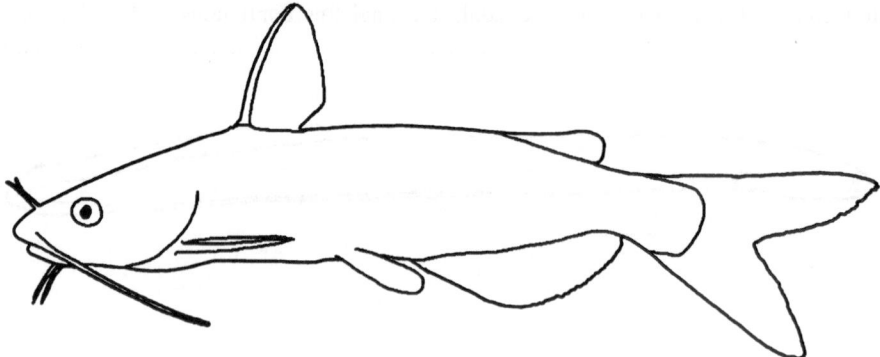

Fig. 5.9 Ictaluridae

 Body with scales...10
10. Pelvic fins thoracic, inferior, or anterior to pectoral fins..11
 Pelvic fins abdominal, well separated from pectoral fins..15
11. Anal spines one or two...12
 Anal spines three or more...13
12. Soft dorsal fin with 23 or more rays; outer pelvic ray 1.25 times longer than the second and third rays, body laterally compressed...Sciaenidae, p. 492

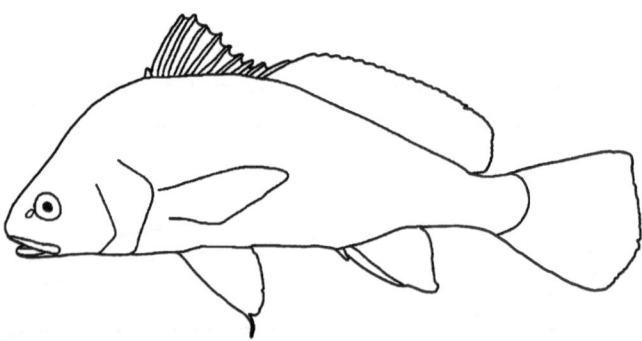

Fig. 5.10 Sciaenidae

 Soft dorsal fin with less than 23 rays; outer pelvic ray as long as the second and third rays; body slender and elongated..Percidae, p. 459

Fig. 5.11 Percidae

13. Dorsal spines 16 or more..Cichlidae, p. 404

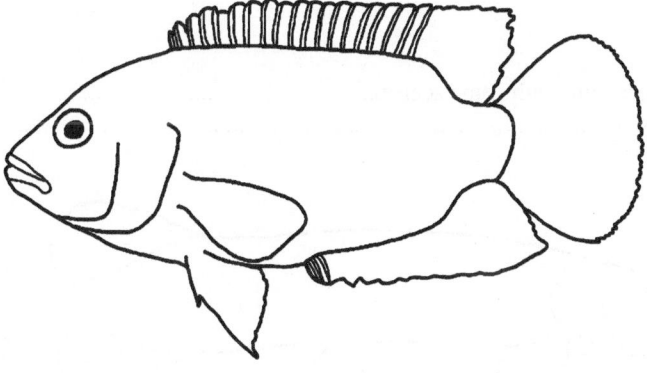

Fig. 5.12 Cichlidae

Dorsal spines 14 or less..14

14. Spinous and soft dorsal fin separate or only slightly connected; posterior edge of the operculum with a well-developed spine directed posteriorly...Moronidae, p. 449

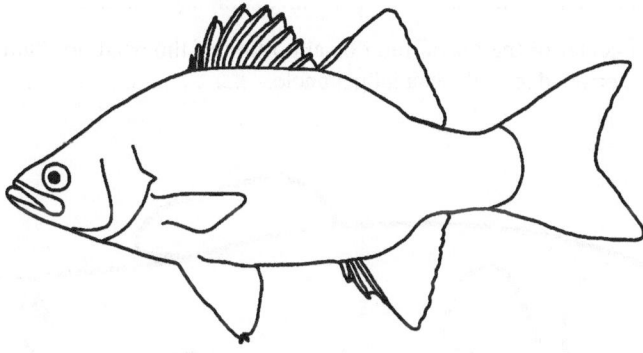

Fig. 5.13 Moronidae

Spinous and soft dorsal fin well connected; posterior edge of the operculum without a well-developed spine directed posteriorly...Centrarchidae, p. 418

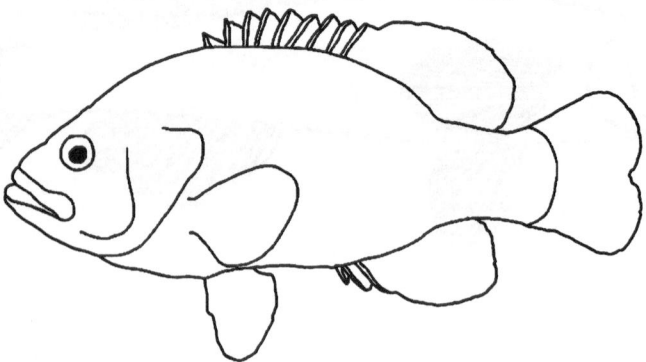

Fig. 5.14 Centrarchidae

15. Top of the head and/or operculum with some scales...16
Head completely lacking scales..18

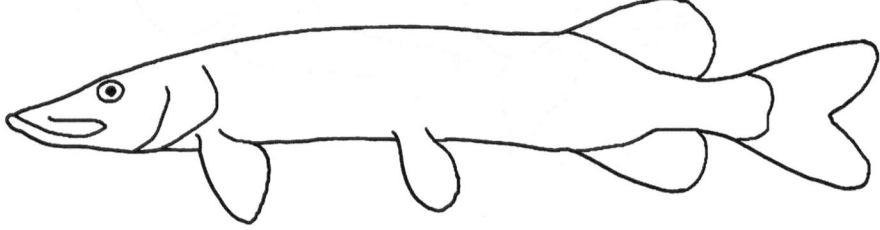

Fig. 5.15 Esocidae

16. Scales 100 or more in the lateral series; jaws resemble a duck's beak with large canine teeth, caudal fin forked..Esocidae, p. 347
Scales 50 or fewer in lateral series; jaws do not resemble a duck's beak with large canine teeth; caudal fin rounded..17

17. Origin of the anal fin at the center of the dorsal fin; dorsal portion of the head not flattened; mouth terminal; distinct vertical black bar on the posterior edge of the caudal peduncle...Esocidae, p. 347

Fig. 5.16 Esocidae

Origin of the anal fin about even with origin of the dorsal fin; dorsal portion of the head flat; mouth superior; no distinct vertical black bar on the posterior edge of the caudal peduncle..Fundulidae, p. 408

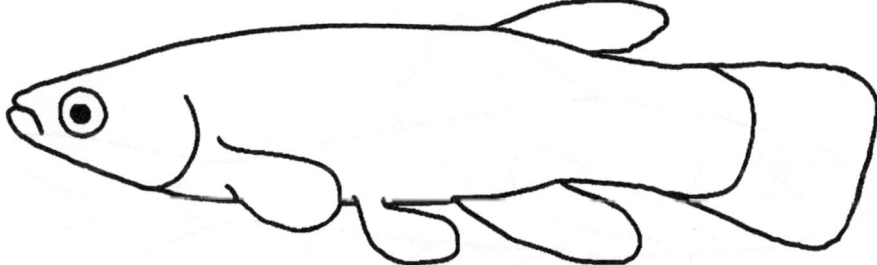

Fig. 5.17 Fundulidae

18. Adipose fin present...19
    Adipose fin absent..21
19. Axillary process present at the base of the pelvic fin...Salmonidae, p. 360

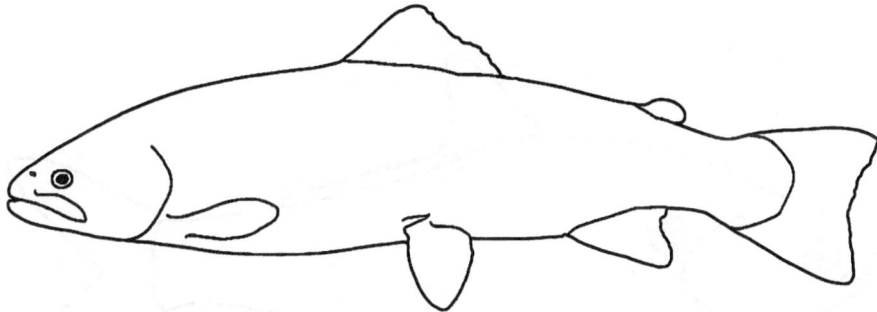

Fig. 5.18 Salmonidae

    Axillary process lacking at the base of the pelvic fin..20
20. Dorsal fin origin at about midbody, approximately equidistant between the snout and the base of the caudal fin; scales cycloid; lateral line absent; 60–72 scales in the lateral series...Osmeridae, p. 392

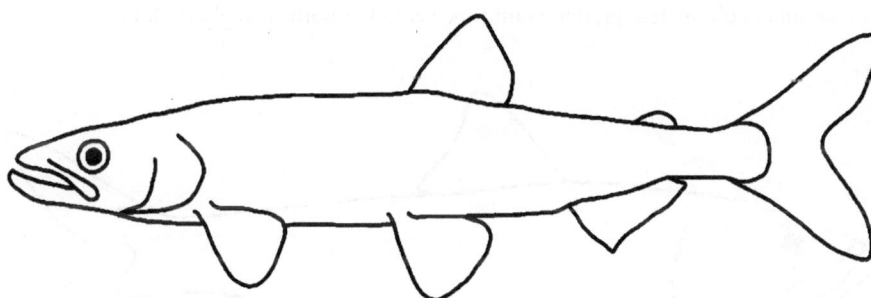

Fig. 5.19 Osmeridae

Dorsal fin origin more anterior, nearer the snout than the caudal fin base; scales ctenoid; lateral line present; 40–55 scales in the lateral series..Percopsidae, p. 396

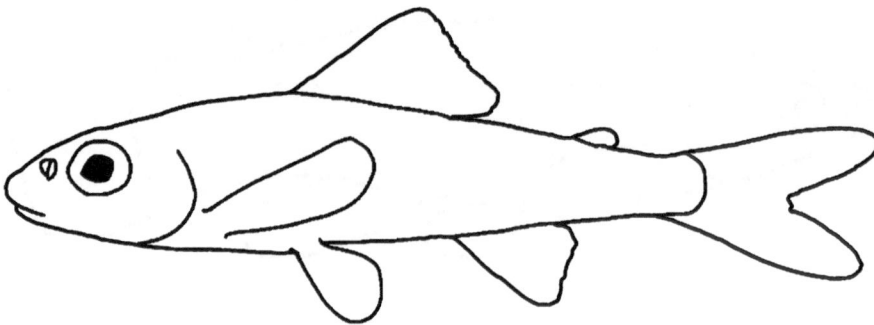

Fig. 5.20 Percopsidae

21. Lateral line absent...22
    Lateral line present..23

22. Last ray of the dorsal fin produced as a long filament; predorsal midline naked...........................Dorosomatidae, p. 137

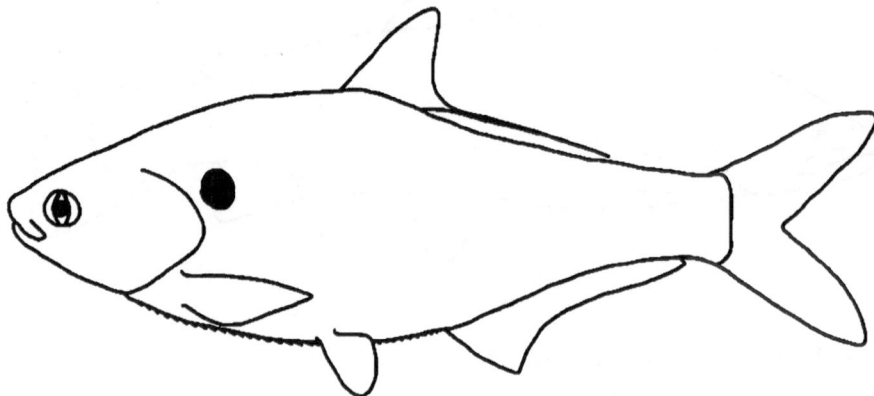

Fig. 5.21 Dorosomatidae

Last ray of the dorsal fin not elongated; predorsal midline scaled; mouth strongly oblique.....................Alosidae, p. 130

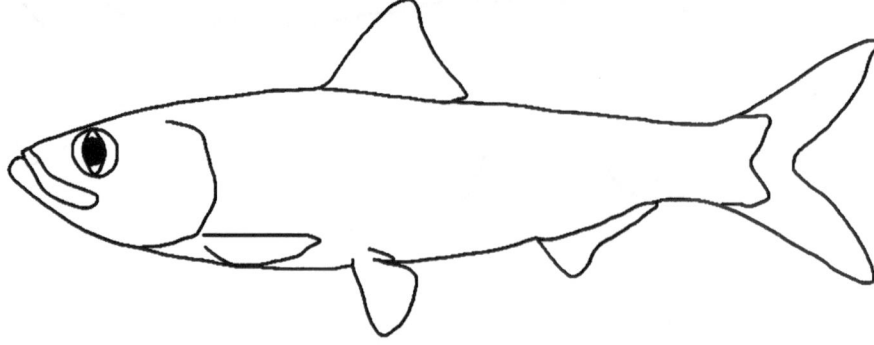

Fig. 5.22 Alosidae

23. Anal fin long, with a combined count of spines and rays totaling 26 or more....................................Hiodontidae, p. 123

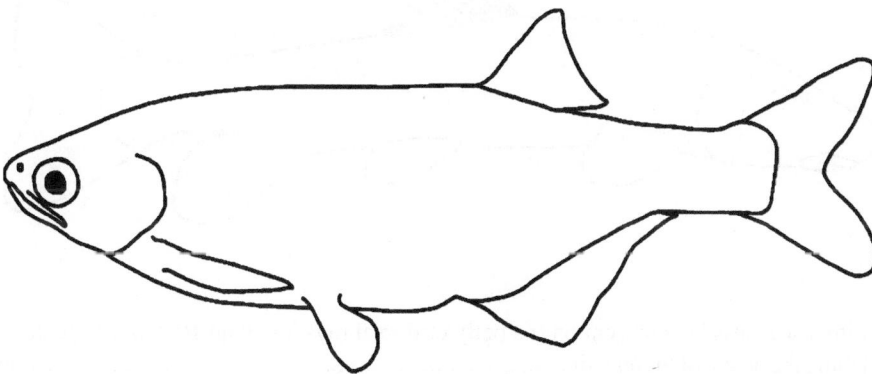

Fig. 5.23 Hiodontidae

Anal fin shorter, with a combined count of spines and rays totaling 17 or fewer...24

24. First ray on the dorsal fin hard and spine-like with posterior serrations...Cyprinidae, p. 190

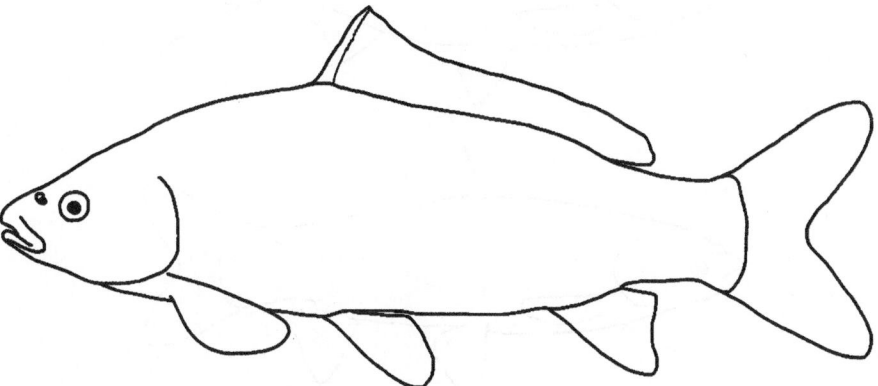

Fig. 5.24 Cyprinidae

First ray on the dorsal fin not hard and spine-like, no posterior serrations..25

25. Dorsal fin with 10 or more soft rays (if 10, no scaleless keel present on the belly); pharyngeal arch with a single row of more than 20 teeth...Catostomidae, p. 144
Dorsal fin with 9 or fewer soft rays or dorsal fin with 10 soft rays and a scaleless keel present on the belly; pharyngeal arch with 1–3 rows with no more than 5 teeth per row...26

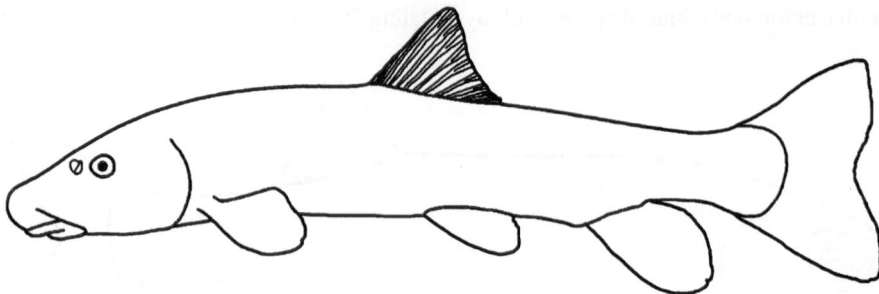

Fig. 5.25 Catostomidae

26. Dorsal rays 10 with a scaleless keel present on the belly or dorsal rays less than 10 with the posterior edge of the erect dorsal fin not reaching the origin of the anal fin..Xenocyprididae, p. 197

Fig. 5.26 Xenocyprididae

   Dorsal fin with 9 or fewer soft rays and the posterior edge of the erect dorsal fin reaches the origin of the anal fin........
   ....Leuciscidae, p. 207

Fig. 5.27 Leuciscidae

5.2 Key to Lampreys of North Dakota and South Dakota

Petromyzontiformes (Family: Petromyzontidae); One Genus: *Ichthyomyzon*

1. Some circumoral teeth bicuspid; 51–54 myomeres (muscle segments) between the last gill cleft and the cloaca; disc length more than 12.5 times the total length..*Ichthyomyzon castaneus* chestnut lamprey, p. 87
   All circumoral teeth unicuspid; 49–52 myomeres between the last gill cleft and the cloaca; disc length 12.5 or fewer times than the total length..*Ichthyomyzon unicuspis* silver lamprey, p. 90

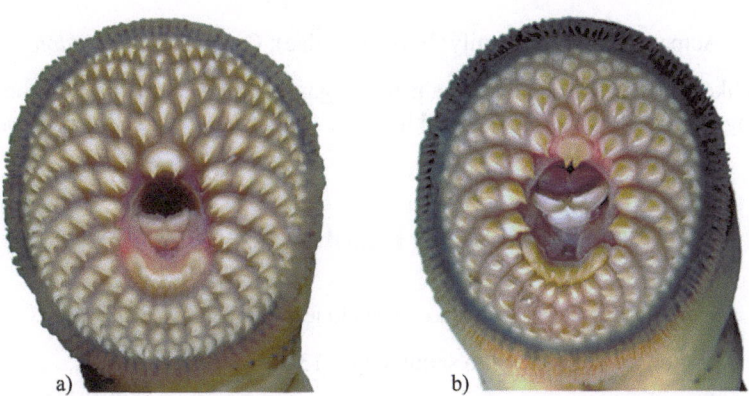

Fig. 5.28 Lamprey buccal funnel morphology: (**a**) chestnut lamprey (*Ichthyomyzon castaneus*) and (**b**) silver lamprey (*Ichthyomyzon unicuspis*) (© Konrad Schmidt)

5.3 Key to Sturgeons of North Dakota and South Dakota

Acipenseriformes (Family: Acipenseridae); Two Genera: *Acipenser* and *Scaphirhynchus*

1. Snout rounded and conical; area posterior to the anal fin with bony plates on the dorsal, lateral, and ventral midlines separated by naked areas; barbels smooth, no lateral projections; caudal peduncle short and partly covered with bony scutes; upper lobe of the caudal fin without a long filament...............................*Acipenser fulvescens* lake sturgeon, p. 94
   Snout flattened and shovel-shaped; area posterior to the anal fin completely covered by bony plates; barbels feathery, with lateral projections; caudal peduncle long and fully covered with bony scutes; upper lobe of the caudal fin with a long filament; genus: *Scaphirhynchus*...2

2. Belly typically covered with small bony plates and feels rough (not easily visible in juveniles); anal fin rays 23 or fewer; dorsal fin rays 36 or fewer; origins of two inner barbels in line with or posterior to the origins of the two outer barbels.. ....*Scaphirhynchus platorynchus* shovelnose sturgeon, p. 100
   Belly naked and feels smooth; anal fin rays 24 or more; dorsal fin rays 37 or more; origins of two inner barbels anterior to the origins of two outer barbels..*Scaphirhynchus albus* pallid sturgeon, p. 97

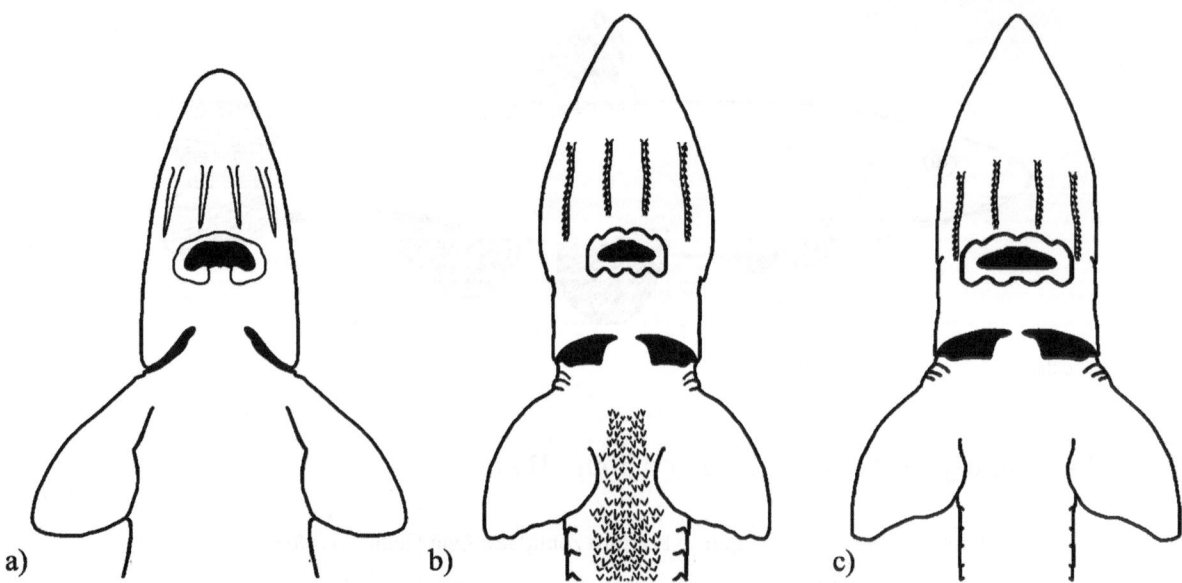

Fig. 5.29 Ventral view of barbels and underside of the (**a**) lake sturgeon (*Acipenser fulvescens*), (**b**) shovelnose sturgeon (*Scaphirhynchus platorynchus*), and (**c**) pallid sturgeon (*Scaphirhynchus albus*)

5.4 Key to Paddlefish of North Dakota and South Dakota

Acipenseriformes (Family: Polyodontidae); One Genus: *Polyodon*

1. Snout long and paddlelike; body naked, except for a patch of ganoid scales at the base of the upper lobe of the caudal fin. Gill cover extending posteriorly as a long, pointed flap......................................*Polyodon spathula*, paddlefish, p. 104

5.5 Key to Gars of North Dakota and South Dakota

Lepisosteiformes (Family: Lepisosteidae); One Genus: *Lepisosteus*

1. Snout long and slender; least snout width fits into snout length 13 or more times..
....*Lepisosteus osseus* longnose gar, p. 108
Snout short and broad; least snout width fits into snout length 10 or fewer times...
....*Lepisosteus platostomus* shortnose gar, p. 111

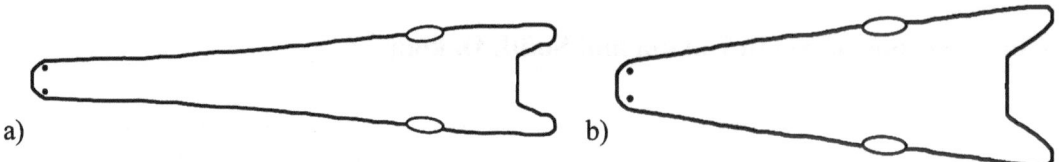

Fig. 5.30 Dorsal view of the head of the (**a**) longnose gar (*Lepisosteus osseus*) and (**b**) shortnose gar (*Lepisosteus platostomus*)

5.6 Key to Bowfin of North Dakota and South Dakota

Amiiformes (Family: Amiidae); One Genus: *Amia*

1. Slightly heterocercal caudal fin; large canine teeth; dorsal fin longer than half of the body length; two barbels on the snout; gular plate present on throat..*Amia ocellicauda* emerald bowfin, p. 115

5.7 Key to Eels of North Dakota and South Dakota

Anguilliformes (Family: Anguillidae); One Genus: *Anguilla*

1. Body snakelike with a small, pointed head; jaws with teeth; pelvic fins absent; pectoral fins present; dorsal fin continuous with the caudal and anal fins extending along the posterior half the body..........*Anguilla rostrata* American eel, p. 119

5.8 Key to Mooneyes of North Dakota and South Dakota

Hiodontiformes (Family: Hiodontidae); One Genus: *Hiodon*

1. Dorsal fin origin slightly posterior to anal fin origin; keel on the belly from the anus to pectoral fin origin; dorsal fin with 9–10 rays..*Hiodon alosoides* goldeye, p. 124
Dorsal fin origin slightly anterior to anal fin origin; keel on the belly from the anus to pelvic fin origin; dorsal fin with 11–12 rays..*Hiodon tergisus* mooneye, p. 127

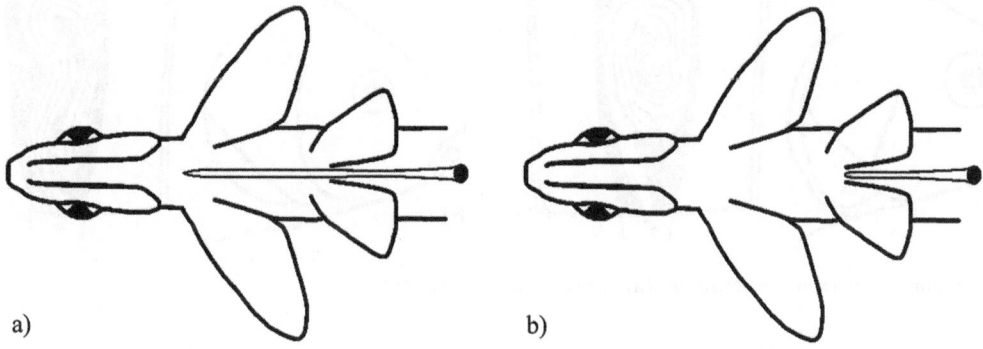

a) b)

Fig. 5.31 Ventral view of the keel of the (**a**) goldeye (*Hiodon alosoides*) and (**b**) mooneye (*Hiodon tergisus*)

5.9 Key to Shad of North Dakota and South Dakota

Clupeiformes (Family: Alosidae); One Genus: *Alosa*

1. Lacking a blue-black spot posterior to the upper edge of the gill cover; the tip of the lower jaw projecting well beyond the tip of the upper jaw; 24 or fewer gill rakers on the lower limb of the first arch..*Alosa chrysochloris* skipjack herring, p. 131
Blue-black spot present posterior to the upper edge of the gill cover; the tip of the lower jaw projecting a little (if any) beyond the tip of the upper jaw; 30 or more gill rakers on the lower limb of the first arch..*Alosa pseudoharengus* alewife, p. 134

5.10 Key to Thread Herrings of North Dakota and South Dakota

Clupeiformes (Family: Dorosomatidae); One Genus: *Dorosoma*

1. Snout bulbous and fleshy, projecting slightly past the upper jaw; anal fin rays 29 or more..*Dorosoma cepedianum* gizzard shad, p. 138
Snout pointed and not projecting past the upper jaw; anal fin rays 17–25..*Dorosoma petenense* threadfin shad, p. 141

5.11   Key to Suckers of North Dakota and South Dakota

Cypriniformes (Family: Catostomidae); Six Genera: *Carpiodes*, *Catostomus*, *Cycleptus*, *Ictiobus*,
Moxostoma, and *Pantosteus*

1. Dorsal fin long with 20 or more rays...2
   Dorsal fin short with 16 or fewer rays...8

2. Body slender; body depth 25% or less of standard length; lateral-line scales 50 or more....................................
   ....*Cycleptus elongatus* blue sucker, p. 160
   Body deep and compressed, depth more than 25% of standard length; lateral-line scales 45 or fewer............3

3. Suboperculum asymmetrical, with its greatest depth anterior to the middle; anal fin rays seven (rarely eight); intestine
   arranged in a whorled pattern; silvery fishes with few or no melanophores on the pelvic fins..4
   Suboperculum symmetrical, with its greatest depth at the middle; anal fin rays 8–11 (rarely 7); intestine arranged in an
   S-shaped pattern; dusky gray fishes with the pelvic fins densely pigmented with melanophores.......................................6

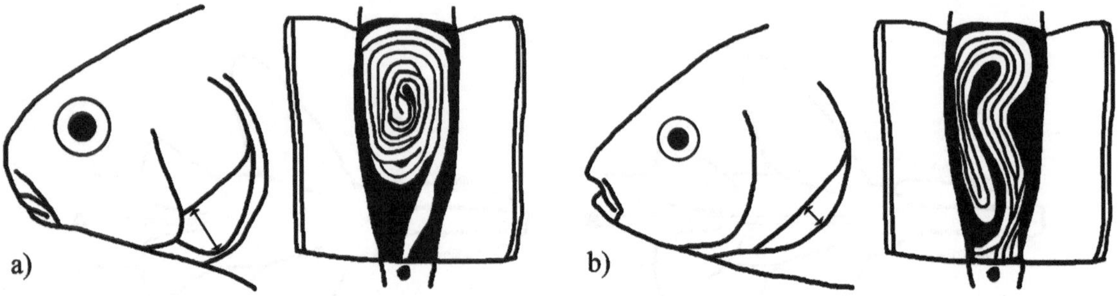

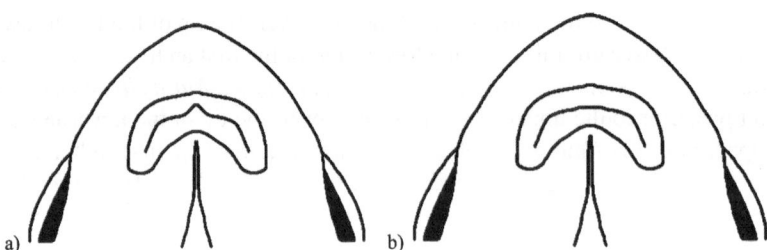

Fig. 5.32 Suboperculum and intestinal pattern of the (**a**) *Carpiodes* and (**b**) *Ictiobus*

4. Distance from the posterior edge of the eye to the gill cleft greater than distance from the anterior edge of the eye snout;
   lateral-line scales 36 or fewer (33–37); nipple-like projection present on the tip of the lower lip; dorsal fin rays usually
   27 or fewer; lips form an obtuse angle (more than 120 degrees)..5
   Distance from the posterior edge of the eye to the gill cleft equal to distance from the anterior edge of the eye snout;
   lateral-line scales 37 or more (36–40); nipple-like projection absent from the tip of the lower lip; dorsal fin rays usually 28 or more; lips form an acute angle (less than 90 degrees)..........................*Carpiodes cyprinus* quillback, p. 148

Fig. 5.33 Mouth (**a**) with a nipple-like projection present on the lower lip of the highfin carpsucker (*Carpiodes velifer*) and river carpsucker
(*Carpiodes carpio*) and (**b**) without a nipple-like projection present on the lower lip of the quillback (*Carpiodes cyprinus*)

5. Anterior rays of the dorsal fin greatly elongated and when depressed reach or nearly reach the end of the dorsal fin base
   ....*Carpiodes velifer* highfin carpsucker, p. 151
   Anterior rays of the dorsal fin not greatly elongated and when depressed reach about two-third the length of the dorsal
   fin base..*Carpiodes carpio* river carpsucker, p. 145

6. Upper lip thinner than lower lip, with shallow grooves; mouth terminal and oblique, with the tip of the upper lip level
   with the lower margin of the eye; gill rakers 40 or more.............................*Ictiobus cyprinellus* bigmouth buffalo, p. 166
   Upper lip as thick as lower lip with grooves; mouth inferior and horizontal, with the tip of the upper lip below the lower
   margin of the eye; gill rakers 35 or fewer..7

7. Extremely deep-bodied, sharply arched/ridged back; steep profile from the head to the dorsal fin, giving a "humped" appearance in lateral view; body depth about 36–42% of standard length; head length about 24–29% of standard length ....*Ictiobus bubalus* smallmouth buffalo, p. 163
   Somewhat deep-bodied but less so and more elongated than other buffalos; back rounded without a distinctive arch or ridge; body depth 29–38% of standard length; head length 26–34% of standard length...
   ....*Ictiobus niger* black buffalo, p. 169

8. Scales in lateral-line series more than 55..9
   Scales in lateral-line series less than 55 (some white sucker (*Catostomus commersonii*) may have scale counts under 55 but have scales crowded anteriorly, whereas the *Moxostoma* spp. do not)..11

9. Lower lip separated from upper lip by lateral notches; moderate median lower lip notch; axillary process of the pelvic fin well developed; cartilaginous scraping ridge inside of the lower lip...............*Pantosteus jordani* plains sucker, p. 187
   Lower lip not separated from upper lip by lateral notches; deep median lower lip notch; lacking axillary process of the pelvic fin; no cartilaginous scraping ridge inside of the lower lip..10

Fig. 5.34 Lateral notch on the lips of the plains sucker (*Pantosteus jordani*)

10. Scales in lateral-line series 53–74...*Catostomus commersonii* white sucker, p. 157
    Scales in lateral-line series 95–120..*Catostomus catostomus* longnose sucker, p. 154

11. Circumpenduncular scale rows usually 16 and lips plicate with halves of the lower lip meeting at an obtuse angle ....*Moxostoma valenciennesi* greater redhorse, p. 184
    Circumpenduncular scale rows usually 12 and lips not as in Fig. 5.35...12

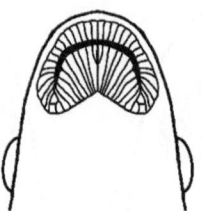

Fig. 5.35 Lips of greater redhorse (*Moxostoma valenciennesi*)

12. Halves of the lower lip meet at an acute angle; majority of the lower lips with plicae broken into papillae (Fig. 5.36); dorsal fin rays usually 14 or more...*Moxostoma anisurum* silver redhorse, p. 172
    Halves of the lower lip in a straight line or at an obtuse angle; lips plicate or plicate with the posterior margin of the lower lip plicae broken into papillae; dorsal fin rays usually 13 or fewer...13

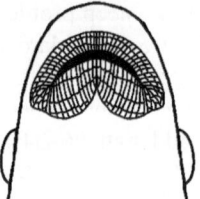

Fig. 5.36 Lips of silver redhorse (*Moxostoma anisurum*)

13. Scales without a dark spot at the base; lateral scale rows 39–42; caudal fin not red in life (except occasionally in young); halves of the lower lips fully plicate (Fig. 5.37)....................................*Moxostoma erythrurum* golden redhorse, p. 178
    Scales with a dark spot at the base; caudal fin red; halves of the lower lips either fully plicate or with posterior margins of the lower lips with plicae broken into papillae...14

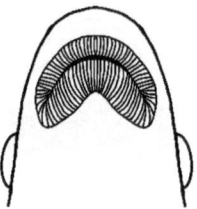

Fig. 5.37 Lips of golden redhorse (*Moxostoma erythrurum*)

14. Posterior margin of the lower lips forming a slight obtuse angle; lower lip plicae not dissected into papillae; pharyngeal arch heavy and triangular, lower teeth large, molariform, increasing in size toward bottom, 6–9 on the lower half of the tooth row; margin of the dorsal fin straight to slightly concave.....................*Moxostoma carinatum* river redhorse, p. 175
    Posterior margin of the lower lips forming a straight line; posterior lower lip plicae dissected into papillae; lower pharyngeal arch weak, not triangular; teeth compressed, comblike, and 12–30 on the lower half of the tooth row, dorsal fin deeply concave..*Moxostoma macrolepidotum* shorthead redhorse, p. 181

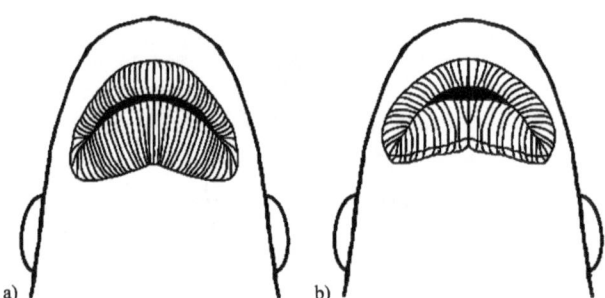

Fig. 5.38 Illustrations of the lips of (**a**) river redhorse (*Moxostoma carinatum*) and (**b**) shorthead redhorse (*Moxostoma macrolepidotum*)

5.12   Key to Barbs and Carps of North Dakota and South Dakota

Cypriniformes (Family: Cyprinidae); Two Genera: *Carassius* and *Cyprinus*

1. Two pairs of barbels present on the upper lip, near the corner of the mouth; pharyngeal teeth molarlike (1,3,3-3,1,1); gill rakers 21–27...*Cyprinus carpio* common carp, p. 194
   Barbels lacking; pharyngeal teeth not molarlike (4-4); gill rakers 37–43.....................*Carassius auratus* goldfish, p. 191

5.13 Key to Sharpbellies of North Dakota and South Dakota

Cypriniformes (Family: Xenocyprididae); Two Genera: *Ctenopharyngodon*
and *Hypophthalmichthys*

1. Distance from the origin of the anal fin to the caudal fin base equal to distance from the anal fin origin to pelvic fin insertion; posterior edge of the erect dorsal fin does not reach the origin of the anal fin; lateral scale rows 34–37..*Ctenopharyngodon idella* grass carp, p. 198
Distance from the origin of the anal fin to the caudal fin base 1.5–2 times that from the anal fin origin to pelvic fin insertion; posterior edge of the erect dorsal fin reaches the origin of the anal fin; lateral scale rows 85 or more......................2

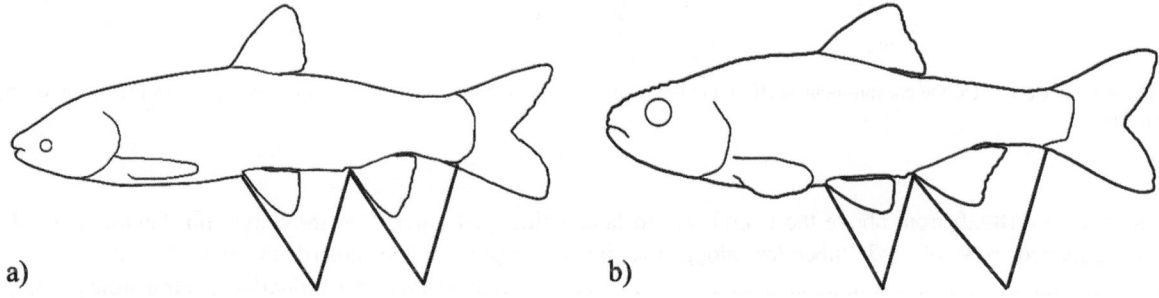

Fig. 5.39 Relative anal fin positions of (**a**) *Ctenopharyngodon* compared to (**b**) other minnows

2. Keel extends from the anus to between the base of the pelvic fins; gill rakers long and slender; body with scattered, irregularly shaped dark blotches; distal tip of the pectoral fin extends well past the origin of the pelvic fin........................
....*Hypophthalmichthys nobilis* bighead carp, p. 204
Keel extends from the anus to the isthmus; gill rakers thin and numerous and covered by a net-like matrix; lacking dark blotches; distal tip of the pectoral fin either extends to the origin of the pelvic fin or does not...
....*Hypophthalmichthys molitrix* silver carp, p. 201.

5.14 Key to Minnows of North Dakota and South Dakota

Cypriniformes (Family: Leuciscidae); 22 Genera: *Alburnops, Campostoma, Chrosomus, Couesius, Cyprinella, Ericymba, Hudsonius, Hybognathus, Luxilus, Macrhybopsis, Margariscus, Miniellus, Nocomis, Notemigonus, Notropis, Paranotropis, Phenacobius, Pimephales, Platygobio, Rhinichthys, Scardinius*, and *Semotilus*

1. Belly with firm keel extending at least from the pelvic fins to anus...2
Belly rounded, without firm keel...3

2. Dorsal rays 9–11; first gill arch with 10–13 gill rakers; lateral scales 36–45; firm keel on the belly from the pelvic fins to anus scaled..*Scardinius erythrophthalmus* rudd, p. 316
Dorsal rays 7–9; first gill arch with 17–19 gill rakers; lateral scales 44–54; firm keel on belly at extending least from pelvic fins to anus scale-less...*Notemigonus crysoleucas* golden shiner, p. 280

3. Lower jaw with a hard cartilaginous ridge that projects anteriad of the tip of the lower jaw, separated from lower lip by groove..4
   Lower jaw lacking a horizontal shelf of cartilage...5

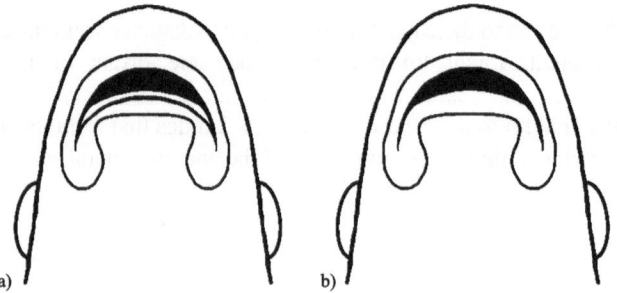

Fig. 5.40 Ventral view of (**a**) the cartilaginous shelf on the lower jaw of *Campostoma* and (**b**) other minnows lacking a cartilaginous shelf on the lower jaw

4. Scale count of 18–20 from above the lateral line to lateral line, just anterior to the dorsal fin; lateral scales 47–55; crescent-shaped row of 1–3 tubercles along the inner margin of the nostril in nuptial males; gill rakers 26–35...*Campostoma anomalum* central stoneroller, p. 211
   Scale count of 13–16 from above the lateral line to the lateral line, just anterior to the dorsal fin; lateral scales 43–47; no crescent-shaped row of 1–3 tubercles along the inner margin of the nostril in nuptial males; gill rakers 19–26..*Campostoma oligolepis* largescale stoneroller, p. 214

5. Mouth sucker-like, lower lip with a fleshy papillose posterior lobe on each side.. ....*Phenacobius mirabilis* suckermouth minnow, p. 298
   Mouth not sucker-like, lower lip without a fleshy papillose posterior lobe on each side...6

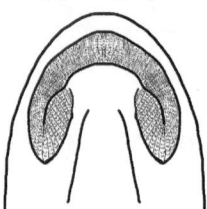

Fig. 5.41 Sucker-like mouth of *Phenacobius*

6. Premaxillae (upper lip) fused to the snout by a smooth continuation of the skin overlying the snout..............................7
   Premaxillae (upper lip) separated from the snout by a continuous groove along the entire posterior margin of the upper jaw 8

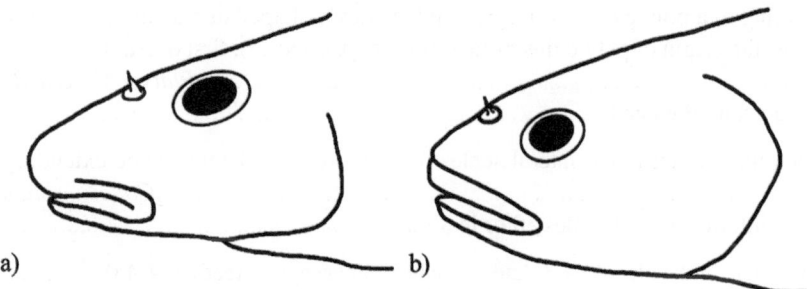

Fig. 5.42 Minnow premaxillaries: (**a**) nonprotractile and (**b**) protractile. This can be checked by either pulling down the lower jaw to force open the mouth or running a probe through the groove of the upper lip

7. Upper jaw scarcely exceeds lower jaw; distance from the tip of the snout to the anterior tip of the lower jaw less than eye diameter...*Rhinichthys obtusus* western blacknose dace, p. 313
   Upper jaw greatly exceeds lower jaw; distance from the tip of the snout to the anterior tip of the lower jaw greater than or equal to eye diameter...*Rhinichthys cataractae* longnose dace, p. 310

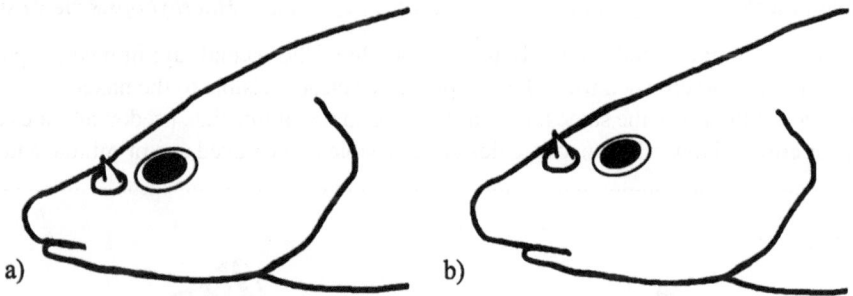

Fig. 5.43 Snout of the (**a**) western blacknose dace (*Rhinichthys obtusus*) and (**b**) longnose dace (*Rhinichthys cataractae*)

8. Large barbels present at the posterior tip of the maxilla (at the corner of mouth); (a transitory fleshy flap that resembles a barbel is present in breeding males of *Pimephales notatus*)...9
   Barbels absent or barbels minute, flat, and concealed in the groove above the maxilla (anterior to posterior tip of the maxilla)...15

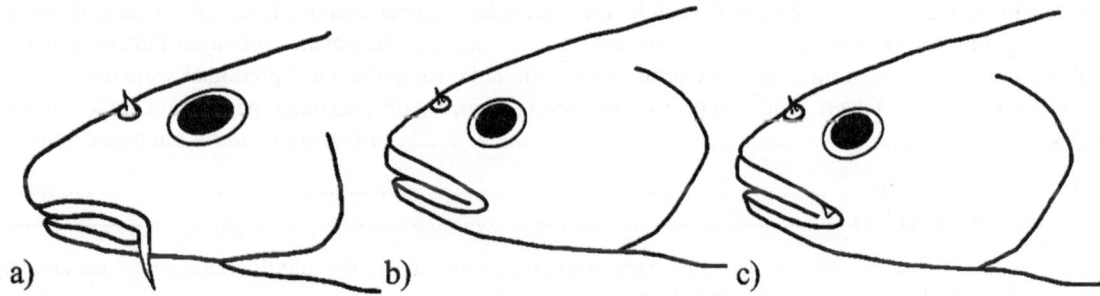

Fig. 5.44 Examples of (**a**) large maxillary barbel, (**b**) no barbel present, or (**c**) barbel minute, flat, and concealed in the groove above the maxilla

9. Anal rays modally 7; mouth terminal; distinct dark caudal spot; breeding males with bright red spot behind the eye (brassy color on females) and large tubercles on top of the head; stout body with 38–45 lateral scale ....
   ....*Nocomis biguttatus* hornyhead chub, p. 277
   Anal rays modally 8..10

10. Broad, flat head tapering to a pointed snout; large, pointed, sickle-shaped dorsal and pectoral fins (tips of the pectoral fins not extending past the origin of pelvic fins as in *Macrhybopsis meeki*); first dorsal fin ray extending beyond the last ray when depressed..*Platygobio gracilis* flathead chub, p. 307
    Head not broad and dorsally flattened..11

11. Mouth slightly subterminal to terminal; lateral scales 53 or more; dark lateral stripe extending from the snout to the caudal peduncle...*Couesius plumbeus* lake chub, p. 226
    Mouth subterminal to inferior; lateral scales 50 or less..12

12. Black spots on the back and side; 32–43 lateral scales; pharyngeal teeth 0,4-4,0..
    ....*Macrhybopsis hyostoma* shoal chub, p. 256
    Body lacking randomly scattered black spots; black lower lobe on the caudal fin..13

13. Longitudinal ridge/keel present on dorsal scales; 39–45 lateral scales; pharyngeal teeth 1,4-4,1......................
    ....*Macrhybopsis gelida* sturgeon chub, p. 253
    Longitudinal ridge/keel absent on dorsal scales...14

14. Dorsal fin origin in front of the pelvic fin origin; 35–48 lateral scales; pharyngeal teeth 1,4-4,1.....................
    ....*Macrhybopsis storeriana* silver chub, p. 262
    Dorsal fin origin over or slightly behind pelvic fin origin; pectoral fins long and sickle-shaped; pharyngeal teeth 0,4-4,0; 43–50 lateral scales..*Macrhybopsis meeki* sicklefin chub, p. 259

15. Leading/rudimentary ray on the dorsal fin nearly half the height of the second ray; immediate predorsal scales larger than nape scales; scales appear crowded toward the nape; scales clearly visible to the naked eye................................16
    Leading ray on the dorsal fin nearly the same length as the second ray; immediate predorsal scales equal to nape scales; scales do not appear crowded toward the nape; scales either visible to the naked eye or minute and barely visible to the naked eye...17

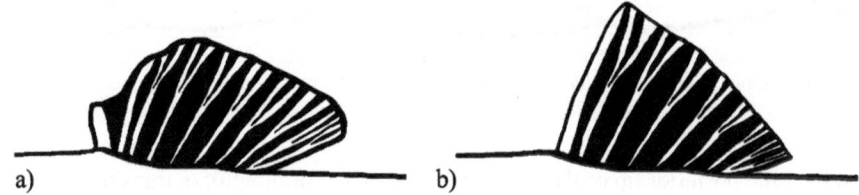

Fig. 5.45 Lateral view of (**a**) the dorsal fin with short rudimentary ray as in *Pimephales* and (**b**) the dorsal fin without shortened rudimentary ray as in many other minnows

16. Lateral line incomplete, not extending to the caudal fin base; sides with herringbone lines; predorsal scale rows usually 23 or more; gill rakers 14 or more; caudal spot absent..............................*Pimephales promelas* fathead minnow, p. 304
    Lateral line complete, extending to the caudal fin; sides without herringbone lines; predorsal scale rows usually 22 or fewer; gill rakers 10 or fewer; caudal spot present; nuptial males with maxillary barbel and three rows of snout tubercles..*Pimephales notatus* bluntnose minnow, p. 301

17. Lateral scale rows 47 or more...18
    Lateral scale rows 46 or less...22

18. Anal rays modally 8; barbels minute, flat, and concealed in the groove above the maxilla (anterior to posterior tip of the maxilla); body cylindrical; scales easily visible without magnification...19
    Anal rays modally 9; no barbels concealed in the groove above the maxilla; body slender and fairly compressed; scales not easily visible without magnification...20

19. Large black spot present on the anterior portion of the dorsal fin base; large terminal mouth reaching past the anterior origin of the eye; nuptial males with pink on lower half of the head and side...
    ....*Semotilus atromaculatus* creek chub, p. 319
    No large black spot present on the anterior portion of the dorsal fin base; small slightly subterminal mouth rarely reaching past the anterior origin of the eye; nuptial males with red along the lower side below the dark lateral stripe............
    ....*Margariscus nachtriebi* northern pearl dace, p. 265

20. Single mid-lateral dark stripe; intestine short, with a single S-shaped loop (less than two times as long as body); ventrolateral surface peppered with melanophores; angle of the mouth extends almost to the front of the pupil...*Chrosomus neogaeus* finescale dace, p. 223
Two lateral dark stripes (one only in young); intestine longer, with two or more loops or coils (more than two times as long as body)...21

21. Mouth upturned, with the chin anterior to the upper lip (reaches less than halfway to the eye); snout rounded...*Chrosomus eos* northern redbelly dace, p. 217
Mouth slightly subterminal (reaches more than halfway to the eye); snout moderately pointed..*Chrosomus erythrogaster* southern redbelly dace, p. 220

22. Intestine long and coiled, greater than twice the standard length when extended; peritoneum black..........................23
Intestine short and with a single S-shaped loop; peritoneum either silvery or black..25

23. Outer margin of the dorsal fin rounded with the first principal ray slightly shorter than the second and third rays; adults with 20 radii on scales; yellowish color in live ....*Hybognathus hankinsoni* brassy minnow, p. 244
Outer margin of the dorsal fin pointed with the first principal ray as long as or longer than the second and third rays; adults with 10 radii on scales; silvery color in live specimens...24

24. Basioccipital process narrow and peg-like, muscles almost touching at the point of attachment to the basioccipital process; head length greater than or equal to five times the eye diameter...*Hybognathus placitus* plains minnow, p. 247
Basioccipital process broad and straight to barely concave, muscles well separated at the point of attachment to the basioccipital process; head length less than five times the eye diameter...*Hybognathus argyritis* western silvery minnow, p. 241

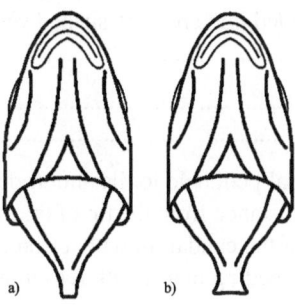

Fig. 5.46 Basioccipital processes of the (**a**) plains minnow (*Hybognathus placitus*) and (**b**) western silvery minnow (*Hybognathus argyritis*). To see the process, make a deep cut at the tip of the isthmus and then carefully bend the head dorsally while holding the chin

25. Dorsal fin origin far posterior (more than 2–3 scale rows) to pelvic fin origin..26
Dorsal fin origin slightly anterior, superior, or slightly posterior to pelvic fin origin..27

26. Snout length greater than eye diameter; dorsal fin rounded with the anterior fin ray not reaching past the posterior fin ray when depressed; lateral stripe continuous to the opercle and dark; melanophores evenly distributed on anterior dorsolateral scales; breeding males with bright red at the bases of fins and on the body...*Notropis percobromus* carmine shiner, p. 289

Snout length not greater than eye diameter; dorsal fin pointed with the anterior fin ray reaching past the posterior fin ray when depressed; lateral stripe continuous to the opercle but starts to fade anterior to the dorsal fin; melanophores only along the distal edges of the anterior dorsolateral scales (creating an outlined appearance); no red color on breeding males..*Notropis atherinoides* emerald shiner, p. 283

27. Basicaudal spot a distinct black spot as large as or larger than the pupil of the eye...
....*Hudsonius hudsonius* spottail shiner, p. 238
Basicaudal spot either absent or smaller than the pupil of the eye...28

28. Peritoneum black to dark brown...29
Peritoneum silvery or silvery with black spots..30

29. Mouth upturned, nearly vertical; dark lateral stripe extending from the caudal peduncle to the snout; scales on the side (front half of the body) as deep as wide; adults without concentrated pigment on some lateral scales causing some scales to appear to be missing..*Miniellus anogenus* pugnose shiner, p. 268
Mouth terminal and oblique; lateral stripe fades anterior to the dorsal fin; scales on the side (front half of body) deeper than wide with the posterior scale edge appearing rounded; adults with concentrated pigment on some lateral scales causing some scales to appear to be missing...*Luxilus cornutus* common shiner, p. 250

30. Upwardly directed eyes (lower margins of the pupils visible when the fish is viewed from above); head elongated with the lower surface distinctly flattened..*Ericymba dorsalis* bigmouth shiner, p. 235
Eyes directed laterally (lower margins of the pupils not visible when the fish is viewed from above); head not elongated, its lower surface not noticeably flattened...31

31. Lateral stripe dark and continuous from the caudal peduncle to the posterior edge of the opercle...............................32
Lateral stripe either absent or present and distinctly fades or constricts anterior to the dorsal fin.................................33

32. Dark borders of lateral-line pores expanded into crescent-shaped vertical bars anteriorly.....................................
....*Notropis heterolepis* blacknose shiner, p. 286
Dark borders of lateral-line pores not expanded into crescent-shaped vertical bars anteriorly.....................................
....*Miniellus topeka* Topeka shiner, p. 274

33. Anal rays modally 7 34
Anal rays modally 8 or more...36

34. Dark lateral stripe extending from the caudal peduncle to the snout; deepest body depth fits into standard length less than 3.5 times; lateral line almost straight; distance from the tip of the snout to dorsal fin origin less than distance from dorsal fin origin to caudal fin origin; base of the caudal fin with a triangular/chevron-shaped black spot (a similar spot sometimes seen in *Miniellus stramineus*); breeding males with red-orange fins and sides of the head............................
....*Miniellus topeka* Topeka shiner, p. 274
Lateral stripe continuous from the caudal peduncle to the eye but fades anterior to the dorsal fin; deepest body depth fits into standard length more than 3.75 times; lateral line slightly decurved; distance from the tip of the snout to dorsal fin origin equal to distance from dorsal fin origin to caudal fin origin; base of the caudal fin without a triangular/chevron-shaped black spot (sometimes seen in *Miniellus stramineus*); red-orange color lacking.......................................35

35. Length of the upper jaw less than eye diameter; mouth does not extend to beneath the eye; mid-dorsal stripe expanded into wedge-shaped spot in front of the dorsal fin and does not surround the fin base; pigment on scales forms a cross-hatch pattern..*Miniellus stramineus* sand shiner, p. 271
Length of the upper jaw greater than eye diameter; mouth extends to beneath the eye; mid-dorsal stripe not expanded into wedge-shaped spot in front of the dorsal fin and surrounds the fin base; pigment on scales does not form a cross-hatch pattern..*Alburnops blennius* river shiner, p. 208

36. Dorsal fin edge rounded; adults with a laterally compressed body (slab-sided); scales on the side (front half of body) with the pigment forming a tall diamond-shaped pattern 37
Dorsal fin edge falcate; adults with an elongated body; scales on the side (front half of the body) without pigment forming a distinct tall diamond-shaped pattern...38

37. Dorsal fin with dark pigment uniformly distributed, not forming a posterior dark spot; body depth less than 3.5 times the standard length; 9 anal rays..*Cyprinella lutrensis* red shiner, p. 229
Dorsal fin with dark pigment concentrated posteriorly forming a dark spot; body depth more than 3.5 times the standard length; 8 anal rays..*Cyprinella spiloptera* spotfin shiner, p. 232

38. Dorsal fin origin slightly anterior to pelvic fin origin; pelvic fin rays 9; anal fin rays 8–9; basicaudal spot absent..........
....*Paranotropis shumardi* silverband shiner, p. 292
      Dorsal fin origin slightly posterior to pelvic fin origin; pelvic fin rays 8; anal fin rays 8; lateral stripe expanded ventrally
      at the base of the caudal fin forming a wedge..39

39. Midline stripe in front of the dorsal fin absent or vague consisting of unconsolidated specks; usually either ovate or
      rectangular dusky blotch at the front of the dorsal fin base without striations; scales over back outlined in black and
      intensifies posteriorly; usually 9 scale rows above the lateral line at 2–3 scales anterior to the dorsal fin insertion.........
....*Paranotropis volucellus* mimic shiner, p. 295
      Midline stripe in front of the dorsal fin 1–3 lines of melanophores (striations); dusky blotch at the front of the dorsal fin
      base usually absent; scales over back not outlined in black but are covered in small melanophores, creating a dusky
      appearance that intensifies posteriorly; usually 11 scale rows above the lateral line at 2–3 scales anterior to the dorsal
      fin insertion..*Paranotropis wickliffi* channel shiner

      Note: No species account is included for the channel shiner as this species is not vouchered from either North Dakota
      or South Dakota. The mimic shiner has been confirmed via voucher specimen from the Red River of the North drainage
      in North Dakota. Reproduction of couplet characters and illustrations courtesy of Robert A. Hrabik, Missouri
      Department of Conservation.

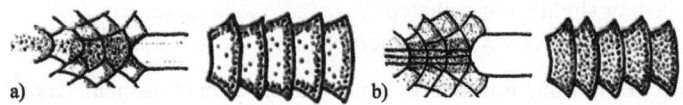

Fig. 5.47 Illustrations of the midline scale pattern anterior to the dorsal fin and posterior to the dorsal fin of the (**a**) mimic shiner (*Paranotropis volucellus*) and (**b**) channel shiner (*Paranotropis wickliffi*)

5.15 Key to Catfish of North Dakota and South Dakota

Siluriformes (Family: Ictaluridae); Four Genera: *Ameiurus*, *Ictalurus*, *Noturus*, and *Pylodictis*

1. Adipose fin fused to the back along its entire length...2
    Adipose fin with the posterior tip not attached to the dorsal surface of body...3

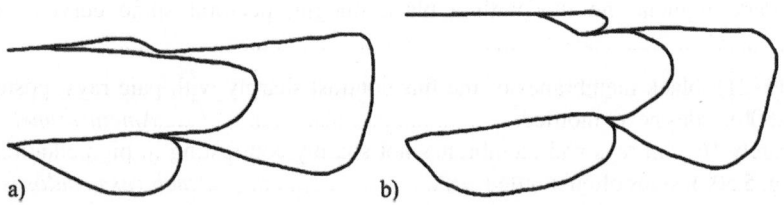

Fig. 5.48 Adipose fins of (**a**) *Noturus* and (**b**) other ictalurids

2. Caudal fin rounded; tooth patch on the upper jaw as a straight bar; lacking pale dorsal areas behind the head and at the
    posterior end of the dorsal fin; posterior basal portion of pectoral spine smooth...
....*Noturus gyrinus* tadpole madtom, p. 341
    Caudal fin truncate; tooth patch on the upper jaw with posterior extensions on each side; pale dorsal areas posterior to
    the head and at the posterior edge of the dorsal fin; posterior basal portion of the pectoral spine serrate......................
....*Noturus flavus* stonecat, p. 338

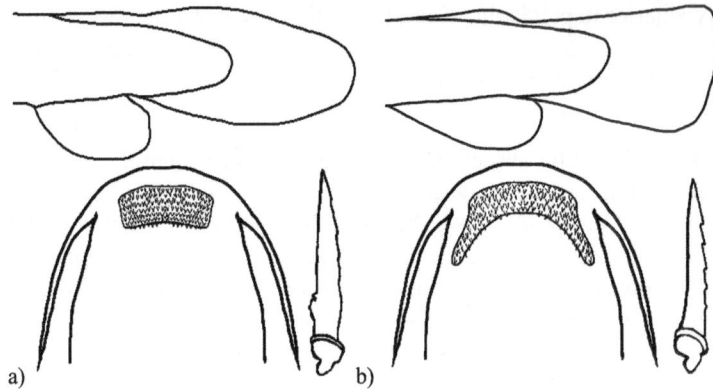

Fig. 5.49 Illustrations of upper jaw tooth patches and pectoral spines of (**a**) tadpole madtom (*Noturus gyrinus*) and (**b**) stonecat (*Noturus flavus*)

3. Anal fin rays fewer than 17; tip of the dorsal lobe of the caudal fin white (juveniles); lower jaw projects out past beyond the upper jaw; head long and flattened..*Pylodictis olivaris* flathead catfish, p. 344
   Anal fin rays 18 or more; tip of the dorsal lobe of the caudal fin never white..4

4. Caudal fin truncated, rounded, or slightly emarginated 5
   Caudal fin deeply forked; anal fin rays always 24 or more..7

5. Chin barbels pale yellowish to white and noticeably lighter than nasal barbels; anal, caudal, dorsal, and pelvic fins typically with a black margin; anal fin rays 24 or more; lower side of the body never marbled or mottled; gill rakers 14 or more; pectoral spine straight and strongly serrated posteriorly (Fig. 5.50a)..

a)   b)   c)

Fig. 5.50 Illustration of pectoral spines of (**a**) yellow bullhead (*Ameiurus natatlis*), (**b**) black bullhead (*Ameiurus melas*), and (**c**) a brown bullhead (*Ameiurus nebulosus*)

   ....*Ameiurus natalis* yellow bullhead, p. 326
   Chin barbels with dark pigment; no fins with a black margin; pectoral spine curved and strongly or weakly serrate..6

6. Gill rakers 16–20 (15–21); black membranes of the fins contrast sharply with pale rays; posterior pectoral fin serrae usually weak (Fig. 5.50b); sides never mottled...*Ameiurus melas* black bullhead, p. 323
   Gill rakers 11–15 (rarely 16); fin rays and membranes not sharply contrasting in pigmentation; posterior pectoral fin serrae prominent (Fig. 5.50c); sides often mottled...*Ameiurus nebulosus* brown bullhead, p. 329

7. Edge of the anal fin straight, anal fin rays 30 or more; never with dark spots on the sides...
   ....*Ictalurus furcatus* blue catfish, p. 332
   Edge of the anal fin rounded, anal fin rays fewer than 30; often with dark spots on the sides..
   ....*Ictalurus punctatus* channel catfish, p. 335

5.16   Key to Pikes and Mudminnows of North Dakota and South Dakota

Esociformes (Family: Esocidae); Two Genera: *Esox* and *Umbra*

1. Caudal fin rounded with a vertical black bar at the caudal fin base; scales 37 or fewer in the lateral series; jaws do not resemble a duck's beak with large canine teeth..*Umbra limi* central mudminnow, p. 357
    Caudal fin forked without a vertical black bar at the caudal fin base; scales 100 or more in the lateral series; jaws resemble a duck's beak with large canine teeth...2

2. .Each side of the lower jaw with 5 or more submandibular pores; branchiostegal rays usually 14–19 per side; opercle with scales only on the upper half 3
    Each side of the lower jaw with 4 or less submandibular pores; branchiostegal rays usually 12 or less per side; opercle fully scaled..*Esox americanus* redfin pickerel, p. 348

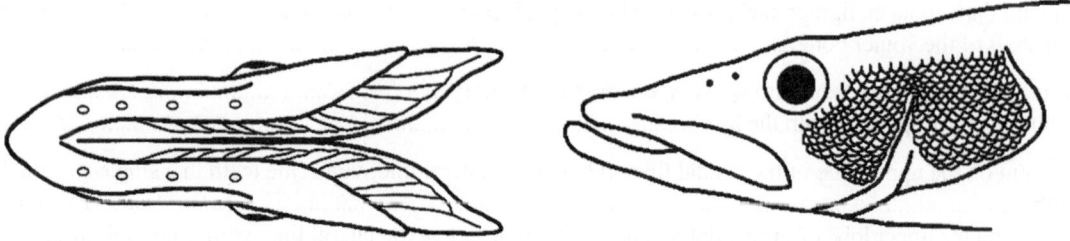

Fig. 5.51 Pattern of submandibular pores, branchiostegal rays, and opercle scalation on the redfin pickerel (*Esox americanus*)

3. Each side of the lower jaw with 5–6 submandibular pores; branchiostegal rays usually 14–16 per side; cheek fully scaled; sides and back green with whitish to yellow spots..*Esox lucius* northern pike, p. 351
    Each side of the lower jaw with 6 or more submandibular pores; branchiostegal rays usually 17–19 per side; cheek with scales only on the upper half; sides and back light yellowish-green with dark spots, blotches, or bars...............................
    ....*Esox masquinongy* muskellunge, p. 354

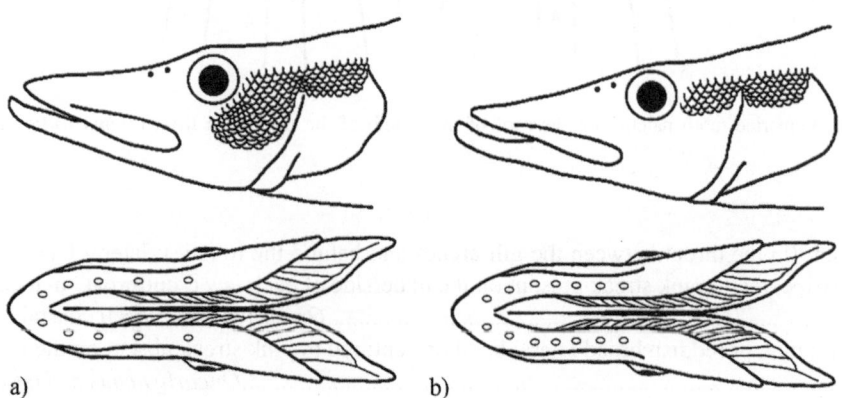

Fig. 5.52 Pattern of submandibular pores, branchiostegal rays, and opercle scalation on the (**a**) northern pike (*Esox lucius*) and (**b**) muskellunge (*Esox masquinongy*)

5.17   Key to Salmonids of North Dakota and South Dakota

Salmoniformes (Family: Salmonidae); Five Genera: *Coregonus, Oncorhynchus, Prosopium, Salmo*, and *Salvelinus*

1. Spots absent on the body and fins; body silvery without patterns or spots; lateral scale rows less than 98...........................2
    Spots present on the body and fins; body color variable with spots or patterns; lateral scale rows more than 100..................4

2. Two flaps of skin between the nostrils; genus *Coregonus*..3
    One flap of skin between the nostrils...*Prosopium gemmifer* Bonneville cisco
    Note: No species account is included as the species is not established in the state. A single specimen exists at the South
    Dakota State University Fish Collection (SDSTATE 3230)

3. Lower jaw extends up to or beyond the tip of the snout; gill rakers 36–64.........................*Coregonus artedi* cisco, p. 361
    Snout overhangs the lower jaw; gill rakers 24–33.....................................*Coregonus clupeaformis* lake whitefish, p. 364

4. Anal rays 12 or less...5
    Anal rays 13 or more 10

5. Caudal fin deeply forked; no black spots in the dorsal fin...................................*Salvelinus namaycush* lake trout, p. 389
    Caudal fin slightly forked, emarginated, or straight-edged; black spots or vermiculations on the dorsal fin.......................6

6. Dorsum with pale spots on dark background; white edge on the pectoral, pelvic, anal, and caudal fins; teeth only on the
    head of the vomer bone...*Salvelinus fontinalis* brook trout, p. 386
    Dorsum with dark spots on lighter background; white edge, if present, only on the anal and pelvic fins; teeth on both the
    head and shaft of the vomer bone...7

7. Caudal fin without dark spots or with spots on the dorsal lobe only..8
    Caudal fin with dark spots on both the lobes..9

8. Spots present on the upper lobe of the caudal fin and below the lateral line; vomerine teeth in a single row on the roof of
    the mouth...*Salmo trutta* brown trout, p. 382
    Spots absent on the upper lobe of the caudal fin and few to none below the lateral line; vomerine teeth in a zigzag pattern
    of two rows on the roof of the mouth...*Salmo salar* Atlantic salmon, p. 379

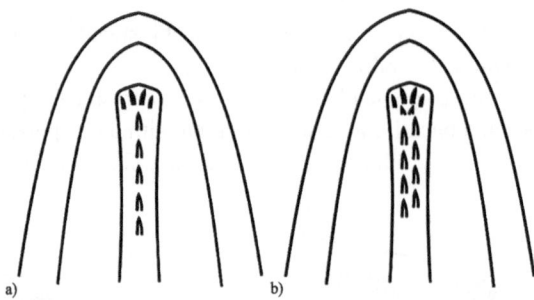

Fig. 5.53 Illustrations of vomerine tooth patches on the roof of the mouth of the (**a**) brown trout (*Salmo trutta*) and (**b**) Atlantic salmon
(*Salmo salar*)

9. Hyoid teeth present (in the throat between the gill arches and behind the tongue); lateral band absent or present as a
    narrow faint red stripe; red to pink streak present on the underside of each jaw ("cutthroat" mark).................................
    ....*Oncorhynchus clarkii* coastal cutthroat trout, p. 367
    Hyoid teeth absent; pink or reddish broad lateral band present; red to pink streak absent on the underside of each jaw..
    ....*Oncorhynchus mykiss* rainbow trout, p. 370

10. Black spots absent on the back and upper lobe of the caudal fin.................*Oncorhynchus nerka*, kokanee salmon, p. 373
    Note: No specimens exist from North Dakota or South Dakota, but the species was stocked in both states. It was
    stocked at the Pactola Reservoir in South Dakota in 2022, so individuals may be collected in the reservoir or Rapid Creek.
    Black spots present on the back and upper lobe of the caudal fin..11

11. Black spots present on the lower lobe of the caudal fin; gums black at the base of teeth...
    ....*Oncorhynchus tshawytscha* Chinook salmon, p. 376
    Black spots absent on the lower lobe of the caudal fin; gums white at the base of the teeth...
    ....*Oncorhynchus kisutch* coho salmon
    Note: No species account is included as the species is not established in the state. A single specimen exists at the South
    Dakota State University Fish Collection (SDSTATE 3215).

5.18 Key to Smelts of North Dakota and South Dakota

Osmeriformes (Family: Osmeridae); One Genus: *Osmerus*

1. Dorsal fin origin at about mid-body, approximately equidistant between the snout and the base of the caudal fin; scales cycloid; lateral line absent; 60–72 scales in the lateral series; adipose fin present; large teeth present; body silvery.........
....*Osmerus mordax* rainbow smelt, p. 393

5.19 Key to Trout-Perch of North Dakota and South Dakota

Percopsiformes (Family: Percopsidae); One Genus: *Percopsis*

1. Dorsal fin origin nearer the snout than the caudal fin base; scales ctenoid; lateral line present; 40–55 scales in the lateral series; adipose fin present; body with brown or black spots on a beige to light brown background
....*Percopsis omiscomaycus* trout-perch, p. 397

5.20 Key to Burbot of North Dakota and South Dakota

Gadiformes (Family: Gadidae); One Genus: *Lota*

1. Two dorsal fins, the first short with 8–16 rays, the second long with 60–80 rays; barbel at the tip of the chin; scales small and embedded...*Lota lota* burbot, p. 401

5.21 Key to Cichlids of North Dakota and South Dakota

Perciformes (Family: Cichlidae); Two Genera: *Rocio* and *Vieja*

1. Black spot present dorsally at the caudal fin base..
....*Rocio octofasciata* Jack Dempsey, p. 405
Black blotch present medially at the caudal fin base and may extend anteriorly to the origin of the anal fin as a wide stripe...*Vieja melanura* redhead cichlid
Note: No species account is included as the species is not established in the state. A single specimen exists at the Florida Museum Fish Collection (UF 96525).

5.22 Key to Topminnows of North Dakota and South Dakota

Cyprinodontiformes (Family: Fundulidae); One Genus: *Fundulus*

1. Dorsal fin origin posterior to anal fin origin; side lacking vertical stripes..
....*Fundulus sciadicus* plains topminnow, p. 415
Dorsal fin origin over or anterior to anal fin origin; side with vertical stripes...2

2. Peritoneum silvery; eye larger, orbit contained 1.3 to 1.7 in postorbital length of head...
....*Fundulus diaphanus* banded killifish p. 409
Peritoneum black; eye smaller, orbit contained 2.0 to 2.5 times in postorbital length of head...
....*Fundulus kansae* northern plains killifish p. 412

5.23   Key to Centrarchids of North Dakota and South Dakota

Perciformes (Family: Centrarchidae); Four Genera: *Ambloplites*, *Lepomis*, *Micropterus*, and *Pomoxis*

1. Anal fin with five or more spines..2
   Anal fin with four or fewer spines...4

2. Dorsal fin with 8 or fewer spines...3
   Dorsal fin with 10 or more spines...*Ambloplites rupestris* rock bass, p. 419

3. Dorsal spines 7–8; dorsal fin soft rays 15–16; side of body randomly mottled (no vertical bars); distance from the origin of the dorsal fin to the posterior edge of the eye equal to the length of the dorsal fin base...
   ....*Pomoxis nigromaculatus* black crappie, p. 446
   Dorsal spines 5–6; dorsal fin soft rays 14; side of body with distinct vertical bars; distance from the origin of the dorsal fin to the posterior edge of the eye greater than the length of the dorsal fin base..
   ....*Pomoxis annularis* white crappie, p. 443

4. Lateral scale rows 53 or more; dorsal fins nearly separate; body elongated and robust; standard length greater than three times body depth..5
   Lateral scale rows 52 or less; dorsal fins continuous; body deep and laterally compressed; standard length less than three times body depth 6

5. Upper jaw does noes extend past the posterior edge of the eye; sides pigmented with vertical bars or spots................
   ....*Micropterus dolomieu* smallmouth bass, p. 437
   Upper jaw extends past the posterior edge of the eye; sides with a continuous black lateral stripe..............................
   ....*Micropterus nigricans* largemouth bass, p. 440

6. Pectoral fin not extending past the front of the eye when bent forward (commonly to the edge for *L. humilis*, but not past it)..7
   Pectoral fin extending past the front of the eye when bent forward...9

7. Pectoral fins short and rounded; mouth large and bass-like with the maxilla extending to or past the middle of the eye; lateral scale rows 41–53; large black spot present in the posterior basal portion of the dorsal fin..................................
   ....*Lepomis cyanellus* green sunfish, p. 422
   Pectoral fins longer and pointed; mouth not large and bass-like, maxilla not reaching the middle of the eye; lateral scale rows 32–46; a large black spot lacking in the posterior basal portion of the dorsal fin...8

8. Short earflap, length shorter than eye diameter; long, thin gill rakers on first gill arch...
   ....*Lepomis humilis* orangespotted sunfish, p. 428
   Long earflap, length longer than eye diameter (may be shorter in juveniles); short, thick gill rakers on first gill arch.....
   ....*Lepomis megalotis* longear sunfish
   Note: No species account is included becacuse this species is not established in the state. A single specimen exists from a single occurence from wolf Creek, McCook County. This voucher is at the Smithsonian museum of Natural History (USNM 366370).

9. Large black spot present in the posterior basal portion of the dorsal fin (may be absent in juveniles); opercular lobe black to the margin; long, thin gill rakers on the first gill arch................................*Lepomis macrochirus* bluegill, p. 431
   No basal dark spot in the posterior dorsal fin; opercular lobe with a pale margin or a spot at the tip; short, thick gill rakers on first gill arch...10

10. Profile of the head pointed; soft dorsal fin lacking distinct wavy lines; pectoral fins extremely long, extending nearly to or beyond the dorsal fin base when angled dorsally; stiff posterior edge of the opercle...
    ....*Lepomis microlophus* redear sunfish, p. 434
    Profile of the head rounded; soft dorsal fin with distinct wavy lines; pectoral fins shorter, extending to about 3–5 scale rows below the dorsal fin base when angled dorsally; flexible posterior edge of the opercle..
    ....*Lepomis gibbosus* pumpkinseed, p. 425

5.24   Key to Temperate Basses of North Dakota and South Dakota

Perciformes (Family: Moronidae); One Genus: *Morone*

1. Spinous and soft dorsal fin connected by a membrane; no patches of teeth on the rear of the tongue; second anal spine as long as the anal fin base when depressed...*Morone mississippiensis* yellow bass, p. 453
   Spinous and soft dorsal fin separate; one to two patches of teeth on the rear of the tongue; second anal spine noticeably shorter than the anal fin base 2

2. Two patches of teeth on the rear of the tongue; body depth less than 33% of the standard length; dorsal fin soft rays usually 12..*Morone saxatilis* striped bass, p. 456
   One patch of teeth on the rear of the tongue; body depth greater than 33% of standard length; dorsal fin soft rays usually 12 or more..*Morone chrysops* white bass, p. 450

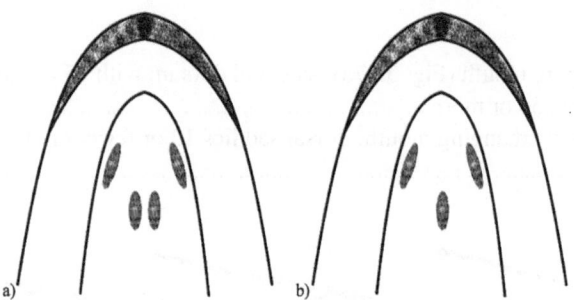

Fig. 5.54 Illustrations of tooth patches on the tongue of (**a**) striped bass (*Morone saxatilis*) and (**b**) white bass (*Morone chrysops*)

5.25   Key to Percids of North Dakota and South Dakota

Perciformes (Family: Percidae); Four Genera: *Etheostoma, Perca, Percina,* and *Sander*

1. Branchiostegal rays 7–8 per side; rear margin of the preopercle strongly serrate...2
   Branchiostegal rays 5–6 per side; rear margin of the preopercle smooth..5

2. Anal fin with 6–8 soft rays; jaws without canine teeth..*Perca flavescens* yellow perch, p. 466
   Anal fin with 11–14 soft rays; jaws with prominent canine teeth..3

3. Opercular spine present..4
   Opercular spine absent...*Sander lucioperca* zander, p. 485

4. Cheek with few or no scales; spinous dorsal fin lacking dark spots on membranes forming longitudinal rows, except for a large black spot present on the posterior basal membranes of the last 2–5 spines of the spinous dorsal fin; pyloric caeca 3..*Sander vitreus* walleye, p. 488
   Cheek with scales abundant; spinous dorsal fin with dark spots on membranes forming longitudinal rows (without a dark spot present on the posterior basal membranes near the base of the last 2–5 spines), pyloric caeca 4 or more.................
   ....*Sander canadensis* sauger, p. 482

5. Sides with pigment on the edge of scales forming W's, M's, or X's...
   ....*Etheostoma nigrum* Johnny darter, p. 463
   Sides without pigment on the edge of scales forming W's, M's, or X's..6

Fig. 5.55 Lateral scale pigment forming W's, M's, or X's on Johnny darter (*Etheostoma nigrum*)

6. No enlarged or modified scales present between the pelvic fins or on the belly; side with pigment forming vertical bars; males with red or blue on the sides..*Etheostoma exile* Iowa darter, p. 460
   One or more enlarged and modified ctenoid scales present between the pelvic fins; midline of the belly either naked or with a row of modified scales (absent in females)..7

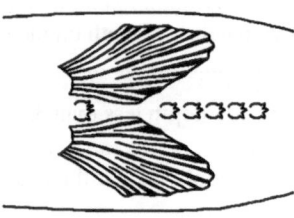

Fig. 5.56 Ventral view of the modified ctenoid scales of *Percina*

7. Conical, fleshy snout overhanging mouth (Fig. 5.57a); side and dorsum with 15 or more narrow vertical bars of various lengths; total dorsal fin elements 30 or more...*Percina caprodes* logperch, p.470
   Snout pointed to blunt and not overhanging mouth; dorsal saddles 13 or fewer or absent; total dorsal fin elements 29 or fewer...8

Fig. 5.57 Head shape of (**a**) logperch (*Percina caprodes*), (**b**) slenderhead darter (*Percina phoxocephala*), and (**c**) river darter (*Percina shumardi*)

8. Lateral blotches usually 9 or fewer and usually wider than tall..........................*Percina maculata* blackside darter, p. 473
   Vertically expanded lateral blotches usually 10 or more and usually taller than wide..9

9. Spinous dorsal fin without an anterior and posterior black basal spot; head shape pointed and elongated (Fig. 5.57b); premaxillary frenum well developed and connecting the upper lip to the snout...
   ....*Percina phoxocephala* slenderhead darter, p. 476
   Spinous dorsal fin with an anterior and posterior black basal spot; head shape curved downward (Fig. 5.57c); premaxillary frenum absent or weakly developed..*Percina shumardi* river darter, p. 479

5.26   Key to Drums of North Dakota and South Dakota

Perciformes (Family: Sciaenidae); One Genus: *Aplodinotus*

1. Pointed caudal fin; spinous dorsal fin with 10 spines; soft dorsal fin with 29–32 rays; extremely long outer pelvic ray; silvery body (gray-black in older individuals); strongly arched body; subterminal mouth..
   ....*Aplodinotus grunniens* freshwater drum, p. 493

5.27   Key to Sticklebacks of North Dakota and South Dakota

Gasterosteiformes (Family: Gasterosteidae); One Genus: *Culaea*

1. Scaleless; narrow caudal peduncle; 4–6 dorsal finlets, followed by a dorsal fin with 14–16 rays; pelvic fin with 1 spine (and 1 ray)...*Culaea inconstans* brook stickleback, p. 497

*Illustrations throughout the keys were either produced by the authors or redrawn from those found in Etnier and Starnes (1993), Ross and Brenneman (2001), or Page and Burr (2011).

References

Etnier DA, Starnes WC (1993) The fishes of Tennessee. University of Tennessee Press, Knoxville

Page LM, Burr BM (2011) Peterson field guide to freshwater fishes of North America North of Mexico, 2nd edn. Houghton Mifflin Harcourt, Boston

Ross ST, Brenneman WM (2001) The inland fishes of Mississippi. University Press of Mississippi, Jackson

Chapter 6
Species Accounts

Kathryn E. Schlafke, Matthew D. Wagner, and Chelsey A. Pasbrig

6.1 Species Accounts

Color Photographs: At least one color photograph of each species is included. When possible, a press box was used to photograph live fish. Otherwise, images of preserved specimens or live fish were obtained from external sources. The images may not capture sexual dimorphism or differences between juveniles and adults in some species.

Distribution Maps: A point distribution map depicts the locations of each species. Each map was created using a 500,000-record database maintained the SDGFP. However, this georeferenced database may not accurately represent species presence and distribution because many areas are under surveyed and some areas, such as the Missouri River reservoirs, lack georeferenced data. In addition to the database, information was included from the South Dakota State University Fish Collection. This collection contains 6000 lots (one or more specimens of a single species collected at a single location at one point in time) from both ND and SD. Because sampling techniques became more standardized in both states in 1990, records on the maps are labeled as either pre-1990 or post-1990.

Etymology: This section provides the meaning of the binomial (*Genus species*) scientific name for each species.

Description: The morphological and anatomical detail in this section can be used as a reference for species identification. Pertinent information is included for the following: body shape, coloration and patterning, scute counts, head shape and size relative to the body, snout shape, eye size and placement, mouth size and orientation, lip description, frenum presence, barbel counts, tooth size and orientation, gill raker size and number, description of all fins, caudal peduncle size, axillary process presence, lateral line description with scale counts, scale type, spawning characteristics, sexual dimorphism, and differences between juveniles and adults.

Similar Species: This section notes other fish species similar in appearance and describes easily observable differences.

Distribution and Habitat: This section expands on the distribution maps by describing the range of the species throughout North America and overall habitat characteristics. Additional information on distribution in the Dakotas, including possible limitations of the maps, is also included. For example, the burbot *Lota lota* distribution map has gaps between observation points in Lake Oahe, even though burbot are known to have a continuous distribution throughout the lake. This is because burbot are not targeted or routinely sampled during annual fish population surveys, leaving some sections of Lake Oahe without an observation point.

Reproduction: This section explains the species reproductive habits, such as the age at sexual maturity, spawning season, spawning behavior and habitat, fecundity, and egg description.

Age and Growth: The longevity and maximum length of the species, in addition to the average age or average growth measurements at each age, are described in this section.

Food and Feeding: This section describes diets and foraging behaviors.

The original version of the chapter has been revised. A correction to this chapter can be found at https://doi.org/10.1007/978-3-031-38040-2_7

K. E. Schlafke
Montana Fish, Wildlife & Parks, Bozeman, MT, USA

M. D. Wagner
U.S. Fish and Wildlife Service, Jackson, MS, USA

C. A. Pasbrig (✉)
South Dakota Game, Fish and Parks, Pierre, SD, USA
e-mail: Chelsey.Pasbrig@state.sd.us

6.2 Petromyzontidae, Lamprey Family

The lamprey family consists of 42 species in eight genera. Members of Petromyzontidae are only found in the cooler waters of North America, Asia, and Europe. Lamprey species can either be entirely freshwater or anadromous, swimming to marine waters after undergoing metamorphosis in freshwater. Two species, the chestnut lamprey *Icthtyomyzon castaneus* and the silver lamprey *Icthtyomyzon unicuspis*, are native to the Hudson Bay, St. Lawrence-Great Lakes, and Mississippi River basins and the Dakotas.

All lamprey species have two main life stages: ammocoetes (larvae) and adults. Ammocoetes are blind, with skin covering their eyes. They also have a toothless hood-shaped mouth. Ammocoetes burrow into sandbars, soft substrates, or beds of organic debris. They filter-feed by projecting their heads out of the water column to catch drifting plankton and bacteria. Metamorphosis varyies by species, with functioning eyes, seven gill openings on each side, and rows of teeth and fringed papillae in a round or oval-shaped sectorial mouth developing over several years. Adult lampreys are entirely cartilaginous and scaleless, with no anal, pelvic, or pectoral fins. They have long, slender, snake-like bodies, and a single dorsal fin continuous with the caudal fin. Lampreys in the Dakotas may be confused with the American eel *Anguilla rostrata*, family Anguillidae. However, eels are easily distinguished by the presence of jaws and pectoral fins.

Lamprey species can be separated into two groups: parasitic and nonparasitic. Parasitic lampreys attach primarily to other ray-finned fish hosts after metamorphosis, using the sectorial disc to feed on host blood, tissue, and other fluids. This parasitism often does not kill the host, but likely weakens its immunity to disease and bacteria. Nonparasitic lamprey do not feed after completing metamorphosis. The chestnut and silver lamprey are both parasitic. After feeding for 1 to 2 years, both species migrate in the late winter and early spring upstream for spawning grounds during the spring and summer. Both chestnut and silver lamprey prefer to spawn over sand or gravel substrate in shallow reaches of smaller, low-order streams with moderate current. Spawning behavior is very similar for all species. Female lampreys attach to a hard and sturdy surface within or over a nest using their mouth. Males attach themselves to the female's head and wrap their bodies around her. Their bodies vibrate together to release eggs and sperm. Like all species of lamprey, chestnut and silver lamprey are semelparous, dying after a single spawning event.

6.2.1   Chestnut Lamprey *Ichthyomyzon castaneus*

Girard, 1858

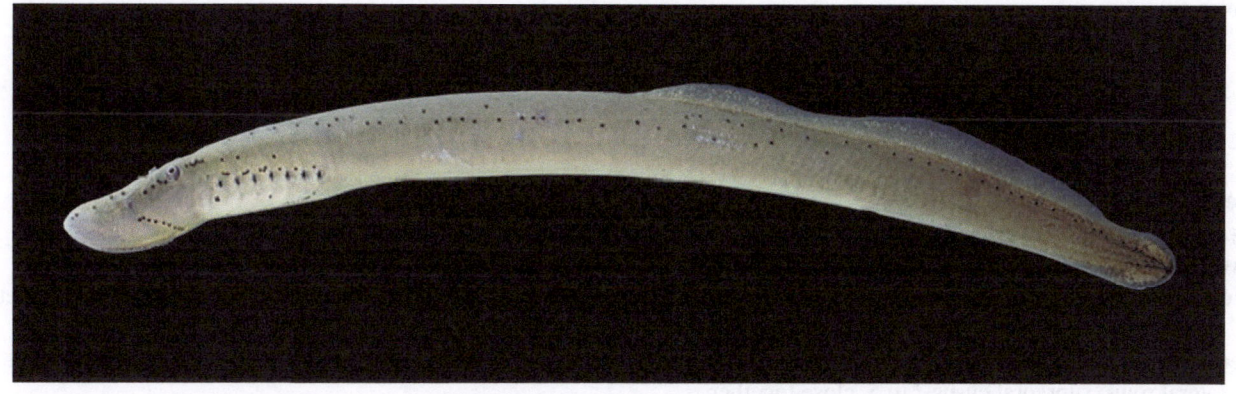

Etymology: *Ichthyomyzon castaneus – Ichthyomyzon* = sucker of fish; *castaneus* = chestnut.

Description:

Body – elongated, cylindrical, laterally compressed posteriorly.
Color –

> *Dorsally* – gray, tan, to olive.
> *Laterally* – tan to olive, sometimes mottled.
> *Ventrally* – light brown to light yellow-cream.
> *Fin* – dorsal: brown tint.

Head – small, cylindrical.
Eyes – small, dorsolaterally on head.
Jaw – absent.
Mouth – disk or oval shaped outlined with fringed papillae; maximum diameter greater than body maximum diameter in adults.
Teeth – circumorals bicuspid (double-pointed), sharp, claw like, 19 to 25 in circumoral row; 3 to 6 in anterior row; 7 to 11 in lateral rows; supraoral cusps 2 to 5, close together.
Gill openings – 7, small, pore-like on each side.
Dorsal fin elongated; shallow notch lacking spines or rays; connected to a protocercal caudal fin; continuous with anal fin.
Lateral line – lateral sides with 50 to 56 myomeres between the last gill opening and vent; small, pigmented lateral pore.
Skin – scaleless.
Juveniles – larvae darker tan to brown; hood-shaped mouth without fringed papillae; lack eyes and sucking disk.

Similar Species: Chestnut lamprey closely resemble silver lamprey. Silver lamprey are more gray, have unicuspid (single-pointed) circumoral teeth, 46 to 53 lateral side myomeres, and pigmented lateral pores in adults. American eel have a snake-like body with jaws and pectoral fins.

Distribution and Habitat: Chestnut lamprey are native to southern Canada and the upper mid-western United States. They range from the Red River of North in eastern North Dakota and western Minnesota to the west, through the St. Lawrence-Great Lakes, upper Mississippi, and Ohio River to the east, and south to the Gulf of Mexico in Louisiana and Texas (Page and Burr 2011). They occur more frequently in smaller waterbodies than the silver lamprey (Becker 1983). Adults are typically found within medium-sized streams, large rivers, and impoundments with moderate flow over gravel, sand, or silt substrate (Cross 1967; Becker 1983; Pflieger 1997). Ammocoetes require backwater areas with moderate current containing sand and silt substrate, beds of organic matter, and light vegetation such as *Chara* (Scott and Crossman 1973).

Reproduction: Upstream spawning migrations to shallow reaches of low-order streams with moderate current begin in late winter and early spring. Spawning dates depend on latitude, but preferred spawning water temperature is less variable (Cochran 2014). Spawning aggregations in Wisconsin occur from May to June, in 15 to 22 °C water at depths of 18 to 46 cm (Cochran 2014). Females attach to a hard surface, such as a rock, using their mouth, and males wrap themselves around the females, instigating a quivering motion to release eggs (Case 1970). Adults construct nests by moving stones with their mouths and excavating a depression roughly 60 cm wide, 100 cm long, and 5 cm deep, within gravel or sand substrate (Case 1970). As many as 50 chestnut lampreys have been observed spawning in the same nest (Case 1970). Fecundity increases with female length. Fecundity ranged from 10,144 to 18,563 eggs per female in Michigan, with 13,000 eggs reported from a 255 mm long female in Wisconsin (Starrett et al. 1960; Becker 1983). Egg diameter has been observed to be 0.6–0.7 mm (Becker 1983). Death occurs shortly after spawning, so no parental care of eggs or larvae occurs. After hatching, larvae (ammocoetes) drift downstream and eventually burrow into sand bars or beds of organic debris (Scott and Crossman 1973). After 5 to 7 years, ammocoetes metamorphosize into the adults beginning in late summer and by the following spring emerge from their burrows to seek out a host (Hall 1963).

Age and Growth: Little information is available on age and growth. Newly morphed adults emerge from their burrows at a total length of approximately 100 mm (Hall 1963). During the adult parasitic phase, most growth takes place during summer months (Cochran 2014). Females tend to achieve greater lengths than males, with maximum lengths of 356 mm (Starrett et al. 1960; Moore and Kernodle 1965; Cochran 2014). Longevity is presumed to be 6 to 7 years.

Food and Feeding: Ammocoetes raise their heads from their burrows to filter feed drifting prey such as diatoms, phytoplankton, detritus, and bacteria (Becker 1983). Adults are parasitic, attaching predominantly to the dorsal side of the host fish by cutting through scales with their teeth and then extracting blood (Hall 1963; Cochran 1986, 2014). The majority of feeding occurs from May to October and peaks in July, with no feeding in winter (Hall 1963; Cochran 2014). Parasitic adults tend to be size selective. They are strongly associated with relatively large hosts having small scales or naked skin, such as paddlefish, northern pike, catfish species, and redhorse species (Cochran 1994, 2014). Adults are parasitic for 1 to 2 years before migrating to spawning grounds, reproducing, and ending their lifecycle (Hall 1963).

References

Becker GC (1983) Fishes of Wisconsin. University of Wisconsin Press, Madison

Case B (1970) Spawning behavior of the chestnut lamprey (*Ichthyomyzon castaneus*). J Fish Res Board Can 27:1872–1874. https://doi.org/10.1139/f70-207

Cochran PA (1986) The daily timing of lamprey attacks. Environ Biol Fish 16:325–329. https://doi.org/10.1007/BF00842989

Cochran PA (1994) Why lampreys and humans "compete" (and when they don't): toward a theory of host species selection by parasitic lampreys. In: Stouder DJ, Fresh KL, Feller RJ (eds) Theory and application in fish feeding ecology, vol 18. Belle W. University of South Carolina Press, Columbia, pp 329–345

Cochran PA (2014) Field and laboratory observations on the ecology and behavior of the chestnut lamprey *Ichthyomyzon castaneus*. J Freshw Ecol 29:491–505. https://doi.org/10.1080/02705060.2014.910477

Cross FB (1967) Handbook of fishes in Kansas. University Press of Kansas, Lawrence

Hall JD (1963) An ecological study of the chestnut lamprey, *Ichthyomyzon castaneus* (Girard), in the Manistee River, Michigan. Dissertation, University of Michigan

Moore GA, Kernodle M (1965) A new size record for the chestnut lamprey, *Ichthyomyzon castaneus* (Girard) in Oklahoma. Proc Okla Acad Sci 1965:68–69

Page LM, Burr BM (2011) Peterson field guide to freshwater fishes of North America North of Mexico, 2nd edn. Houghton Mifflin Harcourt, Boston

Pflieger WL (1997) The fishes of Missouri, revised edn. Missouri Dep Cons, Jefferson City

Scott WB, Crossman EJ (1973) Freshwater fishes of Canada, Bulletin 184. Fisheries Research Board of Canada

Starrett WC, Harth WJ, Smith PW (1960) Parasitic lampreys of the genus *Ichthyomyzon* in the rivers of Illinois. Copeia 1960:337–346. https://doi.org/10.2307/1439761

6.2.2   Silver Lamprey *Ichthyomyzon unicuspis*

Hubbs and Trautman, 1937

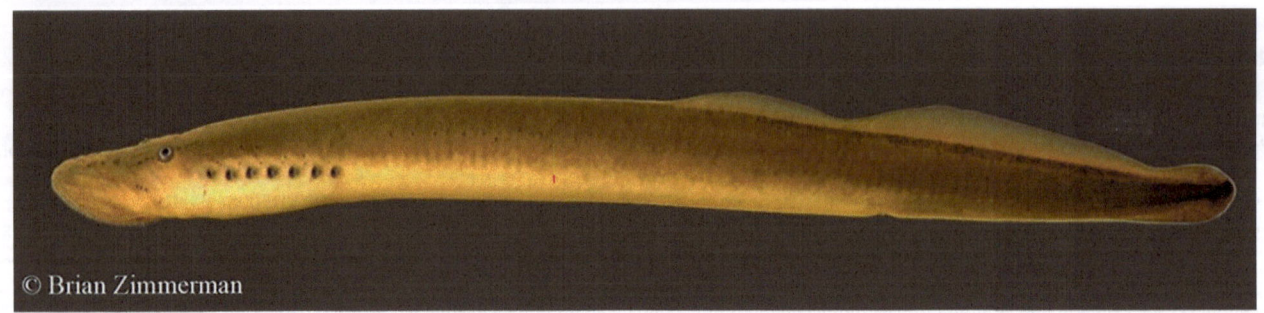

© Brian Zimmerman

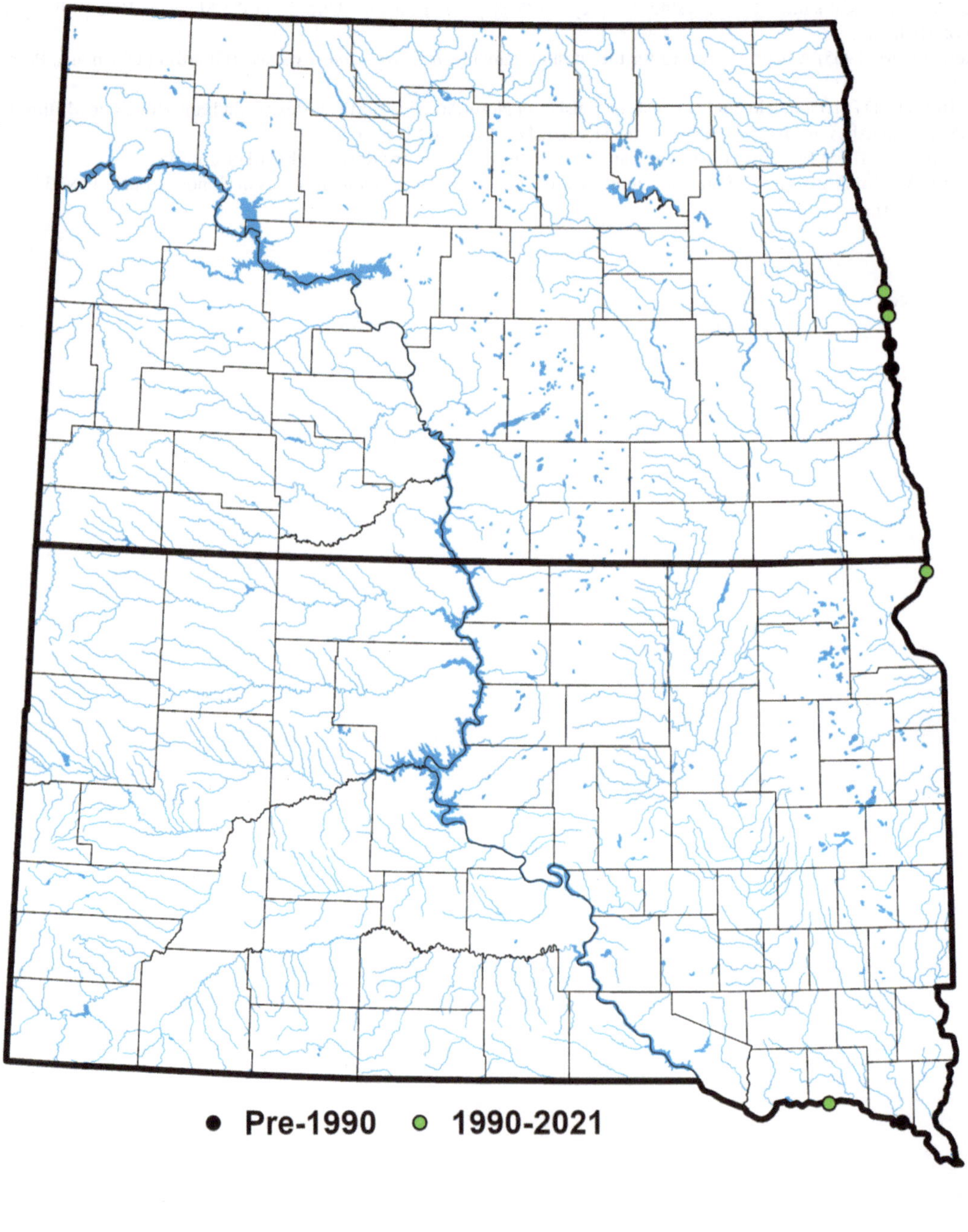

● Pre-1990   ○ 1990-2021

Etymology: *Ichthyomyzon unicuspis* – *Ichthyomyzon* = sucker of fish; *unicuspis* = one elevated tooth part, referencing cirumoral teeth.

Description:

Body – elongated, cylindrical, laterally compressed posteriorly.
Color –

> *Dorsally* – dusky blue to dark gray.
> *Laterally* – fades to bluish gray.
> *Ventrally* – fades to dark gray.
> *Fins* – dorsal – may have a brown tint.

Head – small, cylindrical.
Eyes – small, dorsolaterally on head.
Jaw – absent.
Mouth – disk or oval shaped outlined with fringed papillae; maximum diameter greater than body maximum diameter in adults.
Teeth – circumorals unicuspid (single point), sharp, claw-like; 18 to 25 teeth in circumoral row (in rare cases, they may possess up to two bicuspid circumorals); 2 to 4 in anterior row; 4 to 9 in lateral rows; supraoral cusps 1 to 4, close together.
Gill rakers – 7 gill openings, small, pore like on each side.
Dorsal fin – elongated with shallow notch lacking spines or rays; connected to a protocercal caudal fin; continuous into the anal fin.
Lateral line – lateral sides, 46 to 53 myomeres between the last gill opening and vent; lateral pores small, unpigmented in ammocoetes, pigmented in large adults.
Skin – scaleless.
Juveniles – hood-shaped mouth without fringed papillae; lack eyes and sucking disk.

Similar Species: Silver lamprey closely resemble chestnut lamprey. Chestnut lamprey are tan to brown, with bicuspid (double pointed) circumoral teeth and 50 to 56 myomeres on the lateral sides. Adults have unpigmented lateral pores. American eel have a snake-like body with jaws and pectoral fins.

Distribution and Habitat: The western edge of the silver lamprey native range is the Red River of North in eastern North Dakota and western Minnesota. The range continues east through the St. Lawrence River, Great Lakes, upper Mississippi, Ohio River to New York, and south to central Tennessee. Silver lamprey range expansion has occurred because of canal connection. In the Dakotas, it is found in the Red River of North in North Dakota, the Bois de Sioux River in northeastern South Dakota, the Missouri River near Lewis, and Clark Lake in South Dakota. Adults are found within streams and the mainstems and tributaries of large rivers, as well as lakes and impoundments with gravel substrate. Ammocoetes require sand bars or beds of organic debris within muddy pools or backwaters.

Reproduction: Information on reproductive behavior is scarce. Upstream spawning migrations to areas near run to riffle habitat with gravel, cobble, and boulder substrate begin when water temperature reaches approximately 10 °C (Becker 1983; Cochran and Lyons 2004). Males arrive at the spawning grounds first and may attract females by releasing sexual pheromones (Hardisty 1986; Cochran and Marks 1995). Spawning aggregations occur May through June, in clear to turbid waters at temperatures of 13–23 °C and depths of 23–79 cm (Morman 1979; Cochran and Marks 1995; Cochran and Lyons 2004). As many as ten individuals have been observed spawning in the same nest (Morman 1979). Nests are constructed in sand or gravel substrate in moderate-gradient streams near riffles 8 to 15 cm deep (Trautman 1957; Scott and Crossman 1973; Morman 1979). They range in size from 30 to 122 cm in diameter and 2.5 to 15 cm in depth (Trautman 1957; Scott and Crossman 1973; Morman 1979). Fecundity increases with female length (Schuldt et al. 1987; Cochran and Lyons 2004). Fecundity was reported as 19,012 eggs and 27,400 eggs from 247 and 326 mm (total length) females, respectively (Manion and Hanson 1980; Becker 1983). Egg diameter is 0.8 to 1.0 mm (Manion and Hanson 1980; Becker 1983). Death occurs shortly after spawning. Ammocoetes (larvae) hatch within 5 days at water temperatures of 18 °C (Smith et al. 1968; Scott and Crossman 1973; Becker 1983). They drift downstream and eventually burrow into sand bars or beds of organic debris where they remain for 4 to 7 years (Smith et al. 1968; Scott and Crossman 1973; Becker 1983). Ammocoetes metamorphosize into the adults beginning in August, emerging the following spring to seek a host (Becker 1983).

Age and Growth: Little information is available on age and growth. Newly morphed adults emerging from their burrows range in total length from 89 to 152 mm (Becker 1983). Adult growth occurs mainly during the summer months (Cochran and Marks 1995). Females tend to achieve greater lengths than males, with maximum lengths of 395 mm (Starrett et al. 1960; Becker 1983; Cochran and Marks 1995). Longevity is 6 to 7 years (Cochran and Marks 1995).

Food and Feeding: Ammocoetes raise their heads from their burrows to filter feed drifting prey such as algae, pollen, diatoms, and protozoans (Harlan and Speaker 1951; Becker 1983). Adults are parasitic and typically attach in the springtime to the dorsal side of their host (Cochran and Lyons 2004). They cut through host fish scales with their teeth and then extract blood (Manion and Hanson 1980; Hardisty 1986; Cochran and Lyons 2004). Adults also rely on food they acquired as ammocoetes. Lampreys tend to select relatively large hosts having small scales or naked skin and are strongly associated with paddlefish, lake trout, catfish species, and sturgeon species (Morman 1979; Becker 1983; Cochran 1994; Cochran and Lyons 2004). Although large and numerous, common carp are often not selected because silver lamprey cannot puncture their large, heavy scales (Cochran and Lyons 2004). Adults are parasitic for approximately 1 year before migrating to spawning grounds and ending their lifecycle.

References

Becker GC (1983) Fishes of Wisconsin. University of Wisconsin Press, Madison

Cochran PA (1994) Why lampreys and humans "compete" (and when they don't): toward a theory of host species selection by parasitic lampreys. In: Stouder DJ, Fresh KL, Feller RJ (eds) Theory and application in fish feeding ecology, vol 18. Belle W. University of South Carolina Press, Columbia, pp 329–345

Cochran PA, Lyons J (2004) Field and laboratory observation on the ecology and behavior of the silver lamprey (*Ichthyomyzon unicuspis*) in Wisconsin. J Freshw Ecol 19:245–253. https://doi.org/10.1080/02705060.2004.9664538

Cochran PA, Marks JE (1995) Biology of the silver lamprey, *Ichthyomyzon unicuspis*, in Green Bay and the lower Fox River, with a comparison to the sea lamprey, *Petromyzon marinus*. Copeia 1995:409–421. https://doi.org/10.2307/1446904

Hardisty MW (1986) General introduction to lampreys. In: Holčík J (ed) The freshwater fishes of Europe, vol 1, part I, Petromyzontiformes. AULA, Verlag, Weisbaden, pp 19–83

Harlan JR, Speaker EB (1951) Iowa fish and fishing, 2nd edn. State of Iowa, Des Moines. https://doi.org/10.2307/1439117

Manion PJ, Hanson LH (1980) Spawning behavior and fecundity of lampreys from the upper three Great Lakes. Can J Fish Aquat Sci 37:1635–1640. https://doi.org/10.1139/f80-211

Morman RH (1979) Distribution and ecology of lampreys in the Lower Peninsula of Michigan, 1957–75. Great Lakes Fishery Commission Technical Report No. 33, Ludington, Michigan

Schuldt RJ, Fodale MF, Johnson WJ (1987) Prespawning characteristics of lampreys native to Lake Michigan. J Great Lakes Res 13:264–271. https://doi.org/10.1016/S0380-1330(87)71649-9

Scott WB, Crossman EJ (1973) Freshwater fishes of Canada, Bulletin 184. Fisheries Research Board of Canada

Smith AJ, Howell JH, Piavis GW (1968) Comparative embryology of five species of lampreys of the upper Great Lakes. Copeia 1968:461–469. https://doi.org/10.2307/1442013

Starrett WC, Harth WJ, Smith PW (1960) Parasitic lampreys of the genus *Ichthyomyzon* in the rivers of Illinois. Copeia 1960:337–346. https://doi.org/10.2307/1439761

Trautman MB (1957) The fishes of Ohio. Ohio State University Press, Columbus

6.3 Acipenseridae, Sturgeon Family

The sturgeon family consists of four genera of 25 extant species. The majority of these occur in Europe and Asia, with eight found on the coastal and inland waters of North America. These include five species of the genera *Acipenser* and three *Scaphirhynchus*. Sturgeon are a primitive fish group that evolved during the end of the Triassic period nearly 200 to 250 million years ago; they are the most ancient extant group of the bony fishes (Superclass: Osteichthyes, Class: Actinopterygii). Sturgeons are the largest freshwater fishes. They have a fusiform, elongate, robust, and relatively dorsoventrally flattened body shape, a short dorsal fin far posterior on the body, and a distinguishable heterocercal tail. They also possess two sets of barbels inferior on the head, anterior to the subterminal mouth. Sturgeon bodies are covered in five rows of bony plates, called scutes, that are situated dorsally, laterally, and ventrally.

In North America, sturgeon inhabit benthic habitats of fresh, brackish, and marine waters. While swimming just above the bottom substrate, they use taste buds on sensory barbels to detect food. Sturgeons are carnivorous in all life stages. Larva eat zooplankton, and juveniles and adults consume benthic invertebrates and smaller fishes. Although some species live primarily in brackish or marine waters, all sturgeon migrate and spawn in freshwater rivers. The white sturgeon *Acipenser transmontanus* is the largest North American sturgeon species. It migrates more than 1,000 km from the ocean to spawn in freshwater. The shortnose sturgeon *Acipenser brevirostrum* is the smallest of the North American sturgeon and typically migrates less than 200 km.

Sturgeon populations are dramatically declining because of biological factors and anthropogenic disturbances. They grow slowly and require several years or even decades to sexually mature. Northern populations often take longer to mature, and females frequently mature much later than males. Spawning events are also infrequent. For example, pallid sturgeon *Scaphirhynchus albus* females spawn every 4 to 10 years and males spawn every 2 to 3 years. Sturgeon populations are challenged by slow growth, delayed maturation, and infrequent spawning, and these challenges are compounded when coupled with dam construction and other anthropogenic effects.

Three species of sturgeon occur within the Dakotas: lake sturgeon *Acipenser fulvescens*, shovelnose sturgeon *Scaphirhynchus platorynchus*, and pallid sturgeon. Pallid sturgeon are currently listed as federally endangered.

6.3.1 Lake Sturgeon *Acipenser fulvescens*

Rafinesque, 1817

Etymology: *Acipenser fulvescens* – *Acipenser* = point five, referencing five body scute rows; *fulvescens* = yellowish-brown.

Description:

Body – fusiform but dorsoventrally flattened; 5 rows of large scutes; 9 to 17 dorsal scutes; 29 to 42 lateral scutes; 7 to 12 ventral scutes.

Color –

> *Dorsally* – dark olive to gray.
> *Laterally* – yellow-brown to gray.
> *Ventrally* – white.

Head – elongated, dorsoventrally flattened.
Snout – short, rounded, conical, slightly upturned.
Eyes – small, dorsolaterally on head; spiracle slightly posterior of eye.
Mouth – inferior; fleshy lips; lower lip with two papillose lobes.
Barbels – smooth; four positioned anterior of mouth on ventral side of the snout.
Gill rakers – 25 to 44.
Dorsal fin – 35 to 40 rays; posterior on body.
Caudal peduncle – short, thick; exposed portions of skin present between scutes.
Caudal fin – heterocercal; upper lobe more elongate than lower lobe, but not extending into a long filament.
Anal fin – 25 to 30 rays.
Pelvic fins – posterior; insertions anterior to dorsal fin insertion.
Pectoral fins – large, rounded.
Lateral line – complete, extending through caudal fin upper lobe.
Juveniles – small, dark mottling on body.

Similar Species: Lake sturgeon closely resemble pallid and shovelnose sturgeon. Pallid sturgeon and shovelnose sturgeon have flattened and shovel shaped heads, elongate and slender caudal peduncles without exposed areas of skin, and fringed barbels, but lack spiracles.

Distribution and Habitat: Lake sturgeon are native to eastern North America throughout the Great Lakes, Hudson-James Bay, and Mississippi watersheds, extending as far south as Alabama and northern Mississippi (Lee et al. 1980; Adams Jr 2004). They were historically abundant and widespread throughout their native range, but populations have been declining due to pollution, commercial fishing, habitat fragmentation, and loss of spawning habitat (Pollock et al. 2014). In the Dakotas, lake sturgeon occur in the Red River of the North in North Dakota, the mainstem of the Missouri River below Gavins Point Dam in South Dakota, and Big Stone Lake in the Minnesota River system. The native Big Stone Lake lake sturgeon population was extinct before being re-established through stocking efforts beginning in 2014. In 2021, a lake sturgeon was caught by an angler near Klondike within the Big Sioux River.

Lake sturgeon inhabit large rivers and lakes with sand, gravel, or mud substrate. After hatching, larvae remain burrowed in gravel until yolk sac absorption, after which they drift downstream. Nursery habitats include shallow, low velocity areas with sand substrate and a predominance of dipteran larvae (Benson et al. 2005; Pollock et al. 2014). Juveniles are often found over fine substrate in lower gradient streams at depths ranging from 0.5 to 20 m (Daugherty et al. 2009; Haxton 2011; Pollock et al. 2014). Adults frequent deeper locations during fall and winter in comparison to spring and summer (Briggs et al. 2016).

Reproduction: Upstream spawning migrations begin in early spring shortly after ice-out (Vladykov 1955; Adams Jr 2004). Migrations are often over 129 km and can exceed 402 km (Vladykov 1955; Adams Jr 2004). In the upper Mississippi River basin, lake sturgeon traveled at approximately 0.5 km/day during the spring spawning run (Knights et al. 2002; Adams Jr 2004). Spawning typically occurs between late April and early June over gravel or rocky cobble substrate in the shallow and swift shoals of moderate to high-gradient streams (Becker 1983; Adams Jr 2004; Boschung Jr and Mayden 2004; Daugherty et al. 2009). Males move into spawning sites at water temperatures of 9 to 16 °C, while females arrive later at 19 °C (Bruch and Binkowski 2002; Pollock et al. 2014). Mean age at sexual maturity from Lake of the Woods and the Rainy River system was 16.8 years for males and 25.8 for females (Mosindy and Rusak 1991). Males and females generally only spawn every 1 to 3 years and 4 to 6 years, respectively, and the amount of time between spawning events increases with latitude (Magnin 1966; Adams Jr 2004). Spawning behavior consists of several males swimming alongside a single female and vibrating their bodies against hers to initiate the release of eggs and milt (Becker 1983). Fecundity is highly variable, with females of the same size capable of producing 50,000 to 700,000 eggs (Becker 1983). No nest is constructed, and no parental care is given

(Boschung Jr and Mayden 2004). Eggs are black, demersal, and adhesive. They are approximately 2 to 3 mm in diameter (Adams Jr 2004; Boschung Jr and Mayden 2004). Hatching and incubation time is dependent on water temperature, but typically occurs within 5 to 8 days at 16 or 17 °C (Boschung Jr and Mayden 2004).

Age and Growth: Larvae are approximately 8 mm long at hatching and 125 mm by the first fall (Becker 1983; Boschung Jr and Mayden 2004). Mean total-lengths-at-age of individuals from Rainy Lake were age-15, 1165 mm; age-20, 1233 mm; age-25, 1273 mm; age-30, 1334 mm; age-35, 1442 mm; age-40, 1433 mm (Adams Jr 2004). Females are generally larger than males (Adams Jr 2004). Lake sturgeon are capable of reaching lengths of 2745 mm and can live as long as 154 years (Becker 1983; Etnier and Starnes 1993).

Food and Feeding: Lake sturgeon forage by using barbels to detect prey, and then suction the prey item with their inferior mouth. They are benthic, generalist, opportunistic feeders (Beamish et al. 1998; Nilo et al. 2006; Pollock et al. 2014). Dietary items include blood worms, mayfly nymphs, midge larvae, amphipods, gastropods, crayfish, and other fishes such as gizzard shad (Mosindy and Rusak 1991; Boschung Jr and Mayden 2004; Volkman et al. 2004; Stelzer et al. 2008; Pollock et al. 2014).

References

Adams WE Jr (2004) Lake sturgeon biology in Rainy Lake, Minnesota and Ontario. Dissertation, South Dakota State University

Beamish FWH, Noakes DLG, Rossiter A (1998) Feeding ecology of juvenile lake sturgeon, *Acipenser fulvescens*, in northern Ontario. Can Field Nat 112:459–468

Becker GC (1983) Fishes of Wisconsin. University of Wisconsin Press, Madison

Benson AC, Sutton TM, Elliott RF, Meronek TG (2005) Seasonal movement patterns and habitat preferences of age-0 lake sturgeon in the lower Peshtigo River, Wisconsin. Trans Am Fish Soc 134:1400–1409. https://doi.org/10.1577/T04-179.1

Boschung HT Jr, Mayden RL (2004) Fishes of Alabama. Smithsonian Books, Washington, DC

Briggs AS, Hondorp DW, Quinlan HR, Boase JC, Mohr LC (2016) Electric archival tags provide first glimpse of bathythermal habitat use by free-ranging adult lake sturgeon *Acipenser fulvescens*. J Freshw Ecol 31:477–483. https://doi.org/10.1080/02705060.2016.1152321

Bruch RM, Binkowski FP (2002) Spawning behaviour of lake sturgeon (*Acipenser fulvescens*). J Appl Ichthyol 18:570–579. https://doi.org/10.1046/j.1439-0426.2002.00421.x

Daugherty DJ, Sutton TM, Elliott RF (2009) Suitability modeling of lake sturgeon habitat in five northern Lake Michigan tributaries: implications for population rehabilitation. Restor Ecol 17:245–257. https://doi.org/10.1111/j.1526-100X.2008.00368.x

Etnier DA, Starnes WC (1993) The fishes of Tennessee. University of Tennessee Press, Knoxville

Haxton TJ (2011) Depth selectivity and spatial distribution of juvenile lake sturgeon in a large fragmented river. J Appl Ichthyol 27:45–52. https://doi.org/10.1111/j.1439-0426.2011.01872.x

Knights BC, Vallazza JM, Zigler SJ, Dewey MR (2002) Habitat and movement of lake sturgeon in the upper Mississippi River System, USA. Trans Am Fish Soc 131:507–522. https://doi.org/10.1577/1548-8659

Lee DS, Gilbert CR, Hocutt CH, Jenkins RE, McAllister DE, Stauffer JR (1980) Atlas of North American freshwater fishes. North Carolina Biological Survey Publication, 1980–12. https://doi.org/10.5962/bhl.title.141711

Magnin E (1966) Recherches sur les cycles de reproduction des sturgeons *Acipenser fulvescens* Rafinesque. Annales de la Station Centrale d'Hydrobiologie appliquee 9:8–242

Mosindy T, Rusak J (1991) An assessment of lake sturgeon populations in Lake of the Woods and the Rainy River 1987–90. Ontario Ministry of Natural Resources, Lake of the Woods Fisheries Assessment Unit Report 1991:01, Toronto

Nilo P, Tremblay S, Bolon A, Dodson J, Dumont P, Fortin R (2006) Feeding ecology of juvenile lake sturgeon in the St. Lawrence River system. Trans Am Fish Soc 135:1044–1055. https://doi.org/10.1577/T05-279.1

Pollock MS, Carr M, Kreitals NM, Phillips ID (2014) Review of a species in peril: what we do not know about lake sturgeon may kill them. Environ Rev 23:30–43. https://doi.org/10.1139/er-2014-0037

Stelzer RS, Drecktrah HG, Shupryt MP, Bruch RM (2008) Carbon sources for lake sturgeon in Lake Winnebago, Wisconsin. Trans Am Fish Soc 137:1018–1028. https://doi.org/10.1577/T07-214.1

Vladykov VD (1955) A comparison of Atlantic Sea sturgeon with a new subspecies from the Gulf of Mexico (*Acipenser oxyrhynchus* de sotoi). J Fish Res Board Can 12:754–761. https://doi.org/10.1139/f55-040

Volkman ET, Pangle KL, Rajchel DA, Sutton TM (2004) Hatchery performance attributes of juvenile lake sturgeon fed two natural food types. N Am J Aquac 66:105–112. https://doi.org/10.1577/A03-047.1

6.3.2 Pallid Sturgeon *Scaphirhynchus albus*

(Forbes and Richardson, 1905)

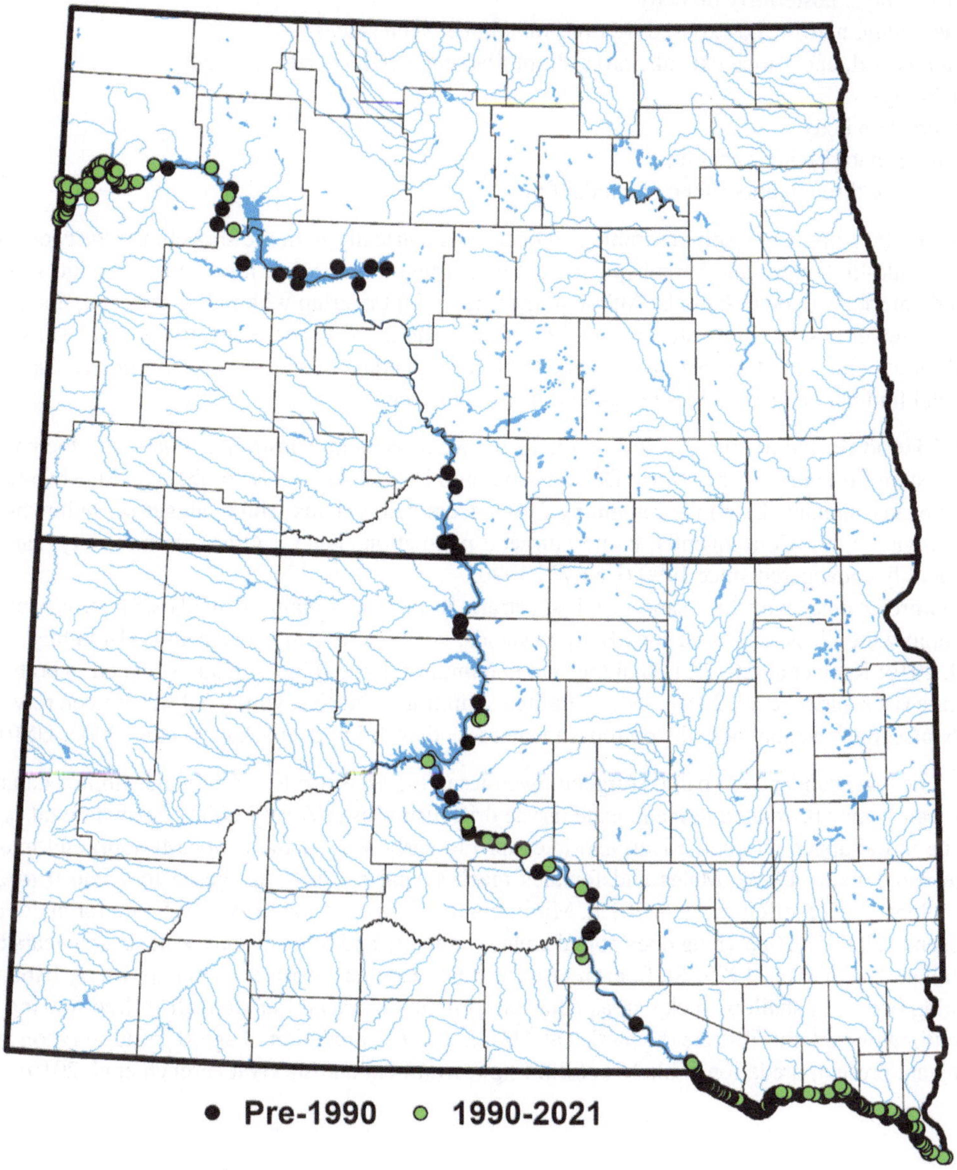

Etymology: *Scaphirhynchus albus* – *Scaphirhynchus* = shovel snout; *albus* = white.

Description:

Body – fusiform, elongated, slightly arched, ventrally flattened, robust dorsally; 5 rows of large scutes; 14 to 18 dorsal scutes, 40 to 48 lateral scutes, 9 to 13 ventral scutes.

Color –

>   *Dorsally* – grayish-blue to light brown.
>   *Laterally* – tan.
>   *Ventrally* – white.

Head – depressed.
Snout – large, elongated, flattened, shovel shaped.
Eyes – small, dorsolateral.
Mouth – inferior; 4 fleshy lobes on lower lip.
Barbels – 4 barbels with accessory fringes; anterior to mouth on ventral side of head with two outer barbels posterior to two inner, thinner, and weakly fringed barbels.
Gill rakers – 13 or 14.
Dorsal fin – 37 to 43 rays; posteriorly on body.
Caudal peduncle – long, narrow; no exposed portions of skin between scutes.
Caudal fin – heterocercal, upper lobe extends into a slight filament; often shortened in adults.
Anal fin – 24 to 29 rays.
Pelvic fins – posterior on body.
Pectoral fins – larger than pelvic or anal fins.
Lateral line – complete through upper lobe of caudal fin.

Similar Species: Pallid sturgeon closely resemble and occur sympatrically with the shovelnose sturgeon. Shovelnose sturgeon have ventral scale-like scutes and barbels positioned in a straight, transverse line, whereas pallid sturgeon have inner barbels positioned anteriorly to outer barbels. Although there is a slight overlap with pallid sturgeon, dorsal and anal fin ray counts are typically lower in shovelnose sturgeon at 39 or fewer, and 25 or fewer, respectively. Lake sturgeon have a short, rounded, conical snout, with smooth barbels, a short and thick caudal peduncle with exposed portions of skin present between scutes, and a caudal fin upper lobe that does not extend into a long filament.

Distribution and Habitat: Pallid sturgeon are native to the Missouri and Mississippi rivers and their major tributaries, including the mainstem Missouri River in North Dakota and South Dakota. Over 50% of their riverine habitat has been channelized, 28% has been impounded, and the remaining 21% occurs below dams which alter flow, sedimentation, and water temperature (Keenlyne 1989). Consequently, pallid sturgeon populations have declined dramatically, and the species has been listed as federally endangered since 1990 (USFWS 1990).

Pallid sturgeon prefer areas with firm sand or silt substrate, in mid- and main-channels with strong current and highly turbid water, around wing dikes, and occasionally near islands, with 1.5 to 4 m water depths (Bramblett and White 2001; Deslauriers et al. 2016; Koch et al. 2012). Lethal thermal maximum for age-0 individuals is related to body mass and acclimation temperature (Deslauriers et al. 2016). Due to limited swimming abilities, young pallid sturgeon rely on water velocities less than 0.6 m/second and stay near the bottom in water less than 1.5 m deep (Adams et al. 2003; USFWS 2003).

Reproduction: The low abundance of pallid sturgeon juveniles in the Missouri River indicates limited natural reproduction and recruitment, which may be related to inadequate larvae drift distances (USFWS 2003). Spawning takes place from late May to early June in the upper Missouri River, and late April to early May in the lower Missouri and middle Mississippi rivers (Keenlyne and Jenkins 1993). Males and females reach sexual maturity at ages 5 to 7 and 9 to 20, respectively (Keenlyne and Jenkins 1993; Steffensen et al. 2013). Minimum fork length-at-maturity is 798 mm for males and 788 mm for females (Steffensen et al. 2013). Spawning does not take place annually and females may wait several years between spawning events (Keenlyne and Jenkins 1993; Steffensen et al. 2013). Detailed information on spawning habitat is scarce, but broadcast spawning occurs in shallow, rocky areas adjacent to main channels. Larvae drift in free-flowing water for 11 to 17 days before settling on bottom substrate (Kynard et al. 2007). Because predation of age-0 pallid sturgeon is low, especially if alternative prey are present, predation is likely not limiting recruitment and survival (French et al. 2010).

Age and Growth: Growth is rapid from 13 to 48 days post-hatch, but then slows until the end of the first growing season (approximately 100 to 110 days) at which time juveniles reach a fork length of 120 to 140 mm (Braaten et al. 2012). Length-at-age may vary depending on time of hatch, water temperature, and growing season duration (Braaten et al. 2012). Optimal growth temperatures for juveniles range from 25 to 28 °C (Chipps et al. 2008). Fork length-at-age for stocked pallid sturgeon was: age-1, 278 mm; age-2, 315 mm; age-3, 351 mm; age-4, 387 mm; age-5, 424 mm; age-6, 460 mm (Shuman et al. 2011; Braaten et al. 2012). Pallid sturgeon are capable of reaching a fork length of 1638 mm and are believed to live at least 50 years (Shuman et al. 2006).

Food and Feeding: As a benthic predator, pallid sturgeon use barbels to detect prey. Wild-feeding young-of-year primarily consume diptera larvae and pupae, ephemeroptera nymphs, and chironomid larvae (USFWS 2003; Wanner et al. 2007; Braaten et al. 2012). Optimal juvenile feeding temperature ranges from 25 to 28 °C (Chipps et al. 2008). Smaller juveniles primarily consume macroinvertebrates but increase their dependency on prey fishes after reaching a fork length of 600 mm, which typically occurs between age-5 and age-7 (Grohs et al. 2009).

References

Adams SR, Adams GL, Parsons GR (2003) Critical swimming speed and behavior of juvenile shovelnose sturgeon and pallid sturgeon. Trans Am Fish Soc 132:392–397. https://doi.org/10.1577/1548-8659

Braaten PJ, Fuller DB, Lott RD, Haddix TM, Holte LD, Wilson RH, Barton ML, Kalie JA, DeHaan PW, Ardren WR, Holm RJ, Jaeger ME (2012) Natural growth and diet of known-age pallid sturgeon (*Scaphirhynchus albus*) early life stages in the upper Missouri River basin, Montana and North Dakota. J Appl Ichthyol 28:496–504. https://doi.org/10.1111/j.1439-0426.2012.01964.x

Bramblett RG, White RG (2001) Habitat use and movements of pallid and shovelnose sturgeon in the Yellowstone and Missouri Rivers in Montana and North Dakota. Trans Am Fish Soc 130:1006–1025. https://doi.org/10.1577/1548-8659

Chipps SR, Klumb RA, Wright EB (2008) Development and application of juvenile pallid sturgeon bioenergetics model. State Wildlife Grant T-24-R. Pierre

Deslauriers D, Heironimus L, Chipps SR (2016) Lethal thermal maxima for age-0 pallid and shovelnose sturgeon: implications for shallow water habitat restoration. River Res Appl 32:1872–1878. https://doi.org/10.1002/rra.3022

French WE, Graeb BDS, Chipps SR, Bertrand KN, Selch TM, Klumb RA (2010) Vulnerability of age-0 pallid sturgeon *Scaphirhynchus albus* to fish predation. J Appl Ichthyol 26:6–10. https://doi.org/10.1111/j.1439-0426.2009.01356.x

Grohs KL, Klumb RA, Chipps SR, Wanner GA (2009) Ontogenetic patterns in prey use by pallid sturgeon in the Missouri River, South Dakota and Nebraska. J Appl Ichthyol 25:48–53. https://doi.org/10.1111/j.1439-0426.2009.01279.x

Keenlyne KD (1989) A report on the pallid sturgeon. U.S. Fish and Wildlife Service, Pierre

Keenlyne KD, Jenkins LG (1993) Age at sexual maturity of the pallid sturgeon. Trans Am Fish Soc 122:393–396. https://doi.org/10.1577/1548-8659

Koch B, Brooks RC, Oliver A, Herzog D, Garvey JE, Hrabik R, Colombo R, Phelps Q, Spier T (2012) Habitat selection and movement of naturally occurring pallid sturgeon in the Mississippi River. Trans Am Fish Soc 141:112–120. https://doi.org/10.1080/00028487.2011.652008

Kynard B, Parker E, Pugh D, Parker T (2007) Use of laboratory studies to develop a dispersal model for Missouri River pallid sturgeon early life intervals. J Appl Ichthyol 23:365–374. https://doi.org/10.1111/j.1439-0426.2007.00908.x

Shuman DA, Willis DW, Krentz SC (2006) Application of a length-categorization system for pallid sturgeon (*Scaphirhynchus albus*). J Freshw Ecol 21:71–76. https://doi.org/10.1080/02705060.2006.9664098

Shuman DA, Klumb RA, Wanner GA, Wilson RH, Steffensen KD, Doyle WJ, Gardner WM, Haddix T, Ruggles M, Horner PT, Jaeger ME, Stukel S (2011) Pallid sturgeon size structure, condition, and growth within the Missouri River basin. J Appl Ichthyol 27:269–281. https://doi.org/10.1111/j.1439-0426.2010.01645.x

Steffensen KD, Pegg MA, Mestl GE (2013) Population characteristics of pallid sturgeon (*Scaphirhynchus albus* (Forbes & Richardson, 1905)) in the Lower Missouri River. J Appl Ichthyol 29:687–695. https://doi.org/10.1111/jai.12196

USFWS (U.S. Fish and Wildlife Service) (1990) Endangered and threatened wildlife and plants; determination of endangered status for the pallid sturgeon. US Fed Reg 55:36641–36647

USFWS (U.S. Fish and Wildlife Service) (2003) Amendment to the 2000 Biological Opinion on the operation of the Missouri River main stem reservoir system, operation and maintenance of the Missouri Riverbank stabilization and navigation project, and operation of the Kansas River reservoir system. Omaha District, Nebraska

Wanner GA, Shuman DA, Willis DW (2007) Food habits of juvenile pallid sturgeon and adult shovelnose sturgeon in the Missouri River downstream of Fort Randall Dam, South Dakota. J Freshw Ecol 22:81–92. https://doi.org/10.1080/02705060.2007.9664148

6.3.3   Shovelnose Sturgeon *Scaphirhynchus platorynchus*

(Rafinesque, 1820)

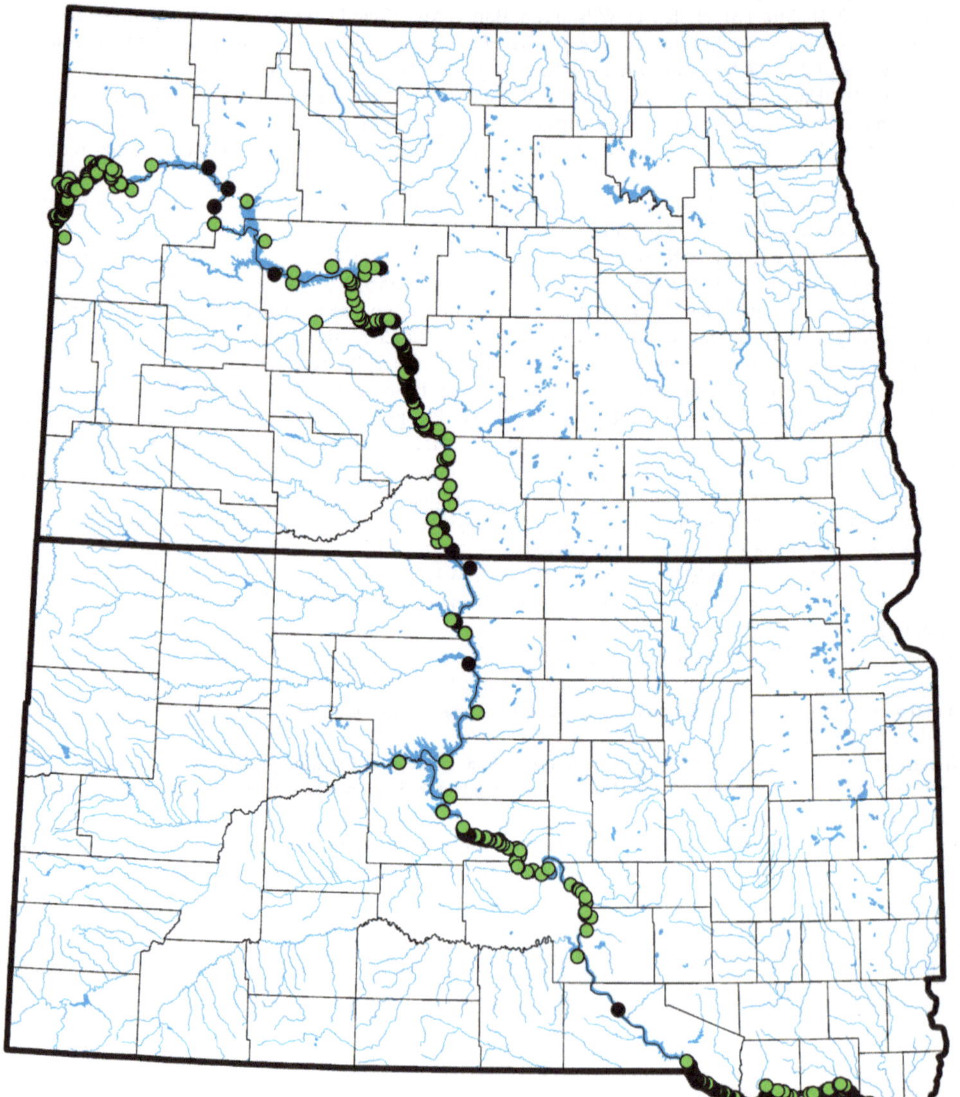

Etymology: *Scaphirhynchus platorynchus* – *Scaphirhynchus* = shovel snout; *platorynchus* = broad or flat snout.

Description:

Body – fusiform, elongated, slightly dorsoventrally flattened; 5 rows of large scutes; 13 to 19 dorsal scutes, 38 to 50 lateral scutes, 9 to 14 ventral scutes.

Color –

> *Dorsally* – light gray, light brown, brown, or olive.
> *Laterally* – light brown to tan.
> *Ventrally* – white.

Head – depressed.
Snout – long, flattened, shovel shaped.
Eyes – small, dorsolaterally on the head.
Mouth – inferior, four fleshy lobes on lower lip.
Barbels – 4 with accessory fringes present anterior in a straight line to mouth; two inner barbels thick and strongly fringed.
Gill rakers – 11 to 14.
Dorsal fin – 29 to 39 rays; posterior.
Caudal peduncle – long, narrow; no exposed portions of skin between scutes.
Caudal fin – heterocercal; upper lobe extends into a slight filament; often shortened in adults.
Anal fin – 18 to 25.
Pelvic fins – posterior on body.
Pectoral fins – larger than pelvic or anal fins.
Lateral line – complete through upper lobe of caudal fin.

Similar Species: Shovelnose sturgeon closely resemble and occur sympatrically with pallid sturgeon. Pallid sturgeon have two sets of barbels with accessory fringes, with the two thin and slightly fringed inner barbels positioned anteriorly to the outer barbels. Pallid sturgeon also lack belly scutes, have a dorsal fin with 37 to 43 rays, and an anal fin with 24 to 29 rays. Lake sturgeon have a short, rounded, conical snout, with smooth barbels, a short and thick caudal peduncle with exposed portions of skin present between scutes, and a caudal fin upper lobe that does not extend into a long filament.

Distribution and Habitat: Shovelnose sturgeon are native to the Missouri and Mississippi rivers, including the mainstem and direct tributaries of the Missouri River throughout North Dakota and South Dakota (Keenlyne 1997; Tripp et al. 2009). Although they are the most widespread sturgeon species in North America, their abundance has been declining because of pollution, habitat alteration, and overharvesting (Keenlyne 1997). Shovelnose sturgeon inhabit swift water over sand, gravel, and cobble substrate in large river main channels and major tributaries (Bramblett and White 2001; Gerrity et al. 2008). They are often found in pools downstream of sandbars, dams, or wing dams (Keenlyne 1997). Juvenile habitat use is very similar to adults (Gerrity et al. 2008).

Reproduction: The shovelnose sturgeon spawning season is protracted. Although it primarily occurs from late April to June, it may be extended into the fall or even become bimodal when fall conditions are similar to the spring (Keenlyne 1997; Tripp et al. 2009). Upstream spawning migrations into smaller tributaries occur at water temperatures of 17 to 21 °C (Becker 1983). Spawning takes place over hard substrates at higher flows and in relatively deeper water (Keenlyne 1997). Shovelnose sturgeon mature earlier than other sturgeon species because of their small size, but maturation may be delayed (Tripp et al. 2009). Males reach sexual maturity at ages five to eight and females at ages seven to nine (Keenlyne 1997; Tripp et al. 2009). Males generally spawn every-other-year, and females every 3 to 4 years (Tripp et al. 2009). Fecundity of females from the middle Mississippi River ranges from 5,733 to 81,842 eggs, with a mean of 29,573 eggs, or 21.7 eggs/g of body weight (Tripp et al. 2009). No nest is constructed, and no parental care is given. Eggs are adhesive, range in color from yellow or white to black, and are approximately 2.5 mm in diameter (Becker 1983; Bryan et al. 2007).

Age and Growth: Shovelnose sturgeon are the smallest North American sturgeon species. Most do not exceed 2.5 kg but individuals from the upper Missouri River can exceed 7 kg (Keenlyne 1997). In the Missouri River, shovelnose sturgeon are long lived and slow growing, unlike individuals from the Mississippi River that are younger and faster growing (Hamel et al. 2014). Differences in growth rates are likely related to environmental and anthropogenic factors (Hamel et al. 2014). Mean fork length at age from the Missouri River were age-1, 213 mm; age-2, 274 mm; age-5, 399 mm; and age-10, 503 mm (Carlander 1969). They can reach 1,200 mm with a longevity of 43 years (Everett et al. 2003).

Food and Feeding: As opportunistic benthivores, shovelnose sturgeon forage by using their barbels to detect prey, which are then captured via suction with their inferior mouth. In the downstream segments of the Missouri River, shovelnose sturgeon do not experience ontogenetic dietary shifts (French et al. 2013). Adults within the Missouri River mainly consume a variety of aquatic arthropods including trichopetra, diptera, and ephemeroptera larvae and odonates, coleopterans, and fish eggs (Modde and Schmulbach 1977; Wanner et al. 2007). Seasonal shifts in the diet are likely influenced by discharge rates and timing (Modde and Schmulbach 1977). Shovelnose sturgeon feeding behavior in the Missouri River has been characterized by benthic foraging in the late spring and summer, feeding on drift in the fall, and consuming a diversity of aquatic and terrestrial invertebrates in the winter months (Modde and Schmulbach 1977). Age-0 juveniles feed primarily on dipteran and ephemeropteran larvae (Braaten et al. 2007).

References

Becker GC (1983) Fishes of Wisconsin. University of Wisconsin Press, Madison

Boschung HT Jr, Mayden RL (2004) Fishes of Alabama. Smithsonian Books, Washington, DC

Braaten PJ, Fuller DB, McClenning ND (2007) Diet composition of larval and young-of-year shovelnose sturgeon in the Upper Missouri River. J Appl Ichthyol 23:516–520. https://doi.org/10.1111/j.1439-0426.2006.00822.x

Bramblett RG, White RG (2001) Habitat use and movements of pallid and shovelnose sturgeon in the Yellowstone and Missouri rivers in Montana and North Dakota. Trans Am Fish Soc 130:1006–1025. https://doi.org/10.1577/1548-8659

Bryan JL, Wildhaber ML, Papoulias DM, DeLonay AJ, Tillitt DE, Annis ML (2007) Estimation of gonad volume, fecundity, and reproductive stage of shovelnose sturgeon using sonography and endoscopy with application to the endangered pallid sturgeon. J Appl Ichthyol 23:411–419. https://doi.org/10.1111/j.1439-0426.2007.00889.x

Carlander KD (1969) Handbook of freshwater fishery biology, vol 1. Iowa State University Press, Ames

Everett SR, Scarnecchia DL, Power GJ, Williams CJ (2003) Comparison of age and growth of shovelnose sturgeon in the Missouri and Yellowstone rivers. N Am J Fish Manag 23:230–240. https://doi.org/10.1577/1548-8675

French WE, Graeb BDS, Bertrand KN, Chipps SR, Klumb RA (2013) Size-dependent trophic patterns of pallid sturgeon and shovelnose sturgeon in a large river system. J Fish Wildl Manage 4:41–52. https://doi.org/10.3996/022012-JFWM-013

Gerrity PC, Guy CS, Gardner WM (2008) Habitat use of juvenile pallid sturgeon and shovelnose sturgeon with implications for water-level management in a downstream reservoir. N Am J Fish Manag 28:832–843. https://doi.org/10.1577/M07-061.1

Hamel MJ, Spurgeon JJ, Pegg MA, Hammen JJ, Rugg ML (2014) Hydrologic variability influences local probability of pallid sturgeon occurrence in a Missouri River tributary. River Res Appl 32:320–329. https://doi.org/10.1002/rra.2850

Keenlyne KD (1997) Life history and status of the shovelnose sturgeon, *Scaphirhynchus platorynchus*. Environ Biol Fish 48:291–298. https://doi.org/10.1023/A:1007349221987

Modde T, Schmulbach JC (1977) Food and feeding behavior of the shovelnose sturgeon, *Scaphirhynchus platorynchus*, in the unchannelized Missouri River, South Dakota. Trans Am Fish Soc 106:602–608. https://doi.org/10.1577/1548-8659

Tripp SJ, Phelps QE, Colombo RE, Garvey JE, Burr BM, Herzog DP, Hrabik RA (2009) Maturation and reproduction of shovelnose sturgeon in the middle Mississippi River. N Am J Fish Manag 29:730–738. https://doi.org/10.1577/M08-056.1

Wanner GA, Shuman DA, Willis DW (2007) Food habits of juvenile pallid sturgeon and shovelnose sturgeon in the Missouri River downstream of Fort Randall Dam, South Dakota. J Freshw Ecol 22:81–92. https://doi.org/10.1080/02705060.2007.9664148

6.4   Polyodontidae, Paddlefish Family

The family Polyodontidae contains only one species, the North American paddlefish *Polyodon spathula*. Chinese paddlefish *Psephurus gladius* historically occurred in the Yangtze-Kiang River in China but were declared extinct in 2020. Paddlefish, along with closely related sturgeon in the family Acipenseridae, belong to Acipenseriformes, an ancient order of bony fishes. The Polyondontidae family is characterized by an extended spatula-like rostrum, cartilaginous skeleton, heterocercal tail, and absence of maxillary and premaxillary bones. Paddlefish are commonly called spoonbills.

North American paddlefish are native to the Mississippi-Missouri River basin in the United States, ranging from New York to Montana, and including the Mobile Bay drainage basin to the Gulf of Mexico. Although paddlefish are relatively numerous throughout most of their range, the Mississippi River and Gulf Coast populations are the most imperiled. They inhabit medium to large, semi-turbid to turbid, rivers and reservoirs, often in lower velocity areas behind sandbars or structures with silt, sand, or gravel substrates. North American paddlefish populations have been negatively impacted by river channelization, dam construction, and other anthropogenic actions that affect migratory patterns. Overharvesting, particularly for roe, has also occurred, although reservoir ranching for paddlefish is allowed in three states for roe production.

Paddlefish use ram ventilation to obtain respiratory oxygen, continuously swimming with an open mouth to force water over their gills. They are a pelagic (midwater) species, with swimming depth to provide ideal plankton for feeding, temperature, and/or dissolved oxygen levels. Paddlefish detect their prey using numerous electrosensory receptors on the dorsal and ventral sides of the rostrum. Spawning occurs in the spring, with fish migrating upstream. Males generally spawn every year, while females spawn every 2 to 3 years. Like sturgeon, paddlefish have delayed maturation. In the Dakotas, paddlefish are considered a game fish with a limited season. Paddlefish can be snagged or targeted with bowfishing gear during designated seasons by tag purchase in North Dakota (ND) or lottery permit in South Dakota (SD).

6.4.1   Paddlefish *Polyodon spathula*

(Walbaum, 1792)

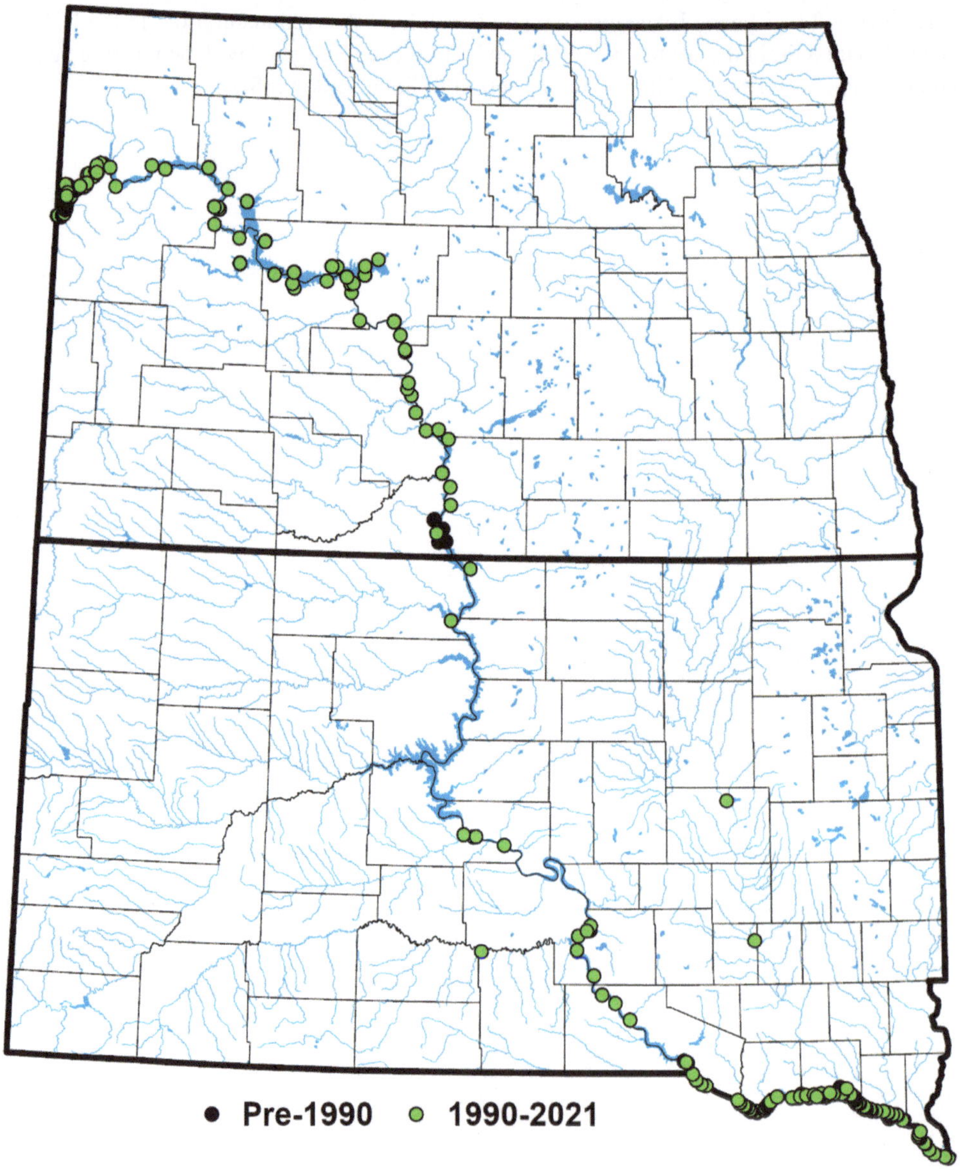

Etymology: *Polyodon spathula* – many teeth, referencing the numerous gill rakers; *spathula* = blade, or paddle, referencing the rostrum shape.

Description:

Body – thick, robust, round in cross section.
Color –

> *Dorsally/Laterally* – blueish-gray to black, often mottled.
> *Ventrally* – cream to white.

Head – broad.
Opercle – extends posteriorly past pectoral fins, tapers into an elongate, pointed flap.
Snout (rostrum) – long, paddle shaped; approximately one-third of adult body length; covered with small sensory pores.
Eyes – small; posterior spiracle.
Mouth – large, subterminal.
Teeth – Juveniles: small, present on basal portions of gill arches, upper and lower jaws.
    Adults: toothless.
Barbels – present but not obvious; one small pair on ventral side of rostrum.
Gill rakers – numerous, long, slender.
Dorsal fin – posterior on body, inserted between pelvic and anal fins.
Caudal peduncle – short, thick.
Caudal fin – heterocercal, strongly forked.
Anal fin – concave distal end.
Pelvic fins – abdominal, fleshy.
Pectoral fins – fleshy.
Lateral line – complete.
Skin – leather-like; scales absent except for single patch of ganoid scales at base of caudal fin upper lobe.

Similar Species: None.

Distribution and Habitat: North American paddlefish are native to the eastern and central United States in the Mississippi-Missouri River basins and all their major tributaries as far south as the Gulf of Mexico. In the Dakotas, they occur within the mainstem and direct tributaries of the Missouri River and have been sampled within the Missouri, White, James, and Big Sioux rivers in SD. Populations still occur over much of the native range, although the species has been negatively affected over time by river channelization, dam construction, dewatering, pollution, and overharvest (Carlson and Bonislawsky 1981; Jennings and Zigler 2000; Paukert and Fisher 2001).

Paddlefish prefer medium to large rivers, reservoirs in open waters near side channels, and areas below structures, sandbars, bays, and eddies with current velocities below 0.3 m/sec (Carlson and Bonislawsky 1981; Southhall and Hubert 1984; Moen et al. 1992). Locations vary seasonally and annually in response to discharge and temperature (Rosen 1976; Southhall and Hubert 1984; Jennings and Zigler 2009). They are most often found over gravel, sand, and silt substrates, and in turbid waters with depths averaging 3.0 m (Rosen 1976; Paukert and Fisher 2001).

Reproduction: Time to sexual maturity is much longer than other freshwater fish species and varies with latitude (Carlson and Bonislawsky 1981). Males typically mature at ages four to nine, with females maturing later at ages 6 to 12 (Carlson and Bonislawsky 1981). Males spawn each year (Russell 1986). Females spawn every 2 to 3 years, likely due to the amount of time needed to acquire the energy resources to produce large egg masses (Carlson and Bonislawsky 1981).

Upstream spawning migrations begin in the spring when flows increase and water temperatures approach 10 °C (Russell 1986; Paukert and Fisher 2001). Migration distances can be several hundred km (Russell 1986; Paukert and Fisher 2001). Spawning occurs in water temperatures from 10 to 16 °C near the water surface over silt free gravel bars and bedrock (Purkett Jr 1961; Russell 1986). If the photoperiod, water temperature, and flow are not suitable, egg resorption occurs (Russell 1986; Jennings and Zigler 2000). Information on the actual spawning behavior is not available.

Fertilized eggs are white, demersal, adhesive, and approximately 2 to 4 mm in diameter (Purkett Jr 1961; Rosen 1976). Fecundity is high but varies among similarly sized females (Russell 1986; Reed et al. 1992). Relative fecundity of paddlefish in Arkansas ranged from 7,794 to 30,247 eggs/kilogram of female body weight (Leone et al. 2012). Hatching success is directly related to water temperature (Jennings and Zigler 2000). In water temperatures of 18 to 21 °C, eggs hatch in 6 to 7 days (Purkett Jr 1961). In temperatures of 11 to 14 °C, eggs hatch in 12 to 14 days (Jennings and Zigler 2000).

Age and Growth: Mean total length of newly hatched larvae is 9 mm (Purkett Jr 1961; Jennings and Zigler 2009). Growth is rapid for the first 5 years and is influenced by food abundance and growing season duration (Russell 1986). Growth rates may be higher in reservoirs than rivers (Paukert and Fisher 2001). Mean eye to fork length (EFL) for 1-year-old paddlefish in Lewis and Clark Reservoir, South Dakota, was 192 mm (Ruelle and Hudson 1977). EFL is measured from the anterior eye to the caudal fin fork and is used instead of total length because rostrum or caudal fins are frequently damaged (Ruelle and Hudson 1977). Adult paddlefish from the Missouri River below Yankton, South Dakota had a mean EFL of 730 mm and a mean weight of 5.9 kg (Rosen 1976). North American paddlefish are capable of reaching lengths of 2.2 m and weights of 72 kg (Allardyce 1992; Epifanio et al. 1996; Jennings and Zigler 2000). Females are longer and heavier than males (Rosen 1976). Mean longevity is 5 to 8 years, but longevity beyond 20 years is not uncommon (Pflieger 1997; Jennings and Zigler 2000).

Food and Feeding: Adults and juveniles are nonselective filter feeders consuming primarily plankton, with small insects, insect larvae, and small fish occasionally eaten (Ruelle and Hudson 1977; Rosen and Hales 1981; Jennings and Zigler 2009). Prey is detected using electrosensory receptors on the rostrum (Wilkins and Hofmann 2007).

References

Allardyce DA (1992) Endangered and threatened wildlife and plants: findings on petition to list the paddlefish. Fed Regist 57:43676–43682

Carlson DM, Bonislawsky PS (1981) The paddlefish (*Polyodon spathula*) fisheries of the Midwestern United States. Fisheries 6:17–27. https://doi.org/10.1577/1548-8446

Epifanio JM, Koppelman JB, Nedbal MA, Philipp DP (1996) Geographic variation of paddlefish allozymes and mitochondrial DNA. Trans Am Fish Soc 125:546–561. https://doi.org/10.1577/1548-8659

Jennings CA, Zigler SJ (2000) Ecology and biology of paddlefish in North America: historical perspectives, management approaches, and research priorities. Rev Fish Biol Fish 10:167–181. https://doi.org/10.1023/A:1016633604301

Jennings CA, Zigler SJ (2009) Biology and life history of paddlefish in North America: an update. Am Fish Soc Symp 66:1–22

Leone FJ, Stoeckel JN, Quinn JW (2012) Differences in paddlefish populations among impoundments of the Arkansas River, Arkansas. N Am J Fish Manag 32:731–744. https://doi.org/10.1080/02755947.2012.686956

Moen CT, Scarnecchia DL, Ramsey JS (1992) Paddlefish movements and habitat use in Pool 13 of the upper Mississippi River during abnormally low river stages and discharges. N Am J Fish Manag 12:744–751. https://doi.org/10.1577/1548-8675

Paukert CP, Fisher WL (2001) Characteristics of paddlefish in a southwestern U.S. reservoir, with comparisons of lentic and lotic populations. Trans Am Fish Soc 130:634–643. https://doi.org/10.1577/1548-8659

Pflieger WL (1997) The fishes of Missouri, revised edn. Missouri Dep Cons, Jefferson City

Purkett CA Jr (1961) Reproduction and early development of the paddlefish. Trans Am Fish Soc 90:125–129. https://doi.org/10.1577/1548-8659(1961)90[125:RAEDOT]2.0.CO;2

Reed BC, Kelso WW, Rutherford DA (1992) Growth, fecundity, and mortality of paddlefish in Louisiana. Trans Am Fish Soc 121:374–384. https://doi.org/10.1577/1548-8659

Rosen RA (1976) Distribution, age and growth, feeding ecology of paddlefish (*Polyodon spathula*) in unaltered Missouri River, South Dakota. Dissertation, South Dakota State University

Rosen RA, Hales DC (1981) Feeding of paddlefish, *Polyodon spathula*. Copeia 1981:441–455. https://doi.org/10.2307/1444235

Ruelle R, Hudson P (1977) Paddlefish (*Polyodon spathula*): growth and food of young of the year and a suggested technique for measuring length. Trans Am Fish Soc 106:609–613. https://doi.org/10.1577/1548-8659

Russell TR (1986) Biology and life history of the paddlefish-a review. In: Dillard JG, Graham LK, Russell TR (eds) The paddlefish: status, management and propagation. American Fisheries Society, special publication 7, Bethesda, pp 2–21

Southhall PD, Hubert WA (1984) Habitat use by adult paddlefish in the upper Mississippi River. Trans Am Fish Soc 113:125–131. doi: https://doi.org/10.1577/1548-8659

Wilkins LA, Hofmann MH (2007) The paddlefish rostrum as an electrosensory organ: a novel adaptation for plankton feeding. Bioscience 57:399–407. https://doi.org/10.1641/B570505

6.5 Lepisosteidae, Gar Family

The gar family consists of seven extant species from two genera, *Lepisosteus* and *Atractosteus*. Gar are an ancient group of fishes, with fossils from the Cretaceous period dating back 75–100 million years from western North America, Europe, Africa, Asia, and South America. The distribution of gar is now restricted to the North American continent, including one species each in Cuba and Central America. While most gar species are freshwater inhabitants, some, like the alligator gar *Atractosteus spatula*, are occasionally found within brackish or marine waters of the Gulf of Mexico. Longnose gar *Lepisosteus osseus* and shortnose gar *Lepisosteus platostomus* occur in the Dakotas.

Gars have elongated, cylindrical, and torpedo-like bodies covered in hard, rhomboid-shaped ganoid scales, a dorsal and anal fin far posterior on the body, and an abbreviate-heterocercal caudal fin. Gar snouts are elongate and lined with multiple sharp teeth along both jaws. Their swim bladder is lung-like, which allows for gulping air. Gar can be bimodal breathers by combining gulping with the normal gill respiration. Gulping occurs more frequently in low oxygenated waters with higher temperatures. This breathing adaptation is a major reason gar have survived fluctuating climatic stages throughout their long evolutionary history.

When foraging, gar stalk their prey for several minutes and exhibit ambush style sit and wait behavior. When attacking, the body forms a rough "S" shape, with the head turning quickly with jaws wide open to grab the prey. Once gar have the prey in their mouth, they shake it back and forth to impair it with their sharp teeth. All species of gar are predators. Adults are primarily piscivorous, consuming other fish, although some species will also consume macroinvertebrates. Gar inhabit quiet, stagnant to low current areas of large rivers, streams, reservoirs, lakes, and ponds with variable substrate. During the late spring and early summer spawning season, gar migrate to shallower areas to spawn over coarse substrate or beds of vegetation. No nest is constructed, but the adhesive and demersal eggs adhere to vegetation or other debris. The eggs also have a gelatinous coating poisonous or toxic to other vertebrates, including humans.

Longnose gar are absent from North Dakota and are uncommon in South Dakota. Shortnose gar occur throughout the Missouri and James rivers in both North Dakota and South Dakota and are more abundant in southeastern South Dakota.

6.5.1 Longnose Gar *Lepisosteus osseus*

(Linnaeus, 1758)

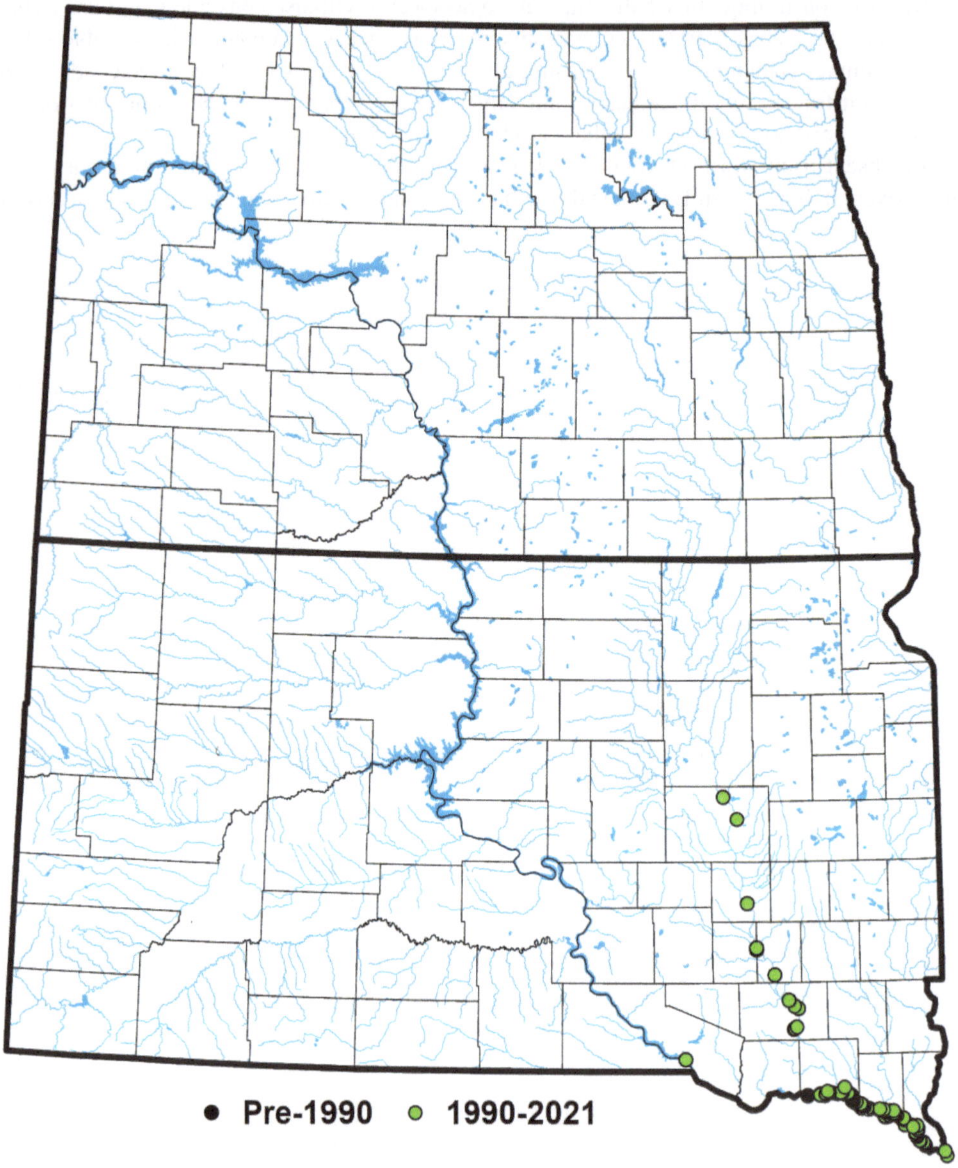

● **Pre-1990**   ○ **1990-2021**

Etymology: *Lepisosteus osseus – Lepisosteus* = scales of bone, *osseus* = bony.

Description:

Body – fusiform, elongated, slender, cylindrical.
Color – varies greatly with habitat.

> *Dorsally* – olive to dark green.
> *Laterally* – olive, brown, or tan; few small dark spots posteriorly.
> *Ventrally* – cream to white.
> *Jaws* – small spots occasionally on ventral sides.
> *Fins* – dorsal/anal/caudal: large, dark spots

Head – elongated.
Snout – long, slender, dorsoventrally flattened; width at nostrils less than diameter of eye; narrowest width fits into snout length 13 or more times.
Eyes – large; directly posterior to opening of mouth.
Mouth – terminal to subterminal.
Teeth – small, numerous, villiform; on both jaws.
Gill rakers – 20 to 31, rudimentary.
Dorsal fin – far posterior.
Caudal peduncle – short, slender.
Caudal fin – heterocercal, fin itself is not bilobed but rather has a rounded distal end.
Anal fin – far posterior; insertion anterior to anterior end of dorsal fin.
Pelvic fins – abdominal.
Pectoral fins – insertion directly posterior to opercle.
Lateral line – 57 to 65 scales; 47 to 55 predorsal scales; 17 to 24 transverse scales.
Skin – thick, ganoid (rhomboid).
Juveniles – distinct brown to black midlateral streak with white underlining from snout to base of caudal fin.
Pectoral air bladder – connected to digestive tract.

Similar Species: Longnose gar closely resemble shortnose gar. Shortnose gar have a shorter, wider, and more blunt snout. The narrowest width of snout fits into snout length ten or fewer times the distance from tip of snout to eye. Shortnose gar are often lighter in color and have 20 to 23 transverse scales.

Distribution and Habitat: Longnose gar occur throughout much of the eastern half of the United States. They are native to the southeastern edge of the Mississippi River basin/Missouri River drainage, and the Ohio and South-Atlantic Gulf. Longnose gar are more common in fresh water but have also been found in brackish waters with salinities up to 31 ppt (Schwartz 2003; McGrath 2010). They are absent from North Dakota and are a rare species in South Dakota, documented only in the James and lower Missouri river. Increased turbidity and siltation levels from impoundment construction has likely impacted their range (Hoagstrom et al. 2006). Longnose gar inhabit quiet, low-current areas of large rivers and streams, reservoirs, and lakes. Adult gar associate with deep open water containing moderate aquatic vegetation or structure over coarse substrate. Juveniles favor shallower water with vegetation during their first summer before moving to deeper waters as they mature (Hasse 1969).

Reproduction: Broad and extensive spawning migrations take place from April through July (Johnson and Noltie 1996; McGrath 2010). Migrations are positively correlated with stream flow and water level and negatively correlated with water temperature (Johnson and Noltie 1996). In lakes and reservoirs, longnose gar migrate to shallower areas and smaller tributary streams to spawn (Hasse 1969; Kelley 2012). Individuals in rivers and streams often migrate upstream to smaller, higher-gradient streams (Yeager and Bryant 1983; Pflieger 1997). Spawning activity occurs over deep, rocky stream reaches, or shallower areas with gravel bars, at peak water temperatures of 20 to 21 °C (Becker 1983; Yeager and Bryant 1983; Pflieger 1997). Males become sexually mature at ages 2 to 4, while females mature at age-6 or age-7 (Hasse 1969; Johnson and Noltie 1996; Ferrara 2001). Males occupy spawning grounds longer than females. One to several young males escort and nudge a single female with their snouts to the spawning ground, where clutches of eggs are released and fertilized (Hasse 1969; Johnson and Noltie 1996). No nest is prepared. Fecundity is related to total body weight and total ovary weight, with estimates ranging from 10,924 to 34,412 eggs per female (Johnson and Noltie 1996). The greenish to gray, demersal, adhesive eggs are 3 mm in diameter (Becker 1983; Ferrara 2001). Hatching is dependent on water temperature, and typically occurs 7 to 10 days post-spawn (Hasse 1969; Simon and Wallus 1989).

Age and Growth: The total length of newly hatched larvae is 9 to 10 mm (Yeager and Bryant 1983). First year growth is rapid, with age-1 juveniles reaching an average length of 400 mm (Ferrara 2001; Sutton et al. 2009; McGrath 2010). Growth significantly slows in both sexes after reaching sexual maturity (Hasse 1969; Ferrara 2001; McGrath 2010). Females grow larger than males, with maximum lengths of 1,829 cm. Average longevity is 11 years in males and 22 years in females (Etnier and Starnes 1993).

Food and Feeding: Longnose gar are a large apex, piscivorous predator. They may stalk prey for several minutes, exhibiting a sit-and-wait behavior before ambushing. Digestion rates are slow (McGrath 2010). Feeding often takes place at night and near the surface. Juveniles feed primarily on small crustaceans, aquatic insects, and insect larvae (Echelle 1968). After reaching a total length greater than 11 cm, they become primarily piscivorous. Longnose gar primarily eat larger forage species such as shad and minnows and secondarily on sport fish (Tyler et al. 1994; McGrath 2010).

References

Becker GC (1983) Fishes of Wisconsin. University of Wisconsin Press, Madison

Echelle AA (1968) Food habits of young-of-year longnose gar in Lake Texacoma, Oklahoma. Southwest Nat 13:45–50. https://doi.org/10.2307/3668814

Etnier DA, Starnes WC (1993) The fishes of Tennessee. University of Tennessee Press, Knoxville

Ferrara AM (2001) Life-history strategy of Lepisosteidae: implications for the conservation and management of alligator gar. Dissertation, Auburn University

Hasse BL (1969) An ecological life history of the longnose gar, *Lepisosteus osseus* (Linnaeus), in Lake Mendota and several other lakes of southern Wisconsin. Dissertation, University of Wisconsin

Hoagstrom CW, Hayer CA, Kral JG, Wall SS, Berry CR Jr (2006) Rare and declining fishes of South Dakota: a river drainage scale perspective. Proc S D Acad Sci 85:171–211

Johnson BL, Noltie DB (1996) Migratory dynamics of stream-spawning longnose gar (*Lepisosteus osseus*). Ecol Freshw Fish 5:97–107. https://doi.org/10.1111/j.1600-0633.1996.tb00041.x

Kelley SW (2012) Age and growth of spawning longnose gar (*Lepisosteus osseus*) in a north central Texas reservoir. West N Am Nat 72:69–77. https://doi.org/10.3398/064.072.0108

McGrath PE (2010) The life history of longnose gar, *Lepisosteus osseus*, an apex predator in the tidal waters of Virginia. Dissertation, College of William and Mary

Pflieger WL (1997) The fishes of Missouri, revised edn. Missouri Dep Cons, Jefferson City

Schwartz J (2003) Longnose gar, *Lepisosteus osseus* (Family Lepisosteidae) in North Carolina, especially the Cape Fear River. J N C Aacd Sci 119:26–32

Simon TP, Wallus R (1989) Contributions to the early life histories of gar (Actinopterygii: Lepidostidae) in the Ohio and Tennessee River basins with emphasis on larval development. Trans Ky Acad Sci 50:59–74

Sutton TM, Grier AC, Frankland LD (2009) Stock structure and dynamics of the longnose gar and shortnose gar in the Wabash River, Indiana-Illinois. J Freshw Ecol 24:657–666. https://doi.org/10.1080/02705060.2009.9664344

Tyler JD, Webb JR, Wright TR, Hargett JD, Mask KJ, Schucker DR (1994) Food habits, sex ratios, and size of longnose gar in southwestern Oklahoma. Proc Okla Acad Sci 74:41–42

Yeager BL, Bryant RT (1983) Larvae of the longnose gar, *Lepisosteus osseus*, from the Little River in Tennessee. J Tenn Acad Sci 58:20–22

6.5.2   Shortnose Gar *Lepisosteus platostomus*

Rafinesque, 1820

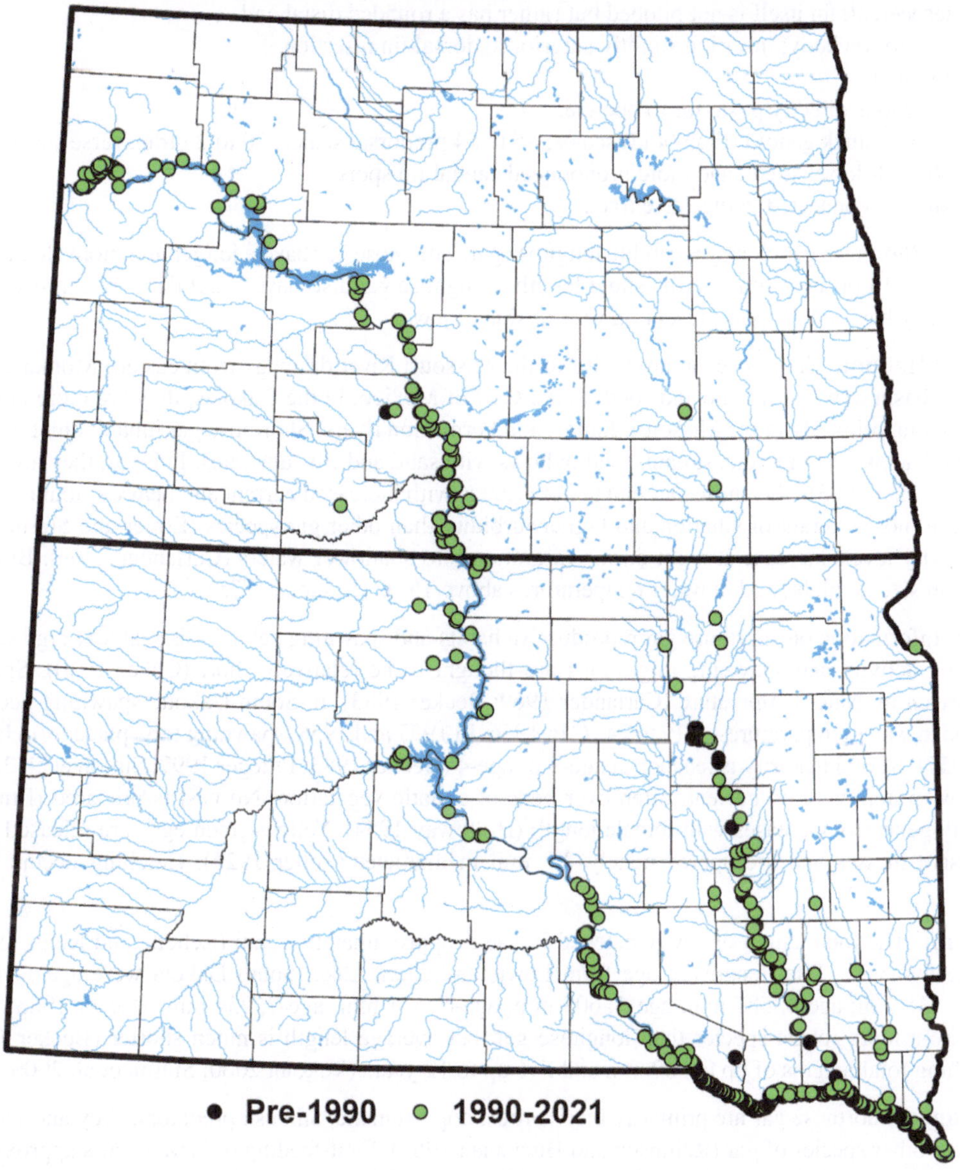

● Pre-1990   ○ 1990-2021

Etymology: *Lepisosteus platostomus* – *Lepisosteus* = scales of bone, *platostomus* = flat mouth.

Description:

Body – fusiform, elongated, slender, cylindrical.
Color –

> *Dorsally* – olive-green.
> *Laterally* – olive silver to tan; small spots faint to absent posteriorly.
> *Ventrally* – cream to white.
> *Fins* – paired fin spots typically absent; dorsal/caudal/anal: a few large dark spots present.

Head – elongated.
Snout – dorsoventrally flattened, short, broad; least snout width ten or fewer times snout length.
Eyes – large; laterally on head, directly posterior to mouth opening.
Mouth – terminal to subterminal.
Teeth – small, numerous, villiform; on both jaws.
Gill rakers – 16 to 25, rudimentary.
Dorsal fin – far posterior.
Caudal peduncle – short, slender.
Caudal fin – heterocercal; fin itself is not bilobed but rather has a rounded distal end.
Anal fin – far posterior on body; insertion slightly anterior to dorsal fin insertion.
Pelvic fins – abdominal.
Pectoral fins – insertions directly posterior to opercle.
Lateral line – 55 to 64 thick ganoid (rhomboid) scales; 45 to 54 predorsal scales; 20 to 23 transverse scales.
Juveniles – darker mid-dorsal streak and more pronounced caudal fin spots.
Pectoral air bladder – connected to digestive tract.

Similar Species: Shortnose gar closely resemble longnose gar. Longnose gar have a longer and more slender snout, and the least snout width fits 13 or more times in the snout length. Longnose gar also have small dark spots present posteriorly on lateral sides and may have small spots present on ventral side of jaw.

Distribution and Habitat: Shortnose gar are native to the Missouri River drainage from eastern Montana, throughout the Mississippi River basin to western Ohio and south to the Gulf of Mexico. In the Dakotas, they are more common in southeastern South Dakota, primarily in the Missouri, James, and Vermillion rivers. Shortnose gar inhabit quiet, low-current areas such as pools and backwaters in rivers, streams, and oxbows with sand and silt substrates. In lakes, they are found over sand bars and shoals (Becker 1983). Shortnose gar tend to avoid areas with increased current and heavy amounts of rooted aquatic vegetation and are more tolerant of siltation and higher turbidity than other gar species (Etnier and Starnes 1993; Pflieger 1997). Their activity levels increase at night during migration into shallower waters (Carlander 1969). Bimodal breathing increases with fish size, at night, and in water temperatures above 15 °C (Saksena 1975).

Reproduction: Information on shortnose gar reproductive habits and spawning behavior is scarce. Large schools form just prior to spawning activity and spawning migrations are thought to be relatively short (Coker 1930). Spawning activity appears to be driven by water temperature (Carlander 1969; Becker 1983). In South Dakota, spawning occurred from late May to mid-June in water temperatures of 19 to 24 °C in 1956. In 1957 and 1958, spawning took place from late June to early July (Shields 1957). Sexual maturity is reached at age-3 or age-4 (Becker 1983; Pflieger 1997; Sutton et al. 2009). Spawning occurs in shallow waters with no current, often over beds of aquatic vegetation. No nest is prepared (Etnier and Starnes 1993). Fecundity generally increases with female length (Holloway 1954). Yellow-green eggs are released in small gelatinous and adhesive masses and are approximately 2.5 mm in diameter (Potter 1926). Hatching occurs in 8 or 9 days (Pflieger 1997).

Age and Growth: The total length of newly hatched larvae is approximately 8 mm (Echelle and Riggs 1972). Growth is rapid during the first year. Total lengths at age from Lewis and Clark Lake, South Dakota were age-1, 417 mm; age-2, 486 mm; age-3, 536 mm; age-4, 587 mm; age-5, 605 mm; age-6, 671 mm; age-7, 734 mm (Becker 1983; Walburg 1964). Shortnose gar are a more robust species than longnose gar, but average length is much shorter (Becker 1983). They are capable of reaching total lengths of up to 900 mm and live up to 12 years (Vokoun 2000; Sutton et al. 2009).

Food and Feeding: Shortnose gar are primarily piscivorous, opportunistic, ambush predators. Prey and feeding habits are more diverse than other species of gar (Robinson and Buchanan 1988). First-feeding of larva occurs approximately 16 days post-hatch (Richardson 1913; Kansas Fishes Committee 2014). Juveniles feed near the surface on insect larvae, small crustaceans, and small fish (Becker 1983; Pflieger 1997; Kansas Fishes Committee 2014). Adults forage throughout the water column and mainly consume other fish species like sunfish and small common carp (Shields 1957; Kansas Fishes Committee 2014). Adults also eat crayfish and insects, such as cicadas, mayflies, and midges (Becker 1983; Etnier and Starnes 1993; Vokoun 2000). Dominance hierarchies and territorial defense during feeding have been reported (Vokoun 2000).

References

Becker GC (1983) Fishes of Wisconsin. University of Wisconsin Press, Madison

Carlander KD (1969) Handbook of freshwater fishery biology. Iowa State University Press, Ames

Coker RE (1930) Studies of common fishes of the Mississippi River at Keokuk, No. 1072. US Government Printing Office, pp 141–225

Echelle AA, Riggs CD (1972) Aspects of the early life history of gars (Lepisosteus) in Lake Texoma. Trans Am Fish Soc 101:106–112. https://doi.org/10.1577/1548-8659

Etnier DA, Starnes WC (1993) The fishes of Tennessee. University of Tennessee Press, Knoxville

Holloway AD (1954) Notes on the life history and management of the shortnose and longnose gar in Florida waters. J Wildl Manag 18:438–449. https://doi.org/10.2307/3797079

Kansas Fishes Committee (2014) Kansas fishes. University Press of Kansas, Lawrence

Pflieger WL (1997) The fishes of Missouri, revised edn. Missouri Dep Cons, Jefferson City

Potter GE (1926) Ecological studies of the short-nose gar-pike (Lepisosteus platostomus). Univ Iowa Stud Nat Hist 11:17–27. https://doi.org/10.21900/j.inhs.v9.390

Richardson RE (1913) Observations on the breeding habitats of fishes at Havana, Illinois, 1910 and 1911. Bull – Ill Nat Hist Surv 9:405–416

Robinson HW, Buchanan TM (1988) Fishes of Arkansas. University of Arkansas Press, Fayetteville

Saksena VP (1975) Effects of temperature and light on aerial breathing of shortnose gar, Lepisosteus platostomus. Ohio J Sci 75:58–62

Shields JT (1957) Report of fisheries investigations during the second year of impoundment of Gavins Point Reservoir, South Dakota, 1956. South Dakota Game, Fish and Parks, Dingell-Johnson Project F-1-R-6

Sutton TM, Grier AC, Frankland LD (2009) Stock structure and dynamics of longnose gar and shortnose gar in the Wabash River, Indiana-Illinois. J Freshw Ecol 24:657–666. https://doi.org/10.1080/02705060.2009.9664344

Vokoun JC (2000) Shortnose gar (Lepisosteus platostomus) foraging on periodical cicadas (Magicicada spp.): territorial defense of profitable pool positions. Am Midl Nat 143:261–265. https://doi.org/10.1674/0003-0031(2000)143[0261:SGLPFO]2.0.CO;2

Walburg CH (1964) Fish population studies, Lewis and Clark Lake, Missouri River, 1956–1962. US Fish and Wildlife Service Special Scientific Report, Fisheries Number 482

6.6 Amiidae, Bowfin Family

During the Jurassic period (145 to 200 million years ago), the family Amiidae contained approximately 30 species. Currently only two species, often referred to as living fossils, exist: bowfin *Amia calva* and emerald bowfin *Amia ocelli- cauda*. Fossil deposits suggest bowfins were once distributed in North America, South America, Europe, and Asia in both marine and freshwater deposits. However, extant bowfin are strictly freshwater. Emerald bowfin are native to eastern North America, primarily in the Mississippi, St. Lawrence, and Great Lakes river basins. They are extremely rare in the Dakotas. Two individuals were collected in 2010 from the Red River of the North, near Wahpeton, North Dakota. These are the only recorded occurrences in the Dakotas, and although not vouchered, they were recorded to be large individuals and are included in the comprehensive species list for fishes of North Dakota.

Bowfin share some primitive morphological characteristics found in extinct bowfin species. These include a primitive skeleton with a highly modified skull and gular plate, a vascularized and subdivided swim bladder allowing them to live in waters with very low oxygen concentrations, and kidney tubules that lead to the body cavity. Of the few primitive fish families (e.g., Acipenseridae, Polyodontidae, Lepisosteidae), bowfin are the only species that build nests and provide parental care for their young.

6.6.1   Emerald Bowfin *Amia ocellicauda*

Richardson, 1836

© Brian Zimmerman

● **Pre-1990**   ● **1990-2021**

Etymology: *Amia ocellicauda* – *Amia* = ancient fish; *ocelli* = eyespot, referencing the large dark spot present at the base of the caudal fin in mature males; cauda = tail.

Description:

Body – cylindrical, elongated.
Color –

> *Dorsally/Laterally* – dark green, olive, or golden olive; darker brown mottling; mottling pattern may appear faded or absent in older adults.
> *Ventrally* – light green, cream, or white.
> *Fins* – dorsal/caudal – dark brown to black bands possible.

Head – moderate, slightly conical; dorsally flat.
Gular plate – bony; on throat.
Snout – short, blunt; nasal tubes.
Eyes – moderately large, laterally on head.
Mouth – terminal, large.
Teeth – on both upper and lower jaws; numerous, canine like.
Barbels – two on snout.
Dorsal fin – more than 45 rays; elongated, extending more than half the body length.
Caudal peduncle – short, very thick.
Caudal fin – slightly heterocercal.
Anal fin – 9 to 12 rays; short, rounded distal end.
Pelvic fin – 7 rays; abdominal, rounded distal end.
Pectoral fin – 16 to 18 rays; rounded distal end.
Lateral line – complete; 62 to 70 scales in series.
Skin – thin, flexible, cycloid; scaleless head.
Juveniles – prominent black and orange mottling pattern on head, dorsal, and lateral sides; black spot at caudal fin base.
Sexual dimorphism – mature males have a prominent black spot outlined in yellow or orange near the caudal fin base, and blue or turquoise on the lips and inside the mouth; mature females lack the distinctive spot near the caudal fin base and tend to have a darker or duller appearance.

Similar Species: Burbot may be misidentified for emerald bowfin but can be easily differentiated because they have a dorsal fin distinctly divided into two sections, a long anal fin, and a single barbel present on the ventral side of the chin.

Distribution and Habitat: Bowfin are native to eastern North America in the St. Lawrence, Great Lakes, and Mississippi river basins from Quebec, Canada, to the Gulf Coast of Texas (Burr and Bennett 2014). Recent research suggests the geographic distribution of *A. calva* and *A. ocellicauda* diverged during the Pleistocene epoch and implies allopatric speciation (Brownstein et al. 2022). Thus, *A. ocellicauda* distribution is suggested to primarily occur throughout the Mississippi, St. Lawrence, and Great Lakes river basins (Brownstein et al. 2022). They are rare in the Dakotas, with no reports from South Dakota and only two records from North Dakota. Two emerald bowfin were sampled from the Red River of the North, Wild Rice River system near Wahpeton, North Dakota in 2010. They were not vouchered but reported to weigh 2.8 and 3.4 kg. Emerald bowfin are found in a wide array of habitats including lakes, wetlands, sloughs, backwaters, oxbows, and medium-to-large creeks and rivers. They are frequently associated with clear to turbid waters and abundant aquatic vegetation (Becker 1983; Burr and Bennett 2014). Emerald bowfin are also seemingly tolerant of pollution and inhabit waters highly modified by anthropogenic factors (Webb 2008; Driver et al. 2009; Burr and Bennett 2014). In waters with lower pH and higher temperatures, Emerald bowfin exhibit facultative air breathing by using a subdivided and highly vascularized swim bladder as a lung (Horn and Riggs 1973; Burr and Bennett 2014). Their critical thermal maximum is approximately 35 °C (Horn and Riggs 1973; Becker 1983; Burr and Bennett 2014).

Reproduction: Sexual maturity occurs between ages 2 to 5 for females and 2 to 4 for males (Carlander 1969; Scott and Crossman 1973; Simon 1990; Koch et al. 2009; Burr and Bennett 2014). Minimum size at maturity is approximately 380 mm TL (Simon 1990). Short spawning migrations into flood plain habitats, backwaters, and sloughs occur in the spring and early summer (Eddy and Underhill 1974; Becker 1983; Simon 1990; Burr and Bennett 2014). Spawning occurs annually and typically at night. Spawning behavior consists of a male biting the snout of a female, and then remaining adjacent to the female until eggs and milt are released (Becker 1983; Burr and Bennett 2014). Males construct nests in shallow waters. They aggressively guard the nest, fertilized eggs, and larval fish until the juveniles reach a total length of approximately 100 mm (Reighard

1903; Breder Jr and Rosen 1966). Fecundity from emerald bowfin in the upper Mississippi River ranged from 9,498 to 110,086 eggs per female (Koch et al. 2009). Mature eggs are roughly 2 to 3 mm in diameter and are adhesive (Simon 1990; Burr and Bennett 2014). Incubation duration and hatching time depend on water temperature, with hatching occurring in 6 days at 16 to 17 °C (Simon 1990; Burr and Bennett 2014).

Age and Growth: Larvae have a total length ranging from 3 to 7 mm at hatch, and growth is known to be rapid during the early stages (Becker 1983; Simon 1990; Burr and Bennett 2014). By the end of year one, juvenile total length may exceed 203 mm (Becker 1983). Total lengths of emerald bowfin ages 2 to 13 from the upper Mississippi River ranged from 392 to 807 mm (Koch et al. 2009). Maximum total length is 1,090 mm (Burr and Bennett 2014). Females live longer than males (Davis 2006; Koch et al. 2009). Longevity is 13 years, but emerald bowfin have lived 30 years in captivity (Carlander 1969; Koch et al. 2009; Burr and Bennett 2014).

Food and Feeding: Emerald bowfin are apex ambush predators with opportunistic foraging behavior. They feed primarily in low light from dusk to dawn on fish, rodents, snakes, frogs, and turtles. They also eat crayfish, aquatic insects, and other invertebrates (Carlander 1969; Becker 1983; Burr and Bennett 2014) and are cannibalistic (Simon 1990). Juveniles feed primarily on phytoplankton (Burr and Bennett 2014).

References

Becker GC (1983) Fishes of Wisconsin. University of Wisconsin Press, Madison

Breder CM, Rosen DE (1966) Modes of reproduction in fishes. TFH Publications, Neptune City

Brownstein CD, Kim D, Orr OD, Hogue GM, Tracy BH, Pugh MW, Singer R, Myles-McBurney C, Mollish JM, Simmons JW, David SR, Watkins-Colwell G, Hoffman EA, Near TJ (2022) Hidden species diversity in a living fossil vertebrate. Biol Lett 18:20220395. https://doi.org/10.1098/rsbl.2022.0395

Burr BM, Bennett MG (2014) Amiidae: bowfins. In: Warren ML Jr, Burr BM (eds) Freshwater fishes of North America, Petromyzontidae to Catostomidae. Johns Hopkins University Press, Baltimore, pp 279–298

Carlander KD (1969) Handbook of freshwater fishery biology, vol 1. Iowa State University Press, Ames

Davis JG (2006) Reproductive biology, life history and population structure of a bowfin *Amia calva* population in southeastern Louisiana. Dissertation, Nicholls State University

Driver LJ, Adams GL, Adams SR (2009) Fish assemblage of a cypress wetland within an urban landscape. Southeast Nat 8:527–536. https://doi.org/10.1656/058.008.0313

Eddy S, Underhill JC (1974) Northern fishes. North Central Publishing, St. Paul

Horn MH, Riggs CD (1973) Effects of temperature and light on the rate of air breathing of the bowfin, *Amia calva*. Copeia 1973:653–657. https://doi.org/10.2307/1443063

Koch JD, Quist MC, Hansen KA, Jones GA (2009) Population dynamics and potential management of bowfin (*Amia calva*) in the upper Mississippi River. J Appl Ichthyol 25:545–550. https://doi.org/10.1111/j.1439-0426.2009.01248.x

Reighard J (1903) The natural history of *Amia calva* Linnaeus. Henry Holt and Company, New York

Scott WB, Crossman EJ (1973) Freshwater fishes of Canada, Bulletin 184. Fisheries Research Board of Canada

Simon TP (1990) Family Amiidae. In: Wallus R, Simon TP (eds) Reproductive biology and early life history of fishes in the Ohio River Drainage, vol 1: Acipenseridae through Esocidae. Tennessee Valley Authority, Chattanooga, pp 89–97. https://doi.org/10.1201/9781420003604-1

Webb PW (2008) The impact of changes in water level and human development on forage fish assemblages in Great Lakes coastal marshes. J Great Lakes Res 34:615–630. https://doi.org/10.1016/S0380-1330(08)71606-X

6.7 Anguillidae, Freshwater Eel Family

The freshwater eel family is one of the most unique and widely dispersed groups of fishes in the world because of their catadromous (migrating from freshwater to saltwater to spawn) life history. The Anguillidae comprise one genus, *Anguilla*, with 16 species occurring worldwide. The American eel *Anguilla rostrata* is the only Anguillidae occurring in North America. Like other eels within the order Anguilliformes, anguillids display an elongate, snake-like body shape. The dorsal, caudal, and anal fins are conjoined into a single continuous soft-rayed fin on the posterior end of the body. Small pectoral fins are also present near the head, but pelvic fins are absent.

Although the exact spawning locations of the American eel remain somewhat mysterious, the Danish biologist Johannes Schmidt discovered in the 1920s that the primary spawning site of the American eel and the European eel *Anguilla anguilla* is the Sargasso Sea near Bermuda and the West Indies. The American eel likely spawns at depths up to 500 m, but the actual spawning act has never been observed. After hatching, it undergoes five main morphological life stages: leptocephalus, glass eel, elver, yellow eel, and silver eel. American eel are semelparous, dying shortly after spawning.

After American eel eggs hatch, the larval stage, leptocephalus, drift in the upper portion of ocean currents for up to a year before reaching North American coastal rivers and streams. At this time, the leptocephalus are transparent and are referred to as glass eels. As the glass eels continue to migrate further into inland freshwater habitats, they completely metamorphosize into darker pigmented elvers. Once elvers reach roughly 127 mm, they turn a yellow-green color and are referred to as yellow eels. This is the longest duration of the life stages. Yellow eels are essentially nonmigratory juveniles. After reaching sexual maturity, they turn a silver color and are called silver eels. The silver eels then begin their long migrations back to the Sargasso Sea to spawn.

Most American eels in the United States occur along the Atlantic Coast. However, the species can travel as far inland as South Dakota. In South Dakota, the American eel occurs in low densities and is rarely documented. It infrequently inhabits the Missouri River below Gavins Point Dam near Yankton.

6.7.1 American Eel *Anguilla rostrata*

(Lesueur, 1817)

● **Pre-1990**   ○ **1990–2021**

Etymology: *Anguilla rostrata – Anguilla* = eel; *rostrata* = beaked or long-nosed.

Description:

Body – elongated, snake like.
Color – Adults (silver eels):

> *Dorsally* – black, dark brown to bronze.
> *Laterally* – brown to tan.
> *Ventrally* – light brown to silver.
> *Fins* – dark brown to bronze; no spots or markings.

Glass eels: pigmented eyes, transparent body.
Elvers: dark.
Yellow eels: yellow-green.
Head – small.
Opercle – single slit gill opening at pectoral fin base.
Snout – moderately pointed.
Eyes – moderate, laterally on head.
Mouth – large, terminal.

> *Jaws* – upper jaw extends to, or just beyond, posterior edge of eye.
Teeth – small, numerous; on both jaws.
Dorsal fin – elongated, insertion far behind head, continuous with caudal and anal fins.
Caudal fin – homocercal; slightly rounded.
Anal fin – insertion posterior from dorsal fin insertion.
Pectoral fins – small, insertion behind gill covers.
Lateral line – complete.
Skin – adults with numerous inconspicuous, embedded cycloid scales not visible to human eye.
Juveniles – larvae, or leptocephalus, are strongly laterally compressed.

Similar Species: American eels may be confused with lamprey species in the Dakotas. Lampreys lack pectoral fins and jaws and have a disk-shaped mouth with multiple small teeth in a circular formation.

Distribution and Habitat: American eel are one of the most widely dispersed fish in North America. Its large geographic range extends from the southern tip of Greenland through North and Central America and south to Venezuela (Tesch 1977). Although widespread, some populations are in decline because of overharvesting, contaminants, barriers to migration, and habitat loss and alteration (Castonguay et al. 1994; Haro et al. 2000). It is rarely collected in South Dakota, with yellow phase eels occasionally documented below Gavins Point Dam (Bailey and Allum 1962). North Dakota has one American eel report in the Red River of the North from 1950, but that is extremely suspect. American eel inhabits a wide variety of habitat types including streams, lakes, marshes, open oceans, and estuaries with freshwater, marine or brackish water and mud, sand, or gravel substrates (Tesch 1977; Helfman et al. 1987). They are catadromous. To elude predation, escape light, or avoid lowering water levels, they will burrow into soft benthic substrates (Tesch 1977). Although American eel are adaptable to a wide variety of temperatures, yellow eel prefer 17 °C (Barila and Stauffer Jr 1980).

Reproduction: A long prespawning migration to saltwater takes place in late summer and early fall and may begin even earlier farther inland (Facey and Van Den Avyle 1987). Spawning in the southern region of the Sargasso Sea occurs during winter and early spring in the upper 500 m section of the water column (Schmidt 1923; Kleckner et al. 1983; McCleave et al. 1987; Braham 2012). American eels are broadcast spawners (Facey and Van Den Avyle 1987). Ages (total lengths) of first reproduction in males and female is 4 to 16 years (228 to 398 mm), and 6 to 43 years (400 to 1,159 mm), respectively (Haro 2014). Fecundity is highly size dependent (Facey and Van Den Avyle 1987). Estimates of fecundity range from 1.8 to 19.9 million eggs for eels 452 to 1,133 mm long (Barbin and McCleave 1997). Eggs range in diameter from less than 0.3 to 1.1 mm (Wenner and Musick 1974; Tesch 1977; Todd 1981). Environmental sex determination occurs in the yellow eel phase, with low densities leading to the development of females and high densities developing males (Krueger and Oliveira 1999; Oliveira 1999). Eggs hatch in 3 or 4 days from February to April, when water temperatures reach 20 °C (Kleckner et al. 1983; Facey and Van Den Avyle 1987; Haro 2014). American eel is a semelparous species, with death occurring shortly after spawning. American eel go through five stages: leptocephalus, glass, elver, yellow, and silver (Tesch 1977; Fahay 1978). After hatching, no parental care is provided and the larvae (leptocephalus) drift with the ocean current in the upper portion of the water column for several months until arriving at coastal rivers and streams (Kleckner and McCleave 1982; Braham 2012). The leptocephalus then turn transparent and are referred to as glass eels as they migrate inland (Dutil et al. 2011; Braham 2012). Glass eels then morph into darker-pigmented elvers, which continue to migrate farther up into freshwater habitats (Facey and Van Den Avyle 1987). After reaching 127 mm, elvers turn yellow-green and enter the yellow phase, which is the longest duration life stage, lasting several years (Tesch 1977; Facey and Van Den Avyle 1987). The silver life stage is reached following sexual maturation when the eels migrate back downriver to spawn in saltwater (Facey and Van Den Avyle 1987).

Age and Growth: American eel length at hatching unknown, but the growth rate of leptocephalus is estimated at 0.2 mm/day (Wippelhauser et al. 1985; Facey and Van Den Avyle 1987). Elver growth is slow, with eels reaching a range of 65 to 127 mm in length after the first year in freshwater (Bigelow and Schroeder 1953; Scott and Crossman 1973; Facey and Van Den Avyle 1987). Most growth occurs during the yellow phase (Tesch 1977; Facey and Van Den Avyle 1987). Growth rates are highly variable within year classes, leading to poor predictability of length at age (Facey and Van Den Avyle 1987). Adult (silver) females grow larger than males, with the mean and maximum female size increasing with latitude (Hurley 1972; Facey and LaBar 1981; Facey and Van Den Avyle 1987; Helfman et al. 1987). Along the North American Atlantic Coast, male and female growth rates are inversely correlated to latitude (Oliveira and McCleave 2002; Jessop 2010). Length ranges from 610 to 914 mm in South Dakota waters (Churchill and Over 1938). American eel can achieve lengths of up to 1,520 mm (Angel and Jones 1974). Male longevity is 16 years, and female longevity is 43 years (Hurley 1972; Facey and LaBar 1981; Haro 2014).

Food and Feeding: American eel are carnivorous, but will rarely consume dead or rotting prey. Foraging primarily occurs at night. Eels use three main modes of feeding: suction, shaking, and spinning (Helfman and Clark 1986; Braham 2012). Prey size increases with eel size (Wenner and Musick 1975). Juvenile life stages primarily feed on larval mayflies, stoneflies, and caddisflies (Ogden 1970). Adults consume a variety of fish, crayfish, invertebrates, and bivalves.

References

Angel NB, Jones WR (1974) Aquaculture of the American eel (*Anguilla rostrata*). International Extension Service School of Engineering, North Carolina State University

Bailey RM, Allum MO (1962) Fishes of South Dakota. University of Michigan, Ann Arbor. https://doi.org/10.3998/mpub.9690435

Barbin GP, McCleave JD (1997) Fecundity of the American eel *Anguilla rostrata* at 45°N in Maine, USA. J Fish Biol 51:840–847. https://doi.org/10.1111/j.1095-8649.1997.tb02004.x

Barila FY, Stauffer JR Jr (1980) Temperature behavioral responses of the American eel, *Anguilla rostrata* (LeSueur), from Maryland. Hydrobiologia 74:49–51. https://doi.org/10.1007/BF00009014

Bigelow HB, Schroeder WC (1953) Fishes of the Gulf of Maine. U.S. Fish and Wildlife Service Fisheries Bulletin 53, Washington DC

Braham MA (2012) Selection of benthic habitat by yellow-phase American eels (*Anguilla rostrata*). Dissertation, West Virginia University, Morgantown, West Virginia

Castonguay M, Hudson PV, Couillard CM, Eckersley MJ, Dutil JD, Verreault G (1994) Why is recruitment of the American eel, *Anguilla rostrata*, declining in the St. Lawrence River and Gulf? Can J Fish Aquat Sci 51:479–488. https://doi.org/10.1139/f94-050

Churchill EP, Over WH (1938) Fishes of South Dakota. Brown & Saenger Printers, Sioux Falls

Dutil JD, Dumont P, Cairns DK, Galbraith PS, Verreault G, Castonguay M, Proulx S (2009) *Anguilla rostrata* glass eel migration and recruitment in the estuary and Gulf of St. Lawrence. J Fish Biol 74:1970–1984. https://doi.org/10.1111/j.1095-8649.2009.02292.x

Facey DE, LaBar GW (1981) Biology of American eels in Lake Champlain, Vermont. Trans Am Fish Soc 110:396–402. https://doi.org/10.1577/1548-8659

Facey DE, Van Den Avyle MJ (1987) Species profiles: life histories and environmental requirements of coastal fishes and invertebrates (north Atlantic): American eel. U.S. Fish and Wildlife Service Biological Report 82(11.74). U.S. Army Corps of Engineers, TR EL-82-4

Fahay MP (1978) Biological and fisheries data on American eel, *Anguilla rostrata* (LeSueue). Technical Series Report No.17, National Marine Fisheries Service, NOAA, Highlands, New Jersey

Haro A (2014) Anguillidae: freshwater eels. In: Warren ML Jr, Burr BM (eds) Freshwater fishes of North America. Johns Hopkins University Press, Baltimore, pp 313–331

Haro AJ, Richkus W, Whalen K, Hoar A, Busch WD, Lary S, Brush T, Dixon D (2000) Population decline of the American eel: implications for research and management. Fisheries 25:7–16. https://doi.org/10.1577/1548-8446

Helfman GS, Clark JL (1986) Rotational feeding: overcoming gape-limited foraging in anguillid eels. Copeia 1986:679–685. https://doi.org/10.2307/1444949

Helfman GS, Facey DE, Hales LS Jr, Bozeman EL Jr (1987) Reproductive ecology of the American eel. Am Fish Soc Symp 1:42–56

Hurley DA (1972) The American eel (*Anguilla rostrata*) in eastern Ontario. Can J Fish Aquat Sci 29:535–543. https://doi.org/10.1139/f72-090

Jessop BM (2010) Geographic effects on American eel (*Anguilla rostrata*) life history characteristics and strategies. Can J Fish Aquat Sci 67:326–346. https://doi.org/10.1139/F09-189

Kleckner RC, McCleave JD (1982) Entry of migrating American eel leptocephali into the Gulf Stream system. Helgoländer Meeresun 35:329–339. https://doi.org/10.1007/BF02006141

Kleckner RC, McCleave JD, Wippelhauser GS (1983) Spawning of American eel, *Anguilla rostrata*, relative to thermal fronts in the Sargasso Sea. Environ Biol Fish 9:289–293. https://doi.org/10.1007/BF00692377

Krueger WH, Oliveira K (1999) Evidence for environmental sex determination in the American eel, *Anguilla rostrata*. Environ Biol Fish 55:381–389. https://doi.org/10.1023/A:1007575600789

McCleave JD, Kleckner RC, Castonguay M (1987) Reproductive sympatry of American and European eel and implications for migration and taxonomy. Am Fish Soc Symp 1:268–297

Ogden JC (1970) Relative abundance, food habits and age of the American eel, *Anguilla rostrata*, in certain New Jersey streams. Trans Am Fish Soc 99:54–59. https://doi.org/10.1577/1548-8659

Oliveira K (1999) Life history characteristics and strategies of the American eel, *Anguilla rostrata*. Can J Fish Aquat Sci 56:795–802. https://doi.org/10.1139/f99-001

Oliveira K, McCleave JD (2002) Sexually different growth histories of the American eel in four rivers of Maine. Trans Am Fish Soc 131:203–211. https://doi.org/10.1577/1548-8659

Schmidt J (1923) The breeding places of the eel. Philos Trans R Soc B 211:179–208. https://doi.org/10.1098/rstb.1923.0004

Scott WB, Crossman EJ (1973) Freshwater fishes of Canada, Bulletin 184. Fisheries Research Board of Canada

Tesch FW (1977) The eel: biology and management of anguillid eels. Chapman and Hall, London. https://doi.org/10.1007/978-94-009-5761-9

Todd PR (1981) Morphometric changes, gonad histology, and fecundity estimates in migrating New Zealand freshwater eels. N Z J Mar Freshw Res 15:155–170. https://doi.org/10.1080/00288330.1981.9515908

Wenner CA, Musick JA (1974) Fecundity and gonad observations of the American eel, *Anguilla rostrata*, migrating from Chesapeake Bay, Virginia. J Fish Res Board Can 31:1387–1391. https://doi.org/10.1139/f74-164

Wenner CA, Musick JA (1975) Food habits and seasonal abundance of the American eel, *Anguilla rostrata*, from the Lower Chesapeake Bay. Chesapeake Sci16:62–66. doi: https://doi.org/10.2307/1351085

Wippelhauser GS, McCleave JD, Kleckner RC (1985) *Anguilla rostrata* leptocephali in the Sargasso Sea during February and March 1981. Dana 4:93–98

6.8 Hiodontidae, Mooneye Family

The mooneye family is composed of only two species, mooneye *Hiodon tergisus* and goldeye *Hiodon alosoides*. Both occur in the Dakotas. This family only occurs in North America and includes several fossilized species. Hiodontids resemble and are often confused with members of the Alosidae (shad) and Dorosomatidae (thread herring) families which also occur in the Dakotas. Hiodontids can easily be distinguished from alosids and dorosomatids by a lateral line, lack of pointed scales, and absence of a serrated keel along the ventral side of the body. In addition, dorsal fin insertion on hiodontids aligns, or nearly aligns, with the insertion of the anal fin, unlike alosids and dorosomatids where the dorsal fin insertion is distinctly positioned anterior to the insertion of the anal fin. Hiodontids also have small but noticeable sharp teeth on both sets of jaws, tongue, and the roof of the mouth.

Both Hiodontidae species inhabit large rivers, small tributaries, reservoirs, and lakes. Mooneye prefer clearer and quieter waters, whereas goldeye are more tolerant of turbid and muddy waters with stronger current. This may explain why mooneye are far less common in South Dakota and are mostly found in the Red River of the North in North Dakota. Spawning takes place during spring, when adults migrate from deeper waters to shallow and quiet areas of rivers, streams, and lakes. Because of the lack of an oviduct, the eggs are released inside the body cavity of females prior to spawning. No parental care is given, and the nonadhesive eggs and larvae develop as they drift with the current. Both species feed opportunistically along the surface at night on small aquatic invertebrates, such as insects, zooplankton, and occasionally small fish.

Although they are not often targeted by anglers in the Dakotas, both goldeye and mooneye are sought after in Canada and elsewhere as a delicacy. They are also a source of cutbait when angling for catfish.

6.8.1   Goldeye *Hiodon alosoides*

(Rafinesque, 1819)

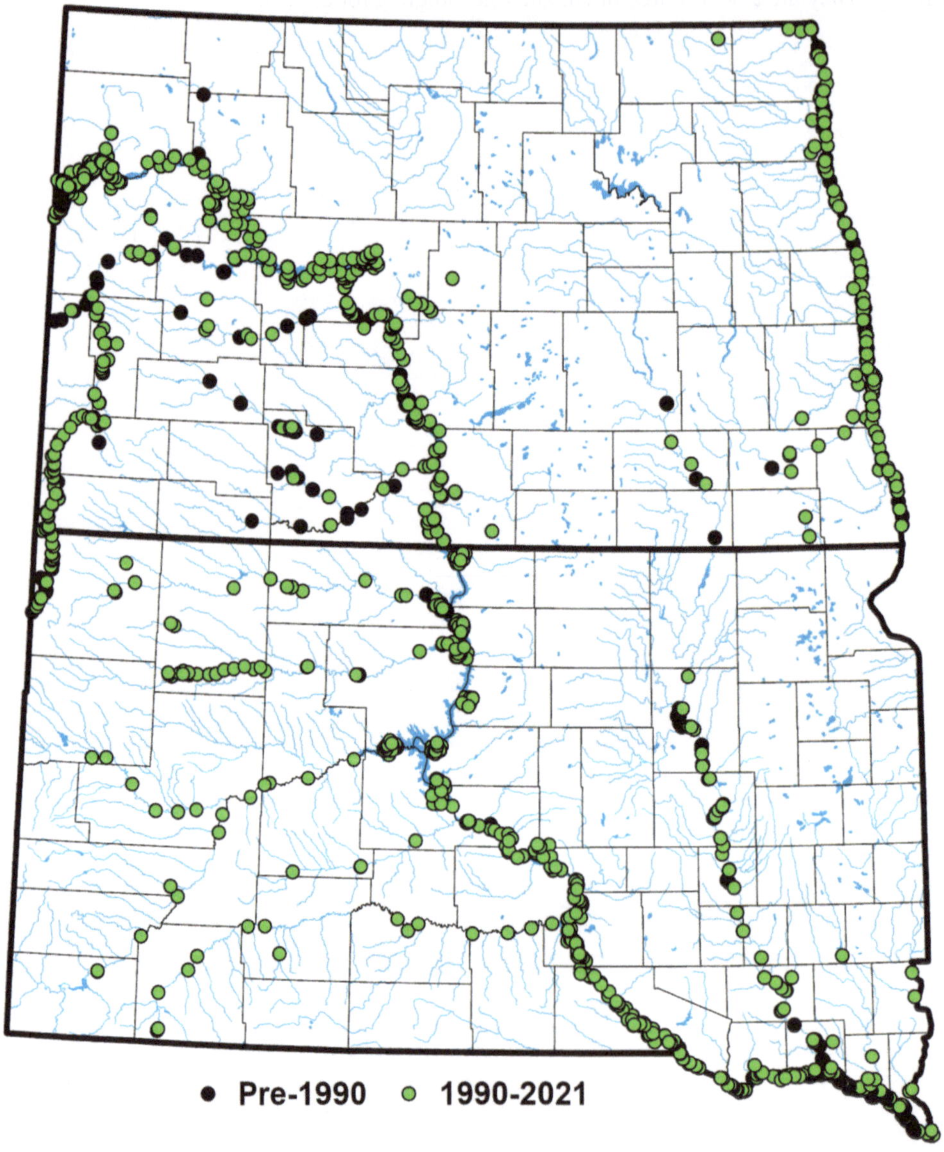

● Pre-1990   ○ 1990-2021

Etymology: ***Hiodon alosoides*** – ***Hiodon*** = toothed tongue; ***alosoides*** = resembling shad, referencing similarity to herrings and shad genus *Alosa*.

Description:

Body – laterally compressed, elongated.
Color –

> *Dorsally* – pale blue-green.
> *Laterally/ventrally* – solid silver to light silver gold.
> *Eyes* – yellow to gold iris.
> *Fins*   no markings.

Head – short.
Snout – blunt.
Eyes – relatively large; adipose eyelid.
Mouth – large, terminal, oblique.
> *Jaws* – upper jaw extends below posterior margin of eye.

Teeth – small, sharp; on both jaws, tongue, and roof of mouth.
Gill rakers – 15 to 17; short, thick.
Dorsal fin – 9 to 10 rays; posteriorly on body.
Caudal peduncle – short, thick.
Caudal fin – forked.
Anal fin – 29 to 34 rays; elongated; insertion slightly anterior to dorsal fin insertion.
Pelvic fins – abdominal, insertion far anterior of dorsal fin insertion; flap-like axillary process.
Pectoral fins – 11 to 12 rays.
Lateral line – complete; 57 to 62 cycloid scales in series.
Keel – on ventral side lacking scales; not serrate; extends from anus to pectoral fins.
Sexual dimorphism – anal fin strongly sickle shaped in males, slightly bowl-shaped in females.

Similar Species: Goldeye and mooneye are very similar. Mooneyes have 11 to 12 dorsal fin rays, with the dorsal fin insertion distinctly anterior of anal fin insertion. The mooneye iris is more silver in color, a maxillary extends to middle of the eye, and a ventral keel extends from the anus to the pelvic fins. Herring, gizzard shad, and alewife have serrated ventral keels, lack a lateral line, and lack teeth on the jaws, tongue, and roof of mouth.

Distribution and Habitat: Goldeye are endemic and widely distributed across North America. Their native range is bordered by the Hudson Bay in the north, the Mississippi River Basin-Missouri River drainage to the south, the Ohio River to the east, and Louisiana and northern Texas to the south. They occur throughout the Missouri River and its major tributaries in Dakotas and the Red River of the North drainage in North Dakota. Goldeye inhabit large rivers, reservoirs, tributaries, and quiet backwaters with turbid, muddy water and little to strong current (Kennedy and Sprules 1967). They are found most frequently just below the surface to midwater depths and overwinter in deep water (Scott and Crossman 1973).

Reproduction: Spawning migrations occur in early spring, with fish moving to the shallow areas of pools, lakes, or streams, within a few hundred meters from the shoreline (Kennedy and Sprules 1967; Scott and Crossman 1973; Donald 1997). Males sexually mature at age-2 or age-3; females typically mature a year later (Kennedy and Sprules 1967). Spawning takes place annually and usually lasts 3 to 6 weeks (Kennedy and Sprules 1967). Spawning activity begins mid-May to early June when water temperatures reach 11 °C, and likely occurs at night in highly turbid water (Scott and Crossman 1973). Eggs are released within the body cavity of the female prior to spawning because of the absence of an oviduct (Kennedy and Sprules 1967). Fecundity from goldeyes in the upper Missouri River ranged from 4,288 to 10,164 eggs/female (Hill 1966). Fertilized eggs are semi-buoyant, nonadhesive, and approximately 2 to 4 mm in diameter. After hatching, larvae drift with the current and wind. Turbulence may increase larval mortality (Donald and Kooyman 1977; Donald 1997).

Age and Growth: Growth varies with location and length of growing season (Kennedy and Sprules 1967). Juveniles of both sexes grow at the same rate, but adult females grow faster than adult males (Kennedy and Sprules 1967). Warm weather during the spring and summer increases growth and development of young, which increases survival rates (Donald 1997). Mean total-lengths-at-age for goldeye from the Missouri River backwaters in North Dakota were age-1, 121 mm; age-2, 189 mm; age-3, 234 mm; age-4, 276 mm; age-5, 298 mm; age-6, 305 mm; and age-7, 314 mm (Moon et al. 1998). Adult total lengths average 381 to 432 mm, with a maximum length of 508 mm (Trautman 1981). Goldeye in South Dakota rarely exceed 907 g.

Females generally live longer and grow larger than males (Neumann and Willis 1994). Longevity is 7 to 9 years but may be greater in northern populations (Hill 1966; Kennedy and Sprules 1967).

Food and Feeding: Goldeye feed on the surface, mostly after dark. Larvae primarily consume daphnia and copepods. Juveniles eat daphnia, copepods, corixids, and aerial insects, depending on food availability (Kennedy and Sprules 1967; Donald and Kooyman 1977). Adults eat zooplankton, a variety of aquatic and terrestrial insects, and occasionally small fish (Donald and Kooyman 1977). Selection of prey varies depending on location and the time of year (Kennedy and Sprules 1967).

References

Donald DB (1997) Relationship between year-class strength for goldeyes and selected environmental variables during the first year of life. Trans Am Fish Soc 126:361–368

Donald DB, Kooyman AH (1977) Food, feeding habits, and growth of goldeye, *Hiodon alosoides* (Rafinesque), in waters of the Peace-Athabasca Delta. Can J Zool 55:1038–1047

Hill WJ (1966) Observations on the life history and movement of the goldeye, *Hiodon alosoides,* in Montana. Dissertation, Montana State University

Kennedy WA, Sprules WM (1967) Goldeye in Canada. Fish Res Board Can Bull 161

Moon DN, Fisher SJ, Willis DW (1998) Goldeye recruitment and growth on two Missouri River backwaters. Proc S D Acad Sci 77:139–144

Neumann RM, Willis DW (1994) Guide to the common fishes of South Dakota. S D Dep Game Fish Parks, Pierre

Scott WB, Crossman EJ (1973) Freshwater fishes of Canada, Bulletin 184. Fisheries Research Board of Canada

Trautman MB (1981) The fishes of Ohio, revised edn. Ohio State University Press, Columbus

6.8.2 Mooneye *Hiodon tergisus*

Lesueur, 1818

© Uland Thomas

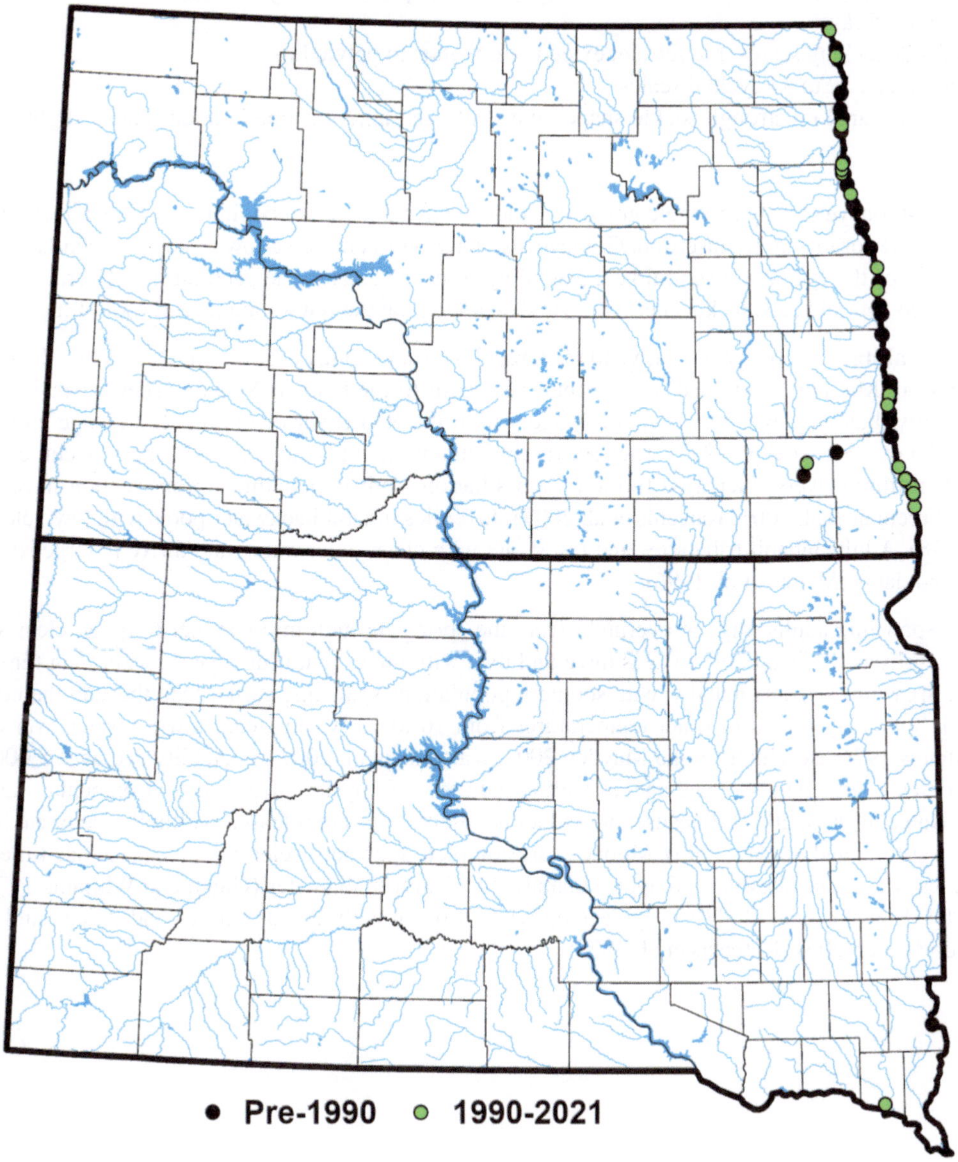

● **Pre-1990** ○ **1990-2021**

Etymology: *Hiodon tergisus* – *Hiodon* = toothed tongue; *tergisus* = polished.

Description:

Body – strongly laterally compressed, elongated.
Color –

> *Dorsally* – pale blue-green.
> *Laterally/Ventrally* – solid silver to light silver gold.
> *Eyes* – iris mainly silver.
> *Fins* – unmarked.

Head – fairly short.
Snout – blunt.
Eyes – relatively large.
Mouth – large, terminal, oblique; maxillary extends to the middle of the eye.
Teeth – small, sharp; on jaws, tongue, and roof of mouth.
Gill rakers – 15 to 17; short, thick, knob like.
Dorsal fin – 11 to 12 rays; posteriorly on body.
Caudal peduncle – short, thick.
Caudal fin – forked.
Anal fin – 26 to 29 rays; elongated, insertion distinctly posterior to dorsal fin insertion.
Pelvic fins – abdominal, insertion anterior of dorsal fin insertion; flap-like axillary process.
Pectoral fins – 13 to 15 rays.
Keel – on ventral side lacking scales, not serrate; extends from anus to pelvic fins
Lateral line – complete; 52 to 57 cycloid scales
Sexual dimorphism – anal fin strongly sickle shaped with enlarged anterior lobe in adult males; slightly bowl shaped in females.

Similar Species: Mooneye closely resemble goldeye. Goldeye have 9 to 10 dorsal fin rays in the dorsal fin, with dorsal fin insertion slightly posterior to anal fin insertion. Goldeye also have an iris more yellow to gold, a maxillary extending below the posterior margin of the eye, and a ventral keel extending from the anus to the pectoral fins. Herring, gizzard shad, and alewife have serrated ventral keels, lack a lateral line, and lack teeth on the jaws, tongue, and roof of mouth.

Distribution and Habitat: Mooneye are native to the Great Lakes and Hudson Bay from Quebec to Alberta, as well as the mainstem Mississippi-Missouri rivers and major tributaries south to the Gulf of Mexico. They occur in the Dakotas, but much less frequently than Goldeye. Mooneye are found within the Red River of the North drainage and Sheyenne River system in North Dakota (Stoner et al. 1993). They are rarely found in South Dakota, but have been documented in the Lower Big Sioux and Vermillion rivers. Increased turbidity and siltation from impoundment construction have likely impacted mooneye range (Freeman et al. 2001; Katechis et al. 2007). Juveniles inhabit backwater pools with low velocity along river-banks (Glenn 1978). Adults inhabit tributaries and reservoirs with clear, quiet water, and thrive in large rivers near the tail-waters of locks and dams (Mettee et al. 1996).

Reproduction: Spawning adults migrate overnight from the spring through mid-summer to quiet, clear streams (Becker 1983). Sexual maturity is reached between ages three and five, with males maturing a year earlier than females (Glenn and Williams 1976). Spawning occurs annually, with southern populations spawning earlier than those in the north (Wallus and Buchanan 1989). Spawning occurs in water temperatures from 10 to 13 °C, in clear, flowing water over coarse substrate (Glenn and Williams 1976; Boschung Jr and Mayden 2004; Katechis et al. 2007). Fecundity averages 5,000 to 9,000 eggs/female and increases with female size and age, as well as a variety of physiological and environmental factors (Glenn and Williams 1976; Katechis et al. 2007). Because an oviduct is absent, eggs are released within the body cavity of the female prior to spawning (Etnier and Starnes 1993). Mooneye are complete spawners, releasing all eggs at one time (Katechis et al. 2007). Eggs are buoyant, nonadhesive, and approximately 2 to 3 mm in diameter (Glenn and Williams 1976; Boschung Jr and Mayden 2004; Katechis et al. 2007). They develop while drifting with the current or wind (Glenn and Williams 1976; Boschung Jr and Mayden 2004; Katechis et al. 2007).

Age and Growth: Newly hatched mooneye larvae are slightly shorter than 7 mm (Snyder and Douglas 1978). Growth is rapid during the first summer and is faster in the northern parts of their range (Glenn 1978; Katechis et al. 2007). During the winter, young of year may lose more than a quarter of their body weight or experience slower or stunted growth (Glenn 1978). Mean total lengths at age of mooneye from the lower Tallapoosa River in Alabama were age-1, 121 mm; age-2, 170 mm; age-3, 206 mm; age-4, 233 mm; age-5, 254 mm; age-6, 269 mm; age-7, 281 mm; and age-8, 290 mm (Katechis et al. 2007). Longevity is 7 to 9 years, with females living longer than males (Glenn and Williams 1976).

Food and Feeding: Adults feed opportunistically on the surface, primarily at night. Although mooneye feed throughout the year, most feeding occurs from the spring through the fall with very little during winter (Glenn 1975, 1978; Glenn and Williams 1976). Diets of young mooneye contain relatively high amounts of zooplankton and immature insects compared to adult diets, indicating that initial feeding takes place further below the surface (Glenn 1978). Individuals over age-1 primarily consume corixids, coleopterans, ephemeropterans, dipterans, odonatans, and other insects (Glenn 1975). Adults also eat zooplankton, crayfish, plant material, and occasionally small fish (Glenn 1975).

References

Becker GC (1983) Fishes of Wisconsin. University of Wisconsin Press, Madison

Boschung HT Jr, Mayden RL (2004) Fishes of Alabama. Smithsonian Books, Washington, DC

Etnier DA, Starnes WC (1993) The fishes of Tennessee. University of Tennessee Press, Knoxville

Freeman MC, Bowen ZH, Bovee KD, Irwin ER (2001) Flow and habitat effects on juvenile fish abundance in natural and altered flow regimes. Ecol Appl 11:179–190. https://doi.org/10.1890/1051-0761(2001)011[0179:FAHEOJ]2.0.CO;2

Glenn CL (1975) Seasonal diets of mooneye, *Hiodon tergisus*, in the Assiniboine River. Can J Zool 53:232–237. https://doi.org/10.1139/z75-029

Glenn CL (1978) Seasonal growth and diets of young-of-the-year mooneye (*Hiodon tergisus*) from the Assiniboine River, Manitoba. Trans Am Fish Soc 107:587–589. https://doi.org/10.1577/1548-8659

Glenn CL, Williams RRG (1976) Fecundity of mooneye, *Hiodon tergisus*, in the Assiniboine River. Can J Zool 54:156–161. https://doi.org/10.1139/z76-016

Katechis CT, Sakaris PC, Irwin ER (2007) Population demographics of *Hiodon tergisus* (mooneye) in the lower Tallapoosa River. Southeast Nat 6:461–470. https://doi.org/10.1656/1528-7092(2007)6[461:PDOHTM]2.0.CO;2

Mettee MF, O'Neil PE, Pierson JM (1996) Fishes of Alabama and the Mobile Basin. Geological Survey of Alabama Monograph 15, Tuscaloosa, Alabama

Snyder DE, Douglas SC (1978) Description and identification of mooneye, *Hiodon tergisus*, protolarvae. Trans Am Fish Soc 107:590–594. https://doi.org/10.1577/1548-8659

Stoner JD, Lorenz DL, Wiche GJ, Goldstein RM (1993) Red River of the North basin, Minnesota, North Dakota, and South Dakota. Am Water Res Assoc 29:575–615. https://doi.org/10.1111/j.1752-1688.1993.tb03229.x

Wallus R, Buchanan JP (1989) Contributions to the reproductive biology and early life ecology of the mooneye in the Tennessee and Cumberland River. Am Midl Nat 122:204–207. https://doi.org/10.2307/2425697

6.9 Alosidae, Shad Family

The shad family consists of approximately 34 species, most of which are anadromous and marine. Some, including the two species present in the Dakotas, can live entirely in freshwater. Alosids were historically included in the family Clupeidae. However, Clupeidae was recently split into the Dorosomatidae (thread herrings) and Alosidae (shads) families. The common names of fishes in the Dakotas makes this somewhat confusing. Gizzard shad and threadfin shad are thread herrings (Dorosomatidae), despite having shad, which is now associated with the Alosidae, as part of their names. Additionally, skipjack herring (*Alosa pseudoharengus*) is an alosid, even though herring is now associated with the family Dorosomatidae.

Alosids are most similar to dorosomatids but can be separated by a scaled predorsal midline along with the lack of a long filament on the last ray of the dorsal fin. Alosids are often mistaken for members of the mooneye family (Hiodontidae) in the Dakotas but can be distinguished by the lack of a lateral line and presence of a ventrally located serrated keel. Shad are rather plain in overall appearance, generally silver and laterally compressed with a black spot behind their opercle. Like members of the trout family (Salmonidae), shad have a small, triangular-shaped flap near the base of the pelvic fins called an axillary process. Shad also have adipose eyelids, which are transparent partial membranes that cover the anterior and posterior corners of the eye.

Shad typically school within deep open water. The two species in the Dakotas occur in the open water of the Missouri River mainstem below Gavin's Point Dam. Their historical range likely extended up the Missouri River, but the dam currently acts as a barrier to upstream migration. Alosids are highly fecund and spawn pelagic eggs near the water surface. Parental care is not provided and mortality during their early life stages is high. Shad diets are primarily plankton, but some species also consume smaller fish. In the Dakotas, alewife *Alosa pseudoharengus* primarily eat algae, plankton, and small aquatic insects, whereas skipjack herring *Alosa chrysochloris* are primarily piscivorous.

6.9.1 Skipjack Herring *Alosa chrysochloris*

(Rafinesque, 1820)

© Uland Thomas

● **Pre-1990**   ○ **1990-2021**

Etymology: *Alosa chrysochloris* – *Alosa* = an old Saxon name for European shad, *Alosa*; *chrysochloris* = golden green, referencing dorsal side coloration.

Description:

Body – fusiform, strongly laterally compressed.
Color –

>   *Dorsally* – olive to dusky dark blue.
>   *Laterally* – silver to bronze; dark, round spot absent near posterior edge of the gill cover.
>   *Ventrally* – silver to white.
>   *Fins* – dorsal/caudal/anal: dusky gray tint; pelvic/pectoral – clear to lightly pigmented.

Head – large, conical.
Snout – pointed, not protruding beyond upper jaw.
Eyes – large, laterally on head.
Mouth – large, terminal.
>   *Jaws* – upper jaw extends to middle of eye; tip of lower jaw protrudes well beyond tip of upper jaw.
Teeth – small, on both jaws and tongue; present in all ages.
Gill rakers – 18 to 30 on first lower arch.
Dorsal fin – 17 to 20 rays; posterior ray short, not extending into a long filament.
Caudal peduncle – short, thick.
Caudal fin – forked.
Anal fin – 17 to 21 rays; straight to slightly concave distal end.
Pelvic fins – abdominal, insertions inferior or slightly posterior to dorsal fin insertion.
Axillary process – present.
Lateral line – absent; 51 to 60 cycloid scales in series.
Ventral keel – 17 to 19 scaled scutes anterior and 14 to 17 posterior to pelvic fin.

Similar Species: Skipjack herring resemble alewife. Alewifes have a slightly deeper body, a dorsal fin with 13 to 14 rays and only 42 to 50 cycloid scales in lateral line series. Threadfin shad have a smaller head, a dark black, round spot similar in size to the pupil posteriorly from the upper edge of the gill cover, a dorsal fin with the last ray extended into a long filament, and 40 to 48 cycloid scales in the lateral line series. Gizzard shad have a more bulbous and fleshy snout that protrudes beyond the upper jaw, a dark black, round spot similar in size to the pupil posteriorly from the upper edge of the gill cover, an anal fin with 29 or more rays, and a dorsal fin with the last ray extended into a long filament.

Distribution and Habitat: Skipjack herring are native to the Mississippi River basin, ranging from eastern South Dakota and central Minnesota in the north, the Ohio River to western Pennsylvania in the east, and the Gulf Slope to the Gulf of Mexico in the south (Lee et al. 1980). They are a very active fish whose migratory patterns have been negatively affected by dams and impoundments (Cross and Huggins 1975; Becker 1983; Pflieger 1997). Although the species is still uncommon within the upper Mississippi and Missouri rivers, they occur more frequently upstream because of the lower turbidities caused by dredging and other modifications in the middle Missouri River (Cross and Huggins 1975; Lee et al. 1980; Pflieger 1997; Neebling and Quist 2008). Skipjack herring are most abundant in the upper Mississippi River below the Ohio River confluence (Pflieger 1997). They have not been reported in North Dakota (Hayer 2014). In South Dakota, they occur as far north as Fort Randall Dam in the Missouri River and in the Whetstone River within the Minnesota River system. Skipjack herring were previously reported from Big Stone Lake but are likely extirpated (Bailey and Allum 1962). They often congregate in large schools in clear to moderately turbid, deep, swift, and open waters of large rivers and occasionally reservoirs (Lee et al. 1980; Trautman 1981; Etnier and Starnes 1993; Pflieger 1997). They are intolerant of high turbidity and generally avoid it (Trautman 1981; Pflieger 1997).

Reproduction: Information on skipjack herring reproductive habits is scarce. The spawning season is prolonged and likely takes place April through July in the upper Mississippi River (Coker 1930). Sexual maturity typically occurs at ages two to three, or when individuals reach a total length of 254 to 305 mm (Coker 1930; Etnier and Starnes 1993). Spawning takes place in the main channel of large rivers over sand and gravel bars (Boschung Jr and Mayden 2004). Although actual spawning behavior has not been observed, it is presumed that individuals do not spawn in large aggregations (Coker 1930; Becker 1983; Boschung Jr and Mayden 2004). Fecundity is dependent on female size but may range from 76,000 to 962,000 eggs (Boschung Jr and Mayden 2004). Eggs are approximately 1 mm in diameter (Becker 1983; Boschung Jr and Mayden 2004). Hatching takes place within 34 to 48 hours at a water temperature of 17 °C (Boschung Jr and Mayden 2004).

Age and Growth: Young of the year skipjack herring generally reach a total length of 75 to 150 mm by the end of the first year (Etnier and Starnes 1993; Pflieger 1997). In Ohio, young of the year were 25 to 100 mm in August and 130 to 200 mm October (Trautman 1981). Maximum total length is 533 mm and longevity is 4 years (Trautman 1981; Boschung Jr and Mayden 2004).

Food and Feeding: Skipjack herring juveniles consume zooplankton, dipterans, and insect larvae (Boschung Jr and Mayden 2004). Fish consumption increases with size, and after reaching 40 mm in total length, skipjack herring primarily forage on small fish (Boschung Jr and Mayden 2004). Adults will consume gizzard shad and threadfin shad near the surface (Etnier and Starnes 1993). Skipjack herring forage by forming big schools in swift waters, and then force large schools of emerald shiner and other minnows toward the surface (Trautman 1981). They then dash into the minnow school and may even leap out of the water to capture jumping minnows (Trautman 1981). This jumping or skipping motion is likely the reason for its common name (Trautman 1981).

References

Bailey RM, Allum MO (1962) Fishes of South Dakota. University of Michigan, Ann Arbor. https://doi.org/10.3998/mpub.9690435

Becker GC (1983) Fishes of Wisconsin. University of Wisconsin Press, Madison

Boschung HT Jr, Mayden RL (2004) Fishes of Alabama. Smithsonian Books, Washington, DC

Coker RE (1930) Studies of common fishes of the Mississippi River at Keokuk. Bulletin of the US Bureau of Fisheries 1072:141–225

Cross FB, Huggins DG (1975) Skipjack herring, *Alosa chrysochloris*, in the Missouri River. Copeia 1975:382–385. https://doi.org/10.2307/1442900

Etnier DA, Starnes WC (1993) The fishes of Tennessee. University of Tennessee Press, Knoxville

Hayer CA (2014) Fish assemblage structure, trophic ecology, and potential effects of invading Asian carps in three Missouri River tributaries, South Dakota. Dissertation, South Dakota State University

Lee DS, Gilbert CR, Hocutt CH, Jenkins RE, McAllister DE, Stauffer JR (1980) Atlas of North American freshwater fishes. North Carolina Biological Survey Publication, 1980–12. https://doi.org/10.5962/bhl.title.141711

Neebling TE, Quist MC (2008) Observations on the distribution and status of western sand darter, spotted gar, and skipjack herring in Iowa rivers. J Iowa Acad Sci 115:24–27

Pflieger WL (1997) The fishes of Missouri, revised edn. Missouri Dep Cons, Jefferson City

Trautman MB (1981) The fishes of Ohio. Ohio State University Press, Columbus

6.9.2 Alewife *Alosa pseudoharengus*

(Wilson, 1811)

● Pre-1990   ○ 1990-2021

Etymology: *Alosa pseudoharengus – Alosa* = old Saxon name for European shad; *pseudoharengus* = false herring.

Description:

Body – fusiform, moderately deep, strongly laterally compressed.
Color –

> *Dorsally* – gray to olive.
> *Laterally* – silver; a dark, diffused blue black spot similar in size to the pupil posteriorly from the upper edge of the gill cover.
> *Ventrally* – silver to white.
> *Fins* – dorsal/caudal/pectoral: gray coloration; anal/pelvic: transparent to white.

Head – small.
Snout – pointed, not protruding beyond upper jaw.
Eyes – large, laterally on head.
Mouth – large, superior, strongly oblique.
> *Jaws* – upper jaw extends to middle of eye; lower jaw protrudes beyond upper jaw.
Gill rakers – 30 to 44 on first arch; fine.
Dorsal fin – 13 to 14 rays; posterior ray short, not extending into a long filament.
Caudal peduncle – short, thick.
Caudal fin – forked.
Anal fin – 17 to 18 rays; straight distal end.
Pelvic fins – 9 rays; abdominal; insertions inferior to slightly posterior of dorsal fin insertion.
Axillary process – present.
Lateral line – absent; 42 to 50 cycloid scales in series; scales present on predorsal midline.
Ventral keel – 17 to 20 scaled scutes anterior and 12 to 15 posterior to pelvic fin.

Similar Species: Alewife resemble skipjack herring. Skipjack herring have a more elongated and conical head and snout, along with small, but visible teeth on jaws, and a dorsal fin with 17 to 20 rays. Gizzard shad have a more bulbous and fleshy snout that protrudes beyond the upper jaw, a dorsal fin with the last ray extended into a long filament, and an anal fin with 29 or more rays. Threadfin shad have fins with a yellow to gold tint, a dorsal fin with the last ray extended into a long filament, and no predorsal midline scales.

Distribution and Habitat: Alewife are an anadromous species native to the Atlantic Coast from Red Bay, Labrador to South Carolina (Lee et al. 1980). Alewife have invaded or been introduced to many landlocked waterbodies including the Great Lakes (Lee et al. 1980; Miller 1957). In the Dakotas, alewife only occur within the Missouri River below Fort Randall Dam in South Dakota. Alewifes inhabit lentic waters of large rivers, streams, lakes, and reservoirs. Larvae and young of the year congregate in large schools in shallow, warmer shoreline areas until early fall before migrating to deeper water (Otto et al. 1976). Adults exhibit diel migrations, moving to deeper depths during daylight and to the surface at night (Pardue 1983). The critical lower lethal temperature of mature alewives is approximately 3 °C (Stanley and Colby 1971; Otto et al. 1976). The critical thermal maximum of the species increases with acclimation, and young-of-the year may be more tolerant of higher water temperatures than adults (Otto et al. 1976). The upper lethal temperature is estimated at 31 to 34 °C, when the fish are acclimated to 27 °C (McCauley and Binkowski 1982). Alewife are tolerant of a wide range of salinity and turbidity (Pardue 1983).

Reproduction: Alewife spawn during late spring and summer in a variety of habitats and substrates (Norden 1967a; Otto et al. 1976; Becker 1983; Pardue 1983). While sexual maturity may be reached at age-1, most spawning individuals are age-2 and age-3 (Norden 1967a). Alewives may spawn only once (Norden 1967a). Adults migrate to shallow, inshore waters for spawning, remaining there until mid-summer migration back to deeper waters (Norden 1967a; Stanley and Colby 1971; Otto et al. 1976). Spawning peaks from midnight to the early morning. It consists of two or more individuals swimming rapidly near the surface in a tight circular motion with lateral sides touching (Edsall 1964; Becker 1983). Females from Lake Michigan produce 11,000 to 22,000 eggs. Anadromous females from Chesapeake Bay can produce 60,000 to 100,000 eggs (Norden 1967a; Foerster and Goodbred 1978; Pardue 1983). The pale-yellow eggs initially are demersal and adhesive but become pelagic and nonadhesive after water hardening (Pardue 1983). Eggs diameter is approximately 1 mm (Norden 1967b). Hatching occurs at water temperatures of 7 to 29 °C, with an optimum hatching temperature of 18 °C (Edsall 1970). Eggs hatch in 15 days at 7 °C, 3.7 days at 21 °C, and 2.1 days at 29 °C (Edsall 1970).

Age and Growth: Larvae have a total length of approximately 4 mm at hatch (Norden 1967a, b). Growth occurs mostly during the first year (Norden 1967a). Total lengths at age of alewife from Lake Michigan were age-0, 96 mm; age-1, 138 mm; age-2, 158 mm; age-3, 173 mm; and age-4, 193 mm (Norden 1967a). They are capable of reaching 350 mm, with a longevity of 10 years (Lee et al. 1980; O'Neill 1980; Pardue 1983).

Food and Feeding: Gizzard shad and alewife have similar diets. Juvenile alewifes primarily consume copepods, cladocerans, and other microcrustacean (Norden 1968). Adults also eat primarily microcrustacea, but also feed on chironomid larvae, filamentous algae, gastropods, and fish eggs (Edsall 1964; Becker 1983; Hrabik et al. 2015). Feeding activity is reduced during the spawning season (Pardue 1983).

References

Becker GC (1983) Fishes of Wisconsin. University of Wisconsin Press, Madison

Edsall TA (1964) Feeding by three species of fishes in eggs of spawning alewives. Copeia 1964:226–227. https://doi.org/10.2307/1440867

Edsall TA (1970) The effect of temperature on the rate of development and survival of alewife eggs and larvae. Trans Am Fish Soc 99:376–380. https://doi.org/10.1577/1548-8659

Foerster JW, Goodbred SL (1978) Evidence for a resident alewife population in the northern Chesapeake Bay. Estuar Coast Mari Sci 7:437–444. https://doi.org/10.1016/0302-3524(78)90120-2

Hrabik RA, Schainost SC, Stasiak RH, Peters EJ (2015) The fishes of Nebraska. University of Nebraska-Lincoln

Lee DS, Gilbert CR, Hocutt CH, Jenkins RE, McAllister DE, Stauffer JR (1980) Atlas of North American freshwater fishes. North Carolina Biological Survey Publication, 1980–12. https://doi.org/10.5962/bhl.title.141711

McCauley RW, Binkowski FP (1982) Thermal tolerance of the alewife. Trans Am Fish Soc 111:389–391. https://doi.org/10.1577/1548-8659

Miller RR (1957) Origin and dispersal of the alewife, *Alosa pseudoharengus*, and the gizzard shad, *Dorosoma cepedianum*, in the Great Lakes. Trans Am Fish Soc 86:97–111. https://doi.org/10.1577/1548-8659

Norden CR (1967a) Age, growth and fecundity of the alewife, *Alosa pseudoharengus* (Wilson), in Lake Michigan. Trans Am Fish Soc 96:387–393. https://doi.org/10.1577/1548-8659(1967)96[387:AGAFOT]2.0.CO;2

Norden CR (1967b) Development and identification of the larval alewife, *Alosa pseudoharengus* (Wilson), in Lake Michigan. Proc Conf Great Lakes Res 10:70–78

Norden CR (1968) Morphology and food habits of the larval alewife, *Alosa pseudoharengus* (Wilson), in Lake Michigan. Proc Conf Great Lakes Res 11:103–110

O'Neill JT (1980) Aspects of the life histories of anadromous alewife and the blueback herring, Marqaree River and Lake Aimsle, Nova Scotia, 1978–1979. Dissertation, Acadia University

Otto RG, Kitchell MA, Rice JO (1976) Lethal and preferred temperatures of the alewife (*Alosa pseudoharengus*) in Lake Michigan. Trans Am Fish Soc 105:96–106. https://doi.org/10.1577/1548-8659

Pardue GB (1983) Habitat suitability index models: alewife and blueback herring. US Dep Inter, Fish Wildl Serv, FWS/OBS-82/10.58

Stanley JG, Colby PJ (1971) Effects of temperature on electrolyte balance and osmoregulation in the alewife (*Alosa pseudoharengus*) in fresh and sea water. Trans Am Fish Soc 100:624–638. https://doi.org/10.1577/1548-8659

6.10   Dorosomatidae, Thread Herring Family

The herring family consists of approximately 116 species, most of which are anadromous and marine. Some, including the two species present in the Dakotas, can live entirely in freshwater. The Dorosomatidae (thread herrings) were historically included in the family Clupeidae. However, Clupeidae was recently split into the Dorosomatidae and Alosidae (shads) families. The common names of fishes in the Dakotas make this somewhat confusing. Gizzard shad and threadfin shad are thread herrings (Dorosomatidae), despite having shad, which is now associated with the Alosidae, as part of their names. Additionally, skipjack herring (*Alosa pseudoharengus*) is an alosid, even though herring is now associated with the family Dorosomatidae.

Dorosomatids are most similar to the alosids but differ by having a long filament on the last ray of the dorsal fin and a naked predorsal midline. Dorosomatids are also often mistaken for hiodontids (mooneye family members) but can be distinguished by the lack a lateral line and presence of a ventrally located serrated keel. Dorosomatid dorsal fin insertion is also distinctly anterior to the insertion of the anal fin, unlike hiodontids where the dorsal and anal fin insertions nearly align. Dorosomatids are rather plain in their overall appearance, generally silver and laterally compressed, with a black spot behind their opercle. Like salmonids (trout family members), dorosomatids have a small, triangular-shaped flap near the base of the pelvic fins called an axillary process. They also have adipose eyelids, which are transparent partial membranes that cover the anterior and posterior corners of the eye.

Dorosomatids typically school within deep open water. The two species in the Dakotas occur primarily in the open water areas of the Missouri River mainstem and major tributaries, as well as several reservoirs. They are highly fecund and spawn pelagic eggs near the water surface. Parental care is not provided and mortality during their early life stages is high. Diets are primarily plankton, but some species also consume smaller fish.

Dorosomatids are economically important and heavily exploited worldwide. They are important forage fish in both marine and freshwater habitats. In the Dakotas, gizzard shad *Dorosoma cepedianum* are an important prey source in the diets of walleye *Sander vitreus* and largemouth bass *Micropterus nigricans*. Large gizzard shad, too big to be eaten, can negatively affect fish populations by reducing zooplankton abundance and recruitment of other sport fish such as bluegill *Lepomis macrochirus*.

6.10.1   Gizzard Shad *Dorosoma cepedianum*

(Lesueur, 1818)

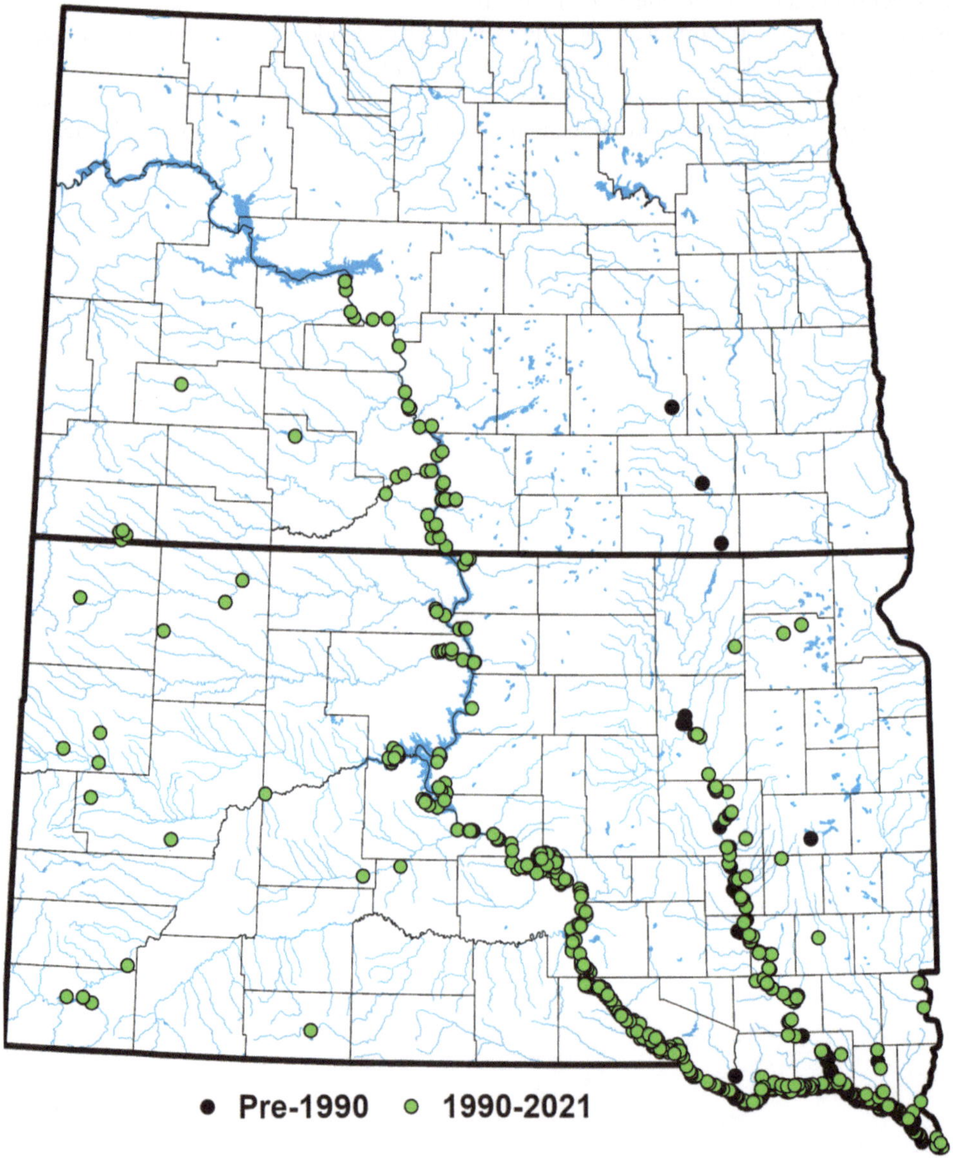

● **Pre-1990** ○ **1990-2021**

Etymology: *Dorosoma cepedianum – Dorosoma:* doris = lance body, referencing head and body shape; *cepedianum* = patronymic for the French ichthyologist Bernard Germain Etienns LaCepede.

Description:

Body – deep, laterally compressed on each side.

Color – dorsal sides silvery-blue fading to silver laterally with a dark purple to black spot larger than the pupil behind the upper portion of gill cover; dark spot may be lacking in older fish.

Head – small, scaleless.

Snout – bulbous, fleshy; protrudes beyond the upper jaw.

Eyes – large; adipose eyelids.

Mouth – small and subterminal; small notch in the middle of the premaxilla; upper jaw extends past lower jaw.

Teeth – absent in adults; present in larvae.

Gill rakers – long, fine; increase in number with age.

Dorsal fin – 10 to 15 rays; last ray elongates into a filament extending halfway back to the base of the caudal fin; insertion in the middle of the body, posterior to pelvic fin insertion.

Caudal fin – forked.

Anal fin – elongate; 29 to 37 rays; straight distal end.

Pelvic fins – abdominal; 7 to 10 rays; axillary process present.

Pectoral fins – 12 to 17 rays.

Lateral line – absent; 52 to 70 cycloid scales in series.

Ventral keel – serrated; 17 to 20 hardened scaled scutes anterior and 10 to 14 posterior of the pelvic fin.

Similar Species: Gizzard shad closely resemble threadfin shad. Threadfin shad have a more pointed snout that does not protrude past the upper jaw, an anal fin with 17 to 25 rays, and a lesser number of ventral keel scaled scutes. Mooneye and goldeye possess a lateral line, a dorsal fin base that is generally over the anal fin, and a toothed tongue. Skipjack herring and alewife have a lower jaw extending beyond the upper jaw when the mouth is closed and a dorsal fin that lacks a long filament. Juvenile gizzard shad may be confused with silver carp and bighead carp, although both carp species have a complete lateral line.

Distribution and Habitat: Gizzard shad are native to North America from New Mexico east to New York, and from southern Ontario and Quebec south to northern Mexico (Wuellner et al. 2009; Fincel et al. 2014). In the Dakotas, the species is primarily found throughout the mainstem and major tributaries of the Missouri River. It has also been widely stocked as a forage fish throughout the Dakotas (Fincel et al. 2017). Gizzard shad are most numerous in lakes, oxbows, impoundments, and low gradient larger streams (Owen et al. 1981). In the Dakotas, gizzard shad are often found over mud and fine silt substrates, with adults in deeper water and juveniles swimming in schools near the surface (Kansas Fishes Committee 2014). Sudden changes in dissolved oxygen and temperature can cause mortality, making gizzard shad prone to winter-kill (Owen et al. 1981).

Reproduction: Most gizzard shad males mature by age-1. While some females mature at that age, the majority mature by age-2 or age-3 (Bodola 1966). Spawning occurs mostly during the day in shallow open water bays and inlets in late spring and early summer at temperatures between 10 and 21 °C (Owen et al. 1981; Heidinger 1983; Michaletz 1998; Kansas Fishes Committee 2014; Greiner et al. 2017; Radigan et al. 2018). Gizzard shad also migrate to smaller streams for spawning (Churchill and Over 1938). Spawning takes place near the surface where males and females swim, roll, and tumble together while expelling eggs and milt (Miller 1960). Fertilized eggs are demersal and attach to vegetation or other debris (Pflieger 1997). Eggs are approximately 1 mm in diameter, and fecundity is approximately 300,000 eggs (Bodola 1966; Etnier and Starnes 1993; Pflieger 1997; Kansas Fishes Committee 2014). Because adults spawn near the surface, there are neither nests nor parental care (Cross 1967; Cross and Collins 1995).

Age and Growth: Gizzard shad total length at hatching is 3 to 5 mm (Miller 1960). Early growth is rapid, with age-1 fish reaching a total length of 101 to 178 mm and age-2 fish reaching approximately 279 mm (Carlander 1969; Williamson and Nelson 1985; Etnier and Starnes 1993; Fincel et al. 2013; Hrabik et al. 2015). After age-2, females grow slightly more rapid than males (Bodola 1966; Greiner et al. 2017). Gizzard shad total length is typically longer in northern waters (Fagan and Fitzpatrick 1978; Wuellner et al. 2009). In South Dakota, gizzard shad from Angostura Reservoir were longer at each age than more-southern fish from Lake Francis Case and Lake Sharpe (Wuellner et al. 2009); Mean total-lengths-at-age from Angostura Reservoir were age-3, 381 mm; age-6, 447 mm; age-10, 441 mm (Wuellner et al. 2009). They are capable of reaching a total length of 521 mm (Trautman 1981). Gizzard shad are a short-lived species with high natural mortality rates (Fincel et al. 2013). Longevity is 10 years (Wuellner et al. 2009).

Food and Feeding: Gizzard shad are an opportunistic omnivore (Yako et al. 1996). Larvae have teeth during the first summer. They feed primarily on zooplankton, but also consume phytoplankton, detritus, and microcrustaceans (Megrey 1980; Kansas Fishes Committee 2014; Hrabik et al. 2015). Larvae only feed during the day, suggesting that they are visual particulate feeders (Dettmers and Stein 1992). Adults consume large amounts of phytoplankton, zooplankton, and detritus. The common name, gizzard shad, stems from the gizzard-like stomach which aids in the breakdown of food (Churchill and Over 1938; Hrabik et al. 2015). Gizzard shad are an important forage species for many South Dakota sportfish (Fincel et al. 2014). However, they are capable of driving zooplankton abundance to low levels, negatively affecting the recruitment of sport fish, such as bluegill and largemouth bass (Michaletz 1998).

References

Bodola A (1966) Life history of the gizzard shad, *Dorosoma cepedianum* (Lesueur), in western Lake Erie. Fish Bull 65:391–425

Carlander KD (1969) Handbook of freshwater fishery biology, vol 1. Iowa State University Press, Ames

Churchill EP, Over WH (1938) Fishes of South Dakota. Brown & Saenger Printers, Sioux Falls

Cross FB (1967) Handbook of fishes in Kansas. University Press of Kansas, Lawrence

Cross FB, Collins JT (1995) Fishes in Kansas, 2nd, revised edn. University Press of Kansas, Lawrence

Dettmers JM, Stein RA (1992) Food consumption by larval gizzard shad: zooplankton effects and implications for reservoir communities. Trans Am Fish Soc 121:494–507. https://doi.org/10.1577/1548-8659

Etnier DA, Starnes WC (1993) The fishes of Tennessee. University of Tennessee Press, Knoxville

Fagan JA, Fitzpatrick LC (1978) Allocation of secondary production to growth and reproduction by gizzard shad *Dorosoma cepedianum* (Clupeidae) in Lewisville, Texas. Southwest Nat 23:247–262. https://doi.org/10.2307/3669773

Fincel MJ, Chipps SR, Graeb BDS, Edwards KR (2013) Larval gizzard shad characteristics in Lake Oahe, South Dakota: a species at the northern edge of its range. J Freshw Ecol 28:17–26. https://doi.org/10.1080/02705060.2012.709887

Fincel MJ, Dembkowski DJ, Chipps SR (2014) Influence of variable rainbow smelt and gizzard shad abundance on walleye diets and growth. Lake Reserv Manag 30:258–267. https://doi.org/10.1080/10402381.2014.914989

Fincel MJ, Smith MJ, Hanten RP, Radigan WJ (2017) Recommendations for stocking gizzard shad in a large upper Midwest reservoir. N Am J Fish Manag 37:599–604. https://doi.org/10.1080/02755947.2017.1300614

Greiner MJ, Fincel MJ, Longhenry CM (2017) The impacts of reduced water temperature on gizzard shad production in two Missouri River reservoirs during a historic flood. Fish Manag Ecol 24:420–425. https://doi.org/10.1111/fme.12238

Heidinger RC (1983) Life history of gizzard shad and threadfin shad as it relates to the ecology of small lake fisheries. In: Bonneau D, Radonski G (eds) Proceedings of the small lakes management workshop: pros and cons of shad. Iowa Conservation Commission, Des Moines, pp 1–18

Hrabik RA, Schainost SC, Stasiak RH, Peters EJ (2015) The fishes of Nebraska. University of Nebraska, Lincoln

Kansas Fishes Committee (2014) Kansas fishes. University Press of Kansas, Lawrence

Megrey BA (1980) *Dorosoma cepedianum* (Lesueur), gizzard shad. In: Lee DS, Gilbert CR, Hocutt CH, Jenkins RE, McAllister DE, Stauffer JR Jr (eds) Atlas of North American freshwater fishes. North Carolina State Museum of Natural History, Raleigh, p 69

Michaletz PH (1998) Population characteristics of gizzard shad in Missouri reservoirs and their relation to reservoir productivity, mean depth, and sport fish growth. N Am J Fish Manag 18:114–123. https://doi.org/10.1577/1548-8675

Miller RR (1960) Systematic and biology of the gizzard shad (*Dorosoma cepedianum*) and related fishes. Fish Bull 173:371–392

Owen JB, Elsen DS, Russell GW (1981) Distribution of fishes in North and South Dakota Basins affected by the Garrison Diversion Unit. University of North Dakota Press, Grand Forks

Pflieger WL (1997) The fishes of Missouri, revised edn. Missouri Dep Cons, Jefferson City

Radigan WJ, Carlson AK, Fincel MJ, Graeb BDS (2018) Otolith chemistry as a fisheries management tool after flooding: the case of Missouri River gizzard shad. River Res Appl 34:270–278. https://doi.org/10.1002/rra.3247

Trautman MB (1981) The fishes of Ohio. Ohio State University Press, Columbus

Williamson KL, Nelson PC (1985) Habitat suitability index models and instream flow suitability curves: gizzard shad, vol 82. Western energy and land use team. US Dep Inter, Fish Wildl Serv, Div Biol Serv, Res Devel

Wuellner MR, Graeb BDS, Ward MJ, Willis DW (2009) Review of gizzard shad population dynamics at the northwestern edge if its range. Am Fish Soc Symp 62:37–653

Yako LA, Dettmers JM, Stein RA (1996) Feeding preferences of omnivorous gizzard shad as influenced by fish size and zooplankton density. Trans Am Fish Soc 125:753–759. https://doi.org/10.1577/1548-8659

6.10.2   Threadfin Shad *Dorosoma petenense*

(Günther, 1867)

© Uland Thomas

● **Pre-1990**   ○ **1990-2021**

Etymology: *Dorosoma petenense* – *Dorosoma* = lanceolate body, referencing juvenile body shape; *petenense* = named after the type locality, Lake Peten, Yucatan.

Description:

Body – fusiform, moderately deep, laterally compressed.
Color – dorsally olive to dusky dark blue; laterally silver with a dark black, round spot similar in size to the pupil posteriorly from the upper edge of the gill cover; ventrally silver to white; all fins except the dorsal fin with yellow to gold tint.
Head – small.
Snout – pointed; does not protrude beyond upper jaw.
Eyes – large, laterally on head.
Mouth – large, terminal; upper jaw does not extend beyond middle of eye.
Gill rakers – long, fine, numerous.
Dorsal fin – 11 to 15 rays, last ray extends into a long filament often reaching to posterior end of the anal fin.
Caudal peduncle – short, thick.
Caudal fin – forked.
Anal fin – elongate; 17 to 25 rays; straight distal end.
Pelvic fins – abdominal; 7 or 8 rays; insertions directly inferior to insertion of dorsal fin; axillary process.
Pectoral fins – 12 to 17 rays.
Lateral line – 40 to 48 cycloid scales in series; scales absent on predorsal midline.
Ventral keel – 16 or 17 scaled scutes anterior and 9 to 11 posterior of pelvic fin.

Similar Species: Threadfin shad closely resemble gizzard shad. Gizzard shad have a more bulbous and fleshy snout protruding beyond the upper jaw, an anal fin with 29 or more rays, and dusky gray fins, lacking any yellow to gold tint. Gizzard shad also grow to a much larger size. Alewife lack a long filament in the dorsal fin and have a scaled predorsal midline.

Distribution and Habitat: Threadfin shad have been widely introduced as a forage species for sport fish outside its native range of the lower Mississippi River and Gulf Slope, south to Central America (DeVries et al. 1991). In the Dakotas, they only occur in the Missouri River as far north as Lewis and Clark Lake in South Dakota. Threadfin shad inhabit reservoirs and big rivers with swift current (Etnier and Starnes 1993). They are intolerant of cold temperatures, and often experience mass winterkill in areas where temperatures suddenly drop or are severe for long-time periods (Parsons and Kimsey 1954; Griffith 1978). In aquaria, lower lethal temperature was approximately 5 °C when fish were exposed to temperatures decreasing at 1 °C every 72 hours (Griffith 1978).

Reproduction: The spawning season for threadfin shad is prolonged. Sexual maturity occurs at less than 130 mm (Griffith 1978). Spawning occurs with schools splashing and swimming quickly back and forth through beds of vegetative matter near the surface and shoreline (Gerdes and McConnell 1963; Lambou 1965). No nest is constructed, and no parental care is given. Spawning lasts for 5 to 10 seconds, after which the fish immediately return to deeper water (Gerdes and McConnell 1963). Fecundity from central Arizona reservoirs ranged from 923 to 8,540 eggs per female, but threadfin shad are capable of producing up to 12,400 eggs (Kilambi and Baglin Jr 1969; Johnson 1971). Fecundity is not related to female length, weight, or age (Johnson 1971). Eggs are adhesive, with a diameter of approximately 0.8 mm (Gerdes and McConnell 1963; Johnson 1971). Water temperatures below 34 °C are needed for successful egg incubation (Hubbs and Bryan 1974). The optimum water temperature for larval survival is 22 °C, with mortality increasing at higher temperatures (Betsill and Van Den Avyle 1997).

Age and Growth: Little information is available on threadfin shad growth rates in the Dakotas. Standard lengths at age of threadfin shad from a Salt River reservoir in central Arizona were age-1, 39 mm; age-2, 63 mm; age-3, 79 mm; age-4, 93 mm (Johnson 1970). Most growth occurs in the spring and summer months (Johnson 1970). Growth of 21-day-old larvae is related to water temperature and prey availability (Betsill and Van Den Avyle 1997). Threadfin shad are capable of reaching a total length of 220 mm, with a longevity of 3 years (Pflieger 1997; Boschung Jr and Mayden 2004).

Food and Feeding: Threadfin shad are opportunistic sight feeders with no dietary differences among fish of different sizes (Miller 1967; Ingram and Ziebell 1983). They are primarily pelagic planktivores because of the long and fine gill filaments used to filter zooplankton (Haskell 1959; Gerdes and McConnell 1963; Miller 1967). They also consume vegetation, organic matter, and other organisms of suitable size (Haskell 1959; Gerdes and McConnell 1963; Miller 1967). In addition, they can shift to feed exclusively on bottom-dwelling prey items, such as chironomids, when other food is lacking (Haskell 1959; Ingram and Ziebell 1983). Feeding behavior consists of quick back-and-forth movements with open mouths for a few seconds at a time (Gerdes and McConnell 1963). Feeding activity decreases at water temperatures below 10 °C (Griffith 1978).

References

Betsill RK, Van Den Avyle MJ (1997) Effect of temperature and zooplankton abundance on growth and survival of larval threadfin shad. Trans Am Fish Soc 126:999–1011. https://doi.org/10.1577/1548-8659

Boschung HT Jr, Mayden RL (2004) Fishes of Alabama. Smithsonian Books, Washington, DC

DeVries DR, Stein RA, Miner JG, Mittelbach GG (1991) Stocking threadfin shad: consequences for young-of-year fishes. Trans Am Fish Soc 120:368–381. https://doi.org/10.1577/1548-8659

Etnier DA, Starnes WC (1993) The fishes of Tennessee. University of Tennessee Press, Knoxville

Gerdes JH, McConnell WMJ (1963) Food habits and spawning of the threadfin shad in a small, desert impoundment. J Ariz Acad Sci 2:113–116. https://doi.org/10.2307/27641798

Griffith JS (1978) Effects of low temperature on the survival and behavior of threadfin shad, *Dorosoma petenense*. Trans Am Fish Soc 107:63–70. https://doi.org/10.1577/1548-8659

Haskell WML (1959) Diet of the Mississippi threadfin shad, *Dorosoma petenense atchafalayae*, in Arizona. Copeia 1959:298–302. https://doi.org/10.2307/1439886

Hubbs C, Bryan C (1974) Maximum incubation temperature of the threadfin shad, *Dorosoma petenense*. Trans Am Fish Soc 103:369–371. https://doi.org/10.1577/1548-8659

Ingram W, Ziebell CD (1983) Diet shifts to benthic feeding by threadfin shad. Trans Am Fish Soc 112:554–556. https://doi.org/10.1577/1548-8659

Johnson JE (1970) Age, growth, and population dynamics of threadfin shad, *Dorosoma petenense* (Günther), in central Arizona reservoirs. Trans Am Fish Soc 99:739–753. https://doi.org/10.1577/1548-8659

Johnson JE (1971) Maturity and fecundity of threadfin shad, *Dorosoma petenense* (Günther), in central Arizona reservoirs. Trans Am Fish Soc 100:74–85. https://doi.org/10.1577/1548-8659

Kilambi RV, Baglin RE Jr (1969) Fecundity of the threadfin shad, *Dorosoma petenense*, in Beaver and Bull Shoals reservoirs. Trans Am Fish Soc 98:320–322. https://doi.org/10.1577/1548-8659(1969)98[320:FOTTSD]2.0.CO;2

Lambou VW (1965) Observations and size distribution and spawning behavior of threadfin shad. Trans Am Fish Soc 94:385–386. https://doi.org/10.1577/1548-8659(1965)94[385:OOSDAS]2.0.CO;2

Miller RV (1967) Food of the threadfin shad, *Dorosoma petenense*, in Lake Chicot, Arkansas. Trans Am Fish Soc 96:243–246. https://doi.org/10.1577/1548-8659(1967)96[243:FOTTSD]2.0.CO;2

Parsons JW, Kimsey JB (1954) A report on the Mississippi threadfin shad. Prog Fish Cult 16:179–181. https://doi.org/10.1577/1548-8659(1954)16[179:AROTMT]2.0.CO;2

Pflieger WL (1997) The fishes of Missouri, revised edn. Missouri Dep Cons, Jefferson City

6.11   Catostomidae, Sucker Family

Nearly all 76 extant fish species in the sucker family are native to North America. Only the Asiatic sucker *Myxocyprinus asiaticus* and longnose sucker *Catostomus* have distributions outside of North America. The Asiatic sucker is native to China and while the longnose sucker is native to North America, its distribution expanded during the Pleistocene glacial period to eastern Siberia.

Catostomids have a variety of body shapes, with several suckers often mistaken for larger, deeper bodied cyprinids such as common carp. Although there are morphological similarities between the two families, a dorsal fin with ten or more soft rays, a lack of barbels, and usually a downward facing, sucker-like mouth differentiates suckers from barbs and carps, sharp-bellies, and minnows. Their mouth is the main morphological feature that also distinguishes the Catostomidae from other fish families in the Dakotas. Sucker lips tend to be enlarged and fleshy, covered with small taste buds and textured with longitudinal grooves (plicate), numerous small bumps (papillose), or a mixture of both grooves and bumps. The variety of sucker lip morphologies is indicative of how they feed. Because of the mouth positioning, suckers can protrude their lips downward to feed along the substrate. The suckermouth minnow *Phenacobius mirabilis* (Leuciscidae) is the only other fish species in the Dakotas with similar mouth morphology. It is differentiated from catastomids by dorsal fin ray count.

Catostomids inhabit a variety of freshwater habitats, including lakes, rivers, and streams. Sucker species in large riverine habitats will either have deep, robust body shapes, or have a more streamlined profile. The body shape of catostomids from smaller streams is typically more tubular and elongate. In the Dakotas, the larger catostomids, like the quillback *carpiodes cyprinus*, river carpsucker *carpiodes carpio*, highfin carpsucker *carpiodes velifer*, blue sucker *Cycleptus elongatus*, redhorse species (*Moxostoma* spp.), and buffalo species (*Ictiobus* spp.), are primarily found in larger rivers and lakes. The smaller species, like the longnose sucker *Catostomus catostomus*, plains sucker *Pantosteus jordani*, and white sucker *Catostomus commersonii*, primarily inhabit smaller tributaries and streams.

A wide variety of catastomids are widely distributed across both North Dakota and South Dakota. Five species of redhorse (*Moxostoma* spp.) primarily occur on the eastern side of both states, with the shorthead redhorse *Moxostoma macrolepidotum* well-distributed west of the Missouri River as well. Bigmouth buffalo *Ictiobus cyprinellus*, smallmouth buffalo *Ictiobus bubalus*, black buffalo *Ictiobus niger*, blue sucker, river carpsucker, and highfin carpsucker all occur within the Missouri River system. Quillback also occur within the Missouri River system and are also found in the Red River of North. White sucker are widespread across the Dakotas. Longnose sucker has a disjunct distribution, occurring in the Missouri River in North Dakota and small areas of the northern Black Hills in South Dakota. The plains sucker range is restricted to streams within the Black Hills of South Dakota.

6.11.1 River Carpsucker *Carpiodes carpio*

(Rafinesque, 1820)

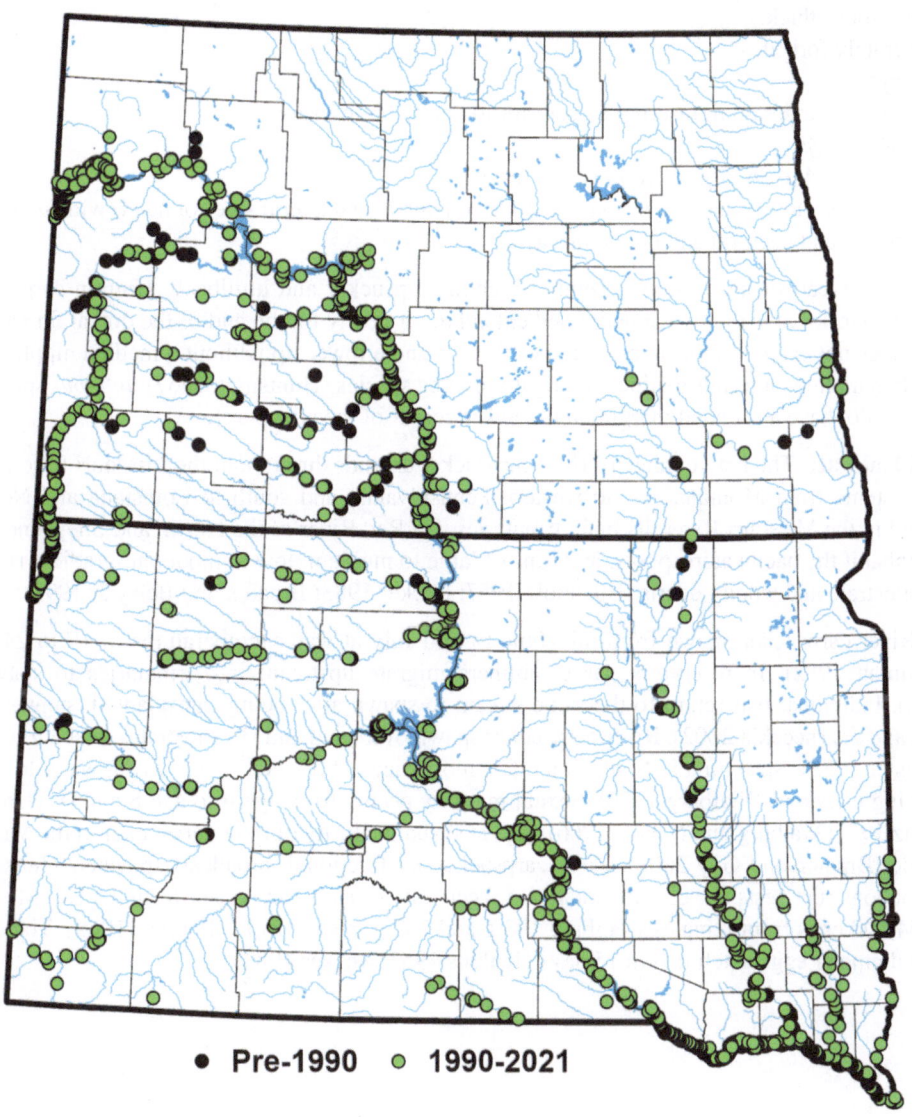

● **Pre-1990** ○ **1990-2021**

Etymology: *carpiodes carpio* – *carpiodes* = carp like; ***carpio*** = carp.

Description:

Body – deep, laterally compressed, highly arched dorsal side.
Color –

> *Dorsally* – olive to brown.
> *Laterally* – silver.
> *Ventrally* – cream to silvery-white.
> *Fins* – dorsal, caudal – light slate gray; ventral – clear to light gray, possible pink-orange near bases.

Head – small.
Opercle – subopercle bone; angled margin on lower corner.
Snout – short, bluntly rounded.
Eyes – large, laterally on head.
Mouth – small, subterminal.

> *Upper jaw* – extends to or beyond anterior end of eye.
> *Lips* – thin; lower lip forms an obtuse angle with a small, median nipple-like projection on tip.

Teeth – pharyngeal numerous, 150 to 180 per side.
Gill rakers – long, fine, narrowly spaced.
Dorsal fin – 23 to 28 rays; elongate and falcate; anterior rays not greatly elongated, when depressed reach roughly two-thirds dorsal fin base length.
Caudal peduncle – short, thick.
Caudal fin – moderately forked.
Anal fin – 7 or 8 rays.
Pelvic fins – 8 to 10 rays; abdominal; insertions posterior to dorsal fin insertion.
Pectoral fins – 14 to 17 rays.
Lateral line – complete; 34 to 36 large cycloid scales in series.
Sexual dimorphism – spawning males have small tubercles on dorsal and lateral sides of head (with an untuberculate region on the opercle bone), and pectoral and pelvic fin rays.

Similar Species: River carpsucker closely resemble highfin carpsucker and quillback. Highfin carpsucker have greatly elongated dorsal fin anterior rays that when depressed extend to, or nearly to, the end of the dorsal fin base. Quillback have an upper jaw that does not extend to the anterior end of the eye and a lower lip without a median nipple-like project on the tip. The nuptial tuberculation pattern is distinct among these three species (Huntsman 1967). *Ictiobus* species, like bigmouth and smallmouth buffalo, have a rounded margin on the lower corner of the subopercle.

Distribution and Habitat: The native range of river carpsucker extends throughout the Missouri River, Mississippi River, and Ohio River systems from Montana east to western Pennsylvania, and south to Louisiana and New Mexico. In the Dakotas, it is found in the Missouri River, its main tributaries, the Red River of the North, and Sheyenne River. River carpsucker primarily inhabit the backwaters, pools, and runs of large to medium, low to moderate-gradient rivers and reservoirs with sand or silt substrate and woody debris (Buchholz 1957; Becker 1983; Braaten and Berry Jr 1997).

Reproduction: River carpsucker spawn from mid-June to mid-July at water temperatures of 19 to 24 °C (Walburg and Nelson 1966; Fuiman 1982). In reservoirs, large numbers migrate upstream into tributaries to spawn (Bonneau and Scarnecchia 2002). Individual fish return to the same stream to spawn after being transplanted, suggesting spawning site fidelity (Bonneau and Scarnecchia 2002). River carpsucker spawn over sand and silt substrates in shallow areas with little to moderate flow (Jester 1972). Spawning in South Dakota is most successful when rising spring water levels flow into flood marshes or low-lying meadows (Walburg 1976). Sexual maturity occurs from age-4 to age-8 and is related more to growth than age (Buchholz 1957; Walburg and Nelson 1966). Age at sexual maturity of river carpsucker in the Missouri River likely varies with latitude (Braaten and Guy 2002). River carpsucker are fractional, broadcast spawners (Jenkins 1953; Behmer 1967). They do not construct nests nor give parental care. Fecundity of river carpsucker from the Des Moines River in Iowa was 4,430 and 154,000 eggs from females weighing 183 g and 737 g, respectively (Behmer 1969). The adhesive, approximately 1 to 2 mm diameter eggs hatch in 8 to 15 days (Jenkins 1953; Becker 1983).

Age and Growth: Male and female river carpsucker growth rates differ little (Walburg and Nelson 1966). In the Missouri River, growth rates at age-1, and at age-6 through age-8, are positively correlated with latitude (Braaten and Guy 2002). River carpsucker from the Des Moines River age-1 growth rates were faster when water levels were low (Keeton 1963). River carpsucker growth is greater in reservoirs than in rivers (Stucky and Klassen 1971). Total lengths at age from Lewis and Clark Reservoir in South Dakota were age-1, 67 mm; age-2, 130 mm; age-3, 184 mm; age-4, 226 mm; age-5, 260 mm; age-6, 286 mm; age-7, 312 mm (Walburg and Nelson 1966). Growth rates from Lewis and Clark Reservoir are slower than other Missouri River reservoirs (Walburg 1976). River carpsucker need at least a 10 °C water temperature to grow (Fuiman 1982). They can reach a total length of 496 mm (Walburg and Nelson 1966). They have a 12-year longevity, which increases from south to north in the Missouri River (Walburg and Nelson 1966; Braaten and Guy 2002).

Food and Feeding: River carpsucker are omnivorous, facultative benthivores that suction feed and selectively filter prey through fine gill rakers and numerous pharyngeal teeth (Eastman 1977; Gido 2001). They eat organic detritus, periphyton, zooplankton, phytoplankton, oligochaetes, mollusks, and aquatic insects associated with benthic habitats (Brezner 1958; Walburg and Nelson 1966; Spiegel et al. 2011). River carpsucker diets vary little seasonally, but they can switch to lower quality prey like detritus during periods of low invertebrate abundance and availability (Buchholz 1957; Walburg and Nelson 1966; Gido 2001).

References

Becker GC (1983) Fishes of Wisconsin. University of Wisconsin Press, Madison

Behmer DJ (1967) Spawning periodicity of river carpsuckers, *carpiodes carpio*. Proc Iowa Acad Sci 72:252–262

Behmer DJ (1969) A method of estimating fecundity; with data on river carpsuckers, *carpiodes carpio*. Trans Am Fish Soc 98:523–524. https://doi.org/10.1577/1548-8659(1969)98[523:AMOEWD]2.0.CO;2

Bonneau JL, Scarnecchia DL (2002) Spawning-season homing of common carp and river carpsucker. Prairie Nat 34:13–20

Braaten PJ, Berry CR Jr (1997) Fish associations with four habitat types in a South Dakota prairie stream. J Freshw Ecol 12:477–489. https://doi.org/10.1080/02705060.1997.9663558

Braaten PJ, Guy CS (2002) Life history attributes of fishes along the latitudinal gradient of the Missouri River. Trans Am Fish Soc 131:931–945. https://doi.org/10.1577/1548-8659

Brezner J (1958) Food habits of the northern river carpsucker. Prog Fish Cult 20:170–174. https://doi.org/10.1577/1548-8659(1958)20[170:FHOTNR]2.0.CO;2

Buchholz M (1957) Age and growth of river carpsucker in Des Moines River, Iowa. Proc Iowa Acad Sci 64:589–600

Eastman JT (1977) The pharyngeal bones and teeth of catostomid fishes. Am Midl Nat 97:68–88. https://doi.org/10.2307/2424686

Fuiman LA (1982) Family Catostomidae, suckers. In: Auer NA (ed) Identification of larval fishes of the Great Lakes basin with emphasis on the Lake Michigan drainage. Great Lakes Fishery Commission, Special Publication 82–3, Ann Arbor, pp 345–435

Gido KB (2001) Feeding ecology of three omnivorous fishes in Lake Texoma (Oklahoma-Texas). Southwest Nat 46:23–33. https://doi.org/10.2307/3672370

Huntsman GR (1967) Nuptial tubercles in carpsuckers (*carpiodes*). Copeia 1967:457–458. https://doi.org/10.2307/1442136

Jenkins RM (1953) Growth histories of the principle fishes in Grand Lake (O' The Cherokees), Oklahoma, through thirteen years of impoundment. Oklahoma Fisheries Research Laboratory Report 34

Jester DB (1972) Life history, ecology and management of the river carpsucker, *carpiodes carpio* (Rafinesque), with reference to Elephant Butte Lake. New Mexico State University, Experiment Station Publication, Research Report 243

Keeton D (1963) Growth of fishes in the Des Moines River, Iowa, with particular reference to water levels. Dissertation, Iowa State University

Spiegel JR, Quist MC, Morris JE (2011) Trophic ecology and gill raker morphology of seven catostomid species in Iowa rivers. J Appl Ichthyol 27:1159–1164. https://doi.org/10.1111/j.1439-0426.2011.01779.x

Stucky NP, Klassen HE (1971) Growth and condition of the carp and the river carpsucker in an altered environment in western Kansas. Trans Am Fish Soc 100:276–282. https://doi.org/10.1577/1548-8659

Walburg CH (1976) Changes in the fish populations of Lewis and Clark Lake, 1956-74, and their relation to water management and the environment. US Dep Inter, Fish Wildl Serv, Res Rep 79

Walburg CH, Nelson WR (1966) carp, river carpsucker, smallmouth buffalo and bigmouth buffalo in Lewis and Clark Lake, Missouri River. US Dep Inter, Fish Wildl Serv, Res Rep 69

6.11.2 Quillback *Carpiodes cyprinus*

(Lesueur, 1817)

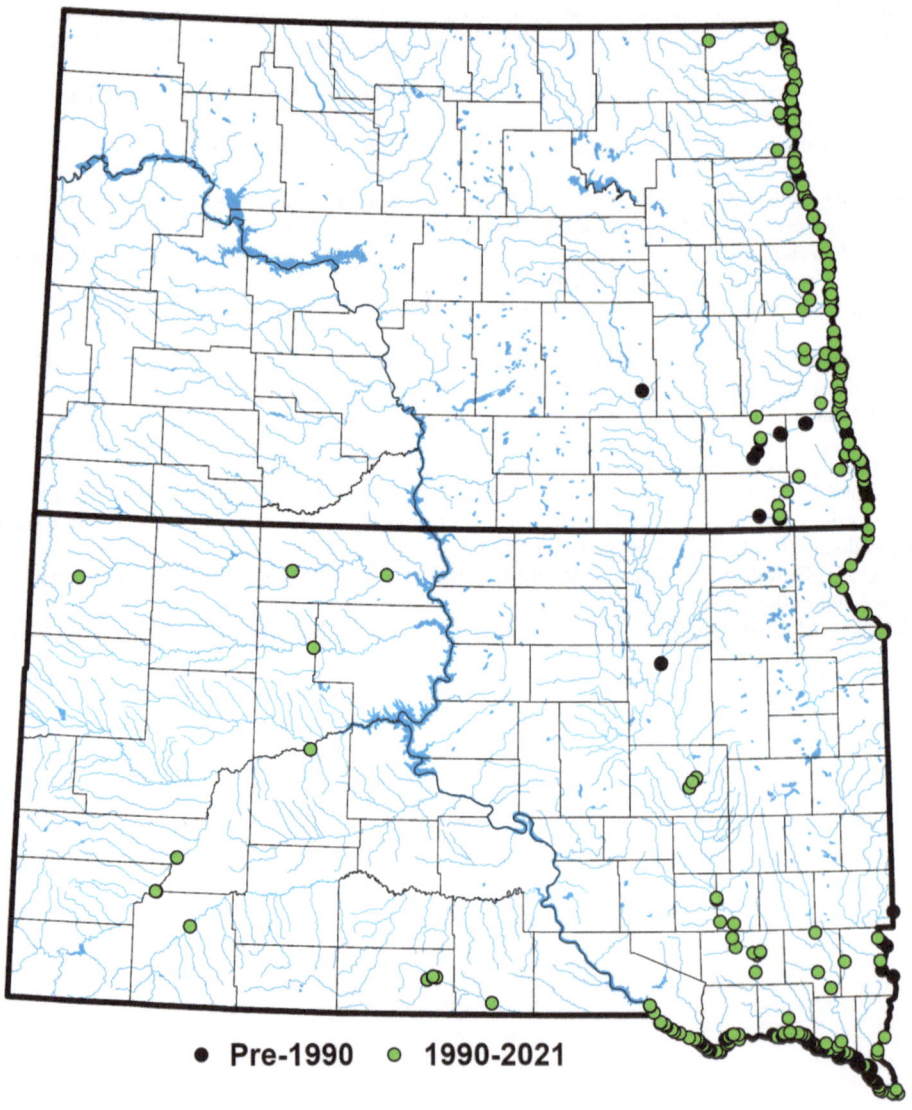

● **Pre-1990**   ● **1990-2021**

Etymology: *carpiodes cyprinus* – *carpiodes* = carplike; *cyprinus* = after Cyprus, the island where the carp was supposedly introduced, also referencing similarity to carp.

Description:

Body – deep, laterally compressed, highly arched dorsal side.
Color –

> *Dorsally* – silvery-gray to light olive.
> *Laterally* – silver.
> *Ventrally* – cream to silvery-white.
> *Fins* – dorsal, caudal – light slate gray; ventral – light olive to gray, possible pink-orange pigmentation near bases.

Head – small.
Snout – short, blunt.
Eyes – large, laterally on head.
Mouth – small, subterminal.

> *Upper jaw* – barely extends to anterior end of eye.
> *Lips* – thin; lower lip forms an acute angle (less than 90°).

Gill rakers – long, thin.
Dorsal fin – 28 to 30 rays; elongated, falcated; anterior ray filamentous, falling between or beyond the middle end of the dorsal fin base when depressed against the body.
Caudal peduncle – short, thick.
Caudal fin – forked.
Anal fin – 7 or 8 rays.
Pelvic fins – 9 to 10 rays; abdominal; insertions posterior to dorsal fin insertion.
Lateral line – complete; 36 to 40 large cycloid scales in series.
Sexual dimorphism – spawning males have small tubercles on the ventral side of the head, lateral sides, and first dorsal fin ray; larger spawning females occasionally have tubercles on the lateral sides of the head.

Similar Species: Quillback closely resemble highfin carpsucker and river carpsucker. Highfin carpsucker have greatly elongated dorsal fin anterior rays that when depressed reach to, or nearly to, the end of the dorsal fin base. Highfin carpsucker also have a lower lip forming an obtuse angle with a small, median nipple-like projection. River carpsucker have dorsal fin anterior rays that when depressed extend about two-thirds the length of the dorsal fin base. River carpsucker also have a lower lip that forms an obtuse angle with a small, median nipple-like projection. The pattern of nuptial tuberculation is distinct among these three species (Huntsman 1967).

Distribution and Habitat: The quillback native range extends from the Great Lakes, St. Lawrence River, Hudson Bay, and Mississippi River systems to the Atlantic seaboard and south to the Gulf of Mexico. In the Dakotas, it is found in the Missouri River, its main tributaries, the Red River of the North, and Sheyenne River (Copes 1965; Peterka 1978; Owen et al. 1981; Hoagstrom 2006). It is rare in mainstem Missouri River reservoirs (Benson 1968; Owen et al. 1981). Quillback inhabit clear to turbid, quiet, minimal-flow waters with little to no vegetation and sand, gravel, silt, mud, or clay substrates (Becker 1983; Kansas Fishes Committee 2014). It is found in medium-to-low gradient large rivers, lakes, creeks, sloughs, and sometimes small streams (Becker 1983). The estimated critical thermal maximum is 39 °C for quillback ranging from 45 to 47 mm (Mundahl 1990; Kansas Fishes Committee 2014).

Reproduction: Tuberculated male quillback have been reported as early as May and as late as September, suggesting a prolonged spawning period (Becker 1983). Males become sexually mature at a minimum total length of 313 mm from age-4 to age-6 (Parker and Franzin 1991). Females mature at 386 mm from age-6 to age-8 (Parker and Franzin 1991). Quillback spawn at water temperatures from 19 to 28 °C (Woodward and Wissing 1976; Becker 1983). In the early springtime, mature adults migrate upstream into small creeks (Trautman 1957; Becker 1983). Quillback are monospawners, releasing only one egg batch per spawning season (Benson 1968; Spiegel et al. 2011). Fecundity and egg size increase with female size, with females producing from 15,235 to 63,779 eggs (Woodward and Wissing 1976; Parker and Franzin 1991; Kansas Fishes Committee 2014). Eggs are deposited over sand, gravel, and mud substrates in quiet waters; no nest is prepared nor parental care provided (Becker 1983). Hatching occurs in 11 to 18 days at 16 °C (Curry and Spacie 1984; Parker and Franzin 1991; Kansas Fishes Committee 2014).

Age and Growth: Male quillback are generally larger than females at each age (Woodward and Wissing 1976; Becker 1983). Quillback average total length is 356 mm but can reach up to 610 mm (Becker 1983; Hrabik et al. 2015). Total lengths at age of quillback from Ohio were age-1, 84 mm; age-2, 156 mm; age-3, 206 mm; age-4, 242 mm; age-5, 280 mm; age-6, 312 mm; age-7, 342 mm; age-8, 368 mm; age-9, 398 mm; age-10, 428 mm; age-11, 448 mm (Woodward and Wissing 1976). Longevity is 11 years (Woodward and Wissing 1976).

Food and Feeding: Quillbacks are generalist, scavenger, benthic feeders that eat vegetation, chironomid and other insect larvae, and a variety of snails, clams, worms, and zooplankton (Cahn 1927; Owen et al. 1981; Becker 1983; Spiegel et al. 2011). Quillback may compete with channel catfish for invertebrate prey (Scott and Crossman 1973; Owen et al. 1981).

References

Becker GC (1983) Fishes of Wisconsin. University of Wisconsin Press, Madison

Benson NG (1968) Review of fishery studies on Missouri River main stem reservoirs, vol 71. Bureau of Sport Fisheries and Wildlife

Cahn AR (1927) An ecological study of the southern Wisconsin fishes: the brook silverside (*Labidesthes sicculus*) and the cisco (*Leucichthys artedi*) in their relations to the region, vol 11. University of Illinois, Champaign. https://doi.org/10.5962/bhl.title.50172

Copes FA (1965) Fishes of the Red River tributaries of North Dakota. Dissertation, University of North Dakota

Curry KD, Spacie A (1984) Differential use of stream habitat by spawning catostomids. Am Midl Nat 111:267–269. https://doi.org/10.2307/2425321

Hoagstrom CW (2006) Fish community assembly in the Missouri River basin. Dissertation, South Dakota State University. https://openprairie.sdstate.edu/etd/471

Hrabik RA, Schainost SC, Stasiak RH, Peters EJ (2015) The fishes of Nebraska. University of Nebraska, Lincoln

Huntsman GR (1967) Nuptial tubercles in carpsuckers (carpiodes). Copeia 1967:457–458. https://doi.org/10.2307/1442136

Kansas Fishes Committee (2014) Kansas fishes. University Press of Kansas, Lawrence

Mundahl ND (1990) Heat death of fish in shrinking stream pools. Am Midl Nat 123:40–46. https://doi.org/10.2307/2425758

Owen JB, Elsen DS, Russell GW (1981) Distribution of fishes in North and South Dakota Basins affected by the Garrison Diversion Unit. Universy of North Dakota Press, Grand Forks

Parker BR, Franzin WG (1991) Reproductive biology of the quillback, *carpiodes cyprinus*, in a small prairie river. Can J Zool 69:2133–2139. https://doi.org/10.1139/z91-298

Peterka JJ (1978) Fishes and fisheries of the Sheyenne River, North Dakota. Proc N D Acad Sci 32:29–44

Scott WB, Crossman EJ (1973) Freshwater fishes of Canada, Bulletin 184. Fisheries Research Board of Canada

Spiegel JR, Quist MC, Morris JE (2011) Trophic ecology and gill raker morphology of seven catostomid species in Iowa rivers. J Appl Ichthyol 27:1159–1164. https://doi.org/10.1111/j.1439-0426.2011.01779.x

Trautman MB (1957) The fishes of Ohio. Ohio State University Press, Columbus

Woodward RL, Wissing TE (1976) Age, growth, and fecundity of the quillback (*carpiodes cyprinus*) and highfin (*C. velifer*) carpsuckers in an Ohio stream. Trans Am Fish Soc 105:411–415. https://doi.org/10.1577/1548-8659

6.11.3   Highfin Carpsucker *Carpiodes velifer*

(Rafinesque, 1820)

© Uland Thomas

● **Pre-1990**   ● **1990-2021**

Etymology: *carpiodes velifer* – *carpiodes* = carplike; ***velifer*** = sailbearer, referencing elongate and filamentous anterior dorsal fin rays that look like a sail when near the surface of the water.

Description:

Body – deep, laterally compressed, highly arched dorsal side.
Color –

> *Dorsally* – olive to brown.
> *Laterally* – silver.
> *Ventrally* – cream to silvery-white.
> *Fins* – clear; may have a slate gray tint.

Head – small.
Opercle – subopercle, angled margin on lower corner.
Snout – short, bluntly rounded.
Eyes – large, laterally on head.
Mouth – small, subterminal.
> *Upper jaw* – extends to or beyond anterior end of eye.
> *Lips* – thin; lower lip forms an obtuse angle with a small, median nipple-like projection on tip.

Gill rakers – long.
Dorsal fin – 22 to 27 rays; elongated, falcated; anterior rays greatly elongated, when depressed reach to, or nearly to, end of dorsal fin base.
Caudal peduncle – short, thick.
Caudal fin – moderately forked.
Anal fin – 8 to 9 rays.
Pelvic fins – 9 to 10 rays; abdominal; insertions posterior to dorsal fin insertion.
Pectoral fins – 14 to 17 rays.
Lateral line – complete; 33 to 35 large cycloid scales in series.
Sexual dimorphism – spawning males have small tubercles on dorsal and lateral sides of head, snout, majority of body, and pectoral and pelvic fin rays, but not on the cheek or the opercle bone.

Similar Species: Highfin carpsucker closely resemble quillback and river carpsucker. River carpsucker have not-greatly elongated dorsal fin anterior rays that when depressed are shorter than the length of the dorsal fin base. Quillback have dorsal fin anterior rays shorter than the dorsal fin base when depressed, an upper jaw that does not extend to the anterior end of the eye, a lower lip forming an acute angle, and lack a median nipple-like projection on the tip of the lower lip. The pattern of nuptial tuberculation is distinct among these three species (Huntsman 1967). *Ictiobus* species, such as the bigmouth and smallmouth buffalo, have a rounded margin on the lower corner of the subopercle.

Distribution and Habitat: The native range of highfin carpsucker extends throughout central North America in the lower Missouri River, Mississippi River, and Ohio River systems from Minnesota to the Gulf of Mexico. It has declined throughout much of its historical range, especially on the western edge, because of water pollution, siltation, and migratory barriers (Cross 1967; Smith 1979; Becker 1983; Hoagstrom et al. 2006). Highfin carpsucker are not found in North Dakota, and their range is restricted to the lower Missouri River below Gavins Point dam and the recreational river section below Ft. Randall dam in South Dakota (Berry Jr and Young 2004). Highfin carpsucker are a riverine specialist species (Lyons 1996). They inhabit medium-to-large rivers with swift velocities, preferring moderately deep to deep pools or areas adjacent to river channels with relatively low turbidity, sand and silt substrates, and little to no aquatic vegetation (Cross 1967; Lee and Platania 1980; Trautman 1981; Becker 1983; Pflieger 1997; Miller and Robinson 2004). When water levels are low, they can typically be found in riffle areas (Becker 1983). Highfin carpsuckers are less tolerant of high turbidity and siltation than other carpsuckers (Becker 1983; Pflieger 1997).

Reproduction: Information on highfin carpsucker spawning and reproductive information is scarce. Spawning migrations to shallow and overflow areas of streams and ponds typically occur in May or early spring (Harlan and Speaker 1951; Becker 1983). They will also spawn in shallow riffles with gravel substrates (Pflieger 1997). In Ohio, highfin carpsucker become sexually mature at a total length of approximately 229 mm (Trautman 1981). In Iowa, sexual maturity occurs at age-3 (Harlan and Speaker 1951). Fecundity was 41,644 eggs from a female with a total length of 275 mm and weight of 247 g, and 62,355 eggs from a 300 mm, 312 g female (Woodward and Wissing 1976). Eggs from highfin carpsucker in Wisconsin averaged 1 mm in diameter (Becker 1983).

Age and Growth: Highfin carpsucker are the smallest carpsucker (Etnier and Starnes 1993). The greatest growth occurs during the first year (Vanicek 1961). Males are typically longer than females at each annulus (Woodward and Wissing 1976). Total lengths at age of highfin carpsucker from the Des Moines River in Iowa were age-1, 89 mm; age-2, 132 mm; age-3, 165 mm; age-4, 196 mm; age-5, 221 mm; age-6, 259 mm; age-7, 279 mm; age-8, 305 mm (Vanicek 1961). Highfin carpsuckers from Ohio and the Illinois River in Oklahoma had higher growth rates than individuals from the Des Moines River, likely because of longer growing seasons (Jenkins et al. 1952; Vanicek 1961; Woodward and Wissing 1976). They can reach a total length of 423 mm and have a longevity of 9 years (Spiegel et al. 2010, 2011; Quist and Spiegel 2012).

Food and Feeding: Highfin carpsucker are benthic omnivores, and have a more specialized diet than the river carpsucker and quillback (Spiegel et al. 2011). They eat detritus, algae, chironomid larvae, oligiochaetes, mollusks, and other aquatic invertebrates associated with benthic habitats (Spiegel et al. 2011).

References

Becker GC (1983) Fishes of Wisconsin. University of Wisconsin Press, Madison

Berry CB Jr, Young B (2004) Fishes of the Missouri national recreational river, South Dakota and Nebraska. Great Plains Res 14:89–114

Cross FB (1967) Handbook of fishes in Kansas. University Press of Kansas, Lawrence

Etnier DA, Starnes WC (1993) The fishes of Tennessee. University of Tennessee Press, Knoxville

Harlan JR, Speaker EB (1951) Iowa fish and fishing. State Conservation Commission, Des Moines. https://doi.org/10.2307/1439117

Hoagstrom CW, Hayer CA, Kral JG, Wall SS, Berry CR Jr (2006) Rare and declining fishes of South Dakota: a river drainage scale perspective. Proc S D Acad Sci 85:171–211

Huntsman GR (1967) Nuptial tubercles in carpsuckers (*carpiodes*). Copeia 1967:457–458. https://doi.org/10.2307/1442136

Jenkins RM, Leonard EM, Hall GE (1952) An investigation of the fishery resources of the Illinois river and pre-impoundment study of Tenkiller Reservoir, Oklahoma. Oklahoma Fisheries Research Laboratory Report No. 26

Lee DS, Platania SP (1980) *carpiodes velifer* (Rafinesque), highfin carpsucker. In: Lee DS, Gilbert CR, Hocutt CH, Jenkins RE, McAllister DE, Stauffer JR Jr (eds) Atlas of North American freshwater fishes. North Carolina Biological Survey No 1980-12, Raleigh, p 369. https://doi.org/10.5962/bhl.title.141711

Lyons J (1996) Effects of flow regulation and restriction of passage due to hydroelectric project operation on the structure of fish and invertebrate communities in Wisconsin's large river systems. 1996 Progress Report, Phase 1.2: Development of an Index of Biotic Integrity. Dep Nat Res, Madison

Miller RJ, Robinson HW (2004) Fishes of Oklahoma, revised edn. University of Oklahoma Press, Norman

Pflieger WL (1997) The fishes of Missouri, revised edn. Missouri Dep Cons, Jefferson City

Quist MC, Spiegel JR (2012) Population demographics of catostomids in large river ecosystems: effects of discharge and temperature on recruitment dynamics and growth. River Res Appl 28:1567–1586. https://doi.org/10.1002/rra.1545

Smith PW (1979) The fishes of Illinois. University of Illinois Press, Urbana

Spiegel JR, Quist MC, Morris JE (2010) Estimating age of highfin carpsucker, quillback carpsucker, and river carpsucker. J Freshw Ecol 25:271–278. https://doi.org/10.1080/02705060.2010.9665077

Spiegel JR, Quist MC, Morris JE (2011) Trophic ecology and gill raker morphology of seven catostomid species in Iowa rivers. J Appl Ichthyol 27:1159–1164. https://doi.org/10.1111/j.1439-0426.2011.01779.x

Trautman MB (1981) The fishes of Ohio, revised edn. Ohio State University Press, Columbus

Vanicek D (1961) Life history of the quillback and highfin carpsuckers in the Des Moines River. Proc Iowa Acad Sci 68:238–246

Woodward RL, Wissing TE (1976) Age, growth, and fecundity of the quillback (*carpiodes cyprinus*) and highfin (*C. velifer*) carpsuckers in an Ohio stream. Trans Am Fish Soc 105:411–415. https://doi.org/10.1577/1548-8659

6.11.4   Longnose Sucker *Catostomus catostomus*

(Forster, 1773)

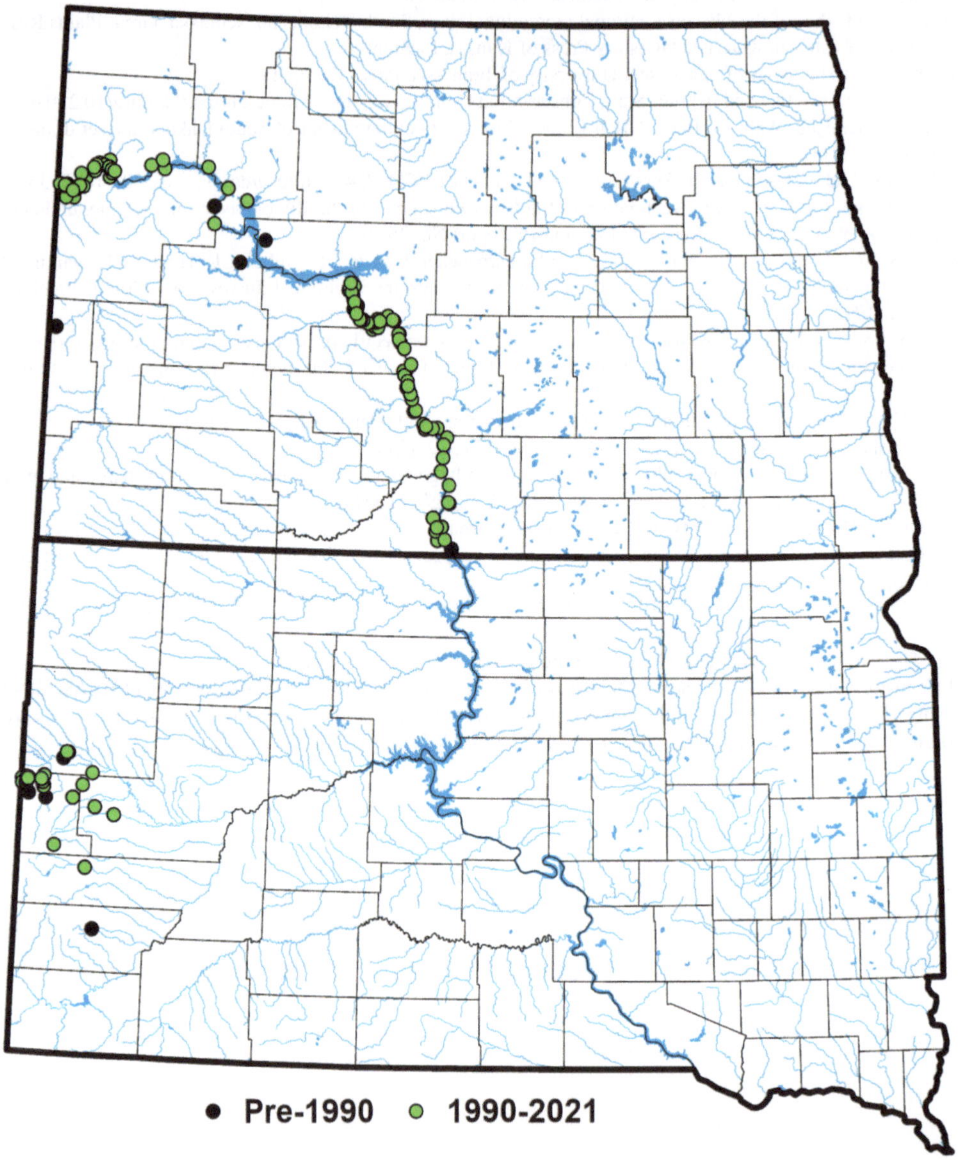

● Pre-1990   ● 1990-2021

Etymology: *Catostomus* – *Catostomus* = inferior mouth; *catostomus* = inferior mouth.

Description:

Body – cylindrical, elongated, slightly dorsoventrally flattened.
Color –

> *Dorsally* – dark gray to black olive.
> *Laterally* – gray to brown.
> *Ventrally* – abruptly changes to silvery-white.
> *Fins* – dorsal, caudal – dark gray; other – light gray to clear.

Head – long, slender, flattened between eyes.
Snout – long, pointed with blunt tip; extends well beyond upper lip.
Eyes – small, high on posterior end of head.
Mouth – inferior, large.

> *Lips* – fleshy, heavily papillose; lower lip completely divided by ventral notch forming an acute angle.
Teeth – pharyngeal comb-like, 44 to 55, in a single row on the fifth gill arch.
Gill rakers – 22 to 27; short; anterior row of first arch rudimentary.
Dorsal fin – 9 to 12 soft rays; slightly concave.
Caudal peduncle – short.
Caudal fin – moderately forked.
Anal fin – 7 rays.
Pelvic fins – 9 rays; insertion posterior to dorsal fin anterior end.
Pectoral fins – rounded.
Lateral line – complete; 95 to 115 cycloid scales crowded anteriorly.
Juveniles – occasional three large black lateral blotches.
Sexual dimorphism – spawning males have a dark red lateral stripe and tubercles on the head, anal fin, and caudal fin.

Similar Species: Longnose sucker closely resemble plains sucker and white sucker. Plains sucker have a lateral line with 75 to 92 cycloid scales, inner cartilaginous ridges on both jaws, lateral notches on each corner of the mouth, and lower lip lobe incompletely divided. White sucker have a slightly pointed snout barely extending past the upper lip and 55 to 75 cycloid scales in lateral line series.

Distribution and Habitat: The native range of longnose sucker extends throughout Alaska and Canada, from Washington to the Great Lakes region and New England, and south to Colorado. Longnose sucker are the only species of North American sucker also native to northern Asia. It occurs throughout the Missouri River in North Dakota, with a likely continuous distribution in Lake Oahe, South Dakota. As a nongame species, longnose sucker are not targeted during surveys or routinely sampled, leaving some sections of the mainstem Missouri River without data points in South Dakota. It is restricted to small areas of the northern Black Hills within the Cheyenne River system in South Dakota (Bailey and Allum 1962). Longnose sucker occupy a wide range of habitats, including small-to-medium-sized rivers, lakes, and reservoirs. It is a benthic dweller, preferring cool, clear water with little to no turbidity and sand or gravel substrates. It is most commonly found at depths of 30 to 40 m in oligotrophic lakes and reservoirs but has been as deep as 183 m in Lake Superior (Scott and Crossman 1973; Walton 1980; Edwards 1983). The optimal water temperature for adult longnose sucker is 10 to 15 °C (Brown and Graham 1953; Edwards 1983). Larvae with total lengths of 11 to 18 mm typically remain in substrate near the spawning grounds for 9 to 14 days before moving to deeper water (Brown 1971; Walton 1980; Childress et al. 2016). Juveniles often school in shallow backwater or lentic areas with dense aquatic vegetation (Swanson 1981).

Reproduction: Water temperature and discharge initiate longnose sucker spawning migrations and spawning behavior (Barton 1980; Childress et al. 2016). Upstream migrations into smaller tributaries occur at water temperatures of 5 to 9 °C (Walton 1980). Males and females become sexually mature at age-4 and age-5, respectively (Rawson and Elsey 1950; Brown 1971). Longnose sucker spawn during the day in slow moving water or pools just downstream of riffles over gravel substrate in water temperatures of 10 to 15 °C (Walton 1980; Swanson 1981; Childress et al. 2016). They do not build nests nor provide parental care. Spawning occurs in groups of up to ten fish, with usually one larger female chased by several smaller males (Swanson 1981; Dion et al. 1993). Males use their snout to nudge females and vibrate their bodies to instigate egg release. The light yellow, demersal, approximately 2 mm diameter eggs adhere to gravel substrate after being broadcast (Becker 1983; Dion et al. 1993). Fecundity depends on female size. Eggs typically hatch in 10 to 20 days. In 12 °C water, hatching occurred in 14 days (Walton 1980).

Age and Growth: Female longnose sucker are typically larger, grow more slowly, and live longer than males. Growth varies greatly after sexual maturity (Swanson 1981). Total lengths at age for both sexes combined from Yellowstone Lake, Wyoming were age-1, 41 mm; age-2, 83 mm; age-3, 147 mm; age-4, 208 mm; age-5, 267 mm; age-6, 313 mm; age-7, 347 mm; and age-8, 370 mm (Swanson 1981). Typically, longnose sucker total lengths range from 254 to 406 mm, with a maximum of 609 mm. Longevity is 8 to 11 years (Barton 1980; Walton 1980).

Food and Feeding: Longnose sucker are specialized, pelagic, benthic omnivores that feed more frequently in the pelagic zone than other sucker species in the Dakotas. They often migrate near shorelines during the night to feed. Fry eat zooplankton and diatoms. Juvenile and adult diets do not drastically differ (Swanson 1981). Adults primarily consume algae and detritus, but also eat amphipods, small aquatic insects, cladocerans, isopods, copepods, and other zooplankton (Weisel 1962). Juveniles primarily consume amphipods, zooplankton, and tendipedidae larvae (Swanson 1981).

References

Bailey RM, Allum MO (1962) Fishes of South Dakota. University of Michigan, Ann Arbor. https://doi.org/10.3998/mpub.9690435

Barton BA (1980) Spawning migrations, age, and growth, and summer feeding of white and longnose suckers in an irrigation reservoir. Can Field Nat 94:300–304

Becker GC (1983) Fishes of Wisconsin. University of Wisconsin Press, Madison

Brown CJD (1971) Fishes of Montana. Montana State University Press, Bozeman

Brown CJD, Graham RJ (1953) Observations on the longnose sucker in Yellowstone Lake. Trans Am Fish Soc 83:38–46. https://doi.org/10.1577/1548-8659(1953)83[38:OOTLSI]2.0.CO;2

Childress ES, Papke R, McIntyre PB (2016) Spawning success and early life history of longnose suckers in Great Lakes tributaries. Ecol Freshw Fish 25:393–404. https://doi.org/10.1111/eff.12220

Dion R, Richardson M, Roy L, Whoriskey FG (1993) Spawning patterns and interspecific matings of sympatric white (*Catostomus commersoni*) and longnose (*C. catostomus*) suckers from the Gouin reservoir system, Quebec. Can J Zool 72:195–200. https://doi.org/10.1139/z94-026

Edwards EA (1983) Habitat suitability index models: longnose sucker. US Dep Int, Fish Wildl Serv, fwsobs82_10_35

Rawson DS, Elsey CA (1950) Reduction in the longnose sucker population of Pyramid Lake, Alberta, in an attempt to improve angling. Trans Am Fish Soc 78:13–31. https://doi.org/10.1577/1548-8659(1948)78[13:RITLSP]2.0.CO;2

Scott WB, Crossman EJ (1973) Freshwater fishes of Canada, Bulletin 184. Fisheries Research Board of Canada

Swanson RD (1981) Some aspects of the biology of the longnose sucker in Yellowstone Lake, Yellowstone National Park, Wyoming. Dissertation, University of Wyoming

Walton BD (1980) The reproductive biology, early life history, and growth of white suckers, *Catostomus commersoni*, and longnose suckers, *C. catostomus*, in the Willow Creek-Chain Lakes System, Alberta. Dissertation, University of Alberta

Weisel GF (1962) Comparative study of the digestive tract of a sucker, *Catostomus catostomus*, and a predaceous minnow, *Ptychocheilus oregonense*. Am Midl Nat 68:334–346. https://doi.org/10.2307/2422739

6.11.5   White Sucker *Catostomus commersonii*

(Lacepède, 1803)

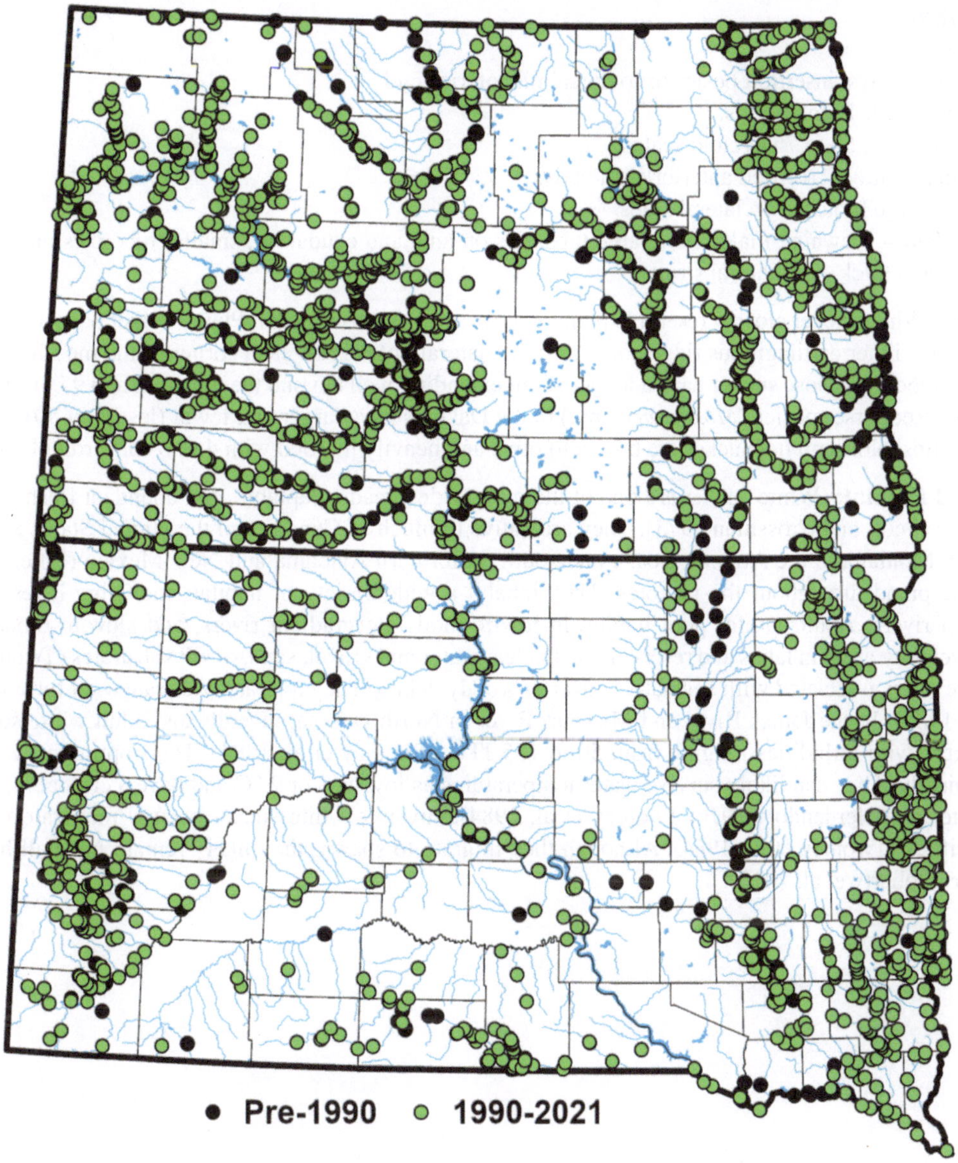

● **Pre-1990**   ○ **1990-2021**

Etymology: *Catostomus commersonii* – *Catostomus* = inferior mouth; *commersonii* = referencing French naturalist Philbert Commerson.

Description:

Body – moderately fusiform, elongated, slightly oval in cross section.
Color –

> *Dorsally/Laterally* – dark gray, brown, or olive; varying small, dark mottling.
> *Ventrally* – silver white.
> *Fins* – dorsal, caudal – light gray; anal, pelvic, pectoral – clear, faint orange.

Head – conical, slightly dorsoventrally flattened, convex between eyes.
Snout – blunt, barely extending past upper lip.
Eyes – small.
Mouth – inferior, large.

> *Lips* – fleshy, heavily papillose; lateral notch absent; lower lip almost twice as thick as upper lip, forms an acute angle.

Teeth – pharyngeal thin, approximately 55 to 58 per arch.
Dorsal fin – 10 to 13 rays; nearly straight; height nearly equal to base length.
Caudal peduncle – short, thick.
Caudal fin – forked.
Anal fin – 7 rays.
Pelvic fins – 10 to 11 rays; insertion posterior to dorsal fin anterior end.
Pectoral fins – rounded, 16 to 19 rays.
Lateral line – complete; 55 to 75 cycloid scales.
Skin – scales smaller, more crowded anteriorly on body.
Juveniles – 3 or 4 large blotches on lateral sides.
Sexual dimorphism – spawning males have large tubercles on anal and caudal fin, smaller tubercles on head and anterior end of body, and a thick, dark lateral stripe.

Similar Species: White sucker closely resemble plains sucker and longnose sucker. Plains sucker have a lateral line with 75 to 92 cycloid scales, inner cartilaginous ridges on both jaws, lateral notches on each corner of the mouth, and a completely divided lower lip lobe. Longnose sucker have a longer snout extending well beyond the upper lip and 95 to 115 cycloid scales in the lateral line. Redhorse species (*Moxostoma* spp.) in the Dakotas have larger and fewer (less than 50) lateral line scales and more plicate lips. Northern hogsucker are brown to olive and heavily mottled with 4 to 6 dark irregular saddles.

Distribution and Habitat: White sucker are one of the most widespread, adaptable, and abundant freshwater fish species in North America (Scott and Crossman 1973). They are native to much of Canada and the United States east of the Rocky Mountains, from Montana to the Atlantic Coast, and south to northern Alabama and New Mexico (Pflieger 1997). White suckers are widespread throughout the Dakotas. They inhabit a wide variety of habitats including lakes, reservoirs, and small-to-medium rivers. Adults prefer pool habitat in low-to-moderate gradient rivers, and slow-to-moderate (less than 40 cm/second) velocity areas in lakes and reservoirs with clear water and gravel, sand, or silt substrates (Twomey et al. 1984). White sucker are often associated with vegetative cover or woody debris. They tolerate wide ranges of turbidity and siltation (Twomey et al. 1984). In the James River and Missouri River in North Dakota, juvenile and adult white sucker have been observed in waters with turbidities ranging from 50 to 135 JTU (Twomey et al. 1984). Their optimum water temperature varies geographically. They can likely survive water temperatures as low as 1 or 2 °C and have a critical thermal maximum of 32 °C (Reutter and Herdendorf 1976; Twomey et al. 1984). Juvenile white sucker occupy moderate-velocity shallow backwaters or riffles (Twomey et al. 1984). Although they migrate to spawn, they likely remain in small home ranges for most of the year (Doherty et al. 2010).

Reproduction: Spawning migrations of up to 40 km typically occur at night in water temperatures of 7 to 10 °C. White sucker spawn in tributaries or shallow areas near riffles with moderate water velocities over sand or gravel substrate (Doherty et al. 2010). They become sexually mature from age-2 to age-4, depending on food availability and location (Becker 1983; Twomey et al. 1984; Chen and Harvey 1995). Males mature 1 or 2 years earlier than females (Scott and Crossman 1973; Becker 1983). Spawning occurs in April and May, with a single female typically wedged between two males until egg release and fertilization (Reighard 1920). White sucker spawning activity increases at dusk and dawn. No nest is prepared, and no parental care is given. The demersal and adhesive eggs are approximately 2 to 3 mm in diameter (Becker 1983). Fecundity from females with total lengths ranging from 406 to 533 mm ranged from 20,000 to 50,000 eggs (Becker 1983). The optimum water temperature for egg incubation is 15 °C, with decreased hatching success at temperatures below 9 °C and over 17 °C (McCormick et al. 1977; Twomey et al. 1984). Spawning may occur multiple times within one season depending on environmental conditions (Doherty et al. 2010).

Age and Growth: Female white sucker are typically larger than males. Growth varies greatly among populations. White sucker typical total lengths range from 203 to 305 mm, with a maximum length of 610 mm. Longevity is 10 to 17 years (Beamish 1973).

Food and Feeding: White suckers are benthic feeders, migrating near shore at night to forage by buccal cavity filtration (Ahlgren 1990). Detritus is an important part of the juvenile diet, along with aquatic invertebrates, cladocerans, and other microcrustaceans (Ahlgren 1990). Adults consume a wide variety of prey items, including plankton, aquatic benthic invertebrates, vegetation, small fish, and fish eggs.

References

Ahlgren MO (1990) Diet selection and the contribution of detritus to the diet of the juvenile white sucker (*Catostomus commersoni*). Can J Fish Aquat Sci 47:41–48. https://doi.org/10.1139/f90-004
Beamish RJ (1973) Determination of age and growth of populations of the white sucker (*Catostomus commersoni*) exhibiting a wide range in size at maturity. J Fish Res Board Can 30:607–616. https://doi.org/10.1139/f73-108
Becker GC (1983) Fishes of Wisconsin. University of Wisconsin Press, Madison
Chen Y, Harvey HH (1995) Growth, abundance, and food supply of white sucker. Trans Am Fish Soc 124:262–271. https://doi.org/10.1577/1548-8659
Doherty CA, Curry RA, Munkittrick KR (2010) Spatial and temporal movements of white sucker: implications for use as a sentinel species. Trans Am Fish Soc 139:1818–1827. https://doi.org/10.1577/T09-172.1
McCormick JH, Jones BR, Hokanson KEF (1977) White sucker (*Catostomus commersonii*) embryo development, and early growth and survival at different temperatures. J Fish Res Board Can 34:1019–1025. https://doi.org/10.1139/f77-154
Pflieger WL (1997) The fishes of Missouri, revised edn. Missouri Dep Cons, Jefferson City
Reighard J (1920) The breeding behavior of the suckers and minnows. I. the suckers. Biol Bull 38:1–32. https://doi.org/10.2307/1536355
Reutter JM, Herdendorf CE (1976) Thermal discharge from a nuclear power plant: predicted effects on Lake Erie fish. Ohio J Sci 76:39–45
Scott WB, Crossman EJ (1973) Freshwater fishes of Canada, Bulletin 184. Fisheries Research Board of Canada
Twomey KA, Williamson KL, Nelson PC (1984) Habitat suitability index models and instream flow suitability curves: white sucker. US Dep Inter, Fish Wildl Serv, FWS/OBS-82/10.64

6.11.6 Blue Sucker *Cycleptus elongatus*

(Lesueur, 1817)

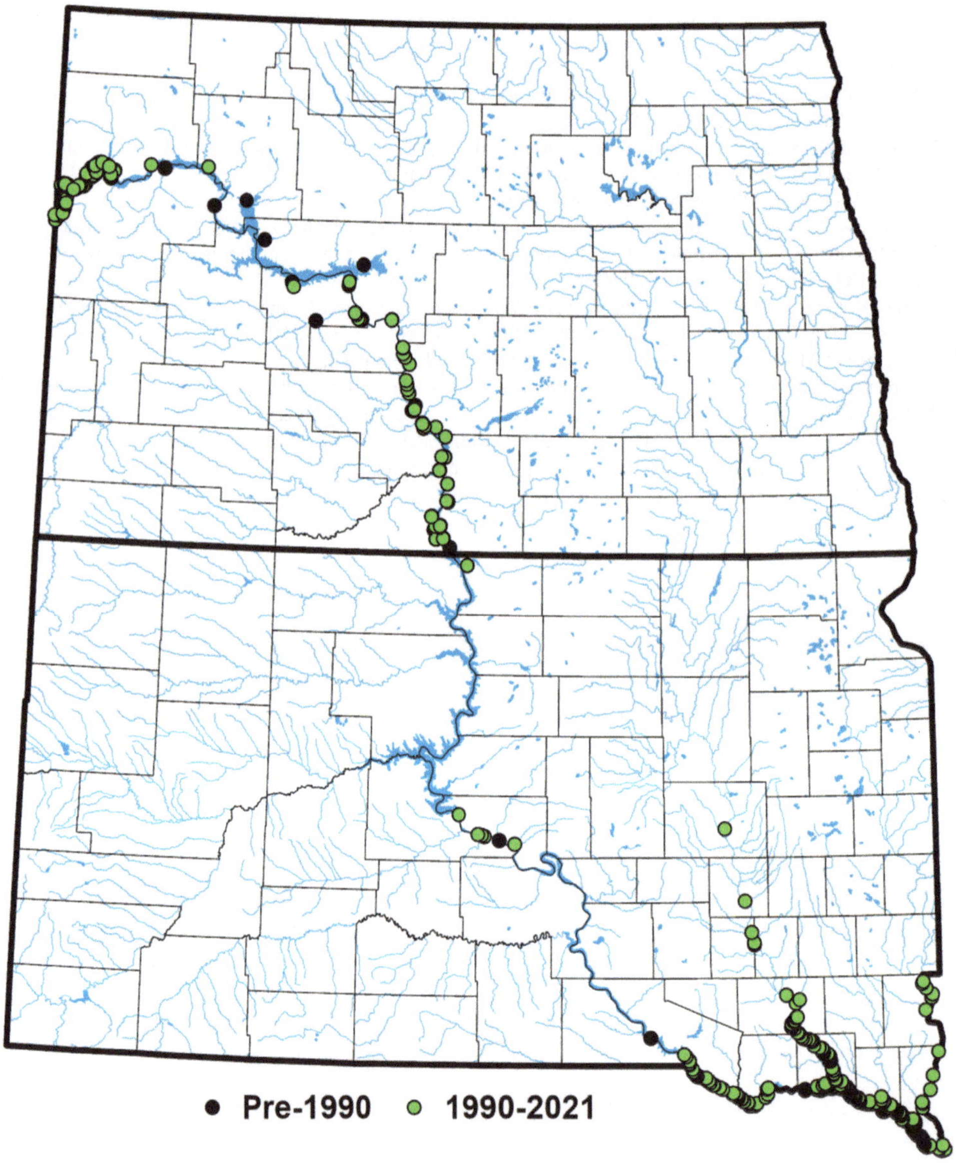

● Pre-1990 ● 1990–2021

Etymology: *Cycleptus elongatus – Cycleptus* = round and thin, referencing the small round mouth; *elongatus* = elongate, referencing body shape.

Description:

Body – elongated, fusiform, terete; anteriorly moderately robust, leading into a slightly laterally compressed posterior end.
Color –

 Dorsally – dark gray or olive; bluish hue.
 Laterally – light bluish-gray.
 Ventrally – light gray, cream, or white.
 Fins – dark, dusky, grayish blue hue.

Head – small, slender.
Snout – elongated, acute; extending well beyond the mouth.
Eyes – small, positioned laterally on head.
Mouth – inferior, protractile, relatively small.

 Lips – thick, heavily papillose.
Teeth – absent on jaws; roughly 35 to 45 long, comb-like pharyngeal teeth per gill arch.
Dorsal fin – 28 to 33 rays; sickle shaped, elongated; anterior rays elongated.
Caudal peduncle – uniformly thick, moderately elongated.
Caudal fin – forked; lobes nearly equal in length, slightly rounded.
Anal fin – 7 to 8 rays, nearly straight distal end.
Pelvic fins – 9 rays; abdominal.
Pectoral fins – large, falcated.
Lateral line – complete; 53 to 58 cycloid scales in series.
Sexual dimorphism – spawning males are blue-black, with small tubercles on the head, body, and fins; spawning females have small tubercles on body and fins.

Similar Species: Blue sucker resemble carpsuckers (*carpiodes* spp.), and buffalo (*Ictiobus* spp.) because of its sickle-shaped dorsal fin. Blue sucker can be easily distinguished by more than 50 lateral line scales and a more elongated body. White sucker have a much less elongated dorsal fin.

Distribution and Habitat: A large river species, blue sucker are native in the mainstem Mississippi River, Missouri River, and the lower sections of their major tributaries. In the Dakotas, they are likely distributed continuously throughout the Missouri River and its reservoirs. In South Dakota, blue sucker also occur in the lower portions of the Big Sioux River and James River systems. They prefer higher velocities in deep channels over hard substrates lacking silt (Steffensen et al. 2015). They are highly mobile and seasonally migrate, using different habitats in the winter, spring, and summer (Neely et al. 2009). The backwaters of large rivers and tributaries and main channel shorelines less than 0.5 m deep are likely crucial nursery habitat for larvae (Brown and Coon 1994; Fisher and Willis 2000; Adams et al. 2006).

Reproduction: Blue sucker spawn in mid-April to May in water temperatures from 20 to 23 °C in tributaries with swift water or riffles over bedrock or cobble substrate (Moss et al. 1983; Vokoun et al. 2003; Neely et al. 2009). The minimum size at maturity varies geographically, but males and females typically become sexually mature at age-4 and age-6, respectively (Rupprecht and Jahn 1980; Moss et al. 1983; Bednarski and Scarnecchia 2006; Daugherty et al. 2008). Spawning behavior information is scarce. Blue suckers near the confluence of the James and Missouri Rivers in South Dakota at total lengths of 669 mm and 755 mm contained 76,880 and 104,564 eggs, respectively. Maximum fecundity was 154,669 eggs for blue suckers in the Missouri River near the James River confluence, and 174,424 eggs for fish from the Big Sioux River. Mean fecundity of female blue suckers from the lower James River, South Dakota, was 90,827 eggs, and ranged from 35,908 to 138,666 eggs (Carlson et al. 2021). Fecundity is positively related to female length and weight (Daugherty et al. 2008; Carlson et al. 2021). The 2-mm diameter, adhesive eggs are opaque with a yellow tint (Moss et al. 1983).

Age and Growth: Larval blue sucker from the Mississippi River estimated to be 12 to 24 days old had total lengths of 13–20 mm, while those 12 to 42 days old had total lengths of 16 to 39 mm (Adams et al. 2006). Initial reports of blue sucker lengths-at-age in South Dakota calculated from pectoral fin rays may have been incorrect because later otolith sectioning indicated substantially older ages (Morey and Berry Jr 2003; Carlson et al. 2021; Radford et al. 2021). At the end of their second growing season, juvenile blue suckers from the James River in South Dakota had a mean total length of 351 mm, with total lengths at age of 545 mm at age-7, 639 mm at age-10, 666 mm at age-20, 687 mm at age-25, 748 mm at age-36, 796 mm at age-42, and 806 mm at age-51 (Carlson 2022). Females are typically longer and heavier than same-age males (Bednarski and Scarnecchia 2006). Blue sucker maximum length is 1,016 mm, with otolith-derived longevity of 66 years from James River and Missouri River fish (Lee et al. 1980; Trautman 1981; Carlson 2022).

Food and Feeding: As a benthic forager, blue sucker use numerous lip papillae to locate and consume caddisflies, midges, hellgrammites, and other immature aquatic insects (Moss et al. 1983; Rupprecht and Jahn 1980; Pflieger 1997). Larvae and early juvenile blue suckers eat chironomid larvae, cyclopods, and cladocerans (Adams et al. 2006). Both aquatic and terrestrial invertebrates are likely very important to blue sucker early life survival (Adams et al. 2006).

References

Adams SR, Flinn MB, Burr BM, Whiles MR, Garvey JE (2006) Ecology of larval blue sucker (*Cycleptus elongatus*) in the Mississippi River. Ecol Freshw Fish 15:291–300. https://doi.org/10.1111/j.1600-0633.2006.00157.x

Bednarski J, Scarnecchia DL (2006) Age structure and reproductive activity of blue sucker in the Milk River, Missouri River drainage, Montana. Prairie Nat 38:167–182

Brown DJ, Coon TG (1994) Abundance and assemblage structure of fish larvae in the lower Missouri River and its tributaries. Trans Am Fish Soc 12:718–732. https://doi.org/10.1577/1548-8659

Carlson T (2022) Population dynamics and seasonal movements of blue suckers (*Cycleptus elongatus*) in the James River, South Dakota. Thesis, University of South Dakota

Daugherty DJ, Bacula TD, Sutton TM (2008) Reproductive biology of blue sucker in a large Midwestern River. J App Ichthyol 24:297–302. https://doi.org/10.1111/j.1439-0426.2007.01042.x

Fisher SJ, Willis DW (2000) Observations of age-0 blue sucker, *Cycleptus elongatus*, utilizing an upper Missouri River backwater. J Freshw Ecol 15:425–427. https://doi.org/10.1080/02705060.2000.9663761

Lee DS, Gilbert CR, Hocutt CH, Jenkins RE, McAllister DE, Stauffer JR (1980) Atlas of north American freshwater fishes. North Carolina Biological Survey Publication 1980-12. https://doi.org/10.5962/bhl.title.141711

Morey NM, Berry CR Jr (2003) Biological characteristics of the blue sucker in the James River and the Big Sioux River, South Dakota. J Freshw Ecol 18:33–41. https://doi.org/10.1080/02705060.2003.9663949

Moss RE, Scanlan JW, Anderson CS (1983) Observations on the natural history of the blue sucker (*Cycleptus elongatus* Le Sueur) in the Neosho River. Am Midl Nat 109:15–22. https://doi.org/10.2307/2425510

Neely BC, Pegg MA, Mestl GE (2009) Seasonal use distributions and migrations of blue sucker in the Middle Missouri River. Ecol Freshw Fish 18:437–444. https://doi.org/10.1111/j.1600-0633.2009.00360.x

Pflieger WL (1997) The fishes of Missouri, revised edn. Missouri Dep Cons, Jefferson City

Radford DS, Lackmann AR, Moody-carpenter CJ, Colombo RE (2021) Comparison of four hard structures including otoliths for estimating age in blue suckers. Trans Am Fish Soc 150:514–527. https://doi.org/10.1002/tafs.10303

Rupprecht RJ, Jahn LA (1980) Biological notes on blue suckers in the Mississippi River. Trans Am Fish Soc 109:323–326. doi: https://doi.org/10.1577/1548-8659

Steffensen KD, Stukel S, Shuman DA (2015) The status of fishes in the Missouri River, Nebraska: blue sucker *Cycleptus elongatus*. Trans Nebr Acad Sci Affili Sci 35:1–11

Trautman MB (1981) The fishes of Ohio, revised edn. Ohio State University Press, Columbus

Vokoun JC, Guerrant TL, Rabeni CF (2003) Demographics and chronology of a spawning aggregation of blue sucker (*Cycleptus elongatus*) in the Grand River, Missouri, USA. J Freshw Ecol 18:567–575. https://doi.org/10.1080/02705060.2003.9663997

6.11.7   Smallmouth Buffalo *Ictiobus bubalus*

(Rafinesque, 1818)

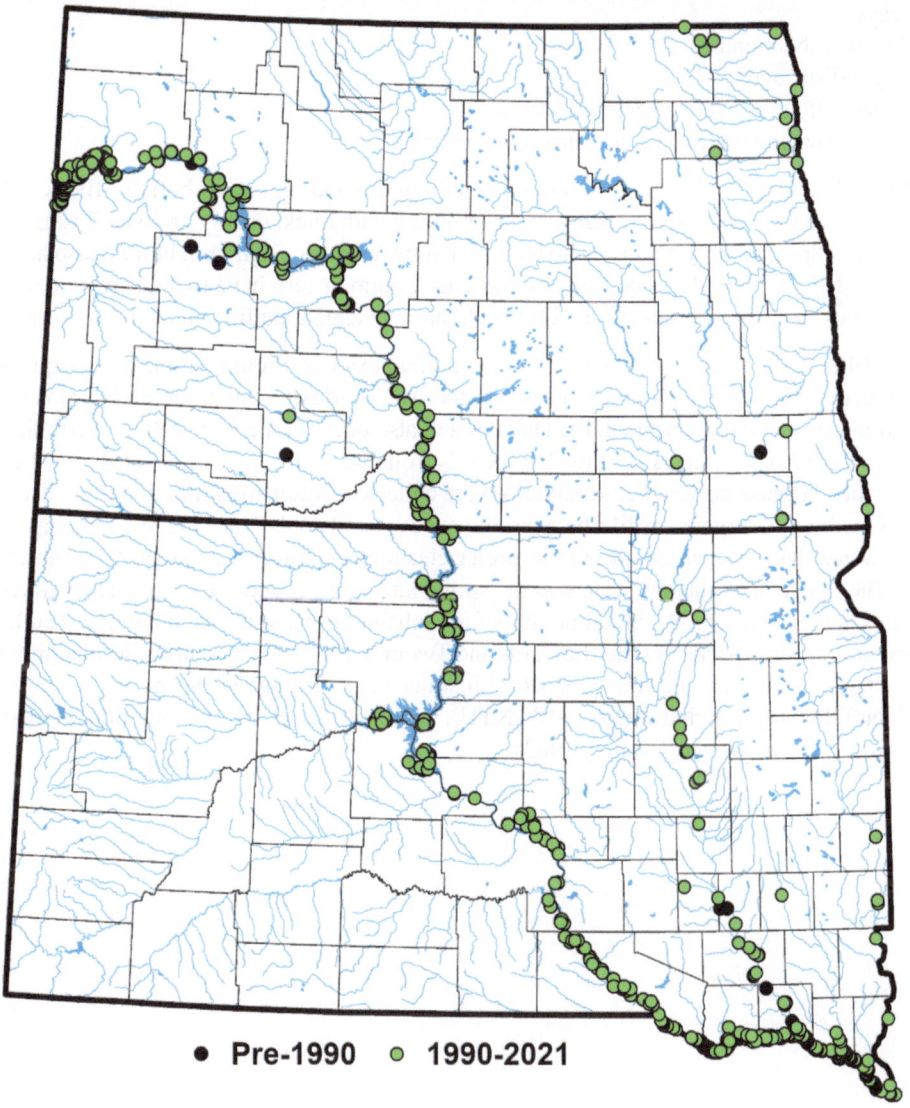

● **Pre-1990**   ● **1990-2021**

Etymology: *Ictiobus bubalus* – *Ictiobus* = fish ox, likely referencing common name or robust appearance; *bubalus* = wild ox or buffalo.

Description:

Body – robust, deep, slightly laterally compressed; steep, humped profile from head to anterior end of dorsal fin.
Color –

> *Dorsally* – bronze, olive-green.
> *Laterally* – light olive.
> *Ventrally* – light yellow-gray to white.
> *Fins* – gray to light brown.

Head – small, conical.
Opercle – subopercle, rounded outline on lower corner.
Snout – blunt.
Eyes – small, positioned anteriorly on head.
Mouth – small, subterminal.

> *Lips* – thick, fleshy, deep grooves.

Teeth – absent on jaws; approximately 130 to 195 small, comb-like pharyngeal teeth per gill arch.
Gill rakers – 30 to 35 on first arch.
Dorsal fin – 25 to 31 rays; long, falcated.
Caudal peduncle – short, thick.
Caudal fin – moderately forked, slightly rounded lobes.
Anal fin – 8 to 10 rays.
Pelvic fins – 9 to 11 rays; abdominal.
Pectoral fins – straight distal end.
Lateral line – complete; 35 to 39 cycloid scales in series.
Sexual dimorphism – spawning males have fine tubercles on head.

Similar Species: Smallmouth buffalo closely resemble black buffalo and bigmouth buffalo. Black buffalo have a more-slender body, large head, and a small, subterminal to inferior mouth with thick, deeply grooved lips. Bigmouth buffalo have a less-deep body and thin lips with only the lower lip having shallow grooves. Bigmouth buffalo also have a large head leading to a less-steep rounded profile to the dorsal fin anterior end. Common carp have two fleshy barbels on each side of the mouth. *Carpiodes* species, such as carpsuckers and quillback, have an angled outline on the lower corner of the subopercle.

Distribution and Habitat: The native range of smallmouth buffalo extends throughout the Mississippi River and Missouri River basin from Montana east to Pennsylvania and south to the Gulf of Mexico. Smallmouth buffalo are primarily found in the Missouri River drainage and its reservoirs in the Dakotas and also occur in the James River, Vermillion River, Big Sioux River, and Red River of the North systems. The twelve records from the Red River of the North may be misidentifications; they could not be verified because no voucher specimens were collected. Adult smallmouth buffalo occur most often in the low velocity areas of medium-to-large rivers like backwaters, oxbows, and pools. They are less frequent in small streams, but also inhabit reservoirs and lakes. Smallmouth buffalo prefer clean, clear, warm water with firm substrates, and abundant aquatic vegetation. They migrate to shallow waters in the spring and summer and retreat to deeper water in fall and winter (Jester 1973; Edwards and Twomey 1982). Smallmouth buffalo inhabit deeper water than bigmouth buffalo. They can withstand moderate current for short periods of time (Edwards and Twomey 1982). Fry and juvenile smallmouth buffalo frequent warm, shallow, vegetated backwaters, or embayments with water velocities under 20 cm/second (Edwards and Twomey 1982). Smallmouth buffalo tolerate turbid waters (>100 NTU), but to a lesser extent than bigmouth buffalo. Optimum water pH ranges from 6.5 to 8.5 (Edwards and Twomey 1982).

Reproduction: Spawning is initiated and is most successful with rising water levels (Jester 1973; Walburg 1976). Smallmouth buffalo typically spawn from March to April but may continue to spawn until September (Jester 1973). Their optimal spawning temperature ranges from 19 to 28 °C, with peak spawning at 22 to 27 °C (Jester 1973). Smallmouth buffalo age at sexual maturity varies and may be a function of size rather than age (Jester 1973). In Lewis and Clark Lake, males become sexually mature from age-7 to age-9, while females mature at age-10 or age-11 (Walburg and Nelson 1966). Prior to spawning, males migrate upstream and congregate with ripe females over or near vegetation in 1.2 to 3.1 m deep areas of shallow shoals, backwaters, or pools (Jester 1973; Edwards and Twomey 1982). Smallmouth buffalo are broadcast spawners that do not create nests nor provide parental care. Fecundity depends on female size. An age-15 female contained 525,500 eggs (Jester 1973). Demersal, approximately 2-mm diameter eggs adhere to bottom substrate or vegetation and hatch in 4 to 14 days depending on water temperature (Jester 1973).

Age and Growth: Newly hatched smallmouth buffalo have a total length of approximately 6 mm (Wrenn and Grinstead 1971). Growth is faster in clear-to-slightly turbid water (Willis 1978; Edwards and Twomey 1982; Pflieger 1997). Seasonal growth begins when water temperatures reach 18 °C (Shields 1957). Total lengths at age of smallmouth buffalo from Wisconsin were age-1, 112 mm; age-2, 245 mm; age-3, 346 mm; age-4, 391 mm; age-5, 428 mm; and age-6, 462 mm (Becker 1983). They can exceed a total length of 762 mm and weight of 18 kg. Smallmouth buffalo longevity is 15 to 20 years (Jester 1973).

Food and Feeding: Smallmouth buffalo are opportunistic planktivores. In Lewis and Clark Lake, fry primarily consume copepods, cladocerans, and other microcrustacean (McComish 1967). Juveniles and adults eat attached algae, cyclops, bosmina, daphnia, and other copepod and cladoceran zooplankton (McComish 1967). Mollusks and insect larvae eggs are also consumed (Forbes and Richardson 1920). A large palatal organ in their pharynx assists with selective food retention (Doosey and Bart 2011; Kansas Fishes Committee 2014). Chlorophyta, insect larvae, detritus, and sand have been found in smallmouth buffalo stomachs, likely indicating foraging along the bottom or near the shoreline (McComish 1967).

References

Becker GC (1983) Fishes of Wisconsin. University of Wisconsin Press, Madison

Doosey MH, Bart HL (2011) Morphological variation of the palatal organ and chewing pad of Catostomidae (Teleostei: Cypriniformes). J Morphol 272:1092–1108. https://doi.org/10.1002/jmor.10966

Edwards EA, Twomey K (1982) Habitat suitability index models: smallmouth buffalo. US Dep Inter, Fish Wildl Serv, FWS/OBS-82/10.13

Forbes SA, Richardson RE (1920) The fishes of Illinois. Illinois State Journal Company, Springfield. https://doi.org/10.5962/bhl.title.5011

Jester DB (1973) Life history, ecology, and management of the smallmouth buffalo, *Ictiobus bubalus* (Rafinesque), with reference to Elephant Butte Lake, New Mexico. New Mexico State University Agricultural Experiment Station Research Report 261

Kansas Fishes Committee (2014) Kansas fishes. University Press of Kansas, Lawrence

McComish TS (1967) Food habits of bigmouth and smallmouth buffalo in Lewis and Clark Lake and the Missouri River. Trans Am Fish Soc 96:70–74. https://doi.org/10.1577/1548-8659(1967)96[70:FHOBAS]2.0.CO;2

Pflieger WL (1997) The fishes of Missouri, revised edn. Missouri Dep Cons, Jefferson City

Shields JT (1957) Report of fisheries investigations during the fourth year of impoundment of Fort Randall Reservoir, South Dakota, 1956. S D Dep Game Fish Parks, Dingell-Johnson Project F-1-R-6:1-60

Walburg CH (1976) Changes in the fish populations of Lewis and Clark Lake, 1956–74, and their relation to water management and the environment. US Dep Inter, Fish Wildl Serv Res Rep 79

Walburg CH, Nelson WR (1966) carp, river carpsucker, smallmouth buffalo, and bigmouth buffalo in Lewis and Clark Lake, Missouri River. US Dep Inter, Fish Wildl Serv Res Rep 69

Willis DW (1978) Investigations of population structure and relative abundance of year-classes of buffalo fishes, *Ictiobus* spp., in Lake Sakakawea, North Dakota. Dissertation, University of North Dakota

Wrenn WB, Grinstead BG (1971) Larval development of the smallmouth buffalo, *Ictiobus bubalus*. J Tenn Acad Sci 46:117–120

6.11.8　Bigmouth Buffalo *Ictiobus cyprinellus*

(Valenciennes, 1844)

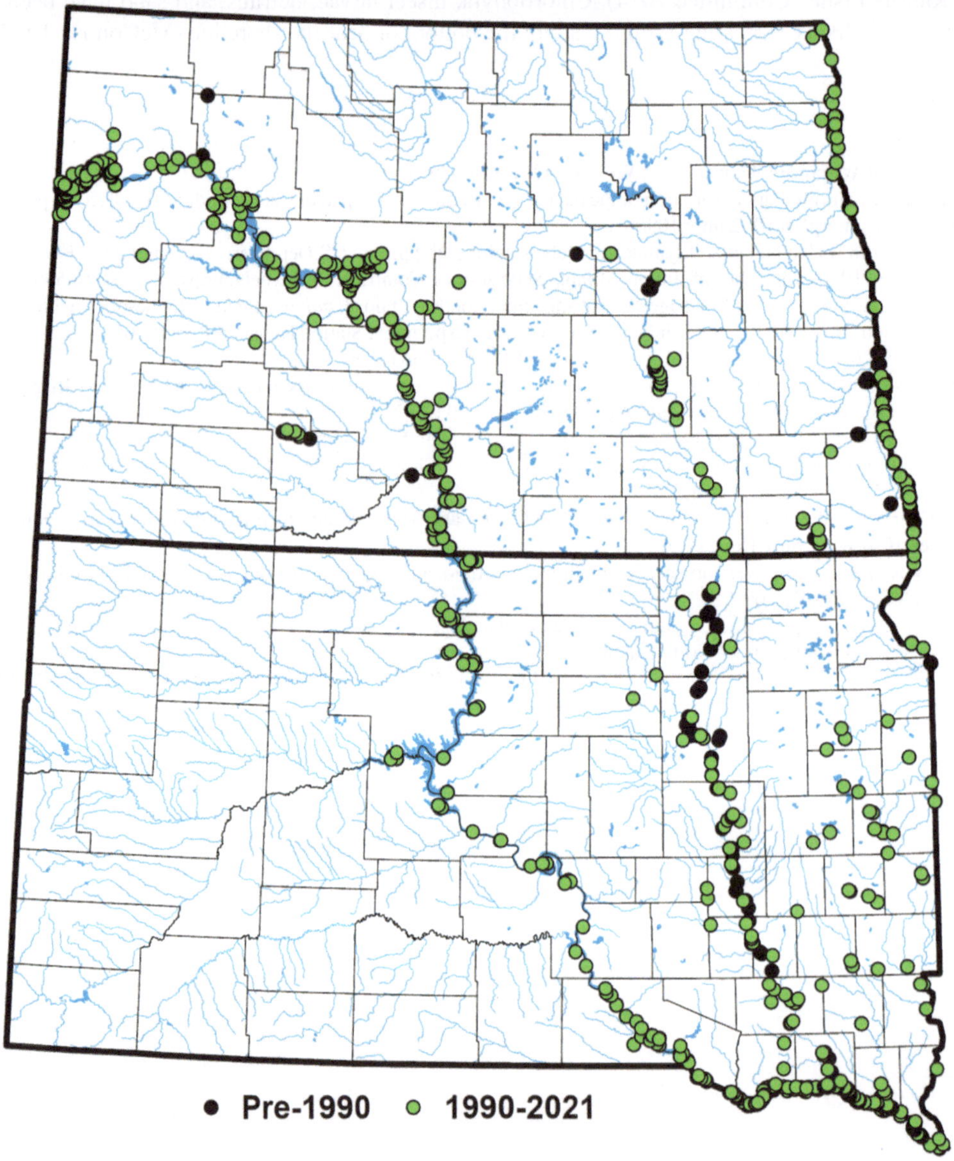

● **Pre-1990**　　○ **1990-2021**

Etymology: ***Ictiobus cyprinellus*** – *Ictiobus* = fish ox, likely referencing common name or robust appearance; *cyprinellus* = carp.

Description:

Body – robust, moderately deep, slightly laterally compressed; distinct rounded profile from head to dorsal fin anterior end.
Color –

> *Dorsally* – dark gray to olive-brown; black and orange markings on dorsal and head more intensified in older individuals.
> *Laterally* – dusky brown to olive.
> *Ventrally* – light gray to white.
> *Fins* – dark gray to black.

Head – large.
Opercle – subopercle; rounded outline on lower corner.
Snout – blunt; slight depression between eyes.
Eyes – small, positioned anteriorly on head.
Mouth – large, terminal, oblique.

> *Lips* – thin, fleshy, smooth; lower lip has shallow grooves.

Teeth – absent on jaws; approximately 130 to 170 short, comb-like pharyngeal teeth per gill arch.
Gill rakers – 40 to 45 on first arch; long.
Dorsal fin – 24 to 32 rays; long, falcated.
Caudal peduncle – short, thick.
Caudal fin – moderately forked, slightly rounded lobes.
Anal fin – 8 to 10 rays.
Pelvic fins – 10 to 11 rays; abdominal.
Pectoral fins – straight distal end.
Lateral line – complete; 34 to 40 cycloid scales in series.
Sexual dimorphism – spawning males have small tubercles on head, nape of neck, and fin rays.

Similar Species: Bigmouth buffalo closely resemble smallmouth buffalo and black buffalo. Smallmouth buffalo have a deeper body, and a small, conical head creating a steep, or humped profile from the head to dorsal fin anterior end. Smallmouth buffalo also have a small, subterminal mouth and thick lips with deep grooves. Black buffalo have a slightly compressed, more slender body, and a small, subterminal to inferior mouth with thick, deeply grooved lips. Common carp have two fleshy barbels on each side of mouth. *Carpiodes* species, such as river carpsucker and quillback, have an angled outline on the lower corner of the subopercle.

Distribution and Habitat: Native throughout the Mississippi River basin, Missouri River basin, lower Great Lakes, and Hudson Bay, bigmouth buffalo are primarily found in and east of the Missouri River in the Dakotas. Bigmouth buffalo inhabit low gradient, medium-to-large rivers in areas with low flow and are well adapted to reservoirs and lakes (Johnson 1963). Bigmouth buffalo are frequently associated with silt, sand, or gravel substrates and abundant vegetation (Edwards 1983). They migrate upstream in the spring to shallow, warm waters, and migrate back in the fall to overwinter in deeper water (Beckman and Elrod 1971; Benson 1980; Edwards 1983). Bigmouth buffalo prefer shallow water from 1.2 to 3.1 m deep and are sometimes seen schooling just below the surface on warm summer days (Johnson 1963; Walburg and Nelson 1966). Their optimum water temperature is 31 to 34 °C (Gammon 1973; Edwards 1983). Fry and juvenile bigmouth buffalo frequent shallow bay areas with little to no flow (Johnson 1963). They can withstand low oxygen levels and warm water temperatures and are more tolerant of turbidity than smallmouth buffalo or black buffalo (Becker 1983; Etnier and Starnes 1993). Bigmouth buffalo prefer water pH ranging from 6.5 to 8.5 (Edwards 1983).

Reproduction: Bigmouth buffalo spawn from April to June at water temperatures from 14 to 18 °C in backwaters or shallow areas near cattails or other aquatic vegetation (Johnson 1963; Benson 1980; Edwards 1983). Spawning ceases when water temperature reaches 27 °C (Edwards 1983). Bigmouth buffalo are annual spawners but may skip spawning if suitable habitat is unavailable (Walburg 1976). In Lewis and Clark Lake, males become sexually mature at age-7 or age-8, with females reaching sexual maturity at age-8 or age-9 (Walburg and Nelson 1966). Spawning occurs when a single female and two or more males splash near the surface until sinking to the bottom where the eggs are released and fertilized (Burr and Heidinger 1983). After spawning, the fish migrate up to 1.6 km back to the mainstream channel or original location (Becker 1983). Bigmouth buffalo do not create nests nor provide parental care. Fecundity is dependent on female size and can reach

750,000 eggs (Johnson 1963). The adhesive eggs are approximately 1 to 2 mm in diameter (Johnson 1963). Hatching occurs in 8 to 14 days depending on water temperature.

Age and Growth: Growth varies by location and is rapid within the first few years of life. Bigmouth buffalo achieve total lengths of 50 to 150 mm during their first summer. Total lengths at age of bigmouth buffalo from Wisconsin were age-1, 131 mm; age-2, 255 mm; age-3, 335 mm; age-4, 393 mm; age-5, 453 mm; age-6, 508 mm; age-7, 561 mm; age-8, 614 mm (Becker 1983). Bigmouth buffalo are the largest *Ictiobus* species. They can reach a total length of 1,250 mm and weigh over 36 kg (Edwards 1983; Lackman et al. 2019). Bigmouth buffalo longevity is 112 years, making them the oldest age-validated freshwater teleost fish (Lackman et al. 2019).

Food and Feeding: Bigmouth buffalo are opportunistic planktivores. They use long gill rakers to filter planktonic prey just above the bottom substrate. Fry feed in shallow bays on a diet of rotifers, copepods, daphnia and other cladocerans, other small crustaceans, and benthic organisms like tendipedid pupae and larvae (McComish 1967; Starostka and Applegate 1970). Juvenile and adult diets primarily eat copepods, cladocerans (especially Daphnia), and midge larvae (McComish 1967; Starostka and Applegate 1970). *Anacytis* and other cyanobacteria can be important food sources from late summer to fall (Starostka and Applegate 1970). Bigmouth buffalo occasionally consume aquatic beetles and small mollusks (Johnson 1963).

References

Becker GC (1983) Fishes of Wisconsin. University of Wisconsin Press, Madison

Beckman LG, Elrod JH (1971) Apparent abundance and distribution of young-of-the-year fishes in Lake Oahe, 1965-69. In: Hall GE (ed) Reservoir fisheries and limnology, Am Fish Soc Special Publ 8, pp 333–347

Benson NG (1980) Effects of post impoundment shore modification on fish populations in Missouri River reservoirs. US Dep Inter, Fish Wildl Serv Res Rep 80

Burr BM, Heidinger RC (1983) Reproductive behavior of the bigmouth buffalo *Ictiobus cyprinellus* in Crab Orchard Lake, Illinois. Am Midl Nat 110:220–221. https://doi.org/10.2307/2425230

Edwards EA (1983) Habitat suitability index models: bigmouth buffalo. US Dep Inter, Fish Wildl Serv, FWS/OBS-82/10.34

Etnier DA, Starnes WC (1993) The fishes of Tennessee. University of Tennessee Press, Knoxville

Gammon JR (1973) The effect of thermal input on the populations of fish and macroinvertebrates in the Wabash River. Purdue University Water Resource Center., Tech Rep 32,, Lafayette

Johnson RP (1963) Studies on the life history and ecology of the bigmouth buffalo, *Ictiobus cyprinellus* (Valenciennes). J Fish Res Board Can 20:1397–1429. https://doi.org/10.1139/f63-095

Lackman AR, Andrews AH, Butler MG, Bielak-Lackman ES, Clark ME (2019) Bigmouth buffalo, *Ictiobus cyprinellus* sets freshwater teleost record as improved age analysis reveals centenarian longevity. Commun Biol 2:197. https://doi.org/10.1038/s42003-019-0452-0

McComish TS (1967) Food habits of bigmouth and smallmouth buffalo in Lewis and Clark Lake and the Missouri River. Trans Am Fish Soc 96:70–74. https://doi.org/10.1577/1548-8659(1967)96[70:FHOBAS]2.0.CO;2

Starostka VJ, Applegate RL (1970) Food selectivity of bigmouth buffalo, *Ictiobus cyprinellus*, in Lake Poinsett, South Dakota. Trans Am Fish Soc 99:571–576. doi: https://doi.org/10.1577/1548-8659

Walburg CH (1976) Changes in the fish populations of Lewis and Clark Lake, 1956–74, and their relation to water management and the environment. US Dept Inter, Fish Wildl Serv Res Rep 79

Walburg CH, Nelson WR (1966) carp, river carpsucker, smallmouth buffalo, and bigmouth buffalo in Lewis and Clark Lake, Missouri River. US Dep Inter, Fish Wildl Serv Res Rep 69

6.11.9 Black Buffalo *Ictiobus niger*

(Rafinesque, 1819)

© Uland Thomas

● **Pre-1990** ○ **1990-2021**

Etymology: *Ictiobus niger* – *Ictiobus* = fish ox, likely referencing common name or robust appearance; ***niger*** = black.

Description:

Body – slightly robust, elongated, slightly compressed; slightly rounded profile from head to dorsal fin anterior end.
Color –

> *Dorsally* – black to dark olive gray.
> *Laterally* – dusky–brown to olive.
> *Ventrally* – light gray to white.
> *Fins* – dark gray to black.

Head – large.
Opercle – subopercle, rounded outline on lower corner.
Snout – blunt.
Eyes – small, positioned anteriorly on head.
Mouth – small, subterminal to inferior.

> *Lips* – thick, with deep grooves.

Teeth – absent on jaws; approximately 195 short pharyngeal teeth per gill arch.
Gill rakers – 30 to 35 on first arch.
Dorsal fin – 25 to 31 rays; long, falcated.
Caudal peduncle – short, thick.
Caudal fin – moderately forked, slightly rounded lobes.
Anal fin – 8 or 9 rays.
Pelvic fins – 9 to 11 rays; abdominal.
Pectoral fins – slightly rounded distal end.
Lateral line – complete; 36 to 39 cycloid scales in series.
Sexual dimorphism – spawning males have small tubercles on head, anterior scales, and pelvic fins.

Similar Species: Black buffalo closely resemble smallmouth buffalo and bigmouth buffalo. Smallmouth buffalo have a deeper, slightly laterally compressed body, and a small conical head creating a humped profile from the head to dorsal fin anterior end. Smallmouth buffalo also have a small, subterminal mouth and thick lips with deep grooves. Bigmouth buffalo have a slightly more robust body, more exaggerated hump from head to dorsal fin-anterior end, and a large, terminal mouth with thin lips. Common carp have two fleshy barbels on each side of the mouth. *Carpiodes* species, such as river carpsucker and quillback, have an angled outline on the lower corner of the subopercle.

Distribution and Habitat: The native range of black buffalo extends throughout the lower Great Lakes, Mississippi River basin, and Missouri River basin from southeastern South Dakota to Michigan and south to Louisiana. It is the least common *Ictiobus* species in its range. Historic occurrences in the Dakotas are from the Missouri River and Lake Mitchell. The North Dakota Missouri River occurrence was likely a prereservoir relict collected near Garrison Dam in 1970. This specimen was vouchered (UMMZ 190381) but subsequently skeletonized, so it cannot be confirmed or denied. This species likely no longer occurs in the Dakotas above Gavins Point Dam. Black buffalo inhabit medium-to-large rivers and reservoirs in sloughs and backwaters near deep, swift riffles. Black buffalo prefer higher water velocities than bigmouth buffalo and smallmouth buffalo and are less tolerant of turbidity than bigmouth buffalo (Cross 1967; Becker 1983).

Reproduction: Little is known about the reproductive biology of the black buffalo, but spawning behavior is thought to be like smallmouth buffalo (Etnier and Starnes 1993). It is an annual spawner. Sexual maturity occurs at age-2 in southern populations (Perry 1976; Becker 1983). Presumable, black buffalo spawn from April to May. In riverine habitat, adults migrate from deeper water within the main channel to shallow backwaters, runs, and pools (Etnier and Starnes 1993; Piller et al. 2003). Spawning occurs when two or more males swim alongside a single female, bumping her near the surface to induce the egg release (Piller et al. 2003). Splashing occurs during this spawning behavior (Piller et al. 2003). Eggs are demersal and adhesive.

Age and Growth: Adult black buffalo total lengths range 381–762 mm and weigh 4.5 to 6.8 kg. Total lengths at age of black buffalo from Kansas were age-1, 310 mm; age-2, 368 mm; age-3, 447 mm; age-4, 543 mm; and age-5, 627 mm (Greer and Cross 1956; Kansas Fishes Committee 2014). They can reach a total length of 800 mm and weigh 13.6 kg. Black buffalo longevity is over 20 years.

Food and Feeding: There is little information on black buffalo diet and feeding habits, but it is thought to be similar to bigmouth buffalo. Black buffalo are opportunistic feeders highly dependent on benthic organisms, stirring up bottom sediments in search of prey (Minckley et al. 1970). Feeding on plankton throughout the water column seems unlikely due to the mouth position (Etnier and Starnes 1993). Adults mainly consume diatoms, cyanobacteria, crustaceans, insects, and detritus (Etnier and Starnes 1993). In Arizona, juvenile Asian clams made up most of the adult diet (Etnier and Starnes 1993). A large palatal organ situated in the pharynx assists black buffalo selectively retain food (Doosey and Bart 2011; Kansas Fishes Committee 2014).

References

Becker GC (1983) Fishes of Wisconsin. University of Wisconsin Press, Madison

Cross FB (1967) Handbook of fishes in Kansas. University Press of Kansas, Lawrence

Doosey MH, Bart HL (2011) Morphological variation of the palatal organ and chewing pad of Catostomidae (Teleostei: Cypriniformes). J Morphol 272:1092–1108. https://doi.org/10.1002/jmor.10966

Etnier DA, Starnes WC (1993) The fishes of Tennessee. University of Tennessee Press, Knoxville

Greer JK, Cross FB (1956) Fishes of El Dorado City Lake, Butler County, Kansas. Trans Kansas Acad Sci 59:358–363. https://doi.org/10.2307/3626610

Kansas Fishes Committee (2014) Kansas fishes. University Press of Kansas, Lawrence

Minckley WL, Johnson JE, Rinne JN, Willoughby SE (1970) Foods of buffalofishes, genus *Ictiobus*, in central Arizona reservoirs. Trans Am Fish Soc 99:333–342. https://doi.org/10.1577/1548-8659

Perry WG (1976) Black and bigmouth buffalo spawn in brackish water ponds. Prog Fish Cult 38:81. https://doi.org/10.1577/1548-8659(1976)38[81:BABBSI]2.0.CO;2

Piller KR, Bart HL Jr, Tipton JA (2003) Spawning in the black buffalo, *Ictiobus niger* (Cypriniformes: Catostomidae). Ichthyol Explor Freshw 14:145–150

6.11.10   Silver Redhorse *Moxostoma anisurum*

(Rafinesque, 1820)

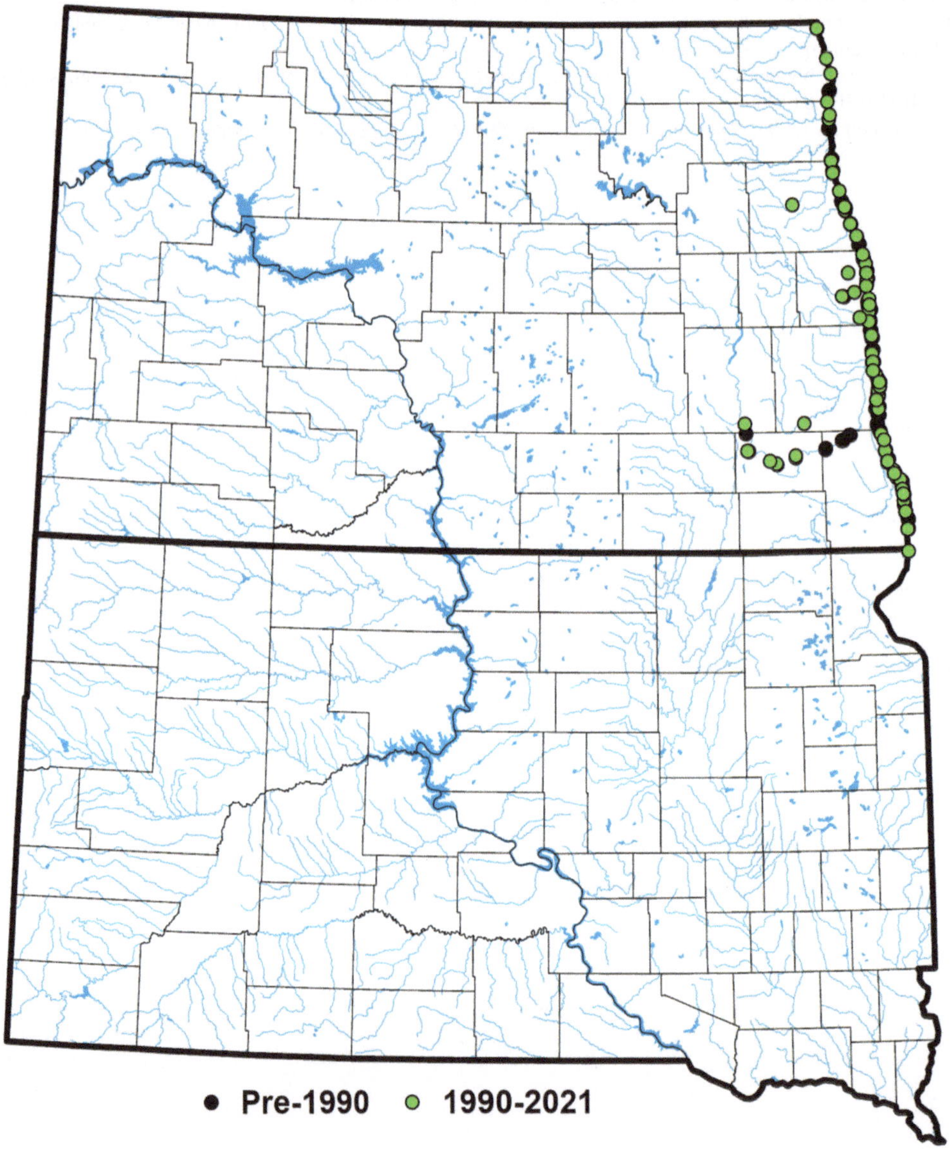

Etymology: *Moxostoma anisurum* – *Moxostoma* = suck mouth; *anisurum* = unequal tail, referencing upper and lower lobes asymmetry.

Description:

Body – fusiform to slightly dorsoventrally flattened, elongated.
Color –

> *Dorsally* – dark gray to olive.
> *Laterally* – silver, scales generally lacking dark spots at anterior base.
> *Ventrally* – white.
> *Fins* – dorsal, caudal – dark gray; anal, pelvic, pectoral – clear to light red orange.

Head – moderately large.
Snout – blunt.
Eyes – moderately large, laterally on upper portion of head.
Mouth – large, inferior; jaws do not extend backward to anterior end of eye.

> *Lips* – plicae; halves of lower lip meet at an acute angle; majority of lower lips with plicae broken into individual papillae; lower thicker than upper.

Teeth – absent on jaws; pharyngeal teeth long, slender.
Gill rakers – 20 to 32.
Dorsal fin – 14 to 17 rays, straight to slightly convex distal end.
Caudal peduncle – short, thick circum peduncular scales usually 12 in a row.
Caudal fin – forked, upper lobe slightly more elongated and pointed.
Anal fin – 7 rays.
Pelvic fins – 9 rays; insertion posterior to dorsal fin insertion.
Pectoral fins – 16 to 19 rays.
Lateral line – complete; 41 to 46 large cycloid scales.
Sexual dimorphism – spawning males have tubercles on head, body, lower lobe of caudal fin, and anal fin.

Similar Species: Silver redhorse closely resemble other redhorse (*Moxostoma*) species. Shorthead redhorse, river redhorse, and greater redhorse all have a red caudal fin and dark spots on the anterior base of scales on the lateral sides. Shorthead redhorse posterior lower lip margin forms a straight line. River redhorse lower lips are plicae, but not broken into individual papillae. Greater redhorse have 13 to 15 dorsal fin rays, typically 16 scales in circum peduncular row, and a posterior lower lip margin that forms an obtuse angle. Golden redhorse have 39–42 lateral line scales in series and a posterior lower lip margin that forms a nearly straight line to obtuse angle.

Distribution and Habitat: The native range of silver redhorse extends from Alberta to Quebec, south to Alabama and the lower Missouri River, east to Vermont, and includes the upper and midcentral Mississippi River, the Great Lakes, St. Lawrence River, and Ohio River basin (Jenkins 1980; Page and Burr 1991; McAllister et al. 2009). Silver redhorse are found in the Red River of the North and Sheyenne River systems in North Dakota and the Bois de Sioux River system of the Red River of North drainage in South Dakota. They inhabit small-to-large rivers, lakes, and impoundments (Scott and Crossman 1973). In rivers, silver redhorse are often found in long, deep pools or backwater habitats with slow current over silty to firm substrate (Meyer 1962; Jenkins 1970; Scott and Crossman 1973). They avoid high-gradient streams with increased turbidity (Becker 1983). During their first year, young silver redhorse inhabit smaller streams with slow water velocity, often near overhanging vegetation and stream banks to avoid predators (Harlan and Speaker 1956; Meyer 1962; Becker 1983).

Reproduction: Information is lacking on silver redhorse spawning behaviors, egg descriptions, and hatching times. They may make small migrations prior to spawning from mid-April to early May (Meyer 1962; Smith 1977). Silver redhorse from the Des Moines River in Iowa reached peak spawning conditions at water temperatures of 13 °C during the first week of May (Meyer 1962). In Alabama, spawning occurs from late March to early April when water temperatures reach 14 °C or 15 °C (Etnier and Starnes 1993). Sexual maturity occurs at age-5 (Meyer 1962). Silver redhorse spawn in river main channels at depths of 0.3 to 0.9 m over coarse substrate (Meyer 1962). In Iowa, fecundity was 14,910 eggs from an age-5 silver redhorse female with a total length of 338 mm and 36,340 eggs from an unknown age female with a total length of 490 mm (Meyer 1962). Eggs are 1 to 2 mm in diameter (Becker 1983).

Age and Growth: The highest silver redhorse growth rates are from late July to early September (Jenkins 1970; Becker 1983). Total lengths at age of silver redhorse from the Des Moines River in Iowa were age-1, 109 mm; age-2, 195 mm; age-3, 271 mm; age-4, 328 mm; age-5, 379 mm; age-6, 432 mm; age-7, 470.2 mm; age-8, 505 mm; and age-9, 513.1 mm (Meyer 1962). They can reach a total length of 675 mm and have a longevity of 25 years (Reid 2007).

Food and Feeding: Information on silver redhorse diet and feeding habits is lacking. Adults eat filamentous algae, detritus, chironomids, ephemeroptera larvae, trichoptera larvae, and other immature aquatic insects (Meyer 1962; Smith 1977).

References

Becker GC (1983) Fishes of Wisconsin. University of Wisconsin Press, Madison

Etnier DA, Starnes WC (1993) The fishes of Tennessee. University of Tennessee Press, Knoxville

Harlan JR, Speaker EB (1956) Iowa fish and fishing. Iowa Conservation Committee, Des Moines. https://doi.org/10.1577/1548-8659(1956)18[177:NOFAF]2.0.CO;2

Jenkins RE (1970) Systematic studies of the catostomid fish Moxostomatini. Dissertation, Cornell University

Jenkins RE (1980) Moxostoma anisurum. In: Lee DS, Gilbert CR, Hocutt CH, Jenkins RE, McAllister DE, Stauffer JR Jr (eds) Atlas of North American freshwater fishes. North Carolina State Museum of Natural History, Raleigh, pp 409–410

McAllister CT, Starnes WC, Robinson HW, Jenkins RE, Raley ME (2009) Distribution of the silver redhorse, *Moxostoma anisurum* (Cypriniformes: Catostomidae), in Arkansas. Southwest Nat 54:514–518. https://doi.org/10.1894/GG-36.1

Meyer WH (1962) Life history of three species of redhorse (*Moxostoma*) in the Des Moines River, Iowa. Trans Am Fish Soc 91:412–419. https://doi.org/10.1577/1548-8659(1962)91[412:LHOTSO]2.0.CO;2

Page LM, Burr BM (1991) Peterson field guide to freshwater fishes of North America North of Mexico. Houghton Mifflin Harcourt, Boston

Reid SM (2007) Comparison of scales, pectoral fin rays, and opercles for age estimation of Ontario redhorse, *Moxostoma*, species. Can Field Nat 121:29–34. https://doi.org/10.22621/cfn.v121i1.389

Scott WB, Crossman EJ (1973) Freshwater fishes of Canada, Bulletin 184. Fisheries Research Board of Canada

Smith CG (1977) The biology of three species of *Moxostoma* (Pisces: Catostomidae) in Clear Creek, hocking and Fairfield counties, Ohio, with emphasis on the golden redhorse, *M. erythrurum* (Rafinesque). Dissertation, Ohio State University

6.11.11 River Redhorse *Moxostoma carinatum*

(Cope, 1870)

© Uland Thomas

● Pre-1990   ○ 1990-2021

Etymology: *Moxostoma carinatum* – *Moxostoma* = suck mouth; *carinatum* = keeled.

Description:

Body – fusiform, elongated, moderately robust.
Color –

   Dorsally – olive to bronze.
   Laterally – golden yellow, olive to silver scales, dark spots at anterior base.
   Ventrally – white.
   Fins – dorsal – dusky red; caudal – red; anal, pelvic, pectoral – orange-red.

Head – moderately large.
Snout – blunt, almost square in shape.
Eyes – moderately large, laterally on upper portion of head.
Mouth – large, inferior; jaws do not extend backward to anterior end of eye.
   Lips – plicate; posterior lower lips margin nearly straight or forming a slightly obtuse angle; lower lip plicae not dissected into individual papillae; lower lip thicker than upper lip.
Teeth – absent on jaws; pharyngeal teeth large, molariform.
Gill rakers – 20 to 31.
Dorsal fin – 13 to 15 rays, straight to slightly concave distal end.
Caudal peduncle – slender, slightly elongated; circum peduncular scales rows usually 12.
Caudal fin – forked, upper lobe slightly more elongated and pointed.
Anal fin – 7 rays.
Pelvic fins – 9 rays; insertion posterior to dorsal fin insertion.
Pectoral fins – 15 to 17 rays.
Lateral line – complete; 41 to 47 large cycloid scales in series.
Sexual dimorphism – spawning males have large tubercles on snout, head, and opercle and small tubercles on body, caudal, and anal fins; larger spawning females have tubercles on the anal fin.

Similar Species: River redhorse closely resemble other redhorse (*Moxostoma*) species. Silver redhorse and golden redhorse have a slate-colored caudal fin and generally lack dark spots on anterior base of scales on lateral sides. Silver redhorse posterior lower lip margin forms an acute angle. Golden redhorse have 39 to 42 lateral line scales and a lower lip posterior margin that forms a nearly straight line or an obtuse angle. Shorthead redhorse have a falcate to concave dorsal fin distal end and a posterior lower lip margin that forms a straight line. Greater redhorse have a slightly convex dorsal fin distal end, a posterior lower lip margin that meets at an obtuse angle and usually 16 scales in circum peduncular row.

Distribution and Habitat: The native distribution of river redhorse is wide but discontinuous. It extends throughout southeastern Canada and the eastern United States, including the Great Lakes, south to central Mississippi, and the Gulf slope (Straight et al. 2015). In the Dakotas, river redhorse have only been found in the lower Missouri River below Gavins Point Dam. Populations within the native historic range have substantially declined over the past century, likely because of habitat loss (Etnier and Starnes 1993; Jenkins and Burkhead 1993; Butler and Wahl 2017). River redhorse inhabit riffle-runs deeper than 1.5 m in medium-to-large rivers and the lower portions of their main tributaries with clear water and current greater than 0.4 m/second (Jenkins 1970; Beckman and Hutson 2012; Butler and Wahl 2017). They may also occupy reservoirs (Jenkins 1970; Beckman and Hutson 2012; Butler and Wahl 2017). River redhorse prefer course substrates like gravel, cobble, or boulder (Butler and Wahl 2017). They move the most in the spring, with the smallest movement occurring in summer (Butler and Wahl 2017). They also prefer faster current areas during winter and spring (Butler and Wahl 2017). River redhorse are intolerant of sedimentation and pollution and negatively impacted by habitat fragmentation and flow regulation (Jenkins 1970; Trautman 1981; Butler and Wahl 2017).

Reproduction: River redhorse spawn from late May through early June at water temperatures of 18 to 19 °C in the northern part of its range (Campbell 2001). Spawning stops at 20 °C (Campbell 2001). River redhorse become sexually mature at approximately age-5 (Ross 2001). Males arrive at the spawning grounds prior to females (Hackney et al. 1967; Campbell 2001; Reid 2006). Redd formation by males sweeping coarse substrate with their bodies and caudal fin has only been observed in Alabama (Hackney et al. 1967; Campbell 2001). Spawning occurs day and night within riffles less than 2 meters deep over gravel substrate in main river channels (Campbell 2001; Straight et al. 2015; Butler and Wahl 2017). Spawning events that last from 3 to 13.5 seconds occur over a 5 to 8-day spawning season (Hackney et al. 1967; Straight et al. 2015). Fecundity is estimated to range from 9,000 to 22,000 eggs (Campbell 2001). Adhesive, yellow to yellow-orange eggs average 3 mm in diameter (Campbell 2001). Laboratory-incubated eggs hatched in 3 to 4 days at a water temperature of 22 °C (Hackney et al. 1967; Becker 1983).

Age and Growth: One of the larger species of *Moxostoma*, river redhorse weights often exceed 5 kg (Jenkins and Burkhead 1993; Beckman and Hutson 2012). Age-at length information is lacking. Most of their growth occurs within the first 5 years, with growth rates slowing around age-8 (Campbell 2001). Females tend to be larger than males (Reid 2006). River redhorse can reach a total length of 798 mm and have a longevity of 28 years (Campbell 2001).

Food and Feeding: River redhorse are benthivores. They use enlarged pharyngeal arches and molariform pharyngeal teeth to eat their primary diet of mussels, snails, and other mollusks (Jenkins 1970). Ephemeropterans, chironomids, trichopterans, and other aquatic insect larvae are also known to be consumed (Hackney et al. 1967).

References

Becker GC (1983) Fishes of Wisconsin. University of Wisconsin Press, Madison

Beckman DW, Hutson CA (2012) Validation of aging techniques and growth of the river redhorse, *Moxostoma carinatum*, in the James River, Missouri. Southwest Nat 57:240–247. https://doi.org/10.1894/0038-4909-57.3.240

Butler SE, Wahl DH (2017) Movements and habitat use of river redhorse (*Moxostoma carinatum*) in the Kankakee River, Illinois. Copeia 105:734–742. https://doi.org/10.1643/CE-17-626

Campbell BG (2001) A study of the river redhorse, *Moxostoma carinatum* (Pisces: Catostomidae), in the tributaries of the Ottawa River near Canada's National Capital and in a tributary of Lake Ontario, the Grand River, near Cayuga, Ontario. Dissertation, University of Ottawa

Etnier DA, Starnes WC (1993) The fishes of Tennessee. University of Tennessee Press, Knoxville

Hackney PA, Tatum WM, Spencer SL (1967) Life history of the river redhorse, *Moxostoma carinatum* (Cope), in the Cahaba River, Alabama, with notes on the management of the species as a sport fish. In: Proceedings of the Conference of the Southeast Association of the Game Fisheries Commission, vol 21, pp 324–332

Jenkins JE, Burkhead NM (1993) Freshwater fishes of Virginia. American Fisheries Society, Bethesda

Jenkins RE (1970) Systematic studies of the catostomid fish tribe Moxostomatini. Dissertation, University of Michigan

Reid SM (2006) Timing and demographic characteristics of redhorse spawning runs in three Great Lakes basin rivers. J Freshw Ecol 21:249–258. https://doi.org/10.1080/02705060.2006.9664993

Ross ST (2001) The inland fishes of Mississippi. University Press of Mississippi, Jackson

Straight CA, Jackson CR, Freeman BJ, Freeman MC (2015) Diel patterns and temporal trends in spawning activities of robust redhorse and river redhorse in Georgia, assessed using passive acoustic monitoring. Trans Am Fish Soc 144:563–576. https://doi.org/10.1080/0002848 7.2014.1001040

Trautman MB (1981) The fishes of Ohio, revised edn. Ohio State University Press, Columbus

6.11.12   Golden Redhorse *Moxostoma erythrurum*

(Rafinesque, 1818)

● Pre-1990   ● 1990-2021

Etymology: *Moxostoma erythrurum* – *Moxostoma* = suck mouth; ***erythrurum*** = red tail.

Description:

Body – fusiform, elongated, moderately laterally compressed.
Color –

> *Dorsally* – olive, bronze or gold.
> *Laterally* – golden yellow to silver, scales without dark spots at anterior base.
> *Ventrally* – white.
> *Fins* – dorsal, caudal – slate; anal, pelvic, pectoral – clear to light orange-red.

Head – moderately large.
Snout – blunt.
Eyes – moderately large, laterally on upper portion of head.
Mouth – large, inferior.

> *Jaws* – do not extend backward to anterior end of eye.
> *Lips* – plicate; posterior margin lower lip plicae broken into individual papillae; halves of lower lip straight line or obtuse angle; lower lip thicker than upper lip.

Teeth – absent on jaws; pharyngeal teeth comb-like, numerous.
Gill rakers – 20 to 32.
Dorsal fin – 12 to 13, slightly concave distal end.
Caudal peduncle – slender, slightly elongated; usually 12 circum peduncular scale rows.
Caudal fin – forked.
Anal fin – 7 rays.
Pelvic fins – 9 rays; insertion posterior to dorsal fin insertion.
Pectoral fins – 16 to 19 rays.
Lateral line – complete, 39 to 42 large cycloid scales in series.
Sexual dimorphism – spawning males have tubercles on snout and dorsal portion of head, and small tubercles on anterior body, and anal and caudal fin rays; spawning females have thickened anal fin rays.

Similar Species: Golden redhorse closely resemble other redhorse (*Moxostoma*) species. Shorthead redhorse, river redhorse, and greater redhorse have a red caudal fin and dark spots on the anterior base of scales on lateral sides. Shorthead redhorse lower lip posterior margin forms a straight line. River redhorse lower lips plicae but are not dissected into individual papillae. Greater redhorse have 13 to 15 dorsal fin rays and typically 16 circum peduncular scale rows. Silver redhorse have 14 to 17 dorsal fin rays and their lower lip halves meet at an acute angle.

Distribution and Habitat: The native range of golden redhorse extends from eastern North Dakota east to the Great Lakes, Hudson Bay, and New York, south to the Atlantic Coast of Alabama and Mississippi, and west to Oklahoma and Kansas. Golden redhorse primarily occur in the Red River of the North and Sheyenne River systems in North Dakota, and the Minnesota River system in South Dakota. It inhabits low gradient rivers, lakes, and reservoirs with warm water temperatures in areas with little aquatic vegetation and coarse substrate (Meyer 1962; Page and Burr 1991; Beckman and Howlett 2013). Golden redhorse migrate downstream during low water periods (Meyer 1962). Optimum water temperature in Indiana is 26 to 28 °C (Gammon 1973). Golden redhorse are tolerant of turbidity and high-water temperatures, with an upper thermal tolerance of approximately 29 °C and a critical thermal maximum of 35 °C (Spiegel et al. 2011). Golden redhorse overwinter in deeper water or pools of larger streams (Trautman 1981). Juveniles prefer low water velocities and are often near stream banks for protection from predators (Meyer 1962).

Reproduction: Golden redhorse spawn during daylight in the spring in water temperatures from 10 to 23 °C in low-to-moderate-flow shoals or riffles 30 to 60 cm deep over gravel and rubble substrate (Smith 1977; Curry and Spacie 1984; Kwak and Skelly 1992). Sexual maturity occurs from age-3 to age-6 (Beckman and Howlett 2013). Male golden redhorse aggressively defend an approximately 0.5 m-diameter territory by ramming or butting other males before and during spawning periods (Meyer 1962; Smith 1977; Kwak and Skelly 1992). Females remain nearby in pools or deeper water until one female is escorted by one or more males into their spawning territory (Kwak and Skelly 1992). Dominant males align themselves on each side of the female and vibrate their bodies against the female for 3 to 6 seconds to induce gamete release (Kwak and Skelly 1992). No nest is constructed, and no parental care is given (Kwak and Skelly 1992). Fecundity of golden redhorse from the Des Moines River in Iowa was 6,100 and 25,350 eggs from females that were age-4 with a total length of 292 mm, and age-6 with a total length of 399 mm, respectively (Meyer 1962). The adhesive eggs are approximately 2 to 3 mm in diameter (Becker 1983).

Age and Growth: After the first year, golden redhorse have an average fork length of 97 mm (Smith 1977). The largest increases in length occur between age-2 and age-4, with growth increments decreasing after age-4 (Smith 1977). Females are generally larger than males (Curry and Spacie 1984). Fork lengths-at-age of golden redhorse from Ohio were age-1, 97 mm; age-2, 126 mm; age-3, 166 mm; age-4, 205 mm; age-5, 236 mm L; age-6, 265 mm; age-7, 290 mm; and age-8, 312 mm (Smith 1977). They can reach a total length of 625 mm and have a longevity of 12 years (Carlander 1969; Becker 1983; Beckman and Howlett 2013).

Food and Feeding: Golden redhorse are benthivores that feed by sucking material off substrate and expelling undesirable prey (Smith 1977). They mostly forage at night (Smith 1977). Adults mainly eat immature insects like chironomids, ephemeropterans, and trichopterans (Meyer 1962; Spiegel et al. 2011). They eat smaller amounts of filamentous algae and detritus, along with worms and small fingernail clams when available (Smith 1977; Etnier and Starnes 1993; Spiegel et al. 2011).

References

Becker GC (1983) Fishes of Wisconsin. University of Wisconsin Press, Madison

Beckman DW, Howlett DT (2013) Otolith annulus formation and growth of two redhorse suckers (Moxostoma: Catostomidae). Copeia 2013:390–395. https://doi.org/10.1643/CG-10-193

Carlander KD (1969) Handbook of freshwater fishery biology. Iowa State University Press, Ames

Curry KD, Spacie A (1984) Differential use of stream habitat by spawning catostomids. Am Midl Nat 111:267–279. https://doi.org/10.2307/2425321

Etnier DA, Starnes WC (1993) The fishes of Tennessee. University of Tennessee Press, Knoxville

Gammon JR (1973) The effects of thermal inputs on the populations of fish and macroinvertebrates in the Wabash River. Purdue Univ Water Res Center Tech Rep, No. 32

Kwak TJ, Skelly TM (1992) Spawning habitat, behavior, and morphology as isolating mechanisms of the golden redhorse, *Moxostoma erythrurum*, and the black redhorse, *M. duquesnei*, two syntopic fishes. Environ Biol Fish 34:127–137. https://doi.org/10.1007/BF00002388

Meyer WH (1962) Life history of three species of redhorse (Moxostoma) in the Des Moines River, Iowa. Trans Am Fish Soc 91:412–419. https://doi.org/10.1577/1548-8659(1962)91[412:LHOTSO]2.0.CO;2

Page LM, Burr BM (1991) Peterson field guide to freshwater fishes of North America North of Mexico. Houghton Mifflin Harcourt, Boston

Smith CA (1977) The biology of three species of *Moxostoma* (Pisces-Catostomidae) in Clear Creek, Hocking, and Fairchild counties, Ohio, with emphasis on the golden redhorse, *M. erythrurum* (Rafinesque). Dissertation, Ohio State University

Spiegel JR, Quist MC, Morris JE (2011) Trophic ecology and gill raker morphology of seven catostomid species in Iowa rivers. J Appl Ichthyol 27:1159–1164. https://doi.org/10.1111/j.1439-0426.2011.01779.x

Trautman MB (1981) The fishes of Ohio, revised edn. Ohio State University Press, Columbus

6.11.13   Shorthead Redhorse *Moxostoma macrolepidotum*

(Lesueur, 1817)

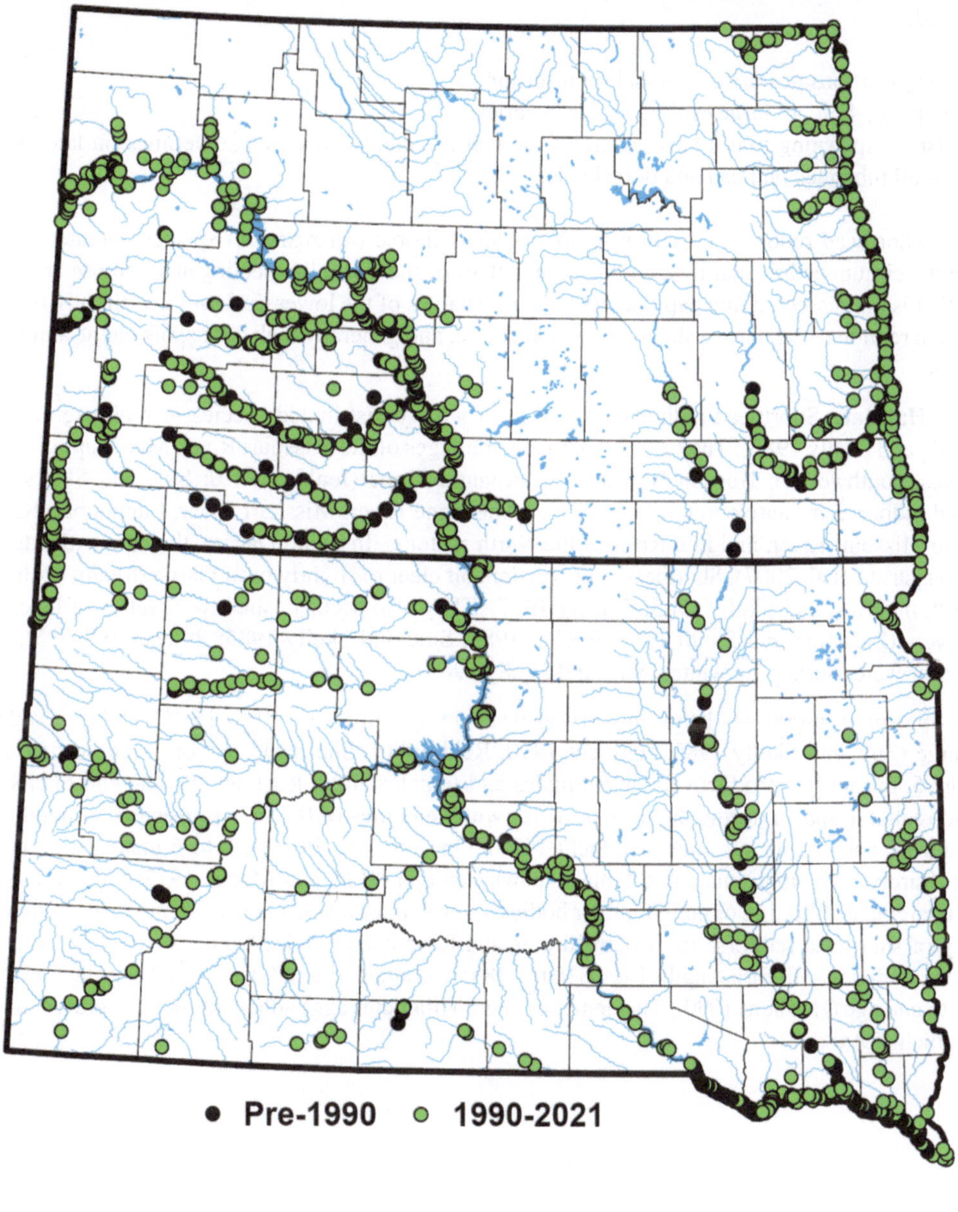

● **Pre-1990**   ○ **1990-2021**

Etymology: *Moxostoma macrolepidotum* – *Moxostoma* = suck mouth; *macrolepidotum* = large scaled.

Description:

Body – fusiform, elongated, moderately laterally compressed.
Color –

> *Dorsally* – olive, bronze, or gold.
> *Laterally* – gold to silver; dark crescent spots at anterior base of scales.
> *Ventrally* – white.
> *Fins* – dorsal, anal, paired fins – faint reddish-orange; caudal – red.

Head – relatively short, small.
Snout – moderately blunt; slightly overhanging the mouth.
Eyes – moderately large, laterally on upper portion of head.
Mouth – small, inferior.

> *Jaws* – do not extend backward to anterior end of eye.
> *Lips* – plicate with the lower lip posterior margin forming a straight line; posterior lower lip plicae dissected into papillae.

Teeth – absent on jaws; pharyngeal teeth compressed, comb-like, 12 to 30 on lower half of tooth row.
Dorsal fin – 12 to 14 rays; deeply concave distal end, anterior rays when depressed do not extend to end of posterior ray.
Caudal peduncle – elongated, moderately thick; circum peduncular scale rows 12 or 13.
Caudal fin – forked.
Anal fin – 7 rays.
Pelvic fins – 9 rays; insertions posterior to dorsal fin insertion.
Lateral line – complete, 41 to 45 large cycloid scales in series.
Sexual dimorphism – spawning males have tubercles on anal fin, caudal fins, and sometimes on lateral sides; spawning females have small tubercles on head and dorsal side.

Similar Species: Shorthead redhorse closely resemble other redhorse (*Moxostoma*) species. Greater redhorse generally have 16 scales in the circum peduncular row, with the two halves of the lower lip meeting at an obtuse angle. River redhorse have a lower lip that is not dissected into papillae, with the two halves of the lower lip forming a slightly obtuse angle. Silver redhorse and golden redhorse both have a slate-colored caudal fin, and generally lack dark spots on the anterior base of scales on the lateral sides.

Distribution and Habitat: Shorthead redhorse are the most widely distributed species of redhorse in the United States (Jenkins 1970; Sule and Skelly 1985). They are native to the drainages of the Missouri River, Mississippi River, St. Lawrence River, Great Lakes, southwestern Hudson Bay, and the Atlantic Slope (Jenkins 1970; Sule and Skelly 1985). In North Dakota, shorthead redhorse are native to the James River, Sheyenne River, Missouri River, Cannonball River, Heart River, Knife River, Little Missouri River, and Red River of the North systems. In South Dakota, they are native to the all the main river systems (Hoagstrom et al. 2007). Shorthead redhorse inhabit clear to slightly turbid large streams with increased velocity and gravel bottoms (Sule and Skelly 1985; Pflieger 1997). They also occur in shallower waters of lakes, reservoirs, and pools of small rivers (Larimore and Smith 1963; Becker 1983; Sule and Skelly 1985; Pflieger 1997). Shorthead redhorse appear to use deeper water during the winter (Sule and Skelly 1985).

Reproduction: Spawning migrations begin in April at water temperatures of 10 to 15 °C, and generally coincide with increased discharge (Sule and Skelly 1985; Harbicht 1990; Reid 2006). Shorthead redhorse can become sexually mature from age-3 to age-5, but most reproductively active males and females are at least age-6 (Sule and Skelly 1985; Harbicht 1990). Shorthead redhorse spawn during daylight in riffles with velocities of 0.3 to 0.7 m/second at depths of 20 to 90 cm over fine sand, gravel, and cobble substrate (Harbicht 1990). One female will have 1 to 6 males behind or alongside her (Harbicht 1990). Rarely do a single male and female spawn together (Burr and Morris 1977; Sule and skelly 1985). The males approach the female, vibrating or nudging their bodies several times against hers to stimulate gamete release (Harbicht 1990). Fecundity of shorthead redhorse from both Illinois and Iowa averages 18,000 eggs (Meyer 1962; Sule and Skelly 1985). In Manitoba, females with fork lengths from 310 to 418 mm contained 12,660 to 44,329 eggs, and fecundity increased with female length and age (Harbicht 1990). The demersal, 1-mm diameter eggs adhere to rocks or gravel substrate (Sule and Skelly 1985; Harbicht 1990).

Age and Growth: Female shorthead redhorse tend to be larger than males (Sule and Skelly 1985; Reid 2006). Total-lengths-at-age of shorthead redhorse from Iowa were age-1, 116 mm; age-2, 224 mm; age-3, 293 mm; age-4, 342 mm; age-5, 418 mm; age-6, 521 mm; age-7, 589 mm; age-8, 655 mm (Meyer 1962). Total lengths at age of shorthead redhorse from Illinois were age-1, 93 mm; age-2, 207 mm; age-3, 295 mm; age-4, 343 mm; age-5, 368 mm; age-6, 390 mm; age-7, 396 mm; and age-8, 402 mm (Sule and Skelly 1985). They can reach a total length of 615 mm and live for 20 years (Reid 2007).

Food and Feeding: Juvenile shorthead redhorse with fork lengths under 100 mm feed throughout the water column on small zooplankton, like cladocerans, copepods, and ostracods (Harbicht 1990). They will also eat chironomids, trichopterans, and other small benthic organisms (Harbicht 1990). Adults and juveniles with fork lengths greater than 100 mm eat a wider variety of predominantly benthic prey, including chironomids, trichopterans, ephemeropterans, and mollusks (Harbicht 1990). Sand, algae, and benthic organic material have also been observed in shorthead redhorse stomachs, likely as a result of feeding along the bottom (Sule and Skelly 1985; Harbicht 1990).

References

Becker GC (1983) Fishes of Wisconsin. University of Wisconsin Press, Madison

Burr BM, Morris MA (1977) Spawning behavior of the shorthead redhorse, (*Moxostoma macrolepidotum*) in Big Rock Creek, Illinois. Trans Am Fish Soc 106:80–82. https://doi.org/10.1577/1548-8659

Harbicht S (1990) Ecology of the shorthead redhorse (*Moxostoma macrolepidotum*, Leseur 1817) in Dauphin Lake, Manitoba. Dissertation, University of Manitoba

Hoagstrom CW, Wall SS, Kral JG, Blackwell BG, Berry CR Jr (2007) Zoogeographic patterns and faunal change of South Dakota fishes. West N Am Nat 67:161–184. https://doi.org/10.3398/1527-0904(2007)67[161:ZPAFCO]2.0.CO;2

Jenkins RE (1970) Systematic studies of the catostomid fish tribe Moxostomatini. Part I and Part II. Dissertation, Cornell University

Larimore RW, Smith PW (1963) The fishes of Champaign County, Illinois, as affected by 60 years of stream changes. Ill Nat His Surv Bull 28:299–382. https://doi.org/10.21900/j.inhs.v28.168

Meyer WH (1962) Life history of three species of redhorse (Moxostoma) in the Des Moines River, Iowa. Trans Am Fish Soc 91:412–419. https://doi.org/10.1577/1548-8659(1962)91[412:LHOTSO]2.0.CO;2

Pflieger WL (1997) The fishes of Missouri, revised edn. Missouri Dep Cons, Jefferson City

Reid SM (2006) Timing and demographic characteristics of redhorse spawning runs in three Great Lakes River basins. J Freshw Ecol 21:249–258. https://doi.org/10.1080/02705060.2006.9664993

Reid SM (2007) Comparison of scales, pectoral fin rays and opercles for age estimation of Ontario redhorse, *Moxostoma*, species. Can Field Nat 12:29–34. https://doi.org/10.22621/cfn.v121i1.389

Sule M, Skelly T (1985) The life history of the shorthead redhorse, *Moxostoma macrolepidotum*, in the Kankakee River drainage, Illinois. Illinois Dep Energ Nat Res, Nat Hist Surv Div, Biol Notes No 123. https://doi.org/10.5962/bhl.title.15169

6.11.14   Greater Redhorse *Moxostoma valenciennesi*

Jordan, 1885

© Uland Thomas

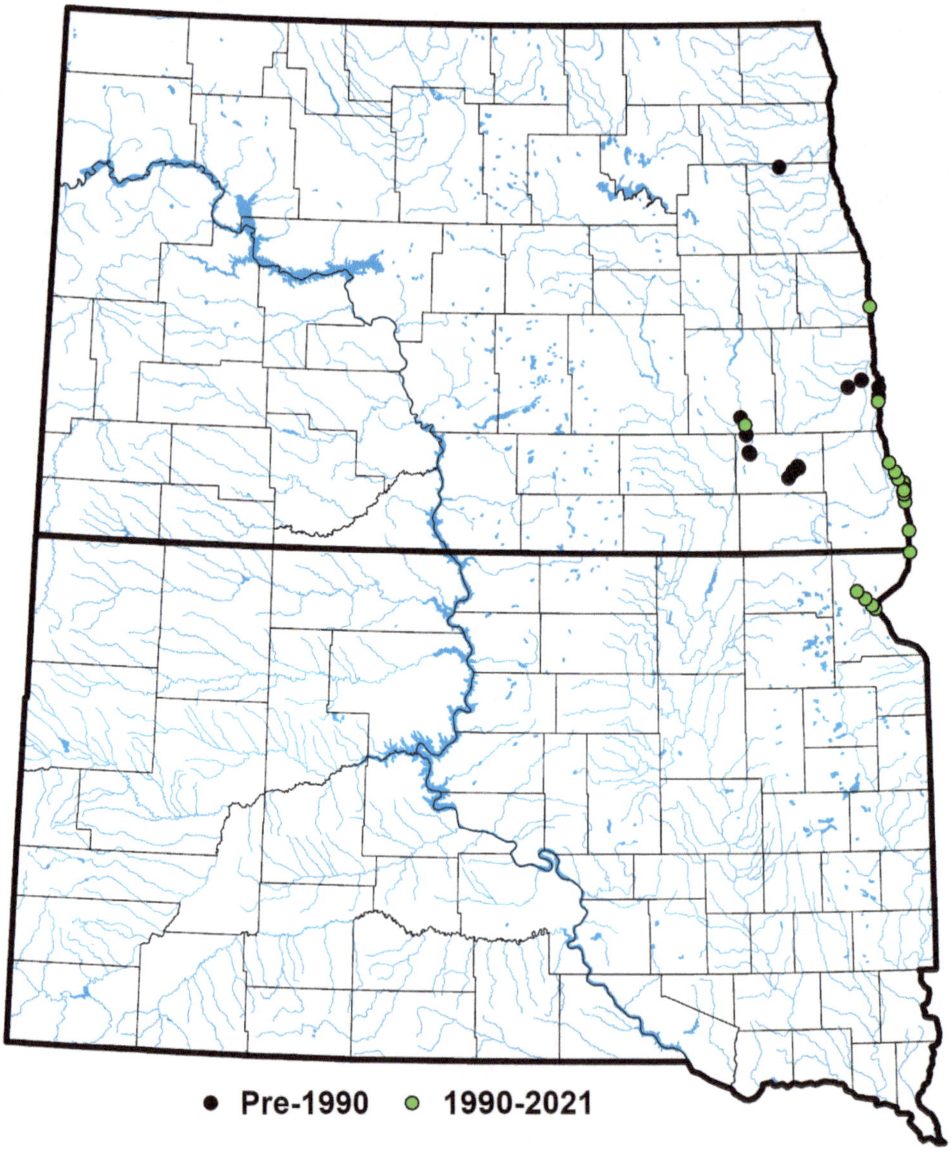

● **Pre-1990**   ● **1990-2021**

Etymology: *Moxostoma valenciennesi – Moxostoma* – suck mouth; *valenciennesi* = referencing M.A. Valenciennes, the French naturalist who first described the species.

Description:

Body – fusiform, elongated, slightly laterally compressed.
Color –

> *Dorsally* – olive, bronze, or gold.
> *Laterally* – golden yellow to silver, scales dark spots at anterior base.
> *Ventrally* – white.
> *Fins* – dorsal, caudal, anal – dark red; pelvic, pectoral – reddish-orange, anterior rays cream.

Head – moderately large.
Snout – moderately blunt.
Eyes – moderately large, laterally on upper portion of head.
Mouth – large, inferior.

> *Jaws* – do not extend backward to anterior end of eye.
> *Lips* – plicate halves of lower lip meeting at obtuse angle; lower lip thicker than upper lip.

Teeth – absent on jaws; pharyngeal teeth thick, numerous.
Dorsal fin – 13 to 15 rays; slightly convex distal end.
Caudal peduncle – slender, slightly elongated circum peduncular scale rows usually 16.
Caudal fin – forked.
Anal fin – 7 rays.
Pelvic fins – 9 rays; insertion posterior to dorsal fin insertion.
Pectoral fins – rounded distal end.
Lateral line – complete, 40 to 45 large cycloid scales in series.
Juveniles – slightly concave dorsal fin.
Sexual dimorphism – spawning males have small tubercles on dorsal and lateral sides of head and body, and caudal fin rays and lower lobe; spawning females have small, fewer tubercles on caudal, anal, pelvic, and pectoral fins.

Similar Species: Greater redhorse closely resemble other redhorse (*Moxostoma*) species. Silver redhorse and golden redhorse both have a slate-colored caudal fin and generally lack dark spots on anterior base of scales on lateral sides. Silver redhorse lower lip posterior margin forms an acute angle. Golden redhorse have 39 to 42 lateral line scales, and the lower lip posterior margin forms a nearly straight line or obtuse angle. Shorthead redhorse have a falcate to concave dorsal fin distal end, with the lower lip posterior margin forming a straight line. River redhorse have a straight to concave dorsal fin distal end, with the lower lip posterior margin forming a straight line or slightly obtuse angle.

Distribution and Habitat: The native range of greater redhorse includes the Great Lakes, St. Lawrence Rivers, northern portion of the Ohio River, upper Mississippi River, Red River of the North, and upper Illinois River (Kay et al. 1994; Bunt and Cooke 2004). They occur within the Red River of the North, Sheyenne River, and Minnesota River systems of eastern North Dakota and South Dakota. Greater redhorse are rare or uncommon in most of its distribution (Becker 1983). They inhabit the low-current areas of clear medium to large rivers with pool-riffle-run habitat, and sand, gravel, or boulder substrate (Trautman 1981; Bunt and Cooke 2001). Greater redhorse will also occupy large lakes and reservoirs (Trautman 1981). After spawning and during the summer, they are found in runs covered with *Cladophora*, cobble or gravel substrate, and surface velocities under 5 cm/second (Bunt and Cooke 2001). They will also occupy riffle and pool habitat to a much lesser extent (Bunt and Cooke 2001). Greater redhorse migrate to overwintering habitat in early fall (Bunt and Cooke 2001). They are sensitive to pollution, siltation, and turbidity, which likely affects their distribution (Becker 1983; Bunt and Cooke 2001).

Reproduction: The reproductive ecology of the greater redhorse is poorly understood (Mongeau et al. 1992; Kay et al. 1994; Bunt and Cooke 2001). Upstream spawning migrations to riffles with increased water velocity and sand, gravel, or cobble substrate occur in early spring (Becker 1983). Males likely become sexually mature at age-5 or age-6 (Jenkins 1970; Becker 1983). Spawning is thought to occur in May and June, but in the Thousand Island region of the St. Lawrence River, greater redhorse spawned from late June through early July in water temperatures of 17 to 19 °C (Scott 1967; Becker 1983). Spawning has been observed in riffles with a mean depth of 34.4 cm and a mean surface velocity of 38 cm/second (Cooke and Bunt 1999). Males defend territories which females periodically visit (Becker 1983). Two males typically spawn with one female (Becker 1983). Greater redhorse do not construct nests nor provide parental care. In the Grand River in Ontario,

Canada, spawning ended in early June, earlier than other larger, and cooler, rivers (Jenkins and Jenkins 1980; Mongeau et al. 1992; Bunt and Cooke 2001). After spawning, greater redhorse move as far as 15.2 km downstream (Bunt and Cooke 2001). The end of spawning is typically associated with a spike in water temperature or heavy rain (Bunt and Cooke 2001). The yellow, nonadhesive, approximately 3-mm diameter, demersal eggs settle into spaces between substrate (Bunt and Cooke 2004). Artificially spawned greater redhorse eggs hatched from 6 to 8 days at 16 to 19 °C (Bunt and Cooke 2004).

Age and Growth: Greater redhorse are the largest redhorse species (Becker 1983). There is little information on their growth rates, which likely vary among populations (Becker 1983). Newly hatched greater redhorse larvae have total lengths of 9 to 14 mm (Bunt and Cooke 2004). Females are often larger than males (Bunt and Cooke 2001). Greater redhorse can reach a total length of 688 mm and live for 20 years (Reid 2007).

Food and Feeding: Like most castostomid benthivores, greater redhorse feed by sucking material off substrate and expelling out undesirable items. Adults mainly consume midge larvae, crustaceans, mollusks, and other aquatic benthic invertebrates (Becker 1983).

References

Becker GC (1983) Fishes of Wisconsin. University of Wisconsin Press, Madison

Bunt CM, Cooke SJ (2001) Post-spawn movements and habitat use by greater redhorse, *Moxostoma valenciennesi*. Ecol Freshw Fish 10:57–60. https://doi.org/10.1111/j.1600-0633.2001.tb00194.x

Bunt CM, Cooke SJ (2004) Ontogeny of larval greater redhorse (*Moxostoma valenciennesi*). Am Midl Nat 151:93–100. https://doi.org/10.1674/0003-0031(2004)151[0093:OOLGRM]2.0.CO;2

Cooke SJ, Bunt CM (1999) Spawning and reproductive biology of greater redhorse, *Moxostoma valenciennesi*, in the Grand River, Ontario. Can Field Nat 113:497–502

Jenkins RE (1970) Systematic studies of the catostomid fish tribe Moxostomatini. Dissertation,. University of Michigan

Jenkins RE, Jenkins DJ (1980) Reproductive behavior of the greater redhorse, *Moxostoma valenciennesi*, in the Thousand Islands Region. Can Field Nat 94:426–430

Kay LK, Wallus R, Yeager BL (1994) Reproductive biology and early life history of fishes in the Ohio River drainage: Catostomidae, vol 2. CRC Press, Boca Raton

Mongeau JR, Dumont P, Cloutier L (1992) La biologie du suceur cuivré (*Moxostoma hubbsi*) compare á celle de quatre autres espèces de *Moxostoma* (*M. anisurum, M. carinatum, M. macrolepidotum* et *M. valenciennesi*). Can J Zool 70:1354–1363. https://doi.org/10.1139/z92-191

Reid SM (2007) Comparison of scales, pectoral fin rays, and opercles for age estimation of Ontario redhorse, *Moxostoma*, species. Can Field Nat 121:29–34. https://doi.org/10.22621/cfn.v121i1.389

Scott WB (1967) Freshwater fishes of eastern Canada. University of Toronto Press, Toronto

Trautman MB (1981) The fishes of Ohio, revised edn. Ohio State University Press, Columbus

6.11.15 Plains Sucker *Pantosteus jordani*

Evermann, 1893

Nuptial Male   Female

● **Pre-1990**   ○ **1990-2021**

Etymology: *Pantosteus jordani – Pantosteus* = all bone, likely referencing the joining of the parietal bones; *jordani* = honoring David Starr Jordan, colleague of Barton Warren Evermann.

Description:

Body – elongated, slender, cylindrical.
Color –

> *Dorsally* – light dusky olive-green to gray; dark mottling.
> *Laterally* – light olive to green.
> *Ventrally* – cream to white.
> *Peritoneum* – black.
> *Fins* – unpigmented; no distinct markings.

Head – short, broad, conical.
Snout – blunt, long.
Eyes – moderately large.
Mouth – inferior, large; deep, lateral notches on each corner.

> *Jaws* – inner cartilaginous ridge.
> *Lips* – fleshy, papillose; lower lip lobes incompletely divided.

Intestine – long, 4 to 6 times body length.
Gill rakers – 23 to 37 external row; 31 to 51 internal row of first gill arch (Baxter and Stone 1995).
Dorsal fin – 8 to 13 rays; usually 10.
Caudal peduncle – short.
Caudal fin – slightly forked.
Anal fin – 7 rays.
Pelvic fins – 9 rays; insertion posterior to dorsal fin anterior end.
Pectoral fins – 15 rays; rounded.
Lateral line – complete; 75 to 92 small cycloid scales, crowded anteriorly.
Sexual dimorphism – spawning males have a well-defined burnt orange lateral stripe above a dark green to gray thin stripe extending from snout tip to caudal fin base and tubercles on entire body including the enlarged anal fin rays; spawning females have a duller lateral stripe extending from operculum to anal fin and tubercles only on the dorsal and lateral sides of the head and body (Hauser 1969).

Similar Species: Plains sucker resemble white sucker and longnose sucker. White sucker have a short snout slightly extending beyond the upper lip and 55 to 75 lateral line scales. They also lack a cartilaginous ridge on inner jaws and notches on each corner of the mouth. Longnose sucker have more than 90 very small cycloid scales in their lateral line and completely divided lower lip lobes. They also lack a cartilaginous ridge on inner jaws.

Distribution and Habitat: *Pantosteus* is one of several genera centered in western North America (Schultz and Bertrand 2012). Mountain sucker *Pantosteus platyrhynchus* were previously recognized as widely distributed throughout the Lahontan and Bonneville basins, Columbia and Snake rivers, and the headwaters of the Missouri drainage and Upper Colorado Basin. However, this genus was recently split into several species, including the plains sucker. Plains sucker are found from the upper Missouri drainage from the Black Hills of South Dakota to western Wyoming, Montana, and Alberta, with Pleistocene fossils from western Kansas (Unmack et al. 2014). They are not present in North Dakota. Plains sucker historically occurred throughout the entire Black Hills region but are now only in the northern and southern portions of the Black Hills within the Cheyenne River system. It is listed as a species of greatest conservation need in South Dakota (Schultz and Bertrand 2012; SDGFP 2006). Although plains sucker occur in large lakes, reservoirs, and rivers throughout its range, in the Black Hills they only inhabit cool, moderate-flow water in low-gradient streams with sand, gravel, or cobble substrates (Baxter and Stone 1995; Schultz and Bertrand 2012). They are often associated with submerged vegetation and canopy cover. Plains sucker density increases with increasing periphyton (Schultz 2011; Schultz et al. 2016). Optimal temperature is from 11 to 19 °C, with a critical thermal maximum in the Black Hills of 32 °C (Hauser 1969; Schultz and Bertrand 2011). Plains suckers can tolerate higher water temperatures than co-occurring salmonids in the Black Hills (Schultz and Bertrand 2011). Juveniles frequently inhabit side channels or deeper pools with little discharge (Hauser 1969). Plains sucker can tolerate moderate turbidity.

Reproduction: In the Black Hills, male plains suckers with breeding colors have been observed from early June through late August, likely indicating protracted spawning (Breeggemann et al. 2014). Spawning occurs in water temperatures from 11 to 19 °C in riffle areas below pools over gravel substrates in water approximately 18 cm deep with velocities of 12 to 15 cm/second (Hauser 1969; Moyle et al. 1995; Wydoski and Wydoski 2002). The light yellow, 2 mm diameter eggs adhere to substrate (Wydoski and Wydoski 2002). Fecundity is strongly correlated with female size (Wydoski and Wydoski 2002). Female plains suckers with a mean total length of 162 mm had a mean fecundity of 2,087 eggs (Wydoski and Wydoski 2002). Males reach sexual maturity at age-2 through age-4, while females mature at age-3 to age-5 (Hauser 1969). In the Black Hills, the smallest mature male and female had total lengths of 95 mm and 101 mm, respectively (Breeggemann et al. 2014).

Age and Growth: Female plains suckers are generally larger and longer lived than males (Hauser 1969). In the Black Hills, plains sucker total length was 100 mm during the fourth growing season, and mean maximum total length was 220 mm (Breeggemann et al. 2014). Total lengths at age of plains suckers from Whitewood and Elk Creeks in the Black Hills were 100 mm at age-3, 111 mm at age-4, 133 mm at age-5, and 151 mm at age-6 (Breeggemann et al. 2014). Growth in the Black Hills is slower than other locations (Breeggemann et al. 2014). Plains sucker can reach a total length of 223 mm and have a longevity of 6 to 9 years (Hauser 1969).

Food and Feeding: Plains suckers are generalist benthic grazers, mainly eating periphyton and diatoms with smaller portions of aquatic insects and larvae. They also consume sand and grit (Moyle et al. 1995). Plains sucker use cartilaginous ridges on the inside of both jaws to scrape algae and benthic material from hard substrates.

References

Baxter GT, Stone MD (1995) Fishes of Wyoming. Wyoming Game and Fish Department, Cheyenne

Breeggemann JJ, Hayer CA, Krause J, Schultz LD, Bertrand KN, Graeb BDS (2014) Estimating the ages of mountain sucker *Catostomus platyrhynchus* from the Black Hills: precision, maturation, and growth. West N Am Nat 74:299–310. https://doi.org/10.3398/064.074.0305

Hauser WJ (1969) Life history of the mountain sucker, *Catostomus platyrhynchus*, in Montana. Trans Am Fish Soc 98:209–215. https://doi.org/10.1577/1548-8659(1969)98[209:LHOTMS]2.0.CO;2

Moyle PB, Yoshiyama RM, Williams JE, Wikramanayake ED (1995) Fish species of special concern in California, 2nd edn. University of California, Davis

Schultz LD (2011) Environmental factors associated with long-term trends of mountain sucker populations in the Black Hills, and the assessment of their thermal tolerance. Dissertation, South Dakota State University

Schultz LD, Bertrand KN (2011) An assessment of the lethal thermal maxima for mountain sucker. West N Am Nat 71:404–411. https://doi.org/10.3398/064.071.0308

Schultz LD, Bertrand KN (2012) Long term trends and outlook for mountain sucker in the Black Hills of South Dakota. Am Midl Nat 167:96–110. https://doi.org/10.1674/0003-0031-167.1.96

Schultz LD, Bertrand KN, Graeb BDS (2016) Factors from multiple scales influence the distribution and abundance of an imperiled fish-mountain sucker in the Black Hills of South Dakota, USA. Environ Biol Fish 99:3–14. https://doi.org/10.1007/s10641-015-0449-6

Scott WB, Crossman EJ (1973) Freshwater fishes of Canada. Fisheries Research Board of Canada, Bulletin 184

SDGFP (South Dakota Department of Game, Fish and Parks) (2006) South Dakota comprehensive wildlife conservation plan. S D Game Fish Parks, Wildl Div Rep 2006–2008, Pierre

Unmack PJ, Dowling TE, Laitinen NJ, Secor CL, Mayden RL, Shiozawa DK, Smith GR (2014) Influence of introgression and geological processes on phylogenetic relationships of Western North American mountain suckers (*Pantosteus*, Catostomidae). PLoS One 9:e90061. https://doi.org/10.1371/journal.pone.0090061

Wydoski RG, Wydoski RS (2002) Age, growth, and reproduction of mountain suckers in Lost Creek Reservoir, Utah. Trans Am Fish Soc 131:320–328. https://doi.org/10.1577/1548-8659

6.12 Cyprinidae, Barb and Carp Family

Until recently, all minnows and carps were included in the Cyprinidae family. However, genetic analysis led to changes in taxonomy, separating Cyprinidae into three distinct families: Cyprinidae, Xenocyprinidae, and Leuciscidae. The Cyprinidae family now only refers to the approximately 1,700 species of barbs and carps native to Eurasia and Africa. Two species have been introduced to North America and can be found in the Dakotas.

Similar to the xenocyprinids and the leuciscids, cyprinids lack teeth on both the upper and lower jaws. Rather, they have well-developed pharyngeal teeth arranged in two to three rows on the gill arches to assist with mastication. Cyprinids have a single continuous dorsal fin but lack an adipose fin. Both of the cyprinids found in the Dakotas have a thick, serrated spine anteriorly on their dorsal and anal fins, unlike the xenocyprinid and leuciscid species also present. Most species of Cyprinidae also have two pairs of barbels. Cyprinids are widely distributed, inhabit a variety of habitats, and possess a diversity of morphological features.

6.12.1   Goldfish *Carassius auratus*

(Linnaeus, 1758)

● **Pre-1990**   ○ **1990-2021**

Etymology: *Carassius auratus – Carassius* = Latinization of the vernacular name Karass or Karausche, of the European Crucian carp; *auratus* = gilded, referencing the golden color.

Description:

Body – deep, slightly laterally compressed, arched back.

Color – greatly variable; generally olive or bronze but may also be solid orange to tangerine, or mottled with white, brown, or black blotches.

Head – short, small, triangular.

Snout – bluntly pointed.

Eyes – moderately large, laterally on head.

Mouth – moderately small, terminal, and oblique.

Lips – thin.

Teeth – pharyngeal tooth pattern 0,4-4,0.

Gill rakers – long, 37 to 43

Dorsal fin – 15 to 19 rays; 1 serrated spine, elongated; soft ray anteriorly.

Caudal peduncle – short, thick.

Caudal fin – moderately forked.

Anal fin – 5 or 6 rays; 1 serrated spine; soft ray anteriorly.

Pelvic fins – 8 or 9 rays; abdominal insertions inferior or slightly posterior to dorsal fin insertion.

Pectoral fins – 15 or 16 rays.

Lateral line – complete; 26 to 31 large cycloid scales in series.

Sexual dimorphism – spawning males develop small tubercles on the head, dorsal side of body, and pectoral fins.

Similar Species: Goldfish resemble common carp. Common carp have two pairs of barbels (one short barbel on each side of the snout, and one longer barbel on each corner of the mouth), 32 to 41 scales in the lateral line, and a more conical, elongate head and snout. Bigmouth buffalo have a large, rounded head, small eyes positioned anteriorly on the head, and pelvic fins with 10 or 11 rays. Smallmouth buffalo have a small, conical head with a humped profile, and a small subterminal mouth with thick lips.

Distribution and Habitat: Native to Asia and Japan, goldfish have been introduced worldwide as an aquarium fish (Blackwell 2007). They are well established and abundant in some parts of North America, but occurrences may be sporadic because of release or escapement (Lee et al. 1980). In North Dakota, a single specimen was vouchered from the Missouri River near Williston. In South Dakota, the distribution is limited to Capitol Lake in Pierre and Fall River in Hot Springs, with prior reports from the Cheyenne River and Cody Lake in Bennett County (Blackwell 2007). In the Dakotas, cold water temperatures limit goldfish distribution (Blackwell 2007). When goldfish are acclimated to 15 °C, critical thermal minimum is 1 °C and critical thermal maximum is 35 °C (Ford and Beitinger 2005). However, when acclimated to 25 °C, critical thermal minimum is 5 °C and critical thermal maximum is 40 °C (Ford and Beitinger 2005). Golfish inhabit heavily vegetated waters with little to no current, including pools in quiet streams, ponds, shallow lakes, and occasionally impoundments (Kansas Fishes Committee 2014). Adults are often found near the bottom substrate, but also school near the surface (Becker 1983). Goldfish are highly tolerant of a wide range of water temperatures, increased salinity, and low dissolved oxygen levels (Lee et al. 1980; Horváth et al. 1992; Imanpoor et al. 2012). They are less tolerant of increased turbidity and moderate or high currents (Trautman 1981).

Reproduction: Goldfish spawning begins when water temperatures reach approximately 16 °C in the spring and extends to late summer (Becker 1983). Spawning generally occurs at dawn (Horváth et al. 1992). Both sexes can become sexually mature at age-1, but in colder climates females may not mature until age-2 (Horváth et al. 1992). During spawning, multiple males chase a single female into dense vegetation, where intense splashing leads to the release of eggs and milt (Horváth et al. 1992). No nest is constructed, and no parental care is given. Goldfish are fractional spawners, with females capable of releasing 2,000 to 4,000 eggs in a single clutch, and ultimately a total of 160,000 to 380,000 eggs (Becker 1983). The adhesive, pale amber eggs are approximately 2 mm in diameter (Innes 1936; Becker 1983; Horváth et al. 1992). Hatching occurs at 46 to 54 hours at 29 °C, 5 days at 20 °C, and 8 to 10 days at 15 °C (Mansueti and Hardy 1967; Becker 1983).

Age and Growth: Growth of young of the year goldfish is highly variable and depends on growing season duration, water temperature, salinity, other environmental conditions, and fish density (Trautman 1981; Moyle 2002; Imanpoor et al. 2012). Total lengths at age of goldfish from Wisconsin were 84 mm at age-0, 120 mm at age-1, 151 mm at age-2, and 188 mm at age-3 (Becker 1983). Goldfish can achieve a total length of 457 mm and live for up to 30 years in ponds not subjected to severe weather (Innes 1936; Becker 1983).

Food and Feeding: Goldfish are omnivorous, with juveniles primarily consuming zooplankton and insect larvae (Lee et al. 1980). Adult diets are similar to those of common carp, consisting of diatoms, phytoplankton, small crustaceans, aquatic insects, green algae, and aquatic vegetation (Churchill and Over 1938; Lee et al. 1980; Horváth et al. 1992; Pflieger 1997). Spawning adults will also often consume their own eggs and newly hatched larvae (Horváth et al. 1992).

References

Becker GC (1983) Fishes of Wisconsin. University of Wisconsin Press, Madison

Blackwell BG (2007) Warm-water fish species. In: Berry C, Higgins K, Willis D, Chipps S (eds) History of fisheries and fishing in South Dakota. S D Dep Game Fish Parks, Pierre, pp 213–238

Churchill EP, Over WH (1938) Fishes of South Dakota. Brown & Saenger Printers, Sioux Falls

Ford T, Beitinger TL (2005) Temperature tolerance in the goldfish, *Carassius auratus*. J Therm Biol 30:147–152. https://doi.org/10.1016/j.jtherbio.2004.09.004

Horváth L, Tamás G, Seagrave C (1992) carp and pond fish culture: including Chinese herbivorous species, pike, tench, zander, wels catfish and goldfish. Halsted Press, New York

Imanpoor MR, Najafi E, Kabir M (2012) Effects of different salinity and temperatures on the growth, survival, haematocrit and blood biochemistry of goldfish (*Carassius auratus*). Aquac Res 43:332–338. https://doi.org/10.1111/j.1365-2109.2011.02832.x

Innes WT (1936) The complete aquarium book: the care and breeding of goldfish and tropical fishes. Halcyon House, New York. https://doi.org/10.5962/bhl.title.5887

Kansas Fishes Committee (2014) Kansas fishes. University Press of Kansas, Lawrence

Lee DS, Gilbert CR, Hocutt CH, Jenkins RE, McAllister DE, Stauffer JR (1980) Atlas of North American freshwater fishes. North Carolina Biological Survey Publication, 1980-12

Mansueti AJ, Hardy JD (1967) Development of fishes of the Chesapeake Bay region: an atlas of egg, larval, and juvenile stages – part 1. University of Maryland, Baltimore

Moyle PB (2002) Inland fishes of California, revised, expanded edn. University of California Press, Berkley

Pflieger WL (1997) The fishes of Missouri, revised edn. Missouri Dep Cons, Jefferson City

Trautman MB (1981) The fishes of Ohio, revised edn. Ohio State University Press, Columbus

6.12.2   Common Carp *Cyprinus carpio*

Linnaeus, 1758

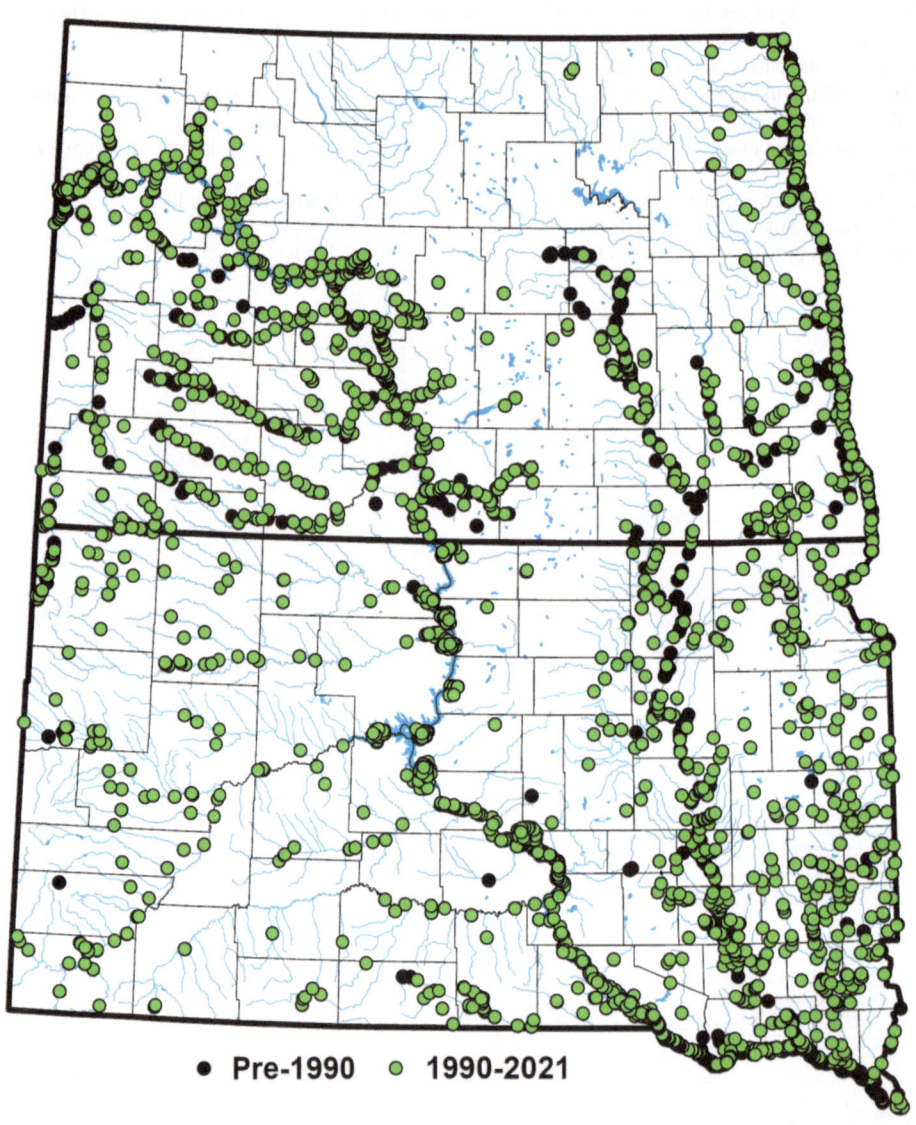

● Pre-1990   ● 1990–2021

Etymology: *Cyprinus carpio* – *Cyprinus* = carp (Greek); *carpio* = carp (Latin).

Description:

Body – moderately fusiform, robust, arched back, deep, laterally compressed.
Color –

> *Dorsally* – dark olive gray to bronze.
> *Laterally* – golden yellow to light yellow tan or olive; scales outlined in darker color appearing cross-hatched.
> *Ventrally* – light yellow to cream.
> *Fins* – pectoral/pelvic/anal/caudal: often yellow-orange on large adults.

Head – conical.
Snout – long.
Eyes – moderately large, laterally on upper portion of head.
Mouth – terminal to slightly subterminal.
> *Jaws* – do not extend backward to anterior edge of eye.
> *Lips* – thin.

Teeth – absent on jaws; pharyngeal teeth molariform, aligned in 3 rows, 1,1,3-3,1,1.
Barbels – 2; 1 short barbel on each side of snout; 1 longer barbel on each corner of mouth.
Gill rakers – 21 to 27.
Dorsal fin 15 to 23 soft rays; 1 stiff serrated spinous ray anteriorly.
Caudal peduncle – short, thick.
Caudal fin – forked.
Anal fin – 4 to 6 rays, including one anterior stiff spinous ray posteriorly serrated.
Pelvic fins – 8 or 9 rays; abdominal.
Pectoral fins – 14 to 17 rays; do not extend to pelvic fin insertions.
Lateral line – complete; 32 to 41 large cycloid scales in series.
Juveniles – deeper body and larger eyes than adults.
Sexual dimorphism – spawning males are slightly darker in color with tubercles on head and pectoral fins.

Similar Species: Common carp may be confused with grass carp, goldfish, carpsuckers (*Carpiodes*), and buffalo (*Ictiobus*). Grass carp lack barbels and have a more elongated, cylindrical, corpulent body than common carp. Goldfish have a similar body shape but lack barbels. Carpsuckers and buffalo lack dorsal fins spines, anal fin spines, and barbels.

Distribution and Habitat: Native to Eastern Europe and central Asia, common carp are one of the most widely distributed fish species in the world. They have been introduced to every continent except Antarctica (McCrimmon 1968). Common carp were stocked in the United States in the 1870s and eastern South Dakota as early as 1885 (Barnes 2007; Blackwell 2007). The United States Bureau of Fish and Fisheries continued stocking common carp into southeastern South Dakota lakes until 1914 (Barnes 2007). Common carp are habitat generalists and are very adaptive to a wide variety of warm water habitats. They often inhabit shallow, highly conductive lakes, reservoirs, streams, ponds, and sloughs with clear to turbid water and abundant aquatic vegetation over soft substrate (Egertson and Downing 2004; Penne and Pierce 2008). They are less frequently found in cold, clear waters. Common carp prefer littoral habitats during spring and summer and move to deeper water during late fall and winter months (Penne and Pierce 2008; Hennen and Brown 2014). In an eastern South Dakota glacial lake, common carp were most active during twilight and least active at dawn (Hennen and Brown 2014). Their feeding and spawning behaviors frequently cause decreased water quality and increased turbidity, negatively impacting native fish communities and aquatic plant growth (Kloskowski 2011; Weber and Brown 2009, 2011b; Kaemingk et al. 2016). Increased common carp density can shift phytoplankton communities from benign green algae to toxic cyanobacteria (Williams and Moss 2003; Kaemingk et al. 2016). Common carp are extremely tolerant of high temperatures, low oxygen levels, and high turbidity, enabling them to frequently become very abundant.

Reproduction: Adult common carp concentrate at spawning grounds well before spawning occurs (Swee and McCrimmon 1966). Spawning is temperature dependent and occurs from late spring through early summer in the shallow shoreline areas of marshes, wetlands, and floodplains where macrophytes are abundant (Swee and McCrimmon 1966; Penne and Pierce 2008; Hennen and Brown 2014). Optimum spawning temperature ranges from 19 to 23 °C, with spawning ceasing at temperatures above 28 °C (Swee and McCrimmon 1966). Several males accompany a single female and splash near the surface during spawning (Etnier and Starnes 1993). Common carp are prolific, fractional, broadcast spawners. Sexual maturity occurs at age-2 or age-3 (Panek 1987). Females may produce over 2,000,000 eggs, with fecundity dependent on female size

(Swee and McCrimmon 1966). No nest is constructed, and no parental care is given. Following spawning, adults broadly disperse (Swee and McCrimmon 1966; Penne and Pierce 2008; Hennen and Brown 2014). Common carp eggs have a diameter of approximately 1 to 2 mm and adhere to submerged macrophytes and soft substrate (Becker 1983). Hatching occurs in 3 to 16 days depending on temperature (Becker 1983). Hatching occurred over 47 days from mid-May to early July in an eastern South Dakota glacial lake (Phelps et al. 2008).

Age and Growth: Growth of common carp is rapid initially but is also variable depending on population densities (Carlander 1969; Shields 1970). Age-0 individuals from South Dakota glacial lakes grew approximately 1 mm/day, with growth rates positively related to water temperatures (Phelps et al. 2008). Earlier hatching juveniles had faster growth rates and achieved greater total lengths during the first growing season than later-hatched fish (Phelps et al. 2008). Common carp size structure decreases and relative abundance increases with increasing water depth (Weber et al. 2010). Average total length at age-1 in South Dakota waters was 132 mm in Lake Francis Case, 97 mm in Lewis and Clark Reservoir, and 90 mm in Lake Oahe (Shields 1970). Common carp can exceed 22.7 kg. Although they rarely live longer than 20 years, a 27-year-old common carp was reported in South Dakota (Weber and Brown 2011a).

Food and Feeding: Common carp juveniles with total lengths less than 100 mm mainly consume large-bodied zooplankton and then shift to benthic prey (Weber and Brown 2009). Adults are opportunistic and omnivorous benthivores with diets of detritus, chironomids, annelids, odonates, amphipods, and other benthic macroinvertebrates. Common carp feed by suctioning sediment into its mouth, where desired food items are retained, and unwanted material ejected (Sibbing 1988). This feeding behavior often uproots submerged aquatic macrophytes, causing sediment resuspension and high levels of turbidity. Large common carp can suction up to 12 cm into the substrate during feeding (Panek 1987).

References

Barnes ME (2007) Fish hatcheries and stocking practices: past and present. In: Berry C, Higgins K, Willis D, Chipps S (eds) History of fisheries and fishing in South Dakota. S D Dep Game Fish Parks, Pierre, pp 267–294

Becker GC (1983) Fishes of Wisconsin. University of Wisconsin Press, Madison

Blackwell BG (2007) Warm-water fish species. In: Berry C, Higgins K, Willis D, Chipps S (eds) History of fisheries and fishing in South Dakota. S D Dep Game Fish Parks, Pierre, pp 213–238

Carlander KD (1969) Handbook of freshwater fishery biology, vol 1. Iowa State University Press, Ames

Egertson CJ, Downing JA (2004) Relationship of fish catch and composition to water quality in a suite of agriculturally eutrophic lakes. Can J Fish Aquat Sci 61:1784–1796. https://doi.org/10.1139/f04-109

Etnier DA, Starnes WC (1993) The fishes of Tennessee. University of Tennessee Press, Knoxville

Hennen MJ, Brown ML (2014) Movement and spatial distribution of common carp in a South Dakota glacial lake system: implications for management and removal. N Am J Fish Manag 34:1270–1281. https://doi.org/10.1080/02755947.2014.959674

Kaemingk MA, Jolley JC, Paukert CP, Willis DW, Henderson K, Holland RS, Lindvall ML (2016) common carp disrupt ecosystem structure and function through middle-out effects. Mar Freshw Res 68:718–731. https://doi.org/10.1071/MF15068

Kloskowski J (2011) Impact of common carp *Cyprinus carpio* on aquatic communities: direct trophic effects versus habitat deterioration. Fundam Appl Limnol 178:245–255. https://doi.org/10.1127/1863-9135/2011/0178-0245

McCrimmon HR (1968) carp in Canada. Fisheries Research Board of Canada Bulletin 165

Panek FM (1987) Biology and ecology of carp. In: Cooper EL (ed) carp in North America. American Fisheries Society, Bethesda, pp 1–15

Penne CR, Pierce CL (2008) Seasonal distribution, aggregation, habitat selection of common carp in Clear Lake, Iowa. Trans Am Fish Soc 137:1050–1062. https://doi.org/10.1577/T07-112.1

Phelps QE, Graeb BDS, Willis DW (2008) First year growth and survival of common carp in two glacial lakes. Fish Manag Ecol 15:85–91. https://doi.org/10.1111/j.1365-2400.2007.00568.x

Shields JT (1970) Changes in the fish population in Lake Francis case in South Dakota in the first 16 years of impoundment. U.S. Bur Sport Fish Wildl, Tech Pap 56, Washington, DC

Sibbing FA (1988) Specialization and limitations in the utilization of food resources by the carp, *Cyprinus carpio*: a study of oral food processing. Environ Biol Fish 22:161–178. https://doi.org/10.1007/BF00005379

Swee UB, McCrimmon HR (1966) Reproductive biology of the carp, *Cyprinus carpio* L., in Lake St. Lawrence, Ontario. Trans Am Fish Soc 95:372–380. https://doi.org/10.1577/1548-8659(1966)95[372:RBOTCC]2.0.CO;2

Weber MJ, Brown ML (2009) Effects of common carp on aquatic ecosystems 80 years after "carp as a dominant": ecological insights for fisheries management. Rev Fish Sci 17:524–537. https://doi.org/10.1080/10641260903189243

Weber MJ, Brown ML (2011a) Comparison of common carp (*Cyprinus carpio*) age estimates derived from dorsal fin spines and pectoral fin rays. J Freshw Ecol 26:195–202. https://doi.org/10.1080/02705060.2011.554218

Weber MJ, Brown ML (2011b) Relationships among invasive common carp, native fishes and physiochemical characteristics in upper Midwest (USA) lakes. Ecol Freshw Fish 20:270–278. https://doi.org/10.1111/j.1600-0633.2011.00493.x

Weber MJ, Brown ML, Willis DW (2010) Spatial variability of common carp populations in relation to lake morphology and physiochemical parameters in the upper Midwest United States. Ecol Freshw Fish 19:555–565. https://doi.org/10.1111/j.1600-0633.2010.00436.x

Williams AE, Moss B (2003) Effects of different fish species and biomass on plankton interactions in a shallow lake. Hydrobiologia 491:331–346. https://doi.org/10.1023/A:1024456803994

6.13   Xenocyprididae, Sharpbelly Family

The Sharpbelly family, Xenocyprinidae, is represented in the Dakotas by three nonnative species, each of which is classified as an Aquatic Invasive/Nuisance Species. This family was recently separated from the Cyprinidae family because of genetic analysis. The 160 species in the Sharpbelly family are native to Asia, with four species introduced to North America. The large size of Xenocyprinids distinguishes them from Leuciscids (minnows). Each of the three Xenocyprinids found in the Dakotas can reach total lengths of over one meter.

Like Cyprinids and Leuciscids, Xenocyprinids lack teeth on both the upper and lower jaws but have well developed pharyngeal teeth arranged on the gill arches for mastication, have a single continuous dorsal fin, and lack an adipose fin. The three species of Xenocyprinids in the Dakotas lack barbels, unlike common carp (Cyprinidae) that has two barbels on each side of the mouth. Each of the four species of Xenocyprinids introduced to North America is long lived with high fecundities.

6.13.1   Grass Carp *Ctenopharyngodon idella*

(Valenciennes, 1844)

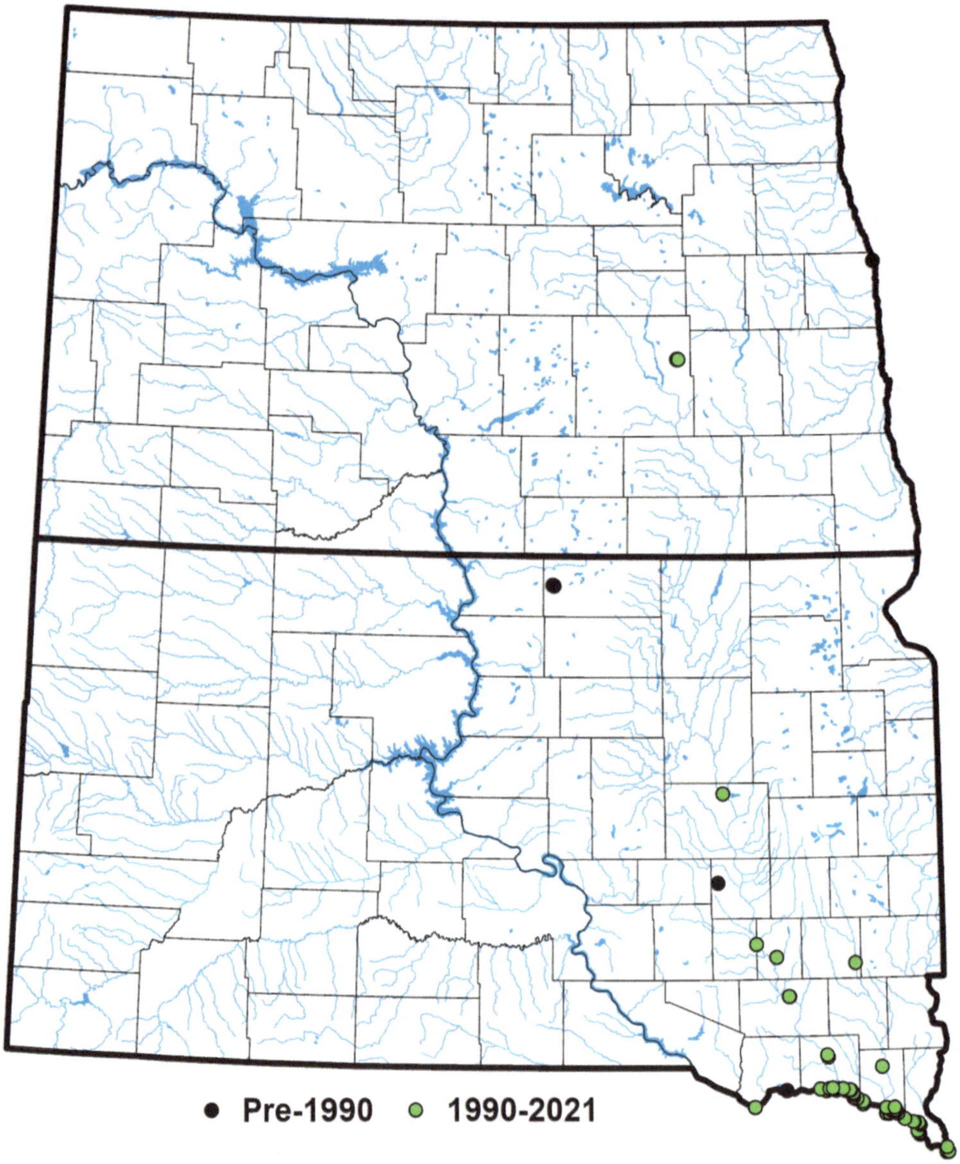

● **Pre-1990**   ○ **1990-2021**

Etymology: *Ctenopharyngodon idella* – *Ctenopharyngodon* = comb throat tooth, referencing the comb-like and sharp pharyngeal teeth; *idella* = distinct.

Description:

Body – slightly elongated, rather cylindrical.
Color –

> *Dorsally* – olive to gray.
> *Laterally* – light olive to silver or bronze; scales outlined in darker color appearing crosshatched.
> *Ventrally* – silver to white.
> *Fins* – dark olive to gray.

Head – broad, small.
Snout – slightly conical, short.
Eyes – moderately large, laterally in middle of head.
Mouth – large, terminal.

> *Jaws* – upper jaw barely extends backward toward anterior edge of eye.
> *Lips* – thin.

Teeth – absent on jaws; pharyngeal teeth deeply grooved, sharp, comb-like; 2,4-5,2 or 2,5-4,2.
Gill rakers – 15 or 16; short, unfused.
Dorsal fin – 7 to 9 rays; short.
Caudal peduncle – slightly elongated, thick.
Caudal fin – forked.
Anal fin – 8 to 10 rays; far posterior on body.
Pelvic fins – 7 or 8 rays; abdominal; insertion slightly posterior to dorsal fin insertion.
Pectoral fins – 18 to 20 rays.
Lateral line – complete; slightly decurved, 34 to 43 large cycloid scales in series.
Juveniles – may be more silver.
Sexual dimorphism – spawning males have minute tubercles on pectoral fins.

Similar Species: Grass carp are easily distinguished from common carp by their more cylindrical body shape and lack of barbels. Bighead carp and silver carp are deeper bodied and more laterally compressed, with much smaller cycloid scales. Buffalo (*Ictiobus*) species are more corpulent and have an elongate dorsal fin.

Distribution and Habitat: Although native to large rivers and lakes from southern Russia to northern China and Vietnam, grass carp have been introduced worldwide (Cudmore and Mandrak 2004). They were imported to the United States in 1963 for aquatic vegetation control and were accidentally released in 1966 by the United States Fish and Wildlife Service in Arkansas (Pflieger 1978; Kelley et al. 2011). Grass carp are now established in the Mississippi River basin and the Trinity River in Texas (Chapman and Hoff 2011). In the Dakotas, they exist east of the Missouri River. In 1984, a single specimen was collected from the Red River of the North in North Dakota. Grass carp inhabit shallow, densely vegetated areas of reservoirs, lakes, ponds, large river backwaters and pools, and occasionally small streams (Nixon and Miller 1978; Shireman and Smith 1983). They are most active during daylight and can negatively impact aquatic ecosystems by overgrazing native macrophytes (Nixon and Miller 1978). Grass carp tolerate water temperatures from 0 to 33 °C, as well as rapid temperature changes (Fedorenko and Fraser 1978; Cudmore and Mandrak 2004). Critical thermal maximum is 39 °C (Chilton III and Muoneke 1992; Cudmore and Mandrak 2004). The lower lethal temperature tolerance of fry is 0 °C, and upper lethal water temperatures range from 33 to 41 °C (Chilton III and Muoneke 1992; Cudmore and Mandrak 2004). Grass carp tolerate oxygen concentrations as low as 0.2 mg/L (Shireman and Smith 1983; Cudmore and Mandrak 2004).

Reproduction: Grass carp are opportunistic broadcast spawners (Stanley 1976). Spawning occurs during spring and early summer in water temperatures from 20 to 30 °C, in turbid and turbulent waters of main river channels and canals during high water (Fedorenko and Fraser 1978; Shireman and Smith 1983; Cudmore and Mandrak 2004). Grass carp have spawned in 15 °C water, but optimum spawning temperatures are from 20 to 22 °C (Shireman and Smith 1983; Cudmore and Mandrak 2004). Sexual maturity occurs at age-4 or age-5 in temperate areas of the United States (Cudmore and Mandrak 2004). Males typically mature one year prior to females (Cudmore and Mandrak 2004). During spawning, a single female is accompanied by two or more males (Cudmore and Mandrak 2004). Average fecundity is one million eggs per female (Aliev and Sukhanova 1974; Stanley 1976). No nest is constructed, and no parental care is given. Semi-buoyant, nonadhesive eggs are approximately 2 to 2.5 mm in diameter (Cudmore and Mandrak 2004). Optimum egg incubation temperatures range from 21 to 26 °C, with hatching occurring within 26 to 60 hours at 17 to 30 °C (Fedorenko and Fraser 1978; Shireman and Smith 1983; Cudmore and Mandrak 2004).

Age and Growth: In hatchery ponds, juvenile grass carp grew 9.8 g/day (Lembi et al. 1978). Under optimal conditions, grass carp often reach 1 kg during their first year (Vietmeyer 1976). In a Florida lake, 90 g grass carp grew 10 g/day the first year after stocking (Shireman et al. 1980). Growth rates of yearling grass carp x bighead carp hybrids in South Dakota was only one-third of similarly sized grass carp (Harberg and Modde 1985). Grass carp can exceed total lengths of 1 m and generally attain weights of from 30 to 50 kg (Cudmore and Mandrak 2004). Lifespans may be longer at higher latitudes with lower water temperatures and shorter growing seasons (Kirk and Socha 2003). Longevity is 15 years, although one triploid grass carp stocked in 1972 was recovered in late summer 2020, and otolith-aging verified as 48 years old (Allen and Wattendorf 1987; Kirk and Socha 2003; Cudmore and Mandrak 2004).

Food and Feeding: Grass carp larvae primarily feed on rotifers and other zooplankton (Stanley et al. 1978). Adult grass carp are mostly opportunistic herbivores, feeding heavily on submerged rooted aquatic macrophytes, benthic invertebrates, filamentous algae, and occasionally flooded terrestrial vegetation (Colle et al. 1978; Shireman and Smith 1981). Active feeding occurs mainly during the morning and evening, with maximum consumption at water temperatures of 22 or 23 °C (Opuszynski 1972; Shireman and Smith 1981). Grass carp may consume 110 to 169% of their body weight each day (Opuszynski 1972; Caldwell 1980). Hybrid daily food consumption rates are approximately one-third that of grass carp (Harberg and Modde 1985). In South Dakota, grass carp x bighead carp hybrids primarily fed diurnally in shallow waters and heavily consumed common water nymph *Najas guadalupensis,* muskgrass *Potamogeton pectinatus,* and sago pondweed *Chara* sp. (Harberg and Modde 1985). Grass carp feeding habits often negatively affect aquatic ecosystems, including native fish and waterfowl (McKnight and Hepp 1995; Cudmore and Mandrak 2004). Sterile triploid grass carp are often used to biologically control aquatic weeds (Allen and Wattendorf 1987; Kelley et al. 2011).

References

Aliev DS, Sukhanova AI (1974) Fecundity of the grass carp *Ctenopharyngodon idella* (Val.) and the silver carp *Hypophthalmichthys molitrix* (Val.) in the Kara-kum canal and its reservoirs. Izv Akad Nauk Turkm SSR Ser Biol Nauk 4:77–83

Allen SK, Wattendorf RJ (1987) Triploid grass carp: status and management implications. Fisheries 12:20–24. https://doi.org/10.1577/1548-8446

Caldwell BA (1980) Ability of grass carp to control aquatic macrophytes in fish culture ponds in Colorado. Dissertation, Colorado State University

Chapman DC, Hoff MH (2011) Introduction. In: Chapman DC, Hoff MH (eds) Invasive Asian carps in North America. American Fisheries Society, Symp 74, Bethesda, pp 1–4. https://doi.org/10.47886/9781934874233.ch1

Chilton EW III, Muoneke MI (1992) Biology and management of grass carp (*Ctenopharyngodon idella*, Cyprinidae) for vegetation control: a North American perspective. Rev Fish Biol Fish 2:283–320. https://doi.org/10.1007/BF00043520

Colle DV, Shireman JV, Rottman RW (1978) Food selection by grass carp fingerlings in a vegetated pond. Trans Am Fish Soc 107:149–152. https://doi.org/10.1577/1548-8659

Cudmore B, Mandrak NE (2004) Biological synopsis of grass carp (*Ctenopharyngodon idella*). Can Manuscr Rep Fish Aquat Sci 2705

Fedorenko AY, Fraser FJ (1978) Review of grass carp biology. Interagency Committee on Transplants and Introductions of Fish and Aquatic Invertebrates in British Columbia, Dep Fish Environ, Fish Marine Serv, Tech Rep No 786

Harberg MC, Modde T (1985) Feeding behavior, food consumption, growth, and survival of hybrid grass carp in two South Dakota ponds. N Am J Fish Manag 5:457–464. https://doi.org/10.1577/1548-8659

Kelley AM, Engle CR, Armstrong ML, Freeze M, Mitchell AJ (2011) History of introduction and governmental involvement in promoting the use of grass, silver, and bighead carps. In: Chapman DC, Hoff MH (eds) Invasive Asian carps in North America. American Fisheries Society, Symp 74, Bethesda, pp 163–174

Kirk JP, Socha RC (2003) Longevity and persistence of triploid grass carp stocked into the same Santee Cooper reservoirs of South Carolina. J Aquat Plant Manag 41:90–92

Lembi CA, Ritenuor BG, Iverson EM, Forss EC (1978) The effects of vegetation removal by grass carp on water chemistry and phytoplankton in Indiana ponds. Trans Am Fish Soc 107:161–171. https://doi.org/10.1577/1548-8659

McKnight SK, Hepp GR (1995) Potential effect of grass carp herbivory on waterfowl foods. J Wildl Manag 59:720–727. https://doi.org/10.2307/3801948

Nixon DE, Miller RL (1978) Movements of grass carp, *Ctenopharyngodon idella*, in an open reservoir system as determined by underwater telemetry. Trans Am Fish Soc 107:146–148. https://doi.org/10.1577/1548-8659

Opuszynski K (1972) Use of phytophagous fish to control aquatic plants. Aquaculture 1:61–74. https://doi.org/10.1016/0044-8486(72)90008-7

Pflieger WL (1978) Distribution and status of the grass carp (*Ctenopharyngodon idella*) in Missouri streams. Trans Am Fish Soc 107:113–118. https://doi.org/10.1577/1548-8659

Shireman JV, Smith CR (1981) Biological synopsis of grass carp (*Ctenopharyngodon idella*). Final Report, contract number 14-16-0009-78-912. U.S. Fish and Wildlife Service, Washington, DC

Shireman JV, Smith CR (1983) Synopsis of biological data on the grass carp *Ctenopharyngodon idella* (Cuvier and Valenciennes, 1844). FAO Fish Synop 135

Shireman JV, Colle DE, Maceina MJ (1980) grass carp growth rates in Lake Wales, Florida. Aquaculture 19:379–382. https://doi.org/10.1016/0044-8486(80)90086-1

Stanley JG (1976) Reproduction of the grass carp (*Ctenopharyngodon idella*) outside its native range. Fisheries 1:7–10. https://doi.org/10.1577/1548-8446

Stanley JG, Miley WW II, Sutton DL (1978) Reproductive requirements and likelihood for naturalization of escaped grass carp in the United States. Trans Am Fish Soc 107:119–128. https://doi.org/10.1577/1548-8659

Vietmeyer ND (1976) grass carp. In: Vietmeyer ND (ed) Making aquatic weeds useful: some perspectives for developing countries. National Academy of Sciences, Washington, DC, pp 15–21

6.13.2   Silver Carp *Hypophthalmichthys molitrix*

(Valenciennes, 1844)

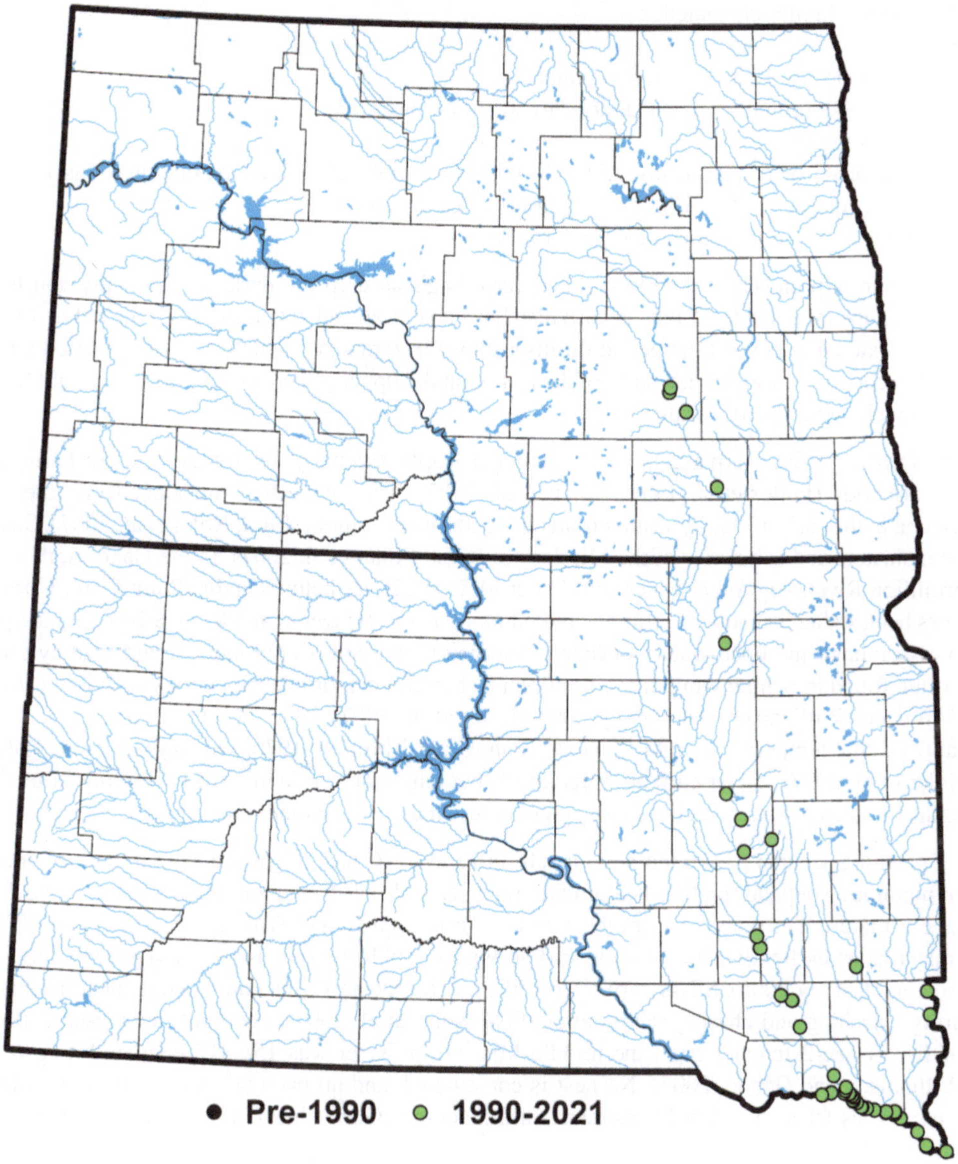

● **Pre-1990**   ○ **1990-2021**

Etymology: *Hypophthalmichthys molitrix* – *Hypophthalmichthys* = fish with eye under or below; *molitrix* = miller or grinder, referencing the pharyngeal apparatus grinding ability.

Description:

Body – fusiform, deep, robust, moderately laterally compressed.
Color –

> *Dorsally* – light gray to olive; head gray to olive.
> *Laterally* – solid silver.
> *Ventrally* – silver to white.
> *Fins* – pelvic: somewhat transparent to white.

Head – large, scaleless.
Snout – short.
Eyes – laterally on lower anterior end of head.
Mouth – large, terminal.

> *Jaws* – lower jaw protrudes forward and beyond upper jaw; upper jaw does not extend to anterior end of eye.
> *Lips* – fleshy.

Teeth – absent on jaws; pharyngeal teeth 0,4-4,0.
Gill rakers – highly modified, fused, sponge like.
Dorsal fin – 1 smooth spinous ray, 8 soft rays.
Caudal peduncle – thick, slightly elongated.
Caudal fin – forked.
Anal fin – 1 smooth spinous ray, 12 to 14 soft rays; falcated.
Pelvic fins – 7 or 8 rays; abdominal; insertion anterior to dorsal fin insertion.
Keel – extends from gills to vent.
Pectoral fins – 1 posterior serrated spinous ray, 15 to 18 soft rays; slightly elongated, extending to insertion of pelvic fins.
Lateral line – decurved, complete; 85 to 100 small cycloid scales in series.
Juveniles – lack spinous rays in dorsal, anal, and pectoral fins.

Similar Species: Silver carp closely resemble bighead carp. Bighead carp are dark gray dorsally and have small, dark, irregular blotches along the lateral sides. Bighead carp also have a ventral keel extending from the pelvic fins to the vent and elongate pectoral fins extending beyond the pelvic fin insertions. Gizzard shad, which may be confused with smaller silver carp, have a long filament on the dorsal fin, a dark to black spot on the upper gill cover, and lack a lateral line. Grass carp and common carp have much larger cycloid scales.

Distribution and Habitat: Silver carp are native to the major Pacific drainages from Russia to northern Vietnam but have been introduced worldwide (Fuller et al. 1999; Kolar et al. 2007). They were first introduced into the United States in Arkansas and Arizona in the early 1970s by commercial aquaculture (Kuronuma 1968; Kolar et al. 2007). Silver carp are now established in the mainstem and tributaries of the Mississippi River (Kolar et al. 2007). In the Dakotas, they are found in the James River, Vermillion River, Big Sioux River, and lower Missouri River below Gavins Point Dam. Upstream dispersion into these rivers has been stalled because of artificial and natural barriers (Hayer et al. 2014). Silver carp are primarily a large river species but also inhabit the open waters of lakes, reservoirs, and ponds (Williamson and Garvey 2005; Kolar et al. 2007). They are often found in behind wing dikes, floodplains, backwaters, or other locations with low velocities and water depths at least 3 m deep (Williamson and Garvey 2005; Kolar et al. 2007). The optimum temperature for larvae is 26 to 34 °C, with an upper lethal temperature of 44 to 47 °C (Panov and Khromov 1970; Opuszynski et al. 1989; Radenko and Alimov 1992; Kolar et al. 2007). Silver carp are tolerant of low temperatures. While overwintering in pools, their activity slows at temperatures less than 4 °C, with little movement at less than 2 °C (Kolar et al. 2007).

Reproduction: Spawning occurs in water temperatures from 17 to 27 °C in spring and early summer (Schofield et al. 2005). Silver carp age of maturation depends on first-year growth rates (Kamilov 1987). In their native range, sexual maturity occurs by age-3, with age-2 mature fish reported from the middle Mississippi River (Williamson and Garvey 2005; Kolar et al. 2007). Females may mature earlier than males in the middle Mississippi River because of rapid growth rates during early-life stages (Williamson and Garvey 2005). Silver carp are prolific batch spawners. Spawning occurs near the surface, with gametes released during splashing and chasing (Kuronuma 1968; Kolar et al. 2007). Fecundity increases with female age and size. Mean fecundity of age-2 females from the middle Mississippi River was 156,312 eggs and ranged from 57,283 to 328,538 eggs (Williamson and Garvey 2005). No nest is constructed, and no parental care is given. Semibuoyant eggs are carried with the current as far as 500 km downstream until hatching (Gorbach and Krykhtin 1989; Laird and Page 1996;

Kolar et al. 2007). The diameter of silver carp eggs is approximately 1 mm before fertilization, and 5 or 6 mm after water hardening (Kolar et al. 2007; Chapman and George 2011). Hatching occurs within 21 to 33 hours in warmer waters (Chapman and George 2011).

Age and Growth: Maximum growth of silver carp occurs in water temperatures of 24 to 34 °C (Mahboob and Sheri 1997; Kolar et al. 2007). Larvae are approximately 5 or 6 mm long at hatching (Chapman and George 2011). They can reach a total length of 400 mm and weigh 270 g by the end of the first growing season and can gain 45 g or more per month (Waterman 1997; Stone et al. 2000; Hayer et al. 2014). Fast initial growth rates are followed by slower growth with increasing age (Williamson and Garvey 2005; Hayer et al. 2014). Growth is primarily influenced by food availability (Kolar et al. 2007). Total lengths at age of silver carp from the middle Mississippi River were age-1, 318 mm; age-2, 531 mm; age-3, 650 mm; age-4, 704 mm; age-5, 723 mm (Williamson and Garvey 2005). Female silver carp are generally heavier than males (Williamson and Garvey 2005). They can reach a total length of 1,260 mm (Kolar et al. 2007). While longevity can be over 15 years for silver carp in their native range, they often reach age-5 in the Mississippi River and its tributaries (Williamson and Garvey 2005; Kolar et al. 2007; Hayer et al. 2014).

Food and Feeding: Silver carp are opportunistic planktivorous filter feeders. Larvae eat rotifers, protozoans, and other small zooplankton, with zooplankton size increasing with larvae size (Marciak and Bogdan 1979; Kolar et al. 2007). Adults primarily consume phytoplankton and zooplankton, with and detritus and bacteria eaten to a lesser extent (Kolar et al. 2007). Silver carp feeding often shifts plankton community composition from larger to smaller species, which can have detrimental effects on native fish (Kolar et al. 2007; Sampson et al. 2009). They can eat smaller particles than bighead carp because their epibranchial organ is smaller (Kolar et al. 2007). Active feeding occurs in water temperatures from 15 to 30 °C, but feeding may continue in temperatures as low as 4 °C (Kolar et al. 2007).

References

Chapman DC, George AE (2011) Developmental rate and behavior of early life stages of bighead and silver carp. U.S. Geological Survey Scientific Investigations Report 2011-5076. https://doi.org/10.3133/sir20115076

Fuller PL, Nico LG, Williams JD (1999) Non-indigenous fishes introduced into inland waters of the United States. American Fisheries Society, Special Publication 27, Bethesda

Gorbach EI, Krykhtin ML (1989) Migration of the white amur, *Ctenopharyngodon idella*, and silver carp, *Hypophthalmichthys molitrix*, in the Amur River Basin. J Ichthyol 28:47–53

Hayer CA, Graeb BDS, Bertrand KN (2014) Adult, juvenile and young-of-year bighead, *Hypophthalmichthys nobilis* (Richardson, 1845) and silver carp, *H. molitrix* (Valenciennes, 1844) range expansion on the northwestern front of the invasion in North America. BioInvasions Records 3:283–289. https://doi.org/10.3391/bir.2014.3.4.10

Kamilov BG (1987) Gonad condition of female silver carp *Hypophthalmichthys molitrix* in relation to growth rate in Uzbekistan. J Ichthyol 27:135–139

Kelley AM, Engle CR, Armstrong ML, Freeze M, Mitchell AJ (2011) History of introduction and governmental involvement in promoting the use of grass, silver, and bighead carps. In: Chapman DC, Hoff MH (eds) Invasive Asian carps in North America. American Fisheries Society, Symposium 74, Bethesda, pp 163–174

Kolar CS, Chapman DC, Courtenay WR Jr, Housel CM, Williams JD, Jennings DP (2007) bighead carps: a biological synopsis and environmental risk assessment. American Fisheries Society, Special Publication 33, Bethesda

Kuronuma K (1968) New systems and new fishes for culture in the Far East. FAO Fish Rep 5:123

Laird CA, Page LM (1996) Non-native fishes inhabiting the streams and lakes of Illinois. Ill Nat Hist Surv Bull 35. https://doi.org/10.5962/bhl.title.50306

Mahboob S, Sheri AN (1997) Growth performance of major, common and some Chinese carps under composite culture system with special reference to pond fertilization. J Aquac Tropics 12:201–207

Marciak Z, Bogdan E (1979) Food requirements of juvenile stages of grass carp, *Ctenopharyngodon idella* Val., silver carp, *Hypophthalmichthys molitrix* Val., and bullhead carp *Aristichthys nobilis* Rich. EMS Special Publication 4:139–157

Opuszynski K, Lirski A, Myszkowski L, Wolnicki J (1989) Upper lethal and rearing temperatures for juvenile common carp, *Cyprinus carpio* L. and silver carp, *Hypophthalmichthys molitrix* (Valenciennes). Aquac Fish Manag 20:287–294. https://doi.org/10.1111/j.1365-2109.1989.tb00354.x

Panov DA, Khromov LV (1970) Why the larvae die. Rybovod i Rybolov 6:12–13

Radenko VN, Alimov IA (1992) Significance of temperature and light for growth and survival of larvae of silver carp, *Hypophthalmichthys molitrix*. J Ichthyol 32:16–27

Sampson SJ, Chick JH, Pegg MA (2009) Diet overlap among two Asian carp and three native fishes in backwater lakes on the Illinois and Mississippi rivers. Biol Invasions 11:483–496. https://doi.org/10.1007/s10530-008-9265-7

Schofield PJ, Williams JD, Nico LG, Fuller P, Thomas MR (2005) Foreign nonindigenous carps and minnows (Cyprinidae) in the United Sates-a guide to their identification, distribution and biology. US Geological Survey Scientific Investigations Report 2005-5041. https://doi.org/10.3133/sir20055041

Stone N, Engle C, Heikes D, Freeman D (2000) bighead carp. Southern Regional Aquaculture Center (SRAC), Publication 428, Stoneville

Waterman MP (1997) Chinese bighead carp continues to draw interest. Aquac Mag 23:73–79

Williamson CJ, Garvey JE (2005) Growth, fecundity, and diets of newly established silver carp in the middle Mississippi River. Trans Am Fish Soc 134:1423–1430. https://doi.org/10.1577/T04-106.1

6.13.3   Bighead Carp *Hypophthalmichthys nobilis*

(Richardson, 1845)

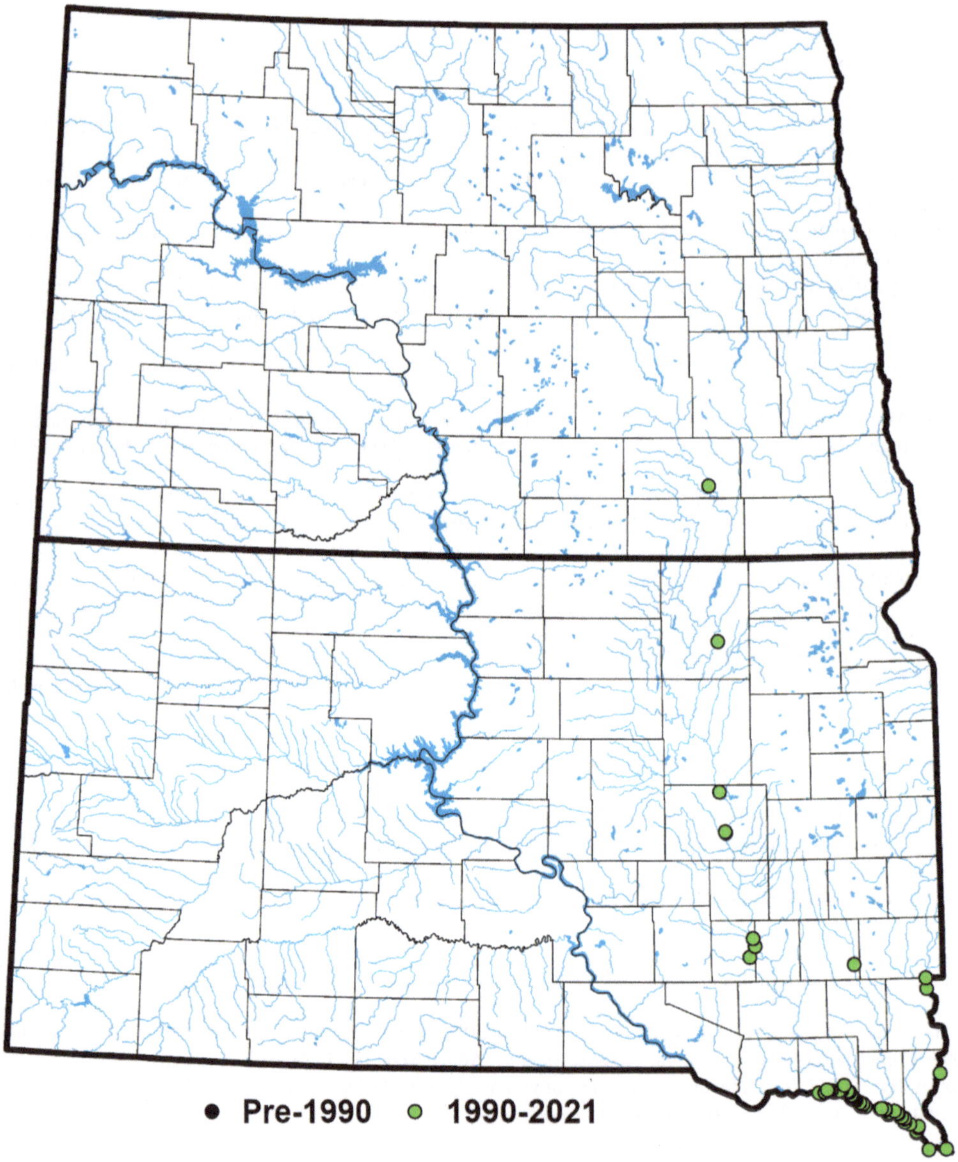

● **Pre-1990**   ○ **1990-2021**

Etymology: *Hypophthalmichthys nobilis* – *Hypophthalmichthys* = fish with eye under or below; *nobilis* = noble.

Description:

Body – fusiform, deep, robust, moderately laterally compressed.
Color –

> *Dorsally* – dark gray; head gray.
> *Laterally* – silver; numerous, small, dark, irregular blotches.
> *Ventrally* – silver to white.
> *Fins* – dark gray.

Head – large, scaleless.
Snout – short.
Eyes – laterally on lower anterior end of head
Mouth – large, terminal.

> *Jaws* – lower jaw protrudes far forward and beyond upper jaw; upper jaw does not extend backward to anterior end of eye.
> *Lips* – fleshy.

Teeth – absent on jaws; pharyngeal teeth long, 0,4-4,0.
Gill rakers – slender, long.
Dorsal fin – 1 smooth spinous ray, 8 to 10 rays.
Caudal peduncle – thick, slightly elongated.
Caudal fin – forked.
Anal fin – 1 smooth spinous ray, 13 or 14 soft rays; falcated.
Pelvic fins – 7 to 9 rays; abdominal; insertion anterior to insertion of dorsal fin.
Keel – smooth, extending from pelvic fins to vent.
Pectoral fins – 1 spinous ray, 16 to 21 soft rays; elongated, extending past pelvic fin insertion.
Lateral line – decurved, complete; 95 to 120 small cycloid scales in series.
Juveniles – more silver dorsally than adults; lack spinous rays in dorsal, anal, and pectoral fins.
Sexual dimorphism – adult males have sharp ridges on several anterior pectoral fin rays.

Similar Species: Bighead carp closely resemble silver carp. Silver carp have solid silver lateral sides without any dark irregular blotches, a well-developed ventral keel extending from the gills to the vent, and pectoral fins that only extend to the pelvic fin insertions. Juvenile bighead carp may be mistaken for golden shiner, which have smaller and fewer (42 to 54) scales in lateral line series. Grass carp and common carp have much larger cycloid scales.

Distribution and Habitat: Originally native to eastern China, eastern Siberia, and extreme eastern North Korea, bighead carp have since been introduced worldwide (Kolar et al. 2007). They were imported into the United States in 1973 for plankton control in catfish ponds, water quality improvement in sewage treatment ponds, and for human consumption (Henderson 1976; Kolar et al. 2007; Shearer 2007; O'Connell et al. 2011). They escaped through flooding or accidental release into the Mississippi River basin and are now established in the Mississippi, Missouri, Ohio, Illinois, and Tennessee rivers (Kolar et al. 2007; Shearer 2007). Bighead carp swam up the Missouri River to South Dakota in the late 1990s and now inhabit the James River, Vermillion River, Big Sioux River, and the Missouri River below Gavins Point Dam (Kolar et al. 2007; Shearer 2007). Upstream dispersion into these rivers has been stalled because of artificial and natural barriers (Hayer et al. 2014). Bighead carp are primarily a large river species but also inhabit lakes, reservoirs, and ponds (Kolar et al. 2007). They are often found behind spur dikes and wing dikes in velocities at or below 0.3 m/second and water depths at least 3 m deep (Kolar et al. 2007). In the Illinois River, water temperature was negatively correlated with movement, suggesting less movement in water exceeding their 26 °C optimum growth temperature (DeGrandchamp et al. 2008). Juveniles are often associated with the backwaters of large rivers rather than main channels (Kolar et al. 2007). Bighead carp are tolerant of water temperature extremes. Young bighead carp prefer water temperatures of 25 to 27 °C and have a critical thermal maximum of 39 °C (Bettoli et al. 1985; Kolar et al. 2007). In the Missouri River, tributary use is highest in the winter. Activity slows at temperatures less than 4 °C, with little movement at less than 2 °C (Kolar et al. 2007).

Reproduction: Bighead carp may migrate more than 80 km upstream to spawning grounds of rapid and turbid areas often found near the confluence of rivers among rocks or sandbars (Jennings 1988; Kolar et al. 2007). Spawning occurs during spring through late summer in the United States (Kolar et al. 2007). Increased water discharge and 22 °C water temperature induced bighead carp spawning activity in the Missouri River (Schrank et al. 2001; Kolar et al. 2007). Sexual maturity begins at least by age-3 in the Missouri River (Schrank and Guy 2002). Bighead carp are prolific spawners. Spawning occurs near the water surface, with two or more males actively chasing and head-butting females (Kolar et al. 2007). Fecundity increases with female size and age. In the lower Missouri River, fecundity averaged 226,213 eggs and ranged from 11,588 to 769,964 eggs (Schrank and Guy 2002). No nest is constructed, and no parental care is given. Egg diameter is typically from 0.5 to 1.2 mm but may reach 1.5 mm from fish in the Missouri River (Schrank and Guy 2002). The semibuoyant eggs are carried with the current until hatching in 24 hours at 28 °C or 30 to 32 hours at 25 °C (Bardach et al. 1972; Kolar et al. 2007).

Age and Growth: Growth rates of bighead carp in the Missouri River are rather fast, peak between age-2 and age-3, and decrease at older ages (Schrank and Guy 2002). Mean lengths are 550 mm at age-3 and 700 mm at age-5 (Schrank and Guy 2002). Bighead carp can reach a total length of 1,100 mm and a weight of 18.1 kg (Howells 2001; Schrank and Guy 2002; Kolar et al. 2007). Longevity is 10 years in the Mississippi River (Garvey et al. 2006).

Food and Feeding: All life stages of bighead carp are opportunistic planktivores. Large quantities of zooplankton are filtered through gill rakers. They have been classified as pump feeders, using buccal pumping to push water through the gill rakers, and ram suspension feeders, using an open mouth while swimming to force water through the gill rakers (Sanderson et al. 1994; Lu and Xie 2001; Kolar et al. 2007). Bighead carp feed near the surface during the evening and night within a wide range of water temperatures and have been observed with full stomachs in the Missouri River at water temperatures as low as 3 °C (Kolar et al. 2007). When zooplankton abundance is low, bighead carp will switch to eating phytoplankton (Kolar et al. 2007). Bighead carp feeding often influences plankton community size structure, potentially negatively affecting paddlefish, gizzard shad, bigmouth buffalo, and other fish species heavily dependent on plankton (Schrank et al. 2003; Sampson et al. 2009).

References

Bardach JE, Ryther JH, McLarney WO (1972) Aquaculture- the farming and husbandry of freshwater and marine organisms. Wiley-Interscience, New York

Bettoli PW, Neill WH, Kelsch SW (1985) Temperature preference and heat resistance of grass carp *Ctenopharyngodon idella* (Valenciennes), bighead carp *Hypophthalmichthys nobilis* (Gray), and their F1 hybrid. J Fish Biol 27:239–247. https://doi.org/10.1111/j.1095-8649.1985.tb04024.x

DeGrandchamp KL, Garvey JE, Colombo RE (2008) Movement and habitat selection by invasive Asian carps in a large river. Trans Am Fish Soc 137:45–56. https://doi.org/10.1577/T06-116.1

Garvey JE, DeGrandchamp KL, Williamson CJ (2006) Life history attributes of Asian carps in the upper Mississippi River system. ANSRP Technical Notes Collection (ERDC/ EL ANSRP-07-1), US Army Corps of Engineers Research and Development Center, Vicksburg. https://doi.org/10.21236/ADA468471

Hayer CA, Graeb BDS, Bertrand KN (2014) Adult, juvenile and young-of-year bighead, *Hypophthalmichthys nobilis* (Richardson, 1845) and silver carp, *H. molitrix* (Valenciennes, 1844) range expansion on the northwestern front of the invasion in North America. Biol Inv Rec 3:283–289. https://doi.org/10.3391/bir.2014.3.4.10

Henderson S (1976) Observations on the bighead and silver carp and their possible application in pond fish culture. Arkansas Game and Fish Commission, Little Rock

Howells RG (2001) Introduced non-native fishes and shellfishes in Texas waters: an updated list and discussion. Texas Parks and Wildlife Department Management Data Series 188

Jennings DP (1988) bighead carp (*Hypophthalmichthys nobilis*): a biological synopsis. US Fish Wildl Serv Biol Rep 88:1–47

Kolar CS, Chapman DC, Courtenay WR Jr, Housel CM, Williams JD, Jennings DP (2007) bighead carps: a biological synopsis and environmental risk assessment. American Fisheries Society, Special Publication 33, Bethesda

Lu M, Xie P (2001) Impacts of filter-feeding fishes on the long-term changes in crustacean zooplankton in a eutrophic subtropical Chinese lake. J Freshw Ecol 16:219–228. https://doi.org/10.1080/02705060.2001.9663806

O'Connell MT, O'Connell AU, Barko VA (2011) Occurrence and predicted dispersal of bighead carp in the Mississippi River system: development of a heuristic tool. In: Chapman DC, Hoff MH (eds) Invasive Asian carps in North America. American Fisheries Society, Symposium 74, Bethesda, pp 51–72

Sampson SJ, Chick JH, Pegg MA (2009) Diet overlap among two Asian carp and three native fishes in backwater lakes on the Illinois and Mississippi river. Biol Invasions 11:483–496. https://doi.org/10.1007/s10530-008-9265-7

Sanderson SL, Cech JJ Jr, Cheer AY (1994) Paddlefish buccal flow velocity during ram suspension feeding and ram ventilation. J Exp Biol 186:145–156. https://doi.org/10.1242/jeb.186.1.145

Schrank SJ, Guy CS (2002) Age, growth, and gonadal characteristics of adult bighead carp, *Hypophthalmichthys nobilis*, in the lower Missouri River. Environ Biol Fishes 64:443–450. https://doi.org/10.1023/A:1016144529734

Schrank SJ, Guy CS, Fairchild JF (2003) Competitive interactions between age-0 bighead carp and paddlefish. Trans Am Fish Soc 132:1222–1228. https://doi.org/10.1577/T02-071

Schrank SJ, Braaten PJ, Guy CS (2001) Spatiotemporal variation in density of larval bighead carp in the lower Missouri River. Trans Am Fish Soc 130:809–814. https://doi.org/10.1577/1548-8659

Shearer JS (2007) Exotic species. In: Berry C, Higgins K, Willis D, Chipps S (eds) History of fisheries and fishing in South Dakota. S D Dep Game Fish Parks, Pierre, pp 253–266

6.14 Leuciscidae, Minnow Family

The minnow family, Leuciscidae, is in the order Cypriniformes, the most diverse order of freshwater fishes. It includes over 650 species native to Europe, Asia, and North America, and northern Africa. It is the largest family of fishes in the Dakotas with 38 species found in the two states. Prior to recent genetic analysis, leuciscids, along with xenocyprinids, used to be part of the family Cyprinidae.

While Leuciscidae is generally referred to as the minnow family, it should not be confused with the lay use of minnow to refer to any small fish. Although most of the family comprises small fishes, typically under 150 mm, leuciscids can be rather sizeable fish. The Colorado pikeminnow *Ptychocheilus Lucius* is the largest leuciscid native to North America and can reach lengths of 1.8 m. Similar to cyprinids and xenocyprinids, leusiscids lack teeth on both the upper and lower jaws and use well-developed pharyngeal teeth arranged on the gill arches in one or two rows for mastication. In addition, they have a single, continuous dorsal fin without spines but lack an adipose fin. Leuciscids also have a Weberian apparatus. This structure, derived from the first few vertebrae and connecting the swim bladder to the inner ear, amplifies sound waves. It enhances hearing and aids in predator evasion.

Leuciscids are widely distributed, inhabit a variety of habitats, and possess diverse morphological features to meet their trophic needs.

6.14.1   River Shiner *Alburnops blennius*

Girard, 1856

© Brian Zimmerman

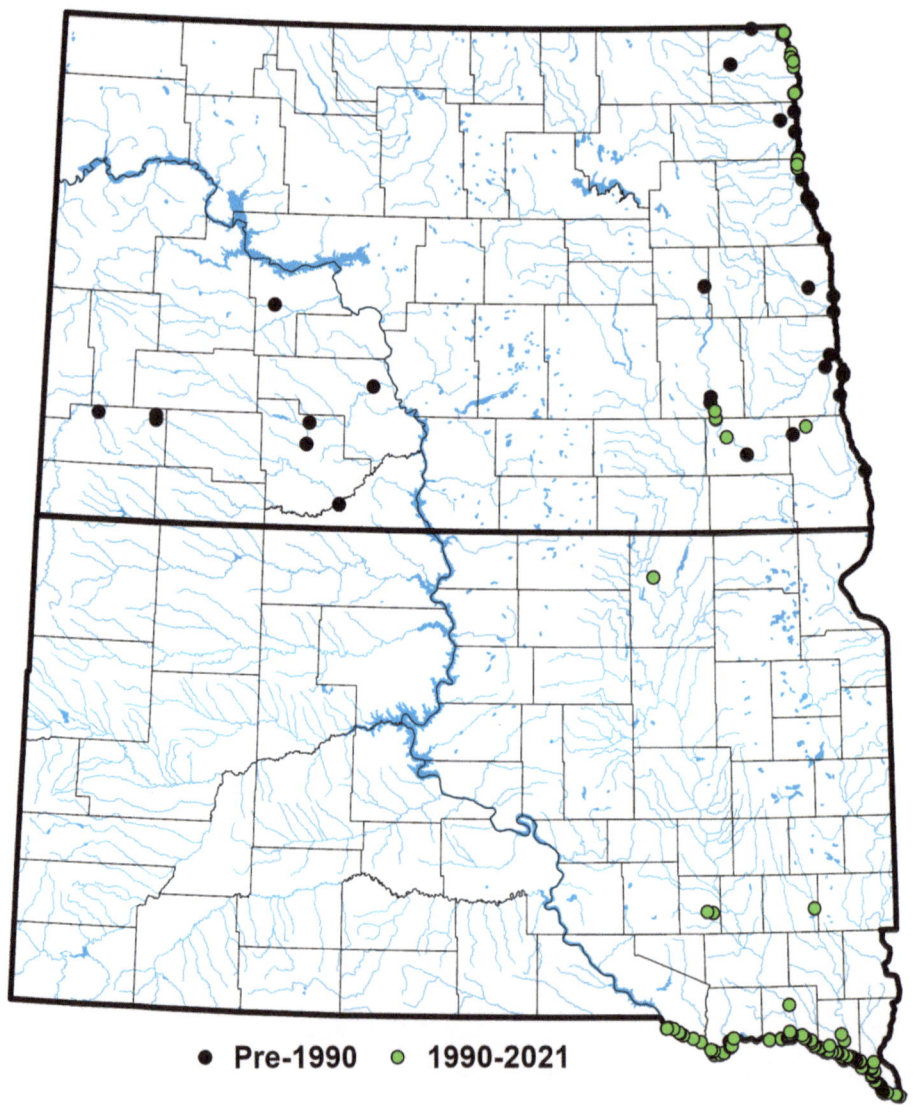

● **Pre-1990**   ○ **1990-2021**

Etymology: *Alburnops blennius – Alburnops* = albu meaning white or pale, ops resembling the European genus *Alburnus*; *blennius* = mucus or slime.

Description:

Body – fusiform, robust, slightly laterally compressed.
Color –

> *Dorsally* – tan to dusky silver; scales faintly outlined in darker pigment; dark, well-defined mid-dorsal stripe surrounding the dorsal fin base.
> *Laterally* – silver; no distinct markings.
> *Ventrally* – silver to white.
> *Fins* – clear.

Head – moderately large.
Snout – bluntly rounded, slightly protruding beyond the mouth.
Eyes – large, laterally on head.
Mouth – large, terminal, slightly oblique.
> *Jaws* – upper jaw extends to or slightly beyond anterior end of eye; greater in length than eye diameter.
Teeth – pharyngeal tooth pattern variable between 1,4-4,1 and 2,4-4,2.
Gill rakers – 6 to 9; short.
Dorsal fin – 8 rays; pointed tip; distal end nearly straight to slightly convex.
Caudal peduncle – slightly elongated, uniform in thickness.
Caudal fin – moderately forked.
Anal fin – 7 rays; straight distal end.
Pelvic fins – 8 rays; abdominal; insertions directly inferior to dorsal fin insertion.
Pectoral fins – 13 to 15 rays.
Lateral line – complete; 32 to 36 cycloid scales in series; slightly decurved.
Sexual dimorphism – spawning males have small tubercles on the dorsal side of the head, pectoral fins, and anterior ends of the dorsal and anal fins.

Similar Species: River shiner closely resemble sand shiner, emerald shiner, bigmouth shiner, channel shiner, and silverband shiner. Sand shiner have small, paired, dark specks, or "mouse tracks" outlining the lateral line and a faint, dusky spot sometimes present at the caudal fin base. Emerald shiner have an anal fin with 10 to 13 rays and a dorsal fin insertion that is distinctly posterior to the pelvic fin insertions. Bigmouth shiner have a large, slightly subterminal mouth with the lower jaw extending to the middle of the eye and an upper jaw length greater than the eye diameter. Channel shiner have a lateral stripe expanding ventrally at the caudal fin base. Silverband shiner have an anal fin with 8 or 9 rays, pelvic fins with 9 rays, and pelvic fin insertions slightly posterior to the dorsal fin insertion.

Distribution and Habitat: The river shiner native range encompasses the Hudson Bay from Manitoba to Alberta, Canada, and the Red River, Mississippi, Middle Missouri, Ohio, and Arkansas rivers (Lee et al. 1980). In North Dakota, river shiner primarily occur in the Red River of North drainage and Sheyenne River system (Owen et al. 1981). In South Dakota, they are confined to the lower Missouri River below Fort Randall dam and the lower reaches of the Vermillion and James river systems (Bailey and Allum 1962). River shiner are typically found schooling in the midwaters of large lakes and large, broad channeled rivers, in addition to the lower sections of main tributaries with clear to turbid water and moderate to swift current over gravel and sand bars (Cross 1967; Trautman 1981; Becker 1983; Pflieger 1997). They can withstand prolonged periods of increased turbidity and typically avoid areas with strong current (Pflieger 1997).

Reproduction: Little information is known about river shiner spawning and reproductive habits. Spawning is prolonged, occurring from June to August (Cross 1967; Becker 1983; Pflieger 1997). Males become sexually mature at age-1 and females at age-2 (Becker 1983). Spawning occurs over sand and gravel substrate (Whitaker Jr 1977). River shiner are fractional spawners, with females producing multiple clutches of eggs over the spawning season (Hatch and Elias 2002). Mature females from the Upper Mississippi river with standard lengths of 48 to 88 mm produced clutches of 436 to 2,754 eggs (Hatch and Elias 2002). Clutch size is positively correlated with female length (Hatch and Elias 2002). Yellow eggs are approximately 1 mm in diameter (Hatch and Elias 2002).

Age and Growth: Total lengths at age of river shiner from the Mississippi River in Wisconsin were 37 mm at age-0, 58 mm at age-1, 77 mm at age-2, 87 mm at age-3, and 97 mm at age-4 (Becker 1983). They can reach lengths of 132 mm and have a longevity of 4 years (Trautman 1981; Becker 1983).

Food and Feeding: River shiner are omnivorous. Adults primarily eat larvae and adult chironomids, trichopterans, corixids, and other aquatic insects, but may also consume algae, microcrustaceans, other plant material, and terrestrial insects (Cross 1967; Whitaker Jr 1977).

References

Bailey RM, Allum MO (1962) Fishes of South Dakota. University of Michigan, Ann Arbor. https://doi.org/10.3998/mpub.9690435

Becker GC (1983) Fishes of Wisconsin. University of Wisconsin Press, Madison

Cross FB (1967) Handbook of fishes in Kansas. University Press of Kansas, Lawrence

Hatch JT, Elias EE (2002) Ovarian cycling, clutch characteristics, and oocyte size of the river shiner *Notropis blennius* (Girard) in the Upper Mississippi River. J Freshw Ecol 17:85–92. https://doi.org/10.1080/02705060.2002.9663871

Lee DS, Gilbert CR, Hocutt CH, Jenkins RE, McAllister DE, Stauffer JR (1980) Atlas of North American freshwater fishes. North Carolina Biological Survey Publication 1980-12. https://doi.org/10.5962/bhl.title.141711

Owen JB, Elsen DS, Russell GW (1981) Distribution of fishes in North and South Dakota Basins affected by the Garrison Diversion Unit. University of North Dakota Press, Grand Forks

Pflieger WL (1997) The fishes of Missouri, revised edn. Missouri Dep Cons, Jefferson City

Trautman MB (1981) The fishes of Ohio, revised edn. Ohio State University Press, Columbus

Whitaker JO Jr (1977) Seasonal changes in food habits of some cyprinid fishes from the White River at Petersburg, Indiana. Am Midl Nat 97:411–418. https://doi.org/10.2307/2425105

6.14.2 Central Stoneroller *Campostoma anomalum*

(Rafinesque, 1820)

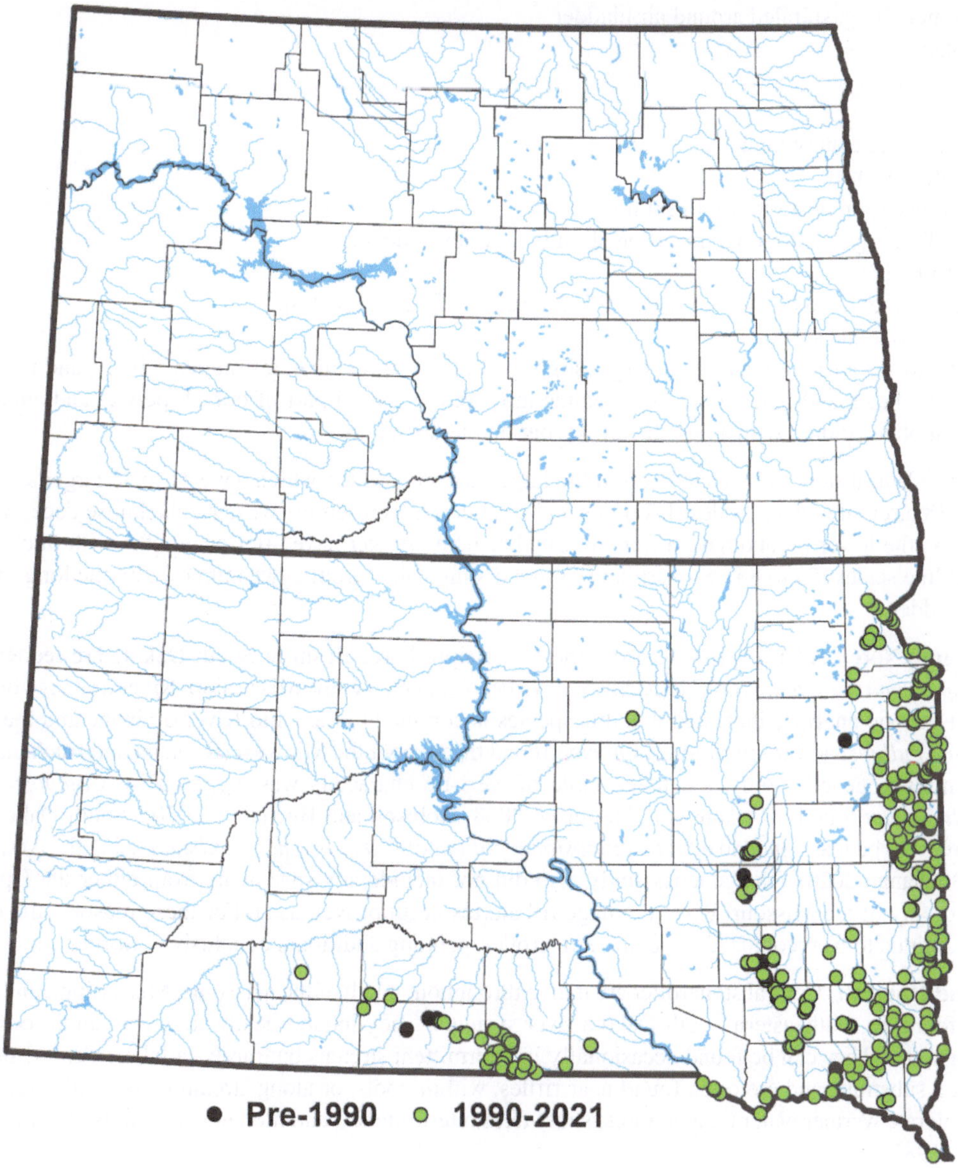

Etymology: *Campostoma anomalum* – ***Campostoma*** = curved mouth; ***anomalum*** = extraordinary.

Description:

Body – elongated, robust, nearly oval to round anteriorly.
Color –

      Dorsally – dark olive to gray.
      Laterally – olive to silver; dark mottling, sometimes faint lateral stripe.
      Ventrally – silver to white.
      Eyes – tan to amber.
      Fins – dorsal/anal: typically clear with dark bands; caudal – dark dusky spot.

Head – short, blunt.
Snout – rounded, slightly protrudes past mouth.
Mouth – subterminal, crescent shaped.
      Jaws – upper jaw short.
      Lips – fleshy; lower lip with inner cartilaginous ridge.

Teeth – pharyngeal teeth 0,4-4,0; slightly hooked.
Intestine – extremely long, spiraled around air bladder.
Peritoneum – black.
Gill rakers – 21 to 33.
Dorsal fin – 8 rays.
Caudal peduncle – moderately thick.
Caudal fin – slightly forked.
Anal fin – 7 rays; insertion posterior to dorsal fin.
Pelvic fins – usually 8 rays; abdominal, insertion slightly anterior to dorsal fin.
Pectoral fins – 15 rays.
Lateral line – complete; 47 to 60 cycloid scales; 38 to 50 circumferential scales.
Juveniles – dark, horizontal stripe on sides.
Sexual dimorphism – spawning males have large, slightly hooked, white tubercles on head, body, and median fins, minor hump behind the head, and black and orange bands through the central part of dorsal, pelvic, and anal fins; spawning females remain olive colored with a possible dusky bar on the dorsal fin.

Similar Species: Central stoneroller closely resemble largescale stoneroller, which have 41 to 48 larger scales in the lateral series and 29 to 38 circumferencial scales. Juveniles and smaller adults resemble western blacknose dace, which lack a cartilaginous ridge on the lower lip and have a frenum. Smaller white sucker have inferior mouths with thick, protruding lips and more lateral line scales (53 to 85). Stonerollers can be distinguished from other minnows by the long intestine spiraled around the air bladder.

Special Taxonomic Note: All *Campostoma* from the Mississippi River Basin in South Dakota are recognized as central stoneroller *Campostoma anomalum* (no *Campostoma* have been collected in the Mississippi River drainage in North Dakota). However, recent genetic analyses indicated that this species is confined to the Ohio River drainage and the Atlantic Slope, with *Campostoma pullum* and *Campostoma plumbeum* found throughout the Midwest (Robison and Buchanan 2020). Based on the aforementioned genetic study, a detailed table of meristic characters was prepared for *Campostoma pullum* and *Campostoma plumbeum* found in Arkansas (Blum et al. 2008; Robison and Buchanan 2020). South Dakota *Campostoma* could not conclusively be reassigned to either *Campostoma pullum* or *Campostoma plumbeum* using the Arkansas characters (Robison and Buchanan 2020). The genetics study also did not include any specimens from either the Upper Mississippi River drainage or any of the Missouri River drainage tributaries north of Nebraska. For these reasons, the *Campostoma* in South Dakota continue to be recognized as central stoneroller, pending additional conclusive information.

Distribution and Habitat: Central stoneroller are native throughout much of the Missouri, Mississippi, and Ohio rivers and are commonly found in southeastern South Dakota river systems. They inhabit headwaters and small-to-medium streams with swift, clear water. They can be found occasionally in intermittent streams but rarely in lakes. Central stoneroller prefer gravel or bedrock substrate, and are often found near riffles, within pools, or along stream banks with overhanging vegetation. They can tolerate warmer water temperatures, short-term intermittency, and moderate amounts of siltation and turbidity (Matthews et al. 1987).

Reproduction: Central stoneroller migrate upstream to a variety of spawning habitats in smaller tributaries (Miller 1962). When water temperatures reach 18 °C in mid-April to May, spawning begins over fine gravel in shallow areas above riffles with swift current, or near deep pools with overhanging vegetation (Owen et al. 1981). Males prepare, and frequently guard, one or more nests by moving large pebble and gravel using their mouths or snout to make a cleared area (Miller 1962). Nests vary in size based on habitats (Miller 1962). During spawning, one or more males escort a female to a nest, where eggs are deposited on the gravel and fertilized. No parental care is given, with both males and females retreating downstream after spawning (Fowler and Taber 1985). The adhesive eggs are bright yellow and have a diameter of approximately 2 mm. Hatching occurs in two or three days. Fecundity ranges from 1,170 to 1,500 eggs per female (McKee and Parker 1982). Central stoneroller become sexually mature at age-1 or age-2 (Pflieger 1997).

Age and Growth: Newly hatched larvae have a total length of approximately 6 mm (Eddy and Underhill 1974). Growth is rapid within the first 2 years, with males longer than females. Total lengths at age for fish from Illinois were age-1, 51 mm; age-2, 79 mm; and age-3, 99 mm (Gunning and Lewis 1956). Total lengths from other populations may range from 100 to 177 mm, and with longer lengths of up to 254 mm in the southern portion of their range (Eddy and Underhill 1974). Longevity is from 3 to 6 years (McKee and Parker 1982).

Food and Feeding: Central stoneroller are omnivorous and feed primarily during the day. Juveniles eat rotifers and small zooplankton, while adults primarily eat filamentous algae, diatoms, and detritus (Owen et al. 1981; Evans-White et al. 2003; Hargrave 2006). Adults also consume macroinvertebrates, small mollusks, and larval fish. Typical feeding behavior consists of side-to-side head movements while scraping attached algae from substrate using the cartilaginous ridge on the inside of the lower lip. They can consume up to 27% body weight in benthic algae per day and can significantly decrease algal biomass (Fowler and Taber 1985; Gelwick and Matthews 1992; Pflieger 1997). Central stoneroller are capable of directly and indirectly affecting the functional and structural properties of stream ecosystems (Matthews et al. 1987; Gido et al. 2010).

References

Blum MJ, Neely DA, Harris PM, Mayden RL (2008) Molecular systematics of the cyprinid genus *Campostoma* (Actinopterygii: Cypriniformes): disassociation between morphological and mitochondrial differentiation. Copeia 2008:360–369. https://doi.org/10.1643/CI-06-093

Eddy S, Underhill JC (1974) Northern fishes with special reference to the upper Mississippi Valley, 3rd edn. University of Minnesota Press, Minneapolis

Evans-White MA, Dodds WK, Whiles MR (2003) Ecosystem significance of crayfishes and stonerollers in a prairie stream: functional differences between co-occurring omnivores. J N Am Benthol Soc 22:423–441. https://doi.org/10.2307/1468272

Fowler JF, Taber CA (1985) Food habits and feeding periodicity in two sympatric central stonerollers (Cyprinidae). Am Midl Nat 113:217–223. https://doi.org/10.2307/2425567

Gelwick FP, Matthews WJ (1992) Effects of an algivorous minnow on temperate stream ecosystem properties. Ecology 73:1630–1645. https://doi.org/10.2307/1940016

Gido KB, Bertrand KN, Murdock JN, Dodds WK, Whiles MR (2010) Disturbance-mediated effects of fishes on stream ecosystem processes: concepts and results from highly variable prairie streams. In: Gido KB, Jackson DA (eds) Community ecology of stream fishes: concepts, approaches, and techniques. American Fisheries Society, Bethesda, pp 593–617. https://doi.org/10.47886/9781934874141.ch31

Gunning GE, Lewis WM (1956) Age and growth of two important bait species in a cold-water stream in southern Illinois. Am Midl Nat 55:118–120. https://doi.org/10.2307/2422325

Hargrave CW (2006) A test of three alternative pathways for consumer regulation of primary productivity. Oecologia 149:123–152. https://doi.org/10.1007/s00442-006-0435-y

Matthews WJ, Stewart AJ, Power ME (1987) Grazing fishes as components of North American stream ecosystems: effects of *Campostoma anomalum*. In: Matthews WJ, Heins DC (eds) Community and evolutionary ecology of North American stream fishes. University of Oklahoma Press, Norman, pp 128–135

McKee PM, Parker BJ (1982) The distribution, biology, and status of the fishes *Campostoma anomalum*, *Clinostomus elongatus*, *Notropis photogenis* (Cyprinidae), and *Fundulus notatus* (Cyprinodontidae) in Canada. Can J Zool 60:1347–1358. https://doi.org/10.1139/z82-182

Miller RJ (1962) Reproductive behavior of the stoneroller minnow, *Campostoma anomallum pullum*. Copeia 1962:407–417. https://doi.org/10.2307/1440909

Owen JB, Elsen DS, Russell GW (1981) Distribution of fishes in North and South Dakota Basins affected by the Garrison Diversion Unit. University of North Dakota Press, Grand Forks

Pflieger WL (1997) The fishes of Missouri, revised edn. Missouri Dep Cons, Jefferson City

Robison HW, Buchanan TM (2020) Fishes of Arkansas. University of Arkansas Press, Fayetteville. https://doi.org/10.2307/j.ctvwh8bnv

6.14.3   Largescale Stoneroller *Campostoma oligolepis*

Hubbs & Greene, 1935

● **Pre-1990**   ● **1990-2021**

Etymology: *Campostoma oligolepis* – *Campostoma* = curved mouth; *oligolepis* = fewer scales.

Description:

Body – elongated, moderately robust, oval in cross section.
Color –

> *Dorsally* – gray, brown to dark olive.
> *Laterally* – olive to silver; dark mottling, sometimes faint lateral stripe.
> *Ventrally* – silver to white.
> *Eyes* – tan to amber.
> *Fins* – typically clear; dorsal: adults dark band; caudal: dark dusky spot.

Head – short, blunt.
Snout – rounded, slightly protrudes past upper lip.
Mouth – subterminal, crescent shaped.

> *Jaws* – upper jaw short.
> *Lips* – fleshy, lower lip with inner cartilaginous ridge.

Teeth – pharyngeal teeth 0,4-4,0; slightly hooked.
Intestine – extremely long, spiraled around air bladder.
Peritoneum – black.
Gill rakers – 20 to 28.
Dorsal fin – 8 rays.
Caudal peduncle – moderately thick.
Caudal fin – slightly forked.
Anal fin – 7 rays.
Pelvic fins – usually 8 rays; abdominal, insertion slightly anterior to dorsal fin insertion.
Pectoral fins – 13 to 17 rays.
Lateral line – complete; 41 to 48 cycloid scales; 29 to 38 circumferential scales.
Juveniles – dark lateral stripe.
Sexual dimorphism – spawning males have large, slightly hooked, light-colored tubercles on head, body, and median fins; small hump behind head, red-orange on ventral scales, pelvic fin, and anal fin, and a light pink spot on each side of the snout; lack inner-nasal tubercles.

Similar Species: Largescale stoneroller closely resemble central stoneroller. Central stoneroller have a more robust body with 47 to 60 smaller scales in lateral series and 38 to 50 circumferencial scales. The adult central stoneroller anal fin has a dark band, and male central stonerollers have inner-nasal tubercles. Juvenile largescale stoneroller and smaller adults resemble blacknose dace, which lack a lower lip cartilaginous ridge and have a frenum. Smaller white sucker have inferior mouths with thick, protruding lips and 53 to 85 lateral line scales. Stonerollers can be distinguished from other minnows by their long intestine spiraled around the air bladder.

Special Taxonomic Note: All *Campostoma* from the Red River of the North drainage in North Dakota are recognized as largescale stoneroller (no *Campostoma* have been collected in the small portion of the Red River of the North drainage in South Dakota). The only vouchered specimens from the Red River of the North drainage in North Dakota are from the Forest River and were morphologically identified as largescale stoneroller based on scale counts (Borgstrom 2010). Although a detailed table of meristic characters was prepared for *Campostoma pullum* and *Campostoma plumbeum* found in Arkansas, North Dakota *Campostoma* could not conclusively be reassigned to either of these likely recognized species using these characters (Rakocinski 1977; Becker 1983; Timms 2017). The population in the Red River of the North drainage was not included in the most recent range-wide genetic analysis of Campostoma (Blum et al. 2008). However, subsequent unpublished genetic analyses identified individuals from the Forest River as *Campostoma pullum* (Blum 2021). Although tentatively genetically assigned to *C. pullum*, because these North Dakota specimens could not be definitively reidentified using meristic characters, this species will be recognized as largescale stoneroller *Campostoma oligolepis* pending additional conclusive evidence.

Distribution and Habitat: The native range of largescale stoneroller extends throughout the Mississippi River basin from northern Wisconsin into eastern Minnesota and Iowa, and southward to the Mobile River and Ozarks. They have been found in eastern North Dakota in the Forest, Park, and Sheyenne Rivers but have not been recorded in South Dakota. The largescale stoneroller is a habitat specialist (Rakocinski 1977; Burr et al. 1979; Fowler and Taber 1985). Adults inhabit headwaters, as well as small-to-medium-sized streams with swift current. It resides near riffles and swift runs over sand, cobble, or gravel substrate, and occasionally pools. Largescale stoneroller prefer larger riffles and faster currents than the central stoneroller (Burr and Smith 1976; Fowler and Taber 1985; Pflieger 1997). Juveniles favor areas with slower current and sand or silt substrates (Burr and Smith 1976). Adults have a high pollution tolerance, but a low silt tolerance (Burr and Smith 1976).

Reproduction: Because the largescale stoneroller is understudied, information regarding reproduction is often derived from other *Campostoma* species. Spawning takes place from March to April and possibly extends into May (Timms 2017). The peak spawning temperature is 12 °C, with spawning activity influenced by seasonal and yearly environmental variations (South and Ensign 2013; Timms 2017). Largescale stoneroller are fractional spawners (Timms 2017). Males create and defend one or more nests by moving pebbles and gravel with their mouths or snout. Nests as large as 30 cm are often in shallow waters of pools or riffles (Timms 2017). During spawning, one or more males escort a single female to a nest where the eggs are deposited and fertilized. After spawning no parental care is given. Fecundity from a 111 mm, 14.7 g, age-2 female was 1,510 eggs (Becker 1983). The adhesive eggs are yellow-amber and have a diameter of approximately 1 mm (Becker 1983). Sexual maturity is reached at age-1 or age-2.

Age and Growth: Average largescale stoneroller length is 102 mm, with a maximum total length of 175 mm (Burr et al. 1979; Mettee et al. 1996). Total lengths at age of fish from the Eau Claire River in Wisconsin were 56 mm at age-1, 90 mm at age-2, and 126 mm at age-3 (Becker 1983). Males are often longer than females (Burr and Smith 1976). Growth is rapid within the first year. Average longevity is 3 to 5 years.

Food and Feeding: Largescale stoneroller are omnivorous, with diets and feeding behavior similar to central stoneroller. Feeding occurs continuously during the day, with adults primarily grazing on filamentous algae (Fowler and Taber 1985). Feeding behavior consists of side-to-side head movements while scraping attached algae from substrate using the cartilaginous ridge on the inside of the lower lip. They are capable of consuming up to 27% body weight in benthic algae per day and can significantly decrease algal biomass (Burr and Smith 1976; Borgstrom 2010; South and Ensign 2013). Like the central stoneroller, largescale stoneroller are capable of changing ecosystems by consuming such large amounts of algal biomass (Burr et al. 1979; Pflieger 1997).

References

Becker GC (1983) Fishes of Wisconsin. University of Wisconsin Press, Madison
Blum MJ (2021) Personal communication regarding unpublished data. Professor at University of Tennessee, Knoxville
Blum MJ, Neely DA, Harris PM, Mayden RL (2008) Molecular systematics of the cyprinid genus Campostoma (Actinopterygii: Cypriniformes): disassociation between morphological and mitochondrial differentiation. Copeia 2008:360–369. https://doi.org/10.1643/CI-06-093
Borgstrom LJ (2010) Fish community assembly in the Forest River, North Dakota, and resolution of Campostoma species presence. Dissertation, South Dakota State University
Burr BM, Smith PW (1976) Status of the largescale stoneroller, *Campostoma oligolepis*. Copeia 1976:521–531. https://doi.org/10.2307/1443370
Burr BM, Cashner RC, Pflieger WL (1979) *Campostoma oligolepis* and *Notropis ozarcanus* (Pisces: Cyprinidae), two additions to the known fish fauna of the Illinois River, Arkansas and Oklahoma. Southwest Nat 24:381–383. https://doi.org/10.2307/3670940
Fowler JF, Taber CA (1985) Food habits and feeding periodicity in two sympatric stonerollers (Cyprinidae). Am Midl Nat 113:217–224. https://doi.org/10.2307/2425567
Metee MF, O'Neil PE, Pierson JM (1996) Fishes of Alabama and the Mobile Basin. Oxmoor House, Birmingham
Pflieger WL (1997) The fishes of Missouri, revised edn. Missouri Dep Cons, Jefferson City
Rakocinski CR (1977) Evolutionary interactions of two sympatric related species of minnows in a creek in Northwestern Illinois: hybridization and isolating mechanisms between *Campostoma oligolepis* and *Campostoma anomalum pullum*. Dissertation, Northern Illinois University
Robison HW, Buchanan TM (2020) Fishes of Arkansas. University of Arkansas Press, Fayetteville. https://doi.org/10.2307/j.ctvwh8bnv
South EJ, Ensign WE (2013) Life history of *Campostoma oligolepis* (largescale stoneroller) in urban and rural streams. Southeast Nat 12:781–789. https://doi.org/10.1656/058.012.0425
Timms DM (2017) Reproductive timing of the largescale stoneroller, *Campostoma oligolepis*, in the Flint River, Alabama. Dissertation, University of Alabama-Huntsville

6.14.4   Northern Redbelly Dace *Chrosomus eos*

Cope, 1861

Nuptial Male   Female

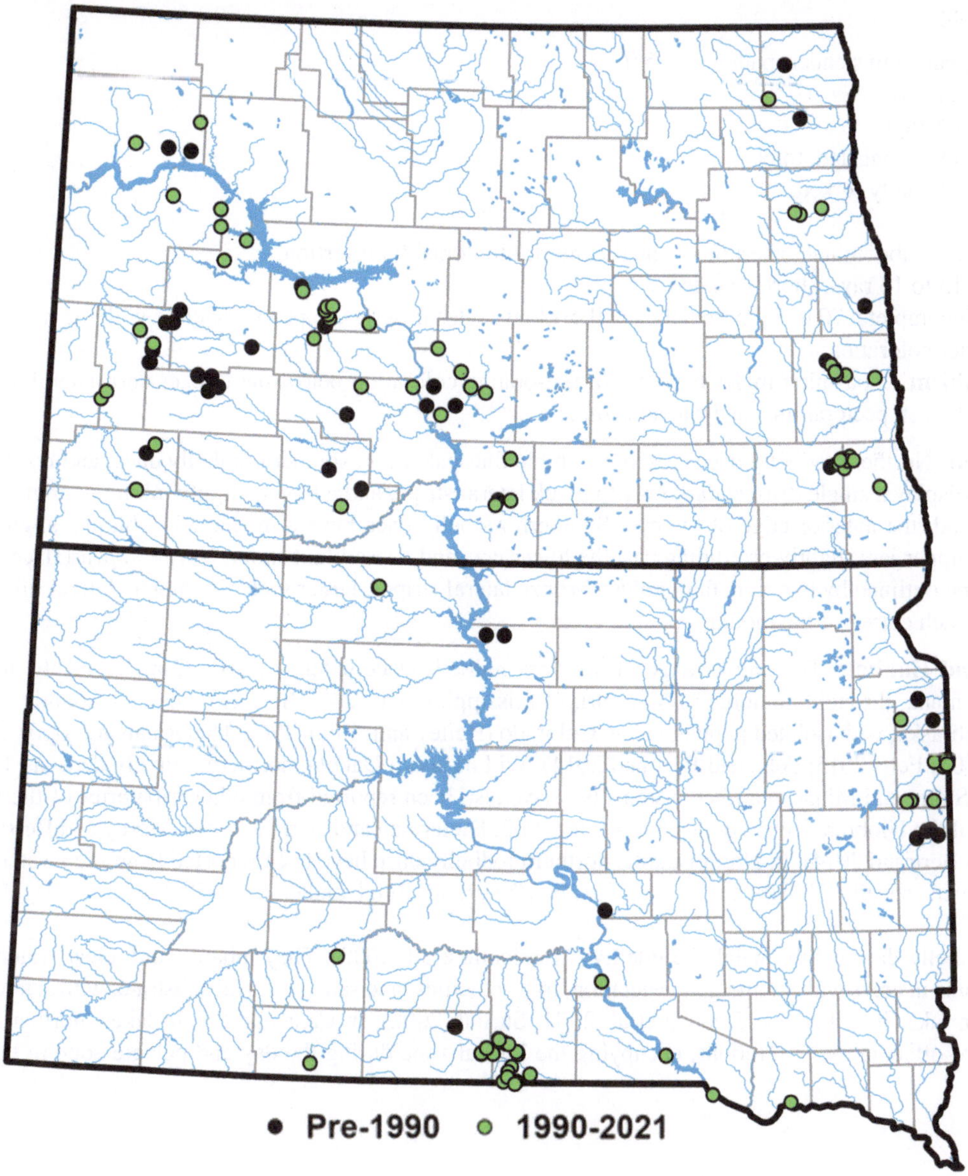

● **Pre-1990**   ● **1990-2021**

Etymology: *Chrosomus eos – Chrosomus* = coloration; *eos* = sunrise, referencing male yellow and orange coloration.
Description:

Body – elongated, cylindrical.
Color –

> *Dorsally* – dark olive to dusky brown.
> *Ventrally* – silvery-white.
> *Laterally* – multiple stripes of color with a thin, narrow dark lateral stripe nearest to dorsal side beginning at posterior edge of operculum and extending to base of caudal fin, a thicker light olive to silvery-tan lateral stripe below, and a thicker dark black lateral stripe beginning at tip of snout and extending to the caudal fin base creating a dark caudal spot).
> *Fins* – generally clear but may have light peach to cream pigmentation.

Head – moderate.
Snout – blunt.
Eyes – moderately large; laterally on head.
Mouth – small, terminal, strongly oblique.

> *Jaws* – lower jaw often protruding past upper jaw; jaws do not extend halfway back toward eye.
> *Lips* – fleshy.

Teeth – pharyngeal teeth slender; pattern 0,5-5,0.
Gill rakers – 10, short.
Dorsal fin – 7 or 8 rays.
Caudal peduncle – elongated, thick.
Caudal fin – moderately forked.
Anal fin – 8 rays.
Pelvic fins – 8 rays; abdominal, insertion distinctly anterior to dorsal fin insertion.
Pectoral fins – 13 to 15 rays; short, rounded.
Lateral line – incomplete; 70 to 87 extremely small cycloid scales in series, appears scaleless with the naked eye.
Juveniles – duller coloration.
Sexual dimorphism – spawning males have bright red-orange coloration below the thicker dark lateral stripe, and small tubercles on breast, pectoral fins, and caudal peduncle.

Similar Species: Northern redbelly dace closely resemble finescale dace, southern redbelly dace, and northern pearl dace. Finescale dace display a single distinct dark gray to black lateral stripe, and a larger, slightly oblique mouth that extends to, or slightly beyond, the anterior edge of the eye. Southern redbelly dace have a moderately oblique, slightly subterminal mouth with the upper jaw slightly protruding past the lower jaw, and a less-blunt snout. Northern pearl dace have one thick, dark lateral stripe outlined by two thin, faint golden bronze lateral stripes, larger scales, complete lateral line, and occasionally a barbel at each corner of the mouth.

Distribution and Habitat: The native range of northern redbelly dace extends across Canada and the northern United States from Montana, through the northern Missouri, Mississippi, Great Lakes-St. Lawrence and Atlantic rivers to Maine, and south to Nebraska, with isolated populations in Colorado (Bailey and Allum 1962; Dieterman and Berry Jr 1994; Morey and Berry Jr 2004; Felts 2013; Felts and Bertrand 2014). Within South Dakota, northern redbelly dace are found primarily within the Big Sioux and Minnesota river systems but have also been reported from other tributaries of the Missouri River and tributaries of the Niobrara River (Bailey and Allum 1962; Dieterman and Berry Jr 1994; Morey and Berry Jr 2004; Felts 2013; Felts and Bertrand 2014). In North Dakota, northern redbelly dace have been reported from the western tributaries of the Missouri River, as well within the Red River of the North system (Reigh and Owen 1978; Stasiak 1980; Morey and Berry Jr 2004).

Northern redbelly dace exhibit natal site fidelity (Massicotte et al. 2008). They inhabit cold, clear waters of first-order spring-fed streams in areas with submerged vegetation, low velocities, and sand or gravel substrate, with schools of 50 to 300 fish observed in lake littoral zones (Dupuch et al. 2004; Stasiak 2006). In Canadian Shield lakes, northern redbelly dace perform onshore-offshore diel migrations, occupying the littoral zone during the day and pelagic zone in the evening, and then migrating back to the littoral zone at sunrise (Naud and Magnan 1988). In littoral zones, they prefer densely covered or structured habitat (Naud and Magnan 1988; Dupuch et al. 2009). Northern redbelly dace may inhabit the littoral zone during the day to avoid predation by pelagic piscivores (Naud and Magnan 1988; Dupuch et al. 2009). When exposed to skin extracts or conspecific chemicals, they exhibit fright reactions such as dashing and freezing behaviors, movement toward substrate, flight from the alarm substance location, and increased school cohesiveness (Dupuch et al. 2004). The fright reaction of individual dace is directly related to the alarm substance concentration, suggesting that the fish may be able to detect different degrees of predation risk (Dupuch et al. 2004).

Reproduction: Northern redbelly dace spawning likely occurs from late spring to early summer (Becker 1983). As many as five males have been observed following a single female into a mass of filamentous algae, where the males vibrate next to the female to initiate gamete release and subsequent fertilization (Cooper 1935; Bertrand-Toline 1994). Sexual maturity occurs at either age-1 or age-2 (Becker 1983). Northern redbelly dace are fractional spawners, with approximately 5 to 30 eggs released during a single spawning event (Cooper 1935). No nest is constructed, and no parental care is given. The non-adhesive, yellow-orange eggs have a diameter of approximately 1 mm (Becker 1983). Hatching occurs in 8 to 10 days in water temperatures of 21 to 27 °C (Hubbs and Cooper 1936; Becker 1983).

Age and Growth: Information on northern redbelly dace age and growth is lacking. They can reach total lengths of 77 mm and live up to 8 years (Scott and Crossman 1973; Becker 1983).

Food and Feeding: Northern redbelly dace are visual omnivores. Adults primarily feed within the water column on green algae, detritus, and small, immature aquatic invertebrates (Cochran et al. 1988). Evening pelagic zone migrations likely increase foraging efficiency on zooplankton (Dupuch et al. 2004).

References

Bailey RM, Allum MO (1962) Fishes of South Dakota. University of Michigan, Ann Arbor. https://doi.org/10.3998/mpub.9690435

Becker GC (1983) Fishes of Wisconsin. University of Wisconsin Press, Madison

Bertrand-Toline CA (1994) Morphometric and genetic differentiation of the northern redbelly dace: investigating microevolutionary processes. Dissertation, University of Toronto

Cochran PA, Lodge DM, Hodgson JR, Knapik PG (1988) Diets of syntopic finescale dace, *Phoxinus neogaeus*, and northern redbelly dace, *Phoxinus eos*: a reflection of trophic morphology. Environ Biol Fish 22:235–240. https://doi.org/10.1007/BF00005384

Cooper GP (1935) Some results of the forage fish investigations in Michigan. Trans Am Fish Soc 65:132–142. https://doi.org/10.1577/1548-8659(1935)65[132:SROFFI]2.0.CO;2

Dieterman D, Berry CR Jr (1994) Fishes in seven streams of the Minnesota River drainage in northeastern South Dakota. Proc S D Acad Sci 73:23–30

Dupuch A, Magnan P, Dill LM (2004) Sensitivity of northern redbelly dace, *Phoxinus eos*, to chemical alarm cues. Can J Zool 82:407–414. https://doi.org/10.1139/z04-003

Dupuch A, Magnan P, Bertolo A, Dill LM, Proulx M (2009) Does predation risk influence habitat use by northern redbelly dace *Phoxinus eos* at different spatial scales? J Fish Biol 74:1371–1382. https://doi.org/10.1111/j.1095-8649.2009.02183.x

Felts EA (2013) Ecology of glacial relict fishes in South Dakota's sandhills region. Dissertation,. South Dakota State University

Felts EA, Bertrand KN (2014) Conservation status of five headwater stream specialists in South Dakota. Am Midl Nat 172:131–159. https://doi.org/10.1674/0003-0031-172.1.131

Hubbs CL, Cooper GP (1936) Minnows of Michigan. Bull (Crankbrook Inst Sci) Cranbrook institute Science), no. 8

Massicotte R, Magnan P, Angers B (2008) Intralacustrine site fidelity and nonrandom mating in the littoral-spawning northern redbelly dace (*Phoxinus eos*). Can J Fish Aquat Sci 65:2016–2025. https://doi.org/10.1139/F08-114

Morey NM, Berry CR Jr (2004) New distributional record of the northern redbelly dace in the Northern Great Plains. Prairie Nat 36:257–260

Naud M, Magnan P (1988) Diel onshore-offshore migrations in northern redbelly dace, *Phoxinus eos* (Cope), in relation to prey distribution in a small oligotrophic lake. Can J Zool 66:1249–1253. https://doi.org/10.1139/z88-182

Reigh RC, Owen JB (1978) Fishes of the western tributaries of the Missouri River in North Dakota. Reprot to the Regional Environmental Assessment Program, contract 6-01

Scott WB, Crossman EJ (1973) Freshwater fishes of Canada. Fisheries Research Board of Canada, Bulletin 184

Stasiak RH (1980) *Phoxinus eos* (Cope), northern redbelly dace. In: Lee DS, Gilbert CR, Hocutt CH, Jenkins RE, McAllister DE, Stauffer JR Jr (eds) Atlas of North American freshwater fishes. North Carolina State Museum of Natural History, Raleigh, p 336

Stasiak RH (2006) Northern redbelly dace (*Phoxinus eos*): a technical conservation assessment. USDA Forest Service, Rocky Mountain Region. https://www.fs.usda.gov/Internet/FSE_DOCUMENTS/stelprdb5206788.pdf. Accessed 24 Apr 2023

6.14.5 Southern Redbelly Dace *Chrosomus erythrogaster*

(Rafinesque, 1820)

Nuptial Male   Female

● **Pre-1990**   ○ **1990–2021**

Etymology: *Chrosomus erythrogaster* – *Chrosomus* = coloration; *erythrogaster* = red belly, referencing the red-orange flank coloration especially evident on males.

Description:

Body – marginally fusiform, elongated, slightly laterally compressed.
Color –

> *Dorsally* – olive to gray; small irregular, dark specks.
> *Ventrally* – cream to white.
> *Laterally* – multiple stripes of color, with a thin, narrow dark lateral stripe nearest to dorsal side beginning at tip of snout and extending to caudal fin base, a thicker silvery-white to cream lateral stripe below, and a thicker dark black lateral stripe beginning at tip of snout and extending to caudal fin base creating a dark caudal spot.
> *Fins* – generally clear, may have light yellow pigmentation.

Head – moderate.
Snout – moderately pointed.
Eyes – moderately large, laterally on head.
Mouth – small, moderately oblique, slightly subterminal.

> *Jaws* – upper jaw protruding past lower jaw; do not extend past anterior end of eye.
> *Lips* – fleshy.

Teeth – pharyngeal teeth slender; pattern 0,5 5,0.
Gill rakers – 7 to 10, short.
Dorsal fin – 8 rays.
Caudal peduncle – elongated, thick.
Caudal fin – moderately forked.
Anal fin – 8 rays.
Pelvic fins – 8 rays; abdominal, insertion distinctly anterior to dorsal fin insertion.
Pectoral fins – 14 to 17 rays; short, rounded.
Lateral line – incomplete; 70 to 95 small cycloid scales in series.
Juveniles – less coloration.
Sexual dimorphism – spawning males have intense, bright red-orange on flanks below lateral stripe, and orange-yellow at base of dorsal, anal, pelvic, and pectoral fins; spawning female coloration is much less and very faint.

Similar Species: Southern redbelly dace closely resemble northern redbelly dace, northern pearl dace, and finescale dace. Northern redbelly dace have a blunt snout, a terminal and strongly oblique mouth, and a lower jaw that often protrudes past the upper jaw. Northern pearl dace have a thick, dark lateral stripe outlined by two thin, faint golden bronze lateral stripes, larger scales, a complete lateral line, and occasionally a barbel at each corner of the mouth. Finescale dace have a single distinct dark gray to black lateral stripe, and a larger, slightly oblique mouth that extends to, or slightly beyond, the anterior edge of the eye.

Distribution and Habitat: Southern redbelly dace are widely distributed in the upper Mississippi, Ohio, and Missouri river systems. They range from southern Minnesota and Wisconsin in the north, Pennsylvania and New York in the east, Arkansas, Oklahoma, and the Tennessee River in the south, and New Mexico and Colorado in the west (Starnes and Starnes 1980; Slack et al. 1997; Stasiak 2007). In the Dakotas, southern redbelly dace have only been reported from two tributaries of the Big Sioux River in Lincoln County in eastern South Dakota: Little Beaver Creek and Ninemile Creek (Springman and Banks 2005; Stasiak 2007). They inhabit small-to-medium low-order streams in waters 0.1–1.5 m deep (Becker 1983; Slack et al. 1997; Stasiak 2007). Southern redbelly dace are abundant and often school in small spring-fed headwaters and slow-flowing pools (velocities of 0.1–0.2 m/sec) in streams with cool, clear waters with substrates of gravel, pebble, sand, and small boulders under dense riparian canopy cover (Phillips 1968; Slack et al. 1997). They also prefer undercut banks and areas with permanent vegetation (Stasiak 2007). Southern redbelly dace avoid substrates with large amounts of clay, silt, mud, and large boulders (Becker 1983; Slack et al. 1997). Their upper critical thermal maximum is 35 °C (Farless and Brewer 2017).

Reproduction: Spawning occurs from May to June, with older adults spawning earlier and younger adults spawning later (Smith 1908; Phillips 1968; Settles and Hoyt 1978). Spawning ceases when water temperatures reach 21 °C (Settles and Hoyt 1978; Hayes 2015). Southern redbelly dace at total lengths under 40 mm are considered immature. Most fish become sexually mature at age-1 at total lengths of approximately 50 mm (Stasiak 2007). Spawning occurs over gravel, pebble, or sand substrate in swift currents (Smith 1908; Hayes 2015). No nest is constructed, but the species is often observed spawning over a nest constructed by creek chub, common shiner, stonerollers, or other minnows (Phillips 1968; Etnier and Starnes 1993; Hayes 2015). Males temporarily defend small territories around one female (Hayes 2015). Spawning begins with several males pursuing a single female, with two males eventually nudging and crowding laterally against the female. One male will clasp the female by placing a pectoral fin under her body and curving his body close to hers (Smith 1908; Phillips 1968; Johnston and Page 1992; Hayes 2015). The male then vibrates rapidly, initiating the release of eggs and milt (Smith 1908). Southern redbelly dace are broadcast spawners (Starnes and Starnes 1981). Fecundity for ages 0, 1, and 2 was 267, 401, and 568 eggs, respectively (Settles and Hoyt 1978; Stasiak 2007). The nearly transparent eggs range from 0.8 to 1.3 mm in diameter (Smith 1908; Becker 1983). Hatching occurs in six days at 20 °C water temperature (Sternburg 1992; Stasiak 2007).

Age and Growth: Female southern redbelly dace are generally smaller than same-age males (Becker 1983; Stasiak 2007). Maximum total length is 89 mm and longevity is 4 years (Etnier and Starnes 1993; Stasiak 2007).

Food and Feeding: Adults are generalist, visual, omnivorous grazers that feed on algae and aquatic invertebrates (Settles and Hoyt 1976; Bertrand and Gido 2007). Moderate densities of southern redbelly dace can temporarily reduce algal filament lengths and particulate organic matter sizes (Bertrand and Gido 2007).

References

Becker GC (1983) Fishes of Wisconsin. University of Wisconsin Press, Madison

Bertrand KN, Gido KB (2007) Effects of the herbivorous minnow, southern redbelly dace (*Phoxinus erythrogaster*), on stream productivity and ecosystem structure. Oecologia 151:69–81. https://doi.org/10.1007/s00442-006-0569-y

Etnier DA, Starnes WC (1993) The fishes of Tennessee. University of Tennessee Press, Knoxville

Farless NA, Brewer SK (2017) Thermal tolerance of fishes occupying groundwater and surface-water dominated streams. Freshwater Sci 36:866–876. https://doi.org/10.1086/694781

Hayes BM (2015) Reproductive biology of syntopic blackside dace and southern redbelly dace in two Kentucky streams. Dissertation, Eastern Kentucky University

Johnston C, Page L (1992) The evolution of complex reproductive strategies in North American minnows (Cyprinidae). In: Mayden R (ed) Systematics, historical ecology, and North American freshwater fishes. Stanford University Press, Stanford, pp 600–621

Phillips GL (1968) *Chrosomus erythrogaster* and *Chrosomus eos* (Osteichthyes: Cyprinidae): taxonomy, distribution, ecology. Dissertation, University of Minnesota

Settles WH, Hoyt RD (1976) Age structure, growth patterns, and food habits of the southern redbelly dace *Chrosomus erythrogaster* in Kentucky. Trans Ky Acad Sci 37:1–10

Settles WH, Hoyt RD (1978) The reproductive biology of the southern redbelly dace, *Chrosomus erythrogaster* Rafinesque, in a spring-fed stream in Kentucky. Am Midl Nat 99:290–298. https://doi.org/10.2307/2424807

Slack WT, O'Connell MT, Peterson TL, Ewing JA III, Ross ST (1997) Ichthyofaunal and habitat associations of disjunct populations of southern redbelly dace. *Phoxinus erythrogaster* (Teleostei: Cyprinidae) in Mississippi. Am Midl Nat 137:251–265. https://doi.org/10.2307/2426844

Smith BG (1908) The spawning habits of *Chrosomus erythrogaster* Rafinesque. Biol Bull 15:9–18. https://doi.org/10.2307/1535560

Springman DJ, Banks RL (2005) Range extension of the southern redbelly dace into South Dakota. Prairie Nat 37:175–176

Starnes LB, Starnes WS (1981) Biology of the blackside dace *Phoxinus cumberlandensis*. Am Midl Nat 106:360–371. https://doi.org/10.2307/2425173

Starnes WC, Starnes LB (1980) *Phoxinus erythrogaster* (Rafinesque), southern redbelly dace. In: Lee DS, Gilbert CR, Hocutt CH, Jenkins RE, McAllister DE, Stauffer JR Jr (eds) Atlas of North American freshwater fishes. North Carolina State Museum of Natural History, Raleigh, p 337

Stasiak RH (2007) Southern redbelly dace (*Phoxinus erythrogaster*): a technical conservation assessment. US Dep Ag, Forest Service, Rocky Mountain Region, golden. https://www.fs.usda.gov/Internet/FSE_DOCUMENTS/stelprdb5206790.pdf. Accessed 24 Apr 2023

Sternburg JG (1992) Spawning of southern and northern redbelly dace. North American Native Fishes Association. http://www.nanfa.org/articles/acredbelly.shtml. Accessed 24 April 2023

6.14.6   Finescale Dace *Chrosomus neogaeus*

(Cope, 1867)

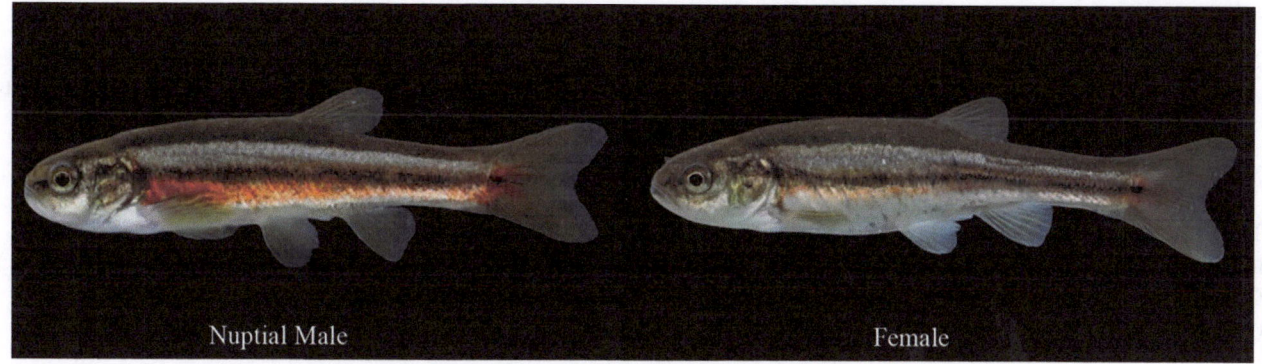

Nuptial Male   Female

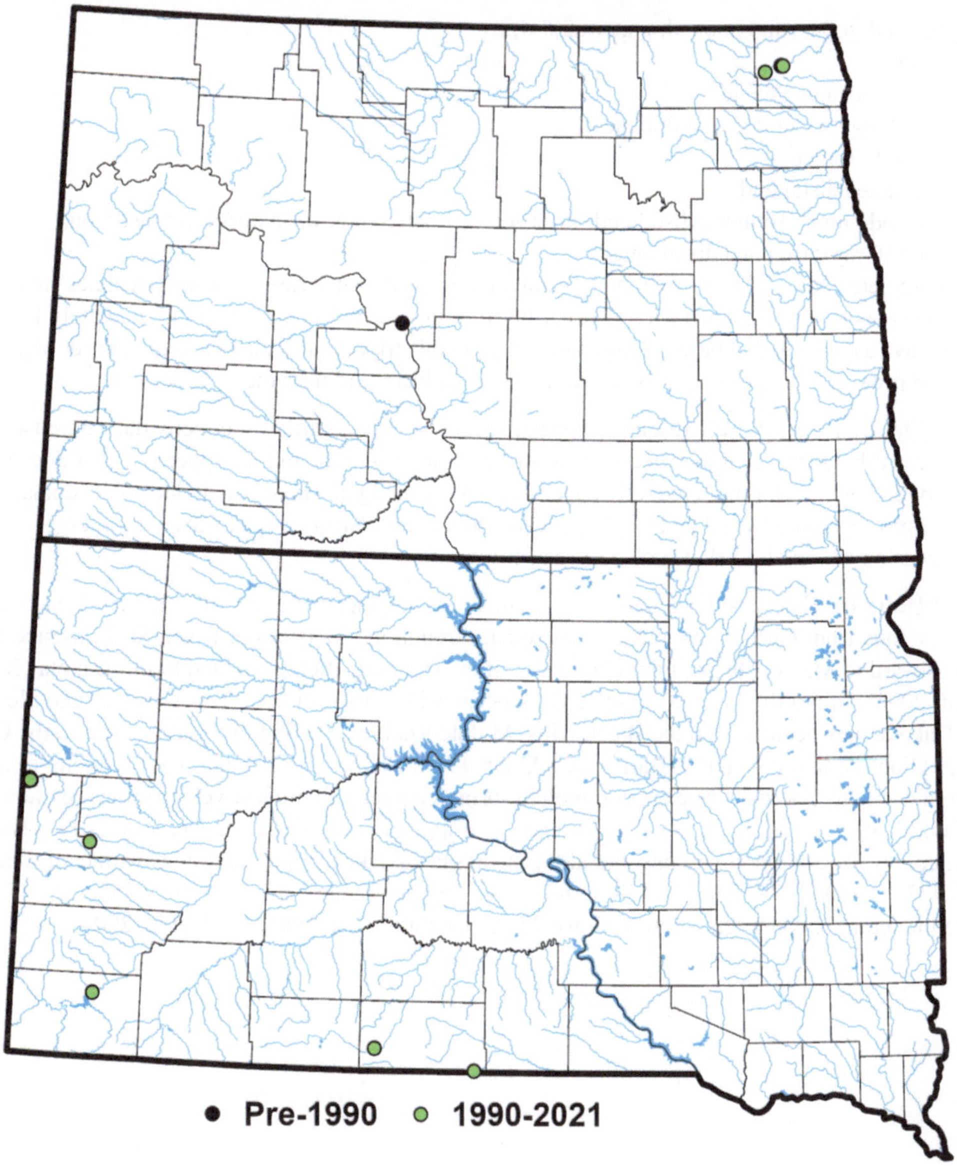

● **Pre-1990**   ○ **1990-2021**

Etymology: *Chrosomus neogaeus* – *Chrosomus* = coloration; *neogaeus* = new world.

Description:

Body – cylindrical anteriorly, laterally compressed posteriorly.
Color –

> *Dorsally* – dusky black to brown.
> *Laterally* – light gray to olive followed by a single distinct dark gray to black lateral stripe extending from tip of snout to caudal fin base, sometimes leading to a dark caudal spot.
> *Ventrally* – light gray fading to white on midventral surface.
> *Fins* – lightly pigmented.

Head – large.
Snout – blunt.
Eyes – moderately large, laterally on head.
Mouth – large, terminal, slightly oblique.

> *Jaws* – upper jaw extends to anterior end or middle of eye.
> *Lips* – fleshy; may slightly protrude past tip of snout.

Teeth – pharyngeal teeth robust and hooked; pattern 2,5-4,2.
Gill rakers – 9; short.
Dorsal fin – 8 rays; rounded distal end.
Caudal peduncle – elongated, uniform in thickness.
Caudal fin – moderately forked.
Anal fin – 8 rays; rounded distal end.
Pelvic fins – 8 rays; abdominal; rounded distal end; insertion distinctly anterior to dorsal fin insertion.
Pectoral fins – 15 or 16 rays; rounded distal ends.
Lateral line – incomplete; short; 70 to 85 extremely small ctenoid scales in series, appearing scaleless to the naked eye.
Sexual dimorphism – spawning males have a distinct burnt orange to red lateral stripe immediately below black lateral stripe; females have a dusky, faded burnt orange to yellow lateral stripe; males also have thickened, highly modified pectoral fin rays and rows of tubercles ventrally, above the anal fin base, and near the opercle.

Similar Species: Finescale dace closely resemble northern redbelly dace. Northern redbelly dace have two black to brown lateral stripes separated by a thicker light olive to gray stripe, a small mouth, and an upper jaw that does not extend to, or beyond, the anterior end of eye. Southern redbelly dace have a more elongate snout, small dark spots dorsally, and two black to brown lateral stripes separated by a thicker light gray to white stripe. Northern pearl dace have larger scales, a complete lateral line, and occasionally one barbel at each corner of the mouth.

Distribution and Habitat: Finescale dace native range includes the glaciated areas of southern Canada and the northern United States. It extends from northern Montana in the west, through the Mississippi River, St. Lawrence River and Great Lakes systems eastward to Maine. Disjunct populations exist in Wyoming, Nebraska, and South Dakota (Bailey and Allum 1962; Isaak et al. 2003). In North Dakota, finescale dace occur within the Pembina River system and along the Missouri River. In South Dakota, they occur within the Black Hills, Middle Rockies and Sandhills ecoregions in the Cheyenne River, White River, and Niobrara River systems (Bailey and Allum 1962; Felts and Bertrand 2014; Amiotte et al. 2015). In the northern portion of its range, finescale dace are common in springfed streams with low velocity and are often associated with beaver ponds (Stasiak 1972; Scholsser 1995; Stasiak and Cunningham 2006; Felts and Bertrand 2014). They are most frequently found in cold water first order streams, but also inhabit glacial lakes, ponds, and bogs with cool water and abundant cover (Stasiak 1972; Stasiak 1978; Felts and Bertrand 2014). Finescale dace are strongly associated with shorelines and structure and are intolerant of water temperature above 25 °C (Cochran et al. 1988; Isaak et al. 2003).

Reproduction: Spawning occurs from late April through late May and may extend into June or early July depending on water temperature (Das and Nelson 1990; Stasiak and Cunningham 2006). In Minnesota, finescale dace spawning in shallow areas 0.5 to 0.9 m deep begins when water temperatures exceed 15 °C and continues until 22 °C (Stasiak 1978; Becker 1983). Sudden fluctuations in water temperature caused by ice-out may initiate spawning (Stasiak 1972). Sexual maturity occurs at age-1 or age-2 (Stasiak 1978). Spawning starts when several males follow one or two ripe females as they move from a school to overhanging vegetation or woody debris (Stasiak 1978). A male uses thickened, modified pectoral fins to grasp a female ventrally just posterior to her pectoral fins (Stasiak 1978). The male then pushes the female against a sturdy object and curves his caudal fin over the females caudal fin which causes the patch of tubercles above the base of his anal fin to rub the vent of the female (Stasiak 1978). This initiates the release of milt and approximately 20 to 30 eggs (Stasiak 1978). Approximately 10 seconds after spawning ceases, females retreat back into schools while the males linger and continue to release milt for several more seconds (Stasiak 1978). No nest is constructed, and no parental care is given (Stasiak 1978). Finescale dace are fractional spawners. Fecundity increases with female length but is rather low for minnow species at 784 to 3,060 eggs per female (Stasiak 1978). Yellow-orange eggs are dermesal, with a diameter ranging from 1.2 to 1.5 mm (Stasiak 1972, 1978; Das and Nelson 1990). Hatching occurs in six days in a water temperature of 20 °C (Stasiak 1978).

Age and Growth: Newly hatched finescale dace larvae have a total length of approximately 4 mm (Stasiak 1978). Standard lengths of both sexes are approximately equal after first year of growth (Stasiak 1978). Maximum total length is 107 mm (Becker 1983). Females tend to live longer than males and have faster growth rates after sexual maturity (Stasiak 1978). Longevity is 6 years (Stasiak 1978).

Food and Feeding: Finescale dace are opportunistic omnivores. Adults feed within shallow shoreline areas on larval chironomids, immature Odonata, Trichoptera larvae, ephemeropterans, coleopterans, and other macroinvertebrates (Cochran et al. 1988). They also consume small mollusks, vegetative matter, green algae, diatoms, and zooplankton (Cochran et al. 1988).

References

Amiotte J, Simpson G, Barnes ME (2015) Re-establishment of finescale dace (*Phoxinus neogaeus*) in Mud Lake, Lawrence County, South Dakota. Proc S D Acad Sci 94:195–200
Bailey RM, Allum MO (1962) Fishes of South Dakota. University of Michigan, Ann Arbor. https://doi.org/10.3998/mpub.9690435
Becker GC (1983) Fishes of Wisconsin. University of Wisconsin Press, Madison
Cochran PA, Lodge DM, Hodgson JR, Knapik PG (1988) Diets of syntopic finescale dace, *Phoxinus neogaeus*, and northern redbelly dace, *Phoxinus eos*: a reflection of trophic morphology. Environ Biol Fish 22:235–240. https://doi.org/10.1007/BF00005384
Das MK, Nelson JS (1990) Spawning time and fecundity of northern redbelly dace, *Phoxinus eos*, finescale dace, *Phoxinus neogaeus*, and their hybrids in Upper Pierre Lake, Alberta. Can Field Nat 104:409–413
Felts EA, Bertrand KN (2014) Conservation status of five headwater stream specialists in South Dakota. Am Midl Nat 172:131–159. https://doi.org/10.1674/0003-0031-172.1.131
Isaak DJ, Hubert WA, Berry CR (2003) Conservation assessment of the lake chub, mountain sucker, and finescale dace in the Black Hills National Forest, South Dakota and Wyoming. USDA Forest Service, Rocky Mountain Region. https://www.fs.usda.gov/Internet/FSE_DOCUMENTS/fsm9_012156.pdf. Accessed 24 Apr 2023
Scholsser IJ (1995) Dispersal, boundary processes, and trophic-level interactions in streams adjacent to beaver ponds. Ecology 76:908–925. https://doi.org/10.2307/1939356
Stasiak RH (1972) The morphology and life history of the finescale dace, *Pfrille neogaea*, in Itasca State Park, Minnesota. Dissertation, University of Minnesota
Stasiak RH (1978) Reproduction, age, and growth of the finescale dace, *Chrosomus neogaeus*, in Minnesota. Trans Am Fish Soc 107:720–723. https://doi.org/10.1577/1548-8659
Stasiak RH, Cunningham GR (2006) Finescale dace (*Phoxinus neogaeus*): a technical conservation assessment. USDA Forest Service, Rocky Mountain Region. https://www.fs.usda.gov/Internet/FSE_DOCUMENTS/stelprdb5206787.pdf. Accessed 24 Apr 2023

6.14.7 Lake Chub *Couesius plumbeus*

(Agassiz, 1850)

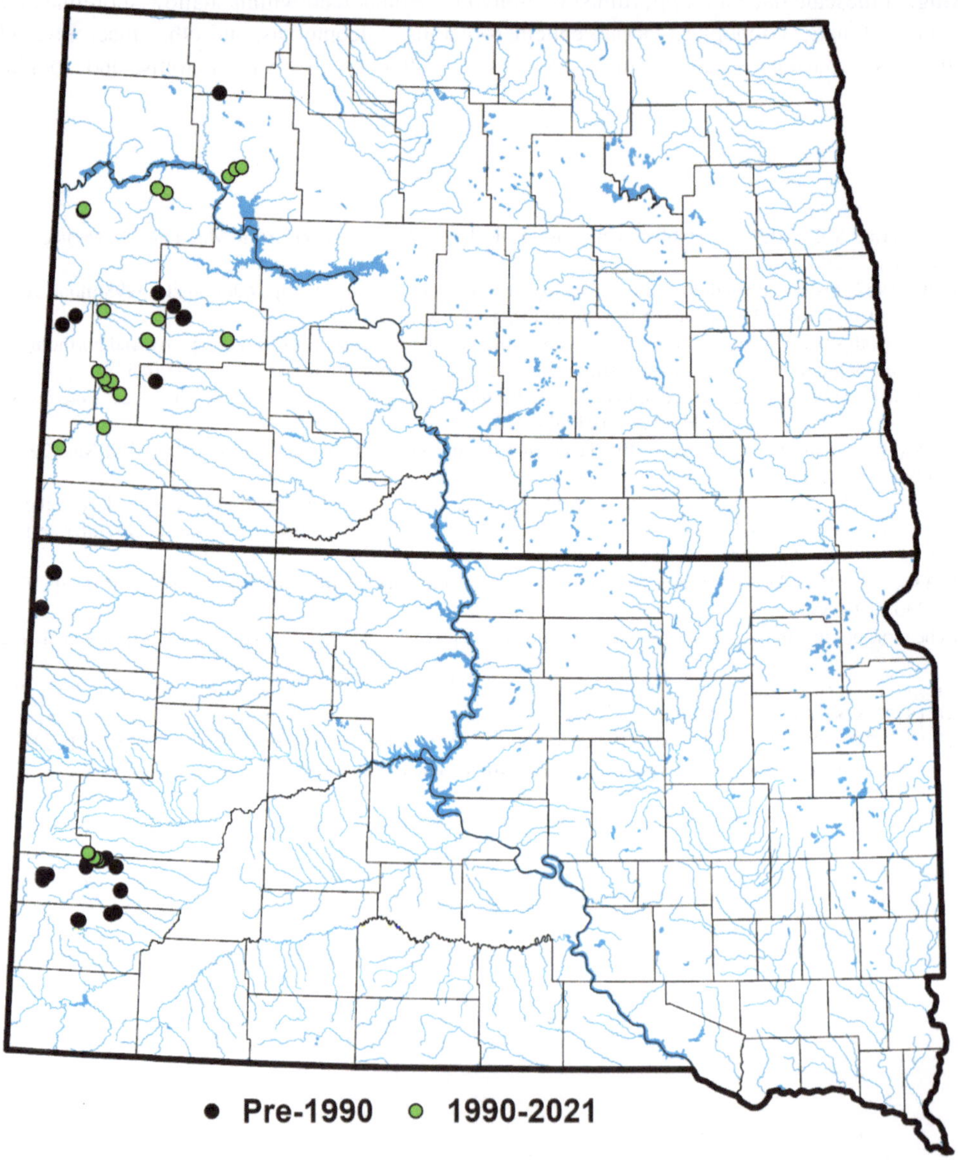

● **Pre-1990**   ○ **1990-2021**

Etymology: *Couesius plumbeus* – *Couesius* = referencing Dr. Elliot Coues, American historian and ornithologist; *plumbeus* = lead colored, referencing dorsal coloration or lateral stripe.

Description:

Body – elongated, rather cylindrical anteriorly; moderately compressed posteriorly.

Color –

> *Dorsally* – blueish-gray to brown.
> *Laterally* – light gray to silver; faint mottling and dusky, faint lead-colored lateral stripe. *Ventrally* – silver to white.
> *Fins* – generally clear to light gray; no distinctive markings.

Head – small, moderately dorsoventrally compressed.

Snout – bluntly pointed, slightly extending beyond the mouth.

Eyes – moderately large, laterally on upper portion of head.

Mouth – relatively small.

> *Jaws* – upper jaw barely reaching to anterior end of eye.
> *Lips* – thin.

Teeth – pharyngeal tooth pattern 2,4-4,2.

Barbels – one on each corner of the mouth; moderately long, prominent.

Gill rakers – 4 to 8; short.

Dorsal fin – 8 rays; nearly straight to slightly concave distal end, rounded tip.

Caudal peduncle – moderately thick.

Caudal fin – forked.

Anal fin – 8 rays; straight distal end.

Pelvic fins – 8 rays; abdominal; insertions anterior to dorsal fin insertion.

Pectoral fins – 15 to 17 rays; short; when depressed do not extend beyond pelvic fin insertions.

Lateral line – complete; 53 to 70 small cycloid scales in series.

Sexual dimorphism – spawning males have red to orange on the bases of the paired fins, red near the mouth and opercle, and a more pronounced lateral stripe; spawning females have faint red on the bases of the pectoral fins; both sexes develop small breeding tubercles on the head, cheeks, dorsal side of the body, and paired fins.

Similar Species: Lake chub closely resemble creek chub. Creek chub have a large mouth with an upper jaw that extends past the anterior edge of the eye and a large dark spot on the anterior basal portion of the dorsal fin. Creek chub barbels are tucked into the upper lip fold away from the corners of the mouth, whereas lake chub barbels are on the corners of the mouth.

Distribution and Habitat: The native range of lake chub extends from the Atlantic Ocean in Canada and the United States west to the Rocky Mountains, north and west to the Pacific Ocean in British Columbia, and to the Arctic Ocean in northwestern Canada, Alaska, and Labrador (Lee et al. 1980; Darveau et al. 2012). In North Dakota it primarily occurs in the upper Missouri, Little Missouri, and Knife rivers, and is considered rare in Lake Sakakawea (Owen et al. 1981). Lake chub are also considered rare in South Dakota, with reports from the Little Missouri and Cheyenne rivers, and most recently in Boxelder Creek, a tributary of the Cheyenne River (Isaak et al. 2003; Hoagstrom et al. 2006). Lake chub inhabit cool, clear waters of lakes and small streams that experience seasonal variations in water temperature from approximately 4 °C to above 20 °C (McPhail and Lindsey 1970; Evermann and Cox 1896; Darveau et al. 2012). They often occur in large schools within shallow waters over sand or gravel substrate and are tolerant of moderate turbidity (McPhail and Lindsey 1970; Becker 1983).

Reproduction: Spawning takes place when water temperatures reach approximately 10 °C, which may be in spring or summer depending on location (Brown et al. 1970; Brown 1971; Becker 1983). Upstream spawning migrations may be up to 3 km (Brown et al. 1970). Sexual maturity occurs at age-3 or upon reaching a total length of approximately 95 mm (Brown et al. 1970; Brown 1971). Males display aggressive behavior during spawning but are not territorial (Brown et al. 1970). Lake chub males court females with a combination of nudging and persistent pursuit movements (Brown et al. 1970). Spawning consists of a single male pressing a single female against substrate or a rock and vibrating his body against hers to initiate eggs and milt release (Brown 1971). This spawning act lasts approximately one second and occurs several times between the same pair with a small number of eggs released each time (Brown 1971). Lake chub are broadcast spawners, with eggs dispersed over silt, gravel, or rocky substrate (Brown et al. 1970; Becker 1983). No nest is constructed, and no parental care is given (Brown et al. 1970). Fecundity generally increases with female size, ranging from 5,290 to 6,630 eggs for females with total lengths of 168 to 181 mm (Brown 1971; Becker 1983). Nonadhesive yellow eggs are approximately 1 to 3 mm in diameter (Brown et al. 1970; Brown 1971; Becker 1983). Hatching occurs in approximately ten days at a water temperature of 13 °C (Brown 1971).

Age and Growth: Information on age and growth of lake chub is scarce, especially in the Dakotas. Total lengths at age of lake chub from Montana were 64 mm at age-1, 114 mm at age-2, and 140 mm at age-3 (Brown 1971). Similar lengths at age were reported from Wisconsin (Becker 1983). Lake chub can reach a total length of 227 mm (Scott and Crossman 1973). Longevity is 7 years, with few individuals surviving beyond age-5 (Becker 1983).

Food and Feeding: Lake chub are likely visual feeders because of the lack of external taste buds (Davis and Miller 1967). Juvenile diets consist primarily of small microcrustaceans such as copepods and cladocerans (Becker 1983). Adults primarily feed on small aquatic invertebrates such as Ephemeroptera, Diptera, and Odonata larvae (Becker 1983). Lake chub have also been known to consume plankton and a small number of crustaceans (Becker 1983).

References

Becker GC (1983) Fishes of Wisconsin. University of Wisconsin Press, Madison

Brown CJD (1971) Fishes of Montana. Montana State University Press, Bozeman

Brown JH, Hammer UT, Koshinsky GD (1970) Breeding biology of the lake chub, *Couesius plumbeus*, at Lac la Ronge, Saskatchewan. J Fish Res Board Can 27:1005–1015. https://doi.org/10.1139/f70-117

Darveau CA, Taylor EB, Schulte PM (2012) Thermal physiology of warm-spring colonists: variation among lake chub (Cyprinidae: *Couesius plumbeus*) populations. Physiol Biochem Zool 85:607–617. https://doi.org/10.1086/665539

Davis BJ, Miller RJ (1967) Brain patterns in minnows of the genus Hybopsis in relation to feeding habits and habitat. Copeia 1967:1–39. https://doi.org/10.2307/1442172

Evermann BW, Cox UO (1896) Report upon the fishes of the Missouri River basin. US Government Printing Office, Washington, DC

Hoagstrom CW, Hayer CA, Kral JG, Wall SS, Berry CR Jr (2006) Rare and declining fishes of South Dakota: a river drainage scale perspective. Proc S D Acad Sci 85:171–211

Isaak DJ, Hubert WA, Berry CR Jr (2003) Conservation assessment for lake chub, mountain sucker, and finescale dace in the Black Hills National Forest, South Dakota and Wyoming. US Dep Ag, Forest Serv, Rocky Moutnain Region. https://www.fs.usda.gov/Internet/FSE_DOCUMENTS/fsm9_012156.pdf. Accessed 24 Apr 2023

Lee DS, Gilbert CR, Hocutt CH, Jenkins RE, McAllister DE, Stauffer JR (1980) Atlas of North American freshwater fishes. North Carolina Biological Survey Publication, 1980-12. https://doi.org/10.5962/bhl.title.141711

McPhail JD, Lindsey CC (1970) Freshwater fishes of northwestern Canada and Alaska. Bulletin 173. Fisheries Research Board of Canada, Ottawa

Owen JB, Elsen DS, Russell GW (1981) Distribution of fishes in North and South Dakota Basins affected by the Garrison Diversion Unit. University of North Dakota Press, Grand Forks

Scott WB, Crossman EJ (1973) Freshwater fishes of Canada. Fisheries Research Board of Canada, Bulletin 184

6.14.8 Red Shiner *Cyprinella lutrensis*

(Baird & Girard, 1853)

Nuptial Male
Female

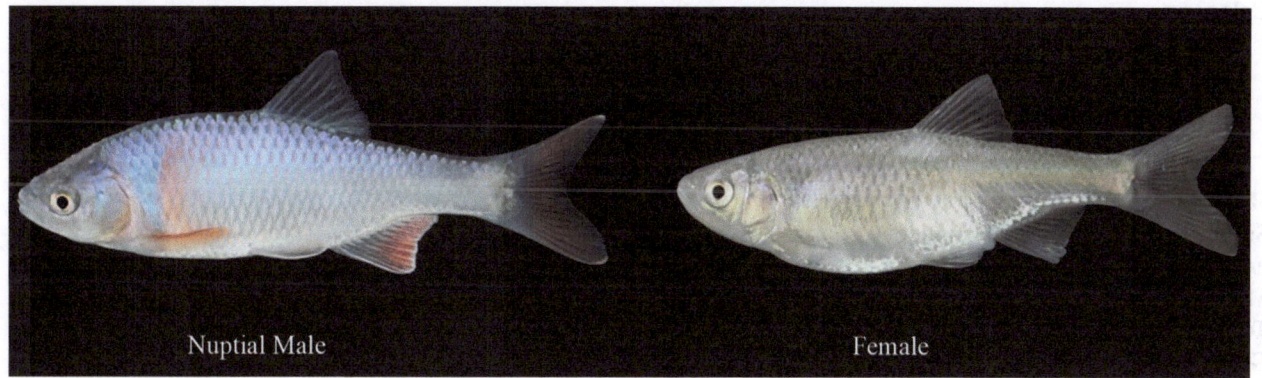

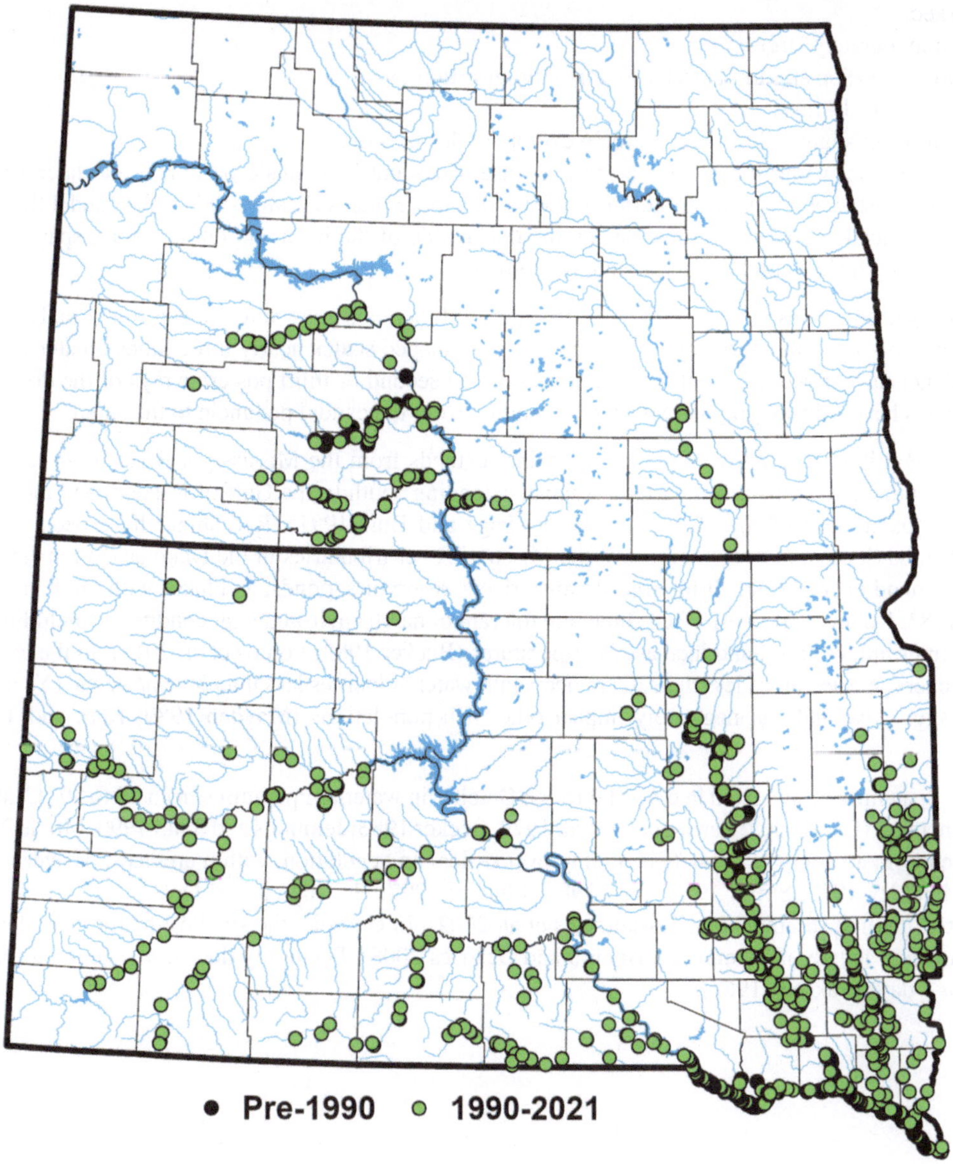
Pre-1990
1990-2021

Etymology: *Cyprinella lutrensis – Cyprinella* = carp; *lutrensis* = Otter, first documented in Otter Creek, Arkansas.

Description:

Body – deep, laterally compressed.
Color –

> *Dorsally* – light olive-green to dusky gray; scales outlined in darker pigment, creating a diamond-shaped pattern.
> *Laterally* – silvery-blue; scales outlined in a darker pigment, creating a diamond-shaped pattern.
> *Ventrally* – silver to white.
> *Fins* – dorsal: clear to light gray; other: clear to faint reddish-orange.

Head – short, deep.
Snout – short, tapering to a blunt point.
Eyes – small, laterally on head.
Mouth – small, terminal, oblique.
Teeth – pharyngeal tooth pattern 0,4-4,0.
Gill rakers – narrow, short.
Dorsal fin – 8 rays; slightly rounded distal end.
Caudal peduncle – short, thick.
Caudal fin – forked.
Anal fin – 9 rays; moderately large.
Pelvic fins – 8 rays; insertion slightly anterior to dorsal fin insertion.
Pectoral fins – generally 14 rays.
Lateral line – complete; slightly decurved, 32 to 36 cycloid scales in series.
Sexual dimorphism – spawning males have a more intense iridescent blue shade on dorsal and lateral sides, a defined, darker blue to purplish crescent mark posterior of the head, a more intense orange-red on the caudal fin, pelvic fins, pectoral fins, and the head dorsally, and tubercles on the dorsal side of the head and on the snout; spawning females have smaller tubercles on the head and much less-intense coloration.

Similar Species: Red shiner resemble common shiner, spotfin shiner, and spawning Topeka shiner males. Common shiner are generally larger in size, with a large eye and large diamond shaped scales nearly three times greater in depth than their width. Spotfin shiner have darker pigment forming a spot on the second or third posterior rays of the dorsal fin. Spawning Topeka shiner males have a dusky gray lateral stripe extending from the caudal peduncle to the snout.

Distribution and Habitat: The native range of red shiner extends from the Mississippi River basin from Wyoming and South Dakota to southern Wisconsin and eastern Indiana, extending south to the Gulf of Mexico, and including Louisiana, Texas, and New Mexico west of the Mississippi River (Page and Burr 1991). Red shiner have been widely introduced throughout the United States and occur in most of the Missouri River tributaries in the Dakotas (Robins et al. 1991). It is a habitat generalist found in creeks, small-to-large streams, rivers, reservoirs, ponds, and lakes with sand, gravel, and silt substrates (Becker 1983; Robins et al. 1991). Red shiner are tolerant of harsh environmental conditions, including high turbidity, siltation, low or intermittent flows, and increased temperatures (Becker 1983; Poulos et al. 2012). In the lower Platte River, Nebraska, red shiners selected areas less than 30 cm deep with water velocities less than 30 cm/second (Yu and Peters 2002). Red shiner are aggressive and can negatively impact other fish populations (Mayden 1989; Karp and Tyus 1990; Nico et al. 2014).

Reproduction: Spawning occurs from late April through October in water temperatures from 16 to 29 °C, allowing populations to rapidly increase (Cross 1967; Farringer III et al. 1979; Becker 1983; Jennings and Saiki 1990; Moyle 2002). Spawning occurs intermittently in a variety of habitats including over riffles, on vegetation, within crevices, or over nests constructed by other fish (Minckley 1972; Smith 1979; Gale 1986; Pflieger 1997; Herrington and DeVries 2008). Sexual maturity is reached as early as the first summer (Marsh-Matthews et al. 2002). In Georgia, the smallest mature females collected from streams had a standard length of 31 mm (Herrington and DeVries 2008). Fecundity does not depend on female size. Eggs hatch within 3 or 4 days (Saksena 1962).

Age and Growth: Rapid growth occurs within the first season (Yildirim and Peters 2006). Larvae in a laboratory reached a total length of 16 mm 34 days after hatching (Saksena 1962). Fork lengths at age of males from the lower Platte River in Nebraska were age-0, 30 mm; age-1, 36 mm; age-2, 46 mm; age-3, 52 mm FL; age-4, 57 mm (Yildirim and Peters 2006). Females in the Platte River were longer and heavier than males in earlier life stages, but in later years, males were longer and heavier (Yildirim and Peters 2006). Red shiner can reach a total length of 90 mm, with 4 years longevity (Pflieger 1997; Yildirim and Peters 2006).

Food and Feeding: Red shiner are sight-feeding, opportunistic, generalist omnivores throughout the water column (Becker 1983). They eat plankton, algae, and the larvae and adults of both terrestrial and aquatic insects (Becker 1983). Red shiner also consume native larval fish in areas where they have been introduced (Ruppert et al. 1993).

References

Becker GC (1983) Fishes of Wisconsin. University of Wisconsin Press, Madison

Cross FB (1967) Handbook of fishes in Kansas. University Press of Kansas, Lawrence

Farringer RT III, Echelle AA, Lehtinen SF (1979) Reproductive cycle of the red shiner, *Notropis lutrensis*, in central Texas and south-central Oklahoma. Trans Am Fish Soc 108:271–276. https://doi.org/10.1577/1548-8659

Gale WF (1986) Intermediate fecundity and spawning behavior of captive red shiners-fractional, crevice spawners. Trans Am Fish Soc 115:429–437. https://doi.org/10.1577/1548-8659

Herrington SJ, DeVries DR (2008) Reproductive and early life history of nonindigenous red shiner in the Chattahoochee River drainage, Georgia. Southwest Nat 7:413–428. https://doi.org/10.1656/1528-7092-7.3.413

Jennings MR, Saiki MK (1990) Establishment of red shiner, *Notropis lutrensis*, in the San Joaquin Valley, California. Vol 76. California Dep Fish Game, pp 46–57

Karp CA, Tyus HM (1990) Behavioral interactions between young Colorado squawfish and six fish species. Copeia 1990:25–34. https://doi.org/10.2307/1445818

Marsh-Matthews E, Matthews WJ, Gido KB, Marsh RL (2002) Reproduction by young-of-year red shiner (*Cyprinella lutrensis*) and its implications for invasion success. Southwest Nat 47:605–610. https://doi.org/10.2307/3672666

Mayden RL (1989) Phylogenetic studies of North American minnows, with emphasis on the genus Cyprinella (Teleostei: Cypriniformes). University of Kansas Museum of Natural History, Lawrence. https://doi.org/10.5962/bhl.title.5480

Minckley WL (1972) Notes on the spawning behavior of red shiner, introduced into Burro Creek, Arizona. Southwest Nat 17:101–103. https://doi.org/10.2307/3669849

Moyle PB (2002) Inland fishes of California. University of California Press, Los Angeles

Nico L, Fuller P, Neilson M (2014) Cyprinella lutrensis. USGS Nonindigenous Aquatic Species Database. https://nas.er.usgs.gov/queries/FactSheet.aspx?SpeciesID=518. Accessed 24 Apr 2023

Page LM, Burr BM (1991) Peterson field guide to freshwater fishes of North America North of Mexico. Houghton Mifflin Harcourt, Boston

Pflieger WL (1997) The fishes of Missouri, revised edn. Missouri Dep Cons, Jefferson City

Poulos HM, Chernoff B, Fuller PL, Butman D (2012) Mapping the potential distribution of the invasive red shiner, *Cyprinella lutrensis* (Teleostei: Cyprinidae) across waterways of the conterminous United States. Aquat Invasions 7:377–385. https://doi.org/10.3391/ai.2012.7.3.009

Robins CR, Bailey RM, Bond CE, Brooker JR, Lachner EA, Lea RN, Scott WB (1991) common and scientific names of fishes from the United States and Canada. American Fisheries Society Special Publication No 20

Ruppert JB, Muth RT, Nesler TP (1993) Predation on fish larvae by adult red shiner, Yampa and Green Rivers, Colorado. Southwest Nat 38:397–399. https://doi.org/10.2307/3671627

Saksena VP (1962) The post-hatching stages of red shiner, *Notropis lutrensis*. Copeia 1962:539–544. https://doi.org/10.2307/1441175

Smith PW (1979) The fishes of Ohio. University of Illinois Press, Urbana

Yildirim A, Peters EJ (2006) Life history characteristics of red shiner, *Cyprinella lutrensis*, in the lower Platte River, Nebraska, USA. J Freshw Ecol 21:307–314. https://doi.org/10.1080/02705060.2006.9665000

Yu S, Peters EJ (2002) Diel and seasonal habitat use by red shiner (*Cyprinella lutrensis*). Zool Stud 41:229–235

6.14.9   Spotfin Shiner *Cyprinella spiloptera*

(Cope, 1867)

Nuptial Male   Female

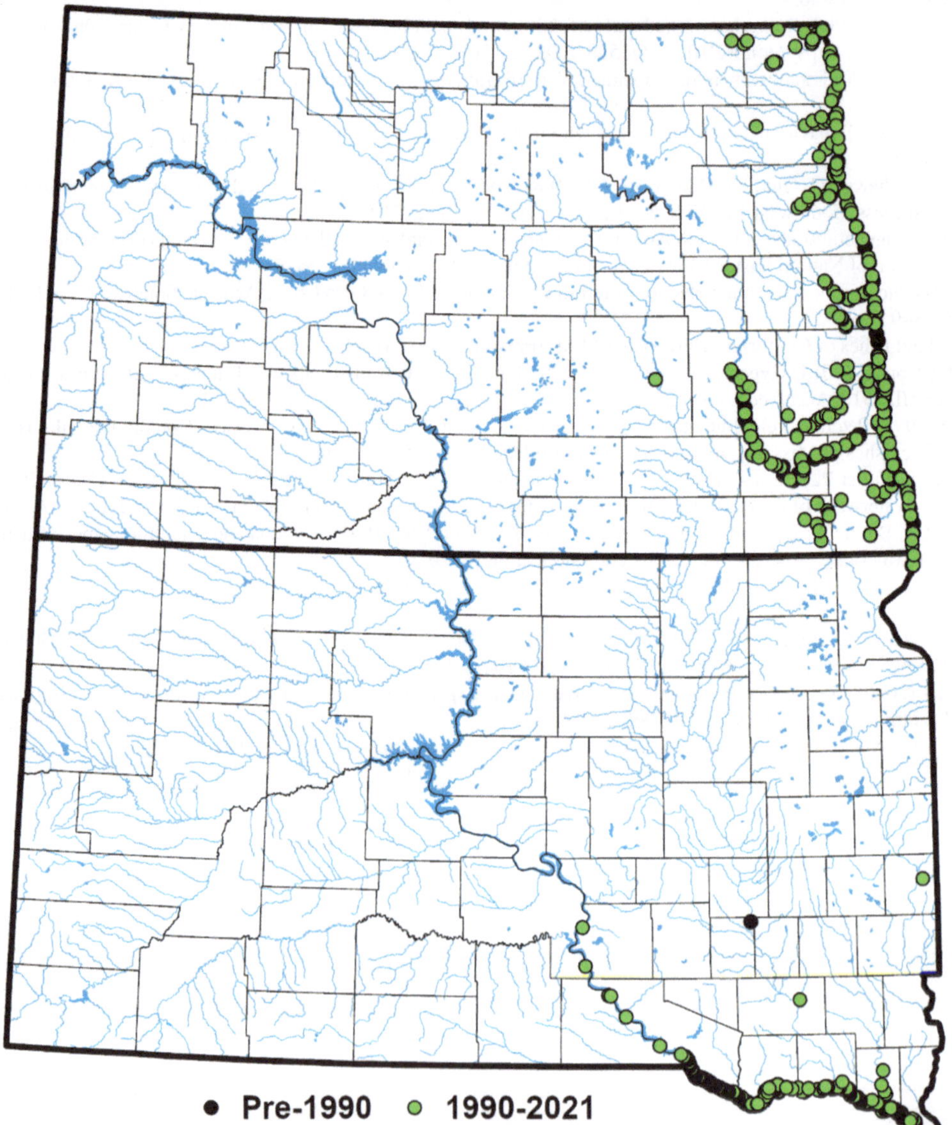

● **Pre-1990**   ○ **1990-2021**

Etymology: *Cyprinella spiloptera* – *Cyprinella* = carp; *spiloptera* = spot, wing, or fin, referencing the dorsal fin dark spot.

Description:

Body – fusiform, moderately deep, laterally compressed; body depth more than 3.5 times standard length.
Color –

> *Dorsally* – dark to dusky gray.
> *Laterally* – silver, sometimes faint iridescent bluish-purple tint.
> *Ventrally* – silver to white.
> *Fins* – transparent; dorsal: single dark spot on second or third posterior rays.

Head – small.
Snout – short, tapering to a blunt point.
Eyes – small, laterally on head.
Mouth – large, terminal, strongly oblique.

> *Jaws* – upper jaw extends below posterior nostril.

Teeth – pharyngeal tooth pattern 1,4-4,1.
Gill rakers – 8 to 10; short, conical.
Dorsal fin – 8 rays.
Caudal peduncle – slightly elongated, uniform thickness.
Caudal fin – forked.
Anal fin – 8 or 9 rays; moderately large.
Pectoral fins – 13 to 15 rays.
Pelvic fins – 8 rays; abdominal; insertion slightly anterior to the dorsal fin insertion.
Lateral line – complete; slightly decurved, 35 to 39 cycloid scales in series.
Sexual dimorphism – spawning males have an intense shade of blue on dorsal and lateral sides, a dusky stripe on posterior lateral sides, yellowish-orange anal, pelvic, and pectoral fins, and tubercles on the snout, dorsal side of the head, lower jaw, and pectoral fins; spawning females have occasional small tubercles on the dorsal side of the head and on mid-dorsal scales.

Similar Species: Spotfin shiner resembles red shiner, common shiner, and sand shiner. Red shiner have a deeper body with a body depth less than 3.5 times standard length and lack a small dark spot of the dorsal fin. Common shiner are generally larger in size, with a large eye and large diamond-shaped scales nearly three times deeper than wide. Sand shiner have small, paired, dark specks (mouse tracks) outlining the lateral line, dorsal scales outlined in darker pigment creating a cross-hatched pattern, and a prominent, narrow mid-dorsal stripe that widens into a wedge-shaped spot at the anterior base of the dorsal fin.

Distribution and Habitat: The native range of spotfin shiner ecompasses the Atlantic Slope and St. Lawrence drainages from New York to eastern North Carolina, throughout the Great Lakes, Hudson Bay and Mississippi River drainages, and eastern North Dakota south to eastern Oklahoma (Page and Burr 1991). In North Dakota, spotfin shiner occur in the Red River of the North and its major tributaries, as well as the Sheyenne, James, Cannonball, Heart, and Knife rivers. They occur in southeastern South Dakota in the Big Sioux, Vermillion, James, and Missouri rivers. Spotfin shiner inhabit shallower riffle or pooled areas of medium-to-large rivers and larger lakes and reservoirs (Becker 1983; Owen et al. 1981). Spotfin shiner are most frequently found over sand, mud, and gravel substrates, and are tolerant of turbid and silt-laden waters (Becker 1983).

Reproduction: Spawning extends from May to September, with activity most frequent in water temperatures of 21 to 24 °C (Becker 1983; Cross and Collins 1995). Spotfin shiners are fractional spawners, using crevices in rocky substrate (Gale and Gale 1977). Males aggressively defend relatively small spawning territories from other males (Pflieger 1965; Gale and Gale 1977; Becker 1983). During spawning territorial encounters, the males, with erect dorsal fins, swim side-by-side, occasionally nudging each other with their snouts, or grabbing the anal or pelvic fins with their mouths (Pflieger 1965; Gale and Gale 1977; Becker 1983). When a female enters a spawning territory, the male makes several prespawning passes before luring the female into a crevice (Gale and Gale 1977; Becker 1983). The pair will then vibrate their bodies together to stimulate the release of gametes (Pflieger 1965; Gale and Gale 1977). Roughly 10 to 97 eggs are released during a single spawn, with as many as 169 to 945 eggs deposited during a single spawning season (Gale and Gale 1977; Becker 1983). The adhesive eggs are approximately 1 mm in diameter, and hatch within approximately 5 days in water temperatures near 22 °C (Hubbs and Cooper 1936; Gale and Gale 1977; Becker 1983).

Age and Growth: Information on spotfin shiner age and growth in the Dakotas is scarce. In Wisconsin, total-lengths-at-age were 53 mm at age-1, 62 mm at age-2, and 75 mm at age-3 (Becker 1983). Males are larger than same-age females, but females live longer than males (Becker 1983). Spotfin shiner can reach a total length of 108 mm and live for 4 years (Stone 1940).

Food and Feeding: Spotfin shiner are opportunistic omnivores. They feed diurnally on a variety of larval and adult aquatic insects, including those in the orders trichoptera and diptera, as well as terrestrial insects, small fish, algae, vegetative matter, and even their own eggs (Pflieger 1965; White and Wallace 1973; Gale and Gale 1977; Becker 1983; Boschung Jr and Mayden 2004).

References

Becker GC (1983) Fishes of Wisconsin. University of Wisconsin Press, Madison

Boschung HT Jr, Mayden RL (2004) Fishes of Alabama. Smithsonian Books, Washington, DC

Cross FB, Collins JT (1995) Fishes in Kansas. University of Kansas, Lawrence

Gale WF, Gale CA (1977) Spawning habits of spotfin shiner (*Notropis spilopterus*) – a fractional crevice spawner. Trans Am Fish Soc 100:170–177. https://doi.org/10.1577/1548-8659

Hubbs CL, Cooper GP (1936) Minnows of Michigan. Cranbrook Institute of Science Bulletin No. 8

Owen JB, Elsen DS, Russell GW (1981) Distribution of fishes in North and South Dakota Basins affected by the Garrison Diversion Unit. University of North Dakota Press, Grand Forks

Page LM, Burr BM (1991) Peterson field guide to freshwater fishes of North America North of Mexico. Houghton Mifflin Harcourt, Boston

Pflieger WL (1965) Reproductive behavior of the minnows *Notropis spilopterus* and *Notropis whipplii*. Copeia 1965:1–8. https://doi.org/10.2307/1441231

Stone UB (1940) Studies on the biology of the satinfin minnows, *Notropis analostanus* and *Notropis spilopterus*. Dissertation, Cornell University

White ST, Wallace DC (1973) Diel changes in the feeding activity and food habits of the spotfin shiner, *Notropis spilopterus* (Cope). Am Midl Nat 90:200–205. https://doi.org/10.2307/2424281

6.14.10   Bigmouth Shiner *Ericymba dorsalis*

(Agassiz, 1854)

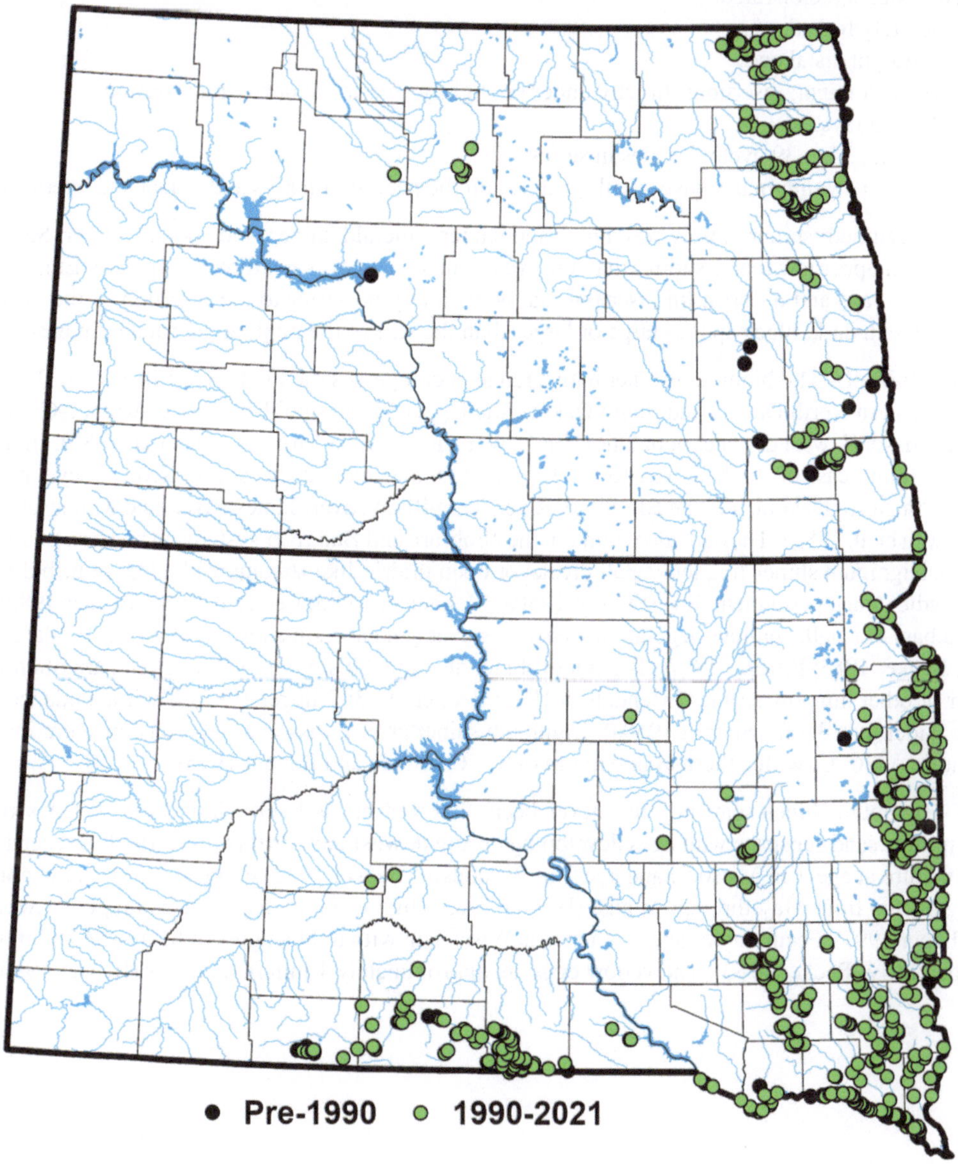

● **Pre-1990**   ○ **1990-2021**

Etymology: *Ericymba dorsalis* – *Ericymba* = ericymba meaning very cavity; referencing externally visible mucous channels on interopercle, suborbital and dentary bones; *dorsalis* = pertaining to the back.

Description:

Body – moderately fusiform, slender, slightly laterally compressed.
Color –

> *Dorsally* – brown, tan, or olive; scales outlined in darker pigment creating a crosshatched pattern.
> *Laterally* – silver; small, paired, dark specks or "mouse tracks" outlining the lateral line.
> *Ventrally* – silver to white.
> *Fins* – little to no pigment; dorsal: prominent, narrow mid-dorsal stripe continuous on each side.

Head – elongated, lower surface distinctly flattened.
Snout – bluntly pointed; slightly overhangs the mouth.
Eyes – large, laterally on head; directed upwards with lower margins of pupils visible when viewed from above.
Mouth – large, slightly subterminal.

> *Jaws* – lower jaw extends to the middle of the eye; length of upper jaw greater than eye diameter.

Teeth – pharyngeal teeth 1,4-4,1.
Gill rakers – approximately 8; short.
Dorsal fin – 8 rays.
Caudal peduncle – slender, elongated.
Caudal fin – moderately forked.
Anal fin – 8 rays; straight distal end.
Pelvic fins – 8 rays; abdominal; insertions directly under or slightly anterior to dorsal fin insertion.
Pectoral fins – 14 or 15 rays.
Lateral line – complete; 32 to 39 cycloid scales in series.
Sexual dimorphism – spawning males have small tubercles on the pectoral fin rays and dorsal and lateral sides of the head.

Similar Species: Bigmouth shiner closely resemble sand shiner, emerald shiner, and spottail shiner. Sand shiner have an anal fin with 7 rays, an upper jaw that is shorter than the eye diameter, and eyes not directed upward. Emerald shiner have an anal fin with 10 to 13 rays and a dorsal fin insertion that is distinctly posterior to the insertion of the pelvic fins. Spottail shiner have a prominent, dark, black spot as large or larger than the size of the pupil at the caudal fin base.

Distribution and Habitat: The bigmouth shiner native range encompasses the Platte River in eastern Wyoming and northeastern Colorado, eastern North Dakota and South Dakota, the Upper Mississippi River, St. Lawrence-Great Lakes, Hudson Bay, Illinois, and Missouri, with disjunct populations in New York and West Virginia (Lee et al. 1980). In the Dakotas, they are predominantly found east of the Missouri River. Bigmouth shiner inhabit the Souris River, Sheyenne River, and Red River of the North in North Dakota, and the Big Sioux River, Vermillion River, James River, and Minnesota River in eastern South Dakota (Owen et al. 1981). They are also found in the Niobrara and Bad river systems in western South Dakota (Bailey and Allum 1962). Bigmouth shiner frequent small creeks or open prairie-like streams with clear to turbid water, slight current, moderate gradient, and permanent flow (Starrett 1950a; Mendelson 1975; Lee et al. 1980; Becker 1983; Pflieger 1997). They prefer sand bars or small, shallow, vegetation-free, areas over sand and silt substrate (Starrett 1950a; Mendelson 1975; Lee et al. 1980; Becker 1983; Pflieger 1997). Bigmouth shiner also inhabit sloughs and lakes and are often found schooling with sand shiner or other minnow species (Mendelson 1975; Becker 1983; Pflieger 1997). Their abundance decreases when streams become wider than 3.1 m (Starrett 1950a). Bigmouth shiner critical dissolved oxygen concentration is 1.02 mg/L at a water temperature of 26 °C, with a thermal maximum of 37 °C (Smale and Rabeni 1995).

Reproduction: Bigmouth shiner spawning has never been observed but is likely similar to other *Notropis* species that spawn midstream over sand substrate with eggs developing during downstream drift (Moore 1944; Becker 1983). Bigmouth shiner sexually mature at age-1 in Missouri and age-2 in Wisconsin (Becker 1983; Pflieger 1997). In the northern part of its range, spawning occurs from May through August (Becker 1983). In the southern part of its range, spawning occurs from June to July (Pflieger 1997). Fecundity of age-2 females in Wisconsin with total lengths of 61 mm and 62 mm was 1,000 and 1,275 eggs, respectively (Becker 1983). The yellow eggs are approximately 1 mm in diameter (Becker 1983).

Age and Growth: Age-0 bigmouth shiner from Missouri were 20 to 41 mm long (Pflieger 1997). Total lengths at age of bigmouth shiner from Wisconsin were 45 mm at age-0, 61 mm at age-1, and 67 mm at age-2 (Becker 1983). Bigmouth shiner can reach a total length of 80 mm with a longevity of 4 years (Lee et al. 1980; Becker 1983).

Food and Feeding: Bigmouth shiner are benthic omnivores, foraging on ephemeropterans, dipterans, and other small aquatic insects, as well as plant material, detritus, and zooplankton (Starrett 1950b; Becker 1983; Pflieger 1997). Sight is presumed to be less important to bigmouth shiners than taste (Pflieger 1997). In an aquarium setting, individuals rapidly swam along the bottom ingesting mouthfuls of sand, sorting the food with their mouths, and then discarding the undesired particles orally or through their gill openings (Pflieger 1997).

References

Bailey RM, Allum MO (1962) Fishes of South Dakota. University of Michigan, Ann Arbor. https://doi.org/10.3998/mpub.9690435

Becker GC (1983) Fishes of Wisconsin. University of Wisconsin Press, Madison

Lee DS, Gilbert CR, Hocutt CH, Jenkins RE, McAllister DE, Stauffer JR (1980) Atlas of North American freshwater fishes. North Carolina Biological Survey Publication 1980-12. https://doi.org/10.5962/bhl.title.141711

Mendelson J (1975) Feeding relationships among species of Notropis (Pisces: Cyprinidae) in a Wisconsin stream. Ecol Monogr 45:199–230. https://doi.org/10.2307/1942422

Moore GA (1944) Notes on the early life history of *Notropis girardi*. Copeia 1944:209–214. https://doi.org/10.2307/1438675

Owen JB, Elsen DS, Russell GW (1981) Distribution of fishes in North and South Dakota Basins affected by the Garrison Diversion Unit. University of North Dakota Press, Grand Forks

Pflieger WL (1997) The fishes of Missouri, revised edn. Missouri Dep Cons, Jefferson City

Smale MA, Rabeni CF (1995) Hypoxia and hyperthermia tolerances of headwater stream fishes. Trans Am Fish Soc 124:698–710. https://doi.org/10.1577/1548-8659

Starrett WC (1950a) Distribution of the fishes of Boone County, Iowa with special reference to the minnows and darters. Am Midl Nat 43:112–127. https://doi.org/10.2307/2421882

Starrett WC (1950b) Food relationships of the minnows of the Des Moines River, Iowa. Ecology 31:216–233. https://doi.org/10.2307/1932388

6.14.11   Spottail Shiner *Hudsonius hudsonius*

(Clinton, 1824)

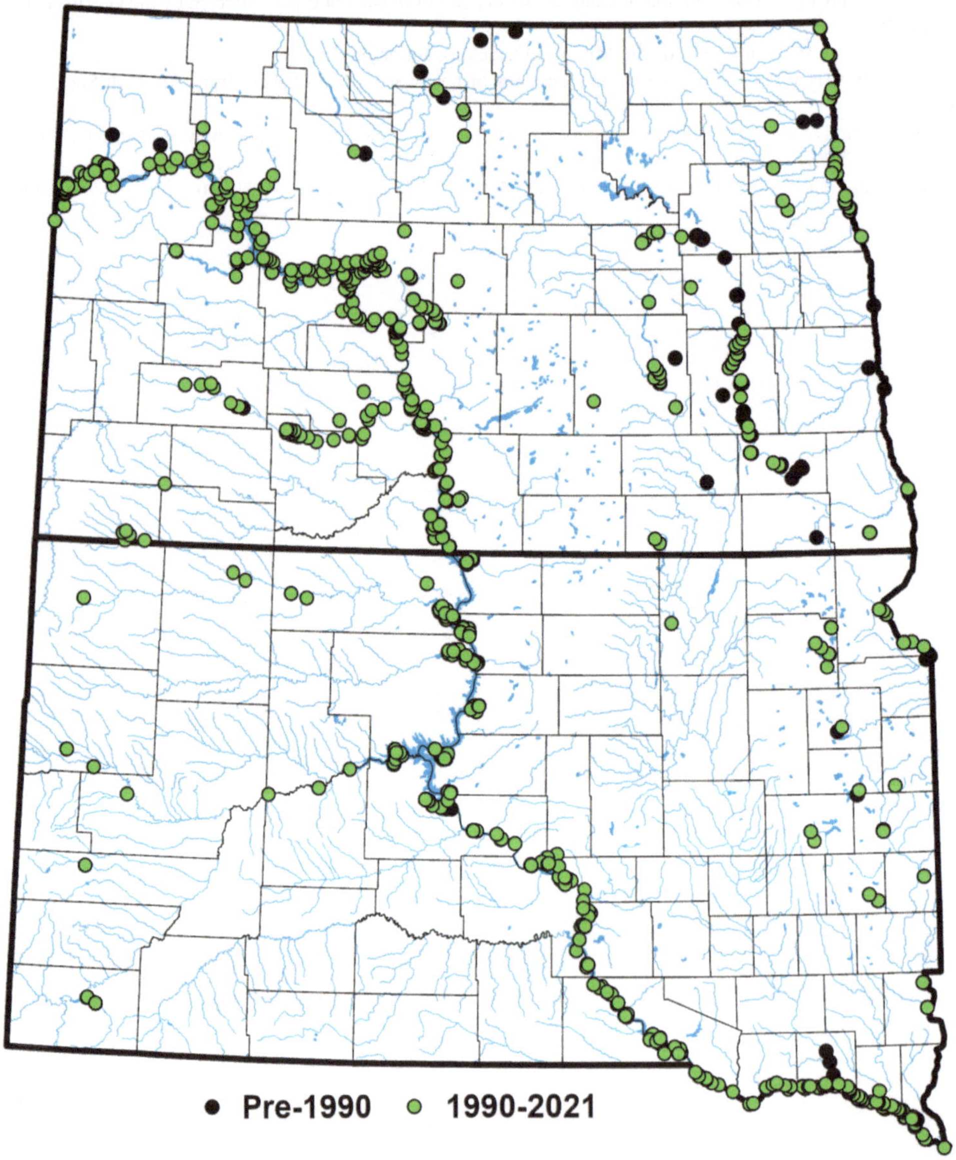

● Pre-1990   ○ 1990-2021

Etymology: *Hudsonius hudsonius* – *Hudsonius* = referencing the Hudson River.

Description:

Body – fusiform, slightly robust, laterally compressed posteriorly.
Color –

> *Dorsally* – olive, tan, or dusky silver.
> *Laterally* – silver with no distinct markings.
> *Ventrally* – silver to white.
> *Fins* – generally clear; prominent, dark, black spot as large or larger than size of pupil at caudal fin base that may be diffuse or absent in larger adults from turbid waters.

Head – small.
Snout – bluntly rounded, slightly protruding beyond mouth.
Eyes – large, laterally on head; diameter greater than upper jaw length.
Mouth – small, terminal to slightly subterminal.
> *Jaws* – upper jaw extending to eye anterior edge.

Teeth – pharyngeal tooth pattern variable between 2,4-4,2 and 0,4-4,0.
Gill rakers – 9; short.
Dorsal fin – 8 rays; pointed tip.
Caudal peduncle – moderately short, slender.
Caudal fin – forked.
Anal fin – 8 rays; slightly concave distal end.
Pelvic fins – 8 rays; abdominal; insertions directly inferior to dorsal fin insertion.
Pectoral fins – 13 to 15 rays.
Lateral line – complete; 36 to 42 cycloid scales in series.
Juveniles – basio-caudal spot sometimes more distinct than adults.
Sexual dimorphism – spawning males have minute tubercles on pectoral fin anterior rays, dorsal side of body, and dorsal and lateral sides of head; spawning females have fewer tubercles on the dorsal sides of the head and body.

Similar Species: Spottail shiner cosely resemble emerald shiner, sand shiner, channel shiner, and river shiner but are easily distinguished by a dark, black spot at their caudal fin base.

Distribution and Habitat: Spottail shiner are widely distributed throughout North America from the Mackenzie River in the Northwest Territories of Canada, through the Hudson Bay and upper eastern Mississippi River basin, to the St. Lawrence-Great Lakes to the Atlantic Slope (Lee et al. 1980). They are also widely distributed throughout the Dakotas (Bailey and Allum 1962; Owen et al. 1981). Spottail shiners were introduced in 1973 into Lake Oahe as a forage fish (Barnes 2007). They inhabit shoal waters of large lakes, glacial lakes, rivers, and large creeks with clear water, little to moderate current, little to moderate aquatic vegetation, and sand or gravel substrate (Trautman 1957; Becker 1983). Spottail shiners prefer the shallow waters of lakes but have been collected as deep as 31 m (Wells and House 1974; Wells 1980). In Lake Oahe, spottail shiners appear to make diel on/offshore movements (Sewell 1993).

Reproduction: Spottail shiners spawn prolifically from late May through early June but may continue spawning into August (Griswold 1963; Sewell 1993). They become sexually mature at a total length of approximately 68 mm, with the age of maturity determined by growth rate (Peer 1966; Wells and House 1974). Males are slightly shorter than females at sexual maturity. Many fish move to shallower water prior to spawning. Fecundity generally increases with increasing female length. Spottail shiners with total lengths of 97 to 131 mm from Lake Michigan had fecundities of 915 to 3,709 eggs, but fecundities of 8,898 eggs have also been reported (Wells and House 1974). Eggs are yellow (Wells and House 1974).

Age and Growth: Females grow faster and are generally longer than males (Wells and House 1974). In Lake Oahe, up to 40% of spottail shiner growth occurs between age-0 and age-1 (Sewell 1993). Total lengths at age of spottail shiners from Lake Oahe in July were 36 mm at age-0, 81 mm at age-1, 107 mm at age-2, and 122 mm at age-3 (Sewell 1993). Total lengths at age of females from southeastern Lake Michigan were 63 mm at age-1, 97 mm at age-2, 114 mm at age-3, 123 mm at age-4, and 131 mm at age-5. Older and longer spottail shiners prefer deeper water (Wells and House 1974). They can reach a total length of 143 mm with a longevity of 5 years (Wells and House 1974; Wells 1980).

Food and Feeding: Spottail shiners are opportunistic omnivores (Griswold 1963; Sewell 1993). In Lake Oahe, *Daphnia*, chironomid larvae, and adult diptera were the most consumed prey items, with insects consumed when zooplankton were scarce (Sewell 1993). In Lake Michigan, spottail shiners primarily consumed immature midges such as chironomids, but also ate *Pontoporeia*, fingernail clams, alewife eggs, trout-perch eggs, copepods, cladocerans, and other zooplankton. In Lake Michigan, they ate midges at shallower depths and *Pontoporeia* in deeper water. Spottail shiner also consume smaller quantities of snails, leeches, ostracods, terrestrial insects, and caddisfly larvae (Wells 1980). In Lake Manitoba, spottail shiner fry consumed small plankton and dipteran larvae (Bernard 1972).

References

Bailey RM, Allum MO (1962) Fishes of South Dakota. University of Michigan, Ann Arbor

Barnes ME (2007) Fish hatcheries and stocking practices: past and present. In: Berry C, Higgins K, Willis D, Chipps S (eds) History of fisheries and fishing in South Dakota. S D Dep Game Fish Parks, Pierre, pp 267–294

Becker GC (1983) Fishes of Wisconsin. University of Wisconsin Press, Madison

Bernard DJ (1972) Ecological divergence between emerald and spottail shiners (Notropis) in Lake Manitoba. Dissertation, University of Manitoba

Griswold BL (1963) Food and growth of spottail shiners and other forage fishes of Clear Lake, Iowa. Proc Iowa Acad Sci 70:215–223

Lee DS, Gilbert CR, Hocutt CH, Jenkins RE, McAllister DE, Stauffer JR (1980) Atlas of North American freshwater fishes. North Carolina Biological Survey Publication 1980-12. https://doi.org/10.5962/bhl.title.141711

Owen JB, Elsen DS, Russell GW (1981) Distribution of fishes in North and South Dakota Basins affected by the Garrison Diversion Unit. University of North Dakota Press, Grand Forks

Peer DL (1966) Relationship between size and maturity in the spottail shiner, *Notropis hudsonius*. J Fish Res Board Can 23:455–457. https://doi.org/10.1139/f66-035

Sewell DC (1993) Food habits and distribution of the emerald shiner, *Notropis atherinoides* Rafinesque, and the spottail shiner, *Notropis hudsonius* (Clinton), in Lake Oahe, South Dakota. Dissertation, South Dakota State University

Trautman MB (1957) The fishes of Ohio. Ohio State University Press, Columbus

Wells L (1980) Food of alewives, yellow perch, spottail shiners, trout-perch, and slimy and fourhorn sculpins in southeastern Lake Michigan. US Dep Inter, Fish Wildl Serv, Tech Pap, vol 98

Wells L, House R (1974) Life history of the spottail shiner (*Notropis hudsonius*) in southeastern Lake Michigan, the Kalamazoo River, and western Lake Erie. US Dep Inter, Fish Wildl Serv, Res Rep vol 78

6.14.12   Western Silvery Minnow *Hybognathus argyritis*

Girard, 1856

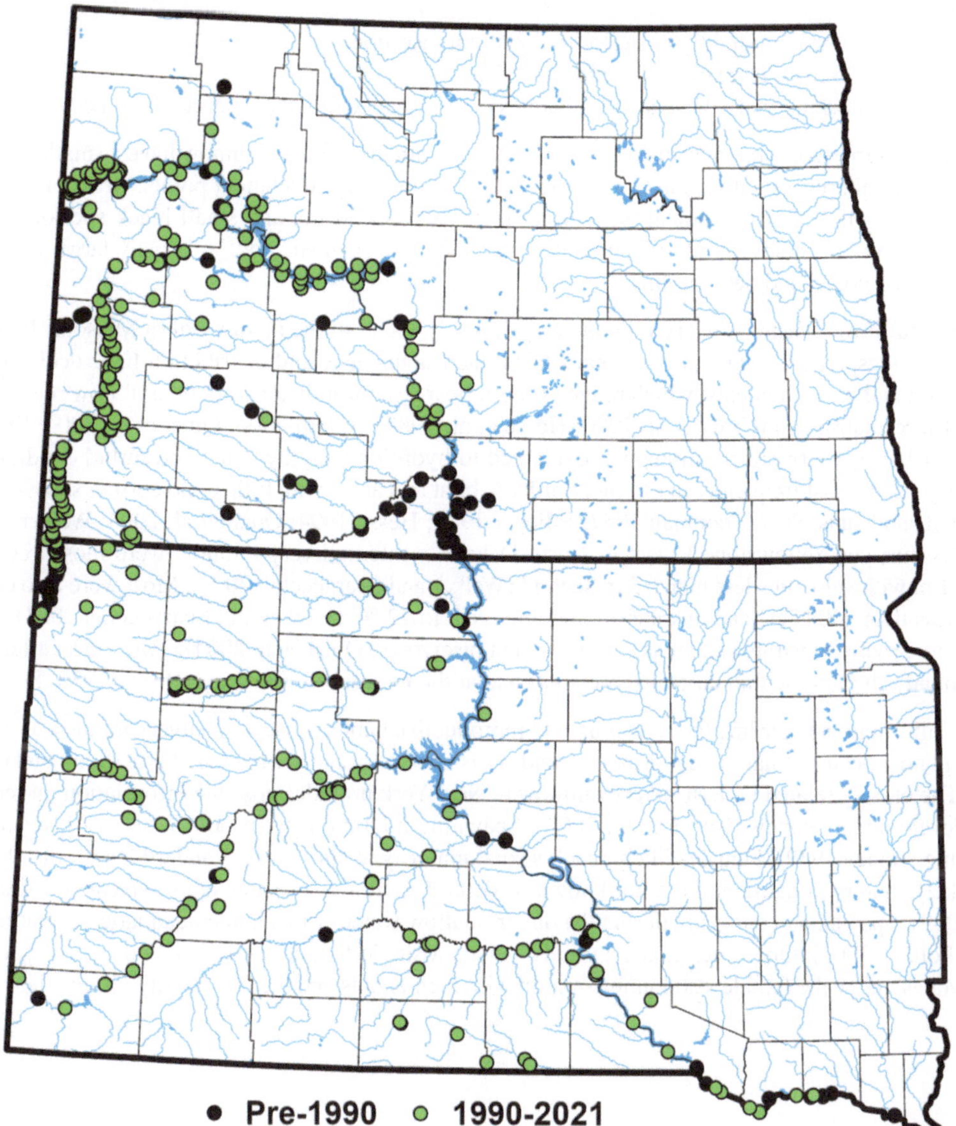

● **Pre-1990**   ○ **1990-2021**

Etymology: *Hybognathus argyritis* – *Hybognathus* = swollen jaw; *argyritis* = unknown.

Description:

Body – moderately fusiform, slightly elongated, moderately laterally compressed.
Color –

> *Dorsally* – tan olive to silver.
> *Laterally* – silver; no obvious markings.
> *Ventrally* – silver to white.
> *Fins* – clear, transparent.

Head – broad; length less than five times eye diameter; basioccipital process internally posterior of head is broad, straight, barely concave, with connecting muscles well separated at attachment points.
Snout – elongated, slightly protrudes past mouth.
Eyes – moderately large, laterally on head.
Mouth – subterminal, small, crescent shaped.
> *Lips* – thin.

Teeth – pharyngeal tooth pattern 0,4-4,0.
Dorsal fin – 8 rays, straight distal end, pointed tip.
Caudal peduncle – slightly elongated, moderately thick.
Caudal fin – forked.
Anal fin – 8 rays.
Pelvic fins – abdominal; insertions slightly posterior to dorsal fin insertion.
Lateral line – complete; 36 to 40 large cycloid scales in series.
Sexual dimorphism – spawning males have small tubercles on dorsal side of head, opercles, and pectoral fins.

Similar Species: Western silvery minnow closely resemble plains minnow. Plains minnow have a smaller eye, and a narrow and peg-like basioccipital process with connecting muscles almost touching attachment points. Brassy minnow have a concave head profile, dusky lateral stripe, dorsal fin with a distinctly rounded tip, and small black specks on the caudal fin. Flathead chub have a broad, dorsoventrally flattened head tapering to a pointed snout, relatively large mouth with barbels, and long, sickle-shaped dorsal and pectoral fins.

Distribution and Habitat: Western silvery minnow are primarily found in the main western Missouri River tributaries in the Dakotas. Once widespread throughout the Great Plains, their abundance and distribution have declined. This is likely because of irrigation and impoundments, which reduce runoff, decrease turbidity, disrupt natural flow regimes, and increase the proportion of large substrates (Cross et al. 1986; Hesse et al. 1989; Pollard 2003; Quist et al. 2004; Smith et al. 2014; Steffensen et al. 2014). Western silvery minnow are adapted to hydrologically dynamic and turbid conditions common in prairie rivers. They occur most frequently in channel border habitat and areas with little to no current such as backwaters and pools with silt and sand substrates (Trautman 1957; Pflieger 1980; Hesse 1994; Quist et al. 2004; Welker and Scarnecchia 2004). In the Missouri and Yellowstone Rivers in North Dakota, western silvery minnow were most frequently found in depths less than 1 m with velocities less than 0.5 m/second (Welker and Scarnecchia 2004). Most were also captured in areas with turbidities less than 250 NTU and in water temperatures 18 to 22 °C (Welker and Scarnecchia 2004). Western silvery minnows avoid large rocky substrate and areas with abundant piscivores (Quist et al. 2004). Two weeks after hatching, juveniles are often found schooling within quiet inshore areas and at the mouths of tributaries (Raney 1939; Pollard 2003).

Reproduction: Information regarding the spawning and reproductive habits of western silvery minnow is extremely scarce or unknown. Spawning occurs in late June in Missouri and is assumed to occur in May in Alberta (Scott and Crossman 1973; Pflieger 1997). The closely related eastern silvery minnow in New York provides possible information on fecundity, spawning, and early development (Raney 1939; Pollard 2003). Spawning migrations to inshore waters of lakes and larger rivers occur in the spring (Raney 1939). Females likely reach sexual maturity at age-1, or a standard length of 50–55 mm (Raney 1939). Males likely mature at age-2 and are not known to defend territories or show aggressiveness toward other mature males (Raney 1939). Spawning occurs only during the day in shallow waters with abundant aquatic vegetation (Raney 1939; Houston 1998; Pollard 2003). One to ten males initially pursue a single female until only two males swim alongside her (Raney 1939). Males nudge the female's abdominal region with their snouts and the fish rapidly vibrate their bodies together to initiate release of eggs and milt (Raney 1939). Females retreat to deeper waters after spawning while males continue to seek other females (Raney 1939). No nest is prepared, and no parental care is given. Fecundity increases with female size (Raney 1939). The nonadhesive, demersal, milky-colored eggs are approximately 1 mm in diameter (Raney 1939). Hatching occurs in 6 or 7 days (Raney 1939).

Age and Growth: Hatched larvae total length is 6 mm (Raney 1939). Total lengths at age from the Moreau River in South Dakota were age-1, 70 mm; age-2, 109 mm; age-3, 119 mm; age-4, 151 mm (Jones 2018). Western silvery minnow can reach a total length of 152 mm, with a longevity of 5.5 years (Cross 1967; Pflieger 1997).

Food and Feeding: Information on the diet of the western silvery minnow is unknown but is likely similar to the plains minnow, eastern silvery minnow, and central silvery minnow (Raney 1939; Cross 1967; Pflieger 1980; Pollard 2003). A long, coiled intestine and a small, crescent-shaped subterminal mouth may allow for the digestion of large amounts of microscopic vegetative matter, diatoms, organic materials, and fungi (Raney 1939; Whitaker Jr 1977).

References

Cross FB (1967) Handbook of fishes in Kansas. University Press of Kansas, Lawrence

Cross FB, Mayden RL, Stewart JD (1986) Fishes in the western Mississippi drainage. In: Hocutt CH, Wiley EO (eds) The zoogeography of North American freshwater fishes. Wiley, Toronto, pp 363–412

Hesse LW (1994) The status of Nebraska fishes in the Missouri River, 5. Selected chubs and minnows (Cyprinidae): sicklefin chub (*M. meeki*), sturgeon chub (*M. gelida*), silver chub (*M. storeriana*), speckled chub (*M. aestivaus*), flathead chub (*Platygobio gracilis*), plains minnow (*Hybognathus placitus*), and western silvery minnow (*H. argyritis*). Trans Nebr Acad Sci Affil Soc 21:99–108

Hesse LW, Schmulbach JC, Carr JM, Keenlyne KD, Unkenholz DG, Robinson JW, Mestl GE (1989) Missouri River fishery resources in relation to past, present, and future stresses. In: Dodge DP (ed) Proceedings of the international large river symposium, Can Spec Publ Fish Aquat Sci, vol 106, pp 352–371

Houston J (1998) Status of the western silvery minnow, *Hybognathus argyritis*, in Canada. Can Field Nat 112:174–153

Jones SJ (2018) Western prairie stream fisheries: an assessment of past and present fish assemblage structure, biotic homogenization, and population dynamics in western South Dakota streams. Dissertation, South Dakota State University

Pflieger WL (1980) *Hybognathus argyritis* Girard, western silvery minnow. In: Lee DS, Gilbert CR, Hocutt CH, Jenkins RE, McAllister DE, Stauffer JR Jr (eds) Atlas of North American freshwater fishes. North Carolina State Museum of Natural History, North Carolina Biological Survey Publication 1980-12, p 174

Pflieger WL (1997) The fishes of Missouri, revised edn. Missouri Dep Cons, Jefferson City

Pollard SM (2003) Status of the western silvery minnow (*Hybognathus argyritis*) in Alberta. Alberta Wildlife Status Report No. 47

Quist MC, Hubert WA, Rahel FJ (2004) Relations among habitat characteristics, exotic species, and turbid-river cyprinids in the Missouri River drainage of Wyoming. Trans Am Fish Soc 133:727–742. https://doi.org/10.1577/T03-124.1

Raney EC (1939) The breeding habits of the eastern silvery minnow, *Hybognathus regius* Girard. Am Midl Nat 21:215–218. https://doi.org/10.2307/2420524

Scott WB, Crossman EJ (1973) Freshwater fishes of Canada. Fisheries Research Board of Canada, Bulletin 184

Smith CD, Fischer JR, Quist MC (2014) Historical changes in Nebraska's lotic fish assemblages: implications of anthropogenic alterations. Am Midl Nat 172:160–184. https://doi.org/10.1674/0003-0031-172.1.160

Steffensen KD, Shuman DA, Stukel S (2014) The status of fishes in the Missouri River, Nebraska: shoal chub (*Macrhybopsis hyostoma*), sturgeon chub (*M. gelida*), sicklefin chub (*M. meeki*), silver chub (*M. storeriana*), flathead chub (*Platygobio gracilis*), plains minnow (*Hybognathus placitus*), western silvery minnow (*H. argyritis*), and brassy minnow (*H. hankinsoni*). Trans Nebr Acad Sci Affil Soc 34:49–67

Trautman MB (1957) The fishes of Ohio. Ohio State University Press, Columbus

Welker TL, Scarnecchia DL (2004) Habitat use and population structure of four native minnows (family Cyprinidae) in the upper Missouri and lower Yellowstone rivers, North Dakota (USA). Ecol Freshw Fish 13:8–22. https://doi.org/10.1111/j.0906-6691.2004.00036.x

Whitaker JO (1977) Seasonal changes in food habits of some cyprinid fishes from the White River at Petersburg, Indiana. Am Midl Nat 90:411–418. https://doi.org/10.2307/2425105

6.14.13   Brassy Minnow *Hybognathus hankinsoni*

Hubbs, 1929

Nuptial Male   Female

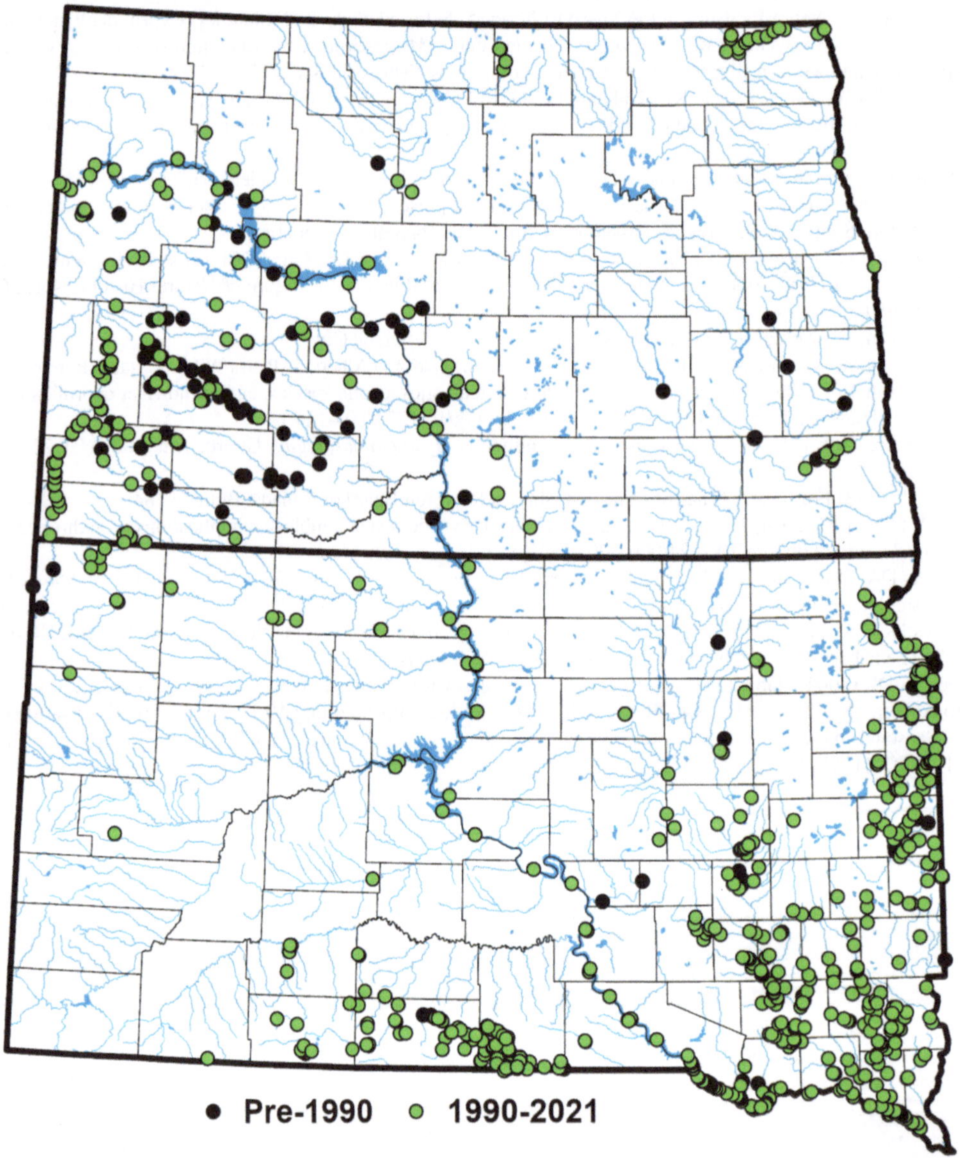

● **Pre-1990**   ● **1990–2021**

Etymology: *Hybognathus hankinsoni – Hybognathus* = swollen jaw; *hankinsoni* = after T.L. Hankinson, Michigan naturalist and ichthyologist.

Description:

Body – moderately fusiform, robust, slightly laterally compressed.
Color –

> *Dorsally* – dark olive to bronze; faint, dark mid-dorsal stripe.
> *Laterally* – olive to bronze or silver; thick, dark lateral stripe, may appear iridescent purple.
> *Ventrally* – cream to white.
> *Fins* – light olive to bronze; caudal: rays of lower lobe have small black specks.

Head – short; concave profile; basioccipital process internally posterior of head is narrow, peg like, and shorter than basioccipital process in plains minnow.
Snout – bluntly rounded, short; slightly protruding past mouth.
Eyes – large, laterally on head.
Mouth – small, terminal, slightly oblique.

> *Jaws* – upper jaw does not extend to eye anterior edge.
> *Lips* – thin.

Teeth – pharyngeal tooth pattern 0,4-4,0.
Dorsal fin – 8 rays; anterior ray slightly shorter than the second or third rays; distal end slightly rounded.
Caudal peduncle – slightly elongated, uniform in thickness.
Caudal fin – moderately forked.
Anal fin – 7 or 8 rays; straight distal end.
Pelvic fins – 8 rays; rounded distal end; insertions slightly posterior to dorsal fin insertion.
Pectoral fins – 13 to 15 rays.
Lateral line – complete; 35 to 39 cycloid scales in series.
Sexual dimorphism – spawning males gold or brassy; small tubercles on pectoral fins.

Similar Species: Brassy minnow closely resemble plains minnow and western silvery minnow. Plains minnow have a smaller eye. Both the plains minnow and western silvery minnow are silver and have a subterminal mouth, straight head profile, shorter snout, and dorsal fin with a straight distal end and pointed tip.

Distribution and Habitat: The native range of brassy minnow is bordered to the north across southern Canada, to the west from Montana to Colorado, to the east to New York, and to the south through northern Missouri and Kansas. It includes the Missouri River, upper Mississippi River, and upper St. Lawrence River basins, and the Great Lakes. In the Dakotas, brassy minnow occur in the Missouri River mainstem and main tributaries, as well as the Red River of the North, Sheyenne, Souris, and Minnesota River systems. Brassy minnow are often found schooling in creeks, headwaters, narrow streams, small rivers, and lakes. They are associated with larger substrates covered in silt or organic debris, and undercut banks in slow runs, pools, backwaters, and other low velocity areas (Bailey 1954; Scheurer and Fausch 2003; Fischer and Paukert 2008; Steffensen et al. 2014). During drought and overwintering periods, they often seek deeper pools (Scheurer and Fausch 2003; Falke et al. 2010). Brassy minnow generally avoid large rivers and lakes (Becker 1983). Rapid dispersal is likely during wet seasons (Scheurer and Fausch 2003). Conductivity levels ranging from 101 to 2,060 µS/cm, high productivity, abundant submerged vegetation, and high nutrient levels are good predictors of brassy minnow presence (Nowosad and Taylor 2013). They can persist in dissolved oxygen levels as low as 0.03 mg/L and in water temperatures as high as 35.5 °C (Scheurer and Fausch 2003). When backwater areas conditions deteriorate, larvae migrate to mid-channel habitats (Falke et al. 2010).

Reproduction: Information on spawning and reproductive habits of brassy minnow is scarce. Spawning occurs during spring and early summer in water temperatures from 16 to 27 °C in quiet, seasonally flooded, shallow, vegetated backwaters and floodplains (Starrett 1951; Copes 1975; Scheurer and Fausch 2003; Falke et al. 2010). Spawning and recruitment are strongly affected by intermittent flows, along with natural and anthropogenic dewatering events (Falke et al. 2010). Sexual maturity occurs at age one or two (Copes 1975; Becker 1983). Spawning occurs in the afternoon, when one or more males pursue a female, with one male ultimately pressing against the female and the pair vibrating together to release eggs and milt (Copes 1975). The adhesive eggs are likely yellow, approximately 0.7 mm in diameter, and hatch within 3 days (Becker 1983; Falke et al. 2010).

Age and Growth: Larvae have a standard length of approximately 4 mm at hatching (Falke et al. 2010). Total lengths at age of brassy minnow from southeastern Wyoming were 24 to 44 mm at age-0, 44 to 68 mm at age-1, 66 to 80 mm at age-2, and 82 to 84 mm at age-3 (Copes 1975). Growth is not significantly different between sexes (Copes 1975). Little growth occurs during winter months, and growth rates are strongly affected by habitat drying (Copes 1975; Falke et al. 2010). Brassy minnow can reach a total length of 97 mm with a longevity of 4 years (Lee et al. 1980; Pflieger 1997).

Food and Feeding: Brassy minnow are opportunistic omnivores (Copes 1975). They often forage in large schools slowly moving upstream, with increased foraging during high water periods (Copes 1975). Brassy minnow feed heavily on phytoplankton, diatoms, algae, and organic debris (Copes 1975). They also consume small aquatic invertebrates, including diptera larvae, and microcrustaceans such as copepods and cladocerans (Copes 1975).

References

Bailey RM (1954) Distribution of the American cyprinid fish *Hybognathus hankinsoni* with comments on its original description. Copeia 1954:289–291. https://doi.org/10.2307/1440044

Becker GC (1983) Fishes of Wisconsin. University of Wisconsin Press, Madison

Copes FA (1975) Ecology of the brassy minnow *Hybognathus hankinsoni* (Cyprinidae). Reports on the Flora and Fauna of Wisconsin 10:46–72

Falke JA, Bestgen KR, Fausch KD (2010) Streamflow reductions and habitat drying affect growth, survival, and recruitment of brassy minnow across a Great Plains riverscape. Trans Am Fish Soc 139:1566–1583. https://doi.org/10.1577/T09-143.1

Fischer JR, Paukert CP (2008) Habitat relationships with fish assemblages in minimally disturbed Great Plains regions. Ecol Freshw Fish 17:597–609. https://doi.org/10.1111/j.1600-0633.2008.00311.x

Lee DS, Gilbert CR, Hocutt CH, Jenkins RE, McAllister DE, Stauffer JR (1980) Atlas of North American freshwater fishes. North Carolina Biological Survey Publication 1980-12. https://doi.org/10.5962/bhl.title.141711

Nowosad DM, Taylor EB (2013) Habitat variation and invasive species as factors influencing the distribution of native fishes in the lower Fraser River Valley, British Columbia, with an emphasis on brassy minnow (*Hybognathus hankinsoni*). Can J Zool 91:71–81. https://doi.org/10.1139/cjz-2012-0177

Pflieger WL (1997) The fishes of Missouri, revised edn. Missouri Dep Cons, Jefferson City

Scheurer JA, Fausch KD (2003) Multiscale processes regulate brassy minnow persistence in a Great Plains river. Trans Am Fish Soc 132:840–855. https://doi.org/10.1577/T02-037

Starrett WC (1951) Some factors affecting the abundance of minnows in the Des Moines River, Iowa. Ecology 32:13–27. https://doi.org/10.2307/1930969

Steffensen KD, Shuman DA, Stukel S (2014) The status of fishes in the Missouri River, Nebraska: shoal chub (*Macrhybopsis hyostoma*), sturgeon chub (*M. gelida*), sicklefin chub (*M. meeki*), silver chub (*M. storeriana*), flathead chub (*Platygobio gracilis*), plains minnow (*Hybognathus placitus*), western silvery minnow (*H. argyritis*), and brassy minnow (*H. hankinsoni*). Trans Nebr Acad Sci Affil Soc 34:49–67

6.14.14   Plains Minnow *Hybognathus placitus*

Girard, 1856

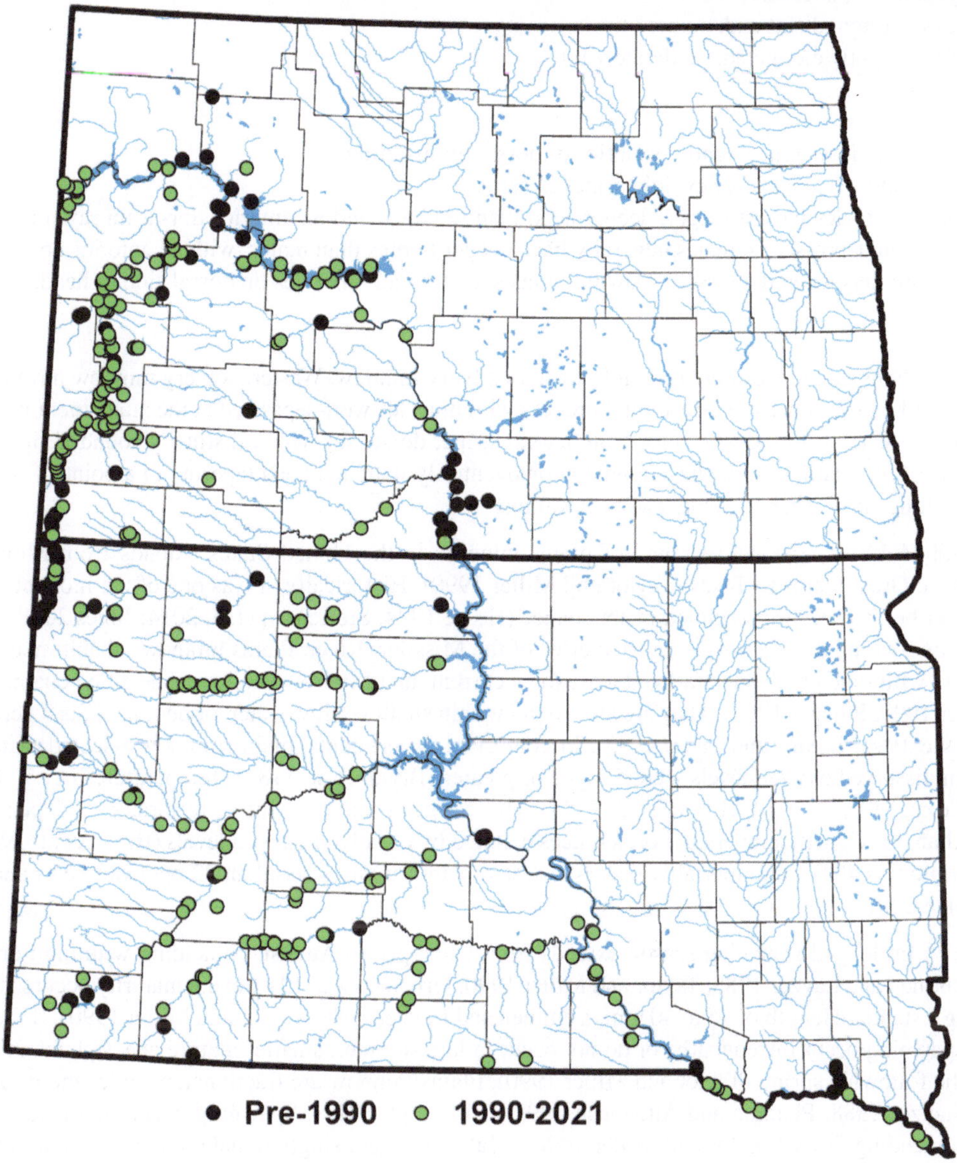

● **Pre-1990**   ● **1990-2021**

Etymology: *Hybognathus placitus* – *Hybognathus* = swollen jaw; ***placitus*** = broad surface, possibly referencing the broad head.

Description:

Body – moderately fusiform, slightly elongated, moderately laterally compressed.
Color –

> *Dorsally* – yellowish-tan to silver; dusky dark olive mid-dorsal stripe.
> *Laterally* – silver; no obvious markings.
> *Ventrally* – silver to white.
> *Fins* – clear, transparent.

Head – broad, slightly ventrally flattened; length greater than or equal to five times eye diameter; basioccipital process internally posterior of head narrow and peg-like with connecting muscles almost touching at point of attachment.
Snout – elongated, slightly protrudes past mouth.
Eyes – moderately small, laterally on head.
Mouth – subterminal, small, crescent shaped.

> *Lips* – thin.

Teeth – pharyngeal tooth pattern 0,4-4,0.
Dorsal fin – 8 rays; straight distal end, pointed tip.
Caudal peduncle – slightly elongated, moderately thick.
Caudal fin – forked.
Anal fin – 8 rays.
Pelvic fins – abdominal; insertions slightly posterior of dorsal fin insertion.
Lateral line – complete; 36 to 40 large cycloid scales in series.
Sexual dimorphism – males have a longer, deeper head than females, and an anterior dorsal fin ray longer than the other dorsal fin rays when depressed; females generally have deeper bodies than males, with an anterior dorsal fin ray equal to the other dorsal fin rays when depressed; spawning males have small tubercles on dorsal side of head, opercles, and pectoral fins.

Similar Species: Plains minnow closely resemble western silvery minnow. Western silvery minnow have a larger eye, and a broad, straight, and barely concave basioccipital process with muscles well-separated at the attachment point. Brassy minnow have a concave head profile, dusky coloration, lateral stripe, dorsal fin with a distinctly rounded tip, and small black specks on the caudal fin. Flathead chub have a broad dorsoventrally flattened head tapering to a pointed snout, wide mouth with barbels, and long sickle-shaped dorsal and pectoral fins.

Distribution and Habitat: The native range of plains minnow in the United States extends from Montana and North Dakota through the Great Plains to Texas (Taylor and Miller 1990). Historically, it was one of the most numerous fishes in the Missouri River but has recently declined in abundance (Hesse 1994; Steffensen et al. 2014; Jones 2018). In the Dakotas, it is primarily found within the main western tributaries of the Missouri River. Plains minnow inhabit medium to large and often turbid rivers with sand or silt substrates, at least some current, and exposed, shallow, sand-filled channels (Cross 1967; Taylor and Miller 1990; Pflieger 1997). They are also found within shallow, slow water, or pool habitats where organic debris accumulate (Hesse 1994; Baxter and Stone 1995). In controlled experiments, plains minnow selected water temperatures near 30 °C when dissolved oxygen levels were 5 mg/L or greater (Bryan et al. 1984). Temperatures of 17 °C were selected when dissolved oxygen was 2 mg/L (Bryan et al. 1984). This suggests that the zone of respiratory independence for plains minnow is less than 30 °C, with temperature preferences dictated by dissolved oxygen levels (Bryan et al. 1984). Mean critical thermal maximum is 40 °C, and mortality occurs in waters with salinities less than 22 ppt (Matthews and Maness 1979; Ostrand and Wilde 2001).

Reproduction: A prolonged spawning season extends from late April to August, coinciding with high or receding flows (Cross 1967; Lehtinen and Layzer 1988; Taylor and Miller 1990; Urbanczyk 2012). Sexual maturity occurs at age-1 or age-2, or upon reaching a standard length of 45 to 50 mm (Lehtinen and Layzer 1988; Taylor and Miller 1990). If the environmental conditions for spawning occur infrequently, or do not occur in late summer, smaller mature age-1 plains minnow may wait to spawn until the following spring (Taylor and Miller 1990). Plains minnow are fractional, pelagic, and broadcast spawners (Lehtinen and Layzer 1988; Platania and Altenbach 1998; Urbanczyk 2012). Spawning occurs in quiet backwaters along sandbars during receding flows (Taylor and Miller 1990). Males pursue a single female, nudge her abdominal region, and eventually wrap their bodies around the female to initiate release of eggs and milt (Platania and Altenbach 1998). No nest is constructed, and no parental care is given. High levels of post-spawning mortality are common (Pflieger 1997). Fecundity increases with size of female (Taylor and Miller 1990). In Oklahoma, fecundity ranged from 417 to 4,134 eggs from females with standard lengths from 51 to 87 mm (Taylor and Miller 1990). The approximately 1-mm diameter, semibuoyant, slightly demersal, and nonadhesive eggs drift downstream until hatching (Sliger 1967; Miller and Robison 1973; Platania and Altenbach 1998). Because eggs and larvae require a certain unimpeded drifting distance to develop, reproductive success may be negatively impacted by stream fragmentation and inconsistent water levels (Urbanczyk 2012). Periods of increased discharge are needed during the spawning season for successful reproduction (Lee et al. 1980).

Age and Growth: Plains minnow total lengths at age from the Moreau River in South Dakota were 62 mm at age-1, 95 mm at age-2, and 104 mm at age-3 (Jones 2018). They can reach a total length of 125 mm and have a longevity of 2 years (Lee et al. 1980; Taylor and Miller 1990).

Food and Feeding: Plains minnow are herbivorous, scraping and scooping up algae off submerged macrophytes and bottom substrates (Cross 1967; Hesse 1994). A long, coiled intestine and small, crescent-shaped subterminal mouth allow it to consume large amounts of vegetative matter and diatoms (Cross 1967).

References

Baxter GT, Stone MD (1995) Fishes of Wyoming. Wyoming Game and Fish Department, Cheyenne

Bryan JD, Hill LG, Neill WH (1984) Interdependence of acute temperature preference and respiration in the plains minnow. Trans Am Fish Soc 113:557–562. https://doi.org/10.1577/1548-8659

Cross FB (1967) Handbook of fishes in Kansas. University Press of Kansas, Lawrence

Hesse LW (1994) The status of Nebraska fishes in the Missouri River, 5. Selected chubs and minnows (Cyprinidae): sicklefin chub (*M. meeki*), sturgeon chub (*M. gelida*), silver chub (*M. storeriana*), speckled chub (*M. aestivaus*), flathead chub (*Platygobio gracilis*), plains minnow (*Hybognathus placitus*), and western silvery minnow (*H. argyritis*). Trans Nebr Acad Sci Affil Soc 21:99–108

Jones SJ (2018) Western prairie stream fisheries: an assessment of past and present fish assemblage structure, biotic homogenization, and population dynamics in western South Dakota streams. Dissertation,. South Dakota State University

Lee DS, Gilbert CR, Hocutt CH, Jenkins RE, McAllister DE, Stauffer JR (1980) Atlas of North American freshwater fishes. North Carolina Biological Survey Publication 1980-12. https://doi.org/10.5962/bhl.title.141711

Lehtinen SF, Layzer JB (1988) Reproductive cycle of the plains minnow, *Hybognathus placitus* (Cyprinidae), in the Cimarron River, Oklahoma. Southwest Nat 33:27–33. https://doi.org/10.2307/3672085

Matthews WJ, Maness JD (1979) Critical thermal maxima, oxygen tolerances and success of cyprinid fishes in a southwestern river. Am Midl Nat 102:374–377. https://doi.org/10.2307/2424665

Miller RJ, Robison HW (1973) The fishes of Oklahoma. Oklahoma State University Press, Stillwater

Ostrand KG, Wilde GR (2001) Temperature, dissolved oxygen, and salinity tolerances of five prairie stream fishes and their role in explaining fish assemblage patterns. Trans Am Fish Soc 130:742–749. https://doi.org/10.1577/1548-8659

Pflieger WL (1997) The fishes of Missouri, revised edn. Missouri Dep Cons, Jefferson City

Platania SP, Altenbach CS (1998) Reproductive strategies and egg types of seven Rio Grande basin cyprinids. Copeia 1998:559–569. https://doi.org/10.2307/1447786

Sliger AS (1967) The embryology, egg structure, micropyle, and egg membranes of the plains minnow *Hybognathus placitus* (Girard). Dissertation, Oklahoma State University

Steffensen KD, Shuman DA, Stukel S (2014) The status of fishes in the Missouri River, Nebraska: shoal chub (*Macrhybopsis hyostoma*), sturgeon chub (*M. gelida*), sicklefin chub (*M. meeki*), silver chub (*M. storeriana*), flathead chub (*Platygobio gracilis*), plains minnow (*Hybognathus placitus*), western silvery minnow (*H. argyritis*), and brassy minnow (*H. hankinsoni*). Trans Nebr Acad Sci Affil Soc 34:49–67

Taylor CM, Miller RJ (1990) Reproductive ecology and population structure of the plains minnow, *Hybognathus placitus* (Pisces: Cyprinidae), in central Oklahoma. Am Midl Nat 123:32–39. https://doi.org/10.2307/2425757

Urbanczyk AC (2012) Reproductive ecology of the plains minnow *Hybognathus placitus* in the Brazos River, Texas. Dissertation, Texas Tech University

6.14.15   Common Shiner *Luxilus cornutus*

(Mitchill, 1817)

Nuptial Male   Female

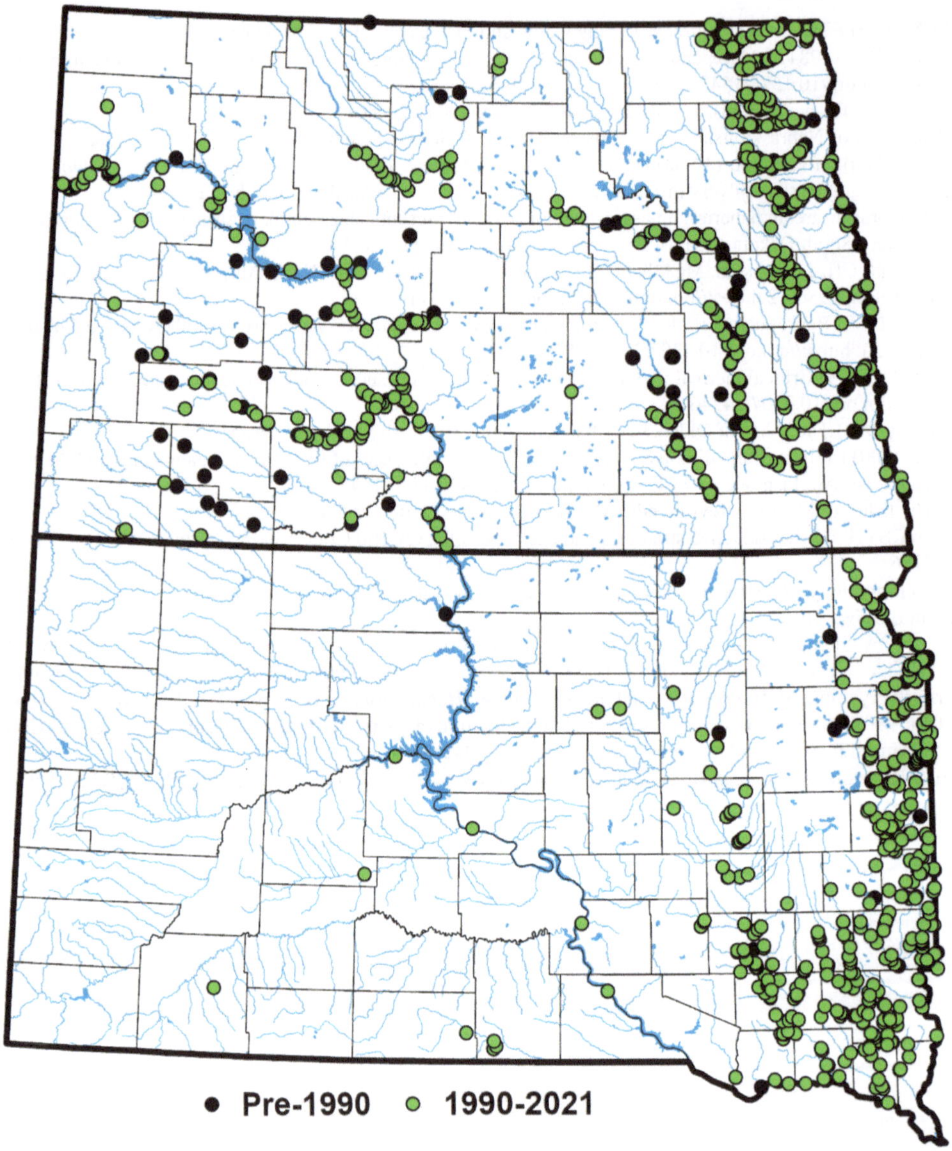

● **Pre-1990**   ● **1990-2021**

Etymology: *Luxilus cornutus – Luxilus* = light; *cornutus* = horned.
Description:

Body – fusiform, deep, laterally compressed.
Color –

> *Dorsally* – dusky gray to olive; wide, prominent, dark mid-dorsal stripe.
> *Laterally* – silver; stripe fading anteriorly to dorsal fin; adults sometimes with small bits of concentrated pigment on some lateral scales.
> *Ventrally* – silver to white.
> *Fins* – clear to lightly pigmented, no distinct markings.

Head – large.
Snout – bluntly pointed.
Eyes – moderately large, laterally on head.
Mouth – large, terminal, oblique.

> *Jaws* – upper jaw extending posteriorly between nostril and anterior edge of eye.

Teeth – pharyngeal tooth pattern 2,4-4,2.
Gill rakers – approximately 9; short.
Dorsal fin – 8 rays, straight distal end, rounded tip.
Caudal peduncle – slender, elongated.
Caudal fin – moderately forked.
Anal fin – 9 rays.
Pelvic fins – 8 rays; abdominal, insertions directly inferior or slightly posterior to dorsal fin insertion.
Pectoral fins – 15 to 17 rays.
Lateral line – complete; 36 to 41 cycloid scales in series, slightly decurved; scales on anterior lateral side of body nearly three times greater in depth than width, posterior edge of scale appears rounded.
Juveniles – decreased body depth.
Sexual dimorphism – spawning males have faint, steel to iridescent blue on head and dorsal side of body, rosy tint on lateral sides of body and fins, and large tubercles on the snout, dorsal side of head, nape, and anterior rays of the dorsal and pectoral fins.

Similar Species: Common shiner resemble golden shiner, red shiner, spotfin shiner, and rudd. Golden shiner have a small, flattened, triangular-shaped head, an elongate and concave anal fin with 11 to 14 rays, and a lateral line with 42 to 54 cycloid scales in series. Red shiner are generally smaller, with a smaller eye and a shorter and thicker caudal peduncle. Spawning male red shiner have iridescent blue hue on their body and red on their caudal, anal, pelvic, and pectoral fins. Spotfin shiner have a moderately deep body and a dark spot on the dorsal fin second or third posterior rays. Rudd have a prominent scaled keel anterior to the vent, 11 to 13 anal fin rays, and red to rosy fins.

Distribution and Habitat: Common shiner are native to North America throughout the upper Atlantic slope, Great Lakes, southern Hudson Bay, and the lower Missouri and upper Mississippi rivers (Lee et al. 1980). They are widespread in the Dakotas but are more prevalent east of the Missouri River in South Dakota (Churchill and Over 1938; Bailey and Allum 1962; Owen et al. 1981). Common shiner inhabit small-to-medium high-gradient streams with sand, gravel, or rubble substrate and clear to slightly turbid, cool water (Cross 1967; Becker 1983; Pflieger 1997). They may also occur in lakes and reservoirs, although this is less common in the Dakotas (Churchill and Over 1938; Becker 1983). They are often found schooling in the midwaters (Pflieger 1997). In Missouri streams, common shiner mean critical dissolved oxygen value was 0.97 mg/L, and the mean critical thermal maximum was 36 °C (Smale and Rabeni 1995).

Reproduction: Spawning occurs from mid-May through late July in water temperatures of 16 to 26 °C (Raney 1940; Starrett 1950; Becker 1983; Pflieger 1997). Males and females become sexually mature at standard lengths of 77 to 120 mm and 75 to 120 mm, respectively (Marshall 1939). Males prefer to spawn in the nests of creek chub, central stoneroller, hornyhead chub, or other species, but if needed will use their snouts to excavate a small depression in shallow riffles over gravel substrate (Raney 1940; Becker 1983). They defend small territories around the nest against male intruders (Becker 1983; Pflieger 1997). During spawning, the male wraps his caudal peduncle around the female and uses his pectoral fins to clasp onto her to initiate release of the gametes (Becker 1983). Females prefer spawning in slower currents (Becker 1983). The orange, demersal, and adhesive eggs are approximately 1.2 to 1.5 in diameter (Becker 1983).

Age and Growth: Growth rates are generally faster in male common shiners than females (Hubbs and Cooper 1936; Fee 1965). Standard lengths at age of males from the Des Moines River in Iowa were age-1, 46 mm; age-2, 67 mm; age-3, 78 mm; age-4, 97 mm; and age-5, 116 mm (Fee 1965). Standard lengths at age of females were 42 mm at age-1, 42 mm and 61 mm at age-2 (Fee 1965). They can reach a length of 208 mm with a longevity of 9 years, although they rarely live past age-6 (Van Oosten 1932; Trautman 1981; Becker 1983).

Food and Feeding: Common shiner are generalized, opportunistic omnivores that feed at the surface or midwater (Pflieger 1997). Adults and juvenile diets are similar (Breder and Crawford 1922; Fee 1965). In Iowa, common shiner ate primarily aquatic and terrestrial insects, with lesser amounts of filamentous algae, bottom ooze, small crustaceans, annelids, and fish (Starrett 1950; Fee 1965). Feeding is greatly influenced by water levels, with more plants than animals consumed during flooding (Fee 1965).

References

Bailey RM, Allum MO (1962) Fishes of South Dakota. University of Michigan, Ann Arbor. https://doi.org/10.3998/mpub.9690435

Becker GC (1983) Fishes of Wisconsin. University of Wisconsin Press, Madison

Breder CM, Crawford DR (1922) The food of certain minnows. Zoologica 2:287–327. https://doi.org/10.5962/p.203790

Churchill EP, Over WH (1938) Fishes of South Dakota. Brown & Saenger Printers, Sioux Falls

Cross FB (1967) Handbook of fishes in Kansas. University Press of Kansas, Lawrence

Fee E (1965) Life history of the northern common shiner, *Notropis cornutus* frontalis, in Boone County, Iowa. Proc Iowa Acad Sci 72:272–281

Hubbs CL, Cooper GP (1936) Minnows of Michigan. Cranbrook institute Science Bulletin 8:1–95

Lee DS, Gilbert CR, Hocutt CH, Jenkins RE, McAllister DE, Stauffer JR (1980) Atlas of North American freshwater fishes. North Carolina Biological Survey Publication 1980-12. https://doi.org/10.5962/bhl.title.141711

Marshall N (1939) Annulus formation in scales of the common shiner, *Notropis cornutus chrysocephalus* (Rafinesque). Copeia 1939:148–154. https://doi.org/10.2307/1436809

Owen JB, Elsen DS, Russell GW (1981) Distribution of fishes in North and South Dakota Basins affected by the Garrison Diversion Unit. University of North Dakota Press, Grand Forks

Pflieger WL (1997) The fishes of Missouri, revised edn. Missouri Dep Cons, Jefferson City

Raney EC (1940) The breeding behavior of the common shiner, *Notropis cornutus* (Mitchill). Zoologica 25:1–14. https://doi.org/10.5962/p.203599

Smale MA, Rabeni CF (1995) Hypoxia and hyperthermia tolerances of headwater stream fishes. Trans Am Fish Soc 124:698–710. https://doi.org/10.1577/1548-8659

Starrett WC (1950) Food relationships of the minnows of the Des Moines River, Iowa. Ecology 31:216–233. https://doi.org/10.2307/1932388

Trautman MB (1981) The fishes of Ohio, revised edn. Ohio State University Press, Columbus

Van Oosten J (1932) The maximum age of fresh-water fishes. Aust Fish 1:3–4

6.14.16   Sturgeon Chub *Macrhybopsis gelida*

(Girard, 1856)

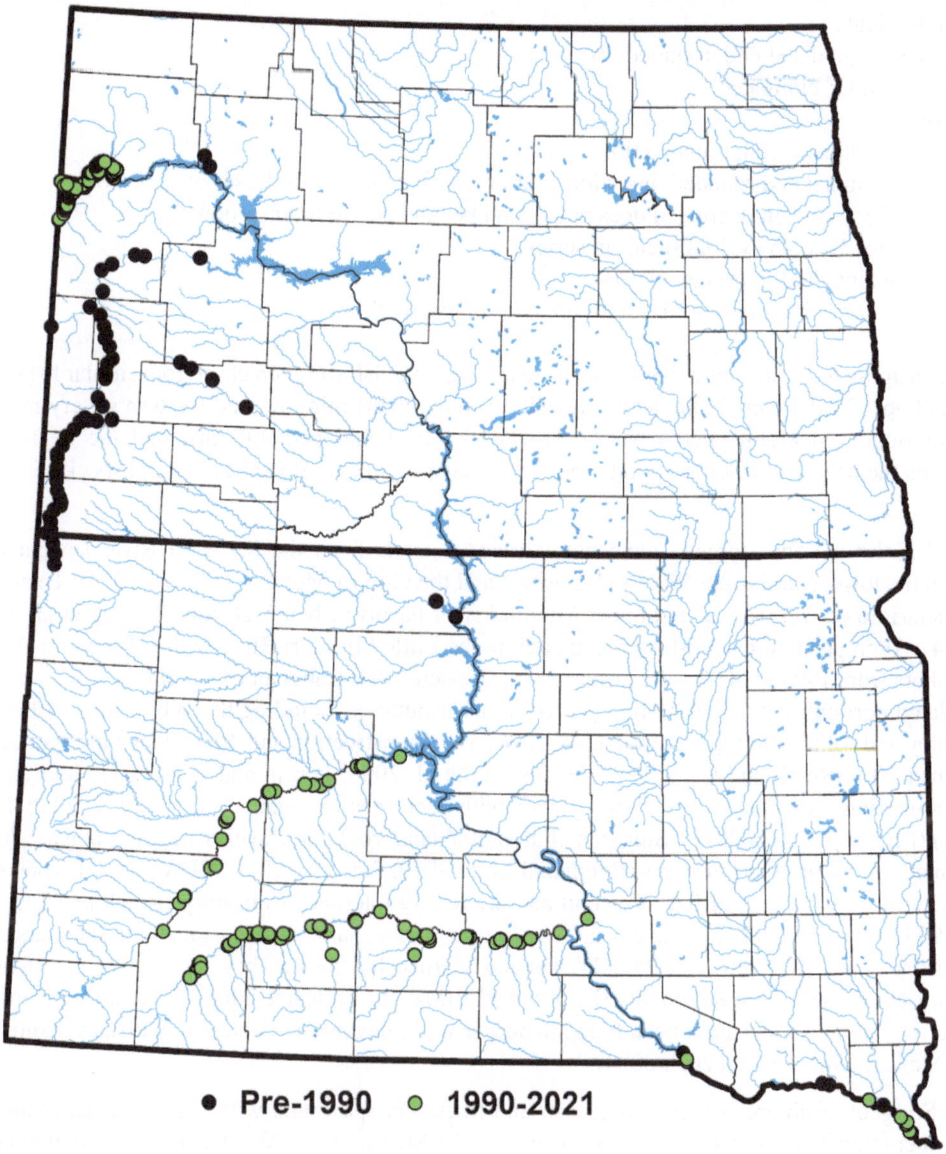

● **Pre-1990**   ○ **1990-2021**

Etymology: *Macrhybopsis gelida – Macrhybopsis* = long blunt snout; *gelida* = ice cold.

Description:

Body – slender, elongated, streamlined.
Color –

   Dorsally – tan to light bronze.
   Laterally – silver; no distinct markings.
   Fins – transparent; caudal: lower lobe dusky gray, white edge.

Head – large, elongated.
Snout – elongated, fleshy, protruding well beyond mouth.
Eyes – moderate, laterally on upper head.
Mouth – large, subterminal.

   Lips – thin.
Teeth – pharyngeal tooth pattern 1,4-4,1.
Rugose folds – on ventral head, extending anteriorly from the middle of the branchial chamber to the posterior third of the
   dentary; end in free, fleshy lobes alternating with transverse, longitudinal, and diagonal grooves.
Barbels – slender, prominent, one on each corner of mouth.
Gill rakers – 3 to 6, blunt.
Dorsal fin – 8 rays, straight distal end, rounded tip.
Caudal peduncle – slender, elongated.
Caudal fin – forked.
Anal fin – 8 rays, straight distal end, rounded tip.
Pelvic fin – 8 rays, abdominal and rounded, insertion over or slightly anterior of dorsal fin insertion.
Pectoral fin – 15 to 17 rays, when depressed does not extend beyond pelvic fin insertion.
Lateral line – complete; 39 to 45 cycloid scales in series.
Scales – on dorsal side, longitudinal ridges (keels).
Sexual dimorphism – spawning males have small tubercles on pectoral fins.

Similar Species: Rugose folds on ventral head surface readily distinguish sturgeon chub from similar fishes (Branson 1966; Hoagstrom et al. 2006). Sicklefin chub, flathead chub, silver chub, plains minnow, and western silvery minnow lack dorsal scale longitudinal ridges. Sicklefin chub have sickle-shaped dorsal and pectoral fins, and the depressed pectoral fin extends well beyond the pelvic fin insertion point. Shoal chub have small dark dorsal and lateral specks. Longnose dace lack a frenum.

Distribution and Habitat: The sturgeon chub native range extends from the Missouri River system in Montana and Wyoming through the lower Mississippi River to Louisiana and the Gulf of Mexico (USFWS 1993). They have been found in the Little Missouri, lower Yellowstone, and upper Missouri river basins in North Dakota (Everett et al. 2004; Welker and Scarnecchia 2004). Attempts to re-establish sturgeon chub in the Little Missouri River within Theodore Roosevelt National Park in North Dakota from 1998 to 2000 were unsuccessful (Kelsch 1994; Rahel and Thel 2004). In South Dakota, sturgeon chub inhabit isolated portions of the lower Missouri River and tributaries including the White River, Cheyenne River, and Grand River (Bailey and Allum 1962; Hoagstrom et al. 2006). Dam construction and dewatering have reduced sturgeon chub range by more than 70% (Reigh and Elsen 1979; Hoagstrom et al. 2006; Perkin and Gido 2011; Steffensen et al. 2014). Sturgeon chub require long stretches of free-flowing river, with decreased abundance for at least 297 km downstream from a dam (184.5 mi) (Perkin and Gido 2011). Sturgeon chub prefer sand, rock, or gravel riffle areas at least 20 cm deep in the benthic zone of medium-to-large turbid rivers (Reigh and Elsen 1979; Stewart 1981; USFWS 1993). They select moderate-to-fast current with mean velocities of 0.6 m/second and are highly tolerant of turbidity (Reigh and Elsen 1979; Stewart 1981; USFWS 1993). Sturgeon chub are found most frequently in main channels, occasionally in sandbars and border channels, and rarely in backwaters (Everett et al. 2004; Welker and Scarnecchia 2004). They are associated with decreasing depth, increasing velocity, and decreasing water quality (Everett et al. 2004). Juvenile and adult sturgeon chub habitat requirements are likely similar, although juveniles have been found in deeper water devoid of vegetation, with sandy substrates, and moderate velocities (Stewart 1981; Everett et al. 2004).

Reproduction: Sturgeon chub are pelagic, broadcast, and fractional spawners, with spawning occurring from mid-June through July at water temperatures of 18 to 22 °C (Lee et al. 1980; Stewart 1981; Werdon 1992). Sexual maturity is at age-2, or at total lengths of 76 to 81 mm for females and 78 or 79 mm for males (Stewart 1981; Hoagstrom et al. 2006; Starks et al. 2016). Fecundity ranges from 2,000 to 3,500 eggs per female (Stewart 1981). Eggs are carried with the current during incubation and larval development (Tibbs and Galat 1997; Hoagstrom et al. 2006). Reproductive success is suspected to be poor (Stewart 1981).

Age and Growth: In North Dakota, sturgeon chub total lengths were 43 mm at age-1, 65 mm at age-2, and 83 mm at age-3 (Everett et al. 2004). Maximum total length is 97 mm and longevity is 4 years (Stewart 1981).

Food and Feeding: Sturgeon chub are benthic insectivores. They use a highly developed cutaneous sensory system of maxillary barbels and numerous external taste buds (including those in rugose folds) to forage for aquatic invertebrates in gravel and sand substrates within highly turbid waters (Branson 1966; Moore 1950; Stewart 1981; Hoagstrom et al. 2006). Juveniles feed primarily on chironomid and other midge larvae (Starks et al. 2016).

References

Bailey RM, Allum MO (1962) Fishes of South Dakota. University of Michigan, Ann Arbor. https://doi.org/10.3998/mpub.9690435

Branson BA (1966) Histological observations on the sturgeon chub, *Hybopsis gelida* (Cyprinidae). Copeia 1966:872–876. https://doi.org/10.2307/1441416

Everett SR, Scarnecchia DL, Ryckman LF (2004) Distribution and habitat use of sturgeon chubs (*Macrhybopsis gelida*) and sicklefin chubs (*M. meeki*) in the Missouri and Yellowstone Rivers, North Dakota. Hydrobiologia 527:183–193. https://doi.org/10.1023/B:HYDR.0000043300.69401.66

Hoagstrom CW, Hayer CA, Kral JG, Wall SS, Berry CR Jr (2006) Rare and declining fishes of South Dakota: a river drainage scale perspective. Proc S D Acad Sci 85:171–211

Kelsch SW (1994) Lotic fish-community structure following transition from severe drought to high discharge. J Freshw Ecol 9:331–341. https://doi.org/10.1080/02705060.1994.9664903

Lee DS, Gilbert CR, Hocutt CH, Jenkins RE, McAllister DE, Stauffer JR (1980) Atlas of north American freshwater fishes. North Carolina Biological Survey Publication 1980-12. https://doi.org/10.5962/bhl.title.141711

Moore GA (1950) The cutaneous sense organs of barbeled minnows adapted to life in the muddy waters of the Great Plains region. Trans Am Microscopy Assoc 69:69–95. https://doi.org/10.2307/3223350

Perkin JS, Gido KB (2011) Stream fragmentation thresholds for a reproductive guild of Great Plains fishes. Fisheries 36:371–383. https://doi.org/10.1080/03632415.2011.597666

Rahel FJ, Thel LA (2004) Sturgeon chub (*Macrhybopsis gelida*): a technical conservation assessment. US Dep Ag, Forest Service, Rocky Mountain Region. https://www.fs.usda.gov/Internet/FSE_DOCUMENTS/stelprdb5206786.pdf. Accessed 18 Aug 2021

Reigh RC, Elsen DS (1979) Status of the sturgeon chub (*Hybopsis gelida*) and sicklefin chub (*H. meeki*) in North Dakota. Prairie Nat 11:49–52

Starks TA, Miller ML, Long JM (2016) Early life history of three pelagic-spawning minnows *Macrhybopsis* spp. in the lower Missouri River. J Fish Biol 88:1335–1349. https://doi.org/10.1111/jfb.12892

Steffensen KD, Shuman DA, Stukel S (2014) The status of fishes in the Missouri River, Nebraska: shoal chub (*Macrhybopsis hyostoma*), sturgeon chub (*M. gelida*), sicklefin chub (*M. meeki*), silver chub (*M. storeriana*), flathead chub (*Platygobio gracilis*), plains minnow (*Hybognathus placitus*), western silvery minnow (*H. argyritis*), and brassy minnow (*H. hankinsoni*). Trans Nebr Acad Sci Affil Soc 34:49–67

Stewart DD (1981) The biology of the sturgeon chub (*Hybopsis gelida* Girard) in Wyoming. Dissertation, University of Wyoming

Tibbs JE, Galat DL (1997) Larval, juvenile, and adult small fish use of Scour Basins connected to the lower Missouri River. Final Report, Missouri Dep Cons, Columbia

USFWS (US Fish and Wildlife Service) (1993) Status report on sturgeon chub (*Macrhybopsis gelida*), a candidate endangered species. US Dept Inter, Fish Wildl Serv, Bismarck

Welker TL, Scarnecchia DL (2004) Habitat use and population structure of four native minnows (family Cyprinidae) in the upper Missouri and lower Yellowstone rivers, North Dakota (USA). Ecol Freshw Fish 13:8–22. https://doi.org/10.1111/j.0906-6691.2004.00036.x

Werdon SJ (1992) Population status and characteristics of *Macrhybopsis gelida*, *Platygobio gracilis*, and *Rhinichthys cataractae* in the Missouri River basin. Dissertation, South Dakota State University

6.14.17 Shoal Chub *Macrhybopsis hyostoma*

(Gilbert, 1884)

● **Pre-1990**   ○ **1990-2021**

Etymology: *Macrhybopsis hyostoma* – *Macrhybopsis* = long blunt snout; *hyostoma* = hogmouth.

Description:

Body – slender, tubular elongated, slightly dorsoventrally flattened.
Color –

> *Dorsally* – tan to bronze; small, prominent, black, scattered specks.
> *Laterally* – silver; small, prominent, black, scattered specks.
> *Ventrally* – silver to white.
> *Fins* – clear; no distinct markings.

Head – elongated.
Snout – fleshy, slightly elongated, extending well beyond mouth.
Eyes – large, laterally on head.
Mouth – small, subterminal; thin lips.
Teeth – pharyngeal tooth pattern 0,4-4,0.
Barbels – 1 or 2 on each corner of mouth; small and conspicuous.
Dorsal fin – 8 rays, nearly straight distal end, rounded tip.
Caudal peduncle – slender, elongated.
Caudal fin – forked.
Anal fin – 7 or 8 rays, straight distal end, rounded tip.
Pelvic fins – abdominal, short, rounded; insertion directly under or slightly anterior to dorsal fin insertion.
Pectoral fins – large, rounded to slightly pointed; when depressed do not extend beyond insertion of pelvic fins.
Lateral line – complete; 32 to 43 cycloid scales, scales typically absent on breast.
Sexual dimorphism – spawning males have small tubercles on the breast, dorsal sides of head and body, and rays of the pectoral and pelvic fins.

Similar Species: The large eyes, small prominent black scattered specks on pale dorsal and lateral sides, and relatively small maximum body size easily distinguish shoal chub from sicklefin chub and sturgeon chub. Sicklefin chub also have sickle shaped and elongate pectoral fins that extend well beyond the insertion of the pelvic fins. Sturgeon chub have a longer snout, dorsal scale longitudinal ridges, and prominent rugose folds on the ventral surface of head. Flathead chub have a dorsoventrally flattened head, a wide terminal mouth, and lack dark specks on their body. Silver chub have more silvery, reflective scales, a slightly laterally compressed body, and reach a much larger size. They also lack dark specks on their dorsal and lateral sides.

Distribution and Habitat: Shoal chub are native throughout much of the south-central United States and Mississippi River basin, ranging from southeastern South Dakota, southern Minnesota and Wisconsin to the north, Texas to the south, and from the Appalachian Mountains to the east, and Rio Grande to the west (Lee et al. 1980). They are absent from North Dakota. Shoal chub occur in South Dakota in the Missouri River below Gavins Point Dam. Extant populations have been experiencing significant declines, likely because of river channelization, impoundments, and other anthropogenic disturbances (Hesse 1994; Steffensen et al. 2014). They inhabit the benthic zone in tributaries and mainstems of small-to-large rivers with moderate-to-strong current over riffles with medium-to-large gravel substrate (Hesse 1994; Boschung Jr and Mayden 2004). In the upper Mississippi River Basin, shoal chubs prefer water temperatures of 21 to 28 °C, moderate current of 0.4 to 0.6 m/second, and relatively shallow depths of 1.0 to 2.7 m (Hesse 1994). They occupy shallow waters during the night (Klutho 1983; Boschung Jr and Mayden 2004). Juveniles prefer sand substrate (Hesse 1994).

Reproduction: Information regarding the spawning and reproductive behavior of shoal chub is limited, but they are likely pelagic broadcast spawners (Perkin and Gido 2011; Williams 2011). Spawning is prolonged, occurring from early spring until late fall (Becker 1983; Etnier and Starnes 1993; Williams 2011). Sexual maturity occurs at age-1 (Williams 2011). Clutch size from females with total lengths of 39 to 70 mm ranged from 34 to 680 eggs per female, with a mean of 199 eggs (Williams 2011). Clutch size is positively related to female total length (Williams 2011). The transparent, semibuoyant, nonadhesive, 2.5 m diameter eggs are carried with the current during incubation and larval development (Platania and Altenbach 1998; Boschung Jr and Mayden 2004; Perkin and Gido 2011). Shoal chub likely require shorter unsegmented stream lengths (103 km) than other pelagic-spawning minnows, but upstream migration is still negatively affected by river fragmentation created because of impoundments (Platania and Altenbach 1998; Perkin and Gido 2011).

Age and Growth: Early growth is rapid, with shoal chub in Texas reaching 45 to 65% of maximum length during the first summer and fall (Williams 2011). They can reach a total length of 76 mm (Lee et al. 1980; Etnier and Starnes 1993). Longevity 2.5 years, with few individuals surviving to age-2 (Williams 2011).

Food and Feeding: Shoal chub are benthic insectivores (Hesse 1994). Adults primarily eat chironomids, dipterans, trichopteran larvae, and other small aquatic invertebrates (Williams 2011; Gaughan 2016). Juveniles consume detritus, sand, aquatic insects, and small crustaceans (Williams 2011). Shoal chub differ than other species of *Macrhybopsis* because they are primarily sight feeders, as evidenced by large eyes, brain morphology, and external taste bud distribution (Davis and Miller 1967).

References

Becker GC (1983) Fishes of Wisconsin. University of Wisconsin Press, Madison

Boschung HT Jr, Mayden RL (2004) Fishes of Alabama. Smithsonian Books, Washington, DC

Davis BJ, Miller RJ (1967) Brain patterns in minnows of the genus *Hybopsis* in relation to feeding habits and habitat. Copeia 1967:1–39. https://doi.org/10.2307/1442172

Etnier DA, Starnes WC (1993) The fishes of Tennessee. University of Tennessee Press, Knoxville

Gaughan S (2016) Habitat use, molecular phylogeny, and population structure of *Macrhybopsis* chubs in the upper Mississippi River basin. Dissertation, University of Nebraska

Hesse LW (1994) The status of Nebraska fishes in the Missouri River, 5. Selected chubs and minnows (Cyprinidae): sicklefin chub (*M. meeki*), sturgeon chub (*M. gelida*), silver chub (*M. storeriana*), speckled chub (*M. aestivaus*), flathead chub (*Platygobio gracilis*), plains minnow (*Hybognathus placitus*), and western silvery minnow (*H. argyritis*). Trans Nebr Acad Sci Affil Soc 21:99–108

Klutho MA (1983) Seasonal, daily, and spatial variation of shoreline fishes in the Mississippi River at Grand Tower, Illinois. Dissertation, Southern Illinois University,

Lee DS, Gilbert CR, Hocutt CH, Jenkins RE, McAllister DE, Stauffer JR (1980) Atlas of North American freshwater fishes. North Carolina Biological Survey Publication 1980-12. https://doi.org/10.5962/bhl.title.141711

Perkin JS, Gido KB (2011) Stream fragmentation thresholds for a reproductive guild of Great Plains fishes. Fisheries 36:371–383. https://doi.org/10.1080/03632415.2011.597666

Platania SP, Altenbach CS (1998) Reproductive strategies and egg types of seven Rio Grande Basin cyprinids. Copeia 1998:559–569. https://doi.org/10.2307/1447786

Steffensen KD, Shuman DA, Stukel S (2014) The status of fishes in the Missouri River, Nebraska: shoal chub (*Macrhybopsis hyostoma*), sturgeon chub (*M. gelida*), sicklefin chub (*M. meeki*), silver chub (*M. storeriana*), flathead chub (*Platygobio gracilis*), plains minnow (*Hybognathus placitus*), western silvery minnow (*H. argyritis*), and brassy minnow (*H. hankinsoni*). Trans Nebr Acad Sci Affil Soc 34:49–67

Williams CS (2011) Life history characteristics of three obligate riverine species and drift patterns of lower Brazos River fishes. Texas State University, Dissertation

6.14.18   Sicklefin Chub *Macrhybopsis meeki*

(Jordan & Evermann, 1896)

● Pre-1990   ● 1990-2021

Etymology: *Macrhybopsis meeki* – *Macrhybopsis* = long blunt snout; *meeki* = after Seth E. Meek, an early American ichthyologist in central North America.

Description:

Body – slender, elongated, somewhat fusiform.
Color –

>   *Dorsally* – tan to light bronze.
>   *Laterally* – silver; no distinct markings.
>   *Ventrally* – silver to white.
>   *Fins* – clear, transparent; caudal: lower lobe dusky gray, white edge.

Head – large, blunt.
Snout – large, blunt; protrudes past the upper jaw.
Eyes – moderate, laterally on upper part of head.
Mouth – large, subterminal; thin lips.
Teeth – pharyngeal tooth pattern 0,4-4,0.
Barbels – short, prominent, one on each corner of mouth.
Gill rakers – 5; blunt.
Dorsal fin – sickle-shaped with pointed tip and 8 rays; insertion over or slightly posterior of pelvic fin insertion.
Caudal peduncle – elongated, relatively thin, uniform in thickness.
Caudal fin – forked.
Anal fin – 8 rays; pointed tip.
Pelvic fins – abdominal, sickle-shaped, pointed.
Pectoral fins – 16 or 17 rays; long, sickle-shaped, when depressed extend beyond pelvic fin insertions.
Lateral line – complete; 43–50 cycloid scales in series; dorsal side scales lack a longitudinal ridge (keel).
Sexual dimorphism – spawning males have small tubercles on all fins except the caudal fin.

Similar Species: Sicklefin chub closely resemble sturgeon chub, flathead chub, silver chub, plains minnow, and western silvery minnow. Sturgeon chub have a longitudinal ridge on each of the dorsal scales, and prominent rugose folds on the ventral surface of the head. Flathead chub have a broad, dorsoventrally flattened head with a snout tapering into a point with a rounded tip, and pectoral fins that when depressed do not extend past the pelvic fin insertions. Silver chub, plains minnow, and western silvery minnow have larger eyes and pectoral fins that when depressed do not extend past the pelvic fin insertions.

Distribution and Habitat: Sicklefin chub are primarily found in the Missouri River main stem and tributaries from Montana to Missouri, and the middle Mississippi River from its confluence with the Ohio River to Missouri (Pflieger 1997; Dieterman and Galat 2004). Its distribution and abundance have declined throughout much of its historical range, likely because of impoundments and channelization reducing turbidity, altering flow regimes, and limiting early life stage habitat (Pflieger and Grace 1987; Steffensen et al. 2014; Panella et al. 2018). In the Dakotas, sicklefin chub are found in the Missouri River drainage. They have been collected most frequently from main channels but are also found near sandbars and in border and side channels (Everett et al. 2004). Sicklefin chub require long free-flowing reaches with natural variability in seasonal flows and turbidity (Dieterman and Galat 2004). Habitat use changes as the fish age (Panella et al. 2018). Post-hatch larvae swim vertically in the water column and rarely rest along the bottom (Albers and Wildhaber 2017). Moderately shallow, approximately 2.1 m deep riverine shoreline areas with relatively slow velocities near 0.3 m/sec provide critical nursery habitat (Dieterman and Galat 2004; Panella et al. 2018). Adults prefer higher water velocities around 0.9 m/sec and mean water depths of 3.1 m (Panella et al. 2018). Sicklefin chub presence increases with increased water depth, decreased velocity, and decreased water clarity (Everett et al. 2004). It rarely occurs in reservoirs and backwaters (Werdon 1993; Hesse 1994; Fisher 1999; Everett et al. 2004).

Reproduction: Spawning of sicklefin chub in the lower Missouri River is protracted, beginning in early June, peaking in mid-July, and subsiding in August (Starks et al. 2016). The minimum spawning water temperature is 20 °C (Braaten and Guy 2002). Sicklefin chub are pelagic broadcast fractional spawners (Albers and Wildhaber 2017). Sexual maturity occurs at age-3 (Dieterman et al. 2006). Spawning occurs at night (Albers and Wildhaber 2017). Spawning occurs when one or two males align their bodies laterally with a single female in head-to-head orientation and nudge her to swim in a circular motion near substrate (Albers and Wildhaber 2017). The male then applies pressure and wraps his caudal fin and caudal peduncle around the female to initiate the release of eggs and milt (Albers and Wildhaber 2017). No nest is constructed, and no parental care is given (Albers and Wildhaber 2017). Sicklefin chub are multiple batch spawners. In aquaria, 104 to 577 eggs were released during a single spawning event, with females producing at least 1,561 eggs over several months (Dieterman et al. 2006; Albers and Wildhaber 2017). The semi-buoyant, nonadhesive eggs average 3 mm in diameter after water hardening and are carried with the current downstream during incubation and larval development (Platania and Altenbach 1998; Albers and Wildhaber 2017). Eggs may sink if water current is not fast enough, leading to reduced survival (Platania and Altenbach 1998; Albers and Wildhaber 2017). Sicklefin chub likely exhibit high post-spawning mortality (Dieterman et al. 2006).

Age and Growth: Sicklefin chub larvae average 5 mm at hatching and require water temperatures of at least 10 °C to grow (Braaten and Guy 2002; Albers and Wildhaber 2017). First year growth rates are significantly greater in higher latitudes than in lower latitudes (Braaten and Guy 2002). Total lengths at age of sicklefin chub from the upper Missouri and lower Yellowstone rivers in western North Dakota were 46 mm at age-1, 75 mm at age-2, 93 mm at age-3, and 107 at age-4 (Everett et al. 2004). Maximum age is significantly related to latitude and increases from low to high latitudes (Braaten and Guy 2002). Sicklefin chub can reach a total length of 128 mm, with a longevity of 4 years (Dieterman et al. 1996; Herman et al. 2008; Steffensen et al. 2014).

Food and Feeding: As benthic insectivores, adult sicklefin chub diets consist primarily of small, benthic, adult, and larval aquatic insects, including dipterans, ephemeropterans, and trichopterans (Werdon 1993; Nocomis 2014). They have unique adaptations, such as elaborate internal and external taste buds, lateral line neuromasts, and numerous small sensory papillae, for foraging in highly turbid waters (Moore 1950; Davis and Miller 1967; Reno 1969; Dieterman and Galat 2004).

References

Albers JL, Wildhaber ML (2017) Reproductive strategy, spawning induction, spawning temperatures and early life history of captive sicklefin chub *Macrhybopsis meeki*. J Fish Biol 91:58–79. https://doi.org/10.1111/jfb.13329

Braaten PJ, Guy CS (2002) Life history attributes of fishes along the latitudinal gradient of the Missouri River. Trans Am Fish Soc 131:931–945. https://doi.org/10.1577/1548-8659

Davis BJ, Miller RJ (1967) Brain patterns in minnows of the genus *Hybopsis* in relation to feeding habits and habitat. Copeia 1967:1–39. https://doi.org/10.2307/1442172

Dieterman DJ, Galat DL (2004) Large-scale factors associated with sicklefin chub distribution in the Missouri and lower Yellowstone rivers. Trans Am Fish Soc 133:577–587. https://doi.org/10.1577/T03-002.1

Dieterman DJ, Roberts E, Braaten PJ, Galat DL (2006) Reproductive development in the sicklefin chub, *Macrhybopsis meeki*, in the Missouri River and Yellowstone River USA. Prairie Nat 38:113–130

Dieterman DJ, Ruggles MP, Wildhaber ML, Galat DL (1996) Population structure and habitat use of benthic fishes along the Missouri and lower Yellowstone rivers. 1996 Annual Report of the Missouri River Benthic Fish Study PD-95-5832 to US Army Corps of Engineers and US Bureau of Reclamation

Everett SR, Scarnecchia DL, Ryckman LF (2004) Distribution and habitat use of sturgeon chubs (*Macrhybopsis gelida*) and sicklefin chubs (*M. meeki*) in the Missouri and Yellowstone Rivers, North Dakota. Hydrobiologia 527:183–193. https://doi.org/10.1023/B:HYDR.0000043300.69401.66

Fisher SJ (1999) Seasonal investigation of native fishes and their habitats in Missouri River and Yellowstone River backwaters. Dissertation, South Dakota State University

Herman P, Plauck A, Utrup N, Hill T (2008) Three-year summary age and growth report for sturgeon chub (*Macrhybopsis gelida*). Report to the US Army Corps of Engineers, Northwestern Division, Omaha

Hesse LW (1994) The status of Nebraska fishes in the Missouri River, 5. Selected chubs and minnows (Cyprinidae): sicklefin chub (*M. meeki*), sturgeon chub (*M. gelida*), silver chub (*M. storeriana*), speckled chub (*M. aestivaus*), flathead chub (*Platygobio gracilis*), plains minnow (*Hybognathus placitus*), and western silvery minnow (*H. argyritis*). Trans Nebr Acad Sci Affil Soc 21:99–108

Moore GA (1950) The cutaneous sense organs of barbeled minnows adapted to life in the muddy waters of the Great Plains region. Trans Am Microscopy Assoc 69:69–95. https://doi.org/10.2307/3223350

Nocomis E (2014) Observations of sicklefin chub diets in the Missouri River. Prairie Nat 46:88–91

Panella MJ, Schainost SC, Mestl GE, Steffensen KD (2018) Listing proposal for four small-bodied fishes in Nebraska: Flathead chub (*Platygobio gracilis*), plains minnow (*Hybognathus placitus*), sicklefin chub (*Macrhybopsis meeki*), and western silvery minnow (*Hybognathus argyritis*). Nebraska Game and Parks Commission, Lincoln

Pflieger WL (1997) The fishes of Missouri, revised edn. Missouri Dep Cons, Jefferson City

Pflieger WL, Grace TB (1987) Changes in the fish fauna of the lower Missouri River, 1940-1983. In: Matthews W, Heins D (eds) Community and evolutionary ecology of North American stream fishes. University of Oklahoma Press, Norman, pp 166–177

Platania SP, Altenbach CS (1998) Reproductive strategies and egg types of seven Rio Grande Basin cyprinids. Copeia 1998:559–569. https://doi.org/10.2307/1447786

Reno HW (1969) Cephalic lateral-line systems of the cyprinid genus *Hybopsis*. Copeia 1969:736–773. https://doi.org/10.2307/1441800

Starks TA, Miller ML, Long JM (2016) Early life history of three pelagic-spawning minnows *Macrhybopsis* spp. in the lower Missouri River. J Fish Biol 88:1335–1349. https://doi.org/10.1111/jfb.12892

Steffensen KD, Shuman DA, Stukel S (2014) The status of fishes in the Missouri River, Nebraska: shoal chub (*Macrhybopsis hyostoma*), sturgeon chub (*M. gelida*), sicklefin chub (*M. meeki*), silver chub (*M. storeriana*), flathead chub (*Platygobio gracilis*), plains minnow (*Hybognathus placitus*), western silvery minnow (*H. argyritis*), and brassy minnow (*H. hankinsoni*). Trans Nebr Acad Sci Affil Soc 34:49–67

Werdon SJ (1993) Status report on sicklefin chub (*Macrhybopsis meeki*), a candidate endangered species. US Fish and Wildlife Service, Bismarck

6.14.19   Silver Chub *Macrhybopsis storeriana*

(Kirtland, 1845)

● Pre-1990   ○ 1990-2021

Etymology: *Macrhybopsis storeriana* – *Macrhybopsis* = long blunt snout; *storeriana* = after David H. Storer, early North American ichthyologist.

Description:

Body – moderately fusiform, slightly stout, laterally compressed body.
Color –

> *Dorsally* – tan to dark gray.
> *Laterally* – silver; no distinct markings.
> *Ventrally* – silver to white.

Head – short.
Snout – blunt, rounded, fleshy; protruding beyond mouth.
Eyes – large, laterally on head.
Mouth – small, subterminal.

> *Lips* – thin, fleshy.

Teeth – pharyngeal tooth pattern 1,4-4,1.
Barbels – 1 at each corner of the mouth; short, conical, inconspicuous.
Dorsal fin – 8 rays, nearly straight distal end, slightly pointed tip.
Caudal peduncle – slender, elongated.
Caudal fin – forked.
Anal fin – 7 or 8 rays, nearly straight distal end and pointed tip.
Pelvic fins – 8 rays; abdominal, short, rounded distal end; insertion slightly posterior to dorsal fin insertion.
Pectoral fins – 17 or 18 rays; when depressed do not extend beyond pelvic fin insertions.
Lateral line – complete; 35 to 48 cycloid scales in series; scales occasionally present only on posterior breast.
Sexual dimorphism – spawning males have large tubercles on pectoral fin rays and smaller tubercles on the dorsal and lateral sides of the head.

Similar Species: Silver chub somewhat resemble shoal chub, sicklefin chub, and sturgeon chub. Shoal chub are much smaller and paler, with small, prominent, black, scattered specks along the dorsal and lateral sides of the body. Sicklefin chub have smaller eyes and sickle-shaped, elongate pectoral fins that extend well beyond pelvic fin insertions. Sturgeon chub have smaller eyes, longitudinal ridges or keels on the dorsal scales, a relatively long snout, and prominent rugose folds on ventral surface of head. Flathead chub have a dorsoventrally flattened head with smaller eyes, long, falcate pectoral and anal fins, and a snout that tapers to a point.

Distribution and Habitat: The native range of silver chub extends throughout the Mississippi River basin from the Missouri River in western Nebraska, to the Monongahela River in Pennsylvania, to the upper Mississippi River along the Minnesota-Wisconsin border, and to the lower Mississippi River in Lousiana (Kinney Jr 1954; Lee et al. 1980; Stauffer Jr et al. 2016; Kočovský 2018; Gaughan 2016). The only lake-dwelling silver chub population is in Lake Erie (Kočovský 2018). In the Dakotas, they primarily occur in the Red River of North and Missouri River drainages (Lee et al. 1980; Harland and Berry Jr 2004; Hayer et al. 2006; Kočovský 2018). Silver chub are macrohabitat generalists, inhabiting large, low gradient, semi-turbid lakes, impoundments, rivers, and headwaters with stronger currents (Kinney Jr 1954; Harlan and Speaker 1987; Steffensen et al. 2014). They rarely occur in small streams, but occasionally inhabit small stream headwaters (Kinney Jr 1954). Silver chub prefer silt, clay, sand, and small gravel substrate, and overwinter in deep pools in the Mississippi River (Wood 1953; Kinney Jr 1954; Becker 1983; Hayer et al. 2006).

Reproduction: Information regarding the spawning and reproductive behavior of the silver chub is limited, especially in lotic habitats. However, they are likely one of several pelagic broadcast spawners in the Great Plains region (Perkin and Gido 2011). Spawning likely occurs in late May through early June when water temperatures reach approximately 20 °C (Kinney Jr 1954; Becker 1983; Harlan and Speaker 1987; Boschung Jr and Mayden 2004). Silver chub sexually mature at age-1 (Kinney Jr 1954). Seasonal upstream migrations have been observed, suggesting possible spring spawning movements (Cross 1967). No nest is constructed, and no parental care is given. Fecundity generally increases with female size (Kinney Jr 1954). Semibuoyant, light orange eggs are likely carried with the current downstream during incubation and larval development (Kinney Jr 1954; Perkin and Gido 2011). The diameter of mature eggs was 1 mm from an age-1 female and 2 mm from an age-3 female (Boschung Jr and Mayden 2004). A minimum of 203 km of unsegmented river is necessary for silver chub reproductive success (Perkin and Gido 2011).

Age and Growth: Information on silver chub age and growth is scarce. Growth occurs mainly during the summer, with little to no growth during the winter (Kinney Jr 1954). Silver chub can reach a total length of 231 mm (Lee et al. 1980). Longevity is 4 years, with females generally living longer than males (Kinney Jr 1954).

Food and Feeding: Silver chub are generalist benthic feeders (Steffensen et al. 2014). Age-0 individuals from western Lake Erie ate copepods, Daphnia, and other microcrustaceans, as well as chironomid larvae and pupae and other small immature insects (Kinney Jr 1954). Adults consume *Hexagenia* and other ephemeroptera, along with small mollusks, *Daphnia* (Kinney Jr 1954; Ross 2001; Gaughan 2016; Kočovský 2018). At water temperatures of 18 to 22 °C, adult silver chub and juveniles can consume between 4 to 5% and 10% of their body weight in food per day, respectively (Kinney Jr 1954). They detect prey with external taste buds on their head and pectoral fins, but unlike sturgeon chub or sicklefin chub, likely also use their large eyes and sight to assist in foraging (Davis and Miller 1967; Hesse 1994; Kansas Fishes Committee 2014).

References

Becker GC (1983) Fishes of Wisconsin. University of Wisconsin Press, Madison

Boschung HT Jr, Mayden RL (2004) Fishes of Alabama. Smithsonian Books, Washington, DC

Cross FB (1967) Handbook of fishes in Kansas. University Press of Kansas, Lawrence

Davis BJ, Miller RJ (1967) Brain patterns in minnows of the genus *Hybopsis* in relation to feeding habits and habitat. Copeia 1967:1–39. https://doi.org/10.2307/1442172

Gaughan S (2016) Habitat use, molecular phylogeny, and population structure of *Macrhybopsis* chubs in the upper Mississippi River Basin. Dissertation,. University of Nebraska

Harlan JR, Speaker EB (1987) Iowa fish and fishing. Iowa Department of Natural Resources, Des Moines

Harland B, Berry CR Jr (2004) Fishes and habitat characteristics of the Kaya Paha River, South Dakota-Nebraska. J Freshw Ecol 19:169–177. https://doi.org/10.1080/02705060.2004.9664529

Hayer CA, Harland BC, Berry CR Jr (2006) Recent range extensions, name changes and status updates for selected South Dakota fishes. Proc S D Acad Sci 85:247–265

Hesse LW (1994) The status of Nebraska fishes in the Missouri River, 5. Selected chubs and minnows (Cyprinidae): sicklefin chub (*M. meeki*), sturgeon chub (*M. gelida*), silver chub (*M. storeriana*), speckled chub (*M. aestivaus*), flathead chub (*Platygobio gracilis*), plains minnow (*Hybognathus placitus*), and western silvery minnow (*H. argyritis*). Trans Nebr Acad Sci Affil Soc 21:99–108

Kansas Fishes Committee (2014) Kansas fishes. University Press of Kansas, Lawrence

Kinney EC Jr (1954) A life history study of the silver chub, *Hybopsis storeriana* (Kirtland), in western Lake Erie with notes on associated species. Dissertation, Ohio State University

Kočovský PM (2018) Diets of endangered silver chub (*Macrhybopsis storeriana*, Kirtland, 1844) in Lake Erie and implications for recovery. Ecol Freshw Fish 2018:1–8. https://doi.org/10.1111/eff.12424

Lee DS, Gilbert CR, Hocutt CH, Jenkins RE, McAllister DE, Stauffer JR (1980) Atlas of North American freshwater fishes. North Carolina Biological Survey Publication 1980-12. https://doi.org/10.5962/bhl.title.141711

Perkin JS, Gido KB (2011) Stream fragmentation thresholds for a reproductive guild of Great Plains fishes. Fisheries 36:371–383. https://doi.org/10.1080/03632415.2011.597666

Ross ST (2001) Inland fishes of Mississippi. University Press of Mississippi, Jackson

Stauffer JR Jr, Criswell RW, Fischer DP (2016) The fishes of Pennsylvania. Cichlid Press, El Paso

Steffensen KD, Shuman DA, Stukel S (2014) The status of fishes in the Missouri River, Nebraska: shoal chub (*Macrhybopsis hyostoma*), sturgeon chub (*M. gelida*), sicklefin chub (*M. meeki*), silver chub (*M. storeriana*), flathead chub (*Platygobio gracilis*), plains minnow (*Hybognathus placitus*), western silvery minnow (*H. argyritis*), and brassy minnow (*H. hankinsoni*). Trans Nebr Acad Sci Affil Soc 34:49–67

Wood KG (1953) Distribution and ecology of certain bottom-living invertebrates of the western basin of Lake Erie. Dissertation, Ohio State University

6.14.20   Northern Pearl Dace *Margariscus nachtriebi*

(Cox, 1896)

Nuptial Male

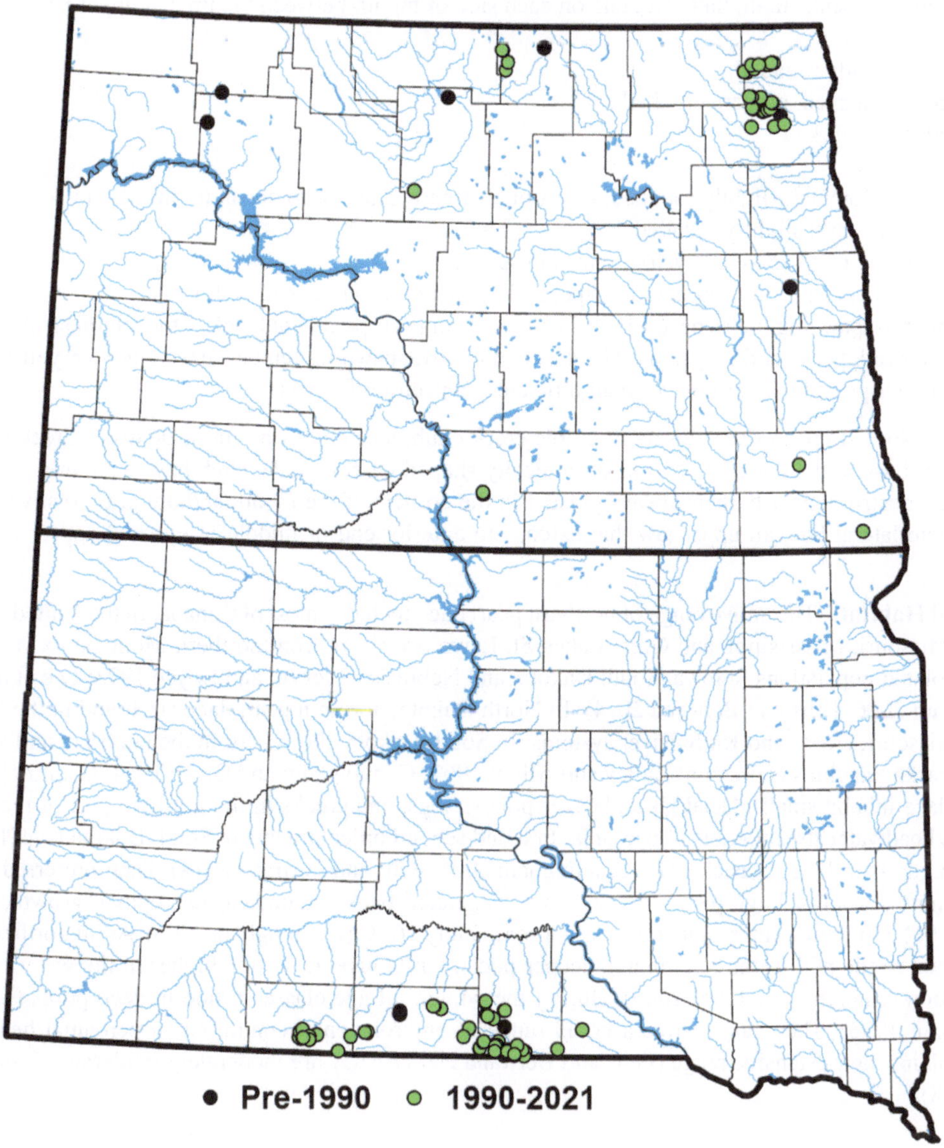

● **Pre-1990**   ○ **1990-2021**

Etymology: *Margariscus nachtriebi* – *Margariscus* = pearl; *nachtriebi* = unknown.

Description:

Body – moderately fusiform, stout, slightly laterally compressed posteriorly.
Color –

> *Dorsally* – dusky brown to dark olive.
> *Laterally* – light bronze to olive; mottled, small irregular dark specks; dark dusky lateral stripe more prominent on posterior end of body, fades with age and growth.
> *Ventrally* – cream to white.
> *Fins* – lightly pigmented.

Head – moderate.
Snout – blunt.
Eyes – moderately large, laterally on head.
Mouth – terminal, slightly oblique, relatively small.

> *Jaws* – extends below nostril.
> *Lips* – fleshy; groove on upper lip continuous over snout.

Teeth – pharyngeal tooth pattern 2,5-4,2.
Barbels – sometimes present; small, flattened, one on each side of mouth between the lip and upper jaw.
Gill rakers – 6 to 8; short.
Dorsal fin – 8 rays; slightly rounded.
Caudal peduncle – elongated, moderately thick.
Caudal fin – moderately forked.
Anal fin – 8 rays.
Pelvic fins – 8 rays; abdominal, slightly rounded; insertion distinctly anterior to dorsal fin insertion.
Pectoral fins – 15 or 16 rays; slightly rounded.
Lateral line – complete; 60 to 75 small cycloid scales in series.
Juveniles – more distinct and prominent lateral stripe, ending in a dark spot at caudal fin base.
Sexual dimorphism – spawning males have a prominent red-orange stripe on flank below the lateral stripe along with small tubercles on the head, pectoral fin rays, lateral sides, and breast; spawning females have a golden yellow stripe on flank below the lateral stripe and may also have small tubercles on pectoral fin rays.

Similar Species: Northern pearl dace closely resemble creek chub, lake chub, and finescale dace. Creek chub have a dark spot at the anterior base of the dorsal fin and more conically shaped barbels. Lake chub have a slightly subterminal mouth extending past the nostril, longer barbels always present in the corners of the mouth and are gray to silver. Finescale dace have an incomplete lateral line, an upper jaw that extends to anterior end or middle of eye and appear scaleless with the naked eye.

Distribution and Habitat: The native range of northern pearl dace includes most of Canada. In the United States, it extends throughout the Missouri, Mississippi, and Great Lakes-St. Lawrence rivers from northern Montana to Maine and south to Virginia, with isolated populations in Iowa, South Dakota, and Nebraska (Menzel and Boyce 1973; Lee et al. 1980; Menzel 1981; Cunningham 2006; Felts and Bertrand 2014). In North Dakota, northern pearl dace are found in the Red River of the North, Souris, Missouri, Heart, and Knife River systems. In South Dakota, they occur in the Niobrara and White River systems of the Nebraska Sandhills ecoregion (Bailey and Allum 1962; Cunningham and Olson 1994; Cunningham et al. 1995). Northern pearl dace inhabit small glacial lakes, beaver ponds, bogs, and creeks in the northern part of their range and the cool, spring-fed, headwater stream pools of perennial first and second order streams in South Dakota (McPhail and Lindsey 1970; Scott and Crossman 1973; Stasiak 1978; Cunningham 2006; Felts and Bertrand 2014). Their preferred habitat includes meandering streams devoid of large piscivores with well-vegetated banks, abundant macrophyte growth, undercut banks (Cunningham 2006). Optimum water temperature is approximately 16 °C (Cunningham 2006). Adult northern pearl dace are generally restricted to the mid-water column and deeper pools, while juveniles often higher in the water column closer to instream vegetation (Cunningham 1995, 2006). During the winter, all age classes reside in deep pools (Tallman and Gee 1982; Cunningham 1995). They can tolerate a broad range of environmental conditions in perennial headwater streams, which likely explains their extensive range (Felts and Bertrand 2014). They are moderately intolerant of turbidity (Whittier and Hughes 1998).

Reproduction: Spawning depends on photoperiod and typically occurs from mid-April to mid-May at water temperatures of 16 to 18 °C (Cunningham 1995, 2006). Both sexes of northern pearl dace become sexually mature at age-1 (Stasiak 1978). Actual spawning has only been described once. In a Michigan stream, males were observed coaxing gravid females into their small and defended territory (Langlois 1929). Males then positioned themselves parallel to the female, grasped her with their tuberculate pectoral fins, and wrapped their caudal fin around her caudal peduncle (Langlois 1929). The bodies of the pair then vibrated together for approximately two seconds to cause gamete release (Langlois 1929). Northern pearl dace are lithophilic spawners, depositing eggs over gravel or rocky substrate in areas with little to no current (Cunningham 2006). No nest is constructed, and no parental care is given (Langlois 1929; Cunningham 2006). Clutch size from northern pearl dace in Nebraska varied, with means of 1,100 and 2,800 eggs for age-1 and age-2 females, respectively (Cunningham 1995, 2006). Eggs are approximately 1 mm in diameter (Fava and Tsai 1974; Becker 1983).

Age and Growth: Standard lengths at age of female northern pearl dace from Nebraska were 56 mm at age-1, 69 mm at age-2, 79 mm at age-3, and 88 mm at age-4 (Stasiak 1978). Females generally grow faster and larger and live longer than males (Stasiak 1978). They can reach a total length of 158 mm, with a longevity of 4 years (Lindsey 1956; Stasiak 1978; Becker 1983).

Food and Feeding: Northern pearl dace are sight-feeding omnivores actively foraging during the day (Tallman and Gee 1982; Cunningham 2006). Individual fish with a total length less than 54 mm primarily ate dipteran larvae, algae, copepods, and cladocerans (Stasiak 1978). Individuals greater than 74 mm have a more diverse diet of snails, fingernail clams, amphipods, and larger insects often found in the benthic zone (Stasiak 1978). Sand, silt, and detritus are also common in the diets of larger northern pearl dace but are likely ingested incidentally while consuming prey along the bottom (Stasiak 1978). Both larvae and adult diptera are the most common dietary item in fish of all sizes (Stasiak 1978).

References

Bailey RM, Allum MO (1962) Fishes of South Dakota. University of Michigan, Ann Arbor. https://doi.org/10.3998/mpub.9690435

Becker GC (1983) Fishes of Wisconsin. University of Wisconsin Press, Madison

Cunningham GR (1995) Life history traits and reproductive ecology of the pearl dace, *Margariscus margarita* (Pisces: Cyprinidae) in the Sandhills of Nebraska. Dissertation, University of Nebraska-Omaha

Cunningham GR (2006) Pearl dace (*Margariscus margarita*): a technical conservation assessment. USDA Forest Service. https://www.fs.usda.gov/Internet/FSE_DOCUMENTS/stelprdb5206789.pdf. Accessed 24 Apr 2023

Cunningham GR, Olson RD, Hickey SM (1995) Fish surveys of the streams and rivers of south-Central South Dakota west of the Missouri River. Proc S D Acad Sci 74:55–64

Cunningham G, Olson R (1994) Fish species collected in streams in West River South Dakota-1994. Unpublished report to South Dakota Game, Fish and Parks. Pierre

Fava JA, Tsai C (1974) The life history of the pearl dace, *Semotilus margarita*, in Maryland. Chesap Sci 15:159–162. https://doi.org/10.2307/1351034

Felts EA, Bertrand KN (2014) Conservation status of five headwater stream specialists in South Dakota. Am Midl Nat 172:131–159. https://doi.org/10.1674/0003-0031-172.1.131

Langlois TH (1929) Breeding habits of the northern dace. Ecology 10:161–163. https://doi.org/10.2307/1940521

Lee DS, Gilbert CR, Hocutt CH, Jenkins RE, McAllister DE, Stauffer JR (1980) Atlas of North American freshwater fishes. North Carolina Biological Survey Publication 1980-12. https://doi.org/10.5962/bhl.title.141711

Lindsey CC (1956) Distribution and taxonomy of fishes in the McKenzie drainage of British Columbia. J Fish Res Board Can 13:759–886. https://doi.org/10.1139/f56-044

McPhail JD, Lindsey CC (1970) Freshwater fishes of Northwestern Canada and Alaska. Fish Res Board Can, Ottawa

Menzel BW (1981) Iowa's waters and fishes: a century and a half of change. Proc Iowa Acad Sci 88:17–23

Menzel BW, Boyce MS (1973) First record of the pearl dace, *Semotilus margarita* (Cope), from Iowa. Iowa State J Res 47:245–248

Scott WB, Crossman EJ (1973) Freshwater fishes of Canada. Fisheries Research Board of Canada, Bulletin 184

Stasiak RH (1978) Food, age and growth of the pearl dace, *Semotilus margarita*, in Nebraska. Am Midl Nat 100:463–466. https://doi.org/10.2307/2424848

Tallman RF, Gee JH (1982) Intraspecific resource partitioning in a headwaters stream fish, the pearl dace, *Semotilus margarita* (Cope). Environ Biol Fish 7:243–249. https://doi.org/10.1007/BF00002499

Whittier TR, Hughes RM (1998) Evaluation of fish species tolerances to environmental stressors in lakes in the northeastern United States. N Am J Fish Manag 18:236–252. https://doi.org/10.1577/1548-8675

6.14.21   Pugnose Shiner *Miniellus anogenus*

(Forbes, 1885)

● Pre-1990   ● 1990-2021

Etymology: *Miniellus anogenus* – *Miniellus* = small, plain; *anogenus* = without chin.

Description:

Body – fusiform, slender, slightly laterally compressed.
Color –

    Dorsally – dusky olive, tan, gold or bronze; scales outlined in slightly darker pigment creating a crosshatched pattern.
    Laterally – silver; prominent black lateral stripe extending from snout through eye to a wedge-shaped spot at the caudal fin base; adults missing concentrated pigment on some lateral scales causing them to appear missing.
    Ventrally – silver to white.
    Fins – generally clear to lightly pigmented.

Head – moderately large.
Snout – blunt.
Eyes – moderately large, greater than the length of snout, laterally on head.
Mouth – small, superior, oblique, nearly vertical.

    Jaws – upper jaw extends posteriorly below the nostril.
Teeth – pharyngeal tooth pattern 0,4-4,0.
Gill rakers – 8; short.
Dorsal fin – 8 or 9 rays, straight distal end.
Caudal peduncle – slightly elongated, thickness uniform.
Caudal fin – moderately forked.
Anal fin – 7 to 10 rays, straight distal end.
Pelvic fins – 8 rays; abdominal; insertions slightly anterior to dorsal fin insertion.
Pectoral fins – 12 to 14 rays.
Lateral line – complete; 34 to 37 cycloid scales in series; scales on the anterior lateral side of body as deep as wide.
Sexual dimorphism – male pelvic fin extends to or slightly beyond cloaca; female pelvic fin does not reach cloaca; spawning males have small tubercles on pectoral fin rays, pelvic fin rays, and dorsal side of head.

Similar Species: A small, superior, oblique, and nearly vertical mouth distinguishes pugnose shiner from other *Notropis*. Pugnose shiner may be confused with blacknose shiner, brassy minnow, and fathead minnow. Blacknose shiner have lateral stripe pore dark borders that expand into crescent-shaped vertical bars anteriorly, a snout that slightly extends beyond the mouth, a lower jaw that extends to the anterior edge of the eye, and an incomplete lateral line. Brassy minnow have a snout that slightly protrudes past the mouth, and pelvic fin insertions slightly posterior to the dorsal fin insertion. Fathead minnow have a dorsal fin with a rounded distal end, pectoral fins with 15 or 16 rays, and an incomplete lateral line with 40 to 54 cycloid scales in series.

Distribution and Habitat: Pugnose shiner are one of the rarest minnows in the northern United States, with small, relatively isolated populations (Bailey 1959; Owen et al. 1981; McCusker et al. 2014). Their native range is fragmented across the Upper Mississippi River, Red River of the North, and Great Lakes drainages, from western New York and eastern Ontario to southeastern North Dakota (Lee et al. 1980; COSEWIC 2002). Pugnose shiner are believed to be extirpated from North Dakota after last being reported in 1964 from the Sheyenne and Turtle Rivers within the Red River of North basin (Owen et al. 1981). They have not been observed in South Dakota. Pugnose shiner may be declining throughout their native range because of extreme sensitivity to turbidity, shoreline erosion and development leading to decreased aquatic vegetation, and the introduction of invasive species (Bailey 1959; Scott and Crossman 1973; Trautman 1981; Lyons 1992; COSEWIC 2002; DFO 2010). They need an abundance of submerged aquatic vegetation for predatory cover, foraging opportunities, and reproductive sites (Becker 1983; COSEWIC 2002). Pugnosed shiner preferred habitat includes clear, quiet, glacial lakes, in addition to stagnant channels of low gradient streams with abundant submerged vegetation and substrates of clean sand or organic debris (Trautman 1981; COSEWIC 2002).

Reproduction: Information on pugnose shiner spawning and reproductive behavior is limited. Spawning in Wisconsin occurs from mid-May through July at water temperatures of 21 to 29 °C (Becker 1983). They are lithophilic batch spawners, releasing eggs more than once during a spawning season (Becker 1983; COSEWIC 2002). No nest is constructed, and no parental care is given. Fecundity ranges from 530 to 1,275 eggs per female. The yellow eggs are approximately 1 mm in diameter (Becker 1983).

Age and Growth: Age-0 pugnose shiner from Ontario, Canada had a total length of 24 mm (COSEWIC 2002; Leslie and Timmins 2002). Total lengths at age of pugnose shiner from Wisconsin were 42 mm at age-1, 46 mm at age-2, and 53 mm at age-3 (Becker 1983). Males and females can reach total lengths of 50 and 60 mm, respectively (COSEWIC 2002). Longevity is 3 years (COSEWIC 2002).

Food and Feeding: Pugnose shiner from Ontario, Canada, ate *Chara*, filamentous green algae, and other minute plant material, as well as *Daphnia*, *Bosmina*, *Chydorus*, other cladocerans, trichopteran larvae, and small leeches (Becker 1983; COSEWIC 2002). Foraging mainly occurred midwater (Becker 1983).

References

Bailey RM (1959) Distribution of the American cyprinid fish (*Notropis anogenus*). Copeia 1959:119–123. https://doi.org/10.2307/1440063

Becker GC (1983) Fishes of Wisconsin. University of Wisconsin Press, Madison

COSEWIC (Canadian Wildlife Service Environment Canada) (2002) COSEWIC assessment and update status report on the pugnose shiner *Notropis anogenus* in Canada. Committee on the Status of Endangered Wildlife in Canada, Ottawa

DFO (Fisheries and Oceans Canada) (2010) Recovery potential assessment of pugnose shiner (*Notropis anogenus*) in Canada. DFO Canadian Science Advisory Secretariat Science Advisory Report 2010/025. https://waves-vagues.dfo-mpo.gc.ca/library-bibliotheque/340942.pdf. Accessed 24 Apr 2023

Lee DS, Gilbert CR, Hocutt CH, Jenkins RE, McAllister DE, Stauffer JR (1980) Atlas of North American freshwater fishes. North Carolina Biological Survey Publication 1980-12. https://doi.org/10.5962/bhl.title.141711

Leslie JK, Timmins CA (2002) Description of age 0 juvenile pugnose minnow *Opsopoedus emiliae* (Hay) and pugnose shiner *Notropis anogenus* Forbes in Ontario. Can Tech Rep Fish Aquat Sci 2397

Lyons J (1992) Using the index of biotic integrity (IBI) to measure environmental quality in warmwater streams of Wisconsin. General technical report NC-149, US Department Agriculture, Forest Service, St. Paul

McCusker MR, Mandrak NE, Egeh B, Lovejoy NR (2014) Population structure and conservation genetic assessment of the endangered pugnose shiner, *Notropis anogenus*. Conserv Genet 15:343–353

Owen JB, Elsen DS, Russell GW (1981) Distribution of fishes in North and South Dakota Basins affected by the Garrison Diversion Unit. University of North Dakota Press, Grand Forks

Scott WB, Crossman EJ (1973) Freshwater fishes of Canada. Fisheries Research Board of Canada, Bulletin 184

Trautman MB (1981) The fishes of Ohio, revised edn. Ohio State University Press, Columbus

6.14.22   Sand Shiner *Miniellus stramineus*

(Cope, 1865)

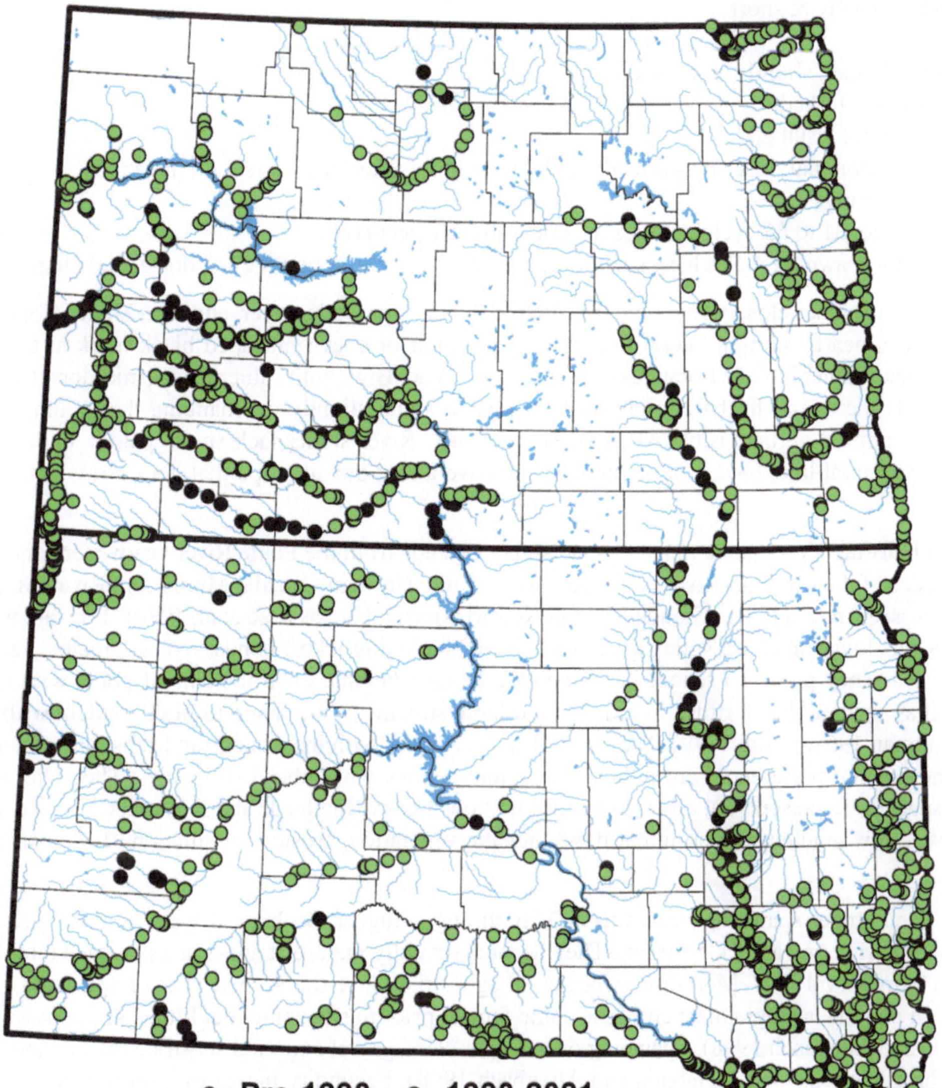

● **Pre-1990**   ● **1990-2021**

Etymology: *Miniellus stramineus* – *Miniellus* = small, plain; *stramineus* = straw, likely reference coloration.

Description:

Body – moderately fusiform, slender, laterally compressed.
Color –

> *Dorsally* – light olive to tan; dorsal scales outlined in darker pigment creating a crosshatched pattern; prominent, narrow mid-dorsal stripe widening into a wedge-shaped spot at the dorsal fin anterior base but not surrounding it.
> *Laterally*: silver; small, paired, dark specks or "mouse tracks" outlining the lateral line; faint stripe continuous from caudal peduncle to eye, fading anterior to dorsal fin.
> *Ventrally* – silver to white.
> *Fins* – little to no pigment; caudal; faint, dusky spot sometimes at base.

Head – short.
Snout – blunt, rounded; does not protrude past the mouth.
Eyes – large, laterally on head.
Mouth – small, subterminal, slightly oblique.

> *Jaws* – lower jaw extending to anterior edge of the eye; length of upper jaw less than eye diameter.

Teeth – pharyngeal teeth 0,4-4,0.
Gill rakers – approximately 8; short.
Dorsal fin – 8 rays.
Caudal peduncle – slender, slightly elongated.
Caudal fin – moderately forked.
Anal fin – 7 rays; straight distal end.
Pelvic fins – 8 rays; abdominal; insertions directly under or slightly anterior to dorsal fin insertion.
Pectoral fins – 13 to 15 rays.
Lateral line – complete; 31 to 35 cycloid scales in series; slightly decurved.
Sexual dimorphism – spawning males have small tubercles on the pectoral fin rays and dorsal and lateral sides of the head.

Similar Species: Sand shiner closely resemble the Topeka shiner, bigmouth shiner, and river shiner. Topeka shiner tend to have a deeper body, a nearly straight lateral line, and a triangular or chevron-shaped black mark at the caudal fin base. Spawning male Topeka shiner have red-orange on their fins and head. Bigmouth shiner have a mid-dorsal stripe that does not widen, but surrounds the dorsal fin base, eight anal fin rays, a snout slightly overhanging the mouth, and eyes directed upwards with lower pupil margins visible when viewed dorsally. River shiner lack small, paired, dark specks or "mouse tracks" outlining the lateral line, have a larger mouth, and have a mid-dorsal stripe that does not widen but surrounds the dorsal fin base.

Distribution and Habitat: The native range of sand shiner extends from the Platte River headwaters in eastern Wyoming, throughout the basins of the Missouri River, lower Red River of the North in Canada, Mississippi River, St. Lawrence River-Great Lakes, Tennessee River, and Rio Grande River in New Mexico and Texas (Lee et al. 1980). They are widely distributed in the Dakotas. Sand shiner inhabit a wide variety of habitats, including lakes, ponds, rivers, streams, and creeks (Becker 1983; Pflieger 1997). They prefer to school in the benthic or near benthic environments of pools with moderate-to-high velocities, as well as sand or gravel riffles in moderate-to-large streams of moderate-to-high gradient with clear to slightly turbid water (Trautman 1957; Smith 1979). Their abundance increases with increasing sand substrate (Mueller Jr and Pyron 2009). Sand shiner habitat preferences are influenced by interactions with other fish species. They chose high flow areas when other fish species are present, but otherwise chose no flow areas (Mueller Jr and Pyron 2009). Sand shiner are absent or rare in areas with low gradients, intermittent flows, aquatic vegetation, erosion, or siltation (Trautman 1957; Mueller Jr and Pyron 2009).

Reproduction: Sand shiner spawning occurs from May through August, peaking at water temperatures of 27 to 37 °C (Starrett 1951; Summerfelt and Minckley 1969). They become sexually mature at age-1 at an average total length of 49 mm (Summerfelt and Minckley 1969). Mortality increases in age-2 fish after the second year of spawning (Summerfelt and Minckley 1969). In aquaria, a single male chases a single female prior to spawning, but little else is known about spawning behavior (Platania and Altenbach 1998). Spawning occurs over clean gravel and sand substrate (Miller and Robinson 2004). Sand shiner are broadcast spawners (Platania and Altenbach 1998). Fecundity increases with female size (Starrett 1951). Fecundity from age-1, age-2, and age-3 sand shiner in Iowa was approximately 250, 1,100 and 1,800 eggs, respectively

(Starrett 1951). The demersal, adhesive, approximately 1-mm diameter eggs quickly settle onto substrate (Summerfelt and Minckley 1969; Platania and Altenbach 1998).

Age and Growth: Total-lengths-at-age of sand shiner in Iowa were 39 mm at age-1, 53 mm at age-2, and 64 mm at age-3 (Smith et al. 2010). They can reach a total length of 87 mm with a longevity of 3 years (Smith et al. 2010).

Food and Feeding: Sand shiner are generalized, opportunistic, benthic omnivores (Starrett 1950; Gillen and Hart 1980). They feed actively throughout the day and are relatively inactive at night (Starrett 1950). Sand shiner primarily consume small mayfly nymphs, stoneflies, midge larvae, and other immature and adult aquatic and terrestrial insects (Gillen and Hart 1980). They also eat detritus and plant material (Starrett 1950).

References

Becker GC (1983) Fishes of Wisconsin. University of Wisconsin Press, Madison

Gillen AL, Hart T (1980) Feeding interrelationships between the sand shiner and the striped shiner. Ohio J Sci 80:71–76

Lee DS, Gilbert CR, Hocutt CH, Jenkins RE, McAllister DE, Stauffer JR (1980) Atlas of North American freshwater fishes. North Carolina Biological Survey Publication 1980-12. https://doi.org/10.5962/bhl.title.141711

Miller RJ, Robinson HW (2004) Fishes of Oklahoma, revised edn. University of Oklahoma Press, Norman

Mueller R Jr, Pyron M (2009) Substrate and current velocity preferences of spotfin shiner (*Cyprinella spiloptera*) and sand shiner (*Notropis stramineus*) in artificial streams. J Freshw Ecol 24:239–245. https://doi.org/10.1080/02705060.2009.9664288

Pflieger WL (1997) The fishes of Missouri, revised edn. Missouri Dep Cons, Jefferson City

Platania SP, Altenbach CS (1998) Reproductive strategies and egg types of seven Rio Grande basin cyprinids. Copeia 1998:559–569. https://doi.org/10.2307/1447786

Smith CD, Neebling TE, Quist MC (2010) Population dynamics of the sand shiner (*Notropis stramineus*) in non-wadeable rivers of Iowa. J Freshw Ecol 25:617–626. https://doi.org/10.1080/02705060.2010.9664411

Smith PW (1979) The fishes of Illinois. University of Illinois Press, Urbana

Starrett WC (1950) Food relationships of the minnows of the Des Moines River, Iowa. Ecology 31:216–233. https://doi.org/10.2307/1932388

Starrett WC (1951) Some factors affecting the abundance of minnows in the Des Moines River, Iowa. Ecology 32:13–27. https://doi.org/10.2307/1930969

Summerfelt RC, Minckley CO (1969) Aspects of the life history of the sand shiner, *Notropis stramineus* (Cope), in the Smoky Hill River, Kansas. Trans Am Fish Soc 98:444–453. https://doi.org/10.1577/1548-8659(1969)98[444:AOTLHO]2.0.CO;2

Trautman MB (1957) The fishes of Ohio. Ohio State University Press, Columbus

6.14.23   Topeka Shiner *Miniellus topeka*

(Gilbert, 1884)

Nuptial Male   Female

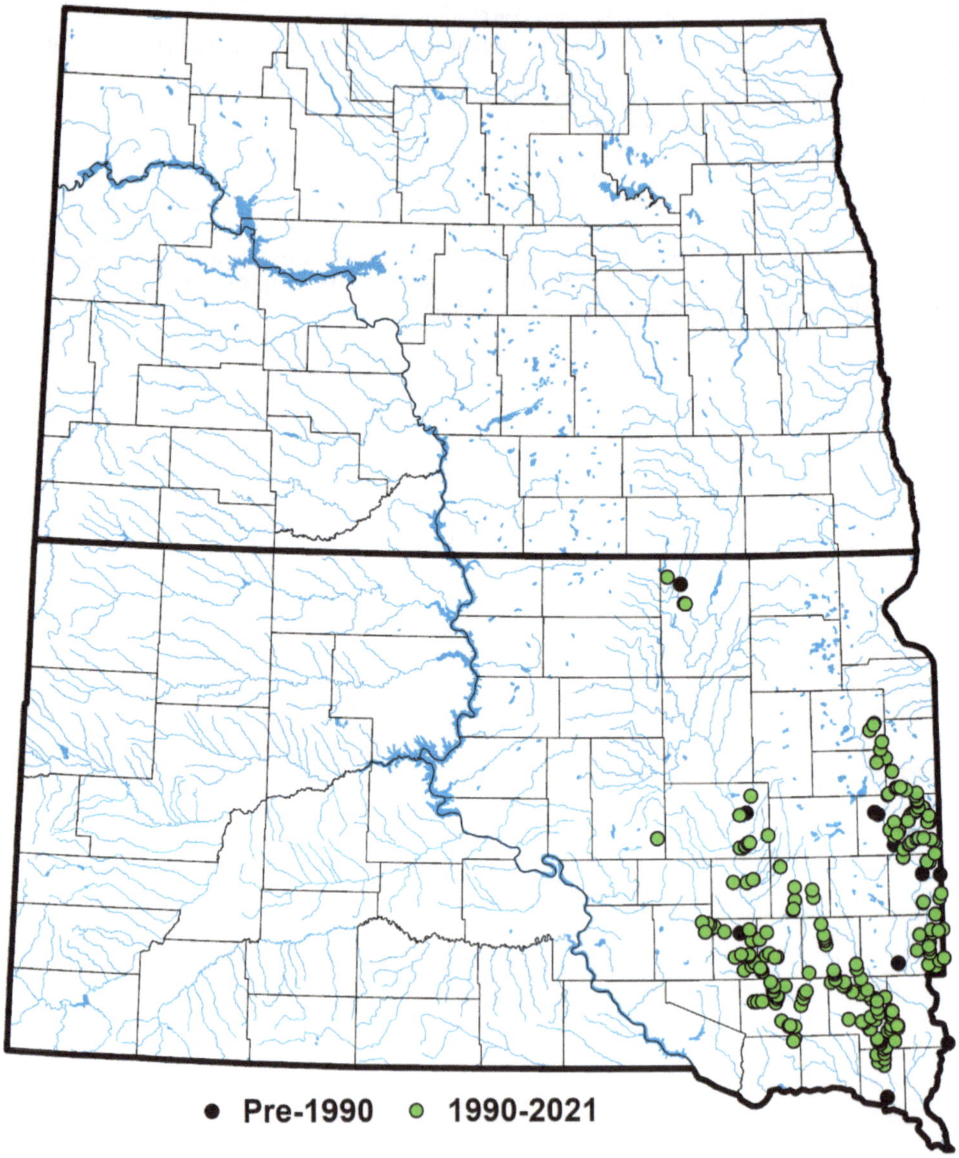

● Pre-1990   ● 1990-2021

Etymology: *Miniellus topeka* – *Miniellus* = small, plain; *topeka* = referencing Topeka, Kansas.

Description:

Body – fusiform, moderately deep, laterally compressed.
Color –

> *Dorsally* – light olive to tan; scales outlined in darker pigment creating a crosshatched pattern; prominent, narrow mid-dorsal stripe slightly widens at, but does not surround, dorsal fin base.
> *Laterally* – silver; small paired dark specks outline the lateral line; dusky lateral stripe from caudal peduncle to snout.
> *Ventrally* – silver to white.
> *Fins* – pigment minimal or absent; caudal: black chevron at base.

Head – short.
Snout – blunt, rounded.
Eyes – large, lateral on head.
Mouth – small, terminal to slightly subterminal, oblique.

> *Jaws* – lower jaw slightly extending beyond anterior edge of eye; upper jaw length less than eye diameter.

Teeth – pharyngeal tooth pattern 0,4-4,0.
Dorsal fin – 8 rays.
Caudal peduncle – slender, slightly elongated.
Caudal fin – moderately forked.
Anal fin – 7 rays, straight distal end.
Pelvic fins – 8 rays; abdominal; insertions directly under or slightly anterior to dorsal fin insertion.
Lateral line – complete; 32 to 37 cycloid scales in series; nearly straight.
Skin – leather like; scaleless except for single patch of ganoid scales at base of caudal fin upper lobe.
Sexual dimorphism – spawning males have bright red-orange fins and side of head, faint blue hue or luminance on lateral sides, and small tubercles on the snout, head, and body.

Similar Species: Topeka shiner resemble sand shiner, bigmouth shiner, and river shiner. Sand shiner have a narrower body and slightly decurved lateral line, with a faint, dusky spot sometimes visible at the caudal fin base. Red or orange coloration on the head or fins does not occur in spawning male sand shiners. Bigmouth shiner have eight anal fin rays, a snout slightly overhanging the mouth, and eyes directed upward with the lower pupil margins visible when viewed from above. Both the bigmouth shiner and river shiner have a uniform-width mid-dorsal stripe surrounding the dorsal fin. River shiner lack small, paired dark specks or "mouse tracks" outlining the lateral line and have a larger mouth.

Distribution and Habitat: Topeka shiner are native to parts of the Mississippi, Missouri, and Arkansas river drainages from central Missouri and Kansas to southern Minnesota, eastern South Dakota, Iowa, and Nebraska (Bailey and Allum 1962). They have not been reported in North Dakota. Previously widespread and abundant, they are now only found in highly fragmented populations in less than 10% of their historical range (Tabor 1998; Blausey 2001). In South Dakota, Topeka shiner are widespread, but not numerous, in the Big Sioux, Vermillion, and James River systems (Blausey 2001; Shearer 2003). They may be more abundant in South Dakota than in any other state (Shearer 2003). Declines in abundance and distribution are likely because of increased sedimentation, nonpoint source pollution, dam construction, and predation by non-native species (Minckley and Cross 1959; USFWS 1998; Adams et al. 2000; Blausey 2001). Topeka shiner inhabit small, quiet, clear, and moderately cool creeks, as well as tributaries and headwaters of larger streams in off-channel habitats with submerged vegetation and sand, gravel or cobble substrate (Pflieger 1997; Blausey 2001; Dahle 2001; Bakevich et al. 2013). They are often found within the limnetic zone of large pools and backwaters (Pflieger 1997; Adams et al. 2000; Blausey 2001). Juveniles prefer shallow, off-channel habitat (Bakevich et al. 2013). Topeka shiner avoid water velocities above 10 cm/second, which can negatively impact reproduction and decrease population sizes (Minckley and Cross 1959; Adams et al. 2000). Most eastern South Dakota Topeka shiner locations have water velocities under 0.20 m/second (Blausey 2001). However, Topeka shiners have been observed swimming at velocities of 30 to 40 cm/second for over 200 minutes, suggesting that fish ways and culverts may facilitate dispersal and recolonization (Pflieger 1997; Adams et al. 2000; Blausey 2001). They have a critical thermal maximum of 39 °C after acclimation to 31 °C and can tolerate oxygen concentrations as low as 1.2 mg/L (Koehle and Adelman 2007).

Reproduction: Spawning occurs from mid-May to early August, peaking in June (Dahle 2001). In aquaria, spawning began when water temperatures reached 24 °C (Katula 1998). Sexual maturity starts at age-1, with most individuals maturing at age-2 (Dahle 2001; Kerns and Bonneau 2002). Mature males are territorial and defend areas under 0.5 m in diameter (Pflieger

1997; Katula 1998; Dahle 2001; Kerns and Bonneau 2002; Stark et al. 2002). In Missouri, Topeka shiners are broadcast spawners and will spawn over green sunfish and orangespotted sunfish nests (Pflieger 1997; Kerns and Bonneau 2002). During spawning, males swim next to females, with both fish vibrating together to cause gamete release (Katula 1998; Stark et al. 2002). Females repeat this process two to four times before resuming normal activity (Katula 1998). Fecundity is more correlated with size than age and has been reported at 351 eggs at age-1, 559 at age-2, and 478 at age-3 (Dahle 2001). The yellow to yellow-orange, adhesive eggs are approximately 1 mm in diameter, and hatch in 5 days at 22 °C (Katula 1998; Dahle 2001; Kerns and Bonneau 2002).

Age and Growth: Growth occurs primarily from May to July, with an optimal temperature of 27 °C (Dahle 2001; Koehle and Adelman 2007). Females are generally shorter than same-age males (Dahle 2001). In Minnesota, standard lengths for females and males were, respectively, 29 and 30 mm at age-1, 41 and 46 mm at age-2, and 47 and 56 mm at age-3 (Dahle 2001). Maximum total length is 75 mm, and longevity is 3 years (Pflieger 1997; Blausey 2001; Dahle 2001). Topeka shiners experience high mortality at the end of summer, prior to young of year recruitment (Dahle 2001).

Food and Feeding: Topeka shiner are generalized, opportunistic, benthic, and nektonic omnivores (Hatch and Besaw 1998, 2001; Dahle 2001). They forage diurnally, with peak feeding from early morning to early afternoon (Kerns and Bonneau 2002). Adult diets consist primarily of both larval and adult aquatic and terrestrial insects, along with microcrustaceans (Dahle 2001; Kerns and Bonneau 2002). Larval fish, snails, worms, algae, vascular plants, and detritus are also consumed (Dahle 2001; Hatch and Besaw 2001).

References

Adams SR, Hover JJ, Killgore KJ (2000) Swimming performance of the Topeka shiner (*Notropis topeka*) an endangered Midwestern minnow. Am Midl Nat 144:178–186. https://doi.org/10.1674/0003-0031(2000)144[0178:SPOTTS]2.0.CO;2

Bailey RM, Allum MO (1962) Fishes of South Dakota. University of Michigan, Ann Arbor. https://doi.org/10.3998/mpub.9690435

Bakevich BD, Pierce CL, Quist MC (2013) Habitat, fish species, and fish assemblage associations of the Topeka shiner in west-central Iowa. N Am J Fish Manag 33:1258–1268. https://doi.org/10.1080/02755947.2013.839969

Blausey CM (2001) The status and distribution of the Topeka shiner *Notropis topeka* in eastern South Dakota. Dissertation, South Dakota State University

Dahle SP (2001) Studies of Topeka shiner (*Notropis topeka*) life history and distribution in Minnesota. Dissertation, University of Minnesota

Hatch JT, Besaw S (1998) Diverse food use in Minnesota populations of the Topeka shiner (*Notropis topeka*). General College and James Ford Bell Museum of Natural History, Minneapolis

Hatch JT, Besaw S (2001) Food use in Minnesota populations of the Topeka shiner (*Notropis topeka*). J Freshw Ecol 16:229–233. https://doi.org/10.1080/02705060.2001.9663807

Katula R (1998) Eureka Topeka! Tropical Fish Hobbyist 47:54–60

Kerns HA, Bonneau JL (2002) Aspects of the life history and feeding habits of the Topeka shiner (*Notropis topeka*) in Kansas. Trans Kans Acad Sci 105:125–142. https://doi.org/10.1660/0022-8443(2002)105[0125:AOTLHA]2.0.CO;2

Koehle JJ, Adelman IR (2007) The effects of temperature, dissolved oxygen, and Asian tapeworm infection on growth and survival of the Topeka shiner. Trans Am Fish Soc 136:1607–1613. https://doi.org/10.1577/T07-033.1

Minckley WL, Cross FB (1959) Distribution, habitat, and abundance of the Topeka shiner *Notropis topeka* (Gilbert) in Kansas. Am Midl Nat 61:210–217. https://doi.org/10.2307/2422352

Pflieger WL (1997) The fishes of Missouri, revised edn. Missouri Dep Cons, Jefferson City

Shearer JS (2003) Topeka shiner (*Notropis topeka*) management plan for the state of South Dakota. S D Dep Game Fish Parks, Wildl Div, Rep No. 2003-10, Pierre

Stark WJ, Luginbill JS, Erbele ME (2002) Natural history of a relict population of Topeka shiner (*Notropis topeka*) in northwestern Kansas. Trans Kans Acad Sci 105:143–152. https://doi.org/10.1660/0022-8443(2002)105[0143:NHOARP]2.0.CO;2

Tabor VM (1998) Final rule to list the Topeka shiner as endangered. Fed Regist 63:69008–69021

USFWS (US Fish and Wildlife Service) (1998) Endangered and threatened wildlife and plants; final rule to list the Topeka shiner as endangered. Fed Regist 63(204):69008–69021

6.14.24   Hornhead Chub *Nocomis biguttatus*

(Kirtland, 1840)

Nuptial Male   Female

● **Pre-1990**   ○ **1990-2021**

Etymology: *Nocomis biguttatus* – *Nocomis* = a Native American name for a group of fishes used by Charles Girard, a French ichthyologist; *biguttatus* = two spotted.

Description:

Body – fusiform, robust, cylindrical anteriorly, laterally compressed posteriorly.
Color –

> *Dorsally* – dark brown to olive.
> *Laterally* – bronze, olive to silver; dark, diffuse lateral stripe ending in a spot at caudal fin base.
> *Ventrally* – silver to white.
> *Fins* – lightly pigmented.

Head – broad.
Snout – bluntly pointed.
Eyes – large, laterally on upper portion of head.
Mouth – large, terminal.

> *Jaws* – upper jaw extends to anterior edge of the eye.
> *Lips* – fleshy.

Teeth – pharyngeal tooth pattern 1,4-4,1.
Barbels – small, inconspicuous; one on each corner of the mouth.
Gill rakers – 9; short.
Dorsal fin – 8 rays, nearly straight to slightly convex distal end.
Caudal peduncle – moderately thick.
Caudal fin – moderately forked.
Anal fin – 7 rays.
Pelvic fins – 8 rays; abdominal; insertions directly under or slightly anterior to dorsal fin insertion.
Pectoral fins – 14 to 17 rays.
Lateral line – complete; 38 to 45 cycloid scales in series.
Juveniles – lateral stripe and caudal spot more pronounced.
Sexual dimorphism – spawning males have a distinct red-orange spot posterior of the eye laterally on the head, large tubercles on the dorsal side of the head, and smaller tubercles on the pectoral fin rays.

Similar Species: Hornyhead chub closely resemble creek chub. Creek chub have an upper jaw extending past the anterior edge of the eye, a black spot on the anterior base of the dorsal fin, an anal fin with eight rays, and pelvic fin insertions anterior to the dorsal fin insertion. Lake chub have a relatively small mouth and an anal fin with eight rays.

Distribution and Habitat: The native range of the hornyhead chub is widespread across the north central glacial regions of the United States, including the Missouri River, upper Mississippi River northern Ohio River, and St. Lawrence-Great Lakes drainages (Lachner and Jenkins 1971). Its distribution has decreased dramatically likely because of agricultural activity, impoundments, stream flow intermittency, siltation, and drought (Lachner and Jenkins 1971; Mammoliti 2002). Hornyhead chub are found throughout the upper Red River of the North in North Dakota and in the Minnesota River system in eastern South Dakota (Bailey and Allum 1962; Lachner and Jenkins 1971). They inhabit backwaters of low-to-moderate gradient streams with aquatic vegetation, clear water, and clean gravel, rubble, or sand substrates (Vives 1990; Lachner and Jenkins 1971; Mammoliti 2002; Hayer et al. 2008). Habitat changes from flow alterations and the presence of large piscivores negatively impacts hornyhead chub abundance (Mammoliti 2002; Hickerson 2018). They have a critical thermal maximum of 36 °C and the critical oxygen concentration of 1.1 mg/L (Smale and Rabeni 1995). Juvenile hornyhead chubs are frequently found within algal and macrophyte beds in stream reaches of little to no current (Lachner 1952).

Reproduction: Hornyhead chub spawning occurs from late May through early June in water temperatures of 16 to 26 °C (Vives 1990; Pflieger 1997). Sexual maturity occurs at age-2 or age-3, and males are longer at maturity than females (Lachner 1952). Flow alterations may reduce spawning fish numbers, thereby negatively affecting or even postponing spawning (Mammoliti 2002). Males use considerable effort constructing nests over gravel substrate in relatively deeper and faster-current open-water areas (Vives 1990; Wisenden et al. 2009). Males carry larger stones and gravel pieces with their mouths or push these materials with the dorsal surface of their head to create a depression approximately 5 to 10 cm deep (Vives 1990). Additional stones and gravel are continuously placed into the depression to make a mound (Vives 1990). On top of the mound, males press and vibrate their bodies against the substrate and use their ventral side and anal fins to form a central depression or cup so that the nest is doughnut shaped (Vives 1990). Spawning begins when a single female approaches a single male. The pair aligns their bodies parallel, and the male wraps his caudal peduncle around the caudal peduncle of the female (Vives 1990). The male then flexes his body and applies pressure to the abdomen of the female, pushing them both into the cup area of the nest where eggs and milt are released (Vives 1990). The male then covers the adhesive eggs with more stones and pebbles (Becker 1983; Vives 1990). A male may occasionally spawn with multiple females (Vives 1990). Fecundity from females with standard lengths of 80 to 89 mm was 460 to 725 eggs (Lachner 1952).

Age and Growth: Total lengths at age of male hornyhead chub from Wisconsin were 36 mm at age-0, 66 mm at age-1, 98 mm at age-2, and 126 mm at age-3 (Becker 1983). Males grow more rapidly than females (Lachner 1952). They can reach a total length of 203 mm, with a longevity of 4 years (Hubbs and Cooper 1936; Lachner 1952).

Food and Feeding: Hornyhead chub are visual omnivores (Davis and Miller 1967; Mammoliti 2002). Adults eat algae, plant material, aquatic insects, detritus, worms, crayfish, snails, and other species of small fish (Scott and Crossman 1973; Pflieger 1997). Juveniles eat much smaller prey items such as algae, zooplankton, diatoms, and small aquatic insect larvae (Scott and Crossman 1973; Lee et al. 1980).

References

Bailey RM, Allum MO (1962) Fishes of South Dakota. University of Michigan, Ann Arbor. https://doi.org/10.3998/mpub.9690435

Becker GC (1983) Fishes of Wisconsin. University of Wisconsin Press, Madison

Davis BJ, Miller RJ (1967) Brain patterns in minnows of the genus *Hybopsis* in relation to feeding habits and habitat. Copeia 1967:1–39. https://doi.org/10.2307/1442172

Hayer CA, Wall SS, Berry CR Jr (2008) Evaluation of predicted fish distribution models for rare fish species in South Dakota. N Am J Fish Manag 28:1259–1269. https://doi.org/10.1577/M07-086.1

Hickerson BT (2018) Conservation and recovery of hornyhead chub. Dissertation, University of Wyoming

Hubbs CL, Cooper GP (1936) Minnows of Michigan. Cranbrook Inst Sci 8:1–95

Lachner EA (1952) Studies of the biology of the cyprinid fishes of the chub genus *Nocomis* of northeastern United States. Am Midl Nat 48:433–466. https://doi.org/10.2307/2422260

Lachner EA, Jenkins RE (1971) Systematics, distribution, and evolution of the *Nocomis biguttatus* species group (Family Cyprinidae: Pisces), with a description of a new species from the Ozark Upland. Smithsonian Contr Zool 91:1–28. https://doi.org/10.5479/si.00810282.91

Lee DS, Gilbert CR, Hocutt CH, Jenkins RE, McAllister DE, Stauffer JR (1980) Atlas of North American freshwater fishes. North Carolina Biological Survey Publication 1980-12. https://doi.org/10.5962/bhl.title.141711

Mammoliti CS (2002) The effects of small watershed impoundments on native stream fishes: a focus on the Topeka shiner and hornyhead chub. Trans Kans Acad Sci 105:219–231. https://doi.org/10.1660/0022-8443(2002)105[0219:TEOSWI]2.0.CO;2

Pflieger WL (1997) The fishes of Missouri, revised edn. Missouri Dep Cons, Jefferson City

Scott WB, Crossman EJ (1973) Freshwater fishes of Canada. Fisheries Research Board of Canada, Bulletin 184

Smale MA, Rabeni CF (1995) Hypoxia and hyperthermia tolerances of headwater stream fishes. Trans Am Fish Soc 124:698–710. https://doi.org/10.1577/1548-8659

Vives SP (1990) Nesting ecology and behavior of hornyhead chub *Nocomis biguttatus*, a keystone species in Allequash Creek, Wisconsin. Am Midl Nat 124:46–56. https://doi.org/10.2307/2426078

Wisenden BD, Unruh A, Morantes A, Bury S, Curry B, Driscoll R, Hussein M, Markegard S (2009) Functional constraints on nest characteristics of pebble mounds of breeding male hornyhead chub *Nocomis biguttatus*. J Fish Biol 75:1577–1585. https://doi.org/10.1111/j.1095-8649.2009.02384.x

6.14.25 Golden Shiner *Notemigonus crysoleucas*

(Mitchill, 1814)

Nuptial Male
Female

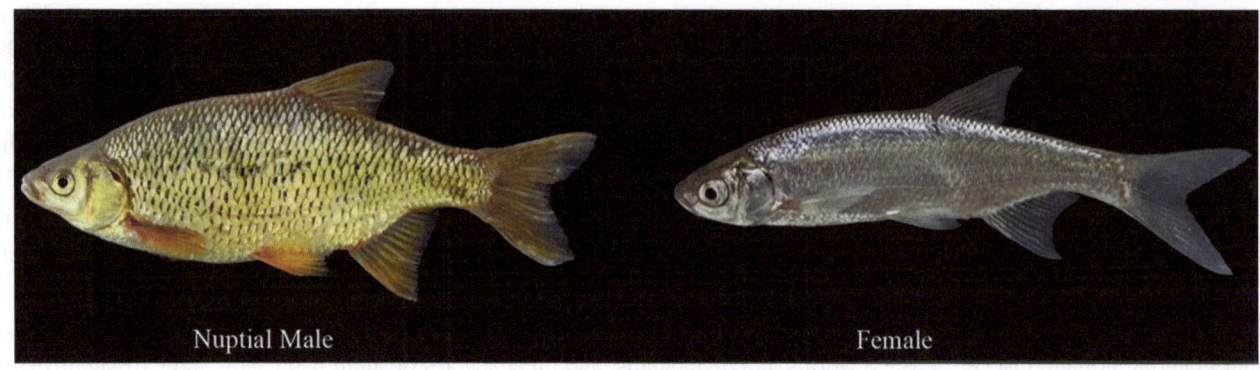

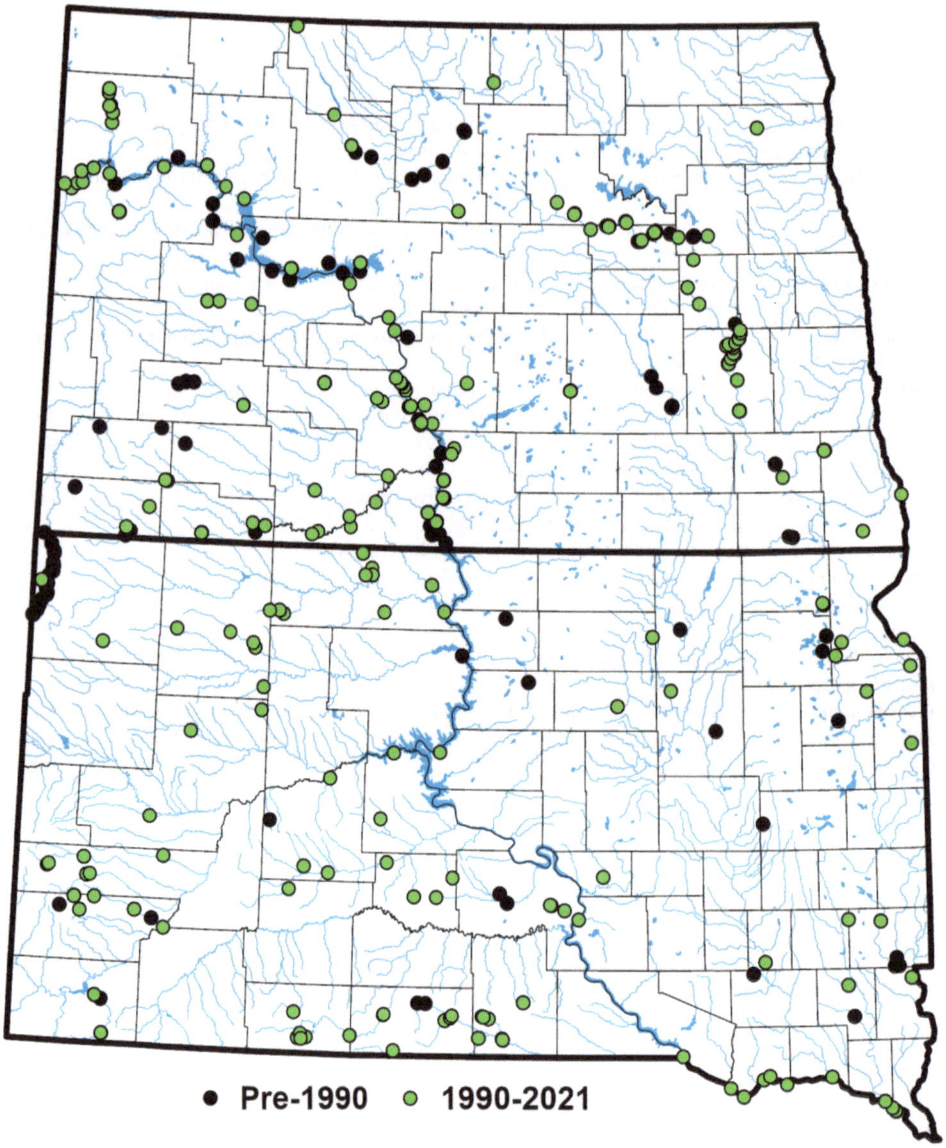
Pre-1990
1990-2021

Etymology: *Notemigonus crysoleucas* – *Notemigonus* = angled back; *crysoleucas* = gold, white.

Description:

Body – deep, fusiform, laterally compressed; moderately arched back; short, single S-shaped loop intestine.
Color – may have an entirely silver body.

> *Dorsally* – iridescent gold to olive.
> *Laterally* – silver to light olive.
> *Ventrally* – silver yellow.
> *Fins* – faint yellow-orange to clear.
> *Peritoneum* – silvery with dark specks.

Head – small, flattened, triangular.
Eyes – large.
Mouth – small, nearly vertical, oblique; upper jaw extends to nostril; snout pointed.
Teeth – pharyngeal tooth pattern 0,5-5,0.
Gill rakers – 16-21; long.
Dorsal fin – 7 to 9 rays; slightly falcate; insertion distinctly posterior to pelvic fin.
Caudal peduncle – short, narrow.
Caudal fin – deeply forked.
Anal fin – 11 to 14 rays; long, concave.
Pelvic fins – abdominal.
Pectoral fins – 15 to 18 rays.
Keel – scaleless from pelvic to anal fin.
Lateral line – complete; 42 to 54 cycloid scales, strongly decurved.
Juveniles – dusky lateral stripe as wide as eye.
Sexual dimorphism – breeding males have orange on caudal, anal, and pelvic fins, and small tubercles on head and lateral sides.

Similar Species: Golden shiner closely resemble rudd, red shiner, common shiner, and goldfish. Rudd have 36 to 45 lateral line scales, 10 to 13 gill rakers, 9 to 11 dorsal fin rays, ventral keel scales from the pelvic to anal fin, and red fins on adults. Red shiner have a blunter snout and a slightly decurved lateral line with 32 to 36 diamond-shaped scales. Common shiner have a slightly decurved lateral line with 36 to 44 diamond-shaped, three times higher than wide scales, and a dorsal fin insertion approximately even with pelvic fin insertions. Goldfish have a hard serrate leading ray on their dorsal and anal fins and 15 to 19 dorsal fin rays.

Distribution and Habitat: Golden shiner are native to southeastern Canada, and the eastern and central United States including the Dakotas (Lee et al. 1980; Hoagstrom et al. 2007; Stone et al. 2016). They have been widely distributed because of their popular use for bait and forage (Hoagstrom et al. 2007). Golden shiner are commonly found in lakes, reservoirs, ponds, creeks, and small-to-medium rivers with little flow, away from fast currents. They are uncommon in intermittent streams or larger rivers with higher flow (Cross 1967). In the Dakotas, golden shiners are abundant in lakes and ponds. They prefer clear water in areas with abundant, submerged aquatic vegetation over mud, sand, or gravel substrates. Golden shiner are commonly found in schools either just below the surface or mid-water column. They can tolerate a wide range of water temperatures and have a thermal maximum from 26 to 40 °C (Stone et al. 2016). They can also tolerate high turbidity and low dissolved oxygen concentrations (Whittier and Hughes 1998; Trebitz et al. 2007; Stone et al. 2016).

Reproduction: Spawning occurs during the late spring and summer in the morning over vegetation beds at water temperatures from 13 to 30 °C (Clemment and Stone 2010; Stone et al. 2016). They become sexually mature between age-1 and age-2 and can mature sooner in the south (Dobie et al. 1948; Stone et al. 2016). Golden shiner are fractional broadcast spawners and can rapidly repopulate after winter kill events. During spawning, one or two males escourt the female to the spawning grounds, often nudging her with their snouts (Etnier and Starnes 1993). The approximately 1 mm diameter eggs adhere to aquatic vegetation, filamentous algae, or substrate. No nest is prepared, and no parental care is given. Golden shiner will occasionally deposit eggs within the active nests of sunfish, bass, and other fish species (Kramer and Smith 1960). Fecundity ranges from 20,602 to 26,079 eggs per female (Clemment and Stone 2010; Stone et al. 2016). Depending on water temperature, eggs hatch within 2 to 5 days (Stone et al. 2016).

Age and Growth: Newly hatched golden shiner larvae have a total length of 4 mm (Buynak and Mohr Jr 1980; Stone et al. 2016). Early growth is rapid at favorable water temperatures. Adult total lengths range from 203 to 305 mm with a maximum of 356 mm. Males grow slower and are smaller than females (Churchill and Over 1938). Adult golden shiner often grow faster in warmer climates, with growth rates highly dependent on water temperature and growing season duration. Their longevity is approximately 8 years.

Food and Feeding: Golden shiners are omnivorous, frequently feeding just below the surface or within the mid-water column. They exhibit diel movement, foraging during the day near shore and within vegetated areas on filamentous algae, small crustaceans (cladocerans, copepods, and ostracods), and terrestrial and aquatic insects and larvae. In the evening, they forage in open and deeper water using their long gill rakers to filter-feed on zooplankton (Hall et al. 1979). Larger golden shiners will also sometimes eat small snails and small fish.

References

Buynak GL, Mohr HW Jr (1980) Larval development of golden shiner and comely shiner from northeastern Pennsylvania. Prog Fish Cult 42:206–211. https://doi.org/10.1577/1548-8659(1980)42[206:LDOGSA]2.0.CO;2

Churchill EP, Over WH (1938) Fishes of South Dakota. Brown & Saenger Printers, Sioux Falls

Clemment T, Stone N (2010) golden shiner egg production during a spawning season. N Am J Aquac 72:272–277. https://doi.org/10.1577/A09-064.1

Cross FB (1967) Handbook of fishes in Kansas. University Press of Kansas, Lawrence

Dobie JR, Meehean OL, Washburn GN (1948) Propagation of minnows and other bait species. US Fish and Wildlife Service Circular 12. https://doi.org/10.1577/1548-8640(1948)10[219:AOMAOB]2.0.CO;2

Etnier DA, Starnes WC (1993) The fishes of Tennessee. University of Tennessee Press, Knoxville

Hall DJ, Werner EE, Gilliam JF, Mittelbach GG, Howard D, Doner CG, Dickerman JA, Stewart AJ (1979) Diel foraging behavior and prey selection in the golden shiner (*Notemigonus crysoleucas*). J Fish Res Board Can 36:1029–1039. https://doi.org/10.1139/f79-145

Hoagstrom CW, Wall SS, Kral JG, Blackwell BG, Berry CR Jr (2007) Zoogeographic patterns and faunal change of South Dakota fishes. West N Am Nat 67:161–184. https://doi.org/10.3398/1527-0904(2007)67[161:ZPAFCO]2.0.CO;2

Kramer RH, Smith LL (1960) Utilization of nests of largemouth bass, *Micropterus salmoides*, by golden shiners, *Notemigonus crysoleucas*. Copeia 1960:73–74. https://doi.org/10.2307/1439868

Lee DS, Gilbert CR, Hocutt CH, Jenkins RE, McAllister DE, Stauffer JR (1980) Atlas of North American freshwater fishes. North Carolina Biological Survey Publication 1980-12. https://doi.org/10.5962/bhl.title.141711

Stone NM, Kelly AM, Roy LA (2016) A fish of weedy waters: golden shiner biology and culture. J World Aquacult Soc 47:152–200. https://doi.org/10.1111/jwas.12269

Trebitz AS, Brazner JC, Brady VJ, Axler R, Tanner DK (2007) Turbidity tolerances of Great Lakes coastal wetland fishes. N Am J Fish Manag 27:619–633. https://doi.org/10.1577/M05-219.1

Whittier TR, Hughes RM (1998) Evaluation of fish species tolerances to environmental stressors in lakes in the northeastern United States. N Am J Fish Manag 18:236–252. https://doi.org/10.1577/1548-8675

6.14.26   **Emerald Shiner** *Notropis atherinoides*

Rafinesque, 1818

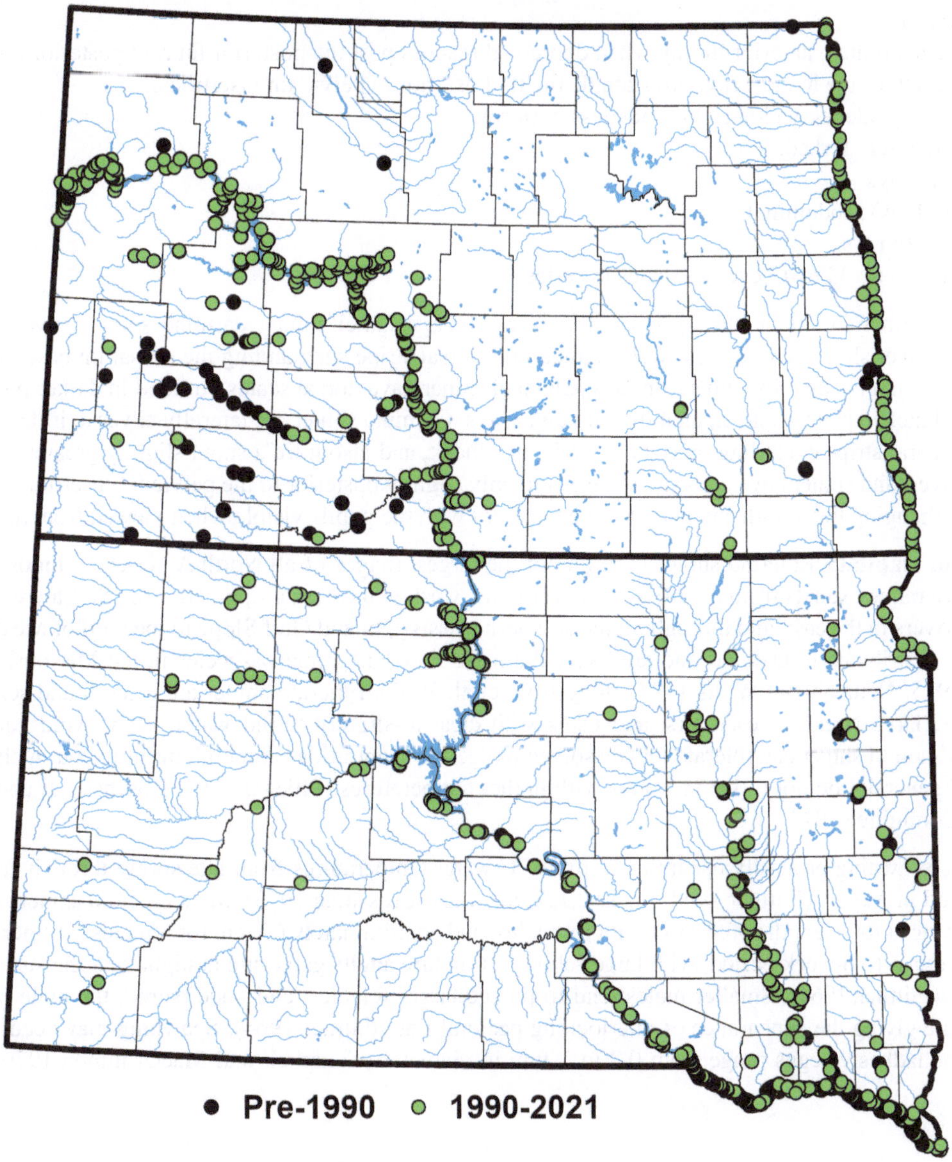

● **Pre-1990**   ○ **1990-2021**

Etymology: *Notropis atherinoides* – *Notropis* = back or keel; ***atherinoides*** = silverside resemblance.

Description:

Body – fusiform, elongated, laterally compressed.
Color –

> *Dorsally* – lightly pigmented; yellowish-olive to iridescent blue-green; melanophores only along distal edges of anterior dorsolateral scales creating an outlined appearance.
> *Laterally* – silver; faint silvery lateral stripe continuous on opercle and fading near dorsal fin anterior end.
> *Ventrally* – silver to white.
> *Mouth* – lips and center of chin deeply pigmented.
> *Fins* – clear to lightly pigmented.

Head – small.
Snout – short, bluntly rounded; length less than two-thirds the distance from the posterior margin of the eye to the posterior margin of the head.
Eyes – large; diameter greater than the snout length; laterally on head.
Mouth – large, oblique, terminal.
Teeth – absent.
Gill rakers – 10 to 12.
Dorsal fin – 8 rays; pointed, anterior fin ray when depressed reaches past the posterior fin ray; posterior fin ray is less than one half the length of the longest ray; insertion distinctly posterior to pelvic fin insertions.
Caudal peduncle – moderately elongated, thickness uniform.
Caudal fin – moderately forked.
Anal fin – 10 to 13 rays.
Pelvic fins – 8 or 9 rays; abdominal.
Pectoral fins – 14-16 rays.
Lateral line – complete; 35 to 43 cycloid scales in series.

Similar Species: Emerald shiner closely resemble river shiner, channel shiner, carmine shiner, silverband shiner, and bigmouth shiner. River shiner have a dark, well-defined mid-dorsal stripe surrounding the dorsal fin base, fewer lateral line scales in the series, and an anal fin with seven rays. Channel shiner have dorsal scales outlined in darker pigment creating a prominent cross-hatched pattern and an anal fin with eight rays. Carmine shiner are laterally silver to iridescent blue, have a faint rosy-pink lateral stripe extending onto the opercle and cheek, and also have a snout slightly greater in length than the eye diameter. Silverband shiner have a dorsal fin insertion only slightly posterior to the pelvic fin insertion. Bigmouth shiner have a snout overhanging the mouth and eyes directed upward with the pupils visible when viewed from above.

Distribution and Habitat: Emerald shiner have one of the largest ranges of all North American minnows. In the United States their native range extends from the Missouri River in the west, to the Mississippi, Hudson, St. Lawrence-Great Lakes, and Mackenzie rivers in the east, and throughout the southern Mississippi and Gulf Slope to the Gulf of Mexico in the south. They are widely distributed in Dakotas and are likely the most abundant minnow species in the Missouri and Mississippi rivers (Russel 1975; Neuman and Willis 1994; Hoagstrom et al. 2007). Emerald shiner are often found schooling near the surface of rivers, large streams, reservoirs, and lakes with clear-to-slightly turbid water. They avoid aquatic vegetation (Brown 1971). Emerald shiner can tolerate low dissolved oxygen concentrations and high turbidity, but with a thermal preference of 25 °C generally do not thrive in waters with higher temperatures (Pflieger 1997; Boschung Jr and Mayden 2004; Kansas Fishes Committee 2014).

Reproduction: Spawning primarily occurs in 22 to 24 °C water temperatures from late spring through early summer but has been observed in August (Flittner 1964; Fuchs 1967; Scott and Crossman 1973). Most emerald shiners sexually mature by age-2, beginning at total lengths of 55 to 60 mm for males, and approximately 65 mm for females (Flittner 1964). Emerald shiner spawn over a 4 to 6-week period, with large schools of mature adults gathering at night near the surface over sand or gravel substrate (Flittner 1964). Smaller males tend to pursue larger females, and once paired, the male presses his body against the female giving the appearance of interlocking pectoral fins (Flittner 1964). Spawning may occur multiple times. The demersal, nonadhesive eggs range from 0.2 to 1 mm in diameter (Campbell and MacCrimmon 1970; Becker 1983).

Fecundity of emerald shiner in Canada ranged from 868 to 8,733 eggs per female, with the number of eggs positively correlated to female length, weight, and gonad weight (Campbell and MacCrimmon 1970). Fecundity of an age-1 female with a total length of 69 mm in Wisconsin was 2,990 eggs, and fecundity of an age-2, 75 mm total length female was 2,040 eggs (Becker 1983). Hatching occurs in 24 to 32 hours depending on water temperature (Becker 1983).

Age and Growth: Initial growth rates are high. In Lewis and Clark Lake, South Dakota, total lengths from first and second annulus formation were 66 mm and 84 mm, respectively (Fuchs 1967). Females may be larger than males after the first year and are larger throughout subsequent years (Fuchs 1967). Maximum emerald shiner growth rates occur at 29 °C (McCormick and Kleiner 1976). Longevity is 4 years (Fuchs 1967; Ross 2001). Most fish only reach age-2, with the older emerald shiners usually female (Fuchs 1967; Ross 2001).

Food and Feeding: Juvenile emerald shiner feed heavily on rotifers and algae (Fuchs 1967). Adults eat primarily cladocerans, copepods, and other large zooplankton, but also consume adult insects, larval insects, annelids, smaller fish, and algae (Fuchs 1967; Boschung Jr and Mayden 2004). Adult emerald shiner have been observed eating drifting invertebrates in the upper water column of rivers, and also rising to the surface to eat insects (Mendelson 1975).

References

Becker GC (1983) Fishes of Wisconsin. University of Wisconsin Press, Madison

Boschung HT Jr, Mayden RL (2004) Fishes of Alabama. Smithsonian Books, Washington, DC

Brown CJD (1971) Fishes of Montana. Montana State University Press, Bozeman

Campbell JS, MacCrimmon HR (1970) Biology of the emerald shiner, *Notropis atherinoides* Raf., in Lake Simcoe, Canada. J Fish Biol 2:259–273. https://doi.org/10.1111/j.1095-8649.1970.tb03284.x

Flittner GA (1964) Morphometry and life history of the emerald shiner *Notropis atherinoides* Rafinesque. Dissertation, University of Michigan

Fuchs EH (1967) Life history of the emerald shiner, *Notropis atherinoides* in Lewis and Clark Lake, South Dakota. Trans Am Fish Soc 96:247–256. https://doi.org/10.1577/1548-8659(1967)96[247:LHOTES]2.0.CO;2

Hoagstrom CW, Wall SS, Kral JG, Blackwell BG, Berry CR Jr (2007) Zoogeographic patterns and faunal change of South Dakota fishes. West N Am Nat 67:161–184. https://doi.org/10.3398/1527-0904(2007)67[161:ZPAFCO]2.0.CO;2

Kansas Fishes Committee (2014) Kansas fishes. University Press of Kansas, Lawrence

McCormick JH, Kleiner CF (1976) Growth and survival of young-of-the-year emerald shiners (*Notropis atherinoides*) at different temperatures. J Fish Board Can 33:839–842. https://doi.org/10.1139/f76-104

Mendelson J (1975) Feeding relationships among species of *Notropis* (Pisces: Cyprinidae) in a Wisconsin stream. Ecol Monogr 45:199–232. https://doi.org/10.2307/1942422

Neuman RM, Willis DW (1994) Guide to the common fishes of South Dakota. Extension Circulars 511

Pflieger WL (1997) The fishes of Missouri, revised edn. Missouri Dep Cons, Jefferson City

Ross ST (2001) The inland fishes of Mississippi. Mississippi Dep Wildl Fish Parks, University Press of Mississippi, Jackson

Russel GW (1975) Distribution of fishes in North Dakota drainages affected by the Garrison Diversion Project. Dissertation, University of North Dakota

Scott WB, Crossman EJ (1973) Freshwater fishes of Canada. Fisheries Research Board of Canada, Bulletin 184

6.14.27   Blacknose Shiner *Notropis heterolepis*

Eigenmann & Eigenmann, 1893

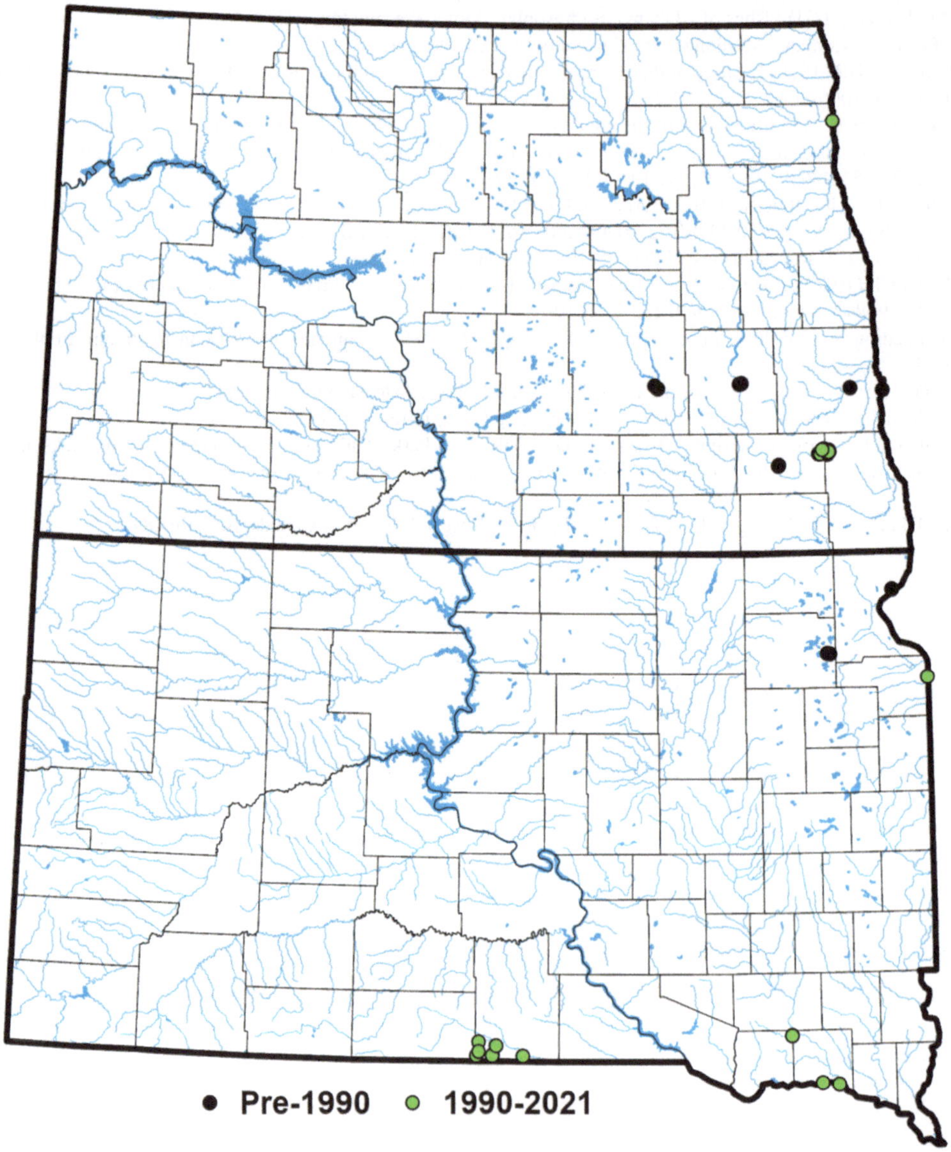

● **Pre-1990**   ○ **1990-2021**

Etymology: *Blacknose shiner – Notropis* = back or keel; ***heterolepis*** = various scale, referencing varying scale shapes.

Description:

Body – fusiform, slender.
Color –

> *Dorsally* – tan to olive; scales outlined in darker pigment creating a crosshatched pattern.
> *Laterally* – olive to bronze; prominent, dark, dusky lateral stripe extending onto snout; dark borders of lateral stripe pores expanded into crescent-shaped vertical bars anteriorly; scales directly above lateral stripe lighter in color.
> *Ventrally* – silver to white.
> *Fins* – little to no-pigment.

Head – moderate.
Snout – bluntly rounded, slightly extending beyond the mouth.
Eyes – moderately large, laterally on head.
Mouth – small, terminal to slightly subterminal, slightly oblique.

> *Jaws* – lower jaw extends to anterior edge of eye.

Teeth – pharyngeal teeth 0,4-4,0.
Dorsal fin – 8 rays.
Caudal peduncle – slender, slightly elongated.
Caudal fin forked.
Anal fin – 7 or 8 rays; straight distal end.
Pelvic fins – 8 rays; abdominal; insertions directly under dorsal fin insertion.
Lateral line – incomplete; 33 to 37 cycloid scales in series.
Sexual dimorphism – spawning males have small tubercles on dorsal side of head.

Similar Species: Blacknose shiner resemble brassy minnow, Topeka shiner, and pugnose shiner. Brassy minnow have a faint, dark mid-dorsal stripe, have pelvic fin insertions that are slightly posterior to the dorsal fin insertion, and lack prominent dorsal side cross-hatching and dark borders on lateral stripe pores. Topeka shiner have a deeper, laterally compressed body with small, paired dark specks outlining the lateral line, and a triangular-shaped black spot present at the caudal fin base. Pugnose shiner have anterior lateral side scales as deep as wide and a lateral stripe without pigment on some of the scales producing a zig-zag appearance.

Distribution and Habitat: Blacknose shiner originally had a broad distribution over the Atlantic, Hudson Bay, Great Lakes, and Mississippi rivers (Page and Burr 1991). They have been extirpated from much of its southern range because of intensive row-crop agriculture, wetland loss, increased turbidity, and siltation from pollution and erosion (Cross and Moss 1987; Hoagstrom et al. 2006; Roberts et al. 2006; Felts and Bertrand 2014). Blacknose shiner are extremely rare in the Dakotas. In North Dakota, they are found in the Red River of the North, Sheyenne River, and James River systems. In South Dakota, they are limited to the perennial and spring-fed streams of the Niobrara River system, with disjunct, peripheral populations in the Big Sioux River, James River, and Minnesota River systems (Owen et al. 1981; Cunningham and Olson 1994; Cunningham et al. 1995; Felts and Bertrand 2014). Blacknose shiner frequently school in the midwaters of clear-to-moderately clear glacial lakes, and small, quiet, prairie streams with pool and run sequences (Page and Burr 1991). They are often associated with considerable aquatic vegetation and organic debris, over sand, gravel, or rock substrate (Page and Burr 1991; Pflieger 1997; Roberts et al. 2006). Vegetation is likely important for foraging and larval growth (Roberts et al. 2006). Blacknose shiner may be intolerant of prolonged high turbidity and siltation (Pflieger 1997).

Reproduction: Little information is known about blacknose shiner reproductive and spawning habits. Sexual maturity generally occurs at standard lengths exceeding 33 mm (Roberts et al. 2006). Spawning in Missouri occurs June through July and in Illinois from late April through late June, with females able to reproduce for 2 to 4 weeks (Pflieger 1997; Roberts et al. 2006). Females are multiple clutch spawners, producing a mean of 167 eggs/clutch with a range of 98 to 330 eggs (Roberts et al. 2006). Clutch size is positively related to female size, but decreases over time, with larger clutches laid earlier in the spawning season (Roberts et al. 2006). It is likely that no parental care is given (Roberts et al. 2006). The yellow eggs have a diameter of 1 mm (Becker 1983; Roberts et al. 2006).

Age and Growth: Total-lengths-at-age of blacknose shiner from Wisconsin were 31 mm at age-0 and 49 mm at age-1 (Becker 1983). Standard lengths-at-age from Michigan were 36 mm at age-0 and 46 mm at age-1 (Emery and Wallace 1974).

Blacknose shiner can reach a standard length of 81 mm (3.2 in) (Hubbs and Lagler 1949; Pflieger 1997). They have a longevity of 2 years, with many likely dying before or during their second winter of life (Emery and Wallace 1974; Becker 1983).

Food and Feeding: Little information is known about the feeding habits of the blacknose shiner. A subterminal mouth suggests it is primarily a benthic feeder (Becker 1983). In Illinois, blacknose shiner foraged during the mornings in spring, and throughout the morning and night during the summer (Roberts et al. 2006). They generally consumed less food in the summer and fall compared to spring (Roberts et al. 2006). Blacknose shiner primarily eat chydorid and bosminid cladocerans, ostracods, copepods, ephemeropterans, dipterans, plant material, arachnids, and possibly fish eggs (Becker 1983; Roberts et al. 2006).

References

Becker GC (1983) Fishes of Wisconsin. University of Wisconsin Press, Madison

Cross FB, Moss RE (1987) Historic changes in fish communities and aquatic habitat in plains streams of Kansas. In: Matthews WJ, Heins DC (eds) Community and evolutionary ecology of North American stream fishes. University of Oklahoma Press, Norman,Oklahoma, p 155–165

Cunningham G, Olson R (1994) Fish species collected in streams in West River South Dakota-1994. Unpublished report to South Dakota Game, Fish and Parks, Pierre

Cunningham GR, Olson RD, Hickey SM (1995) Fish surveys of the streams and rivers of South-Central South Dakota west of the Missouri River. Proc S D Acad Sci 74:55–64

Emery L, Wallace DC (1974) The age and growth of the blacknose shiner, *Notropis heterolepis* Eigenmann and Eigenmann. Am Midl Nat 91:242–243. https://doi.org/10.2307/2424526

Felts EA, Bertrand KN (2014) Conservation status of five headwater stream specialists in South Dakota. Am Midl Nat 172:131–159. https://doi.org/10.1674/0003-0031-172.1.131

Hoagstrom CW, Hayer CA, Kral JG, Wall SS, Berry CR (2006) Rare and declining fishes of South Dakota: a river drainage scale perspective. Proc S D Acad Sci 85:171–211

Hubbs CL, Lagler KF (1949) Fishes of Isle Royale, Lake Superior, Michigan. Univ Michigan Acad Sci 33:73–133

Owen JB, Elsen DS, Russell GW (1981) Distribution of fishes in North and South Dakota Basins affected by the Garrison Diversion Unit. University of North Dakota Press, Grand Forks

Page LM, Burr BM (1991) Peterson field guide to freshwater fishes of North America North of Mexico. Houghton Mifflin Harcourt, Boston

Pflieger WL (1997) The fishes of Missouri, revised edn. Missouri Dep Cons, Jefferson City

Roberts ME, Burr BM, Whiles MR (2006) Reproductive ecology and food habits of the blacknose shiner, *Notropis heterolepis*, in northern Illinois. Am Midl Nat 155:70–83. https://doi.org/10.1674/0003-0031(2006)155[0070:REAFHO]2.0.CO;2

6.14.28 Carmine Shiner *Notropis percobromus*

(Cope, 1871)

Nuptial Male

● **Pre-1990**   ○ **1990-2021**

Etymology: *Notropis percobromus – Notropis* = back or keel; ***percobromus*** = unknown.

Description:

Body – fusiform, slender, elongated.
Color –

> *Dorsally* – olive-green; evenly distributed melanophores; dorsal scales outlined in slightly darker pigment creating a faint and less prominent crosshatched pattern.
> *Laterally* – silver to irredescent blue; faint rosy to pink lateral stripe extending onto opercle and cheek.
> *Ventrally* – silver to white.
> *Fins* – generally clear.

Head – moderate, slightly elongated.
Snout – blunt, conical; length slightly greater than eye diameter.
Eyes – large, laterally on head.
Mouth – large, terminal, slightly oblique.

> *Jaws* – upper jaw extends to or slightly beyond anterior edge of eye.
> *Lips* – pigmented, not extending onto chin.

Teeth – pharyngeal tooth pattern 2,4-4,2.
Gill rakers – 5 to 7; short.
Dorsal fin – 8 rays, rounded tip; anterior fin ray when depressed does not extend past posterior fin ray.
Caudal peduncle – slender, slightly elongated.
Caudal fin – moderately forked.
Anal fin – 9 or 10 rays; straight distal end.
Pelvic fins – 8 rays; abdominal; insertions distinctly anterior to dorsal fin insertion.
Pectoral fins – 12 or 13 rays.
Lateral line – slightly decurved; 36 to 40 cycloid scales in series.
Sexual dimorphism – spawning males have numerous minute tubercles on head and pectoral fin rays, and red, pink, or orange on the head, cheeks, body, and base of the fins; spawning females have less intense coloration and fewer tubercles.

Similar Species: Carmine shiner closely resemble emerald shiner. Emerald shiner have a more bluntly rounded snout shorter than their eye diameter, 10 to 12 gill rakers, and a pointed-tip dorsal fin. Carmine shiner are closely related and morphologically similar to the rosyface shiner and were previously referred to as rosyface shiner in South Dakota (Bailey and Allum 1962; Hayer et al. 2008). However, the two species are genetically distinct and have different geographical distributions (Wood et al. 2002; Hayer et al. 2008).

Distribution and Habitat: The native range of carmine shiner encompasses an area including southern Manitoba, Canada, eastern North Dakota, eastern South Dakota, Minnesota, Wisconsin, Indiana, Arkansas, and Oklahoma (Pandit et al. 2017). They are native to the Sheyenne River system and Red River of North drainage in North Dakota, as well as the Minnesota and Bois de Sioux River systems in eastern South Dakota (Bailey and Allum 1962; Hoagstrom et al. 2006). Specific carmine shiner habitat information is minimal and has typically been extrapolated from that of rosyface shiners outside of the carmine shiner range (Fisheries and Oceans Canada 2007). Carmine shiner exist in varied climates (Pandit et al. 2017). They are associated with the headwaters of streams with perennial flow in the Minnesota River basin of South Dakota (Hayer et al. 2008). In Manitoba, they are generally found in the mid-waters of fast-flowing clear to semi-turbid creeks or small rivers with clean gravel or rubble substrate (Fisheries and Oceans Canada 2007). Carmine shiner are often found in or near riffles and pools and may move into deeper areas during the winter (Fisheries and Oceans Canada 2007; Pandit et al. 2017). In aquaria, the minimum, preferred, and maximum temperatures were 18 °C, 24 °C, and 29 °C, respectively (Stol et al. 2013).

Reproduction: Although little is known about carmine shiner reproduction, it is likely similar to rosyface shiner (Fisheries and Oceans Canada 2007). They become sexually mature in southern Manitoba at a minimum fork length of 46 mm (Carr et al. 2015). Spawning individuals from Manitoba were sampled in water ranging from 19 to 23 °C, velocities up to 0.53 m/second, and depths of 0.2 to 1.4 m (Fisheries and Oceans Canada 2007). In the more southern part of their range, spawning occurs in water temperatures of 20 to 29 °C, with peak activity from late April or early May through June (Cross 1967; Becker 1983; Pflieger 1997; Carmine shiner Recovery Team 2007). Spawning occurs in riffles over sand, cobble, boulder, or bedrock substrates (Fisheries and Oceans Canada 2007). Fecundity of carmine shiner females from Manitoba ranged from 694 to 2,806 eggs and increased with increasing female size and age (Becker 1983; Fisheries and Oceans Canada 2007). The yellow, demersal, and adhesive eggs are approximately 2 mm in diameter (Pfeiffer 1955; Becker 1983; Fisheries and Oceans Canada 2007). Hatching occurs at 57 to 59 hours at 21 °C water temperature (Becker 1983; Fisheries and Oceans Canada 2007).

Age and Growth: Total-lengths-at-age from carmine shiner in Wisconsin were 52 to 60 mm at age-1, 58 to 75 mm at age-2, and 69 to 76 mm at age-3 (Becker 1983). They can reach a length of 89 mm and have a longevity of 3 years (Becker 1983; Pflieger 1997).

Food and Feeding: Carmine shiner are benthic, visual omnivores (Pfeiffer 1955; Fisheries and Oceans Canada 2007). They consume a variety of immature and mature aquatic and terrestrial invertebrates, and especially eat dipterans during the summer (Fisheries and Oceans Canada 2007). Young of the year carmine shiners primarily eat diatoms and algae; as they age, they consume more insects (Pflieger 1997).

References

Bailey RM, Allum MO (1962) Fishes of South Dakota. University of Michigan, Ann Arbor. https://doi.org/10.3998/mpub.9690435

Becker GC (1983) Fishes of Wisconsin. University of Wisconsin Press, Madison

Fisheries and Oceans Canada (2007) Recovery strategy for the carmine shiner (*Notropis percobromus*) in Canada [Proposed]. Species at Risk Act Recovery Strategy Series, Fisheries and Oceans Canada, Ottawa

Carmine shiner Recovery Team (2007) Recovery strategy for the carmine shiner (*Notropis percobromus*) in Canada [Proposed]. Species at Risk Act Recovery Strategy Series, Fisheries and Oceans Canada, Ottawa

Carr M, Watkinson DA, Svedsen JC, Enders EC, Long JM, Lindenschmidt KE (2015) Geospatial modeling of the Birch River: distribution of carmine shiner (*Notropis percobromus*) in geomorphic response units (GRU). Int Rev Hydrobiol 100:129–140. https://doi.org/10.1002/iroh.201501789

Cross FB (1967) Handbook of fishes in Kansas. University Press of Kansas, Lawrence

Hayer CA, Wall SS, Berry CR Jr (2008) Evaluation of predicted fish distribution models for rare fish species in South Dakota. N Am J Fish Manag 28:1259–1269. https://doi.org/10.1577/M07-086.1

Hoagstrom CW, Wall SS, Kral JG, Blackwell BG, Berry CR Jr (2006) Zoogeographic patterns and faunal change of South Dakota fishes. West N Am Nat 67:161–184. https://doi.org/10.3398/1527-0904(2007)67[161:ZPAFCO]2.0.CO;2

Pandit SN, Maitland BM, Pandit LK, Poesch MS, Enders EC (2017) Climate change risks, extinction debt, and conservation implications for a threatened freshwater fish: carmine shiner (*Notropis percobromus*). Sci Total Environ 598:1–11. https://doi.org/10.1016/j.scitotenv.2017.03.228

Pfeiffer RA (1955) Studies on the life history of the rosyface shiner, *Notropis rubellus*. Copeia 1955:95–104. https://doi.org/10.2307/1439311

Pflieger WL (1997) The fishes of Missouri, revised edn. Missouri Dep Cons, Jefferson City

Stol JA, Svendsen JC, Enders EC (2013) Determining the thermal preferences of carmine shiner (*Notropis percobromus*) and lake sturgeon (*Acipenser fulvescens*) using an automated shuttlebox. Can Tech Rep Fish Aquat Sci 3038

Wood RM, Mayden RL, Matson RH, Kuhajda BR, Laymen SR (2002) Systematics and biogeography of the *Notropis rubellus* species group (Teleostei: Cyprinidae). Bull Alabama Mus Nat Hist 22:37–80

6.14.29   Silverband Shiner *Paranotropis shumardi*

Girard, 1856

● **Pre-1990**   ○ **1990-2021**

Etymology: *Paranotropis volucellus – Paranotropis* = near back or keel; *shumardi* = after George C. Shumard, a naturalist on the Mexican Boundary Survey.

Description:

Body – fusiform, slender, elongate.
Color –

>   *Dorsally* – tan to silver; scales faintly outlined in darker pigment.
>   *Laterally* – silver; no distinct markings.
>   *Ventrally* – silver to white.
>   *Fins* – generally clear.

Head – small.
Snout – bluntly rounded, not protruding beyond mouth.
Eyes – moderately large, laterally on head.
Mouth – large, terminal, oblique.

>   *Jaws* – upper jaw extends to anterior end of eye.
Teeth – pharyngeal tooth pattern variable between 1,4-4,1 and 2,4-4,2.
Gill rakers – 7 to 9.
Dorsal fin – 8 rays; pointed tip; distal end straight to slightly falcate.
Caudal peduncle – moderately short, slender.
Caudal fin – forked.
Anal fin – 8 or 9 rays; straight distal end.
Pelvic fins – 9 rays; abdominal; insertions slightly posterior to dorsal fin insertion.
Pectoral fins – 13 to 16 rays.
Lateral line – complete; 34 to 37 cycloid scales in series.
Sexual dimorphism – spawning males have small tubercles on the pectoral fin rays, dorsal and lateral sides of the head, and nape and breast body regions.

Similar Species: Silverband shiner closely resemble emerald shiner, sand shiner, channel shiner, and river shiner. Emerald shiner have an anal fin with 10 to 13 rays, and a dorsal fin insertion distinctly posterior to the insertions of the pelvic fins. Sand shiner have small, paired, dark specks, or "mouse tracks" outlining the lateral line and sometimes a faint, dusky spot at the caudal fin base. Channel shiner have eight pelvic fin rays and a lateral stripe expanding ventrally at the caudal fin base. River shiner have eight pelvic fin rays and seven anal fin rays.

Distribution and Habitat: The native range of silverband shiner encompasses the area including the Missouri River up to South Dakota, the Mississippi River, Illinois River, lower Ohio River, Arkansas River, and Red River to the Gulf of Mexico, along with several isolated populations near the Gulf slope in Texas (Lee et al. 1980). It has not been reported in North Dakota. It is considered rare in South Dakota, the northernmost part of its range (Miller 1972; Madsen 1985; Berry Jr and Young 2004). Silverband shiner occur primarily in the Missouri River drainage including Fort Randall Reservoir, Lake Francis Case, Lake Oahe, and Lewis and Clark Lake (Underhill 1959; Harlan and Speaker 1987; Berry Jr and Young 2004; Jones 2018). It inhabits large turbid rivers with moderate-to-strong current over sand, gravel, silt, shale, or mud substrate in the main channel, embayments, and backwater pools downstream of point bars (Lee et al. 1980; Pflieger 1997).

Reproduction: Little is known about silverband shiner spawning and reproductive habits, especially in its northern range. The spawning season is prolonged, from late May through mid-August in Tennessee, and from June to August in Lousiana (Etnier and Starnes 1993; Pflieger 1997). Sexual maturity occurs at age-1 (Williams 2011). Silverband shiners are fractional, broadcast spawners, with multiple egg clutches throughout the season (Suttkus 1980; Williams 2011). Clutch sizes averaged 387 eggs and ranged from 144 to 750 eggs from 47 to 76 mm total length females from the Brazos River, Texas (Williams 2011). Clutch size is positively correlated to female total length (Williams 2011).

Age and Growth: Age-0 silverband shiner exhibit rapid growth (Williams 2011). Total lengths at age of fish from Texas were 41 mm at age-0, 67 mm at age-1, and 86 mm at age-2 (Williams 2011). Silverband shiner can reach a total length of 91 mm and have a longevity of 2.5 years (Pflieger 1997; Williams 2011).

Food and Feeding: Silverband shiner are invertivores. The little information available suggests their diet primarily consists of trichoptera, ephemeroptera, diptera, and other aquatic insects, and may also include smaller portions of detritus, terrestrial insects, crustaceans, and algae. Aquatic insects become more important in adult diets (Williams 2011).

References

Berry CR Jr, Young B (2004) Fishes of the Missouri national recreational river, South Dakota and Nebraska. Great Plains Res 14:89–114

Etnier DA, Starnes WC (1993) The fishes of Tennessee. University of Tennessee Press, Knoxville

Harlan JR, Speaker EB (1987) Iowa fish and fishing. Iowa Dep Nat Res, Des Moines

Jones SJ (2018) Western prairie stream fisheries: an assessment of past and present fish assemblage structure, biotic homogenization, and population dynamics in western South Dakota streams. Dissertation, South Dakota State University

Lee DS, Gilbert CR, Hocutt CH, Jenkins RE, McAllister DE, Stauffer JR (1980) Atlas of North American freshwater fishes. North Carolina Biological Survey Publication 1980–12

Madsen TI (1985) The status and distribution of the uncommon fishes in Nebraska. M.S. Dissertation, University of Nebraska-Omaha

Miller RR (1972) Threatened freshwater fishes of the United States. Trans Am Fish Soc 101:239–252

Pflieger WL (1997) The fishes of Missouri, revised edn. Missouri Dep Cons, Jefferson City

Suttkus RD (1980) *Notropis candidus*, a new cyprinid fish from the Mobile Bay basin, and a review of the nomenclatural history of *Notropis shumardi* (Girard). Bull Alabama Mus Nat Hist 5:1–15

Underhill JC (1959) Fishes of the Vermillion River, South Dakota. Proc S D Acad Sci 38:96–102

Williams CS (2011) Life history characteristics of three obligate riverine species and drift patterns of lower Brazos River fishes. Texas State University, Dissertation

6.14.30   Mimic Shiner *Paranotropis volucellus*

(Cope, 1865)

© Uland Thomas

● **Pre-1990**   ● **1990-2021**

Etymology: *Paranotropis volucellus* – *Paranotropis* = near back or keel; *volucellus* = swift or winged.

Description:

Body – fusiform, slightly deep, laterally compressed.
Color –

> *Dorsally* – dusky gray, tan, or olive; scales outlined in black creating a crosshatched pattern intensifying posteriorly; midline stripe anterior of dorsal fin of unconsolidated specks vague or absent.
> *Laterally* – silver; small, paired, faint darker specks or "mouse tracks" outlining the lateral line; stripe expanding ventrally at the base of the caudal fin to form a wedge faint or absent.
> *Ventrally* – silver to white.
> *Fins* – clear to lightly pigmented.

Scales – cycloid.
Head – elongated, slightly conical.
Snout – blunt, slightly protruding beyond mouth.
Eyes – large, laterally on head; diameter greater than snout length.
Mouth – small; terminal to slightly subterminal; oblique.

> *Jaws* – upper jaw extends to posterior end of eye.

Teeth – absent on jaws; pharyngeal tooth pattern 0,4-4,0.
Dorsal fin – 8 rays; straight distal end.
Caudal peduncle – slender, slightly elongated.
Caudal fin – moderately forked.
Anal fin – 8 rays; straight distal end.
Pelvic fin – 8 rays, abdominal; insertions inferior or slightly anterior to dorsal fin insertion.
Pectoral fin – 14 to 16 rays.
Lateral line – complete; 32 to 39 scales in series.
Sexual dimorphism – spawning males have small tubercles on head and pectoral fins.

Similar Species: Mimic shiner appear very similar to channel shiner, creating taxonomic confusion. However, channel shiner do not occur in the Dakotas. Mimic shiner may be confused with bigmouth shiner, emerald shiner, river shiner, and sand shiner in the Dakotas. Bigmouth shiner have a large, slightly subterminal mouth with the lower jaw extending to the middle of the eye, and an upper jaw the length greater than eye diameter. Emerald shiner have an anal fin with 10 to 13 rays, and a dorsal fin insertion distinctly posterior to the insertion of the pelvic fins. River shiner have a dark, well-defined mid-dorsal stripe that surrounds the dorsal fin base, an upper jaw extending or slightly beyond the anterior end of eye, and an upper jaw length greater than the eye diameter. Sand shiner have an anal fin with seven rays, and an upper jaw shorter than the eye diameter.

Distribution and Habitat: The native range of the mimic shiner extends from the Hudson Bay, St. Lawrence River/Great Lakes, and Mississippi River basins southward through the Atlantic Slope drainages to the Gulf of Mexico. A single vouchered specimen was collected in 2018 near Fargo, North Dakota from the Red River of the North drainage (UT 44.14073). Mimic shiner are not known to occur in South Dakota. Little habitat information exists. They school in moderately clear, mid-to-deep waters of lakes, large pools, or backwaters in rivers or creeks over sand and gravel substrates (Trautman 1981; Becker 1983). Mimic shiner are moderately intolerant to turbidity (Trebitz et al. 2007). They have been reported to move inshore nocturnally to areas with little to no aquatic vegetation during the summer months (Hanych et al. 1983).

Reproduction: Information on mimic shiner reproduction is scarce. In Minnesota, sexually mature mimic shiners had total lengths over 42 mm (Moyle 1973). Spawning occurs in the spring in water depths under 5 m over sand and gravel substrates with high levels of both submergent and emergent aquatic vegetation (Lane et al. 1996). No nest is constructed, and no parental care is given. Mimic shiner are broadcast spawners (Lane et al. 1996). Fecundity from females in Wisconsin with total lengths of 62 and 65 mm was 635 and 960 eggs, respectively (Becker 1983). Fecundity from females in Indiana with total lengths from 36 to 63 mm averaged 367 eggs and ranged from 67 to 920 eggs (Black 1945). The orange eggs are approximately 1 mm in diameter (Becker 1983).

Age and Growth: Mimic shiners have indeterminate growth, and females are typically larger than males (Becker 1983; Middleton et al. 2013). Total lengths at age from mimic shiners in Wisconsin were 43 to 52 mm at age-1 and 52 to 65 mm at age-2 (Becker 1983). The total lengths of age-1 females and males from Lake Erie ranged from 29 to 51 mm and 30 to 46 mm, respectively, while age-2 females and males ranged from 57 to 61 mm and 54 to 56 mm, respectively (Middleton et al. 2013). Maximum total length is 76 mm and longevity is 2 to 3 years (Trebitz et al. 2007).

Food and Feeding: Mimic shiner are active insectivores within the littoral zone (Olmstead et al. 1979). They primarily forage in the early morning and early evening, eating primarily aquatic and terrestrial insects (Black 1945; Moyle 1973). Green algae, blue-green algae, and other vegetative matter are also consumed (Black 1945; Moyle 1973).

References

Becker GC (1983) Fishes of Wisconsin. University of Wisconsin Press, Madison

Black JD (1945) Natural history of the northern mimic shiner, *Notropis volucellus volucellus* Cope. Invest Indiana Lakes Stream 2:449–466

Hanych DA, Ross MR, Magnien RE, Suggars AL (1983) Nocturnal inshore movement of the mimic shiner (*Notropis volucellus*): a possible predator avoidance behavior. Can J Fish Aquat Sci 40:888–894. https://doi.org/10.1139/f83-115

Lane JA, Portt CB, Minns CK (1996) Spawning habitat characteristics of Great Lakes fishes. Can Manuscript Rep Fish Aquat Sci 2368

Middleton S, Perello M, Simon TP (2013) Length-weight relationships of the mimic shiner Notropis volucellus (Cope 1865) in the western basin of Lake Erie. Ohio J Sci 112:44–50

Moyle PB (1973) Ecological segregation among three species of minnows (Cyprinidae) in a Minnesota lake. Trans Am Fish Soc 102:794–805. https://doi.org/10.1577/1548-8659

Olmstead LR, Krater S, Williams GE, Jaegar RG (1979) Foraging tactics of the mimic shiner in a two-prey system. Copeia 1979:437–441. https://doi.org/10.2307/1443219

Trautman MB (1981) The fishes of Ohio. Ohio State University Press, Columbus

Trebitz AS, Brazner JC, Brady VJ, Axler R, Tanner DK (2007) Turbidity tolerances of Great Lakes coastal wetland fishes. N Am J Fish Manag 27:619–633. https://doi.org/10.1577/M05-219.1

6.14.31 Suckermouth Minnow *Phenacobius mirabilis*

(Girard, 1856)

© Uland Thomas

● **Pre-1990**   ○ **1990-2021**

Etymology: *Phenacobius mirabilis* – *Phenacobius* = deceptive life, likely referencing the long intestine suggesting herbivory for this insectivorous species; *mirabilis* = strange or wonderful.

Description:

Body – slender, elongated, cylindrical anteriorly, slightly laterally compressed posteriorly.
Color –

> *Dorsally* – dark gray to olive.
> *Laterally* – silver to tan or light olive; dark, dusky stripe ending in a prominent, dark spot at the caudal fin base; scales above stripe outlined in darker pigment creating a herringbone or crosshatched pattern.
> *Ventrally* – silver to white.
> *Fins* – pectoral/dorsal/caudal: olive to tan, lightly pigmented; anal/pelvic: typically white to clear.

Head – short, slightly dorsoventrally compressed.
Snout – elongate, bluntly rounded, protruding beyond mouth.
Eyes – moderately large, laterally on upper portion of head.
Mouth – small, subterminal, sucker like.

> *Jaws* – upper jaw does not extend to anterior edge of eye.
> *Lips* – fleshy; lower lip: fleshy posterior lobe on each side.

Teeth – pharyngeal tooth pattern 0,4-4,0.
Gill rakers – 9; short.
Dorsal fin – 8 rays; straight to slightly concave distal end.
Caudal peduncle – moderately short, uniform in thickness.
Caudal fin – moderately forked.
Anal fin – 6 or 7 rays; straight distal end.
Pelvic fins – 8 rays; abdominal; insertions slightly posterior to dorsal fin insertion.
Pectoral fins – 14 to 16 rays.
Lateral line – complete; 40 to 51 cycloid scales in series; slightly decurved; predorsal scale rows typically 18 to 21.
Sexual dimorphism – spawning males have small tubercles on dorsal side of head, anterior end of body, and pectoral fins; females may have fewer small tubercles on the dorsal side of head.

Similar Species: Suckermouth minnow closely resemble longnose dace, western blacknose dace, central stoneroller, and shoal chub. Both longnose dace and western blacknose dace have a frenum, barbels, pelvic fin insertions slightly anterior to the dorsal fin insertion, and more than 60 small cycloid scales in lateral line series. Central stoneroller have a more robust body with mottling on the lateral sides, and a cartilaginous ridge on the inner lower lip. Shoal chub have barbels and small, prominent, black, scattered specks on the dorsal and lateral sides.

Distribution and Habitat: The suckermouth minnow native range extends from South Dakota, eastern Wyoming, and Colorado east through the Mississippi River basin to Ohio and south through the Tennessee river valleys to Texas and northeastern New Mexico (Lee et al. 1980). Its limited distribution and low abundance is likely because of habitat alteration, fragmentation, groundwater withdrawal, and impoundments (Bestgen et al. 2003). Suckermouth minnow are not found in North Dakota. In South Dakota, they are found at low levels with a restricted distribution in the Missouri River, Big Sioux River, and Niobrara River systems. Suckermouth minnow prefer any size streams of any gradient with permanent flow, gravel substrate, and deep pools, riffles, and runs (Pflieger 1997; Bestgen et al. 2003). They are likely tolerant of water level and turbidity fluctuations as long as the gradient and continuous flow are sufficient to maintain silt-free gravel substrate (Trautman 1981; Pflieger 1997). Young-of-the-year suckermouth minnow inhabit riffles over gravel substrate like adults, but also frequent deep, low velocity pools or backwaters with sand and silt substrate (Haas 1977; Bestgen et al. 2003).

Reproduction: Suckermouth minnow spawning is prolonged, extending from March through August (Bestgen et al. 2003; Brewer et al. 2006). Sexual maturity occurs at age-2 (Becker 1983). Suckermouth minnows are fractional spawners, with females likely releasing only a small batch of ripe eggs during each of the two or more spawning events each season (Becker 1983; Bestgen and Compton 2007). Smaller females spawn earlier than larger females (Haas 1977). While suckermouth minnow likely spawn within their year-round habitat of riffles with gravel or cobble substrate, they also spawn in shallower habitats before migrating to deeper areas postspawning (Haas 1977; Pflieger 1997; Brewer et al. 2006; Bestgen and Compton 2007). In a laboratory setting, spawning was successful in water temperatures from 17 to 23 °C (Bestgen et al. 2003). In Missouri, spawning began at a 9 °C water temperature and peaked at water temperatures from 12 to 20 °C (Brewer et al.

2006). During spawning, the fish rapidly chase each other in circles near the substrate. Males position themselves perpendicular to the female, nudging their vent areas, and rapidly shudder to likely initiate gamete release (Bestgen et al. 2003; Bestgen and Compton 2007). Fecundity is approximately 200 to 500 eggs per female and is generally positively related to female size (Haas 1977; Bestgen and Compton 2007). A female suckermouth minnow from Wisconsin with a total length of 90 mm produced 640 eggs (Becker 1983). The dark yellow to gold, adhesive eggs are approximately 2 mm in diameter one hour after fertilization (Bestgen et al. 2003; Bestgen and Compton 2007). In aquaria, hatching began approximately 3 days at a water temperature of 23 °C and 4 days at 17 to 19 °C (Bestgen et al. 2003).

Age and Growth: The development and growth of suckermouth minnow embryos and larvae are much faster at warmer water temperatures (Bestgen et al. 2003). Larvae total lengths average 5 mm at hatching (Bestgen and Compton 2007). In aquaria, 10 days post-hatch larvae grew 0.4 mm/day at 17 °C and 0.5 mm/day at 23 °C (Bestgen et al. 2003; Bestgen and Compton 2007). Up to 72% of their maximum length is achieved during the first growing season (Haas 1977). Suckermouth minnow can reach a length of 122 mm and have a longevity of 3 years (Haas 1977; Trautman 1981).

Food and Feeding: Suckermouth minnow are benthic insectivores. They primarily eat diptera, trichoptera, ephemeroptera, and other aquatic insects (Haas 1977). They may be selective and sensitive feeders, using numerous sensory organs and fleshy lips to feed by taste and touch, rather than by sight (Haas 1977; Bestgen et al. 2003).

References

Becker GC (1983) Fishes of Wisconsin. University of Wisconsin Press, Madison

Bestgen KR, Compton RI (2007) Reproduction and culture of suckermouth minnow. N Am J Aquac 69:345–350. https://doi.org/10.1577/A06-077.1

Bestgen KR, Zelasko K, Compton R (2003) Environmental factors limiting suckermouth minnow *Phenacobius mirabilis* populations in Colorado. Larval Fish Laboratory 136, Dep Fish Wildl Biol, Colorado State University

Brewer SK, Papoulias DM, Rabeni CF (2006) Spawning habitat associations and selection by fishes in a flow-regulated prairie river. Trans Am Fish Soc 135:763–778. https://doi.org/10.1577/T05-021.1

Haas MA (1977) Some aspects of the life history of the suckermouth minnow, *Phenacobius mirabilis* (Girard). Dissertation, University of Missouri

Lee DS, Gilbert CR, Hocutt CH, Jenkins RE, McAllister DE, Stauffer JR (1980) Atlas of North American freshwater fishes. North Carolina Biological Survey Publication 1980-12. https://doi.org/10.5962/bhl.title.141711

Pflieger WL (1997) The fishes of Missouri, revised edn. Missouri Dep Cons, Jefferson City

Trautman MB (1981) The fishes of Ohio. Ohio State University Press, Columbus

6.14.32   Bluntnose Minnow *Pimephales notatus*

(Rafinesque, 1820)

Nuptial Male

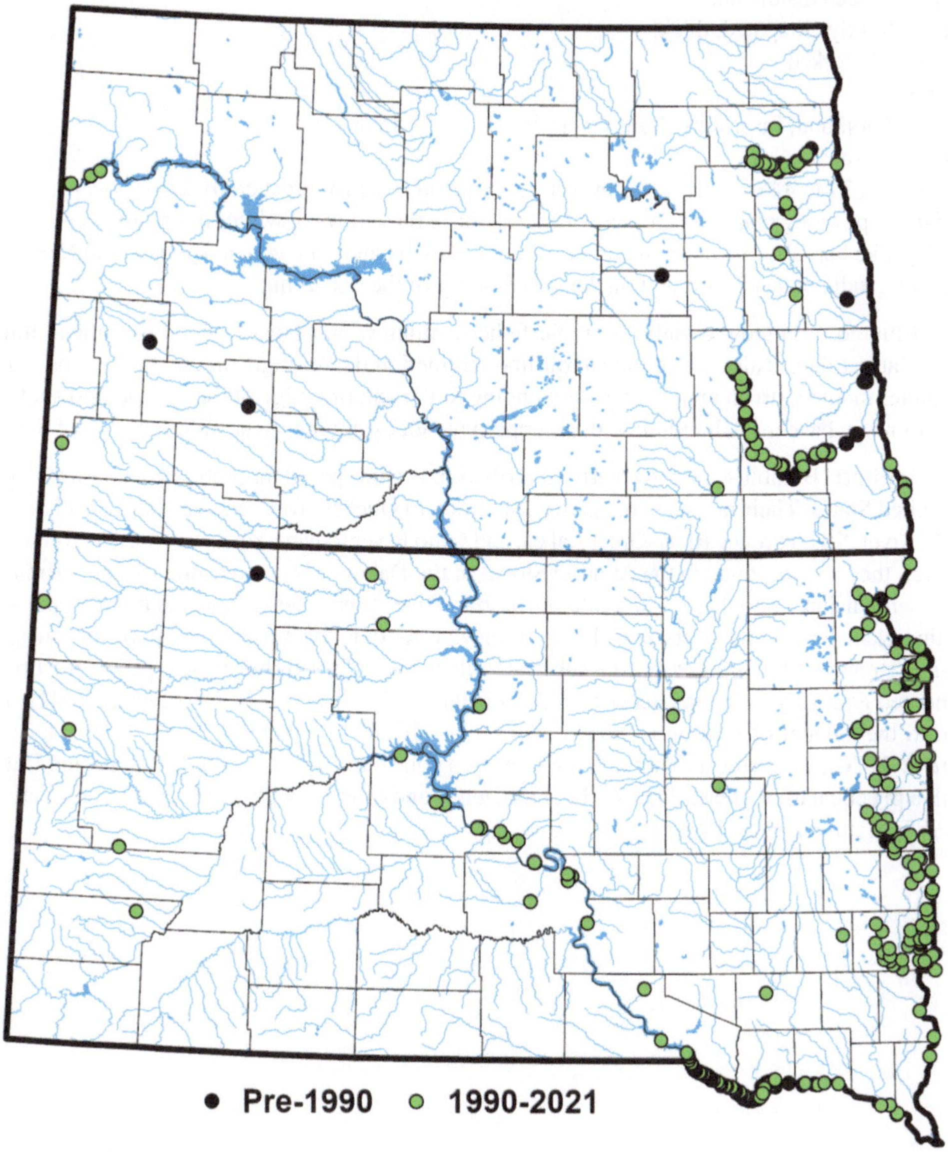

● **Pre-1990**   ○ **1990-2021**

Etymology: *Pimephales notatus* – *Pimephales* = fat head; *notatus* = marked or spotted.

Description:

Body – robust, rather cylindrical anteriorly; laterally compressed posteriorly.
Color –

> *Dorsally* – dark olive to brown.
> *Laterally* – brown to tan; dark stripe extending onto the opercle and snout, leading into a caudal spot that fades with age.
> *Ventrally* – cream to white.
> *Fins* – lightly pigmented; dorsal: small, dusky, sometimes anterior base dark spot on larger fish.

Head – broad.
Snout – bluntly rounded, slightly protruding beyond the mouth.
Eyes – moderately large, laterally on head.
Mouth – subterminal, moderately large.

> *Jaws* – upper jaw extends to anterior edge of eye.
> *Lips* – thin, uniform in thickness.

Teeth – pharyngeal tooth pattern 0,4-4,0.
Gill rakers – 7 to 10; short.
Dorsal fin – 8 rays; rounded distal end.
Caudal peduncle – slightly elongated, thick.
Caudal fin – moderately forked.
Anal fin – 7 or 8 rays.
Pelvic fins – 8 rays; abdominal; insertions slightly anterior to dorsal fin insertion.
Pectoral fins – 15 or 16 rays.
Lateral line – complete; 37 to 44 cycloid scales in series; predorsal scale rows generally 22 or fewer.
Sexual dimorphism – spawning males are dark gray to black, have three rows of white, sharp tubercles on the snout, a single pair of small, flap-like barbels, one on each corner of the mouth, a patch of swollen, spongy skin (dorsal pad) anterior to the dorsal fin, and a rudimentary (short and thickened) first ray of the dorsal fin.

Similar Species: Bluntnose minnow closely resemble fathead minnow and brassy minnow. Fathead minnows generally have an incomplete lateral line, scales above the lateral line outlined in darker pigment creating a crosshatched or herringbone pattern, a smaller mouth with an upper jaw not extending to the anterior edge of the eye, and no spot at the base of the caudal fin. Brassy minnow have a small mouth with an upper jaw not extending to the anterior edge of the eye.

Distribution and Habitat: Bluntnose minnow native distribution is widespread throughout southern Canada and the northeast and central United States. Their range encompasses Manitoba to Quebec, from the Dakotas to Oklahoma, east throughout the Mississippi River, St. Lawrence River-Great Lakes, and Ohio River to New York, and south to Tennessee and the Gulf of Mexico. However, they are not found in the Atlantic Slope. In the Dakotas, bluntnose minnow are primarily found east of the Missouri River with disjunct populations in western Missouri River tributaries. Bluntnose minnow inhabit a wide variety of habitats including glacial lakes and reservoirs. However, they prefer quiet backwaters and pools of low gradient, medium-to-moderately large streams with sand or gravel substrate, clear-to-slightly turbid water, persistent flow, and moderate amounts of submerged aquatic vegetation (Becker 1983; Pflieger 1997; Boschung Jr and Mayden 2004). Their presence is positively related to the amount of floating overhead cover (Gatz Jr 2008). Bluntnose minnow have a critical upper-level water temperature of 37 °C and a critical lower dissolved oxygen concentration of 1.1 mg/L (Smale and Rabeni 1995). They prefer waters with a pH equal to or greater than 7.8 but has been found in water with a pH equal to or greater than 5.6 (Rahel and Magnuson 1983; Matuszek et al. 1990).

Reproduction: Bluntnose minnow spawning occurs from May through August, beginning at water temperatures of approximately 21 °C (Carlander 1969; Moyle 1973; Pot 1985). The age of sexual maturity may vary between populations, but typically occurs at age-1 or age-2 (Westman 1938). Bluntnose minnow construct cavity-like nests under rocks, logs, or other structures along the bottom substrate, with a single layer of eggs adhering to the ceiling of the structure (Westman 1938; Scott and Crossman 1973; Pot 1985). Nests are generally at water depths up to 1 m but may be as deep as 2.4 m (Westman 1938; Pot 1985). If necessary, the male may dig out sediments (Pot 1985). Females prefer spawning with larger males, likely increasing the safety of their young (Pot 1985). Spawning generally occurs at night (Westman 1938; Gale 1983). Bluntnose minnow are fractional spawners, with 2 to 14 days between spawning (Gale 1983). Females may spawn more than 10 times during a single season (Gale 1983; Pot 1985). Eleven bluntnose minnow pairs from Pennsylvania produced 7 to 19 clutches, with an average clutch size of 93 to 239 eggs, for a total of 1,112 to 4,195 eggs per spawning season (Gale 1983). The first clutch of eggs generally contains fewer eggs than later clutches (Gale 1983). Fecundity is unrelated to the length and weight of the female (Gale 1983). Eggs are approximately 2 mm in diameter (Westman 1938). Males provide parental care, guarding and cleaning the eggs until hatching (Pot 1985). Males use their dorsal pad to clean the eggs, which helps keep the eggs free of silt, snails, or parasites (Pot 1985). Hatching occurs within 6 or 7 days (Pot 1985).

Age and Growth: One-day post-hatch larvae have a total length of approximately 5 mm (Westman 1938). They are sexually dimorphic, with males growing faster and becoming considerably longer than females (Moyle 1973; Pflieger 1997). Bluntnose minnow can reach a total length of 109 mm and have a longevity of 3 years (Trautman 1957; Moyle 1973).

Food and Feeding: Bluntnose minnow are opportunistic benthic omnivores. They frequently forage during the day in large schools along the bottom, eating a mixture of algae, large diatoms, aquatic insect larvae, and microcrustacean (Moyle 1973; Boschung Jr and Mayden 2004). The tendency to feed in aquatic vegetation and along bottom substrate likely increases with fish size (Moyle 1973). Females will consume embryos in the nests when males are distracted by other females or defending the nest (Pot 1985).

References

Becker GC (1983) Fishes of Wisconsin. University of Wisconsin Press, Madison

Boschung HT Jr, Mayden RL (2004) Fishes of Alabama. Smithsonian Books, Washington, DC

Carlander KD (1969) Handbook of freshwater fishery biology, vol 1. Iowa State University Press, Ames

Gale WF (1983) Fecundity and spawning frequency of caged bluntnose minnows-fractional spawners. Trans Am Fish Soc 112:398–402. https://doi.org/10.1577/1548-8659

Gatz AJ Jr (2008) The use of floating overhead cover by warmwater stream fishes. Hydrobiologia 600:307–310. https://doi.org/10.1007/s10750-007-9252-5

Matuszek JE, Goodier J, Wales DL (1990) The occurrence of cyprinidae and other small fish species in relation to pH in Ontario lakes. Trans Am Fish Soc 119:850–861. https://doi.org/10.1577/1548-8659

Moyle PB (1973) Ecological segregation among three species of minnows (Cyprinidae) in a Minnesota lake. Trans Am Fish Soc 102:794–805. https://doi.org/10.1577/1548-8659

Pflieger WL (1997) The fishes of Missouri, revised edn. Missouri Dep Cons, Jefferson City

Pot W (1985) Competition for nests, parental care, and female choice in the bluntnose minnow, *Pimephales notatus*. Dissertation, University of Guelph

Rahel FJ, Magnuson JJ (1983) Low pH and the absence of fish species in naturally acidic Wisconsin lakes: inferences for cultural acidification. Can J Fish Aquat Sci 40:3–9. https://doi.org/10.1139/f83-002

Scott WB, Crossman EJ (1973) Freshwater fishes of Canada. Fisheries Research Board of Canada, Bulletin 184

Smale MA, Rabeni CF (1995) Hypoxia and hyperthermia tolerances of headwater stream fishes. Trans Am Fish Soc 124:698–710. https://doi.org/10.1577/1548-8659

Trautman MB (1957) The fishes of Ohio. Ohio State University Press, Columbus

Westman J (1938) Studies on the reproduction and growth of the bluntnose minnow, *Hyborhynchus notatus* (Rafinesque). Copeia 1938:57–61. https://doi.org/10.2307/1435690

6.14.33   Fathead Minnow *Pimephales promelas*

Rafinesque, 1820

Nuptial Male

Female

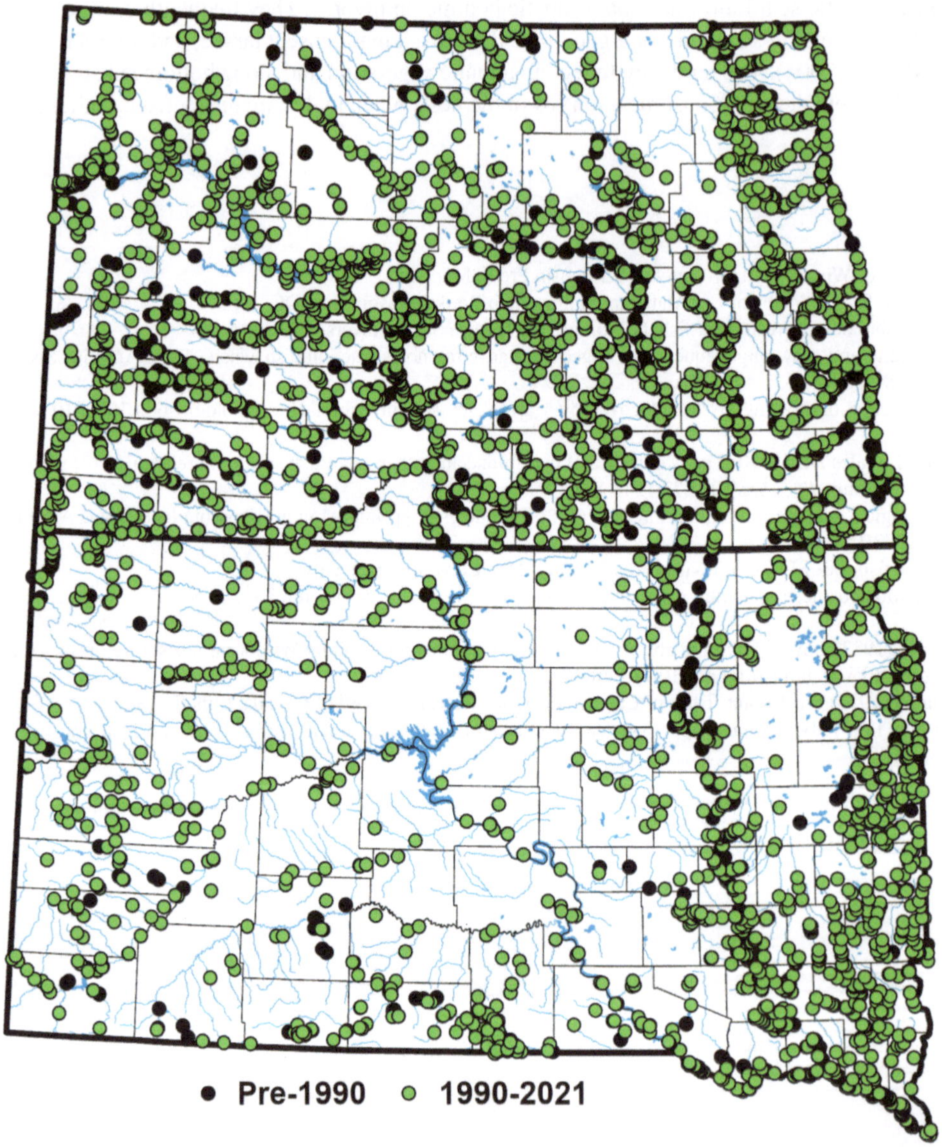

● Pre-1990   ● 1990-2021

Etymology: *Pimephales promelas* – *Pimephales* = fat head; *promelas* = before black, likely referencing the dark head of spawning males.

Description:

Body – moderately fusiform, robust, moderately deep.
Color –

> *Dorsally* – dark olive to brown.
> *Laterally* – tan, olive to silver; a dark lateral stripe fades with age; scales above lateral line outlined in darker pigment creating a crosshatched or herringbone pattern.
> *Ventrally* – cream to white.
> *Fins* – lightly pigmented; small, dusky, dorsal: sometimes anterior base dark spot on larger fish.

Head – broad.
Snout – bluntly rounded, not extending beyond the upper lip.
Eyes – moderate, laterally on head.
Mouth – very small, terminal, oblique.

> *Jaws* – upper jaw does not extend to anterior edge of the eye.
> *Lips* – thin.

Teeth – pharyngeal tooth pattern 0,4-4,0.
Gill rakers – 14 to 16.
Dorsal fin – 8 rays; rounded distal end.
Caudal peduncle – slightly elongated, thick.
Caudal fin – moderately forked.
Anal fin – 7 rays.
Pelvic fins – 8 rays; abdominal; insertions slightly anterior to dorsal fin insertion.
Pectoral fins – 15 or 16 rays.
Lateral line – incomplete; 40 to 54 cycloid scales in series; generally 23 or more predorsal scale rows.
Sexual dimorphism – spawning males are often darker, especially on the head; have a fleshy dorsal pad, rudimentary (short and thickened) first ray of the dorsal fin, and possibly large breeding tubercles on the head, snout, and pectoral fins rays.

Similar Species: Fathead minnow closely resemble bluntnose minnow and brassy minnow. Bluntnose minnow have a complete lateral line with scales above not outlined in darker pigment, a small subterminal mouth with an upper jaw extending beyond the anterior edge of the eye, and usually a caudal spot. Brassy minnow have a complete lateral line with 35 to 39 cycloid scales in series, and pelvic fin insertions slightly posterior to the dorsal fin insertion.

Distribution and Habitat: The fathead minnow native range includes most of North America, with a northern boundary in Canada from the Northern Territories to Quebec. The western boundary in the United States extends from Montana to New Mexico. The range includes the basins of the Missouri River, Mississippi River, St. Lawrence River/Great Lakes, and Ohio River east of the Appalachian Mountains from Vermont to Texas. Fathead minnow have been widely distributed outside their native range because of their popularity as bait fish and tolerance of a wide variety of habitats. It is found throughout the Dakotas. Fathead minnow are well adapted to shallow, pooled areas in wetland and stream habitats, but also occur in creeks, ponds, and lakes (Pflieger 1997; Herwig and Zimmer 2007). They tolerate water quality extremes common to wetlands of the northern prairie region, including high alkalinity, turbidity, ammonia, and temperatures (McCarraher and Thomas 1968; Brungs 1971; Thurston et al. 1986; Pflieger 1997). Acclimation to higher water temperature generally increases the upper lethal temperature, which can be as high as 34 °C (Brungs 1971). Fathead minnow are highly tolerant of hypoxic conditions, although not for prolonged periods (Klinger et al. 1982; Danylchuk and Tonn 2003). They may take advantage of physiological tolerance differences to reduced dissolved oxygen to avoid predation (Abrahams and Sloan 2012).

Reproduction: Spawning occurs from April to August at surface water temperatures ranging from 14 to 18 °C (McCarraher and Thomas 1968; Gale and Buynak 1982). Fathead minnows will not spawn in water temperatures over 30 °C (Brungs 1971). Under optimal conditions, they may become sexually mature 4 to 5 months after hatching, but typically will not spawn until age-1 (Markus 1934; Brungs 1971). While males and females may be the same size during the spawning period, females often spawn with larger males (Markus 1934). Fathead minnows generally begin spawning before dawn and continue until midmorning (Gale and Buynak 1982). They are fractional spawners and spawn every 2 to 16 days (Gale and Buynak 1982). Female spawning frequency and fecundity gradually decrease in water temperatures over 24 °C (Brungs

1971). Six fathead minnow pairs from Pennsylvania produced 16 to 26 clutches with 9 to 1,136 eggs per clutch yielding a total of 6,803 to 10,164 eggs per female per spawning season (Gale and Buynak 1982). The first clutch generally has fewer eggs than later clutches (Gale and Buynak 1982). Fathead minnow construct cavity-like nests under rocks, logs, or other structures along the bottom substrate, with one or two layers of adhesive eggs adhering to the ceiling or underside of the structure (Markus 1934). Males actively guard and clean the eggs until hatching (Markus 1934; Smith and Murphy 1974). They keep the eggs silt-free by rubbing the eggs with their dorsal pad while also coating the eggs with a mucus layer, which may increase egg survival (Smith and Murphy 1974). If water pH exceeds 5.9, egg production, egg quality, and survival to hatch are reduced (Mount 1973). Eggs are approximately 1 mm in diameter and hatch at 25 °C in 4 or 5 days (Markus 1934; Devlin et al. 1996).

Age and Growth: At hatching, fathead minnow larvae have a total length of approximately 5 mm (Markus 1934). They experience rapid growth, with juveniles reaching total lengths of 45 to 50 mm at 90 days (Held and Peterka 1974; Herwig and Zimmer 2007). Total lengths at age of fathead minnows in North Dakota were 41 mm at age-1, 41 mm and 58 mm at age-2 (Held and Peterka 1974). Males generally grow faster and are longer than females (Held and Peterka 1974). Fathead minnows can reach a total length of 101 mm and have a longevity of 3 years, with most only surviving until age-2 (McCarraher and Thomas 1968; Held and Peterka 1974).

Food and Feeding: Fathead minnows are opportunistic omnivores (Held and Peterka 1974; Herwig and Zimmer 2007). Adults display both particulate and filter feeding behaviors (Hambright and Hall 1992). They consume a wide variety of prey items, including detritus, zooplankton, and macroinvertebrates (Held and Peterka 1974; Price et al. 1991; Duffy 1998; Herwig and Zimmer 2007). Juveniles and adults select invertebrate prey over detritus when invertebrate prey is available and intra-specific competition is low (Herwig and Zimmer 2007). Detritus is an important food for all fathead minnow sizes because it allows for survival when more nutritional food is lacking (Held and Peterka 1974; Price et al. 1991; Herwig and Zimmer 2007).

References

Abrahams MV, Sloan J (2012) Risk of predation, variation in dissolved oxygen, and their impact upon habitat selection decisions by fathead minnow. Trans Am Fish Soc 141:580–584. https://doi.org/10.1080/00028487.2012.683470

Brungs WA (1971) Chronic effects of elevated temperature on the fathead minnow (*Pimephales promelas* Rafinesque). Trans Am Fish Soc 100:659–664. https://doi.org/10.1577/1548-8659

Danylchuk AJ, Tonn WM (2003) Natural disturbances and fish: local and regional influences on winterkill of fathead minnows in boreal lakes. Trans Am Fish Soc 32:289–298. https://doi.org/10.1577/1548-8659

Devlin EW, Brammer JD, Puyear RL, McKim JM (1996) Prehatching development of the fathead minnow *Pimephales promelas* Rafinesque. US Environmental Protection Agency, Office of Research Development, National Health and Environmental Effects Research Laboratory

Duffy WG (1998) Population dynamics, production, and prey consumption of fathead minnows (*Pimephales promelas*) in prairie wetlands: a bioenergetics approach. Can J Fish Aquat Sci 54:15–27. https://doi.org/10.1139/f97-204

Gale WF, Buynak GL (1982) Fecundity and spawning frequency of the fathead minnow- a fractional spawner. Trans Am Fish Soc 111:35–40. https://doi.org/10.1577/1548-8659

Hambright KD, Hall RO (1992) Differential zooplankton feeding behaviors, selectivities, and community impacts of two planktivorous fishes. Environ Biol Fish 35:401–411. https://doi.org/10.1007/BF00004992

Held JW, Peterka JJ (1974) Age, growth, and food habits of the fathead minnow, *Pimephales promelas*, in North Dakota saline lakes. Trans Am Fish Soc 103:743–757. https://doi.org/10.1577/1548-8659

Herwig BR, Zimmer KD (2007) Population ecology and prey consumption by fathead minnows in prairie wetlands: importance of detritus and larval fish. Ecol Freshw Fish 16:282–294. https://doi.org/10.1111/j.1600-0633.2006.00220.x

Klinger SA, Magnuson JJ, Gallepp GW (1982) Survival mechanisms of the central mudminnow (*Umbra limi*), fathead minnow (*Pimephales promelas*), and brook stickleback (*Culaea inconstans*) for low oxygen in winter. Environ Biol Fish 7:113–120. https://doi.org/10.1007/BF00001781

Markus HC (1934) Life history of the blackhead minnow (*Pimephales promelas*). Copeia 1934:116–122. https://doi.org/10.2307/1436755

McCarraher DB, Thomas R (1968) Some ecological observation on the fathead minnow, *Pimephales promelas*, in the alkaline waters of Nebraska. Trans Am Fish Soc 97:52–55. https://doi.org/10.1577/1548-8659(1968)97[52:SEOOTF]2.0.CO;2

Mount DI (1973) Chronic effect of low pH on fathead minnow survival, growth and reproduction. Water Res 7:987–993. https://doi.org/10.1016/0043-1354(73)90180-2

Pflieger WL (1997) The fishes of Missouri, revised edn. Missouri Dep Cons, Jefferson City

Price CJ, Tonn WM, Paszkowski CA (1991) Intraspecific patterns of resource use by fathead minnows in a small boreal lake. Can J Zool 69:2109–2115. https://doi.org/10.1139/z91-294

Smith RJF, Murphy BD (1974) Functional morphology of the dorsal pad in fathead minnows (*Pimephales promelas* Rafinesque). Trans Am Fish Soc 103:65–72. https://doi.org/10.1577/1548-8659

Thurston RV, Russo RC, Meyn EL, Zajdel RK, Smith CE (1986) Chronic toxicity of ammonia to fathead minnows. Trans Am Fish Soc 117:196–207. https://doi.org/10.1577/1548-8659

6.14.34   Flathead Chub *Platygobio gracilis*

(Richardson, 1836)

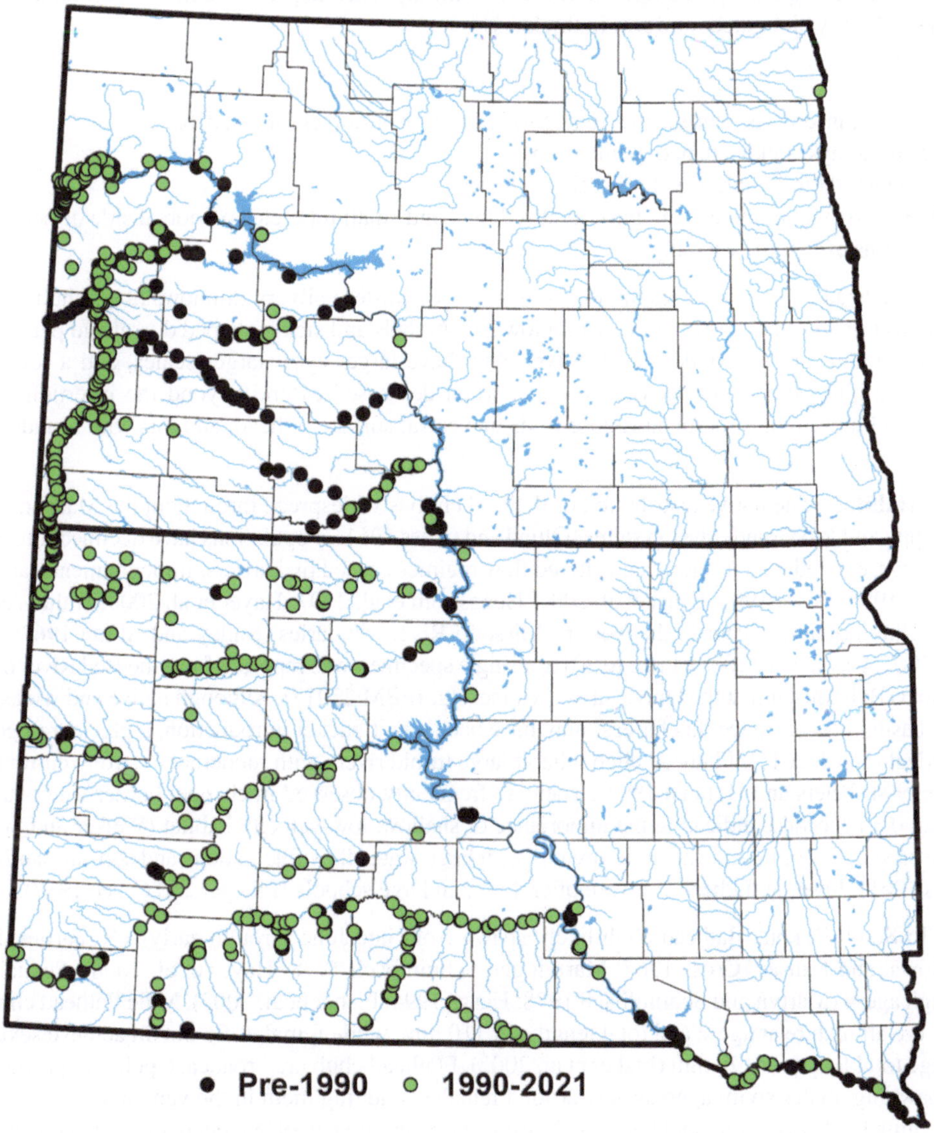

● **Pre-1990**   ○ **1990-2021**

Etymology: *Platygobio gracilis* – *Platygobio* = flat; *gracilis* = slender, possibly referencing body shape.

Description:

Body – slender, elongated; slightly laterally compressed.
Color –

>   *Dorsally* – tan to light bronze.
>   *Laterally* – silver; no distinct markings.
>   *Ventrally* – silver to white.
>   *Fins* – clear, transparent; caudal: lower lobe dusky gray, white edge.

Head – broad, short, dorsoventrally flattened; wider than deep.
Snout – tapered to a point; rounded tip protruding past upper jaw.
Eyes – moderately small, laterally on head.
Mouth – subterminal, large, slightly oblique; extending to anterior end of eye.

>   *Lips* – thin.

Teeth – pharyngeal tooth pattern 2,4-4,2.
Barbels – one on each corner of mouth; short, rounded.
Gill rakers – 4 to 6; short.
Dorsal fin – 8 rays; sickle shaped, pointed tip; anterior dorsal fin ray when depressed extends beyond last ray.
Caudal peduncle – slightly elongated, uniform in thickness.
Caudal fin – forked.
Anal fin – 8 rays.
Pelvic fins – abdominal; insertions nearly equal to slightly posterior of dorsal fin insertion.
Pectoral fins – 15 to 18 rays; sickle shaped, pointed tip.
Lateral line – complete; 44 to 59 large cycloid scales in series.
Sexual dimorphism – spawning males may have faint red fins, and small tubercles on dorsal side of head and body, caudal peduncle, and all fins except the caudal.

Similar Species: Flathead chub closely resemble plains minnow, western silvery minnow, silver chub, sicklefin chub, and sturgeon chub. Plains minnow and western silvery minnow lack barbels and sickle-shaped dorsal and pectoral fins. They also have much smaller mouths and a rounded head. Silver chub have larger eyes, larger scales, and a rounded, short snout. Sicklefin chub have a blunt snout and pectoral fins that when depressed extend beyond the insertion of the pelvic fins. Sturgeon chub have a blunt snout, prominent rugose folds on ventral surface of head, and fins with rounded tips lacking any sickle shape.

Distribution and Habitat: The native distribution of flathead chub is widespread throughout the Mackenzie, Saskatchewan, Missouri-Mississippi, and Rio Grande river systems (Olund and Cross 1961). Populations have declined throughout the historical range, likely because of reduced turbidity and altered flow regimes caused by barriers, impoundments, and channel modifications (Cross et al. 1986; Hesse 1994; Quist et al. 2004; Hoagstrom et al. 2007; Hayer et al. 2008; Walters et al. 2014). In the Dakotas, flathead chub primarily occur in the western Missouri River tributaries (Bailey and Allum 1962; Quist et al. 2004; Hoagstrom et al. 2007; Hayer et al. 2008). Prior to 2016, a single specimen was reported from the Red River of the North drainage in North Dakota (Bell Museum at the University of Minnesota, JFBM 22917). Although native and widespread throughout the Nelson River basin, an occurrence this far east may have been a bait bucket introduction, a glacial refuge, or just a strong swimming individual (Hatch 2015). Flathead chub inhabit large turbid rivers with moderate to strong current over sand or silt substrates (Hesse 1994; Peters and Holland 1994). They are frequently observed in channel border waters under 1-m deep in velocities less than 0.3 m/second, indicating the importance of shallow, low-velocity habitat (Welker and Scarnecchia 2004). Flathead chub are especially common in alkaline streams with large water-flow fluctuations and shifting sandy-loam substrates (Olund and Cross 1961). Juvenile flathead chub are often found in large schools (Pflieger and Grace 1987).

Reproduction: In South Dakota, flathead chub likely spawn from late June through early July, earlier than observations from Montana, Iowa, and Kansas (Cross 1967; Martyn and Schmulbach 1978; Hayer et al. 2008). Sexual maturity occurs from age-2 through age-4 (Martyn and Schmulbach 1978; Hesse 1994; Fisher et al. 2002). Male flathead chub from the upper Missouri River were all mature at age-2 and total lengths of 110 mm, while females did not all achieve sexual maturity until age-3 at an average total length of 170 mm (Fisher et al. 2002). Flathead chub are broadcast, pelagic spawners (Bestgen et al. 2016). During spawning, males swim aggressively around females, nudging them in the vent and abdominal areas (Bestgen et al. 2016). Spawning likely occurs in pool habitats (Martyn and Schmulbach 1978). Fecundity ranged from 360 to 735 eggs for flathead chub in Montana, but mature females in Iowa averaged 4,974 eggs/ovary (Martyn and Schmulbach 1978; Gould

1985). The number of eggs increases with female standard lengths up to 130 mm and then slightly declines (Martyn and Schmulbach 1978). The semi-buoyant, nonadhesive, approximately 1 to 2 mm diameter, orange eggs are carried with the current downstream during incubation and larval development (Gould 1985; Platania and Altenbach 1998; Durham and Wilde 2005; Bestgen et al. 2016). Hatching occurs in 6 or 7 days at water temperatures of 20 to 22 °C (Bestgen et al. 2016).

Age and Growth: The total length of newly hatched flathead chub larvae is approximately 6 mm (Haworth 2015). Growth rates are determined by hatching date, with fish spawning later during the season having slower growth rates than fish that spawn earlier (Durham and Wilde 2005). Total lengths at age of flathead chub from the Moreau River, South Dakota were age-1, 60 mm; age-2, 99 mm; age-3, 133 mm; age-4, 154 mm; age-5, 169 mm; age-6, 182 mm (Jones 2018). Total lengths increased more rapidly after age-4 in flathead chub from the White River system in South Dakota, possibly because of the availability of large terrestrial insects (Harland 2003). Flathead chub in the Missouri River near border of North Dakota and Montana have reached a total length of 275 mm (10.8 in), while those in Canada have reached 370 mm (Kristensen 1980; Fisher et al. 2002; Hayer et al. 2008). Longevity is 10 years (Bishop 1975).

Food and Feeding: Adult flathead chub primarily eat plant material and terrestrial insects drifting with the current (Hesse 1994). In high-flow years, fish with total lengths under 60 mm ate hemipterans and copepods, but in years of average flow they ate predominantly coleopterans (Fisher et al. 2002). Turbidity does not affect food consumption because of barbels and other morphological adaptations, and a combination of visual and nonvisual cues (Bonner and Wilde 2002).

References

Bailey RM, Allum MO (1962) Fishes of South Dakota. University of Michigan, Ann Arbor. https://doi.org/10.3998/mpub.9690435

Bestgen KR, Crockett HJ, Haworth MR, Fitzpatrick RM (2016) Production of nonadhesive eggs by flathead chub and implications for downstream transport and conservation. J Fish Wildl Manage 7:434–443. https://doi.org/10.3996/022016-JFWM-018

Bishop FG (1975) Observation on the fish fauna of the Peace River in Alberta. Can Field Nat 89:423–430

Bonner TH, Wilde GR (2002) Effects of turbidity on prey consumption by prairie stream fishes. Trans Am Fish Soc 131:1203–1208. https://doi.org/10.1577/1548-8659

Cross FB (1967) Handbook of fishes in Kansas. University Press of Kansas, Lawrence

Cross FB, Mayden RL, Stewart JD (1986) Fishes in the western Mississippi drainage. In: Hocutt CH, Wiley EO (eds) The zoogeography of North American freshwater fishes. Wiley, Toronto, pp 363–412

Durham BW, Wilde GR (2005) Relationship between hatch date and first-summer growth of five species of prairie-stream cyprinids. Environ Biol Fish 72:45–54. https://doi.org/10.1007/s10641-004-4186-5

Fisher SJ, Willis DW, Olson MM, Krentz SC (2002) Flathead chubs, *Platygobio gracilis*, in the upper Missouri River: the biology of a species at risk in an endangered habitat. Can Field Nat 116:26–41

Gould W (1985) Aspects of the biology of the flathead chub (*Hybopsis gracilis*) in Montana. Great Basin Nat 45:332–336

Harland BC (2003) Survey of the fishes and habitat of western South Dakota streams. Dissertation, South Dakota State University

Hatch JT (2015) Minnesota fishes: just how many species are there anyway? Am Curr 40:10–21

Haworth MR (2015) Reproduction and recruitment dynamics of flathead chub *Platygobio gracilis* relative to flow and temperature regimes in Fountain Creek, Colorado. Dissertation, Colorado State University

Hayer CA, Ahrens NL, Berry CR Jr (2008) Biology of flathead chub, *Platygobio gracilis* in three Great Plains rivers. Proc S D Acad Sci 87:185–196

Hesse LW (1994) The status of Nebraska fishes in the Missouri River, 5. Selected chubs and minnows (Cyprinidae): sicklefin chub (M. meeki), sturgeon chub (M. gelida), silver chub (M. storeriana), speckled chub (M. aestivaus), flathead chub (Platygobio gracilis), plains minnow (Hybognathus placitus), and western silvery minnow (H. argyritis). Trans Nebr Acad Sci Affil Soc 21:99–108

Hoagstrom CW, Wall SS, Kral JG, Blackwell BG, Berry CR (2007) Zoogeographic patterns and faunal change of South Dakota fishes. N Am Nat 67:161–184. https://doi.org/10.3398/1527-0904(2007)67[161:ZPAFCO]2.0.CO;2

Jones SJ (2018) Western prairie stream fisheries: an assessment of past and present fish assemblage structure, biotic homogenization, and population dynamics in western South Dakota streams. Dissertation, South Dakota State University

Kristensen J (1980) Large flathead chub (*Platygobio gracilis*) from the Peace-Athabasca Delta, Alberta, including a Canadian record. Can Field Nat 94:342

Martyn HA, Schmulbach JC (1978) Bionomics of the flathead chub, *Hybopsis gracilis* (Richardson). Proc Iowa Acad Sci 85:62–65

Olund LJ, Cross FB (1961) Geographic variation in the North American cyprinid fish, *Hybopsis gracilis*, vol 13. University of Kansas, pp 323–348

Peters EJ, Holland RS (1994) Biological and economic analyses of the fish communities in the Platte River: modifications and tests of habitat suitability criteria for fishes in the Platte River. Completion report, Federal Aid in Fish Restoration Project Number F-78-R, Study III: Job III-2. University of Nebraska-Lincoln

Pflieger WL, Grace TB (1987) Changes in the fish fauna of the lower Missouri River, 1940-1983. In: Matthews W, Heins D (eds) Community and evolutionary ecology of North American stream fishes. University of Oklahoma Press, Norman, pp 166–177

Platania SP, Altenbach CS (1998) Reproductive strategies and egg types of seven Rio Grande Basin cyprinids. Copeia 1998:559–569. https://doi.org/10.2307/1447786

Quist MC, Hubert WA, Rahel FJ (2004) Relations among habitat characteristics, exotic species, and turbid-river cyprinids in the Missouri River drainage of Wyoming. Trans Am Fish Soc 133:727–742. https://doi.org/10.1577/T03-124.1

Walters DM, Zuellig RE, Crockett HJ, Bruce JF, Lukacs PM, Fitzpatrick RM (2014) Barriers impede upstream spawning migration of flathead chub. Trans Am Fish Soc 143:17–25. https://doi.org/10.1080/00028487.2013.824921

Welker TL, Scarnecchia DL (2004) Habitat use and population structure of four native minnows (family Cyprinidae) in the upper Missouri and lower Yellowstone rivers, North Dakota (USA). Ecol Freshw Fish 13:8–22. https://doi.org/10.1111/j.0906-6691.2004.00036.x

6.14.35   Longnose Dace *Rhinichthys cataractae*

(Valenciennes, 1842)

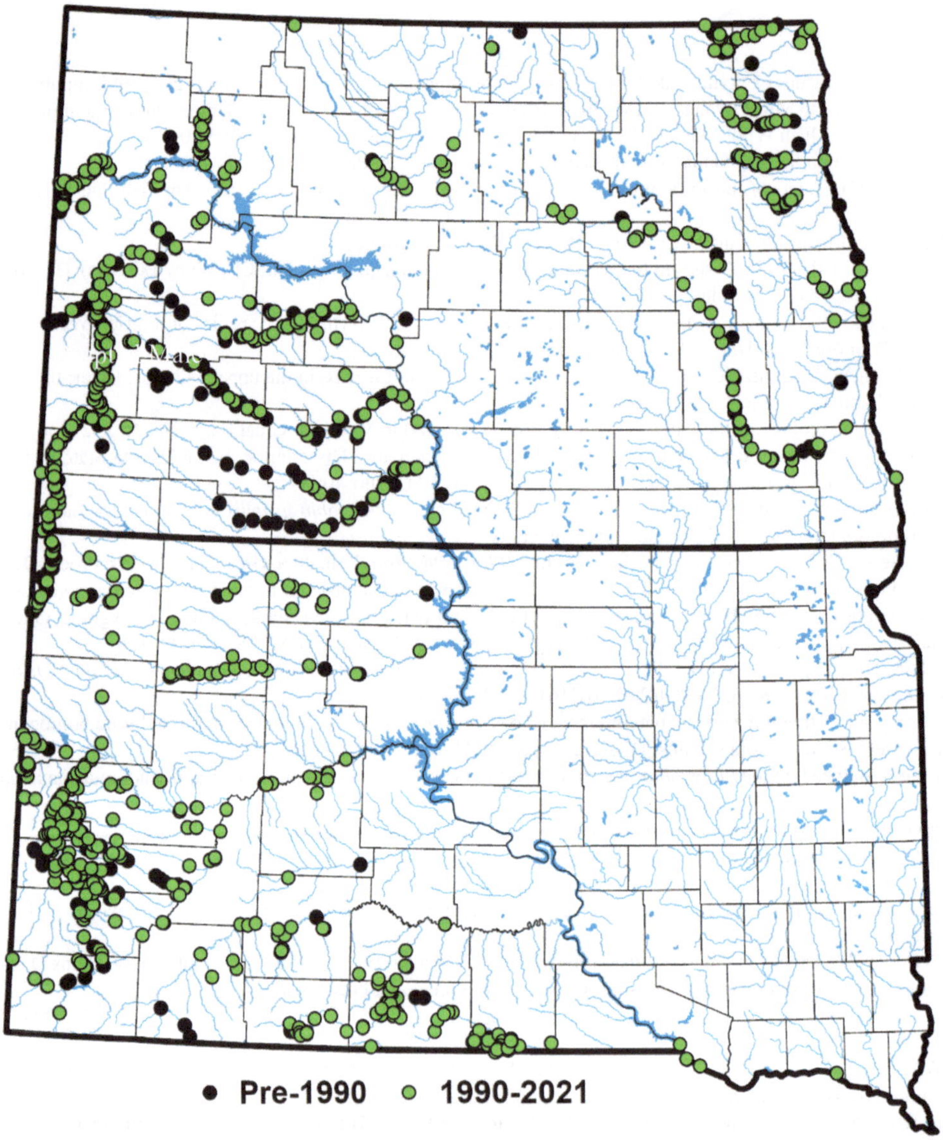

Etymology: *Rhinichthys cataractae – Rhinichthys* = snout fish; *cataractae* = of the cataract, referencing the type locality, Niagara Falls.

Description:

Body – elongated, rather cylindrical anteriorly, slightly laterally compressed posteriorly.
Color –

>   *Dorsally* – dark brown to dark olive; mottled.
>   *Laterally* – olive to gray; dark mottled specks often forming a weak stripe and a more-defined nasal stripe anterior of eye in adults.
>   *Ventrally* – cream to silvery-white.
>   *Fins* – lightly pigmented; slight cream to light golden orange tint.

Head – slightly dorsoventrally flattened.
Snout – elongated, protrudes far beyond mouth.
Eyes – moderate, laterally on upper portion of head.
Mouth – small, subterminal.

>   *Jaws* – upper jaw greatly exceeds lower jaw and extends to posterior end of nostril; frenum. *Lips* – fleshy, thick.
Teeth – pharyngeal tooth pattern 2,4-4,2.
Barbels – sometimes present; small, inconspicuous; one on each corner of mouth.
Gill rakers – 6 to 10; short.
Dorsal fin – 8 rays.
Caudal peduncle – elongated, uniform in thickness.
Caudal fin – moderately forked.
Anal fin – 7 rays.
Pelvic fins – 8 rays; abdominal; insertions slightly anterior to dorsal fin insertion; axillary process sometimes present above origin of pelvic fins.
Pectoral fins – 13 to 15 rays.
Lateral line – complete; 61 to 76 small cycloid scales in series.
Juveniles – dark spot at caudal fin base; more distinct, dark lateral stripe on body fades with age.
Sexual dimorphism – spawning males have small tubercles on dorsal side of head and body, larger tubercles on pelvic fin rays, orangish-red on lips, cheeks, all fin bases, and ventral side of body; spawning females have light orange to yellow nuptial coloration, but to a lesser extent than males.

Similar Species: Longnose dace closely resemble western blacknose dace and sturgeon chub. Western blacknose dace have an upper jaw that barely exceeds the lower jaw and a snout that only slightly overhangs the mouth, with the distance from the snout tip to the anterior tip of the lower jaw less than the eye diameter. Sturgeon chub lack a frenum, have prominent rugose folds on ventral surface of head, tend to be less mottled, and have a longitudinal ridge or keel on the dorsal scales.

Distribution and Habitat: The native range of longnose dace extends as far as the Mackenzie River near the Arctic Circle, to the northern Pacific Coast and Rocky Mountains to the Rio Grande River, throughout the northern glaciated regions of the Missouri River, Mississippi River, St. Lawrence River-Great Lakes, the Atlantic slope, and Appalachian Mountains (Bartnik 1972; Lee et al. 1980; Edwards et al. 1983). In the Dakotas, they primarily occur west of the Missouri River, but are also found in the Red River of the North, Souris River, and Sheyenne River in North Dakota. Longnose dace inhabit coldwater streams and northern lakes and are most abundant in steep gradient headwaters or tributaries of larger rivers (Kuehn 1949; Edwards et al. 1983; Etnier and Starnes 1993; Roberts and Grossman 2001). They prefer riffles with overhead cover and avoids pools and other low-flow areas (McPhail and Lindsey 1970; Bartnik 1973; Becker 1983). Within 6 weeks after hatching, longnose dace migrate to riffles, where they remain for the rest of their lives (Gee and Northcote 1963; Gee and Machniak 1972; Mullen and Burton 1995). Both juveniles and adults prefer higher water velocities and boulder substrate, which is used as shelter from the fast current (Gee and Northcote 1963; McPhail and Lindsey 1970; Culp 1989; Mullen and Burton 1995). Longnose dace generally select water depths of 10 to 19 cm; juveniles avoid depths greater than 20 cm and adults avoid depths less than 10 cm and greater than 30 cm (Mullen and Burton 1995). Their home range is limited, and they may only use 10 to 15 m of stream during their entire lives (Hill and Grossman 1987; Etnier and Starnes 1993). Longnose dace fry inhabitat areas with overhead cover (Bartnik 1973). Although longnose dace are intolerant of long-term turbidity, they will temporarily reside in swift, turbid streams (Lee et al. 1980; Edwards et al. 1983).

Reproduction: Longnose dace spawn from late spring to early summer at water temperatures of 12 to 19 °C (Brown 1971; Brazo et al. 1978; Edwards et al. 1983). In lotic habitats, they spawn over pits or riffles with loose, coarse substrate and water velocities of 45 to 60 cm/second near vegetative cover (Bartnik 1970; Roberts and Grossman 2001). In lentic habitats, they spawn near inshore, wave-swept areas (Brazo et al. 1978; Edwards et al. 1983). Sexual maturity occurs at age-2 (McPhail and Lindsey 1970; Edwards et al. 1983). Males actively defend territories before and during spawning by biting and butting any intruders, and court females to the spawning site by nudging and quivering movements (Bartnik 1970). Longnose dace are fractional spawners, with females capable of more than six clutches of eggs per year (Roberts and Grossman 2001). Fecundity is positively correlated with female size (Roberts and Grossman 2001). Eggs are deposited in crevices in coarse substrate and are likely guarded by one parent (McPhail and Lindsey 1970). The approximately 1 mm diameter unfertilized eggs increase to 2 mm after post-fertilization water absorption (Cooper 1980). The transparent, demersal, and adhesive fertilized eggs hatch in 3 or 4 days at a water temperature of 21 °C, and 7 to 10 days at 16 °C (McPhail and Lindsey 1970; Cooper 1980).

Age and Growth: The total length of newly hatched longnose dace is 5 or 6 mm (Cooper 1980). By age-3, females are usually longer than males and will also live to older ages (Kuehn 1949; Gerald 1966). Longnose dace can reach a total length of 178 mm and have a longevity of 5 years (Kuehn 1949; Gerald 1966; Becker 1983; Etnier and Starnes 1993).

Food and Feeding: Longnose dace are opportunistic benthivores that mostly eat aquatic invertebrates and periphyton (Gerald 1966; Edwards et al. 1983). Fry initially consume algae and as they grow, transition to mayflies and chironomids (Gerald 1966; Edwards et al. 1983). Adult longnose dace primarily consume baetids, tendipedids, chironimids, simulids, and other ephemeropterans and dipterans, with annelids, crustaceans, mollusks, and fish eggs eaten to a lesser extent (Gerald 1966; Becker 1983; Edwards et al. 1983).

References

Bartnik VG (1970) Reproductive isolation between two sympatric dace, *Rhinichthys atratulus* and *Rhinichthys cataractae*, in Manitoba. J Fish Res Board Can 27:2125–2141. https://doi.org/10.1139/f70-242

Bartnik VG (1972) Comparison of the breeding habits of two subspecies of longnose dace, *Rhinichthys cataractae*. Can J Zool 50:83–86. https://doi.org/10.1139/z72-015

Bartnik VG (1973) Behavioral ecology of the longnose dace, *Rhinichthys cataractae* (Pisces, Cyprinidae), significance of the dace social organization. Dissertation, University of British Columbia

Becker GC (1983) Fishes of Wisconsin. University of Wisconsin Press, Madison

Brazo DC, Liston CR, Anderson RC (1978) Life history of the longnose dace, *Rhinichthys cataractae*, in the surge zone of eastern Lake Michigan near Ludington, Michigan. Trans Am Fish Soc 107:550–556. https://doi.org/10.1577/1548-8659

Brown CJD (1971) Fishes of Montana. Montana State University Press, Bozeman

Cooper JE (1980) Egg, larval and juvenile development of longnose dace, *Rhinichthys cataractae*, and river chub, *Nocomis micropogon*, with notes on their hybridization. Copeia 1980:469–478. https://doi.org/10.2307/1444524

Culp JM (1989) Nocturnally constrained foraging of a lotic minnow (*Rhinichthys cataractae*). Can J Zool 67:2008–2012. https://doi.org/10.1139/z89-285

Edwards EA, Li H, Schreck CB (1983) Habitat suitability index models: longnose dace. US Dep Int, Fish Wildl Serv, FWS/OBS-82/10.33

Etnier DA, Starnes WC (1993) The fishes of Tennessee. University of Tennessee Press, Knoxville

Gee JH, Machniak K (1972) Ecological notes on a lake-dwelling population of longnose dace (*Rhinichthys cataractae*). J Fish Res Board Can 29:330–332. https://doi.org/10.1139/f72-054

Gee JH, Northcote TG (1963) Comparative ecology of two sympatric species of dace (*Rhinichthys*) in the Fraser River system, British Columbia. J Fish Res Board Can 20:105–118. https://doi.org/10.1139/f63-010

Gerald JW (1966) Food habits of the longnose dace, *Rhinichthys cataractae*. Copeia 1966:478–485. https://doi.org/10.2307/1441069

Hill J, Grossman GD (1987) Home range estimates for three North American stream fishes. Copeia 1987:376–380. https://doi.org/10.2307/1445773

Kuehn JH (1949) A study of a population of longnose dace (*Rhinichthys c. cataractae*). Proc Minn Acad Sci 17:81–87

Lee DS, Gilbert CR, Hocutt CH, Jenkins RE, McAllister DE, Stauffer JR (1980) Atlas of North American freshwater fishes. North Carolina Biological Survey Publication 1980-12. https://doi.org/10.5962/bhl.title.141711

McPhail JD, Lindsey CC (1970) Freshwater fishes of northwestern Canada and Alaska. Bull Fish Res Board Can 173:1–381

Mullen DM, Burton TM (1995) Size-related habitat use by longnose dace (*Rhinichthys cataractae*). Am Midl Nat 133:177–183. https://doi.org/10.2307/2426359

Roberts JH, Grossman GD (2001) Reproductive characteristics of female longnose dace in the Coweeta Creek drainage, North Carolina, USA. Ecol Freshw Fish 10:184–190. https://doi.org/10.1034/j.1600-0633.2001.100308.x

6.14.36   Western Blacknose Dace *Rhinichthys obtusus*

Agassiz, 1854

Nuptial Male   Female

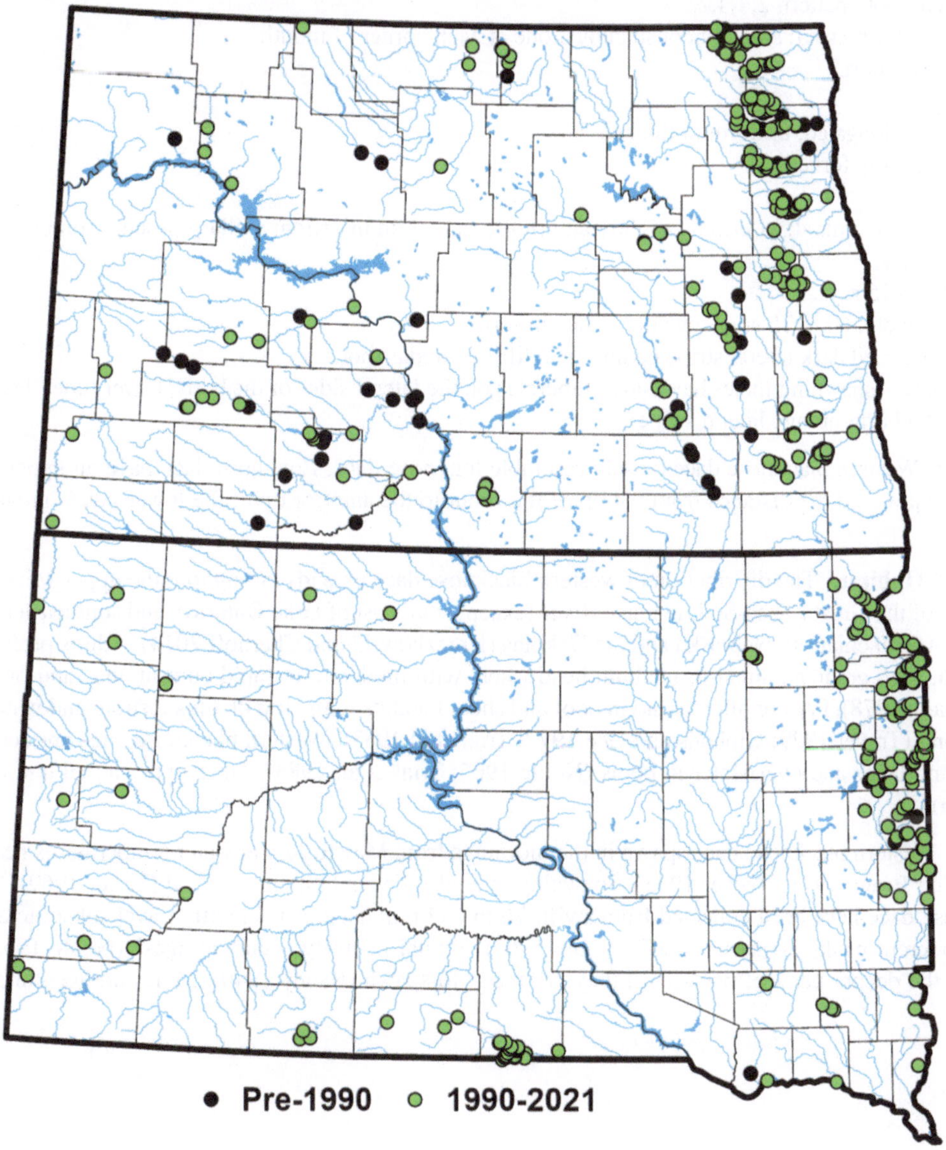

● **Pre-1990**   ○ **1990-2021**

Etymology: *Rhinichthys obtusus* – *Rhinichthys* = snout fish; *obtusus* = blunt, referencing the bluntly pointed snout.

Description:

Body – fusiform, elongated, cylindrical anteriorly, slightly laterally compressed posteriorly.
Color –

> *Dorsally* – blackish-brown to dark olive; small dark specks or mottling.
> *Laterally* – gray to olive-brown; small dark specks and mottling, sometimes forming a diffuse dark stripe extending through the eye onto the snout.
> *Ventrally* – cream to silvery-white.
> *Fins* – lightly pigmented; light gray to cream tint.

Head – slightly dorsoventrally flattened.
Snout – snout elongated, bluntly pointed, slightly protruding past mouth.
Eyes – moderate, laterally on upper portion of head.
Mouth – small, subterminal, slightly oblique.

> *Jaws* – upper jaw scarcely exceeds lower jaw and extends to posterior end of nostril; frenum.
> *Lips* – fleshy, thick.

Teeth – pharyngeal tooth pattern 2,4-4,2.
Barbels – occasionally present; small, inconspicuous, one at each corner of mouth.
Gill rakers – 6 or 7; short.
Dorsal fin – 8 rays.
Caudal peduncle – elongated, uniform in thickness.
Caudal fin – moderately forked.
Anal fin – 7 rays.
Pelvic fins – 8 rays; abdominal; insertions slightly anterior to dorsal fin insertion; axillary process sometimes present above origin of pelvic fins.
Pectoral fins – 13 to 15 rays.
Lateral line – complete; 60 to 75 small cycloid scales in series.
Juveniles – more defined dark lateral stripe ending in a diffuse caudal spot.
Sexual dimorphism – spawning males have small tubercles on the lateral sides of the body, larger tubercles on the head and pelvic fins, and a burnt orange lateral stripe.

Similar Species: Western blacknose dace closely resemble longnose dace. Longnose dace have an upper jaw that greatly exceeds the lower jaw, and the distance from the tip of the snout to the anterior tip of the lower jaw is greater than or equal to eye diameter.

Distribution and Habitat: The distribution of western blacknose dace extends throughout the upper Mississippi and Ohio rivers, tributaries of the Great Lakes-St. Lawrence River (except southeast of Lake Ontario), and southeastern Atlantic drainages west of the Appalachians to Nebraska and the Dakotas (Kraczkowski and Chernoff 2014). Adults prefer shallow, clear-to-slightly turbid, high gradient, small-to-medium headwaters with moderate-to-rapid current and sand or gravel substrate (Noble 1965; Bragg 1978). Fry are often found in pool and shoal habitats with sand or silt substrate and water velocities less than 15 cm/second (Traver 1929; Gibbons and Gee 1972; Trial et al.1983). Western blacknose dace are occasionally found in larger rivers and lakes and overwinter in pools (Noble 1965; Trial et al. 1983). Their upper lethal temperature is 29 °C (Trial et al. 1983).

Reproduction: Western blacknose dace spawn in the mornings from May to July at water temperatures near 21 °C (Traver 1929; Noble 1965; Bragg 1978; Becker 1983). Spawning may also depend on photoperiod (Tarter 1969). Sexually mature western blacknose dace have been collected from swift, 76 to 152 mm deep, waters with gravel substrate and overhanging vegetation or undercut banks (Trautman 1957; Noble 1965). Sexual maturity occurs at age-2 (Noble 1965; Bragg 1978). Females stay in the deeper parts of pools until they are ready to spawn (Raney 1940). Males are territorial towards other males, but not toward females (Phillips 1967). Two or more males court a single female to the spawning grounds (Raney 1940). One male places his caudal peduncle over the female, and the pair vibrates together for approximately two seconds at a time to initiate gamete release (Raney 1940; Becker 1983). Females retreat back to deeper waters post-spawning, while males continue defending the spawning site (Raney 1940). Fecundity increases with female size (Noble 1965). In Iowa, the fecundity of western blacknose dace females with standard lengths of 38 to 61 mm SL ranged from 375 to 2,500 (Noble 1965). Eggs are approximately 1 to 2 mm in diameter (Noble 1965; Bragg 1978).

Age and Growth: At hatching, western blacknose dace have a total length of approximately 5 mm (Traver 1929; Trial et al. 1983). In Iowa, age-0 individuals total lengths were 24 to 27 mm at the end of their first year (Noble 1965). The fastest growth rates for age-1 and age-2 fish occurred from May through July (Noble 1965). Standard lengths at age from Nebraska were 22 to 44 mm at age-1, 40 to 72 mm at age-2, and 68 to 81 mm at age-3 (Bragg 1978). Mature females are heavier and longer than males (Bragg 1978). Western blacknose dace can reach a total length of 104 mm and have a longevity of 3 years (Tarter 1969; Gibbons and Gee 1972).

Food and Feeding: Western blacknose dace consume large quantities of diptera larvae and adults, as well as smaller amounts of amphipods, isopods, and ephemeroptera larvae (Noble 1965; Tarter 1970; Bragg 1978). Juveniles feed in quiet, shallow waters over soft substrates, while adults often feed over riffles with some vegetation or within deep eddying pools (Tarter 1970). They mostly eat in the morning, with limited feeding activity at night (Tarter 1970).

References

Becker GC (1983) Fishes of Wisconsin. University of Wisconsin Press, Madison

Bragg RJ (1978) The distribution and ecology of the blacknose dace, *Rhinichthys atratulus* (Herman) in Nebraska. Dissertation, University of Nebraska-Omaha

Gibbons JRH, Gee JH (1972) Ecological segregation between longnose and blacknose dace (genus *Rhinichthys*) in the Mink River, Manitoba. J Fish Res Board Can 29:1245–1252. https://doi.org/10.1139/f72-190

Kraczkowski ML, Chernoff B (2014) Molecular phylogenetics of the Eastern and Western blacknose dace, *Rhinichthys atratulus* and *R. obtusus* (Teleostei: Cyprinidae). Copeia 2014:325–338. https://doi.org/10.1643/CG-14-002

Noble RL (1965) Life history and ecology of Western blacknose dace, Boone County, Iowa, 1963-1964. Iowa Acad Sci 72:282–293

Phillips GL (1967) Sexual dimorphism in the Western blacknose dace, *Rhinichthys atratulus meleagris*. J Minn Acad Sci 34:11–13

Raney EC (1940) Comparison of the breeding habits of two subspecies of black-nosed dace, *Rhinichthys atratulus* (Hermann). Am Midl Nat 23:399–403. https://doi.org/10.2307/2420673

Tarter DC (1969) Some aspects of reproduction in the Western blacknose dace, *Rhinichthys atratulus meleagris* Agassiz, in Doe Run, Meade County, Kentucky. Trans Am Fish Soc 98:454–459. https://doi.org/10.1577/1548-8659(1969)98[454:SAORIT]2.0.CO;2

Tarter DC (1970) Food and feeding habits of the Western blacknose dace, *Rhinichthys atratulus* meleagris Agassiz, in Doe Run, Meade County, Kentucky. Am Midl Nat 83:134–159. https://doi.org/10.2307/2424012

Trautman MB (1957) The fishes of Ohio. Ohio State University Press, Columbus

Traver JR (1929) The habits of the blacknosed dace, *Rhinichthys atronasus* (Mitchell). J Elisha Mitchell Sci Soc 45:101–129

Trial JG, Stanley JG, Batcheller M, Gebhart G, Maughan OE, Nelson PC (1983) Habitat suitability information: blacknose dace. US Dep Int, Fish Wildl Serv, FWS/OBS-82/10.41

6.14.37   Rudd *Scardinius erythrophthalmus*

(Linnaeus, 1758)

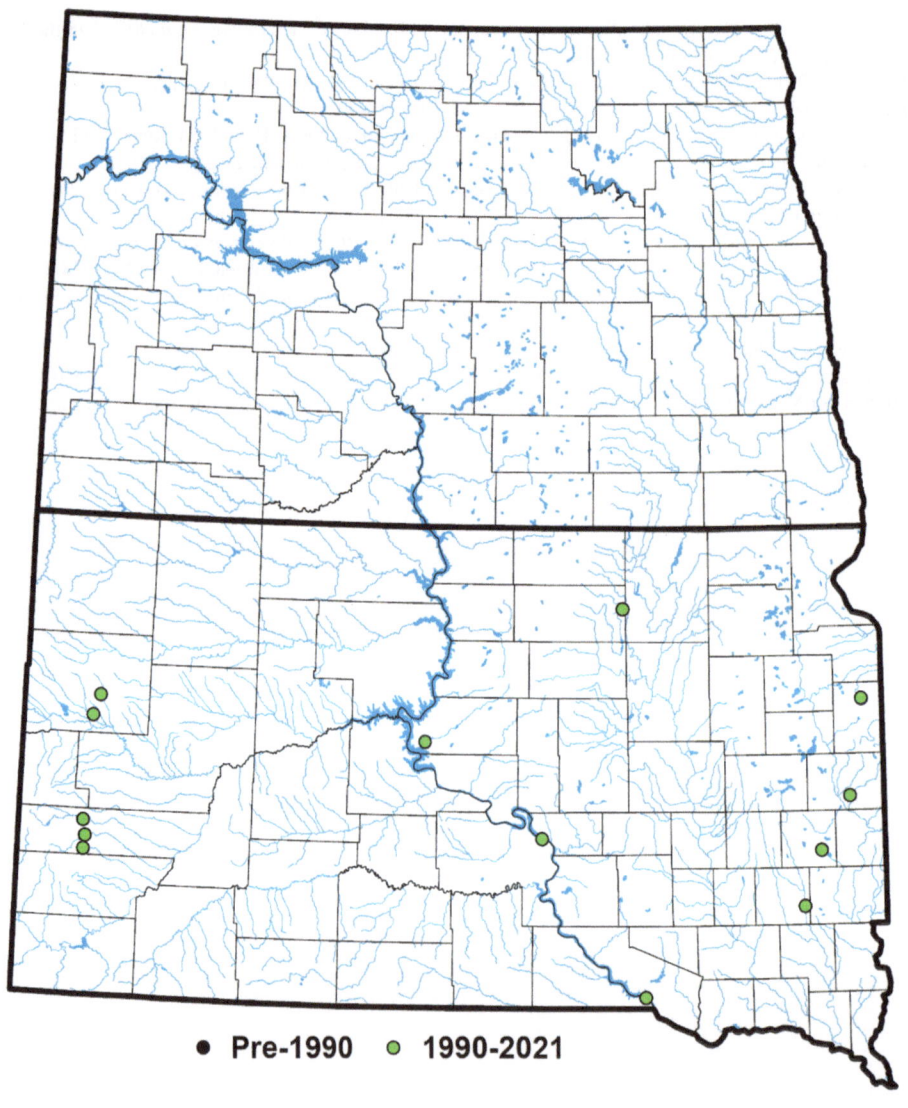

● **Pre-1990**   ○ **1990-2021**

Etymology: *Scardinius erythrophthalmus – Scardinius* = referencing the Scardus mountain range between Moesia and Macedonia; *erythrophthalmus* = red eye, referencing the slight fleck of red in the eye.

Description:

Body – deep, laterally compressed; arched back
Color –

> *Dorsally* – gold to brassy olive.
> *Laterally* – silver to gold; no distinct markings.
> *Ventrally* – silver to white.
> *Eyes* – golden with slight fleck of red.
> *Fins* – red to rosy orange tint.

Head – small, triangular.
Snout – bluntly pointed.
Eyes – moderately large; laterally on head.
Mouth – small, terminal to superior.

> *Jaws* – lower jaw slightly protruding beyond the upper jaw.

Teeth – pharyngeal tooth pattern 3,5-5,3.
Gill rakers – 10 to 13, on first gill arch.
Dorsal fin – 9 to 11 rays; slightly falcate.
Caudal peduncle – short, thick.
Caudal fin – forked; lower lobe slightly more elongate than upper lobe.
Anal fin – 11 to 13 rays; elongated, moderately falcate.
Pelvic fins – 8 or 9 rays; abdominal; insertions distinctly anterior to dorsal fin insertion.
Pectoral fins – 15 to 17 rays.
Keel – scaled; extends from pelvic fins to vent.
Lateral line – decurved, complete; 36 to 45 large cycloid scales rounded on the distal end in series.
Juveniles – less fin coloration than adults.
Sexual dimorphism – spawning males have bright red fins with distinct pale margins and small tubercles on head, anterior end of body, and pectoral, dorsal, and anal fin rays.

Similar Species: Rudd closely resemble golden shiner. Golden shiner have a scaleless keel, faint yellow-orange to clear fins, and a dorsal fin with seven to nine rays. Common shiner have a larger head, a large terminal and oblique mouth, a dorsal fin with eight rays, a slenderer and more elongate caudal peduncle, and an anal fin with only nine rays.

Distribution and Habitat: Native to the Europe, Asia Minor, and interior Russia, rudd were introduced to the United States in 1916 (Cahn 1927; Burkhead and Williams 1991). They were detected in South Dakota during the early 1990s but have not been observed in North Dakota (Burkhead and Williams 1991; Blackwell et al. 2009). South Dakota waters with rudd include Interstate Lake, Lake Alice, Lake Madison, East Vermillion Lake, Mina Lake, Lake Francis Case, Belle Fourche Reservoir, Newell Lake, Sheridan Lake, Pactola Reservoir, Canyon Lake, and Angostura Reservoir (Hoagstrom et al. 2007; Blackwell et al. 2009). Rudd inhabit a variety of habitats, including rivers, streams, lakes, and reservoirs, in littoral areas with submerged aquatic vegetation and quiet, slow waters like pools and backwaters (Haberlehner 1988; Schofield et al. 2005; Hrabik et al. 2015). Rudd tolerate a wide variety of habitats and environmental conditions, likely contributing to their wide distribution across the United States (Schofield et al. 2005).

Reproduction: Rudd spawn in shallow vegetated waters during late spring to summer when water temperatures exceed 16 °C (Hrabik et al. 2015). They do no construct nests nor give any parental care. Fecundity ranged from 12,827 to 156,775 eggs from rudd weighing 138 to 1,701 g collected in and near Lake Erie (Kapuscinski et al. 2012). The colorless to pale yellow, adhesive eggs are approximately 1 mm in diameter (Haberlehner 1988; Patimar et al. 2010; Hrabik et al. 2015). Hatching occurs 4 to 20 days after fertilization depending on water temperature (Hrabik et al. 2015).

Age and Growth: Total-lengths-at-age of rudd from Lake Alice in eastern South Dakota were age-1, 64 mm; age-2, 166 mm; age-3, 258 mm; age-4, 315 mm; age-5, 354 mm; age-6, 373 mm; age-7, 391 mm; age-8, 405 mm; age-9, 407 mm; age-10, 421 mm (Blackwell et al. 2009). Total-lengths-at-age of rudd from Newell Lake in western South Dakota were age-1, 57 mm; age-2, 128 mm; age-3, 217 mm; age-4, 278 mm; age-5, 306 mm; age-6, 319 mm; age-7, 325 mm; and age-8, 339 mm (Blackwell et al. 2009). Growth of rudd in South Dakota appears to be relatively fast, likely because of the high productivity of eutrophic waters and the low population densities typical of newly introduced populations (Blackwell et al. 2009). Rudd can reach a total length of 617 mm and live for 30 years (Schofield et al. 2005; Šprem et al. 2010).

Food and Feeding: Adult rudd are primarily herbivorous, eating aquatic macrophytes during the summer, with filamentous algae and fish, like emerald shiner, consumed in the spring and fall (Kapuscinski et al. 2012). They undergo an ontogenetic diet shift, with young-of-the-year rudd eating cladocerans and chironomid larvae, and juveniles consuming benthic invertebrates and some aquatic macrophytes (Hicks 2003; Kapuscinski et al. 2012). After reaching 150 mm, juveniles eat primarily macrophytes (Hicks 2003; Kapuscinski et al. 2012). Rudd with fork lengths over 200 mm have diets of over 80% plant material (Hicks 2003). Feeding activity is significantly reduced during the spawning season and is positively related to water temperature (Kapuscinski et al. 2012).

References

Blackwell BG, Kaufman TM, Miller WH (2009) Occurrence of rudd (*Scardinius erythrophthalmus*) and dynamics of three populations in South Dakota. J Freshw Ecol 24:285–291. https://doi.org/10.1080/02705060.2009.9664294

Burkhead NM, Williams JD (1991) An intergeneric hybrid of a native minnow, the golden shiner, and an exotic minnow, the rudd. Trans Am Fish Soc 120:781–795. https://doi.org/10.1577/1548-8659

Cahn AR (1927) The European rudd (*Scardinius*) in Wisconsin. Copeia 162:5. https://www.jstor.org/stable/1436212

Haberlehner E (1988) Comparative analysis of feeding and schooling behaviour of the Cyprinidae *Alburnus alburnus*, *Rutilus rutilus* and *Scardinius erythrophthalmus*. Int Rev Hydrobiol 73:537–546. https://doi.org/10.1002/iroh.19880730504

Hicks BJ (2003) Biology and potential impacts of rudd (*Scardinius erythrophthalmus* L.) in New Zealand. In: Munro R (ed) Managing invasive freshwater fish in New Zealand. Proceedings of workshop hosted by Dep Cons, Wellington, New Zealand, 10–12 May 2001, pp 49–58

Hoagstrom CW, Wall SS, Kral JG, Blackwell BG, Berry CR Jr (2007) Zoogeographic patterns and faunal change of South Dakota fishes. West N Am Nat 67:161–184. https://doi.org/10.3398/1527-0904(2007)67[161:ZPAFCO]2.0.CO;2

Hrabik RA, Schainost SC, Stasiak RH, Peters EJ (2015) The fishes of Nebraska. University of Nebraska, Lincoln

Kapuscinski KL, Farrell JM, Wilkinson MA (2012) Feeding patterns and population structure of an invasive cyprinid, the rudd *Scardinius erythrophthalmus* (Cypriniformes, Cyprinidae), in buffalo Harbor (Lake Erie) and the upper Niagara River. Hydrobiologia 693:169–181. https://doi.org/10.1007/s10750-012-1106-0

Patimar R, Nadjafypour E, Yaghouby M, Nadjafy M (2010) Reproduction characteristics of a stunted population of rudd, *Scardinius erythrophthalmus* (Linnaeus, 1758) living in the Anzali Lagoon (the southwest Caspian Sea, Iran). J Ichthyol 50:1060–1065. https://doi.org/10.1134/S0032945210110111

Schofield PJ, Williams JD, Nico LG, Fuller P, Thomas MR (2005) Foreign nonindigenous carps and minnows (Cyprinidae) in the United States- a guide to their identification, distribution, and biology. US Geol Surv Sci Invest Rep 2005-5041. https://doi.org/10.3133/sir20055041

Šprem N, Matulić D, Treer T, Aničič I (2010) Short communication, a new maximum length and weight for *Scardinius erythrophthalmus*. J Appl Ichthyol 26:618–619. https://doi.org/10.1111/j.1439-0426.2010.01391.x

6.14.38   Creek Chub *Semotilus atromaculatus*

(Mitchill, 1818)

Nuptial Male   Female

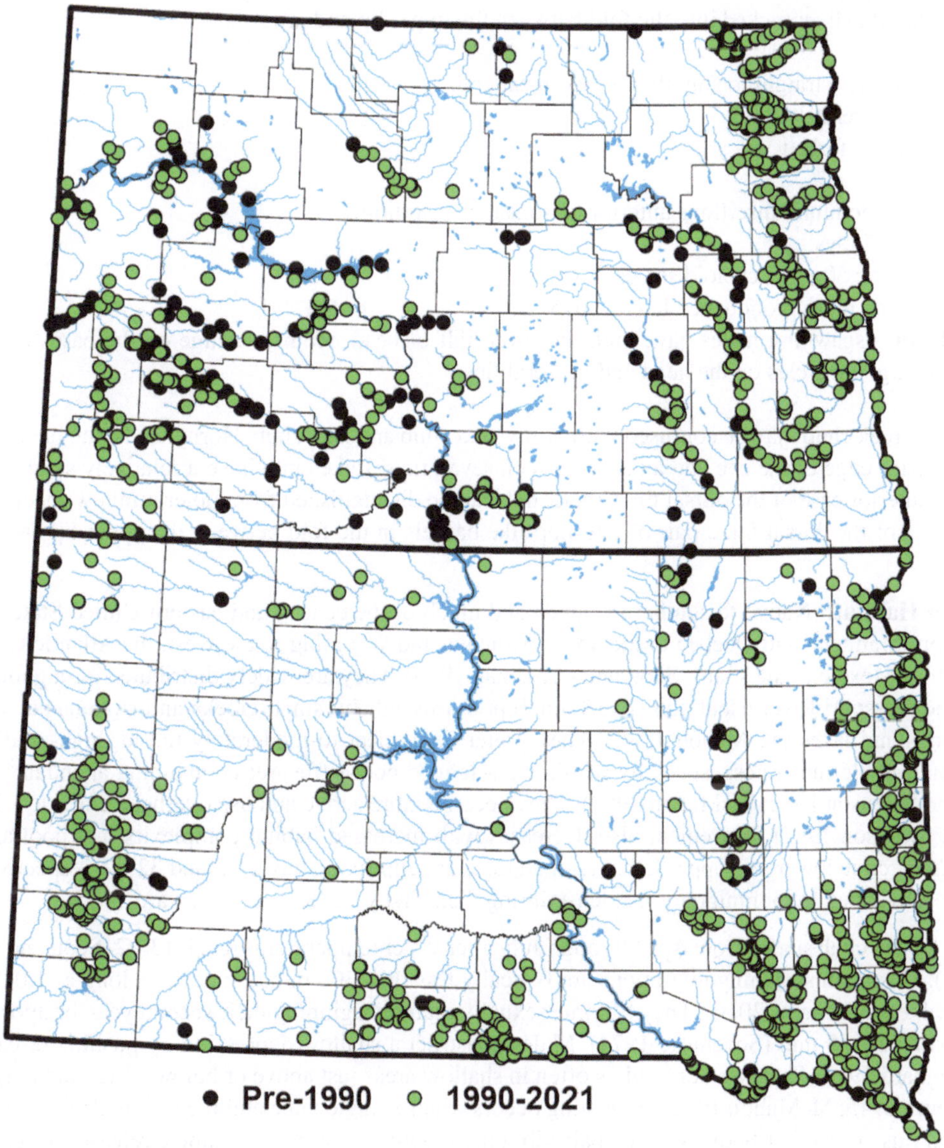

● **Pre-1990**   ○ **1990-2021**

Etymology: *Semotilus atromaculatus* – *Semotilus* = spotted banner, referencing the black spot at the anterior dorsal fin base; *atromaculatus* = black spot.

Description:

Body – fusiform, robust, cylindrical anteriorly, laterally compressed posteriorly.
Color –

> *Dorsally* – dark brown to olive.
> *Laterally* – bronze to silver; dark, diffuse stripe ending in a spot at the caudal fin base.
> *Ventrally* – silver to white.
> *Fins* – black spot on the dorsal fin anterior base; remaining fins lightly pigmented

Head – broad.
Snout – bluntly pointed.
Eyes – large, laterally on upper portion of head.
Mouth – large, terminal.

> *Jaws* – upper jaw extends past the anterior edge of the eye.
> *Lips* – fleshy.

Teeth – pharyngeal tooth pattern 2,5-4,2.
Barbels – small, flat, flap like; tucked into the fold between the upper lip and jaw.
Gill rakers – 8; short.
Dorsal fin – 8 rays; nearly straight to slightly convex distal end.
Caudal peduncle – moderately thick.
Caudal fin – moderately forked.
Anal fin – 8 rays.
Pelvic fins – 8 rays; abdominal; insertions anterior to dorsal fin insertion.
Pectoral fins – 16 or 17 rays.
Lateral line – complete; 50 to 65 cycloid scales in series.
Juveniles – more silver, more pronounced lateral stripe.
Sexual dimorphism – spawning males have pink on lower half of head, reddish-orange on the bases of the pectoral and pelvic fins, and large tubercles on the head and pectoral fins.

Similar Species: Creek chub may be confused with hornyhead chub and lake chub. Hornyhead chub have an upper jaw that extends to the anterior edge of the eye and an anal fin with seven rays. Lake chub have a relatively small mouth and lack a black spot on the anterior base of the dorsal fin. Creek chub can be distinguished from other minnow species by a black spot on the anterior base of the dorsal fin, a pair of flat, flap-like barbels in the fold between the upper lip and jaw, and a large terminal mouth.

Distribution and Habitat: Native to southern Canada and much of the central and eastern United States, the creek chub range extends from Manitoba and Quebec to the north, Montana and Wyoming to the west, the Atlantic Coast drainages to the east, and Gulf of Mexico to the south (Scott and Crossman 1973). They are widely distributed throughout the Dakotas in small streams and Missouri River tributaries. Creek chub primarily inhabit small creeks and streams but will occasionally occur within ponds and lakes (McMahon 1982). They prefer cool, clear-to-moderately-turbid, high-gradient waters with gravel, cobble, and rubble substrates, especially within well-defined pool-riffle areas containing abundant cover (Dinsmore 1962; Gradall and Swenson 1982; McMahon 1982). They become more active and use overhead cover less in higher turbidity (Gradall and Swenson 1982). Movement also decreases with increased habitat complexity and pool area (Walker and Adams 2016). Creek chub lower and upper lethal temperatures are approximately 2 °C and 32 °C, respectively (Brett 1944; McMahon 1982). Juveniles are nonmigratory, at least during their first year (Katz and Howard 1955).

Reproduction: Creek chub spawn from April through July after water temperatures reach 13 °C (Scott and Crossman 1973; McMahon 1982). Upstream spawning migrations are typically less than 300 m but may be as long as 1,069 m (Storck and Momot 1981; Walker and Adams 2016). They become sexually mature beginning their second year, or upon reaching a total length of approximately 63 mm (Schemske 1974). Males construct a shallow depression by pushing or carrying gravel in their mouth (Reighard 1910). This nest, or redd, is often in shallow areas just above or below riffles and may be of considerable size (Washburn 1945; McMahon 1982). Spawning occurs when a single male embraces a single female by wrapping his head and pectoral fins around her body, with the pair vibrating together to stimulate gamete release (Reighard 1910). After the eggs settle into the nest, females retreat to cover or continue to spawn at another nest (Reighard 1910). Males guard the nest only until the eggs have been completely covered (Reighard 1910). Fecundity generally increases with female size (Schemske 1974; Reash and Berra 1986). Fecundity from creek chub in Michigan with total lengths of 127 to 152 mm ranged from 3,500 to 5,000 eggs, compared to 4,193 to 4,671 eggs for females with total lengths of 114 to 121 mm in Ohio (Clark 1943; Washburn 1945; Schemske 1974). The yellow-orange eggs range in diameter from 1 to 2 mm, with egg size generally increasing with female size (Schemske 1974). Eggs hatch in 10 days at 13 °C (Washburn 1945).

Age and Growth: At hatching, creek chub larvae have a total length of approximately 6 or 7 mm (Boschung Jr and Mayden 2004). Growth of age-0 juveniles is rapid, but highly variable depending on food availability, stream flow, dissolved oxygen concentration, and pollution (Dinsmore 1962; Katz and Howard 1955). Growth is also related to the amount of woody debris, likely because woody debris provides quality nutrients for the aquatic invertebrates that constitute a large portion of juvenile creek chub diets (Dinsmore 1962; Quist and Guy 2001). Total-lengths-at-age of creek chub from the Des Moines River in Iowa were 58 mm at age-1, 95 mm at age-2, and 128 mm at age-3 (Dinsmore 1962). Slightly longer lengths-at-age were observed in northern Kansas (Quist and Guy 2001). Creek chub can reach a total length of 305 mm and have a longevity of 8 years (Pflieger 1997).

Food and Feeding: Creek chub are opportunistic, visual, generalist omnivores (Evans 1952). Juveniles consume small terrestrial and aquatic insects during the day (Barber and Minckley 1971; Copes 1978). Adults primarily consume plant material, and aquatic and terrestrial insects, including ephemeropterans and coleopteran adults (Dinsmore 1962). They also eat annelids, crayfish, gastropods, and mussels (Dinsmore 1962; Barber and Minckley 1971; Quist et al. 2006). Larger creek chub will prey on smaller fish (Dinsmore 1962). In intermittent, reduced-resource environments, creek chub gut lengths increase to facilitate survival and maintain body condition (Christian and Adams 2014).

References

Barber WE, Minckley WL (1971) Summer foods of the cyprinid fish *Semotilus atromaculatus*. Trans Am Fish Soc 100:283–289. https://doi.org/10.1577/1548-8659

Boschung HT Jr, Mayden RL (2004) Fishes of Alabama. Smithsonian Books, Washington, DC

Brett JR (1944) Some lethal temperature relations of Algonquin Park fishes. University of Toronto Studies Biological Series, No 52, Ontario, Canada

Christian JM, Adams GL (2014) Effects of pool isolation on trophic ecology of fishes in a highland stream. J Fish Biol 85:752–772. https://doi.org/10.1111/jfb.12453

Clark CF (1943) Creek chub minnow propagation. Ohio Cons Bull 7:12–13

Copes F (1978) Ecology of the creek chub, *Semotilus atromaculatus* (Mitchill), in northern waters. University of Wisconsin, Fauna Flora Wisconsin Rep No 12, Stevens Point

Dinsmore JJ (1962) Life history of the creek chub, with emphasis on growth. Iowa Acad Sci 69:296–301

Evans HE (1952) The correlation of brain pattern and feeding habits in four species of cyprinid fishes. J Comp Neurol 97:133–142. https://doi.org/10.1002/cne.900970108

Gradall KS, Swenson WA (1982) Responses of brook trout and creek chubs to turbidity. Trans Am Fish Soc 111:392–395. https://doi.org/10.1577/1548-8659

Katz M, Howard WC (1955) The length and growth of 0-year class creek chubs in relation to domestic pollution. Trans Am Fish Soc 84:228–238. https://doi.org/10.1577/1548-8659(1954)84[228:TLAGOY]2.0.CO;2

McMahon TE (1982) Habitat suitability index models: creek chub. US Dep Int, Fish Wildl Serv, FWS/OBS-82/10.4

Pflieger WL (1997) The fishes of Missouri, revised edn. Missouri Dep Cons, Jefferson City

Quist MC, Guy CS (2001) Growth and mortality of prairie stream fishes: relations with fish community and instream habitat characteristics. Ecol Freshw Fish 10:88–96. https://doi.org/10.1034/j.1600-0633.2001.100203.x

Quist MC, Bower MR, Hubert WA (2006) Summer food habits and trophic overlap of roundtail chub and creek chub in muddy creek, Wyoming. Southwest Nat 51:22–27. https://doi.org/10.1894/0038-4909(2006)51[22:SFHATO]2.0.CO;2

Reash RJ, Berra TM (1986) Fecundity and trace-metal content of creek chubs from a metal-contaminated stream. Trans Am Fish Soc 115:346–351. https://doi.org/10.1577/1548-8659

Reighard J (1910) Methods of studying the habits of fishes, with an account of the breeding habits of the horned dace. Bull US Bureau Fish 28:1111–1136

Schemske DW (1974) Age, length and fecundity of the creek chub, *Semotilus atromaculatus* (Mitchill), in central Illinois. Am Midl Nat 92:505–509

Scott WB, Crossman EJ (1973) Freshwater fishes of Canada. Fisheries Research Board of Canada, Bulletin 184

Storck T, Momot WT (1981) Movements of the creek chub in a small Ohio stream. Ohio Acad Sci 81:9–13

Walker RH, Adams GL (2016) Ecological factors influencing movement of creek chub in an intermittent stream of the Ozark Mountains, Arkansas. Ecol Freshw Fish 25:190–202. https://doi.org/10.1111/eff.12201

Washburn GN (1945) Propagation of the creek chub in ponds with artificial raceways. Trans Am Fish Soc 75:336–350. https://doi.org/10.1577/1548-8659(1945)75[336:POTCCI]2.0.CO;2

6.15 Ictaluridae, North American Catfish Family

Over 2,000 marine and freshwater catfish species in 34 different families exist worldwide. Ictaluridae, the only catfish family in North America, contains 46 species. The native distribution of Ictaluridae is the freshwaters from the Atlantic-Pacific Continental Divide east to the Atlantic Coast, but several species have been widely introduced elsewhere.

Ictalurids can easily be identified from other fish families in the Dakotas by their scaleless body, flap-like adipose fin (sometimes connected to the caudal fin), four sets of conspicuous, long barbels (commonly called whiskers, creating the image of a cat-like fish), and spines in the dorsal and pectoral fins. Like all fish barbels, those in catfish have numerous tiny tastebuds which use chemical cues to locate prey. Pectoral and dorsal spines can inject a small amount of nonlethal venom, creating a painful sting that may be intensely painful and, because of an anticoagulant, can induce bleeding. The Ictaluridae family is often associated with larger species like channel catfish *Ictalurus punctatus*, flathead catfish *Pylodictis olivaris*, blue catfish *Ictalurus furcatus*, and the somewhat smaller bullheads (*Ameiurus*). However, the family also includes smaller species like the tadpole madtom *Noturus gyrinus* and stonecat *Noturus flavus*, the only two species of *Noturus* that occur in the Dakotas, which are typically less than 250 mm in total length.

In the Dakotas, blue catfish, flathead catfish, and channel catfish inhabit large rivers, major tributaries, and reservoirs. Channel catfish can also occasionally be found in larger creeks. Bullheads, highly tolerant of turbidity and low dissolved oxygen, primarily inhabit lakes, ponds, wetlands, and streams. Tadpole madtoms prefer streams, sloughs, ponds, and lakes with silt, sand, pebble, or gravel substrates. Stonecats are highly adaptable and can be found in the Missouri River as well as smaller streams. Ictalurids are generally most active during the night, migrating from deeper to shallower waters to forage. The numerous taste buds on their body and barbels make the catfish family well adapted to foraging at night and in turbid water. Large-bodied catfish in the genera *Ictalurus* and *Pylodictis* are considered opportunistic, nocturnal predators and undergo an ontogenetic diet shift to piscivory at large sizes. Bullheads are also opportunistic nocturnal foragers, with adults consuming a wide variety of prey items including zooplankton, benthic aquatic invertebrates, mollusks, and small fish. Madtoms are also nocturnal and primarily feed along the bottom substrate on organic debris, zooplankton, and aquatic insect larvae. Males and females of certain species, construct nests or depressions within the substrate, or cavities within submerged structures, where they provide parental care by fanning the eggs until they hatch. Some members of Ictaluridae will guard and swim alongside their young, which form tight schools called a ball.

Channel catfish, flathead catfish, and blue catfish are important sportfish species in the Dakotas. Walleye *Sander vitreus*, northern pike *Esox Lucius*, and other sportfish frequently eat bullheads and madtoms.

6.15.1   Black Bullhead *Ameiurus melas*

(Rafinesque, 1820)

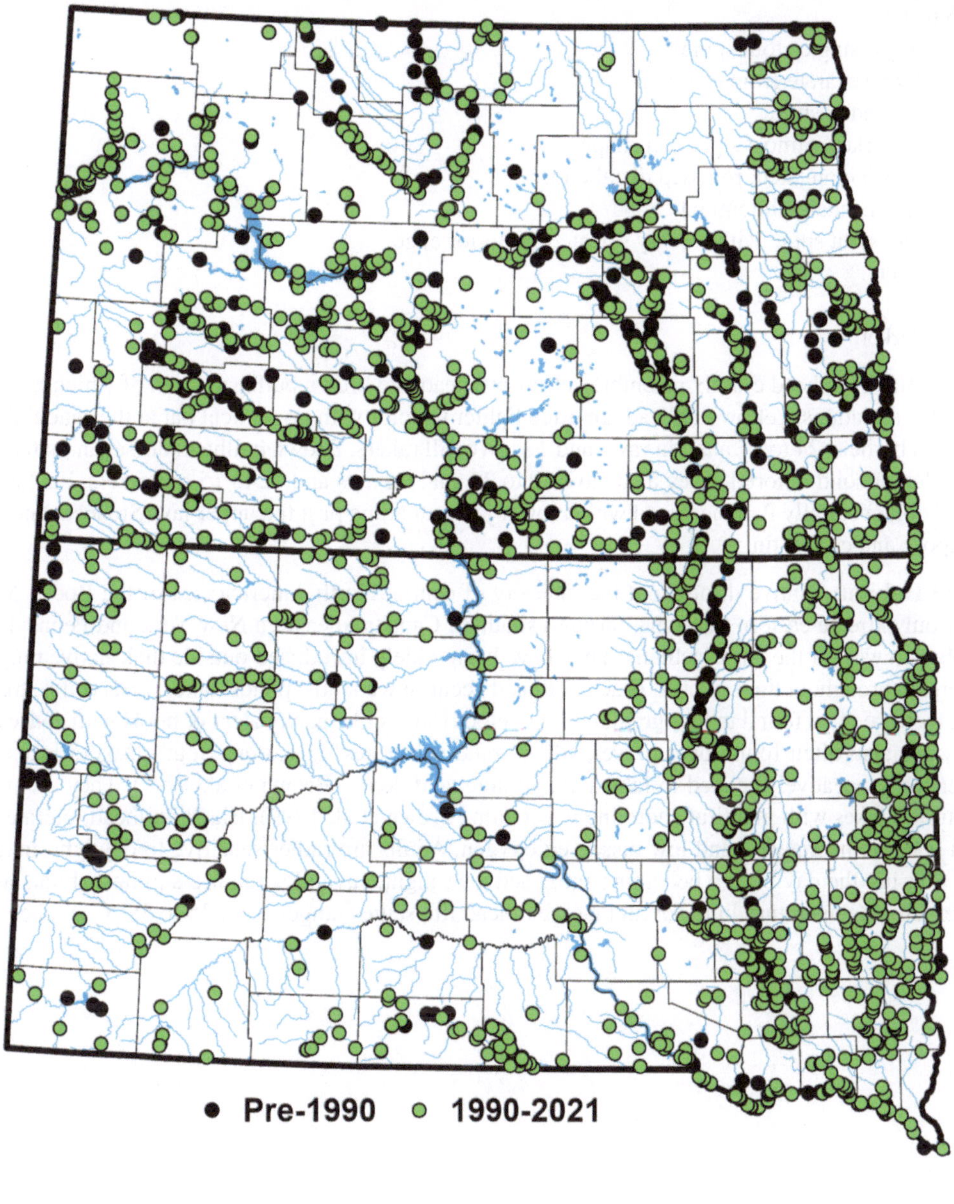

● **Pre-1990**   ● **1990-2021**

Etymology: *Ameiurus melas* – *Ameiurus* = without tail, referencing the unforked caudal fin; *melas* = black.

Description:

Body – robust, stout, rounded anteriorly, laterally compressed posteriorly.
Color –

> *Dorsally* – solid black, dark brown, or dark olive-green.
> *Laterally* – gray, tan-brown.
> *Ventrally* – yellow-cream to white.
> *Barbels* – dark gray, brown to black.
> *Fins* – uniformly dark in color; caudal – faint yellow-cream vertical bar at base.

Head – slightly dorsoventrally flattened.
Snout – blunt.
Eyes – small, positioned dorsolaterally on head.
Mouth – terminal, wide.

> *Jaws* – nearly equal length; upper jaw occasionally extending slightly past lower jaw.

Teeth – patches on jaws; numerous sharp, small teeth.
Barbels – four pairs; one pair top of snout, one long pair corners of upper jaw; two pairs chin in transverse line.
Gill rakers – 15 to 19.
Dorsal fin – 5 or 6 rays; single stout spine.
Adipose fin – flap like; separate from caudal fin.
Caudal peduncle – slender, short.
Caudal fin – weakly forked, almost square in shape.
Anal fin – 16 to 21 rays; elongated, rounded distal end.
Pelvic fins – 7 or 8 rays; small, abdominal.
Pectoral fins – single, stout spine with weakly serrated posterior edge.
Lateral line – complete.
Skin – scaleless.
Juveniles – much darker.

Similar Species: Black bullhead closely resemble yellow bullhead and brown bullhead, both of which have strong pectoral spine posterior-edge serrations. Yellow bullhead are often lighter in color with no vertical bar at the caudal fin base, and have cream to white chin barbels, 24 to 27 anal fin rays, and 12 to 16 gill rakers. Brown bullhead dorsal and lateral sides are often brown-black mottled or solid-colored. They also have 20 to 24 anal fin rays and 12 to 15 gill rakers. Flathead catfish have a more pronounced dorsoventrally flattened head with a lower jaw extending past the upper jaw. Stonecat and tadpole madtom have a fused adipose and caudal fin.

Distribution and Habitat: Native throughout the majority of central North America east of the Rocky Mountains, black bullhead range from extreme eastern Montana, through southern Canada to eastern New York and Pennsylvania, and south to the Gulf of Mexico west of the Appalachians. They have been widely introduced outside their native range and are abundant and widespread throughout the Dakotas. Black bullhead occur in wetlands, ponds, lakes, reservoirs, streams, and rivers. In lotic habitats, they are often found in low-gradient, low-velocity areas like backwaters or pools with dense cover or aquatic vegetation. They are also often found over gravel, silt, or mud substrates. In eastern South Dakota lakes, black bullhead abundance is generally negatively related to lake area, volume, and depth (Brown et al. 1999). They are most abundant in small, shallow, turbid lakes with plentiful shoreline vegetation (Brown et al. 1999). Black bullhead tolerate a wide range in turbidities, high water temperatures, and low dissolved oxygen. Adults are rather sensitive to daylight, inactive and hiding under vegetation during the day before becoming most active at night. Juveniles require similar habitats as adults (Stuber 1982). The upper lethal temperature limit for all black bullhead life stages ranges from 35 to 39 °C (Stuber 1982).

Reproduction: Black bullhead spawn from mid-May to early July when water temperature reaches 20 °C (Dennison and Bulkley 1972; Scott and Crossman 1973; Stuber 1982). Sexual maturity occurs at a total length of approximately 160 mm and varies by age. Females construct saucer-shaped nests approximately 15 to 35 cm in diameter using their snouts, pelvic fins, and anal fins (Wallace 1967; Pflieger 1997). Nests are near vegetative cover or woody debris in sand, silt, or mud substrates at a depth of 0.6 to 1.2 m (Wallace 1967). The yellow eggs are approximately 1 mm in diameter (Becker 1983). Following fertilization, both sexes fan the gelatinous egg masses. Hatching occurs in 5 to 10 days depending on water temperature. Fry form tight schools, or ball, near the surface and are guarded by the adults for 2 weeks or longer (Becker 1983).

Age and Growth: Growth is density dependent and highly variable among black bullhead populations (Brown et al. 1999; Hanchin et al. 2002). In South Dakota, black bullhead grow larger in deeper lakes with relatively small populations than from shallow lakes with larger populations (Hanchin et al. 2002). Stunting occurs in dense populations with fewer predators (Hanchin et al. 2002). In the James River, growth is faster in years with high spring flows (Tol 1976). Although survival rates may increase in turbid water, growth rates increase in clear waters (Hanchin et al. 2002). Total-lengths-at-age of black bullhead from eastern South Dakota lakes were age-1, 101 mm; age-2, 155 mm; age-3, 203 mm; age-4, 232 mm; age-5, 262 mm; and age-6, 289 mm (Hanchin et al. 2002). Adult total lengths are typically less than 305 mm, but black bullheads total length can exceed 381 mm. Longevity is 10 years.

Food and Feeding: Black bullheads are opportunistic, selective zooplanktivores. Diet varies little with size, with all life stages selective of the largest available cladocerans (Repsys 1972). Young-of-year black bullhead primarily eat limnetic cladocerans and copepods (Repsys 1972). Adults primarily eat zooplankton, but also scavenge on insect larvae, crayfish, snails, and small fish. Black bullhead mostly feed near the bottom at night in shallow waters.

References

Becker GC (1983) Fishes of Wisconsin. University of Wisconsin Press, Madison
Brown ML, Willis DW, Blackwell BG (1999) Physiochemical and biological influences on black bullhead populations in eastern South Dakota glacial lakes. J Freshw Ecol 14:47–60. https://doi.org/10.1080/02705060.1999.9663654
Dennison SG, Bulkley RV (1972) Reproductive potential of the black bullhead, *Ictalurus melas*, in Clear Lake, Iowa. Trans Am Fish Soc 101:483–487. https://doi.org/10.1577/1548-8659
Hanchin PA, Willis DW, Hubers MJ (2002) Black bullhead growth in South Dakota waters: limnological and community influences. J Freshw Ecol 17:65–73. https://doi.org/10.1080/02705060.2002.9663869
Pflieger WL (1997) The fishes of Missouri, revised edn. Missouri Dep Cons, Jefferson City
Repsys AJ (1972) Food selectivity of the black bullhead (*Ictalurus melas*, Rafinesque) in Lake Poinsett, South Dakota. Dissertation, South Dakota State University
Scott WB, Crossman EJ (1973) Freshwater fishes of Canada. Fisheries Research Board of Canada, Bulletin 184
Stuber RJ (1982) Habitat suitability index models: black bullhead. US Dept Inter, Fish Wildl Serv, FWS/OBS-82/10.14
Tol D (1976) An evaluation of the fishery resources in a portion of the James River, South Dakota scheduled for channel modification. Dissertation, South Dakota State University
Wallace CR (1967) Observations on the reproductive behavior of the black bullhead (*Ictalurus melas*). Copeia 1967:852–853. https://doi.org/10.2307/1441904

6.15.2   Yellow Bullhead *Ameiurus natalis*

(Lesueur, 1819)

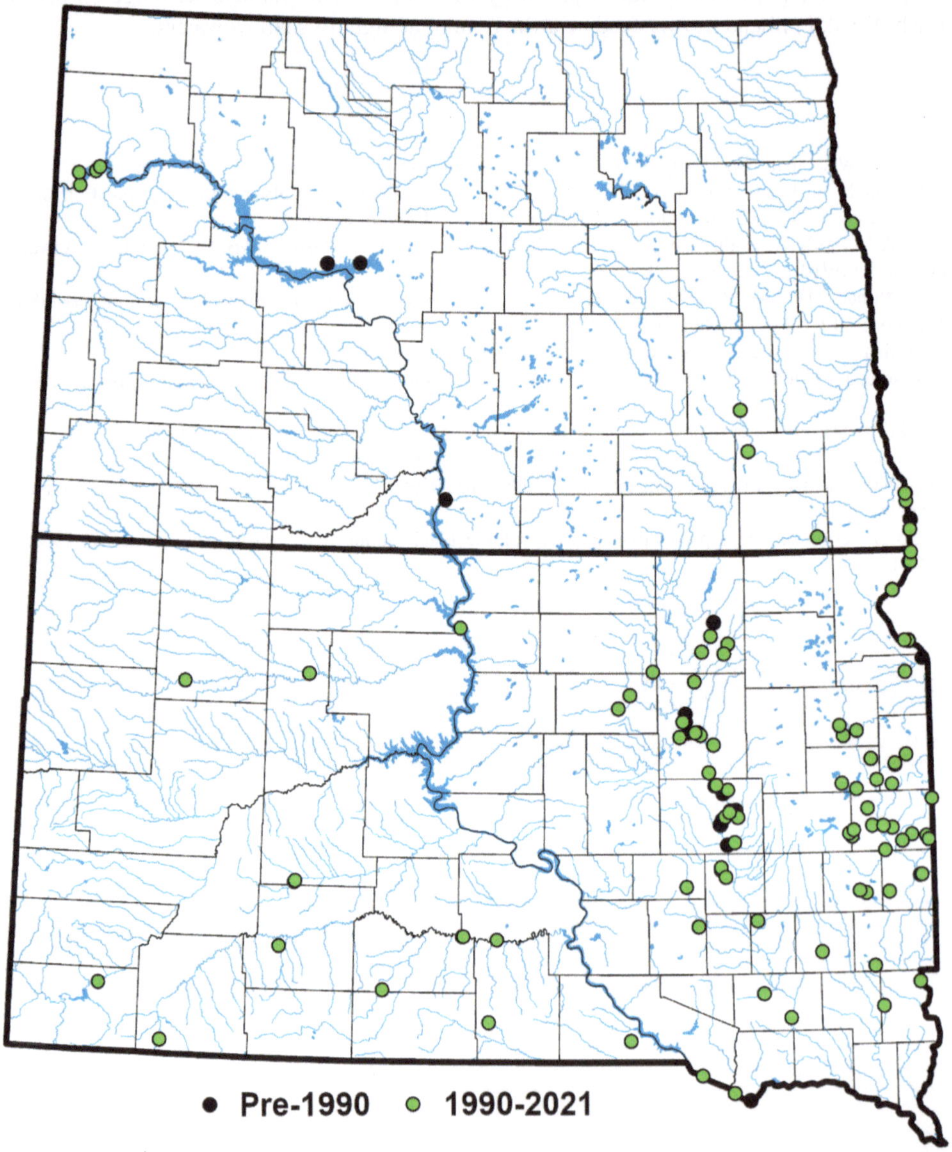

● **Pre-1990**   ● **1990-2021**

Etymology: *Ameiurus natalis – Ameiurus* = without tail, referencing the unforked caudal fin; *natalis* = large buttocks.

Description:

Body – robust, stout, rounded anteriorly, laterally compressed posteriorly.
Color –

> *Dorsally* – solid yellow-olive to dark brown-black.
> *Laterally* – yellow-olive to brown.
> *Ventrally* – yellow-cream to white.
> *Barbels* – cream to white.
> *Fins* – dusky dark olive.

Head – slightly dorsoventrally flattened.
Snout – blunt.
Eyes – small, positioned dorsolaterally on head.
Mouth – terminal, wide.

> *Jaws* – nearly equal in length; upper jaw occasionally extending slightly past lower jaw.

Teeth – tooth patches; numerous sharp, small teeth.
Barbels – four pairs; one pair top of snout, one long pair corners of upper jaw, two chin pairs in transverse line.
Gill rakers – 12 to 16.
Dorsal fin – 5 or 6 rays; single stout spine.
Adipose fin – flap-like; separated from caudal fin.
Caudal peduncle – slender, short.
Caudal fin – weakly forked, almost square in shape.
Anal fin – 24 to 27 rays; elongated, nearly straight distal end.
Pelvic fins – 7 or 8 rays; small, abdominal.
Pectoral fins – single, stout spine, serrated on posterior edge.
Lateral line – complete.
Skin – scaleless.
Juveniles – less yellow.

Similar Species: Yellow bullhead closely resemble black bullhead and brown bullhead. Black bullhead have weak serrations on the pelvic spine posterior edge, dark gray or brown to black chin barbels, 16 to 21 anal fin rays, and 15 to 19 gill rakers. Brown bullhead dorsal and lateral sides are often brown-black mottled or solid-colored. They also have 20 to 24 anal fin rays and 12 to 15 gill rakers. Flathead catfish have a more pronounced dorsoventrally flattened head with a lower jaw extending past the upper jaw. Stonecat and tadpole madtom have fused adipose and caudal fins.

Distribution and Habitat: Native throughout the eastern United States, the range of the yellow bullhead includes southeastern Canada and the Great Plains, south to the Gulf of Mexico. They are less abundant in the Dakotas than black bullhead. In the Dakotas, yellow bullhead occur throughout the Missouri River and its major tributaries, as well as the Red River of the North drainage. Yellow bullhead inhabit wetlands, ponds, lakes, reservoirs, streams, and rivers, but are most common in shallow, low-flow areas of lakes and streams with clear water and abundant aquatic vegetation. They are often found 0.5 to 1.5 m deep over gravel, sand, or mud substrates. Yellow bullhead are more abundant in clearer streams than turbid waters. They tolerate low dissolved oxygen and high temperatures, with an upper lethal limit of 40 °C (Carveth et al. 2004).

Reproduction: Yellow bullhead spawn from May to July, typically earlier than black bullhead or brown bullhead. They become sexually mature at age-2 or age-3, or around a total length of 140 mm (Scott and Crossman 1973). Their spawning behavior is like black bullhead, with the male and female laying side-by-side in opposing directions, vibrating their bodies together (Scott and Crossman 1973). One or both adults construct saucer-shaped nests, often near submerged structure. Yellow bullhead will also spawn in natural cavities within rocks or submerged trees. The adhesive, yellow-cream eggs are approximately 2 to 3 mm in diameter (Becker 1983). After fertilization, one or both adults fan the gelatinous egg masses. Yellow bullhead fecundity ranges from 1,650 to 4,300 eggs, with approximately 300 to 700 eggs per nest (Becker 1983). Eggs hatch in 5 to 10 days depending on water temperature. Several hundred fry will form a tight school near the surface. Fry are guarded by an adult, usually the male, until they reach a total length of 50 mm (Becker 1983).

Age and Growth: Information on yellow bullhead growth is scarce, but they are generally smaller than black bullhead or brown bullhead. Growth is highly variable among populations and is dependent on environmental conditions. Total-lengths-at-age of yellow bullhead from Wisconsin were 206 mm at age-2, 226 mm at age-3, and 272 mm at age-4 (Becker 1983). Yellow bullhead can reach a total length of 298 mm and live for 12 years (Murie et al. 2009).

Food and Feeding: Yellow bullhead are voracious, opportunistic omnivores. Their diet is like black bullhead, but with a greater amount of aquatic vegetation and benthic detritus. Yellow bullhead can locate prey using the taste buds on their barbels and elsewhere on their bodies. Adults eat small fish, benthic invertebrates, mollusks, and macrophytes (Pflieger 1997). They mostly feed near the bottom at night in shallow waters (Pflieger 1997). Adult yellow bullhead will also scavenge on dead fish after winterkill (Schneider 1998).

References

Becker GC (1983) Fishes of Wisconsin. University of Wisconsin Press, Madison

Carveth CJ, Widmer A, Bonar SC, Matter W (2004) Estimation of acute upper lethal water temperature tolerances of native Arizona fishes. Report submitted to the Water Resources Research Center, University of Arizona

Murie DJ, Parkyn DC, Loftus WF, Nico LG (2009) Variable growth and longevity of yellow bullhead (*Ameiurus natalis*) in the Everglades of South Florida, USA. J Appl Ichthyol 25:740–745. https://doi.org/10.1111/j.1439-0426.2009.01300.x

Pflieger WL (1997) The fishes of Missouri, revised edn. Missouri Dep Cons, Jefferson City

Schneider JC (1998) Fate of dead fish in a small lake. Am Midl Nat 140:192–196. https://doi.org/10.1674/0003-0031(1998)140[0192:FODFI A]2.0.CO;2

Scott WB, Crossman EJ (1973) Freshwater fishes of Canada. Fisheries Research Board of Canada, Bulletin 184

Wallace CR (1967) Observations on the reproductive behavior of the black bullhead (*Ictalurus melas*). Copeia 1967:852–853. https://doi.org/10.2307/1441904

6.15.3 Brown Bullhead *Ameiurus nebulosus*

(Lesueur, 1819)

© Uland Thomas

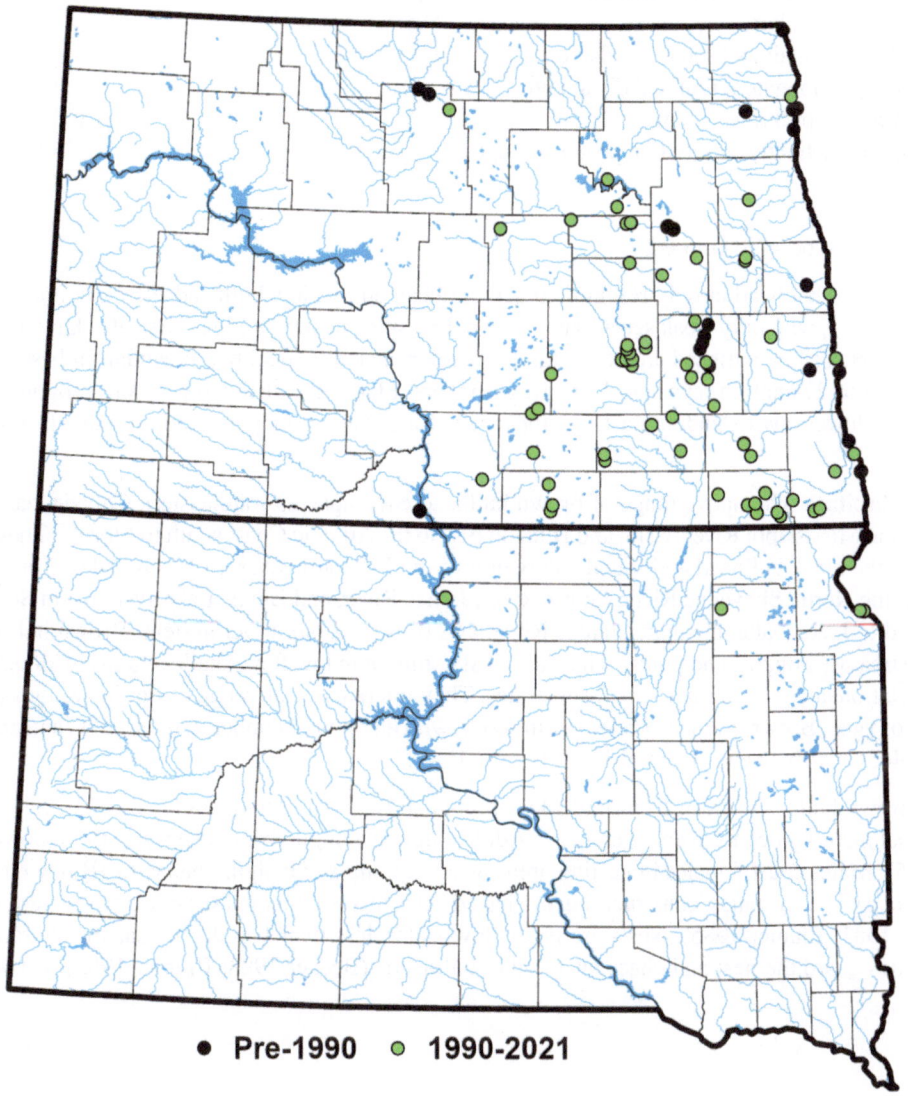

● **Pre-1990**   ● **1990–2021**

Etymology: *Ameiurus nebulosus* – *Ameiurus* = without tail, referencing the unforked caudal fin; *nebulosus* = clouded, referencing mottled coloring.

Description:

Body – robust, stout, rounded anteriorly; laterally compressed posteriorly.
Color –

> *Dorsally* – mottled; dark brown to black; occasionally solid color.
> *Laterally* – mottled; light, and dark brown.
> *Ventrally* – cream to white.
> *Barbels* – black to yellow-brown.
> *Fins* – dark, may be mottled.

Head – slightly dorsoventrally flattened.
Snout – blunt.
Eyes – small, positioned dorsolaterally on head.
Mouth – terminal, wide.

> *Jaws* – nearly equal in length; upper jaw occasionally extending slightly past lower jaw.

Teeth – patches present on jaws; numerous, sharp, small.
Barbels – four pairs; one pair top of snout, one long pair near corners of upper jaw, two chin pairs on transverse line.
Gill rakers – 12 to 15.
Dorsal fin – 6 or 7 rays; 1 stout spine.
Adipose fin – flap like; separated from caudal fin.
Caudal peduncle – slender, short.
Caudal fin – weakly forked, almost square in shape.
Anal fin – 20 to 24 rays; elongated, rounded distal end.
Pelvic fins – abdominal; 7 or 8 rays; small.
Pectoral fins – 1 spine; distinctly serrated on posterior edge.
Lateral line – complete.
Skin – scaleless.
Juveniles – more likely to be solid colored.

Similar Species: Brown bullhead closely resemble black bullhead and yellow bullhead. Black bullhead are solid black or dark brown to olive-green, with only weak serrations on the pelvic spine posterior edge. They also have 16 to 21 anal fin rays and 15 to 19 gill rakers. Yellow bullhead are often lighter in color, lack a vertical bar at caudal fin base, and have cream to white chin barbels, 24 to 27 anal fin rays, and 12 to 16 gill rakers. Flathead catfish have a more pronounced dorsoventrally flattened head with a lower jaw extending past the upper jaw. Stonecat and tadpole madtom have fused adipose and caudal fins.

Distribution and Habitat: The native range of brown bullhead encompasses southeastern Canada, eastern North Dakota and South Dakota, the Mississippi River basin, and Hudson Bay to the Atlantic Coast south to Florida. They are less abundant in the Dakotas than the black bullhead and found solely east of the Missouri River. In North Dakota, brown bullhead occur in the Red River of the North, Sheyenne River, Souris River, James River, and Devils Lake systems. In South Dakota, brown bullhead are limited to Lake Oahe, the upper James River, and Bois de Sioux River system. Brown bullhead inhabit lakes, reservoirs, rivers, streams, sloughs, and ponds. They are most common in low flow, shallow areas of sloughs and ponds with dense aquatic vegetation and gravel, sand, or silt substrates. They tolerate a wide range of turbidity, high water temperature, and low dissolved oxygen. Brown bullhead prefer warm water, with an optimal temperature of 26 °C and an upper lethal limit of 38 °C (Brett 1944; Crawshaw and Hammel 1974; Becker 1983).

Reproduction: Brown bullhead spawn from mid-May through July at water temperatures of 21 to 25 °C (Mansueti and Hardy 1967). Sexual maturity occurs at age-2 or age-3. Adults migrate at night from deep water to shallower water spawning sites (Blumer 1985). Males court females in nesting areas using nipping and budging behavior (Blumer 1985). Brown bullhead females, and occasionally males, construct saucer-shaped nests near submerged structure or vegetation in sand or gravel substrates (Becker 1983; Blumer 1985). Nests are in shallow water, less than 1-m deep, near the shoreline (Becker 1983; Blumer 1985). Males guard the nest and chase away other bullheads (Becker 1983; Blumer 1985). Brown bullhead, black bullhead, and yellow bullhead have similar spawning behavior. Spawning occurs in the early morning to midafternoon, with a male and female laying side-by-side in opposing directions vibrating their bodies together (Breder Jr and Rosen 1966; Becker 1983). Fecundity increases with female size. Fecundity ranged from 2,000 to 13,000 eggs from females with total lengths ranging from 202 to 230 mm (Etnier and Starnes 1993). The adhesive, cream-colored eggs are approximately 3 mm in diameter (Breder Jr and Rosen 1966; Becker 1983). One or both adults use their caudal, anal, and pectoral fins to fan the gelatinous egg masses (Blumer 1985). At water temperatures of 21 to 23 °C, hatching typically occurs within 6 to 9 days, but may take up to 13 days (Becker 1983; Blumer 1985). Fry remain in the nest for 7 to 10 days after hatching, and then form tight schools guarded by adults from potential predators (Blumer 1982; Becker 1983). Parental care continues until fry reach a total length of approximately 50 mm (Etnier and Starnes 1993).

Age and Growth: There is little information on brown bullhead growth. Newly hatched larvae total length ranges from 4 to 8 mm (Mansueti and Hardy 1967). Growth rates are highly variable among populations and are dependent on environmental conditions. Maximum growth rates occur at water temperatures of 20 to 25 °C (Hartman 2017). Brown bullhead can reach a total length of 532 mm and live for 11 years (Carlander 1969).

Food and Feeding: Brown bullhead are opportunistic omnivores who use the tastes buds on their barbels and elsewhere on their body to locate prey. Fry and young-of-year brown bullhead primarily eat zooplankton and chironomids (Carlander 1969). Adults eat benthic invertebrates, small fish, aquatic vegetation, and mollusks. Feeding occurs both day and night, primarily along the bottom. Brown bullhead metabolic rates increase with water temperature (Keast 1984). Food consumption increases with increasing temperature from 10 to 25 °C, begins declining at 30 °C, and ceases at 35 °C (Hartman 2017).

References

Becker GC (1983) Fishes of Wisconsin. University of Wisconsin Press, Madison

Blumer LS (1982) Parental care and reproductive ecology of the North American catfish, *Ictalurus nebulosus*. Dissertation, University of Michigan

Blumer LS (1985) Reproductive natural history of the brown bullhead, *Ictalurus nebulosus* in Michigan. Am Midl Nat 114:318–330. https://doi.org/10.2307/2425607

Breder CM Jr, Rosen DE (1966) Modes of reproduction in fishes. Natural History Press, Garden City

Brett JR (1944) Some lethal temperature relations of Algonquin Park fishes. University of Toronto Studies, Biol Ser No, p 52

Carlander KD (1969) Handbook of freshwater fishery biology. Iowa State University Press, Ames

Crawshaw LI, Hammel HT (1974) Behavioral regulation of internal temperature in the brown bullhead, *Ictalurus nebulosus*. Comp Bioch Physiol, Part A: Physiol 47:51–60. https://doi.org/10.1016/0300-9629(74)90050-4

Etnier DA, Starnes WC (1993) The fishes of Tennessee. University of Tennessee Press, Knoxville

Hartman KJ (2017) Bioenergetics of brown bullhead in a changing climate. Trans Am Fish Soc 146:634–644. https://doi.org/10.1080/00028487.2017.1293563

Keast A (1984) Growth responses of the brown bullhead (*Ictalurus nebulosus*) to temperature. Can J Zool 63:1510–1515. https://doi.org/10.1080/00028487.2017.1293563

Mansueti AJ, Hardy JD (1967) Development of fishes of the Chesapeake Bay region. University of Maryland, Baltimore

6.15.4   Blue Catfish *Ictalurus furcatus*

(Valenciennes, 1840)

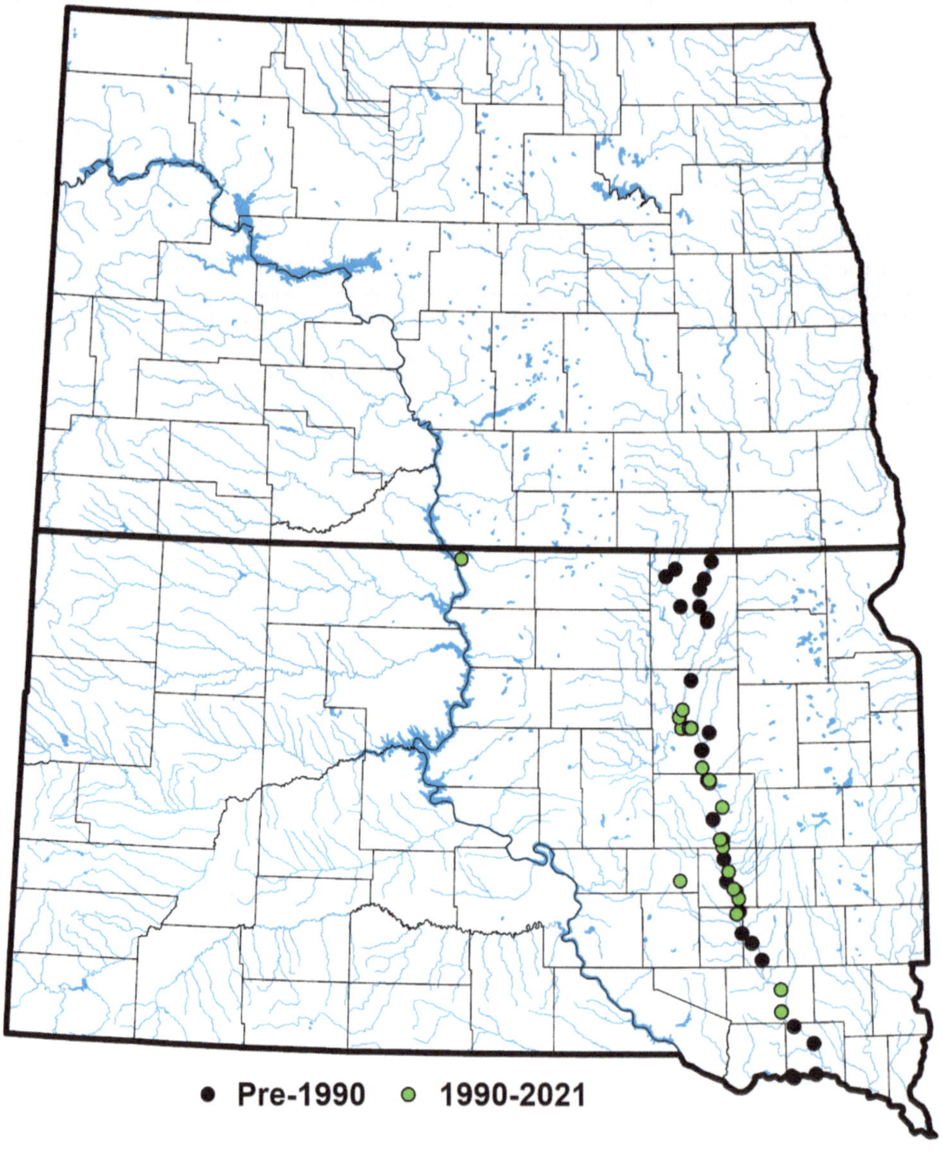

● **Pre-1990**   ● **1990–2021**

Etymology: *Ictalurus furcatus* – *Ictalurus* = fishcat; *furcatus* = forked, referencing the tail.

Description:

Body – deep; robust.
Color –

> *Dorsally* – blue-silver; larger fish gray.
> *Laterally* – blue-silver; larger fish blue.
> *Ventrally* – white.
> *Barbels* – white.

Head – broad, wedge-shaped; virtually straight profile from snout tip to anterior part of dorsal fin.
Mouth – upper jaw extends further than lower jaw.
Teeth – on inner margin of upper jaw.
Swim bladder – 2 chambers.
Barbels – 2 sets; 2 chin pairs.
Dorsal fin – 6 rays; 1 spine.
Adipose fin – present.
Caudal peduncle – concise.
Caudal fin – deeply forked, separate from adipose fin.
Anal fin – 30 to 35 rays; long, nearly straight.
Pelvic fins – 8 rays.
Pectoral fins – 9 rays; 1 spine.
Lateral line – complete, scaless.
Juveniles – silver-white.
Sexual dimorphism – male genital orifice circular, prominent papillae; female genital orifice slit-like, less-prominent papillae.

Similar Species: Blue catfish appear like larger channel catfish that develop a bluish hue and lose spots during the breeding season. Channel catfish have a rounded anal fin, a rounder profile from snout tip to the anterior part of the dorsal fin, and a single chamber swim bladder.

Distribution and Habitat: Native to the Mississippi River, Missouri River, and Ohio River basins, blue catfish are also found in Gulf Coast streams from Alabama to Mexico and northern Central America (Glodek 1980; Graham 1999). They are not found in North Dakota (Graham 1999). Blue catfish are native to South Dakota with a range restricted to the Missouri River and James River basins (Bailey and Allum 1962). Blue catfish are primarily a big-river species. They occur in large rivers with deep, swift channels or flowing pools, with normal turbidity, and a mud or silt substrate (Graham 1999). They are also found in open waters of large reservoirs, with large individuals often residing in dam tailwaters (Mettee et al. 1996; Graham 1999). Blue catfish can withstand moderately high turbidity (Graham 1999). They will move into tributaries and backwaters during high water periods (Hrabik et al. 2015). Blue catfish are the most migratory of the catfish family, migrating several hundred kilometers upstream in the spring and downstream in the fall in response to water temperature changes (Pflieger 1997; Graham 1999).

Reproduction: Spawn timing and reproductive habits of blue catfish are relatively unknown, but likely similar to channel catfish (Lagler 1961; Harlan et al. 1987; Pflieger 1997). Blue catfish spawn in late spring or early summer in protected areas with minimal current. Larger fish often migrate both upstream and downstream to seek spawning sites (Graham 1999). Both sexes mature earlier in the southern portion of their range (Graham 1999). In Lake of the Ozarks, Missouri, blue catfish became sexually mature at total lengths ranging from 42 to 48 cm and age-6 to age-7 (Graham and DeiSanti 1999). Egg diameter is approximately 3 mm (Graham 1999). Hatching occurs in 7 or 8 days in water temperatures of 21 to 24 °C (Henderson 1972; Pflieger 1997; Graham 1999). Like other ictalurid catfish species, males guard eggs and fry.

Age and Growth: Blue catfish are the largest catfish in the United States. They grow rapidly, especially after becoming piscivorous (Graham 1999). Growth is similar for both sexes. Adult blue catfish in the Midwest are typically longer than 122 cm and can reach a length of 165 cm and weigh 57 kg (Kansas Fishes Committee 2014). Blue catfish grow faster in southern regions because of a longer growing season, warmer water temperatures, and a more diverse forage base (Graham 1999). Blue catfish can live up to 25 years (Hale and Timmons 1990; Graham 1999).

Food and Feeding: Blue catfish are opportunistic omnivores. Smaller blue catfish eat mostly small invertebrates and some small fish, while larger fish eat mostly fish and larger invertebrates (Pflieger 1997). Blue catfish also eat fingernail clams and freshwater mussels (Graham 1999). They feed along the bottom substrate using taste and smell more than sight (Pflieger 1997; Graham 1999).

References

Bailey RM, Allum MO (1962) Fishes of South Dakota. University of Michigan, Ann Arbor. https://doi.org/10.3998/mpub.9690435

Glodek GS (1980) *Ictalurus furcatus* (LeSueur) blue catfish. In: Lee DS, Gilbert CR, Hocutt CH, Jenkins RE, McAllister DE, Stauffer JR (eds) Atlas of North American freshwater fishes. North Carolina Biological Survey Publication 1980-12, Raleigh, p 439

Graham K (1999) A review of the biology and management of blue catfish. In: Irwin ER, Hubert WA, Rabeni CF, Schramm HL Jr, Coon T (eds) Catfish 2000: proceedings of the international ictalurid symposium, Am Fish Soc Symp 24, pp 37–49

Graham K, DeiSanti K (1999) The population and fishery of blue catfish and channel catfish in the Harry S Truman Dam tailwater, Missouri. In: Irwin ER, Hubert WA, Rabeni CF, Schramm HL Jr, Coon T (eds) Catfish 2000: proceedings of the international ictalurid symposium, Am Fish Soc Symp 24, pp 361–376

Hale RS, Timmons TJ (1990) Growth of blue catfish in the lacustrine and riverine areas of the Tennessee portion of Kentucky Lake. J Tenn Acad Sci 65:86–90

Harlan JR, Speaker EB, Mayhew J (1987) Iowa fish and fishing, 5th edn. Iowa Dep Nat Res, Des Moines

Henderson GG (1972) Rio Grande blue catfish study. Texas Park Wildl Dep, Fed Aid Fish Restor Proj F-18-R-7, Job 11, Prog Rep, Austin

Hrabik RA, Schainost SC, Stasiak RH, Peters EJ (2015) The fishes of Nebraska. University of Nebraska, Lincoln

Kansas Fishes Committee (2014) Kansas fishes. University Press of Kansas, Lawrence

Lagler KF (1961) Freshwater fishery biology. William C Brown Co, Dubuque

Mettee MF, O'Neil PE, Pierson JM (1996) Fishes of Alabama and the mobile basin. Oxmoor House Inc, Birmingham

Pflieger WL (1997) The fishes of Missouri, revised edn. Missouri Dep Cons, Jefferson City

6.15.5 Channel Catfish *Ictalurus punctatus*

(Rafinesque, 1818)

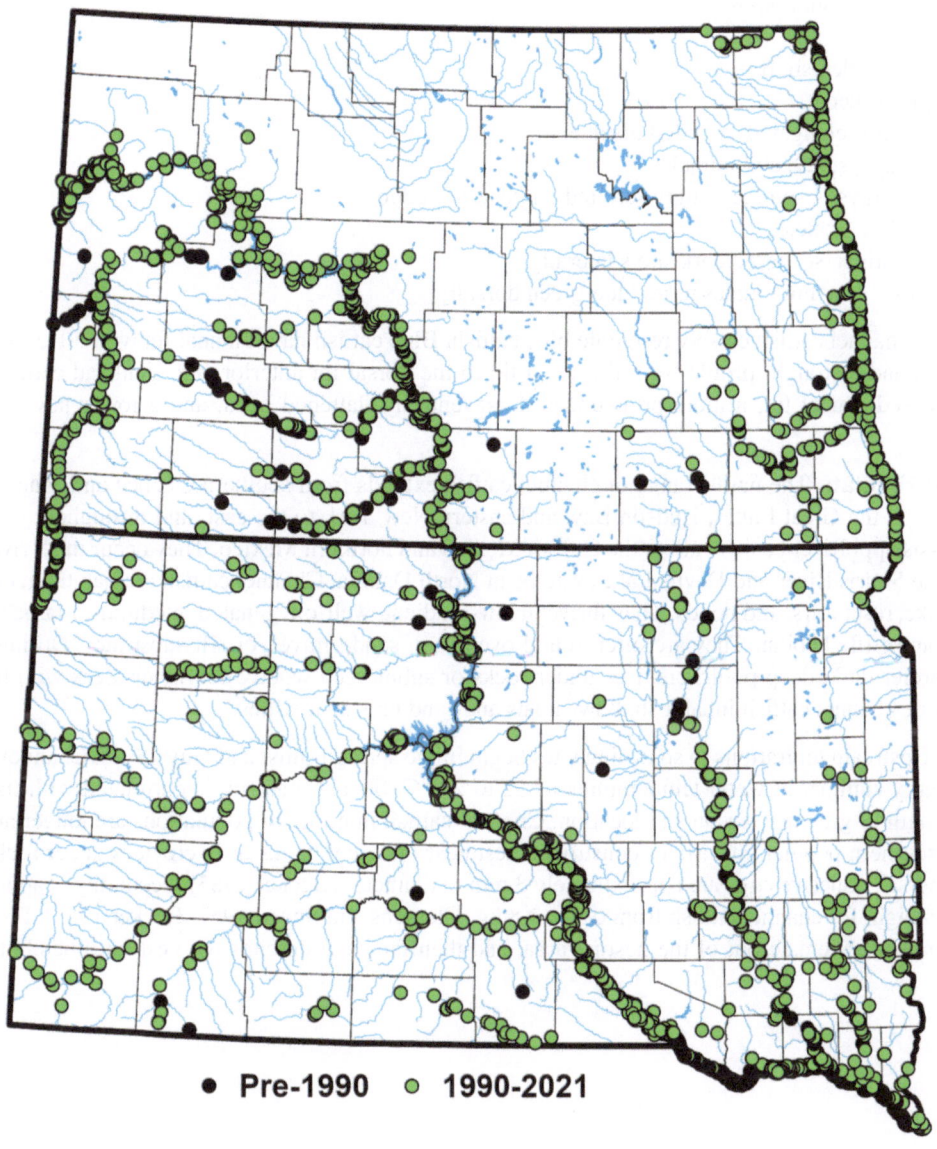

● **Pre-1990** ○ **1990-2021**

Etymology: *Ictalurus punctatus* – *Ictalurus* = fish cat, referencing barbels resembling cat whiskers; *punctatus* = spotted, referencing lateral spots on juveniles and young adults.

Description:

Body – slender, elongated, laterally compressed posteriorly.
Color –

> *Dorsally* – dusky blue or pale to slate olive-gray.
> *Laterally* – slate to light gray.
> *Ventrally* – white.
> *Barbels* – upper jaw – dark brown; ventral – lightly pigmented.
> *Fins* – lightly pigmented.

Head – small; slightly dorsoventrally flattened; profile slightly rounded from snout tip to dorsal fin anterior end.
Snout – blunt.
Eyes – small, positioned dorsolaterally on head.
Mouth – wide, subterminal.

> *Jaws* – upper jaw extends past lower jaw.

Teeth – small, sharp; premaxillary tooth patch on upper jaw without backward extensions.
Barbels – 4 pairs; one pair top of snout, one long pair near corners of upper jaw, two chin pairs (inner pair shorter than outer).
Gill rakers – 14–18.
Dorsal fin – 6 or 7 rays; 1 short spine.
Adipose fin – flap-like.
Caudal peduncle – slender, short.
Caudal fin – deeply forked.
Anal fin – 24 to 29 rays; elongated, rounded distal end.
Pelvic fins – 8 or 9 rays; small, abdominal.
Pectoral fins – 7 to 9 rays; 1 spine, strongly serrated on posterior edge.
Lateral line – complete, scaleless.
Juveniles – small, distinct, scattered dark spots laterally.
Sexual dimorphism – spawning males more blue-green dorsally.

Similar Species: Channel catfish closely resemble blue catfish. Blue catfish have an anal fin with 30–35 rays with a nearly straight distal end, and a straight profile from the snout tip to the dorsal fin anterior end. Flathead catfish have an almost square slightly forked caudal fin, a more pronounced dorsoventrally flattened head, and a lower jaw extending past the upper jaw.

Distribution and Habitat: The native range of channel catfish extends from southern Quebec and Manitoba in the north, Montana in the west, the Great Lakes, Hudson Bay, and eastern New York to the east, and throughout the Missouri River drainage and Mississippi River basin south to the Gulf of Mexico and northern Mexico. They occur in all river systems in the Dakotas except the Souris River and Devils Lake systems in North Dakota. Channel catfish are habitat generalists, living in rivers, streams, lake, reservoirs, and ponds. They thrive in low-gradient, well-oxygenated, medium-to-large turbid rivers with alternating pool and riffle habitat. They are often found over mud, sand, gravel, or silt substrates. During daylight hours, channel catfish prefer dark, deep pools, crevices under rocks or submerged woody debris, or areas with moderate aquatic vegetation. Juvenile channel catfish inhabit shallow shoals and sand bars near riffles.

Reproduction: Upstream migrations to spawning sites begin in the spring (Quist and Guy 1998). Channel catfish spawn in late spring and early summer at water temperatures of 22 to 28 °C. Sexual maturity is correlated with fish size, which is determined by length of growing season and location. Channel catfish in northern populations are longer at sexual maturity than those from southern populations. Males construct a nest near undercut banks or in crevices under rocks and trees. The light yellow, demersal, adhesive eggs are approximately 3 mm in diameter (Becker 1983). Fecundity is approximately 8,800 eggs/kg of body weight for channel catfish from 0.5 to 1.8 kg (Clemens and Sneed 1957; Becker 1983). Following fertilization, the male chases the female out of the nesting area, and then fans and defends the nest until hatching in 7 or 8 days (Pflieger 1997).

Age and Growth: Newly hatched channel catfish larvae have a total length of approximately 6 mm (Becker 1983). Growth rates vary among populations and may be influenced by localized environmental conditions, climate, and resource availability (Kirby 2001; Pegg and Pierce 2001; Stevens 2013). Low-flow conditions increase shallow, warm, and low velocity habitat that facilitate young channel catfish growth (Hogberg et al. 2016). Total-lengths-at-age of channel catfish from the upper segment of the Big Sioux River were age-2, 216 mm; age-3, 307 mm; age-4, 391 mm; age-5, 456 mm; age-6, 495 mm; age-7, 623 mm; age-8, 631 mm; age-9, 678 mm; age-10, 672 mm (Kirby 2001). Channel catfish can exceed total lengths of 1,000 mm (Siddons and Pegg 2016). While average longevity is 10 to 15 years, they can live as long as 27 years (Siddons and Pegg 2016).

Food and Feeding: Adult channel catfish are generalist, opportunistic omnivores which allows them to adjust to changing conditions within and among years (Stevens 2013). They migrate to shallow waters near rifles at night to feed. Larvae primarily eat zooplankton. Juveniles consume macroinvertebrates, with the size of macroinvertebrates eaten increasing with increasing juvenile size (Bailey and Harrison 1948; Stevens 2013). Juvenile channel catfish undergo an ontogenetic diet shift to piscivory at a total length of approximately 280 mm (Hill et al. 1995; Stevens 2013). However, benthic macroinvertebrates remain in the diet of individuals of all sizes (Hill et al. 1995; Stevens 2013). Invertebrate availability influences channel catfish growth and condition (Hampton and Berry Jr 1997; Quist and Guy 1998). Adults primarily eat small fish, but also consume dipterans, annelids, crayfish, mollusks, immature ephemeroptera, immature trichoptera, nematodes, filamentous algae, birds, and rodents (Stevens 2013; Peterson 2017).

References

Bailey RM, Harrison HM (1948) Food habits of the southern channel catfish (*Ictalurus Lacustris Punctatus*) in the Des Moines River, Iowa. Trans Am Fish Soc 75:110–138. https://doi.org/10.1577/1548-8659(1945)75[110:FHOTSC]2.0.CO;2

Becker GC (1983) Fishes of Wisconsin. University of Wisconsin Press, Madison

Clemens HP, Sneed KE (1957) The spawning behavior of the channel catfish *Ictalurus punctatus*. US Dep Inter, Fish Wildl Serv, Special Sci Rep, Fish No 219

Hampton DR, Berry CR Jr (1997) Fishes of mainstem Cheyenne River in western South Dakota. Pro S D Acad Sci 76:11–25

Hill TD, Duffy WG, Thompson MR (1995) Food habits of channel catfish in Lake Oahe, South Dakota. J Freshw Ecol 10:319–323. https://doi.org/10.1080/02705060.1995.9663454

Hogberg NP, Hamel MJ, Pegg MA (2016) Age-0 channel catfish *Ictalurus punctatus* growth related to environmental conditions in the channelized Missouri River, Nebraska. River Res Appl 32:744–752. https://doi.org/10.1002/rra.2890

Kirby DJ (2001) An assessment of the channel catfish population in the Big Sioux River, South Dakota. Dissertation, South Dakota State University

Pegg MA, Pierce CL (2001) Growth rate responses of Missouri and lower Yellowstone River fishes to a longitudinal gradient. J Fish Biol 59:1529–1543. https://doi.org/10.1111/j.1095-8649.2001.tb00218.x

Peterson E (2017) Invertebrate prey selectivity of channel catfish (*Ictalurus punctatus*) in western South Dakota prairie streams. Dissertation, South Dakota State University

Pflieger WL (1997) The fishes of Missouri, revised edn. Missouri Dep Cons, Jefferson City

Quist MC, Guy CS (1998) Population characteristics of channel catfish from the Kansas River, Kansas. J Freshw Ecol 13:351–359. https://doi.org/10.1080/02705060.1998.9663628

Siddons SF, Pegg MA (2016) Age, growth, and mortality of a trophy channel catfish population in Manitoba, Canada. N Am J Fish Manag 36:1368–1374. https://doi.org/10.1080/02755947.2016.1224783

Stevens TM (2013) Feeding ecology and factors influencing growth and recruitment of channel catfish in South Dakota reservoirs. South Dakota State University, Dissertation

6.15.6   Stonecat *Noturus flavus*

Rafinesque, 1818

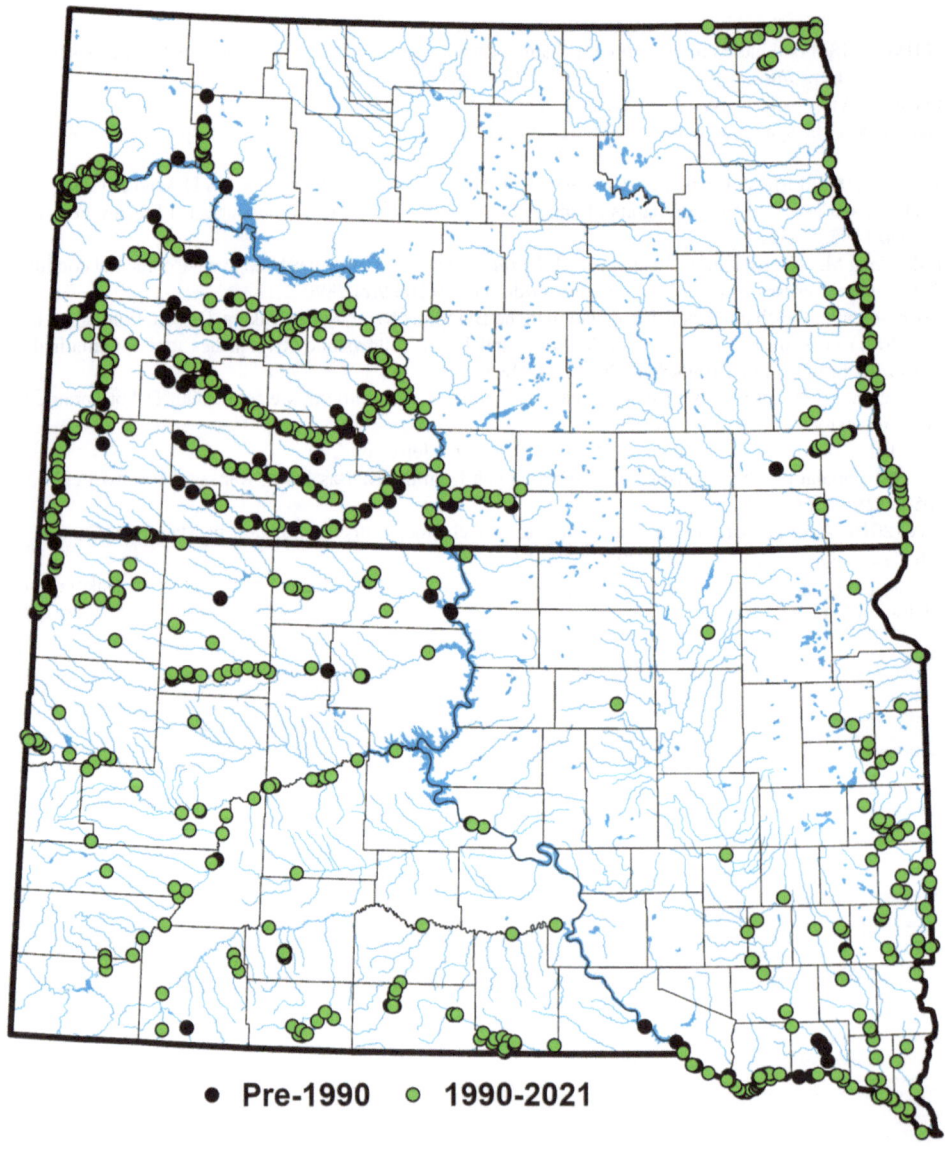

Etymology: *Noturus flavus* – *Noturus* = back tail, referencing the fused adipose and caudal fins; *flavus* = yellow.

Description:

Body – elongated; cylindrical anteriorly; laterally compressed posteriorly
Color –

  Dorsally – tan to dark brown.
  Laterally – brown to tan.
  Ventrally – cream to yellow.
  Fins – dorsal – cream-white spot at rear; ventral – dark base, fading to pale yellow on distal ends; caudal – often horizontal stripe at base

Head – broad, slightly dorsoventrally compressed.
Snout – bluntly pointed.
Eyes – small, positioned on dorsal side of head.
Mouth – subterminal, small, wide.

  Jaws – upper jaw protrudes past lower jaw.
  Lips – fleshy; upper lip overhangs lower.

Teeth – small, sharp, numerous on upper and lower jaws; tooth patch on upper jaw.
Barbels – 8; on head surrounding mouth and posterior nostrils.
Gill rakers – 6 to 7.
Dorsal fin – 6 or 7 rays; 1 short venomous spine.
Adipose fin – long, short, fused with caudal fin and separated by a shallow notch.
Caudal fin – square-shaped, nearly straight distal end.
Anal fin – 15 to 18 rays.
Pelvic fins – 8 to 10 rays; insertions posterior to dorsal fin insertion.
Pectoral fins – 10 rays; 1 short, venomous spine, serrae on anterior edge, smooth on posterior edge.
Lateral line – incomplete, scaleless.

Similar Species: Stonecat closely resemble tadpole madtom. Tadpole madtom also have a fused adipose and caudal fin, but the caudal fin has a more rounded distal end. Tadpole madtom have jaws that are nearly equal in length and eyes laterally on the head. They lack a dorsal spot. Tadpole madtom also prefer stagnant water, whereas stonecat prefer rivers. Stonecat can be differentiated from other ictalurids in the region by the fused adipose and caudal fins.

Distribution and Habitat: The native range of stonecat extends from Montana and Wyoming eastward through the Missouri River, upper Mississippi River, St. Lawrence River, Great Lakes, Ohio River systems to New York, and south to Tennessee and Oklahoma (Scott and Crossman 1973; Jenkins and Burkhead 1993; Pollard 2004). In the Dakotas, it occurs in the Missouri River western tributaries, as well as the Red River of North, James River, Vermillion River, and Big Sioux River systems. Stonecat inhabit a wide variety of fluvial cool and warm-water systems, including medium-to-large streams and lakes with clear to slightly turbid water and sand, pebble, gravel, or rocky substrates (Trautman 1981; Walsh and Burr 1985; Brewer and Rabeni 2008). They are often associated with shallow, 0.03 to 0.5 m deep, stream shorelines with moderate water velocities of 0.0 to 0.7 m/second in run and riffle sequences (Banks and DiStefano 2002; Brewer et al. 2006). Stonecat generally avoid areas up to 5 m away from the nearest shoreline (Brewer and Rabeni 2008). Stonecat use shorelines to feed on drifting aquatic invertebrates that become available with increased flows (Brewer et al. 2006). Stonecat are nocturnal, selecting the shallowest water at night, and likely use crevices between coarse substrates as cover during the day (Carlson 1966; Burr and Stoeckel 1999; Brewer et al. 2006; Brewer and Rabeni 2008). They are susceptible to dewatering and avoid intermittent and high-gradient streams (Carlson 1966; Burr and Stoeckel 1999; Brewer et al. 2006; Brewer and Rabeni 2008). Similar to adults, juveniles are often associated with coarse substrates. However, they tend to avoid shoreline areas and use the midchannel to migrate to shallower habitats (Brewer and Rabeni 2008).

Reproduction: Stonecat spawn from May to July when water temperatures reach 17 °C (Walsh and Burr 1985; Brewer et al. 2006). Spawning stops when water temperatures are consistently 25 to 27 °C (Walsh and Burr 1985; Brewer et al. 2006). They likely move from shallow, warm, shorelines areas to deeper waters for spawning (Brewer et al. 2006). Warmer water temperatures near shorelines may increase metabolism and growth and subsequently enhance reproduction (Brewer et al. 2006). Females sexually mature at age-3 or age-4 at an average standard length of 119 mm (Walsh and Burr 1985). Males likely become sexually mature at 87 mm (Walsh and Burr 1985). Male stonecat construct nests in pools, or between crevices under rocks, with aquatic vegetation, at depths of 53 to 117 cm by moving small particles with their mouths (Cochran 1996; Walsh and Burr 1985). Fecundity increases with female size, and ranges from 189 to 570 eggs, with approximately 104 to 306 eggs per clutch (Walsh and Burr 1985). The demersal, opaque-yellow, approximately 3 to 4 mm diameter eggs are held together by an adhesive jelly-like mixture (Becker 1983; Walsh and Burr 1985; Simon and Burr 2004). Males guard and defend the nests until larvae can swim away after reaching a total lengths of 12 to 15 mm (Burr and Stoeckel 1999).

Age and Growth: Newly hatched stonecat larvae have total lengths of 7 to 8 mm (Walsh and Burr 1985). Growth is fastest during the first year (Walsh and Burr 1985). Total-lengths-at-age of stonecat from South Dakota were age-1, 79 mm; age-2, 97 mm; age-3, 114 mm; age-4, 138 mm; age-5, 155 mm; age-6, 176 mm; age-7, 193 mm (Carlson 1966). Stonecat in lentic environments typically live longer and achieve greater lengths than those in lotic environments (Gilbert 1953; Carlson 1966; Scott and Crossman 1973). They can reach a total length of 312 mm and live up to 9 years (Gilbert 1953; Scott and Crossman 1973).

Food and Feeding: Stonecat are nocturnal, bottom dwelling, opportunistic predators. They use their barbels and senses of smell and taste when foraging along the bottom substrate (Pflieger 1997). Stonecat eat a diversity of prey, with some shift relative to body size (Walsh and Burr 1985; Pollard 2004). Smaller stonecat eat large numbers of aquatic insect larvae, including mayfly, stonefly, and caddisfly. Adults eat mollusks, minnows, crayfish, and aquatic vegetation, in addition to smaller quantities of aquatic insect larvae (Scott and Crossman 1973; Walsh and Burr 1985; Pollard 2004).

References

Banks SM, DiStefano RJ (2002) Diurnal habitat association of the madtoms, *Noturus albater, N. exilis, N. flavater* and *N. flavus* in Missouri Ozarks streams. Am Midl Nat 148:138–145. https://doi.org/10.1674/0003-0031(2002)148[0138:DHAOTM]2.0.CO;2

Becker GC (1983) Fishes of Wisconsin. University of Wisconsin Press, Madison

Brewer SK, Rabeni CF (2008) Seasonal and diel habitat shifts by juvenile ictalurids in a flow-regulated prairie river. Am Midl Nat 159:42–54. https://doi.org/10.1674/0003-0031(2008)159[42:SADHSB]2.0.CO;2

Brewer SK, Papoulias DM, Rabeni CF (2006) Spawning habitat associations and selection by fishes in a flow-regulated prairie river. Trans Am Fish Soc 135:763–778. https://doi.org/10.1577/T05-021.1

Burr BM, Stoeckel JN (1999) The natural history of madtoms (genus Noturus), North American's diminutive catfishes. In: Irwin ER, Hubert WA, Rabeni CF, Schramm HL Jr, Coon T (eds) Catfish 2000: proceedings of the international ictalurid symposium, Am Fish Soc Symp 24, pp 51–101

Carlson DR (1966) Age and growth of the stonecat, *Noturus flavus* Rafinesque, in the Vermillion River. Proc S D Acad Sci 45:131–137

Carlson T, Schall BJ, Lucchesi D (2021) Biological characteristics and seasonal use of blue sucker in the James River, South Dakota. South Dakota Game, Fish and Parks Interim Report. State Wildlife Grant T-86-R-1. Pierre, SD

Cochran PA (1996) Cavity enhancement by madtoms (genus Noturus). J Freshw Ecol 11:521–522. https://doi.org/10.1080/02705060.1996.9664480

Gilbert CR (1953) Age and growth of the yellow stone catfish, *Noturus flavus* (Rafinesque). Dissertation, Ohio State University

Jenkins RE, Burkhead NM (1993) Freshwater fishes of Virginia. American Fisheries Society, Bethesda

Pflieger WL (1997) The fishes of Missouri, revised edn. Missouri Dep Cons, Jefferson City

Pollard SM (2004) Status of the stonecat (*Noturus flavus*) in Alberta. Alberta Sustain Res Devel, Fish Wildl Divi, Alberta Conserv Assoc Wildl Status Rep No 52, Edmonton

Scott WB, Crossman EJ (1973) Freshwater fishes of Canada. Fisheries Research Board of Canada, Bulletin 184

Simon TP, Burr BM (2004) Description of developmental stages of the stonecat, *Noturus flavus* and the slender madtom, *Noturus exilis* (siluriformes: ictaluridae). Proc Indiana Acad Sci 113:123–132

Trautman MB (1981) The fishes of Ohio, revised edn. Ohio State University Press, Columbus

Walsh SJ, Burr BM (1985) Biology of the stonecat, *Noturus flavus* (Siluriformes, Ictaluridae), in central Illinois and Mississippi streams, and comparison with Great Lakes populations and congeners. Ohio J Sci 85:85–96

6.15.7   Tadpole Madtom *Noturus gyrinus*

(Mitchill, 1817)

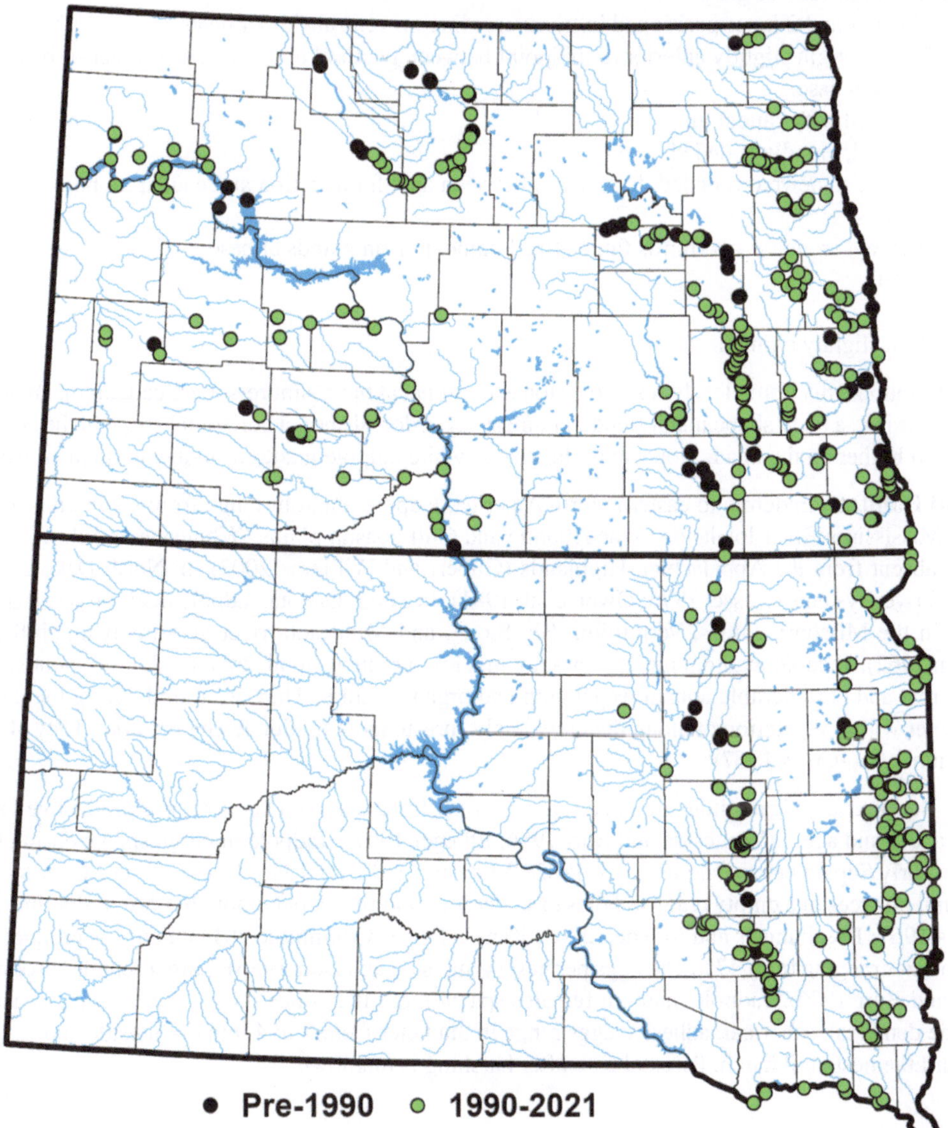

Etymology: *Noturus gyrinus* – *Noturus* = back tail, referencing the fused adipose and caudal fins; ***gyrinus*** = tadpole, referencing the body shape like a tadpole.

Description:

Body – stocky, rounded anteriorly, compressed posteriorly; posterior end angled downward giving curved or bent appearance.
Color –

>   *Dorsal* – caudal peduncle dark brown to olive gray; thin skin sometimes transparent.
>   *Laterally* – dark brown to olive gray.
>   *Ventrally* – light yellow-brown to tan; narrow mid-lateral stripe extending to caudal fin.
>   *Barbels* – dark colored.
>   *Fins* – dark brown; paired fins light-to-medium tan.

Head – wide; somewhat depressed dorsally.
Eyes – small, positioned lower on head.
Mouth – wide.

>   *Jaws* – about equal in length.
>   *Lips* – fleshy.

Teeth – small, sharp, on inside margin jaws.
Barbels – 2 pairs; 2 outer and 2 inner positioned in transverse line on ventral side of head.
Dorsal fin – 6 to 7 rays; origin slightly anterior to midpoint between pectoral and pelvic fins, anterior base swollen; 1 short spine, poison glands at base.
Adipose fin – short, fused with caudal fin.
Caudal fin – homocercal, broadly rounded.
Anal fin – 13 to 18 rays; somewhat rounded, little space between anal fin insertion and caudal fin base.
Pelvic fins – 8 to 10 rays.
Pectoral fins – 7 to 9 rays; roughly two-thirds fin length; 1 spine, poison glands at base.
Lateral line – complete.
Skin – scaleless.
Juveniles – caudal fin slightly pointed.

Similar Species: Tadpole madtom closely resemble stonecat. Stonecat have a more-square caudal fin, an upper jaw extending beyond the lower jaw, a body shape much slenderer anteriorly, a dorsal spot, a horizontal stripe in the center of the caudal fin, and eyes placed higher on the head. Stonecat prefer rivers, while tadpole madtom frequents stagnant waters.

Distribution and Habitat: Endemic to eastern North America, tadpole madtom range from southern Canada to the Great Lakes along the Mississippi River lowlands to the Atlantic and Gulf coastal plains (Whiteside and Burr 1986; Hrabik et al. 2015). They are absent from the Appalachian Highlands (Gilbert and Williams 2002). In North Dakota, tadpole madtom occur in nearly all river systems (Russel 1975; Owen et al. 1981). In South Dakota, tadpole madtom occur in the eastern half of the state within the Missouri, James, Vermillion, Big Sioux, and Minnesota river systems (Churchill and Over 1938; Bailey and Allum 1962). Tadpole madtom prefer slow-moving, low-gradient small streams, sloughs, ponds, and lakes with a silt or mud substrate and considerable aquatic vegetation and organic debris. They are found during the day in moderately clear to turbid water within vegetation or along undercut banks, presumably to hide from predators. They also inhabit empty bottles or aluminum cans (Cross 1967).

Reproduction: Because tadpole madtom are nocturnal and wary, little is known about their reproductive behaviors. Sexual maturity occurs age-1 and age-2 (Whiteside and Burr 1986; Pflieger 1997). Tadpole madtom spawn during the summer and occasionally into early September (Lindquist et al. 1982; Whiteside and Burr 1986; Kansas Fishes Committee 2014). Nests are made in natural crevices like empty crayfish holes or under logs and rocks, but empty bottles or aluminum cans have also been used (Cross 1967; Boschung Jr and Mayden 2004; Kansas Fishes Committee 2014). Females produce from 48 to 323 eggs, with clutch sizes ranging from 47 to 150 eggs per nest (Whiteside and Burr 1986). Larger females produce more mature ova (Whiteside and Burr 1986). Tadpole madtom females spawn more than once during the breeding season (Menzel and Raney 1973). The demersal, spherical, adhesive eggs range in diameter from 3 to 4 mm (Whiteside and Burr 1986). One or both parents guard the nest until hatch. Parental care after hatching is unknown.

Age and Growth: Adult tadpole madtom typically have total lengths ranging from 75 to 102 mm but can be up to 130 mm. Tadpole madtom in Minnesota had total lengths of 15 to 35 mm at age-0; 43 to 85 mm at age-1, and 78 to 104 at age-2 (Hooper 1949; Case 1970). They typically live for 2 to 3 years, with some surviving to age-4.

Food and Feeding: Tadpole madtom feed primarily at night along the substrate or within aquatic vegetation on organic debris, chironomids, and crustaceans, like amphipods, cladocerans, and ostracods. They will also eat small fishes, dragonflies, and worms. Smaller tadpole madtom rely heavily on chironomids, oligiochates, and small fly larvae.

References

Bailey RM, Allum MO (1962) Fishes of South Dakota. University of Michigan, Ann Arbor. https://doi.org/10.3998/mpub.9690435

Boschung HT Jr, Mayden RL (2004) Fishes of Alabama. Smithsonian Books, Washington, DC

Case BE (1970) An ecological study of the tadpole madtom *Noturus gyrinus* (Mitchell), with special reference to movements and population fluctuations. Dissertation, University of Manitoba

Churchill EP, Over WH (1938) Fishes of South Dakota. Brown & Saenger Printers, Sioux Falls

Cross FB (1967) Handbook of fishes in Kansas. University Press of Kansas, Lawrence

Gilbert CR, Williams JD (2002) National Audubon Society field guide to fishes: North America, revised edn. Knopf Doubleday Publishing Group, New York

Hooper FF (1949) Age analysis of a population of the ameurid fish *Schilbeodes mollis* (Hermann). Copeia 1:34–38. https://doi.org/10.2307/1437660

Hrabik RA, Schainost SC, Stasiak RH, Peters EJ (2015) The fishes of Nebraska. University of Nebraska, Lincoln

Kansas Fishes Committee (2014) Kansas fishes. University Press of Kansas, Lawrence

Lindquist DG, Shute PW, Shute JR (1982) Spawning and nest site selection by the broadtail and tadpole madtoms: utilization of experimental spawning cover in Lake Waccamaw, North Carolina. Paper presented at the 62nd annual meeting of American Society of Icthyologists and Herpetologists Annual Meeting, Northern Illinois University, DeKalb

Menzel B, Raney E (1973) Hybrid madtom catfish, *Noturus gyrinus* x *Noturus miurus*, from Cayuga Lake, New York. Am Midl Nat 90:165–176. https://doi.org/10.2307/2424276

Owen JB, Elsen DS, Russell GW (1981) Distribution of fishes in North and South Dakota Basins affected by the Garrison Diversion Unit. University of North Dakota Press, Grand Forks

Pflieger WL (1997) The fishes of Missouri, revised edn. Missouri Dep Cons, Jefferson City

Russel GW (1975) Distribution of fishes in North Dakota drainages affected by the Garrison Diversion Project. Dissertation, University of North Dakota

Whiteside LA, Burr BM (1986) Aspects of the life history of the tadpole madtom, *Noturus gyrinus* (Siluriformes: Ictaluridae), in Southern Illinois. Ohio J Sci 86:153–160

6.15.8   Flathead Catfish *Pylodictis olivaris*

(Rafinesque, 1818)

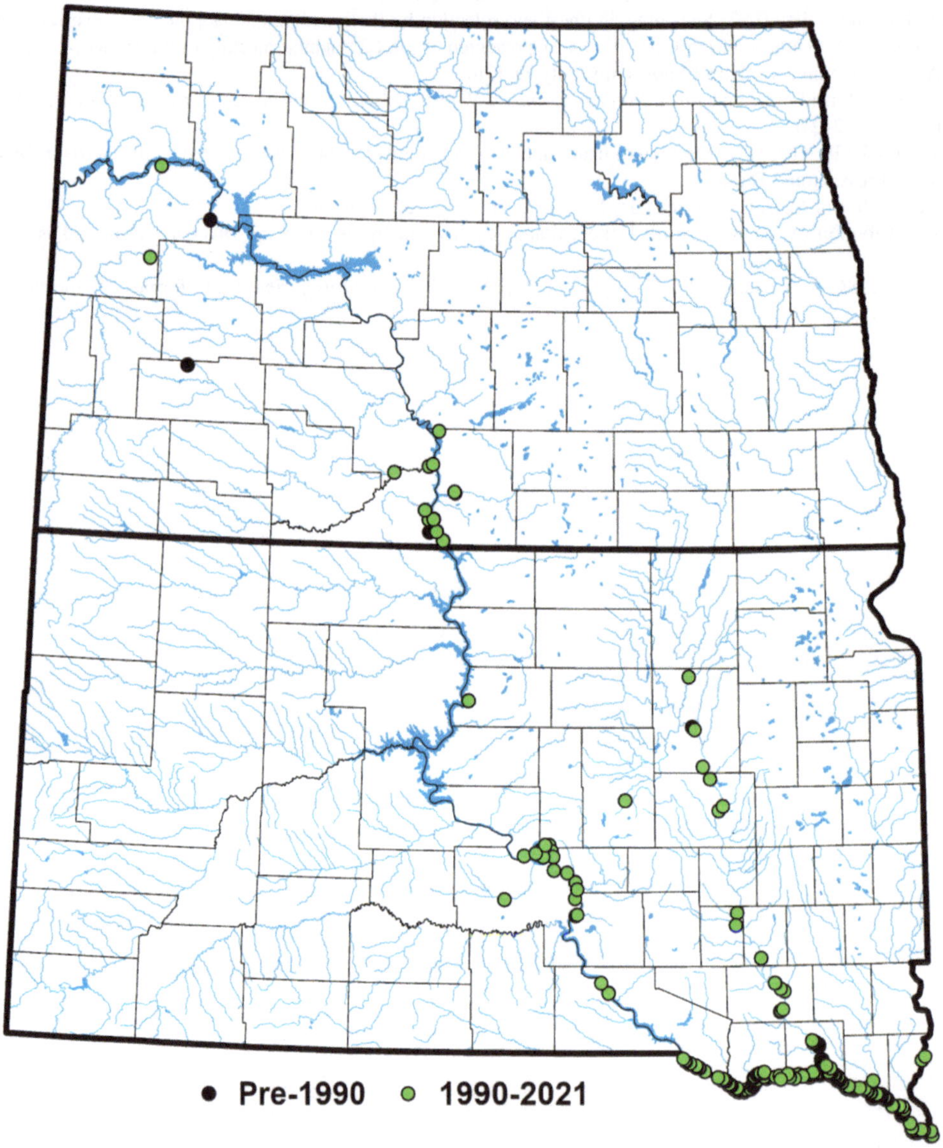

● **Pre-1990**   ● **1990-2021**

Etymology: *Pylodictis olivaris* – *Pylodictis* = mud fish, referencing habitat preference; *olivaris* = olive coloration.

Description:

Body – elongated, moderately dorsoventrally compressed anteriorly.
Color – varies widely with habitat and size.

> *Dorsally* – light brown, olive, or light green-yellow; dark black or brown mottling.
> *Laterally* – light brown, olive, or light green-yellow; dark black or brown mottling.
> *Ventrally* – yellow, tan, or cream.
> *Barbles* – light to dark brown olive.
> *Fins* – dark, dusky; caudal upper lobe – white to cream, mottling and white edge that fades with age.

Head – broad, dorsoventrally flattened.
Snout – broad, blunt.
Eyes – small, positioned dorsally on head.
Mouth – wide, superior.

> *Jaws* – lower jaw extends past upper jaw.

Teeth – small, sharp, on upper and lower jaws; premaxillary tooth patch on upper jaw with backwards extensions.
Barbels – 4 pairs; one pair top of snout; one long pair near corners of upper jaw; two chin pairs, inner pair shorter than outer.
Gill rakers – 9 to 13.
Dorsal fin – 6 rays; 1 short spine.
Adipose fin – elongated, flap like.
Caudal peduncle – slender, short.
Caudal fin – slightly forked, almost square in shape.
Anal fin – 14 to 17 rays; rounded distal end; directly under adipose fin.
Pelvic fins – 8 or 9 rays; small, abdominal.
Pectoral fins – 1 spine, strongly serrated on anterior and posterior edges; over 1/2 fin length.
Lateral line – complete; scaleless.
Juveniles – prominent white or cream upper caudal lobe.

Similar Species: Juveniles and young adult flathead catfish closely resemble bullheads, stonecat, and tadpole madtom. Bullheads have an anal fin with more than 17 rays. Stonecat lack distinct mottling and have a fused adipose and caudal fin. Tadpole madtom also have a fused adipose and caudal fin which is broadly rounded. Channel catfish and blue catfish have deeply forked caudal fins and upper jaws extending past the lower jaw.

Distribution and Habitat: Flathead catfish are native to the southern Great Lakes region, and the Mississippi River, Mobile River, and Rio Grande River drainages. They have been widely introduced outside their native range. Flathead catfish are likely continuously distributed throughout the Missouri River and its major tributaries in North Dakota and South Dakota. They are primarily found in large rivers, but also inhabit creeks, lakes, and reservoirs. They prefer slow water velocities in dark areas with abundant submerged cover, like rocks, riprap, or woody debris (Lucchesi et al. 2017). Flathead catfish over-winter in deeper main channel pools with riprap and reduced flows (Daugherty and Sutton 2005). In Michigan, flathead catfish migrated an average of 1,146 m in the spring to summer habitats from fall to wintering habitats when water temperatures reached 10 °C (Daugherty and Sutton 2005). They are primarily nocturnal with little activity during the day. Young flathead catfish inhabit shallow areas near gravel riffles (Becker 1983).

Reproduction: Flathead catfish spawn in June and July at water temperatures of 22 to 24 °C (Minckley and Deacon 1959; Becker 1983). They become sexually mature at age-3 or age-4 (Becker 1983). Both sexes use their mouths and tails to create nests, which are substrate depressions often near banks or cavities within submerged structures (Becker 1983). The golden-yellow eggs are adhesive (Becker 1983). Fecundity increases with female size, and ranges from 3,520 to 3,783 eggs/kg of body weight (Summerville and Crawley 1970; Colehour 2009). Following fertilization, the male chases the female out of the nesting area, and then continues to fan and defend until hatching in 6 to 9 days (Minckley and Deacon 1959; Becker 1983).

Age and Growth: Flathead catfish are the second largest ictalurid in North America. Larvae are approximately 4 to 11 mm long at hatch. Flathead catfish grow faster in lakes than rivers (Arterburn 2001; Goble 2011; Lucchesi et al. 2017). In rivers, growth rates are highest in the shallow, turbid, lower segments (Minckley and Deacon 1959). Total-lengths-at-age of flathead catfish from the James River were age-1, 83 mm; age-2, 186 mm; age-3, 294 mm; age-4, 385 mm; age-5, 468 mm; age-6, 539 mm; age-7, 599 mm; age-8, 650 mm; age-9, 707 mm; age-10, 755 mm; age-11, 793 mm (Arterburn 2001). They can reach total lengths exceeding 1,000 mm and 50 kg, respectively. Flathead catfish longevity is 28 years, although fish over age-13 are rare in most populations (Kwak et al. 2006).

Food and Feeding: Feeding mostly occurs at night when flathead catfish move from deep to shallow areas near riffles. Adults are primarily piscivorous, opportunistic, apex predators that primarily eat gizzard shad, freshwater drum, yellow perch, and common carp. Juvenile flathead catfish eat primarily invertebrates. In Lake Mitchell, South Dakota, flathead catfish with total lengths under 300 mm mainly consumed crayfish. Flatheads with total lengths from 300 to 600 mm consumed nearly equal amounts of crayfish and smaller prey fish (centrarchids), while large adults with total lengths over 600 mm consumed only prey fish (Lucchesi et al. 2017). A diet shift to piscivory occurs at total lengths of approximately 400 mm for flathead catfish in Lake Mitchell and from 200 to 350 mm in the Missouri River (Minckley and Deacon 1959; Hogberg and Pegg 2016; Lucchesi et al. 2017). The size at which the fish shift diets may depend on prey quantity and availability (Minckley and Deacon 1959; Hogberg and Pegg 2016; Lucchesi et al. 2017).

References

Arterburn JE (2001) Population characteristics and sampling methods of catfish for the James and Big Sioux rivers. Dissertation, South Dakota State University

Becker GC (1983) Fishes of Wisconsin. University of Wisconsin Press, Madison

Colehour JD (2009) Fecundity of flathead catfish and blue catfish from the Mississippi River between Hannibal and Cape Girardeau, Missouri. Dissertation, University of Central Missouri

Daugherty DJ, Sutton TM (2005) Seasonal movement patterns, habitat use, and home range of flathead catfish in the Lower St. Joseph River, Michigan. N Am J Fish Manag 25:256–269. https://doi.org/10.1577/M03-252.2

Goble CW (2011) Ecology and management of channel catfish *Ictalurus punctatus* and flathead catfish *Pylodictis olivaris* populations in the Missouri River, NE. Dissertation, University of Nebraska-Lincoln

Hogberg NP, Pegg MA (2016) Flathead catfish *Pylodictis olivaris* diet composition during extreme flow events in a large river. J Freshw Ecol 31:431–441. https://doi.org/10.1080/02705060.2016.1172523

Kwak TJ, Pine WE III, Waters DS (2006) Age, growth, and mortality of introduced flathead catfish and a review of other populations. N Am J Fish Manag 26:73–87. https://doi.org/10.1577/M04-144.1

Lucchesi DO, Wagner MD, Stevens TM, Graeb BDS (2017) Population dynamics of introduced flathead catfish in Lake Mitchell, South Dakota. J Freshw Ecol 32:323–336. https://doi.org/10.1080/02705060.2017.1292963

Minckley WL, Deacon JE (1959) Biology of flathead catfish in Kansas. Trans Am Fish Soc 88:344–355. https://doi.org/10.1577/1548-8659(1959)88[344:BOTFCI]2.0.CO;2

Summerville VC, Crawley HD (1970) Egg production of flathead catfish. Prog Fish Cult 32:191. https://doi.org/10.1577/1548-8640(1970)32[191:EPOFC]2.0.CO;2

6.16   Esocidae, Pike and Mudminnow Family

The Esocidae family traditionally held only pike and pickerel of the genus *Esox*. Because of recent genetic and paleontological information, it now also contains the fish species formerly in Umbridae, the mudminnow family. Worldwide, four of the five *Esox* species are native to North America, with the circumpolar range of northern pike *Esox lucius* also including Asia and Europe. Amur pike *Esox reichertii* are endemic only in the Amur River basin of China, Mongolia, and Russia. Northern pike, muskellunge *Esox masquinongy*, and redfin pickerel *Esox americanus* occur within the Dakotas.

Species of the genus *Esox* have elongate, cylindrical, torpedo-like bodies and elongate, dorsoventrally flattened, duckbill-like snouts. They also have large, sharp teeth, a complete lateral line, and dorsal and anal fins far posterior on the body. Although *Esox* species are commonly identified by coloration and markings, these vary enough to make differentiating species difficult. *Esox* species can more easily be distinguished by opercle scaling and mandibular pores. For example, the opercle is only scaled on the top half of muskellunge and northern pike but is fully scaled on redfin pickerel. Northern pike have five or six mandibular pores on each side of the jaw, muskellunge have six to nine, and redfin pickerel have four.

Pikes occur within lakes, reservoirs, sloughs, streams, and rivers, preferring little to no flow. They are visual, diurnal, ambush predators, and although primarily piscivorous, will opportunistically feed on other prey. Spawning occurs in the spring, with a single female spawning with one or more males. Pikes do not construct nests nor provide parental care. Northern pike and muskellunge can reach large sizes and are sought-after game species in the Dakotas.

The mudminnows are a small group of northern-hemisphere freshwater fishes represented by seven species within three genus *Dallia*, *Novumbra*, and *Umbra*. Of the seven mudminnow species, four are native to North America and three are native to Europe and Northern Asia. The central mudminnow *Umbra limi* is the only species in the Dakotas.

Mudminnows are small-bodied fishes under 200 mm in total length with a rounded caudal fin, no lateral line, and a dorsal fin far posterior on the body. They often occur in isolated, small populations in underutilized habitats with rather harsh conditions. Most mudminnow species perform supplemental aerial respiration, a physiological trait allowing them to gulp air from the surface of waters with low dissolved oxygen levels and even use air pockets under ice cover in the winter months. Mudminnows are also very tolerant of colder water temperatures.

6.16.1 Redfin Pickerel *Esox americanus*

Gmelin, 1789

Etymology: *Esox americanus* – *Esox* = old European vernacular name for pike; *americanus* = America. Commonly referred to as grass pickerel.

Description:

Body – elongated, cylindrical.
Color –

> *Dorsally* – dark green to olive.
> *Laterally* – olive-green; darker green patches forming wavy, oblique vertical bars; dark brown to black bar below eye, slanting slightly posterior.
> *Ventrally* – cream to white.
> *Fins* – generally clear, no markings, slight gold to cream possible.

Head – elongated.
Opercle – fully scaled; cheek fully scaled.
Snout – elongated, flattened dorsoventrally.
Eyes – relatively large, laterally on upper portion of head.
Mouth – terminal, large.

> *Jaws* – upper jaw does not extend past middle of eye; mandibular pores 4 or less per ventral side of the lower jaw; branchiostegal rays 12 or less per side of jaw.

Teeth – on lower jaws; large, canine like.
Gill rakers – sharp.
Dorsal fin – 12 or 13 rays; far posterior on body; nearly rounded distal end.
Caudal peduncle – elongated, thick.
Caudal fin – moderately forked, slightly rounded lobes.
Anal fin – 11 or 12 rays; insertion directly inferior to dorsal fin insertion.
Pelvic fins – 9 to 11 rays; abdominal.
Pectoral fins – 14 or 15 rays; rounded distal ends.
Lateral line – complete; 97 to 110 cycloid scales in series.
Juveniles – distinct light, pale lateral band, that breaks up and blends as fish grows.

Similar Species: Although redfin pickerel are much smaller, they closely resemble northern pike and muskellunge. They can easily be distinguished by a brown to black, dark, vertical bar underneath the eye. Northern pike have an opercle only scaled on the top half, approximately eight horizontal rows of small, light yellow to cream bean-shaped spots on the lateral sides, teeth on both jaws, and five or six mandibular pores on each ventral side of the lower jaw. Muskellunge have an opercle only scaled on the top half, six to nine mandibular pores on each ventral side of the lower jaw, and faint, dark vertical bars or irregular spots on the lateral sides.

Distribution and Habitat: Redfin pickerel are native west of the Appalachian Mountains in the Mississippi River and Great Lakes basins (Crossman 1966). In the Dakotas, they have only been documented in the Missouri River drainage below Lake Francis Case in the southeastern corner of the South Dakota. Redfin pickerel inhabit locations in sloughs, ponds, small streams, lakes, and shallower, sluggish bays of larger water bodies that are clear to slightly turbid, generally under 2-m deep, with little to no-flow, and silt and mud substrates (Kleinert and Mraz 1966; Becker 1983; COSEWIC 2005). They are very reliant on abundant aquatic vegetation (COSEWIC 2005).

Reproduction: Increasing water levels and new vegetative growth in the spring spur redfin pickerel spawning in sloughs, near shorelines, and within bays near vegetation (Kleinert and Mraz 1966; COSEWIC 2005). Both sexes may become sexually mature within the first year of life (Kleinert and Mraz 1966). Redfin pickerel are broadcast spawners (COSEWIC 2005). The demersal, clear to yellowish-amber, approximately 2-mm-diameter eggs adhere to vegetation and hatch in 11 to 15 days in 8 to 9 °C water temperatures (Becker 1983; COSEWIC 2005).

Age and Growth: Redfin pickerel are the smallest members of the genus *Esox* in the Dakotas. Total-lengths-at-age of redfin pickerel from Pony Lake in the Nebraska were 166 mm at age-2, 172 mm at age-3, 178 mm at age-4, and 224 mm at age-5 (Jolley and Willis 2008). However, these growth rates may be relatively slow because of a high abundance of common carp and a reduced amount of submerged aquatic vegetation (Kleinert and Mraz 1966; Carlander 1969; Jolley and Willis 2008). Total-lengths-at-age from multiple lakes in Wisconsin were 145 mm at age-0, 208 mm at age-1, 251 mm at age-2, 287 mm at age-3, and 356 mm at age-4 (Becker 1983). Female redfin pickerel are generally longer than males at each age (Kleinert and Mraz 1966). Redfin pickerel can reach a total length of 381 mm and live up to 5 years (Trautman 1957; Serns and McKight 1977; Becker 1983).

Food and Feeding: Like other esocid species, redfin pickerel are opportunistic ambush predators feeding by sight primarily in the afternoon and evening (Becker 1983). They undergo an ontogenetic diet shift, seeking larger prey items as they grow longer (Kleinert and Mraz 1966; Weinman and Lauer 2007). In Indiana, at total lengths from 57 to 95 mm, redfin pickerel ate mostly fish, at total lengths from 96 to 150 mm ate mostly fish and crayfish, and at total lengths over 150 mm ate mostly crayfish (Weinman and Lauer 2007). In Wisconsin, redfin pickerel at total lengths from 10 to 15 mm primarily consumed cladocerans, copepods, ostracods, and other zooplankton, while at total lengths of 15 to 40 mm, diets included copepods, cladocerans, tendipedid larvae, odonata nymphs, and fish (Kleinert and Mraz 1966). For their size, redfin pickerel can eat relatively large fish. Stomachs from 198-mm and 244-mm total-length redfin pickerel contained a 76-mm yellow perch and a 102-mm bluegill, respectively (Kleinert and Mraz 1966). They are cannibalistic, with smaller individuals found in the stomachs of juvenile redfin pickerel (Crossman 1962; Kleinert and Mraz 1966; COSEWIC 2005). Redfin pickerel also eat coleopterans, ephemeropterans, annelids, and frogs (Weinman and Lauer 2007).

References

Becker GC (1983) Fishes of Wisconsin. University of Wisconsin Press, Madison

Carlander KD (1969) Handbook of freshwater fishery biology. Iowa State University Press, Ames

COSEWIC (Committee on the Status of Endangered Wildlife in Canada) (2005) COSEWIC assessment and status report on the grass pickerel *Esox americanus vermiculatus* in Canada. https://www.canada.ca/en/environment-climate-change/services/species-risk-public-registry/cosewic-assessments-status-reports/grass-pickerel.html. Accessed 24 Apr 2023

Crossman EJ (1962) The grass pickerel *Esox americanus vermiculatus* LeSueur in Canada. Royal Ontario Museum, Life Sciences Contributions, No 55. https://doi.org/10.5962/bhl.title.52172

Crossman EJ (1966) A taxonomic study of *Esox americanus* and its subspecies in eastern North America. Copeia 1966:1–20. https://doi.org/10.2307/1440756

Jolley JC, Willis DW (2008) Characteristics of a grass pickerel (*Esox americanus vermiculatus*) population in Pony Lake, Nebraska. J Freshw Ecol 23:497–499. https://doi.org/10.1080/02705060.2008.9664234

Kleinert SJ, Mraz D (1966) Life history of the grass pickerel (*Esox americanus vermiculatus*) in southeastern Wisconsin. Wis Conserv Dep Tech Bull 37, Madison, Wisconsin

Serns SL, McKight TC (1977) The occurrence of northern pike x grass pickerel hybrids and an exceptionally large grass pickerel in a northern Wisconsin stream. Copeia 1977:780–781. https://doi.org/10.2307/1443191

Trautman MB (1957) The fishes of Ohio. Ohio State University Press, Columbus

Weinman ML, Lauer TE (2007) Diet of grass pickerel (*Esox americanus vermiculafus*) in Indiana streams. J Freshw Ecol 22:451–460. https://doi.org/10.1080/02705060.2007.9664175

6.16.2   Northern Pike *Esox lucius*

Linnaeus, 1758

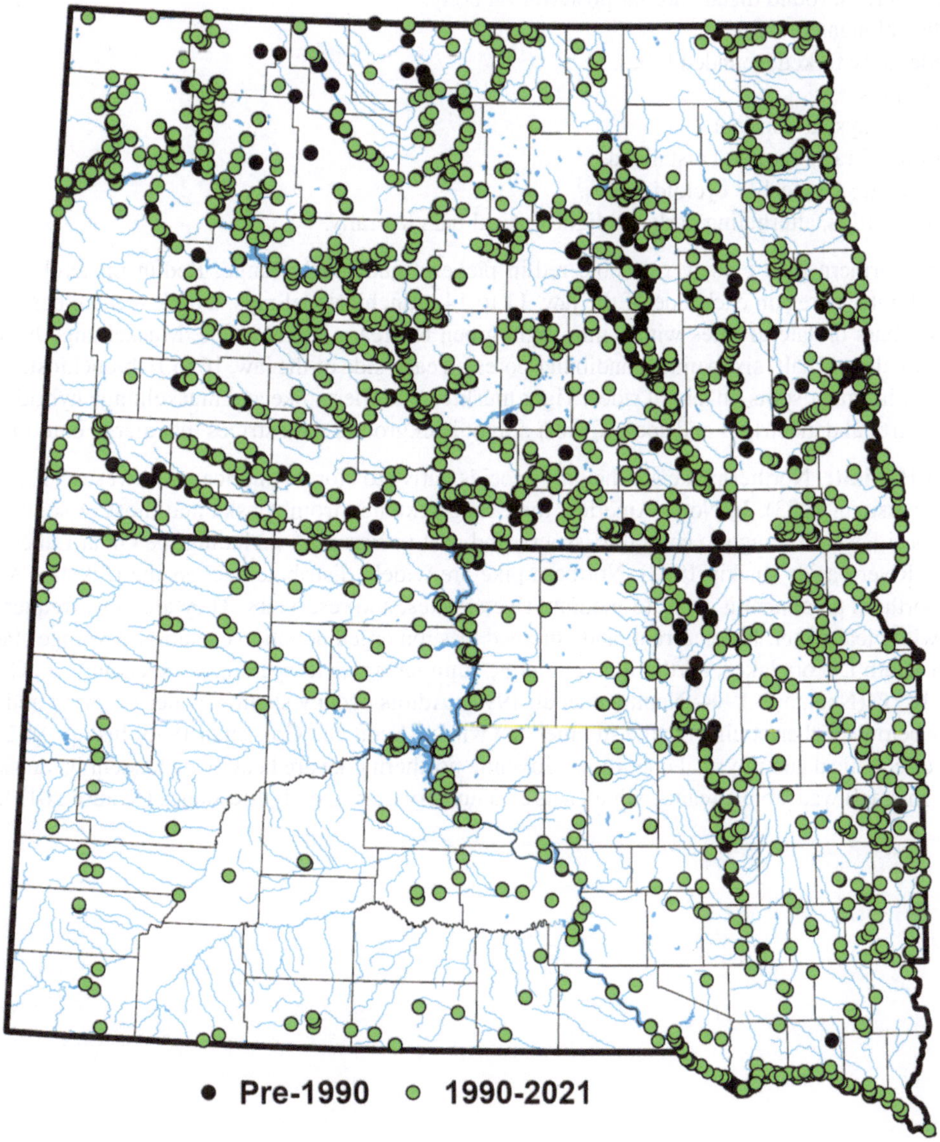

● **Pre-1990**   ● **1990-2021**

Etymology: *Esox lucius* – *Esox* = an old European vernacular name for pike; *lucius* = light.

Description:

Body – elongated, cylindrical.
Color –

> *Dorsally* – dark green to olive.
> *Laterally* – olive; 8 horizontal rows of small, bean-shaped light yellow to cream spots.
> *Ventrally* – cream to white.
> *Fins* – gold to amber, dark spots or blotches.

Head – elongated.
Opercle – scaled on top half only; cheek fully scaled.
Snout (rostrum) – elongated, flattened dorsoventrally.
Eyes – high on head.
Mouth – terminal, large; mandibular pores 5 per ventral side of lower jaw; branchiostegal rays 14 to 16 per jaw.
Teeth – on jaws, roof of mouth, tongue, branchial bones; lower jaw large, sharp teeth; tongue and roof of mouth short, comb-like teeth.
Gill rakers – sharp.
Dorsal fin – 15 to 19 rays; round distal end, far posterior on body.
Caudal peduncle – elongated, thick.
Caudal fin – moderately forked, rounded lobes.
Anal fin – 12 to 15 rays.
Pelvic fins – 10 or 11 rays; abdominal.
Pectoral fins – 14 to 17 rays; rounded distal end.
Lateral line – complete; 105 to 148 cycloid scales.
Juveniles – faintly mottled, alternating dark and light vertical bars laterally.

Similar Species: Northern pike closely resemble redfin pickerel and muskellunge. Redfin pickerel have a fully scaled opercle, four mandibular pores on each side of the jaw, 11 to 14 branchiostegal rays, a dark vertical bar below the eye, and wavy, vertical dark bars on lateral sides with a thin, light green to cream lateral stripe. Muskellunge have an opercle and cheek scaled only on the top half, six to nine mandibular pores on each side of the jaw, 16 to 19 branchiostegal rays, and faint, dark vertical bars or irregular spots on lateral sides. Tiger muskie, a northern pike and muskellunge hybrid, have an irregular pattern of narrow vertical dark stripes and spots against a light background with stripes that merge onto the dorsal side.

Distribution and Habitat: Northern pike are the only esocids native to North America, northwestern Europe, and northern Asia (Scott and Crossman 1973). In North America, their range extends from Alaska and most of southern Canada south through the Missouri River drainage to central Missouri, and east to Vermont, including the Great Lakes, Ohio River, and upper Mississippi River basins (Inskip 1982). Northern pike are widely distributed across the Dakotas. As a mesothermal, cool-water fish, northern pike inhabit small-to-medium rivers, lakes, and reservoirs. They are most frequent in lakes, pools, and backwaters with clear water, little current, and emerged or submerged aquatic vegetation coverage (Paukert and Willis 2003). In eastern South Dakota lakes, northern pike tolerate summer water temperatures exceeding 25 °C, well above their optimum of 8 to 18 °C (Neumann 1994; Neumann et al. 1994). Adults occupy shallow water in spring and fall, and then go deeper, but rarely below the thermocline, in the summer and winter (Scott and Crossman 1973; Inskip 1982). They can tolerate a wide range of chemical and physical conditions. Juvenile northern pike are heavily dependent on dense aquatic vegetation, particularly in small wadeable streams, which acts as a nursery during early life stages (Krause 2013).

Reproduction: Adult northern pike migrate in spring from deeper to shallow areas or calm backwaters to spawn over submerged vegetation in water temperatures from 8 to 12 °C (Inskip 1982). Poor reproductive success occurs without submerged vegetation (Hassler 1970). Males enter spawning grounds first, with one or two escorting a single female just prior to egg release and fertilization. Adults return to deeper waters after spawning. Males become sexually mature at age-1 or age-2, with females maturing at age-2 or age-3. No nest is constructed, and no parental care is given. Fecundity may exceed 100,000 eggs and depends on female size. Demersal, 3-mm diameter, amber eggs adhere to submerged vegetation or substrate and hatch in 10 to 14 days (Inskip 1982). Optimal temperature for successful egg incubation is 6 to 10 °C (Bondarenko et al. 2015).

Age and Growth: Growth is highly variable among and within northern pike populations (Inskip 1982). In South Dakota, optimum growth is in water temperatures from 8 to 18 °C (Neumann 1994; Neumann et al. 1994). Growth is rapid during the first 2 years and females grow faster and are often larger than males. Northern pike can achieve total lengths up to 914 mm (Churchill and Over 1938). They typically live 8 years but can reach a maximum age of 30 years.

Food and Feeding: Northern pike are visual, opportunistic, ambush predators, actively feeding during the day. Adults are primarily piscivorous, but feed opportunistically on a variety of prey items, including leopard frogs (Chapman et al. 1989; Sammons 1993; Krause 2013). In Lake Thompson, South Dakota, adult northern pike ate fathead minnows all years, with common carp, black crappie, and white crappie consumed seasonally (Sammons 1993). Age-0 northern pike in small streams eat a wide range of vertebrate and invertebrate prey items, including diptera, ephemeroptera, larval common carp, and amhipods (Krause 2013). Smaller northern pike from Pactola Lake, South Dakota, ate rainbow smelt, rock bass, bluegill, and other centrarchids (Scheibel 2015).

References

Bondarenko V, Drozd B, Policar T (2015) Effect of water temperature on egg incubation time and quality of newly hatched larvae of northern pike (*Esox lucius* L., 1758). J Appl Ichthyol 31:45–50. https://doi.org/10.1111/jai.12851

Chapman LJ, Mackay WC, Wilkinson CW (1989) Feeding flexibility in northern pike (*Esox lucius*): fish versus invertebrate prey. Can J Fish Aquat Sci 46:666–669. https://doi.org/10.1139/f89-085

Churchill EP, Over WH (1938) Fishes of South Dakota. Brown & Saenger Printers, Sioux Falls

Hassler TJ (1970) Environmental influences in early development and year-class strength of northern pike in Lakes Oahe and Sharpe, South Dakota. Trans Am Fish Soc 99:369–375. https://doi.org/10.1577/1548-8659

Inskip PD (1982) Habitat suitability index models: northern pike. US Dep Inter, Fish Wildl Serv, FWS/OBS-82/10.17

Krause JR (2013) Biotic integrity and northern pike ecology in Eastern South Dakota. Dissertation, South Dakota State University

Neumann RM (1994) Growth, distribution, and movement of northern pike in a South Dakota natural lake, with sampling considerations. Dissertation, South Dakota State University

Neumann RM, Willis DW, Sammons SM (1994) Seasonal growth of northern pike (*Esox lucius*) in a South Dakota glacial lake. J Freshw Ecol 9:191–196. https://doi.org/10.1080/02705060.1994.9664886

Paukert CP, Willis DW (2003) Population characteristics and ecological role of northern pike in shallow natural lakes in Nebraska. N Am J Fish Manag 23:313–322. https://doi.org/10.1577/1548-8675

Sammons SM (1993) Annual food habits of northern pike in an eastern South Dakota natural lake. Dissertation, South Dakota State University

Scheibel NC (2015) Age, growth, and trophic interactions of lake trout and northern pike in Pactola Reservoir: implications for lake trout management. Dissertation, South Dakota State University

Scott WB, Crossman EJ (1973) Freshwater fishes of Canada. Fisheries Research Board of Canada, Bulletin 184

6.16.3 Muskellunge *Esox masquinongy*

Mitchill, 1824

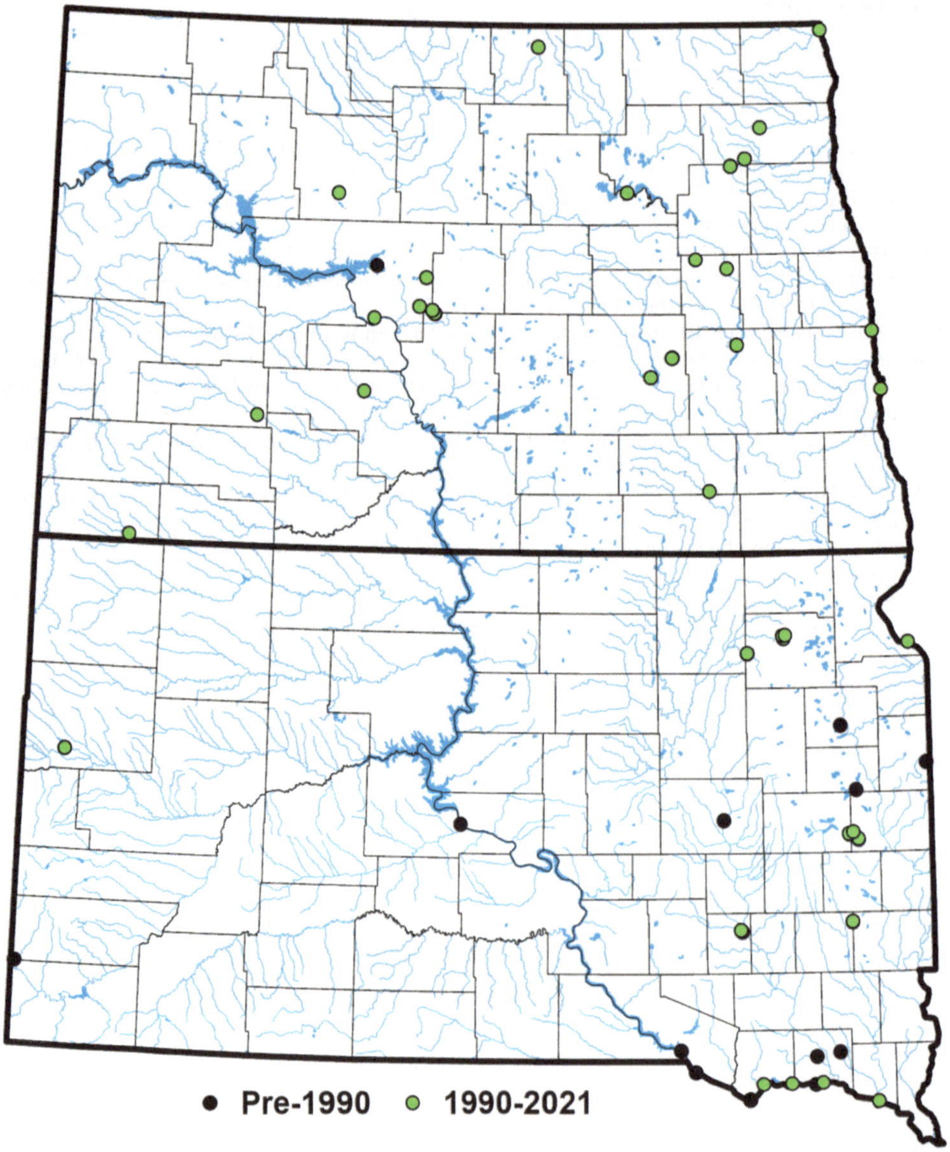

● **Pre-1990** ○ **1990–2021**

Etymology: *Esox masquinongy* – *Esox* = old European vernacular name for the pike; *masquinongy* = named by Ojibwae Native Americans.

Description:

Body – elongated, cylindrical.
Color –

> *Dorsally* – dark olive.
> *laterally* – olive to silver; dark, diffused vertical bars or spots.
> *Ventrally* – cream to white.
> *Fins* – yellow to amber; dorsal, anal, caudal – large, dark, diffused spots.

Head – elongated.
Opercle – scaled on top half.
Snout – elongated, flattened dorsoventrally.
Eyes – high on head.
Mouth – terminal, large.

> *Jaws* – upper jaw extending to mid or posterior edge of pupil; 6 to 9 mandibular pores per ventral side of lower jaw; 16 to 19 branchiostegal rays per jaw.

Teeth – on both jaws, roof of mouth, tongue, and branchial bones; lower jaw large, sharp canines; tongue and roof of mouth with short, comb-like teeth.
Gill rakers – sharp, tooth like.
Dorsal fin – 15 to 19 rays; slightly straight distal end, far posterior on body.
Caudal peduncle – elongated, thick.
Caudal fin – moderately forked, slightly pointed lobes.
Anal fin – 14 to 16 rays.
Pelvic fins – 11 or 12 rays; abdominal.
Pectoral fins – 14 to 19 rays; slightly straight distal end.
Lateral line – complete; 147 to 155 cycloid scales in series.
Juveniles – broken dark vertical bars appearing as spots in irregular columns.

Similar Species: Muskellunge closely resemble northern pike. Northern pike have an opercle that is scaled on only the top half but have a fully scaled cheek. Northern pike also have five or six mandibular pores on each side of the jaw, 14 to 16 branchiostegal rays, and roughly eight horizontal rows of small, bean-shaped light yellow to cream spots on the lateral sides. Tiger muskie, a hybrid between a northern pike and muskellunge, have an irregular pattern of narrow vertical dark stripes and spots against a light background with stripes that merge onto the dorsal side. Tiger muskie also have a more rounded caudal fin than muskellunge.

Distribution and Habitat: Endemic to North America, muskellunge are native to the Upper Mississippi River, Great Lakes, Ohio River, and Hudson Bay drainages. Because of poor reproductive success and spawning habitat destruction, muskellunge are declining throughout their native range (Dombeck et al. 1984; Cook and Solomon 1987; Crane et al. 2015). Many populations are maintained by stocking, including those in eastern North Dakota and South Dakota. Muskellunge inhabit lakes, rivers, impoundments, and ponds. In lakes, they typically occur at depths less than 5 m but are sometimes found between 12 and 15 m (Becker 1983). In rivers, they occur in low gradient pools with little current. Muskellunge are often associated with submerged vegetation, rock reefs, trees, or other structure. Their optimal water temperature is 25 to 26 °C (Scott and Crossman 1973). Especially in the summer, muskellunge establish home ranges of 20 to 500 ha (Diana et al. 2015). Daily distance traveled increases with increasing temperature (Younk 1982). They can withstand low oxygen levels (Becker 1983). Juvenile muskellunge depend on submerged aquatic vegetation to escape predators.

Reproduction: Muskellunge spawn in spring, typically late April or early May, in water temperatures of 9 to 16 °C (Scott and Crossman 1973). Spawning habitat depends on location, but generally is in water from 1 to 3 m deep with abundant debris and dead aquatic vegetation, and gravel, sand, or silt substrates (Scott and Crossman 1973; Zorn et al. 1998; Crane et al. 2015). Eggs and larvae need high dissolved oxygen concentrations to successfully develop (Dombeck et al. 1984). Males sexually mature at age-3 or age-4, with females maturing at age-4 or age-5. Males arrive first at spawning grounds. A single female is escorted over the spawning ground by one or two males, vibrating their bodies together to release and fertilize the eggs (Scott and Crossman 1973; Cook and Solomon 1987). Muskellunge are fractional spawners, with multiple irregular spawning intervals over several days (Scott and Crossman 1973; Cook and Solomon 1987). Nonadhesive, 3-mm diameter, amber eggs are broadcast over vegetated substrate and debris (Scott and Crossman 1973). No nest is built, and no parental care is given. Fecundity ranges from 6,000 to 265,000 eggs and increases with female size (Cook and Solomon 1987). Eggs hatch in 8 to 14 days in water temperatures of 12 to 17 °C. Larvae remain on the bottom substrate for another 10 days before moving to deeper water (Becker 1983; Zorn et al. 1998).

Age and Growth: Growth is highly variable among muskellunge populations because of differences in prey fish abundance (Scott and Crossman 1973; Cook and Solomon 1987). Females grow faster, live longer, and are typically larger than males at any given age (Scott and Crossman 1973; Cook and Solomon 1987). Muskellunge also live longer and grow slower in the northern part of their range (Scott and Crossman 1973; Cook and Solomon 1987). Growth is most rapid during the first 3 years (Becker 1983). Typical muskellunge total lengths are from 762 to 1168 mm (Scott and Crossman 1973). Maximum total length and weight are 1829 mm and 31,752 g, respectively. Muskellunge live from 10 to 15 years but can live as along as 30 years.

Food and Feeding: Muskellunge are large, apex, visual, ambush predators. Adults mainly feed during the day on a wide variety of fish including sunfish, minnows, yellow perch, northern pike, and smaller muskellunge. Crayfish, frogs, mice, and small waterfowl are also eaten. Muskellunge fry mainly eat larger zooplankton like cladocerans, and switch to piscivory upon reaching a total length of approximately 30 to 40 mm (Cook and Solomon 1987).

References

Becker GC (1983) Fishes of Wisconsin. University of Wisconsin Press, Madison

Cook MF, Solomon RC (1987) Habitat suitability index models: muskellunge. U.S. Dep Inter, Fish Wildl Serv, Biol Rep 82(10.148)

Crane DP, Miller LM, Diana JS, Casselman JM, Farrell JM, Kapuscinski KL, Nohner JK (2015) Muskellunge and northern pike ecology and management: important issues and research needs. Fisheries 40:258–267. https://doi.org/10.1080/03632415.2015.1038382

Diana JS, Hanchin P, Popoff N (2015) Movement patterns and spawning sites of muskellunge Esox masquinongy in the Antrim chain of lakes, Michigan. Environ Biol Fish 98:833–844. https://doi.org/10.1007/s10641-014-0319-7

Dombeck MP, Menzel BW, Hinz PN (1984) Muskellunge spawning habitat and reproductive success. Trans Am Fish Soc 113:205–216. https://doi.org/10.1577/1548-8659

Scott WB, Crossman EJ (1973) Freshwater fishes of Canada. Fisheries Research Board of Canada, Bulletin 184

Younk JA (1982) Distribution, movement, and temperature selection of adult walleye and muskellunge in a power plant cooling reservoir. Dissertation, South Dakota State University

Zorn SA, Margenau TL, Diana JS, Edwards CJ (1998) The influence of spawning habitat on natural reproduction of muskellunge in Wisconsin. Trans Am Fish Soc 127:995–1005. https://doi.org/10.1577/1548-8659

6.16.4   Central Mudminnow *Umbra limi*

(Kirtland, 1840)

● Pre-1990   ● 1990-2021

Etymology: *Umbra limi* – *Umbra* = shadow, shade, or dark, possibly referencing preferred habitat; *limi* = mud.

Description:

Body – fusiform to cylindrical; moderately elongated.
Color –

> *Dorsally* – dark brown to olive-green.
> *Laterally* – tan to black-brown; approximately 14 irregular dark brown vertical bars.
> *Ventrally* – tan to cream.
> *Fins* – uniformly pale brown; caudal – prominent dark vertical bar at base.

Head – scaled dorsally, laterally.
Snout – blunt, short.
Eyes – relatively large, laterally on head.
Mouth – terminal.
> *Lower jaw* – slightly protrudes past upper jaw.
> *Upper jaw* – extends to anterior edge of pupil.
> *Lips* – noncontinuous groove on upper lip.

Teeth – small, villiform; on premaxilla, roof of mouth, and mandible.
Gill rakers – 13 to 15.
Dorsal fin – 13 to 15 rays; rounded; insertion slightly posterior to insertion of pelvic fins.
Caudal peduncle – thick.
Caudal fin – homocercal, rounded.
Anal fin – 7 to 9 rays.
Pelvic fins – 6 or 7 rays.
Pectoral fins – 14 to 16 rays.
Lateral line – absent; 34 to 37 cycloid scales in lateral line series.
Sexual dimorphism – spawning males have an elongated, iridescent purple to blue-green anal fin almost extending to caudal fin base; females have a short anal fin.

Similar Species: Central mudminnow most closely resemble banded killifish. Banded killifish have a more elongated body, 12 to 20 dark greenish-brown vertical bars laterally, a continuous upper lip groove, 10 to 12 anal fin rays, and a dorsal fin insertion distinctly posterior to the pelvic fins. They also lack a dark vertical bar at the base of the caudal fin.

Distribution and Habitat: Central mudminnow are common in northern glacial regions, with a native range encompassing the upper Mississippi River drainage from Quebec to Manitoba, Great Lakes tributaries, headwaters of the Hudson Bay (Red River of North), and south to northwestern Tennessee (Etnier and Starnes 1993; Fuller et al. 1999; Schilling et al. 2006). Isolated populations occur within the Missouri River drainage in east-central South Dakota, eastern Nebraska, and western Iowa (Fuller et al. 1999; Schilling et al. 2006). In the Dakotas, central mudminnow primarily occur east of the Missouri River in the Souris River, Red River of the North, Minnesota River, and Big Sioux River systems. Four central mudminnow were vouchered in 1976 from the Green River, a major tributary to the Heart River in North Dakota, but have since been lost. This is the only central mudminnow occurrence documented west of the Missouri River (Reigh and Owen 1979). Central mudminnow are hardy, habitat specialists that inhabit lake shorelines, wetlands, and pools in small streams less than 1.0 m deep with little to no flow. They are less common in deeper waters with greater flow. Central mudminnow prefer moderate to dense submerged aquatic vegetation or detritus over gravel, sand, silt, and mud substrates. As facultative air breathers, they tolerate hypoxic, isolated, stagnant conditions and high temperatures. The swim bladder connected to the pharynx assists in respiration, and all sizes of central mudminnow will gulp air from the surface (Gee 1980; Martin-Bergmann and Gee 1985). They also use air bubbles under the ice (Magnuson et al. 1983; Martin-Bergmann and Gee 1985).

Reproduction: Spawning migrations upstream, or laterally to adjacent flooded areas with abundant vegetation, begin in early spring, typically March and April (Peckham and Dineen 1957). Central mudminnow spawn in water temperatures from 3 to 16 °C (Becker 1983). They sexually mature at age-1 or age-2. Adhesive, 1-mm diameter, yellow-orange eggs are deposited onto beds of vegetation (Peckham and Dineen 1957). Although no nest is built, the female provides parental care and will consume any undeveloped eggs (Westman 1938; Becker 1983). Fecundity increases with female size and ranges from 220 to 1,489 eggs (Peckham and Dineen 1957; Martin-Bergmann and Gee 1985). Eggs hatch within a week (Peckham and Dineen 1957; Martin-Bergmann and Gee 1985). Juveniles move away from the breeding site upon reaching a total length of approximately 30 mm.

Age and Growth: Newly hatched central mudminnow larvae have a total length of approximately 5 mm (Peckham and Dineen 1957). They grow the fastest within the first year, reaching an average age-1 total length of 43 mm (Becker 1983). In Minnesota, central mudminnow total-lengths-at-age were 62 to 122 mm at age-2, 91 to 138 mm at age-3, and 110–143 mm at age-4 (Jones 1973). They rarely exceed a total length of 140 mm and have a longevity of 7 to 9 years (Schilling et al. 2006).

Food and Feeding: Central mudminnow are primarily benthic carnivores. Juveniles primarily consume chironomids, ostracods, copepods, and cladocerans (Peckham and Dineen 1957). Chironomids are the most important food for age-0 and age-1 central mudminnow (Martin-Bergmann and Gee 1985). Adults rely less on small crustaceans and eat a variety of prey including aquatic insects (ostracods, amphipods, ephemeropterans, odonates, and dipterans), terrestrial insects (oligochaetes, dipterans, araneaens, and coleopterans), small mollusks, and small fish (Jones 1973; Martin-Bergmann and Gee 1985). Vegetative matter is eaten but is likely consumed accidentally (Cahn 1927, Peckham and Dineen 1957). In aquariums, central mudminnow lay motionless before ambushing prey (Etnier and Starnes 1993). They are active winter feeders and can rapidly digest food in cold temperatures (Chilton et al. 1984; Martin-Bergmann and Gee 1985).

References

Becker GC (1983) Fishes of Wisconsin. University of Wisconsin Press, Madison

Cahn AR (1927) An ecological study of southern Wisconsin fishes; the brook silversides (*Labidesthes sicculus*) and the cisco (*Leucichthys artedi*) in their relations to the region. Dissertation, University of Illinois. https://doi.org/10.5962/bhl.title.50172

Chilton G, Martin KA, Gee JH (1984) Winter feeding: an adaptive strategy broadening the niche of the central mudminnow, *Umbra limi*. Environ Biol Fish 10:215–219. https://doi.org/10.1007/BF00001129

Etnier DA, Starnes WC (1993) The fishes of Tennessee. University of Tennessee Press, Knoxville

Fuller PL, Nico LG, Williams JD (1999) Nonindigenous fishes introduced into inland waters of the United States. American Fisheries Society, Special Publ 27, Bethesda

Gee JH (1980) Respiratory patterns and antipredator responses in the central mudminnow, *Umbra limi*, a continuous, facultative, air-breathing fish. Can J Zool 58:819–827. https://doi.org/10.1139/z80-114

Jones JA (1973) The ecology of the mudminnow, *Umbra limi*, in Fish Lake (Anoka County, Minnesota). Dissertation, Iowa State University

Magnuson JJ, Keller JW, Beckel AL, Gallepp GW (1983) Breathing gas mixtures different from air: an adaptation for survival under the ice of a facultative air-breathing fish. Science 220:312–314. https://doi.org/10.1126/science.220.4594.312

Martin-Bergmann KA, Gee JH (1985) The central mudminnow, *Umbra limi* (Kirtland), a habitat specialist and resource generalist. Can J Zool 63:1753–1764. https://doi.org/10.1139/z85-264

Peckham RS, Dineen CF (1957) Ecology of the central mudminnow, *Umbra limi* (Kirtland). Am Midl Nat 58:222–231. https://doi.org/10.2307/2422370

Reigh RC, Owen JB (1979) Fishes of the western tributaries of the Missouri River in North Dakota. Report No 79-2, North Dakota Region Environ Assess Program, Bismarck

Schilling EG, Halliwell DB, Gullo AM, Markowsky JK (2006) First records of *Umbra limi* (central mudminnow) in Maine. Northeast Nat 13:287–290. https://doi.org/10.1656/1092-6194(2006)13[287:FROULC]2.0.CO;2

Westman JR (1938) Studies on the reproduction and growth of the bluntnose minnow, *Hyborhynchus notatus* (Rafinesque). Copeia 1938:57–61. https://doi.org/10.2307/1435690

6.17   Salmonidae, Salmon, Trout, and Char Family

The Salmonidae family contains approximately 170 species of salmon, trout, and char. Salmonids are native to the Northern Hemisphere, but have been introduced worldwide. They are not native to North Dakota nor South Dakota but have been widely introduced in each state. Eight salmonid species now regularly occur in the Dakotas.

Salmonids are adapted to colder water temperatures and thrive in waters below 18 °C. Some species are anadromous, spawning in freshwater with progeny migrating to sea to spend most of their life before returning to freshwater for spawning. Other salmonids complete their entire lifecycle in freshwater; there are no entirely marine salmonid species. Salmnonids in the Dakotas prefer cool, clear streams and lakes with high dissolved oxygen levels, submerged vegetation, and sand, gravel, or rocky substrate.

Salmonids are characterized by an adipose fin, a pelvic fin axillary process, soft-rayed fins, and numerous small, embedded cycloid scales. Coloration varies widely, ranging from more colorfully patterned fish like rainbow trout *Oncorhynchus mykiss* and brown trout *Salmo trutta* to the more solid silver cisco *Coregonus artedi* and lake whitefish *Coregonus clupeaformis*.

Rainbow smelt *Osmerus mordax,* family Osmeridae, occurs sympatrically in the Dakotas with some salmonid species. While they look similar, rainbow smelt are smaller and lack an axillary process. Trout-perch *Percopsis omiscomaycus*, family Percopsidae, have trout in their common name but differ from salmonids by having ctenoid scales and soft spines in the dorsal and anal fins.

Salmonids were introduced in 1886 into the Black Hills of western South Dakota, with brown trout and brook trout *Salvelinus fontinalis* quickly becoming self-sustaining. Rainbow trout also naturally reproduce in a few creeks in the Black Hills. Numerous ponds, lakes, and streams also contain salmonids, particularly rainbow trout, but require annual stockings and are frequently only seasonal fisheries. Construction of the Missouri River reservoirs in the 1950s and 1960s created new coldwater habitat in both North Dakota and South Dakota. Chinook salmon *Oncorhynchus tshawytscha*, Atlantic salmon *Salmo salar*, cisco, lake whitefish, and other salmonids were subsequently introduced into the cold, well-oxygenated, deep pelagic waters.

Spawning migrations of salmonids can be strenuous and long. Anadromous salmonids may swim hundreds or thousands of miles from the ocean back to the exact stream where they hatched. Chinook salmon in the Missouri River reservoirs still mimic this behavior in their landlocked freshwater habitats. During the fall they return to locations like Whitlock's Bay area on Lake Oahe for spawning by fisheries staff. Fish hatcheries incubate the eggs, grow the fish, and restock the reservoirs to artificially complete the salmon life cycle.

6.17.1   Cisco *Coregonus artedi*

Lesueur, 1818

● Pre-1990   ● 1990-2021

Etymology: *Coregonus artedi* – *Coregonus* = angle-eye; *artedi* = referencing Petrus Artedi, the Father of Ichthyology, who was an associate of Linnaeus. Commonly referred to as lake herring, even though it is not a member of the thread herring family, Dorosomatidae.

Description:

Body – elongated, fusiform, laterally compressed, deepest near dorsal fin insertion.
Color –

> *Dorsally* – gray, olive to blue-green.
> *Laterally* – solid silver; faint blue, purple, or pink iridescence.
> *Ventrally* – silver to cream.
> *Fins* – creamy white to light gray.

Head – small, conical.
Snout – pointed.
Eyes – moderately large, laterally on head.
Mouth – terminal, oblique.

> *Jaws* – equal in length, extending to middle of eye.

Teeth – small, on both jaws.
Gill rakers – 36 to 64; long, comb-like.
Dorsal fin – 8 to 11 rays.
Adipose fin – small, flap-like.
Caudal peduncle – short, moderately thick.
Caudal fin – strongly forked.
Anal fin – 10 to 13 rays.
Pelvic fins – 10 to 12 rays; abdominal, axillary process.
Pectoral fins – 14 to 17 rays.
Lateral line – complete; 70 to 94 cycloid scales in series.
Sexual dimorphism – spawning males have small tubercles on head and body; females have fewer, less-developed tubercles.

Similar Species: Cisco resemble lake whitefish. Lake whitefish have a blunt, extended snout overhanging an inferior to subterminal mouth. Skipjack herring have a slightly superior mouth and lack an adipose fin. Goldeye and mooneye also lack adipose fins.

Distribution and Habitat: Cisco are native to the north-central and eastern United States, the Great Lakes, and the majority of Canada to the Arctic Ocean (Scott and Crossman 1973). Population declines have occurred because of rising water temperatures, land-use changes, and invasive species like the rainbow smelt (Lawrie and Rahrer 1973; Hartman 1973; Bryan 1995). In the Dakotas, cisco are found primarily in Lake Sakakawea and Lake Oahe, but they may entrain into other Missouri River waters (Hartman 1973; Lawrie and Rahrer 1973; Bryan 1995). Cisco that escaped from Blue Dog Lake State Fish Hatchery in Day County, South Dakota in the early 1990s have become established in Waubay Lake. Cisco are most frequent in the hypolimnion pelagic zone in well-oxygenated, cold, clear water of deep lakes and reservoirs. They form schools during the day and disperse at night (Milne et al. 2005). Cisco are intolerant of warm water temperatures and low dissolved oxygen concentrations, with adults less tolerant than juveniles (Frey 1955; Edsall and Colby 1970). Upper lethal temperatures are 20 °C and 26 °C for adults and young-of-year, respectively (Frey 1955; Colby and Brooke 1969; Edsall and Colby 1970). Young-of-year and adult lower lethal temperature is 0 °C (Frey 1955; Edsall and Colby 1970). Larvae inhabit shallow bays near the surface for the first month after hatching before moving offshore to deeper waters, where they remain near the surface (MacKay 1963; Clady 1976).

Reproduction: Prior to spawning, cisco move to near-shore, shallower areas of lakes, or upstream large rivers. This movement and subsequent spawning transports energy from the pelagic zone to the shoreline, resulting in an important resource subsidy to nearshore egg predators (Stockwell et al. 2014). Spawning occurs at night during the fall near the surface over a variety of substrates, peaking when water temperature reaches 4 °C (John 1956; Becker 1983). Age at sexual maturity varies among populations. Most cisco mature at age-1 or age-2, with maturity up to age-5 also possible (Hile 1936; Smith 1956; Smith 1957; Becker 1983). Males arrive at spawning grounds 2 to 4 days prior to females (Cahn 1927; Scott and Crossman 1973). Initially, up to 12 males may escort a single female to the spawning grounds, but later during the spawn, one or two males is more common (Cahn 1927; Becker 1983). Cisco are broadcast spawners. Fecundity varies with female size and was 4,314 to 10,250 eggs from females with total lengths of 269 to 356 from Lake Superior (Dryer and Beil 1964; Becker 1983).

Eggs are slightly adhesive and average 2 mm in diameter (Becker 1983). Eggs incubate overwinter under the ice until hatching in early spring. Hatching occurs from 37 to 49 days at 10 °C (Hinrichs and Booke 1975; Becker 1983).

Age and Growth: Information on cisco age and growth is limited in the Dakotas. The optimum temperature range for normal cisco development is approximately 2 to 8 °C (Colby and Brooke 1969; Becker 1983). Growth rates are highest in summer and early fall (Hile 1936). In the Missouri River reservoirs, adults have total lengths from 254 to 381 mm but can reach a total length 559 mm. Longevity is 10 to 15 years.

Food and Feeding: Cisco are primarily offshore pelagic zone zooplanktivores. They use long gill rakers to strain large numbers of small copepods, cladocerans, and other zooplankton from the water column. They also eat chironomids and cannibalize fish eggs (Dryer and Beil 1964; Becker 1983; Aku and Tonn 1999). Cisco stomachs are more full during the day than night, suggesting that night feeding is less efficient (Milne et al. 2005). Stomach fullness also increases with increased school size, suggesting that schooling increases foraging efficiencies (Milne et al. 2005).

References

Aku PMK, Tonn WM (1999) Effects of hypolimnetic oxygenation on the food resources and feeding ecology of cisco in Amisk Lake, Alberta. Trans Am Fish Soc 128:17–30. https://doi.org/10.1577/1548-8659

Becker GC (1983) Fishes of Wisconsin. University of Wisconsin Press, Madison

Bryan SD (1995) Bioenergetics of walleye in Lake Oahe, South Dakota. Dissertation, South Dakota State University

Cahn AR (1927) An ecological study of the southern Wisconsin fishes. The brook silverside (*Labidesthes sicculus*) and the cisco (*Leucichthys artedi*) in their relations to the region. Ill Biol Monograph 11:1–151. https://doi.org/10.5962/bhl.title.50172

Clady MD (1976) Distribution and abundance of larval ciscoes, *Coregonus artedii*, and burbot, *Lota lota*, in Oneida Lake. J Great Lakes Res 2:234–247. https://doi.org/10.1016/S0380-1330(76)72288-3

Colby PJ, Brooke LT (1969) Cisco (*Coregonus artedii*) mortalities in a southern Michigan lake. Limn Oceano 14:958–960. https://doi.org/10.4319/lo.1969.14.6.0958

Dryer WR, Beil J (1964) Life history of the lake herring in Lake Superior. US Dept Inter, Fish Wildl Serv, Fish Bullet 63:493–530

Edsall TA, Colby PJ (1970) Temperature tolerance of young-of-the-year cisco, *Coregonus artedii*. Trans Am Fish Soc 99:526–531. https://doi.org/10.1577/1548-8659

Frey DG (1955) Distributional ecology of the cisco (*Coregonus artedii*) in Indiana. Invest Indiana Lakes Streams 4:177–228

Hartman WL (1973) Effects of exploitation, environmental changes, and new species on the fish habitats and resources of Lake Erie. Great Lakes Fish Comm Rep 22

Hile R (1936) Age and growth of the cisco, *Leucichthys artedi* (LeSueur), in the lakes of the northeastern highlands, Wisconsin. Great Lakes Science Center, Bull Bur Fish 19:211–317

Hinrichs MA, Booke HE (1975) Egg development and larval feeding of the lake herring, *Coregonus artedii* LeSueur. University of Wisconsin-Stevens Point, Rep Fauna Flora Wis 10:75–86

Hoagstrom CW (2006) Fish community assembly in the Missouri River basin. Dissertation, South Dakota State University

John KR (1956) Onset in spawning activities of the shallow water cisco, *Leucichthys artedi* (LeSueur), in Lake Mendota, Wisconsin, relative to water temperatures. Copeia 1956:116–118. https://doi.org/10.2307/1440428

Lawrie AH, Rahrer JF (1973) Lake Superior: case history of the lake and its fisheries. Great Lakes Fish Comm Tech Rep 19

MacKay HH (1963) Fishes of Ontario. Ontario Department of Lands and Forest, Toronto

Milne SW, Shuter BJ, Sprules WG (2005) The schooling and foraging ecology of lake herring (*Coregonus artedi*) in Lake Opeongo, Ontario, Canada. Can J Fish Aquat Sci 62:1210–1218. https://doi.org/10.1139/f05-030

Scott WB, Crossman EJ (1973) Freshwater fishes of Canada. Fisheries Research Board of Canada, Bulletin 184

Smith SH (1956) Life history of lake herring of Green Bay, Lake Michigan. US Dep Inter, Fish Wildl Serv, Fish Bull 57:87–138

Smith SH (1957) Evolution and distribution of the coregonids. J Fish Res Board Can 14:599–604. https://doi.org/10.1139/f57-018

Stockwell JD, Yule DL, Hrabik TR, Sierszen ME, Isaac EJ (2014) Habitat coupling in a large lake system: delivery of an energy subsidy by an offshore planktivore to the nearshore zone of Lake Superior. Freshw Biol 59:1197–1212. https://doi.org/10.1111/fwb.12340

6.17.2 Lake Whitefish *Coregonus clupeaformis*

(Mitchill, 1818)

© Konrad Schmidt

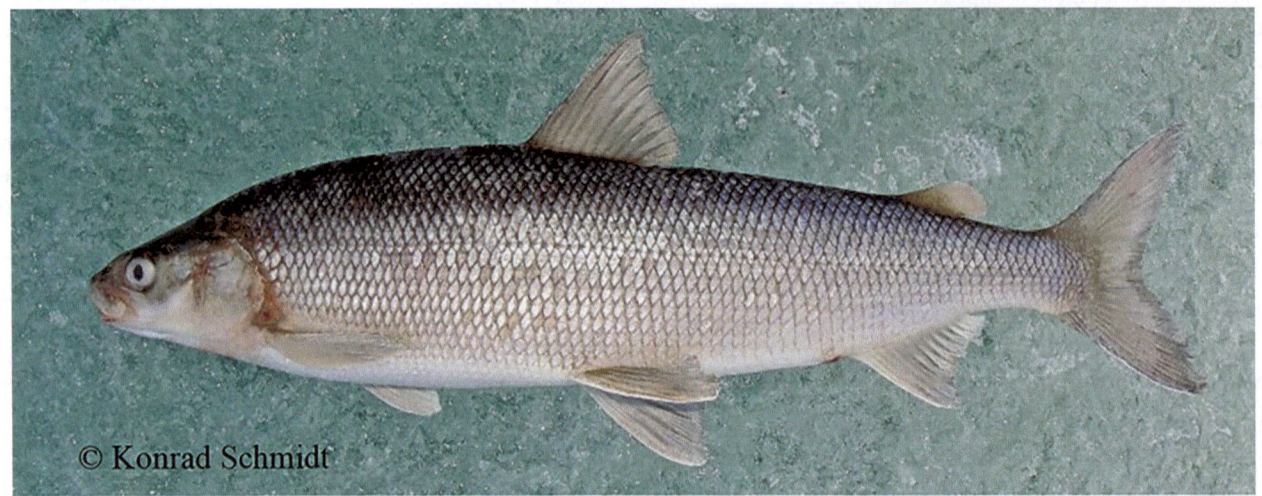

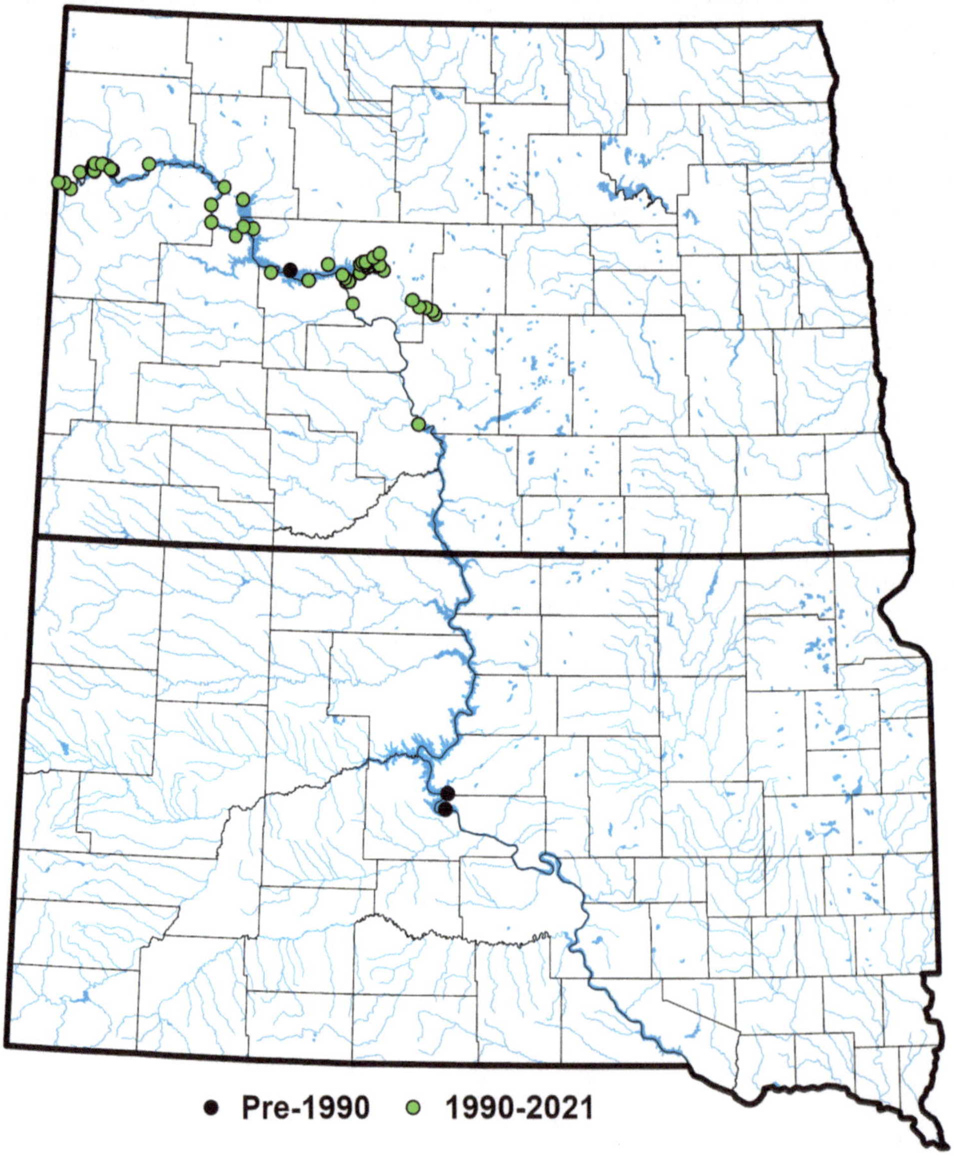
● Pre-1990 ● 1990-2021

Etymology: *Coregonus clupeaformis* – *Coregonus* = angle-eye; *clupeaformis* = herring-shaped.

Description:

Body – elongated, fusiform, laterally compressed.
Color –

   Dorsally – dark, pale olive-green, bronze, or silvery-gray.
   Laterally – solid silver, bluish sheen.
   Ventrally – cream to white; dorsal side of head and upper jaw dark gray; lower jaw silvery-white.
   Fins – light to dark gray.

Head – small, conical.
Snout – blunt, overhanging mouth.
Eyes – moderately large, laterally on head.
Mouth – inferior to subterminal, slightly oblique.

   Upper jaw – rarely extends past anterior edge of pupil.
Teeth – small, on both jaws.
Gill rakers – 24 to 33; rather short.
Dorsal fin – 10 to 12 rays.
Adipose fin – small, flap-like.
Caudal peduncle – slightly elongated, moderately thick.
Caudal fin – strongly forked.
Anal fin – 10 to 12 rays.
Pelvic fins – 10 to 12 rays; abdominal, axillary process.
Pectoral fins – 14 to 17 rays.
Lateral line – complete; 74 to 93 cycloid scales in series.
Sexual dimorphism – spawning males have small tubercles on head and body; females have fewer, less-developed tubercles.

Similar Species: Lake whitefish closely resemble cisco. Cisco have a more pointed snout and terminal mouth. Skipjack herring have a slightly superior mouth and lack an adipose fin. Goldeye and mooneye lack an adipose fin.

Distribution and Habitat: The native range of lake whitefish extends throughout the majority of Canada, north to the Arctic Ocean, and south throughout the north-central and eastern United States including the Great Lakes (Hubbs and Lagler 2004). They are not native to the Dakotas but were stocked into Lake Sakakawea and Lake Oahe (Owen et al. 1981). Lake whitefish distribution and recruitment is negatively affected by rainbow smelt competition and larval predation (Crowder 1980; Loftus and Hulsman 1986). Adult lake whitefish school 15 to 50 m deep in cold, oligotrophic lakes, reservoirs, and rivers (Walden 1964). They vertically migrate daily, especially in the spring and fall, occupying deeper depths during the day and moving up in the water column at night (Gorsky 2011). Their preferred water temperatures range from 10 to 14 °C (Christie and Regier 1988; Gorsky 2011). Lake whitefish migrate to deeper waters in the fall to overwinter (Gorsky 2011). Larvae inhabit surface waters immediately after hatching (Ihssen et al. 1981). Lake whitefish maximum oxygen consumption and critical swimming speed are reached at 12 °C (Bernatchez and Dodson 1985). The highest larval survival is at 10 °C, while juveniles prefer water temperatures of 16 to 20 °C (Edsall 1999a, b). They are intolerant of warm water, with an adult upper lethal temperatures of 27 °C (Esdall and Rottiers 1976).

Reproduction: Lake whitefish begin migrating in the late fall to shallow spawning grounds. Spawning occurs at night from October through early December in water temperatures under 6 °C and lasts for approximately a week (Hart 1930; Nester and Poe 1984; Anras et al. 1999). Spawning sites are generally 1 to 4 m deep with clean, sandy, rocky, cobble or boulder substrate in the inlets of lakes, streams, or windswept shorelines (Anras et al. 1999; Gorsky 2011). Lake whitefish reach sexual maturity from age-3 to age-6 (Gorsky 2011). Males arrive at spawning grounds prior to females. One to two males escort a female to the water surface where gametes are released during a period of splashing and jumping (Becker 1983). Lake whitefish are broadcast spawners. Adults migrate back to deeper waters after spawning. Fecundity and egg size increase with female size and fecundity varies substantially among populations (Ihssen et al. 1981). The semi-buoyant, yellow-orange eggs are approximately 3 mm in diameter (Booke 1970; Ihssen et al. 1981). Eggs incubate over winter and hatch in spring. Time to hatch is inversely related to water temperature, ranging from 41.7 days at 10 °C to 182 days at 0 °C (Brooke 1975). Optimum water temperature for egg incubation is from 3 to 8 °C (Brooke 1975).

Age and Growth: Information on lake whitefish age and growth in the Dakotas is limited. Lake whitefish are the largest of the *Coregonus* species, but growth varies widely among populations. Lake whitefish larvae have a total length of approximately 13 to 14 mm at hatching (Gorsky 2011). Larval growth is fastest at 18 °C and is highly correlated with food availability (Edsall 1999a; Gorsky 2011). Adults typically have total lengths ranging from 254 to 381 mm but can reach 737 mm (Dryer 1963). Lake whitefish longevity is 12 years (Gorsky 2011).

Food and Feeding: Adult lake whitefish are primarily opportunistic benthivores, consuming large amounts of benthic invertebrates like insects, crustaceans, and mollusks (Reckahn 1970; Pothoven and Nalepa 2006). They also occasionally consume zooplankton, *Mysis*, oligochaetes, and small fish like sticklebacks and alewives (Reckahn 1970). While benthic prey size increases with increased fish size, pelagic prey size is not related to fish size (Pothoven and Nalepa 2006). Larvae eat large-bodied copepods, cladocerans, rotifers, and other zooplankton (Teska and Behmer 1981; Pothoven and Nalepa 2006).

References

Anras MLB, Cooley PM, Bodaly RA, Anras L, Fudge RJP (1999) Movement and habitat use by lake whitefish during spawning in a boreal lake: integrating acoustic telemetry and geographic information systems. Trans Am Fish Soc 128:939–952. https://doi.org/10.1577/1548-8659

Becker GC (1983) Fishes of Wisconsin. University of Wisconsin Press, Madison

Bernatchez L, Dodson JJ (1985) The influence of current speed and temperature on the swimming capacity of lake whitefish (*Coregonus clupeaformis*) and cisco (*C. artedii*). Can J Fish Aquat Sci 42:1522–1529. https://doi.org/10.1139/f85-190

Booke HE (1970) Speciation parameters in Coregonine fishes: part I. Egg size. Part II. Karyotype. In: Linsey CC, Woods CS (eds) Biology of coregonid fishes. University of Manitoba Press, Winnipeg, pp 61–66

Brooke LT (1975) Effect of different constant incubation temperatures on egg survival and embryonic development in lake whitefish (*Coregonus clupeaformis*). Trans Am Fish Soc 104:555–559. https://doi.org/10.1577/1548-8659

Christie GC, Regier HA (1988) Measures of optimal thermal habitat and their relationship to yields of four commercial fish species. Can J Fish Aquat Sci 45:301–314. https://doi.org/10.1139/f88-036

Crowder LB (1980) Alewife, rainbow smelt and native fishes in Lake Michigan: competition or predation? Environ Biol Fish 5:225–233. https://doi.org/10.1007/BF00005356

Dryer WR (1963) Age and growth of the whitefish in Lake Superior. Fish Bull 63:77–95

Edsall TA (1999a) The growth-temperature relation of juvenile lake whitefish. Trans Am Fish Soc 128:962–964. https://doi.org/10.1577/1548-8659

Edsall TA (1999b) Preferred temperature of juvenile lake whitefish. J Great Lakes Res 25:583–588. https://doi.org/10.1016/S0380-1330(99)70761-6

Esdall TA, Rottiers DV (1976) The temperature tolerance of young-of-the-year lake whitefish, *Coregonus clupeaformis*. J Fish Res Board Can 33:177–180. https://doi.org/10.1139/f76-021

Gorsky D (2011) Lake whitefish (*Coregonus clupeaformis*) habitat utilization, early life history, and interactions with rainbow smelt (*Osmerus mordax*) in northern Maine. Dissertation, University of Maine

Hart JL (1930) The spawning and early life history of the whitefish, *Coregonus clupeaformis* (Mitchill) in the Bay of Quinte, Ontario. Contr Can Biol Fish 6:165–214. https://doi.org/10.1139/f31-007

Hubbs C, Lagler K (2004) Fishes of the Great Lakes region. University of Michigan Press, Ann Arbor. https://doi.org/10.3998/mpub.17658

Ihssen PE, Evans DO, Christie WJ, Reckahn JA, DesJardine RL (1981) Life history, morphology, and electrophoretic characteristics of five allopatric stocks of lake whitefish (*Coregonus clupeaformis*) in the Great Lakes region. Can J Fish Aquat Sci 38:1790–1807. https://doi.org/10.1139/f81-226

Loftus DH, Hulsman PF (1986) Predation on larval lake whitefish (*Coregonus clupeaformis*) and lake herring (*C. artedii*) by adult rainbow smelt (*Osmerus mordax*). Can J Fish Aquat Sci 43:812–818. https://doi.org/10.1139/f86-100

Nester RT, Poe TP (1984) Predation on lake whitefish eggs by longnose suckers. J Great Lakes Res 10:327–328. https://doi.org/10.1016/S0380-1330(84)71846-6

Owen JB, Elsen DS, Russell GW (1981) Distribution of fishes in North and South Dakota Basins affected by the Garrison Diversion Unit. University of North Dakota Press, Grand Forks

Pothoven SA, Nalepa TF (2006) Feeding ecology of lake whitefish in Lake Huron. J Great Lakes Res 32:489–501. https://doi.org/10.3394/0380-1330(2006)32[489:FEOLWI]2.0.CO;2

Reckahn JA (1970) Ecology of young lake whitefish (*Coregonus clupeaformis*) in South Bay, Manitoulin Island, Lake Huron. In: Linsey CC, Woods CS (eds) Biology of coregonid fishes. University of Manitoba Press, Winnipeg, pp 437–460

Teska JD, Behmer DJ (1981) Zooplankton preference of larval lake whitefish. Trans Am Fish Soc 110:459–461. https://doi.org/10.1577/1548-8659

Walden H (1964) Familiar freshwater fishes of America. Harper and Row, New York

6.17.3   Coastal Cutthroat Trout *Oncorhynchus clarkii*

(Richardson, 1836)

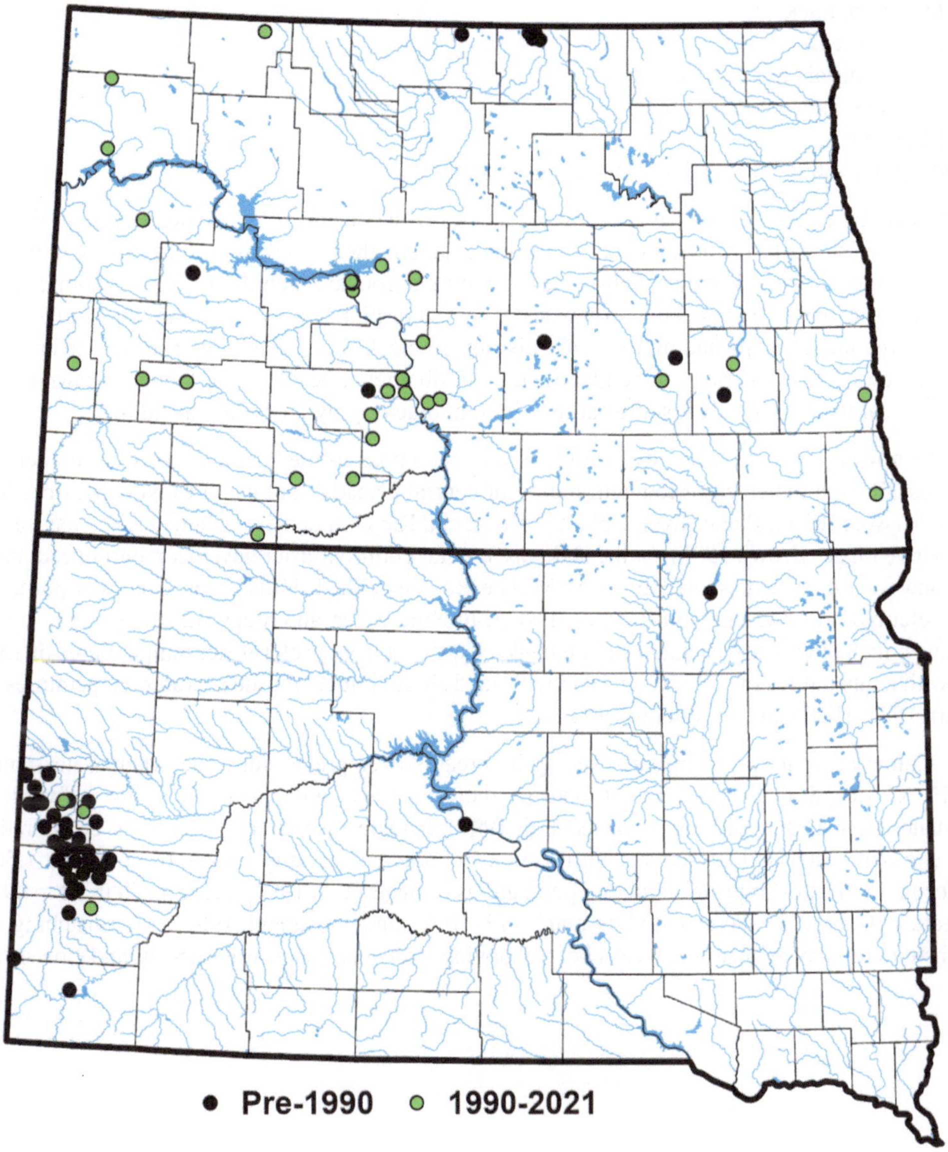

● **Pre-1990**   ◯ **1990-2021**

Etymology: *Oncorhynchus clarkii* – *Oncorhynchus* = hooked snout, referencing the elongated hooked jaw, or kype, of a spawning male; *clarkia* = in honor of William Clark, from the Lewis and Clark expedition.

Description:

Body – slender, elongated, slightly laterally compressed.
Color –

  Dorsally – greenish-blue to dusky-gray; dark gray to black round spots, more numerous posteriorly.
  Laterally – yellow to silver; dark gray to black round spots, more numerous posteriorly.
  Mouth – reddish-orange slash mark on each underside of lower jaw.
  Fins – dorsal, adipose, caudal – heavily spotted; pectoral, pelvic, anal – brown to red, possible spots.

Head – small to moderate.
Snout – short, blunt.
Eyes – moderately large, sometimes spotted.
Mouth – small, terminal.

  Upper jaw – extending beyond posterior edge of eye.
Teeth – small, sharp, on posterior end of tongue and jaws.
Gill rakers – 17 to 21.
Dorsal fin – 8 to 11 rays.
Adipose fin – small, slender.
Caudal peduncle – short, thick.
Caudal fin – slightly forked.
Anal fin – 8 to 12 rays; distal end rounded.
Pelvic fins – 9 or 10 rays; small axillary process.
Pectoral fins – 12 to 15 rays.
Lateral line – complete; 130 to 240 small cycloid scales in series.

Similar Species: Coastal cutthroat trout have several subspecies with considerable genetic diversity (Hickman and Raleigh 1982). The red-orange slash marks on the underside of the lower jar and the numerous posterior black spots easily distinguish them from other trout species in the Dakotas. Subspecies of cutthroat trout are identified by the location and size of the spots. Brook trout lack red slash marks on underside of jaw and have red or pink spots outlined in blue on lateral sides. Rainbow trout have a pinkish-red lateral stripe and more-evenly distributed black spots. Rainbow trout will hybridize with cutthroat trout to produce cutbows. These hybrids can be identified by basibranchial teeth on the tongue. Brown trout generally lack dark spots on the caudal fin and have dark spots outlined in light blue or gray along lateral sides.

Distribution and Habitat: The native range of cutthroat trout encompasses the Pacific coast from southern Alaska to northern California, west through the Rocky Mountain region of central Canada, and south to New Mexico. It is native to the headwaters of the Missouri River, Platte River, Colorado River, and Rio Grande River drainages and has been widely distributed outside its native range (Brown 1971). In the Dakotas, coastal cutthroat trout have been introduced and, although rare, are primarily found west of the Missouri River, the Middle Rockies (Black Hills) ecoregion, and put-and-take fisheries. Cutthroat trout prefer river headwaters with clear, cold, swift-moving water, and rocky substrates. They are often associated with abundant cover, including well-vegetated stream banks. Cutthroat trout prefer water temperatures of 15 to 16 °C, rarely occur in waters with temperatures over 22 °C, and can withstand short periods of time in water temperatures as high as 26 °C (Behnke and Zarn 1976; Hickman and Raleigh 1982).

Reproduction: Cutthroat trout spawn in early spring in streams over gravel substrate when water temperatures reach approximately 10 °C. Length determines sexual maturity more than age, with males maturing at fork lengths of 110 to 160 mm and females from 150 to 180 mm (Downs et al. 1997). Females use their caudal fin to create redds, small gravel depressions for egg deposition and subsequent fertilization by males. Fecundity ranges from 500 to 1,500 eggs (Brown 1971). Length-fecundity relationships differ among populations (Downs et al. 1997). Eggs hatch in 28 to 49 days depending on water temperature (Cope 1957; Scott and Crossman 1973; Hickman and Raleigh 1982). After hatching, fry reside in the rocky, gravel substrate for approximately 2 weeks before moving to deeper, faster water (Scott and Crossman 1973).

Age and Growth: The average length of adult cutthroat trout ranges from 304 to 406 mm, with older individuals reaching up to 508 mm. Their temperature for optimal growth is 15 °C. Cutthroat trout typically live 4 to 7 years but can live up to 8 years.

Food and Feeding: Juvenile and adult cutthroat trout eat all life stages of aquatic and terrestrial insects, including mayflies, stoneflies, caddisflies, and dipterans. Larger fish are more opportunistic, and will also eat smaller fish, fish eggs, snails, small rodents, and annelids. Fry eat zooplankton. Cutthroat trout will defend feeding territories in, or downstream of, riffles.

References

Behnke RJ, Zarn M (1976) Biology and management of threatened and endangered western trout. US Forest Serv General Tech Rep, RM-28

Brown CJD (1971) Fishes of Montana. Montana State University Press, Bozeman

Cope OB (1957) The choice of spawning sites by cutthroat trout. Proc Utah Acad Sci Art Lett 34:73–79

Downs CC, White RG, Shepard BB (1997) Age at sexual maturity, sex ratio, fecundity, and longevity of isolated headwater populations of West-slope cutthroat trout. N Am J Fish Manag 17:85–92. https://doi.org/10.1577/1548-8675

Hickman T, Raleigh RF (1982) Habitat sustainability index models: cutthroat trout. US Dept Inter, Fish Wildl Serv, FWS/OBS-82/10.5

Scott WB, Crossman EJ (1973) Freshwater fishes of Canada. Fisheries Research Board of Canada, Bulletin 184

6.17.4   **Rainbow Trout** *Oncorhynchus mykiss*

(Walbaum, 1792)

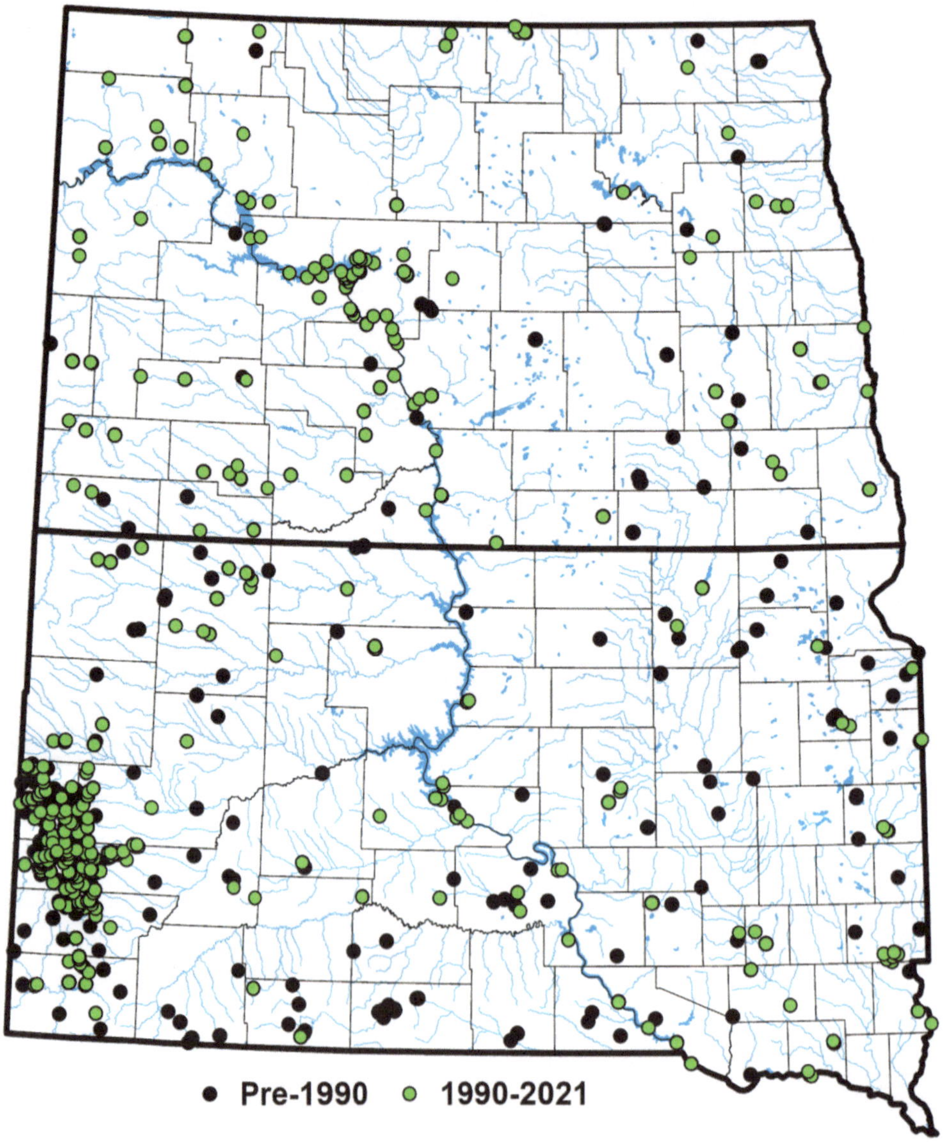

● Pre-1990   ● 1990-2021

Etymology: *Oncorhynchus mykiss* – *Oncorhynchus* = hooked snout, referencing the elongated hooked jaw, or kype, of a spawning male; *mykiss* = likely derived from a Kamchatkan word for trout.

Description:

Body – slender, elongated, slightly laterally compressed.
Color – heavily spotted with small black spots on head, body, and fins.

> *Dorsally* – dark olive to blue-green.
> *Laterally* – silver to light green; distinct bright pink lateral stripe.
> *Ventrally* – white to silver.
> *Fins* – dorsal – cream colored tip possible; pelvic, anal – white distal end.

Head – short.
Snout – rounded.
Eyes – moderately large.
Mouth – large, terminal.
Teeth – small; on jaws and tongue.
Gill rakers – 16 to 22.
Dorsal fin – 10 to 12 rays; nearly straight distal end.
Adipose fin – present.
Caudal peduncle – stout, thick.
Caudal fin – moderately forked.
Anal fin – 8 to 12 rays, relatively large.
Pelvic fins – 9 or 10 rays; abdominal, insertion slightly posterior to dorsal fin insertion; small axillary process.
Pectoral fins – 11 to 17 rays.
Lateral line – complete; 100 to 150 small cycloid scales in series.
Juveniles – 5 to 10 dark, vertical bars, or parr marks on lateral sides.
Sexual dimorphism – spawning males have a hook, or kype on lower jaw, lack intensely red lateral stripe.

Similar Species: Rainbow trout may be confused with brown trout, brook trout, or coastal cutthroat trout, but can be easily distinguished by their bright pink lateral stripe. Brown trout generally lack caudal fin spots and have black and red spots outlined in light blue. Brook trout lack a pink lateral stripe and have well-defined red or pink spots outlined in blue. Coastal cutthroat trout have red to orange slash marks on each underside of the lower jaw.

Distribution and Habitat: Rainbow trout are native to the Pacific Coast drainages of northeastern Asia, Canada, and the United States from Alaska to California (Becker 1983). They have been introduced widely outside of their native range for sport fishing. Rainbow trout are stocked throughout the Black Hills, eastern South Dakota urban waters, and tailwaters below Missouri River dams. In the Dakotas, natural reproduction only occurs in a few Black Hills streams; periodic stocking maintains all other populations. Rainbow trout can be anadromous (known as steelhead), reside entirely in freshwater streams, or dwell in freshwater lakes and reservoirs. They thrive in clear, well oxygenated water with optimal temperatures from 12 to 18 °C (Baxter and Stone 1995). They are often found over gravel and rocky substrates, near riffles, and other areas rich in vegetative cover (Baxter and Stone 1995).

Reproduction: Rainbow trout begin to spawn in early spring or summer over gravel substrates at water temperatures of 6 to 7 °C (Behnke 1979; Bjornn and Reiser 1991). Stream-resident and lake- and reservoir-dwelling rainbow trout migrate short distances to spawning grounds in smaller inlets and tributary streams (Behnke 1979). They become sexually mature at age-2 or age-3, with males often reaching maturity before females (Baxter and Stone 1995). With their tails, females excavate one or more redds, depressions in the gravel streambed often downstream of a pool or near the head of a series of riffles, where eggs are deposited (Bjornn and Reiser 1991; Baxter and Stone 1995). After fertilization, the female covers the eggs with gravel. Hatching occurs in 3 to 4 weeks at water temperatures of 10 to 15 °C (Davis 2012).

Age and Growth: Wild rainbow trout grow slower than those raised in a hatchery (Kansas Fishes Committee 2014: Kientz 2016). Growth varies within and among populations because of differing diets, temperatures, and other factors (Lynott et al. 1995; Koth 1980; Kansas Fishes Committee 2014). Total-lengths-at-age of rainbow trout from Deerfield Reservoir in South Dakota were 40 mm at age-0, 60 mm at age-1, 125 mm at age-2, 175 mm at age-3, and 210 mm at age-4 (Marcogliese and Casselman 1998). Rainbow trout can reach a total length of 678 mm in the Dakotas. While few rainbow trout live past age-6, some can live to age-11 (Nelitz et al. 2007).

Food and Feeding: Adult rainbow trout eat all life stages of aquatic and terrestrial insects, leeches, fish eggs, and small crustaceans such as zooplankton and scuds. In Lake Oahe and other populations, rainbow trout undergo an ontogenetic diet shift and may consume smaller fishes (Raleigh et al. 1984; Railsback and Rose 1999). Juveniles mainly consume zooplankton and aquatic macroinvertebrates such as mayflies, caddisflies, and stoneflies (Railsback and Rose 1999; Yard et al. 2011).

References

Baxter GT, Stone MD (1995) Fishes of Wyoming. Wyoming Game and Fish Department, Cheyenne

Becker GC (1983) Fishes of Wisconsin. University of Wisconsin Press, Madison

Behnke RJ (1979) Monograph of the native trouts of the genus *Salmo* of western North America. US Dep Ag, Fish Wildl Serv, Lakewood, Denver. https://doi.org/10.5962/bhl.title.110986

Behnke RJ (2002) Trout and salmon of North America. Simon and Schuster, New York

Bjornn TC, Reiser DW (1991) Habitat requirements of salmonids in streams. In: Meehan D (ed) Influences of forest and rangeland management on salmonid fishes and their habitats. American Fisheries Society, Spec Publ 19, Bethesda, pp 83–138

Davis JL (2012) Contribution of natural recruitment to the rainbow trout *Oncorhynchus mykiss* sport fishery in Deerfield Reservoir. Dissertation, South Dakota State University

Kansas Fishes Committee (2014) Kansas fishes. University Press of Kansas, Lawrence

Kientz JL (2016) Survival, abundance, and relative predation of wild rainbow trout in the Deerfield Reservoir system, South Dakota. Dissertation, South Dakota State University

Koth RM (1980) Food habits and growth of rainbow trout in a prairie pond. Dissertation, South Dakota State University

Lynott ST, Bryan SD, Hill TD, Duffy WG (1995) Monthly and size-related changes in the diet of rainbow trout in Lake Oahe, South Dakota. J Freshw Ecol 10:399–340. https://doi.org/10.1080/02705060.1995.9663463

Marcogliese LA, Casselman JM (1998) Scale methods for discriminating between Great Lakes stocks of wild and hatchery rainbow trout, with a measure of natural recruitment in Lake Ontario. N Am J Fish Manag 18:253–268. https://doi.org/10.1577/1548-8675

Nelitz MA, MacIsaac EA, Peterman RM (2007) A science-based approach for identifying temperature-sensitive streams for rainbow trout. N Am J Fish Manag 27:405–424. https://doi.org/10.1577/M05-146.1

Railsback SF, Rose KA (1999) Bioenergetics modeling of stream trout growth: temperature and food consumption effects. Trans Am Fish Soc 128:241–256. https://doi.org/10.1577/1548-8659

Raleigh RF, Hickman T, Solomon RC, Nelson PC (1984) Habitat suitability information: rainbow trout. US Dep Inter, Fish Wildl Serv, FWS/OBS-82/10.60

Yard MD, Coggins LG Jr, Baxter CV, Bennett GE, Korman J (2011) Trout piscivory in the Colorado River, Grand Canyon: effects of turbidity, temperature, and fish prey availability. Trans Am Fish Soc 140:471–486. https://doi.org/10.1080/00028487.2011.572011

6.17.5   Kokanee Salmon, *Oncorhynchus nerka*

(Waldbaum, 1792)

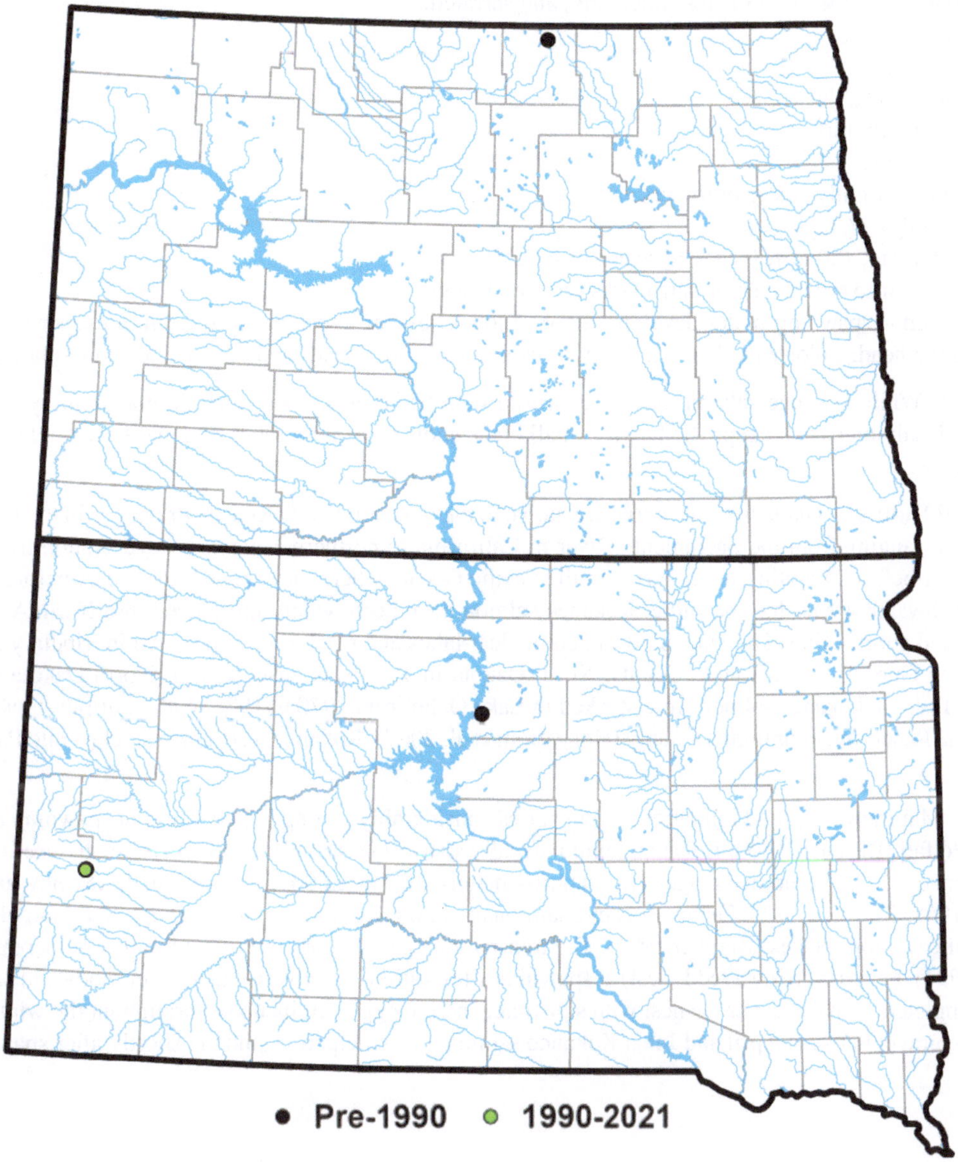

● Pre-1990   ● 1990-2021

Etymology: *Oncorhynchus nerka-Oncorhynchus* = hooked snout, referencing the elongated hooked jaw, or kype, of a spawning male; *nerka* = vernacular name from the Koryak languages of Kamchatka. The common name sockeye is the saltwater, or anadromous form, derived from various Native American words. The common name kokanee is the landlocked, freshwater, non-anadromous form, and means red fish, named by the Kootenai Native Americans.

Description:

Body – fusiform, elongate, slightly laterally compressed.
Color – varies depending on age and habitat type.

> *Dorsally* – dark green and silvery-blue with a few very fine black specks.
> *Laterally* – silver.
> *Ventrally* – silvery-white.

Head – short, conical.
Snout – short, bluntly pointed in nonspawning individuals.
Eyes – moderate, laterally on head.
Mouth – moderately large, terminal.

> *Upper jaw* – extending past posterior edge of eye.
Teeth – small, well developed, present on both jaws.
Gill rakers – 30 to 40 on the first gill arch, fine, long, and serrated.
Dorsal fin – 10 to 12 rays.
Adipose fin – present.
Caudal peduncle – short, thick.
Caudal fin – moderately forked.
Anal fin – 12 to 18 rays.
Pelvic fins – 11 rays, slightly posterior to dorsal fin.
Pectoral fins – roughly 16 rays.
Lateral line – complete; 120 to 150 cycloid scales in series.
Juveniles – 6 to 12 dark vertical bars or parr marks on lateral sides.
Sexual dimorphism – spawning males develop a hooked lower jaw, or kype, enlarged teeth, a deeper body depth, and a hump behind the head. Color will also turn dusky olive-green on their heads and dorsal side and red laterally.

Similar Species: While kokanee salmon may be confused with Chinook salmon, they do not occur sympatrically in the Dakotas. Chinook salmon have a deeper body with small black spots present on the dorsal and lateral sides, as well as both caudal fin lobes.

Distribution and Habitat: *Oncorhynchus nerka* are native to the northern Pacific coastal drainages in the United States and northeastern Asia, ranging from the Sacramento River in California to northern Japan. Sockeye salmon live the first 1 to 2 years in freshwater before migrating in the spring to the Pacific Ocean, where they spend 2 to 3 years before migrating back to their natal freshwater stream for spawning. Kokanee salmon have been widely introduced into North America for commercial harvest and sportfishing. They inhabit deep, cold, clear lakes and reservoirs, and spawn in tributary streams. A one-time stocking into Gordon Lake, Rolette County, North Dakota in 1955 was unsuccessful (North Dakota Game and Fish 1994). In South Dakota, kokanee salmon were stocked in Lake Oahe from 1970 to 1976, with multiple stockings from 1970 to 1984 in Pactola Reservoir, Pennington County (Hoagstrom et al. 2007). Kokanee were again stocked into Pactola Reservoir in 2022.

Reproduction: Maturation age and migration timing varies considerably among populations but spawning typically occurs in late summer to fall (Burgner 1991). Kokanee salmon usually spawn in streams but will occasionally use lake bottoms and shorelines (Behnke 2002). Fecundity ranges from approximately 300 to 2,000 eggs per female (Seeley and McCammon 1966; Foerster 1968; Burgner 1991). Female sockeye salmon are known to have the smallest egg sizes and highest fecundity rates of the Pacific salmon species (Burgner 1991). The transparent eggs are dark orange-red. Egg size increases with female size (Burgner 1991). Females create a redd (nest) at spawning sites within the substrate using their body and caudal fin. When spawning is completed, females cover the nest with substrate, and the eggs will incubate throughout the winter. Fry begin to emerge from the gravel between April and June. Kokanee salmon are semelparous and die shortly after spawning.

Age and Growth: Newly emerged fry have a total length of 25 to 31 mm, and growth rates vary greatly among populations (Burgner 1991). Kokanee salmon can reach a total length of 660 mm, with a longevity of 4 years (Behnke 2002).

Food and Feeding: Fry consume cladocerans, copepods, and ostracods. Juvenile kokanee salmon forage in the limnetic zone on zooplankton and aquatic insect larvae (Burgner 1991). Adults use their gill rakers to strain large numbers of zooplankton, and also eat aquatic insects and freshwater shrimp (Burgner 1991). Like other species of Pacific salmon, feeding ends during spawning migrations.

References

Behnke R (2002) Trout and salmon of North America. Simon and Schuster, New York

Burgner RL (1991) Life history of sockeye salmon (*Oncorhynchus nerka*). In: Groot C, Margolis L (eds) Pacific salmon life histories. UBC Press, Vancouver

Foerster RE (1968) The sockeye salmon, *Oncorhynchus nerka*. Fish Res Board Can, Bull 16(162)

Hoagstrom CW, Wall SS, Kral JG, Blackwell BG, Berry CR (2007) Zoogeographic patterns and faunal change of South Dakota fishes. West N Am Nat 67:161–184. https://doi.org/10.3398/1527-0904(2007)67[161:ZPAFCO]2.0.CO;2

North Dakota Game and Fish (1994) Fishes of the Dakotas. Brochure. N D Game Fish Dep, Bismark, North Dakota

Seeley CM, McCammon GW (1966) Kokanee. Inland fisheries management. Calif Dep Fish Game, Sacramento, California, pp 274–294

6.17.6 Chinook Salmon *Oncorhynchus tshawytscha*

(Walbaum, 1792)

Nuptial Male   Female

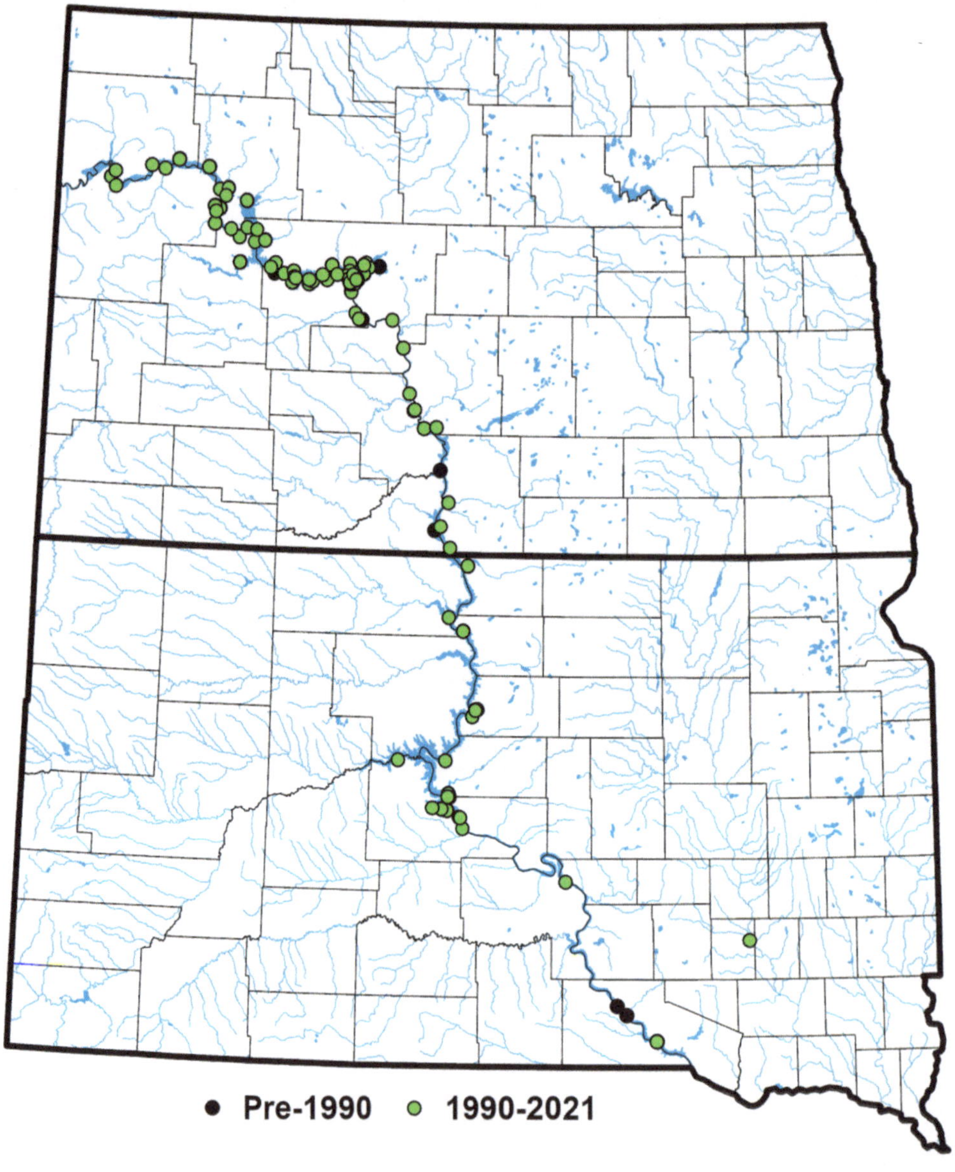

● Pre-1990   ○ 1990-2021

Etymology: *Oncorhynchus tshawytscha-Oncorhynchus* = hooked snout, referencing the elongated hooked jaw, or kype, of a spawning male; *tshawytscha* = vernacular name in Kamchatka. The common name derives from the Chinook Native American tribe of northwestern North America.

Description:

Body – fusiform, laterally compressed.
Color – few small dark spots present on dorsal, lateral, and fins.
>   *Dorsally* – dark gray to blue-green.
>   *Laterally* – silver to silvery-olive-blue.
>   *Ventrally* – silver to white.
>   *Mouth* – gum of lower jaw black; tongue gray.

Head – elongated.
Snout – elongated.
Eyes – moderate, laterally on head.
Mouth – large, terminal, slightly oblique.

>   *Upper jaw* – extending past posterior edge of eye.
Teeth – large, sharp; present on both jaws.
Gill rakers – 16 to 26.
Dorsal fin – 10 to 14 rays.
Adipose fin – present.
Caudal peduncle – short, thick.
Caudal fin – shallowly forked.
Anal fin – 14 to 19 rays.
Pelvic fins – 10 or 11 rays, axillary process.
Pectoral fins – 14 to 17 rays.
Lateral line – 130 to 165 cycloid scales in series.
Juveniles – 6 to 12 dark vertical bars or parr marks on lateral sides.
Sexual dimorphism – spawning males have a hooked lower jaw, or kype.

Similar Species: Black gums on the lower jaw and a gray tongue distinguish Chinook salmon from other salmonids in the Dakotas.

Distribution and Habitat: Chinook salmon are native to the Arctic and Pacific drainages in North America from Alaska to California. They inhabit cold, clear, well-oxygenated pelagic waters. In the Dakotas, they only occur in the coldwater habitat of the Missouri River reservoirs and tailraces (Cordes 2007). Chinook salmon were stocked into Lake Sakakawea in 1976, with some migrating downstream to Lake Oahe by 1979 (Warnick 1987; Marrone and Stout 1996; Hill 1997). Chinook salmon were subsequently stocked into Lake Oahe in 1982 (Warnick 1987). Dam entrainment occurs, with salmon occasionally present in dam tailwaters. In an unusual case, on November 9, 2019, an angler caught a male that had been entrained through four Missouri River dams and swam over 150 river miles up the James River to Firesteel Creek, about one mile east of Lake Mitchell. As the Missouri River reservoirs warm in the summer and begin to thermally stratify, Chinook salmon go deeper (Aaland 1987; Hill 1997). The preferred water temperatures of adults and juveniles are 11 °C and 14 °C, respectively (Brett 1952; Haynes and Keleher 1986). Lake Oahe Chinook salmon have poor homing ability (Lott et al. 1997).

Reproduction: Natural reproduction of Chinook salmon does not occur in the Missouri River because suitable spawning habitat is nonexistent (Marrone and Stout 1996). Chinook salmon populations in the Dakotas are maintained entirely by hatchery production (Lott et al. 1997). Artificial spawning of Lake Oahe Chinook salmon occurs in October at Whitlock Bay Spawning Station near Gettysburg, South Dakota. The size of spawning fish varies yearly, but fish from age-1 to age-4 typically ascend the fish ladder, with most of the fish spawned age-3 (Marrone and Stout 1996; Lott et al. 1997; Barnes 2007). Female size at spawning and fecundity are considerably less than Chinook salmon in their native range, likely because of dietary differences (Barnes et al. 2000). Mean spawning female total length is 660 mm, with a fecundity of 2,708 eggs and a relative fecundity of 3.6 eggs/mm (Barnes et al. 2000; Becket et al. 2015). Lake Oahe Chinook salmon eggs are pea-sized and egg size is not related to female size (Barnes et al. 2000). Egg survival is relatively poor during hatchery incubation, but ranges from zero to nearly 100% depending on the female (Barnes et al. 2000, 2003). In their native range, Chinook salmon are anadromous, spending 3 to 4 years in the ocean before migrating to their natal freshwater rivers to spawn over clean gravel substrate. Females deposit eggs within a nest, or redd, which they construct using their body and caudal fin. Shortly after fertilization, females cover the redd with gravel before moving upstream to repeat the process multiple times. Chinook salmon are semelparous and die shortly after spawning.

Age and Growth: In Lake Oahe, male Chinook salmon are slightly longer than females at each age (Hill 1997). Their greatest growth increment is from age-1 to age-2 (Hill 1997). Total lengths at age for male Chinook salmon from Lake Oahe in 1995 were 373 mm at age-1, 373 mm, 572 mm at age-2, and 672 mm at age-3 (Hill 1997). Chinook salmon from age-1 to age-2 grow faster in Lake Oahe than in Lake Sakakawea (Aadland 1987; Hill 1997). Few Chinook salmon live past age-3 in Lake Oahe but have obtained age-5 in Lake Sakakawea (Aadland 1987; Marrone and Stout 1996; Hill 1997). In Lake Oahe, they can reach a total length of 914 mm and a weight of 13.8 kg but can exceed 45 kg in their native range. Longevity is 9 years (Becker 1983).

Food and Feeding: In Lake Oahe, Chinook salmon are heavily dependent on the abundance of rainbow smelt (Hill 1997). Juveniles eat zooplankton, aquatic and terrestrial insects, and both juvenile and adult rainbow smelt (Hill 1997). After juvenile Chinook salmon are stocked in late May, they immediately eat invertebrates but within about 2 months switch to eating rainbow smelt (Hill 1997). Age-1 Chinook salmon primarily consume age-0 and adult rainbow smelt, while age-3 salmon eat only adult rainbow smelt (Hill 1997). The length of rainbow smelt consumed increases with increased Chinook salmon length (Hill 1997). In years with relatively low rainbow smelt abundance, adult Chinook salmon diets include a variety of aquatic invertebrates, zooplankton, and lesser rainbow smelt amounts (Barnes 2007). In their native range, Chinook salmon consume fish, euphausiids, crab larva, squid, and pelagic amphipods (Higgs et al. 1995).

References

Aadland LPE (1987) Food habits, distribution, age and growth of Chinook salmon, and predation on newly stocked Chinook salmon in Lake Sakakawea, North Dakota. Dissertation, University of North Dakota

Barnes ME (2007) Fish hatcheries and stocking practices: past and present. In: Berry C, Higgins K, Willis D, Chipps S (eds) History of fisheries and fishing in South Dakota. S D Game Fish Park, Pierre, pp 267–293

Barnes ME, Hanten RP, Cordes RJ, Sayler WA, Carreiro J (2000) Reproductive performance of inland fall Chinook salmon. N Am J Aquacult 62:203–211. https://doi.org/10.1577/1548-8454

Barnes ME, Sayler WA, Cordes RJ, Hanten RP (2003) Potential indicators of egg viability in landlocked fall Chinook salmon spawn with or without the presence of overripe eggs. N Am J Aquacult 65:49–55. https://doi.org/10.1577/1548-8454

Becker GC (1983) Fishes of Wisconsin. University of Wisconsin Press, Madison

Becket KH, Barnes ME, Durben DJ, Parker TM (2015) Landlocked fall Chinook salmon ovarian fluid turbidity and egg survival. N Am J Aquacult 77:18–21. https://doi.org/10.1080/15222055.2014.951811

Brett JR (1952) Temperature tolerance in young Pacific salmon, genus *Oncorhynchus*. J Fish Res Board Can 9:265–323. https://doi.org/10.1139/f52-016

Cordes R (2007) Cold-water fish species. In: Berry C, Higgins K, Willis D, Chipps S (eds) History of fisheries and fishing in South Dakota. S D Game Fish Park, Pierre, pp 201–211

Haynes JM, Keleher CJ (1986) Movements of Pacific salmon in Lake Ontario in spring and summer: evidence of wide spatial dispersal. J Freshw Ecol 3:289–297. https://doi.org/10.1080/02705060.1986.9665120

Higgs DA, Macdonald JS, Levings CD, Dosanjh BS (1995) Nutrition and feeding habits in relation to life history stage. In: Grott C, Margolis L, Clarke WC (eds) Physiological ecology of Pacific salmon. University of British Columbia Press, Vancouver, pp 159–316

Hill TD (1997) Life history and bioenergetics of Chinook salmon in Lake Oahe, South Dakota. Dissertation, South Dakota State University

Lott J, Marrone G, Stout D (1997) Influence of size-and-date at stocking, imprinting attempts and growth on initial survival, homing ability, maturation patterns and angler harvest of Chinook salmon in Lake Oahe, South Dakota. S D Dep Game Fish Park, Prog Rep 97-20, Pierre

Marrone GM, Stout DA (1996) Whitlocks Bay spawning and imprinting station annual report 1995. S D Dep Game Fish Park, Wildl Divi, Pierre

Warnick DC (1987) The introduction of selected fish species into the Missouri River system, 1987-1985. S D Dep Game Fish Park, Wildl Divi, Prog Rep 87-10, Pierre

6.17.7   Atlantic Salmon *Salmo salar*

Linnaeus, 1758

● **Pre-1990**   ● **1990-2021**

Etymology: *Salmo salar* – *Salmo* = salmon; *salar* = the leaper, referencing the ability to leap over low waterfalls and rapids to reach the spawning habitat.

Description:

Body – fusiform, laterally compressed.
Color – varies depending on age and habitat type; further described as an anadromous form.

> *Migrating young* -

>> *Dorsally* – dark gray to black.
>> *Laterally* – silver.
>> *Ventrally* – white.

> *Nonmigrating adults (ocean)* – bright silvery-blue, dark x-shaped spots on body above lateral line, head, opercle, and fins.

>> *Migrating adults (freshwater)* – darken to bronze or black color.

Head – small.
Snout – blunt.
Eyes – relatively small, laterally on head.
Mouth – relatively small.

> *Upper jaw* – extends to posterior edge of eye or slightly beyond eye.

Teeth – well developed, on both jaws; small, conical.
Gill rakers – 15 to 20 on the first arch, slender.
Dorsal fin – 10 to 12 rays.
Adipose fin – present.
Caudal peduncle – narrow.
Caudal fin – shallowly forked.
Anal fin – 8 to 11 rays.
Pelvic fins – insertion posterior to dorsal fin insertion, axillary process.
Lateral line – complete; 109 to 121 rather large cycloid scales in series.
Juveniles – referred to as parr in freshwater – bronze to brown appearance; laterally – small red and black spots, 8 to 11 dark vertical bars or parr marks.
Sexual dimorphism – spawning males have a kype and may even develop a slight red or olive-green hue.

Similar Species: Atlantic salmon may be confused with Chinook salmon in Lake Oahe, South Dakota. Chinook salmon have a lower jaw black gum and an anal fin with 14–19 rays.

Distribution and Habitat: The native range of Atlantic salmon encompasses the east and west coasts of the North Atlantic Ocean, including North America, Iceland, Greenland, Europe, and Russia. Within the United States, Atlantic salmon are native to Maine, New Hampshire, Vermont, Massachusetts, Rhode Island, Connecticut, and New York. Although not native to the Dakotas, Atlantic salmon were introduced into the South Dakota portion of Lake Oahe in 2018. They likely occur as a continuous distribution throughout Lake Oahe and downstream because of entrainment. Atlantic salmon can be both anadromous and non-anadromous, and therefore can occupy a wide range of habitats (Elliot et al. 1998). Entirely landlocked populations exist because of barriers created during a land mass uplift following Pleistocene glaciation (Power 1958; Berg 1985; Fleming 1996).

Reproduction: Age and size at sexual maturity in anadromous Atlantic salmon increases with increasing freshwater migration distance (Fleming 1996). Within-population size-at-sexual-maturity variability is likely the greatest of all salmonids, which may help maximize survival and population stability (Scott and Crossman 1973; Fleming 1996; Klemetsen et al. 2003). Like other salmonids, anadromous Atlantic salmon have high spawning-site fidelity, migrating from the ocean to spawn in the same freshwater rivers where they developed as juveniles (Fleming 1996). Non-anadromous, lacustrine populations complete their entire lifecycle in freshwater and migrate into tributaries to spawn. Females produce uniformly sized eggs, but the size and weight of the eggs may vary by female among and within populations (Fleming 1996). In general, fecundity is lower in larger females and higher in smaller females (Fleming 1996). Egg size increases with female size, with larger females having larger but fewer eggs compared to smaller females with smaller but more numerous eggs (Fleming 1996). Females build redds, nests in gravel beds upstream of riffles and gravel bars or near the tail end of pools where water velocity increases (Fleming 1996). After egg release and fertilization, females cover the redd and retreat to a nearby pool or mainstem (Webb and Hawkins 1989; Fleming 1996). Atlantic salmon are iteroparous and spawn multiple years, unlike semelparous Chinook salmon that die after spawning.

Age and Growth: Lengths at age are highly variable among Atlantic salmon populations. Average parr total length in the United States is 150 mm (Hutchings and Jones 1998). Non-anadromous Atlantic salmon from Maine and Lake Ontario can weigh up to 12.2 kg and 20.4 kg, respectively, similar to anadromous populations (Scott and Crossman 1973; Warner and Harvey 1985; Klemetsen et al. 2003).

Food and Feeding: Fry and juvenile Atlantic salmon eat plankton, blackfly larvae, stoneflies, caddisflies, chironomids, and other small aquatic invertebrates. Adults are active pelagic predators, eating plankton and aquatic invertebrates, with larger adults also eating herring, alewives, rainbow smelt and other fishes (Sayers Jr et al. 2011).

References

Berg OK (1985) The formation of non-anadromous populations of Atlantic salmon, *Salmo salar* L., in Europe. J Fish Biol 27:805–815. https://doi.org/10.1111/j.1095-8649.1985.tb03222.x

Elliot SR, Coe TA, Helfield JM, Naiman RJ (1998) Spatial variation in environmental characteristics of Atlantic salmon (*Salmo salar*) rivers. Can J Fish Aquat Sci 55:267–280. https://doi.org/10.1139/d98-001

Fleming IA (1996) Reproductive strategies of Atlantic salmon: ecology and evolution. Rev Fish Biol Fish 6:379–416. https://doi.org/10.1007/BF00164323

Hutchings JA, Jones MEB (1998) Life history variation and growth rate thresholds for maturity in Atlantic salmon, *Salmo salar*. Can J Fish Aquat Sci 55:23–47. https://doi.org/10.1139/d98-004

Klemetsen A, Amundsen PA, Dempson JB, Jonsson B, Jonsson N, O'Connell MF, Mortensen E (2003) Atlantic salmon *Salmo salar* L., brown trout *Salmo trutta* L., and Arctic charr *Salvelinus alpinus* (L.): a review of aspects of their life histories. Ecol Freshw Fish 12:1–59. https://doi.org/10.1034/j.1600-0633.2003.00010.x

Power G (1958) The evolution of the freshwater races of the Atlantic salmon (*Salmo salar* L.) in eastern North America. Arctic 11:86–92. https://doi.org/10.14430/arctic3735

Sayers RE Jr, Moring JR, Johnson PR, Roy SA (2011) Importance of rainbow smelt in the winter diet of landlocked Atlantic salmon in four Maine lakes. N Am J Fish Manag 9:298–302. https://doi.org/10.1577/1548-8675

Scott WB, Crossman EJ (1973) Freshwater fishes of Canada. Fisheries Research Board of Canada, Bulletin 184

Warner K, Harvey KA (1985) Life history, ecology, and management of Maine landlocked salmon. Maine Dep Inland Fish Wildl 64. https://digitalmaine.com/ifw_docs/64

Webb JH, Hawkins AD (1989) The movements and spawning behaviour of adult salmon in the Girnock Burn, a tributary of the Aberdeenshire Dee, 1986. Dep Agricult Fish Scotl, 40

6.17.8   Brown Trout *Salmo trutta*

Linnaeus, 1758

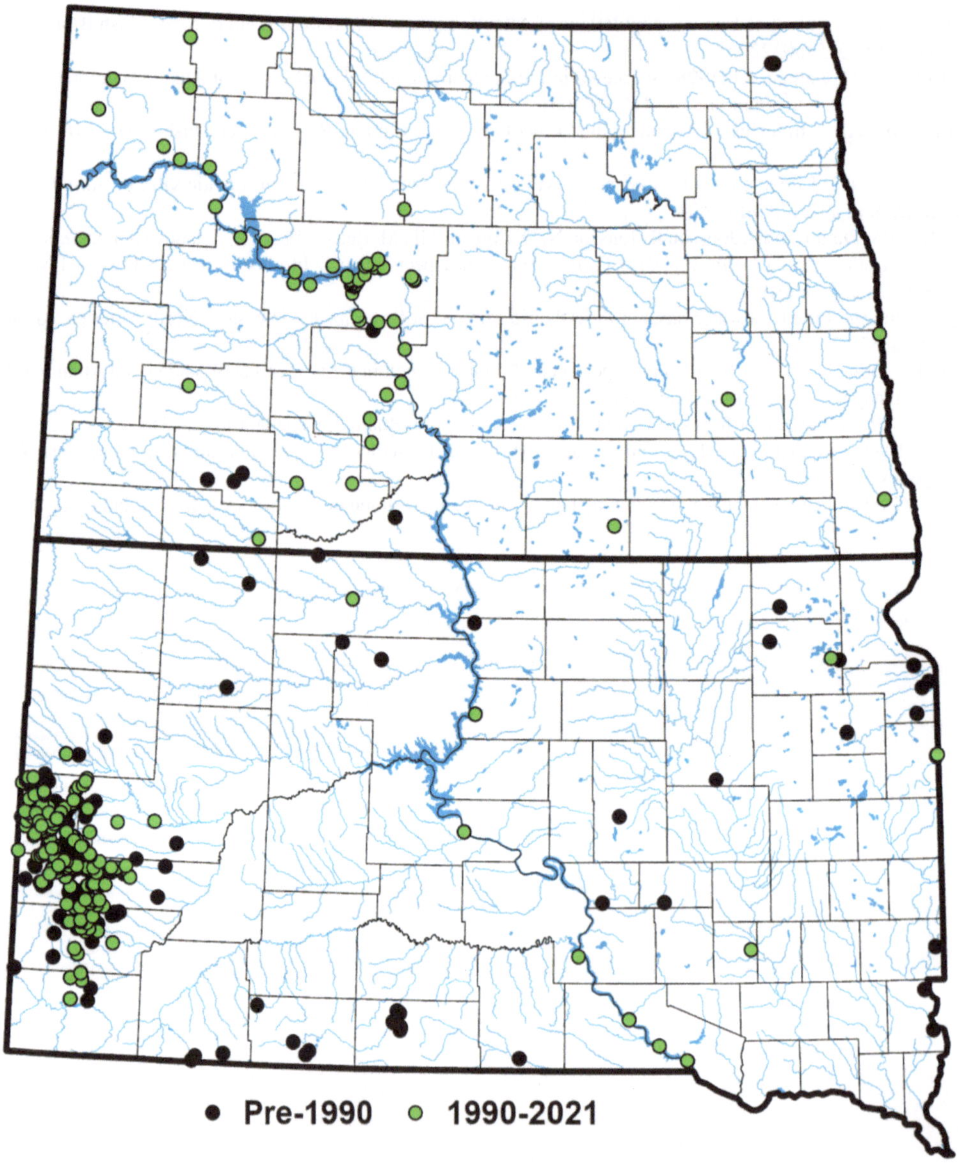

● Pre-1990   ● 1990-2021

Etymology: *Salmo trutta – Salmo* = salmon *trutta* = trout.

Description:

Body – slender, elongated, slightly laterally compressed.
Color –

   Dorsally – brown to olive; large dark brown to black spots as large as pupil, outlined in tan to light blue, leading onto lateral sides.
   Laterally – light olive to tan; continued large, round to irregularly shaped dark spots outlined in light tan or blue; less numerous small, prominent red spots, also outlined in light tan or blue.
   Ventrally – cream to white.
   Fins – dorsal – numerous small dark brown spots; adipose – orange to red on distal end; caudal – little to no spots; anal – white edge on distal end.

Snout – rounded.
Eyes – moderately large, laterally on head.
Mouth – large, terminal, slightly oblique.

   Upper jaw – extending to or slightly beyond posterior edge of eye.
Teeth – present on both jaws, tongue, vomer, and palatines.
Gill rakers – 14 to 20.
Dorsal fin – 12 to 14 rays.
Adipose fin – present.
Caudal peduncle – stout.
Caudal fin – square to slightly forked.
Anal fin – 10 to 12 rays.
Pelvic fins – 9-10 rays; abdominal, small axillary process.
Pectoral fins – 13 to 14 rays.
Lateral line – complete; 120 to 130 cycloid scales in series.
Juveniles – less numerous spots on lateral sides, 7 to 11 dark, oval shaped blotches, or parr marks on lateral sides.
Sexual dimorphism – spawning males have a hooked lower jaw or kype.

Similar Species: Brown trout resemble brook trout, coastal cutthroat trout, and rainbow trout. Brook trout have well-defined small pink or red spots outlined in blue against a dark olive background on lateral sides, and well-defined black stripes and white edging on lower lobe distal ends of the caudal, anal, pelvic, and pectoral fins. Coastal cutthroat trout have numerous small black spots on the body extending onto the caudal fin and reddish-orange slash marks on each underside of the lower jaw. Rainbow trout have a bright pink to reddish lateral stripe and numerous small black spots on the body.

Distribution and Habitat: Brown trout are native to Europe, northern Africa, and western Asia (MacCrimmon et al. 1970). They have been widely introduced outside their native range and now have a world-wide distribution (Elliott 1994). Brown trout were introduced into the United States in 1883 (Behnke 2002). In the Dakotas, they occur mainly west of the Missouri River, but are also stocked elsewhere to create temporary put-and-take fisheries. Adult brown trout inhabit deeper areas with vegetative cover and little flow in streams with gravel or stony substrate (Klemetsen et al. 2003). In the Black Hills of South Dakota, runs and pools are used more often than riffles in fall and winter, with runs used more in the spring (James et al. 2007). They are rather sedentary with small home ranges and migrations (Bachman 1984; James et al. 2007). Female brown trout typically move more than males (Campbell 1977; Elliott 1994; Klemetsen et al. 2003). Movements are induced by water temperature and flow changes (Jonsson 1991; Klemetsen et al. 2003). Annual movements of brown trout in the Black Hills averaged 23 m and ranged from 2 to 150 m, with the longest moves during fall and spring (James et al. 2007). When water temperature drops below 8 °C, activity decreases and the fish migrate to deeper overwintering habitat (Heggenes and Dokk 2001; Klemetsen et al. 2003; James et al. 2007). Fry often move downstream after hatching to shallow, slow-flowing areas under 30 cm deep near shorelines or banks with water velocities from 0 to 20 cm/second and sand or gravel substrate (Elliott 1986; Roussel and Bardonnet 1999; Armstrong et al. 2003; Klemetsen et al. 2003).

Reproduction: Spawning migrations begin in the fall. Brown trout in the Black Hills of South Dakota migrate up to 1091 m upstream to spawn, and after spawning often return within a few meters of their prespawning localities (James et al. 2007). Spawning occurs in fall or early winter at water temperatures of 3 to 13 °C in water velocities from 10.8 to 80.2 cm/second at depths around 25.5 cm over gravel substrate near instream cover (Witzel and MacCrimmon 1983). The timing of brown trout sexual maturity is highly variable and can occur from age-1 to age-10, with males maturing earlier than females

(Klemetsen et al. 2003). Brown trout mature later in colder water typically found in northern rivers and mountain lakes (Jonsson et al. 1991; Klemetsen et al. 2003). Several males court a single female, but only one large male will likely fertilize most of the eggs (Jones and Ball 1954; Largiander et al. 2001; Klemetsen et al. 2003). Females construct redds (nests) at least 14 cm deep in gravel or sand substrate (Witzel and MacCrimmon 1983). Fecundity depends on female size and age. Brown trout are fractional spawners, with females depositing only a portion of their eggs in each redd (Klemetsen et al. 2003). After egg deposition, females cover the eggs with stones and gravel to incubate for several months. Females then abandon the redd and let the male continue spawning with other females (Klemetsen et al. 2003). Egg size increases with the female size (Klemetsen et al. 2003). Hatching occurs in spring. Egg mortality is typically less than 5% in water temperatures of 9 to 10 °C. Mortality increases with rising temperatures, becoming 50% at 12 °C and 100% at temperatures over 16 °C (Jowett 1992; Crisp 1993; Armstrong et al. 2003).

Age and Growth: Differences in temperature and food availability make growth highly variable among populations (Klemetsen et al. 2003). Growth is density-dependent in both streams and lakes (Jensen 1977; Jenkins et al. 1999; Klemetsen et al. 2003). Brown trout larvae have a total length of approximately 20 mm at hatch, but larval size may decrease with increased incubation temperatures (Klemetsen et al. 2003). The optimal growth temperature for brown trout is from 13 to 18 °C (Klemetsen et al. 2003). Their upper and lower critical temperatures for growth are 25 to 26 °C and 3 to 6 °C, respectively (Klemetsen et al. 2003). Brown trout often reach total lengths from 254 to 381 mm but can reach 762 mm in the Dakotas. Longevity is 20 years (Brown 1957; Klemetsen et al. 2003).

Food and Feeding: Brown trout are opportunistic, voracious, visual predators. Larvae and juveniles consume aquatic insect larvae near the shorelines and water surface (Klemetsen et al. 2003). Diets vary by season, habitat, fish age, and fish size (Bridcut and Giller 1995; Klemetsen et al. 2003). Prey size increases with fish size. Adults primarily feed during the day near the surface and shoreline on ephemeroptera larvae, plecopteran larvae, trichoptera larvae, and other aquatic invertebrates. Adults also eat terrestrial insects, worms, crayfish, and small fish. In lakes, zoobenthos and zooplankton are important food sources (Klemetsen et al. 2003). Brown trout optimum feeding temperature is 16 °C (Becker 1983).

References

Armstrong JD, Kemp PS, Kennedy GJA, Ladle M, Milner NJ (2003) Habitat requirements of Atlantic salmon and brown trout in rivers and streams. Fish Res 62:143–170. https://doi.org/10.1016/S0165-7836(02)00160-1

Bachman RA (1984) Foraging behavior of free-ranging wild and hatchery brown trout in a stream. Trans Am Fish Soc 113:1–32. https://doi.org/10.1577/1548-8659

Becker GC (1983) Fishes of Wisconsin. University of Wisconsin Press, Madison

Behnke RJ (2002) Trout and salmon of North America. Simon and Schuster, New York

Bridcut EE, Giller PS (1995) Diet variability and foraging strategies on brown trout (*Salmo trutta*): an analysis from subpopulations to individuals. Can Fish Aqua Sci 52:2543–2552. https://doi.org/10.1139/f95-845

Brown ME (1957) Experimental studies on growth. In: Brown ME (ed) The physiology of fishes. Academy Press, New York, pp 361–400. https://doi.org/10.1016/B978-1-4832-2817-4.50015-9

Campbell JS (1977) Spawning characteristics of brown trout and sea trout *Salmo trutta* L. in Kirk Burn, River Tweed, Scotland. J Fish Biol 11:217–229. https://doi.org/10.1111/j.1095-8649.1977.tb04115.x

Crisp DT (1993) The environmental requirements of salmon and trout in fresh water. Freshwater Forum 3:176–202

Elliott JM (1986) Spatial distribution and behavioural movements of migratory trout *Salmo trutta* in a Lake District stream. J Anim Ecol 55:907–922. https://doi.org/10.2307/4424

Elliott JM (1994) Quantitative ecology and the brown trout. Oxford University Press, Oxford

Heggenes J, Dokk JG (2001) Contrasting temperatures, waterflows, and light: seasonal habitat selection by young Atlantic salmon and brown trout in a boreonemoral river. Regul Rivers: Res Manage 17:623–635. https://doi.org/10.1002/rrr.620

James DA, Erickson JW, Barton BA (2007) Brown trout seasonal movement patterns and habitat use in an urbanized South Dakota stream. N Am J Fish Manag 27:978–985. https://doi.org/10.1577/M06-091.1

Jenkins TM, Diehl S, Kratz KW, Cooper SD (1999) Effects of population density on individual growth of brown trout in streams. Ecology 80:941–956. https://doi.org/10.1890/0012-9658(1999)080[0941:EOPDOI]2.0.CO;2

Jensen KW (1977) On the dynamics and exploitation of the population of brown trout, Salmo trutta L., in Lake Øvre Heimdalsvatn, southern Norway. Rep Inst Freshw Res Drottningholm 56:18–69

Jones JW, Ball JN (1954) The spawning behavior of brown trout and salmon. Br J Anim Behav 2:103–114. https://doi.org/10.1016/S0950-5601(54)80046-3

Jonsson N (1991) Influence of water flow, water temperature and light on fish migration in rivers. Nordic J Freshwater Res 66:20–35

Jonsson B, L'Abèe-Lund JH, Heggberget TG, Jensen AJ, Johnsen BO, Næsje TF, Sættem LM (1991) Longevity, body size and growth in anadromous brown trout. Can J Fish Aquat Sci 48:1838–1845. https://doi.org/10.1139/f91-217

Jowett IG (1992) Models of abundance of large brown trout in New Zealand rivers. N Am J Fish Manag 12:417–432. https://doi.org/10.1577/1548-8675

Klemetsen A, Amundsen PA, Dempson JB, Jonsson B, Jonsson N, O'connell MF, Mortensen E (2003) Atlantic salmon *Salmo salar* L., brown trout *Salmo trutta* L. and Arctic charr *Salvelinus alpinus* (L.): a review of aspects of their life histories. Ecol Freshw Fish 12:1–59. https://doi.org/10.1034/j.1600-0633.2003.00010.x

Largiander CR, Estoup A, Lecerf F, Champigneulle A, Guyomard R (2001) Microsatellite analysis of polyandry and spawning site competition in brown trout (*Salmo trutta* L.). Gene Select Evol 33:205–222. https://doi.org/10.1186/BF03500881

MacCrimmon HR, Marshall TL, Gotos BL (1970) World distribution of brown trout, *Salmo trutta*: further observations. J Fish Res Board Can 27:811–818. https://doi.org/10.1139/f70-085

Roussel JM, Bardonnet A (1999) Ontogeny of diel pattern of stream-margin habitat use by emerging brown trout, *Salmo trutta*, in experimental channels: influence of food and predator presence. Environ Biol Fish 56:253–262. https://doi.org/10.1023/A:1007504402613

Witzel L, MacCrimmon H (1983) Redd-site selection by brook trout and brown trout in southwestern Ontario streams. Trans Am Fish Soc 112:760–771. https://doi.org/10.1577/1548-8659

6.17.9   Brook Trout *Salvelinus fontinalis*

(Mitchill, 1814)

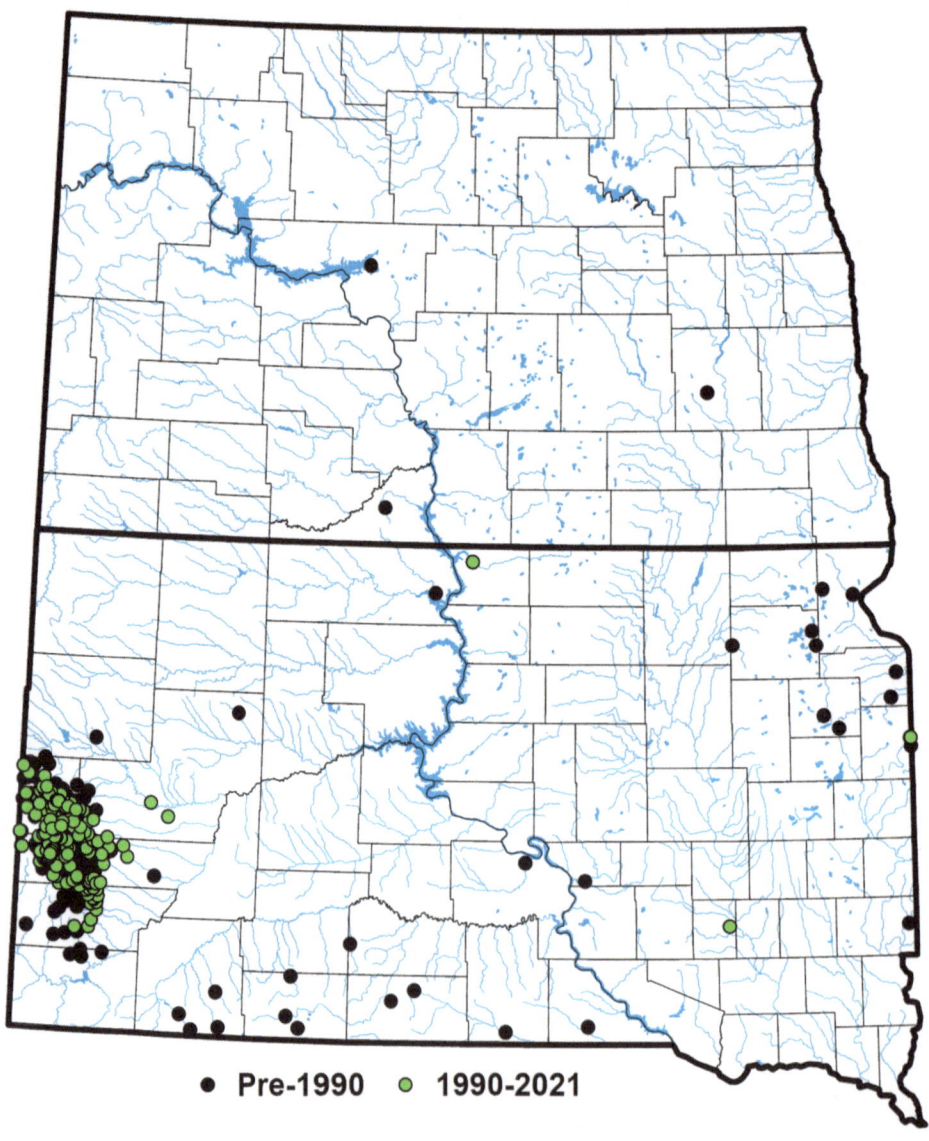

Etymology: *Salvelinus fontinalis* – *Salvelinus* = a deviation of the old vernacular name for char; *fontinalis* = of springs, likely referencing habitat.

Description:

Body – slender, elongated, slightly laterally compressed.
Color –

    Dorsally – olive to dark bluish-gray; irregular light gray blotches or vermiculations.
    Laterally – light, well-defined red or pink spots outlined in light blue against a dark olive background, light cream to gray irregular blotches.
    Ventrally – light yellow, cream to white.
    Fins – dorsal, caudal – dark brown to black spots; lower lobe caudal, anal, pelvic, pectoral – light red to orange leading to distinct black stripe, followed by white edge on distal end.

Head – small.
Snout – rounded.
Eyes – moderately large, laterally on head.
Mouth – large, terminal, slightly oblique.

    Upper jaw – extends beyond posterior edge of eye.
Teeth – on both jaws and tongue.
Gill rakers – 9 to 12.
Dorsal fin – 10 to 14 rays.
Adipose fin – present.
Caudal peduncle – stout.
Caudal fin – squared, slightly forked.
Anal fin – 9 to 14 rays.
Pelvic fins – 8 to 10 rays; abdominal, small axillary process.
Pectoral fins – 11 to 14 rays.
Lateral line – complete; 210 to 240 cycloid scales in series.
Juveniles – dorsally darker; laterally 7 to 9 dark, oval shaped blotches, or parr marks.
Sexual dimorphism – spawning males have black gum lines, a hooked lower jaw or kype, and bright coloration; lateral spots become more intense; lateral-ventrally bright gold to red; black pigmented striped on each side of the belly; pectoral, pelvic, and anal fins bright red.

Similar Species: Well-defined pink or red spots outlined in blue against a dark olive background on their lateral sides and black stripes on the lower lobe of the caudal, anal, pelvic, and pectoral fins distinguish brook trout from other trout species in the Dakotas. Brown trout have black to brown spots outlined in blue on light olive to tan lateral sides, and generally lack numerous caudal fin dark spots. Rainbow trout have a pinkish-red lateral stripe and numerous small black spots. Lake trout have a slate-gray to emerald-green head and body heavily marked with creamy white spots, and a forked caudal fin. Splake (a hybrid resulting from crossing a female lake trout and a male brook trout) body coloration more closely resembles that of a lake trout.

Distribution and Habitat: Brook trout are native to most of eastern Canada and the northeastern United States from eastern Minnesota and the upper Mississippi River basin to the Great Lakes, Hudson Bay, and Atlantic drainages from New York south through the Appalachian Mountains to northern Georgia. Brook trout have been introduced to much in the western United States. Most brook trout in the Dakotas reproduce naturally and are abundant in the Black Hills of South Dakota, but they have also been stocked in temporary put-and-take fisheries across both states. Brook trout inhabit cool, clear, spring-fed ground waters of headwater streams and oligotrophic lakes with gravel substrate and aquatic vegetation or submerged structure (Hoxmeier and Dieterman 2013). They migrate downstream after spawning to overwinter in deeper, quiet waters, such as eddies or under banks (Benson 1955). Brook trout are rather sedentary, with little annual movement outside of spawning. They prefer water temperatures from 11 to 16 °C (Coutant 1977). Brook trout fry have an optimum water temperature of 8 to 12 °C and inhabit shallow areas near shorelines with aquatic vegetation for cover (Raleigh 1982; Peterson et al. 1979). Brook trout are sensitive to moderate turbidity and warm water temperatures.

Reproduction: Upstream migrations to headwater streams or tributaries occur in the late summer and fall. Male brook trout sexually mature earlier than females. Brook trout typically mature early in life, but maturation timing is variable within populations (Raleigh 1982). Spawning occurs in late fall in water temperatures of 4 to 10 °C (Raleigh 1982). Females use their body and caudal fin to make redds (depressions in coarse or gravel substrate) in water from 36 to 61 cm deep with surface velocities of approximately 13 to 24 cm/second (Hazzard 1932; Essington et al. 1998; Bernier-Borugault and Magnan 2002). Females prefer to spawn with larger males and may delay spawning if paired with smaller males (Blanchfield and Ridgway 1999). Fecundity increases with female size and age. In Wisconsin, a female with a total length of 127 mm produced less than 100 eggs, while a female with a total length of 356 mm produced approximately 1,200 eggs (Becker 1983). After egg deposition and fertilization, females cover the redd with gravel and then depart, leaving the male to provide parental care for up to 2 weeks (Hazzard 1932). Hatching occurs in late winter to early spring depending on water temperature. Brook trout eggs hatch in 45 days at 10 °C and 165 days at 3 °C (Brasch et al. 1958; Raleigh 1982).

Age and Growth: The optimal water temperature for juvenile brook trout growth and survival is 12 to 15 °C (McCormick et al. 1972). Growth rates are fastest in spring and summer (Hoxmeier and Dieterman 2013). Optimal water temperatures for adult growth are from 9 to 17 °C (Baldwin 1956; Raleigh 1982). Brown trout grow more quickly in populations with fewer conspecifics (Carlson et al. 2007; Hoxmeier and Dieterman 2013). Annual growth rates vary depending on water temperatures, densities, and prey availability (Hoxmeier and Dieterman 2013). Density-dependent growth often occurs with high water temperatures and limited prey resources (Utz and Hartman 2009; Xu et al. 2010). Male brook trout growth rates are faster than females. Brook trout total length can exceed 508 mm. Average longevity in streams is 3 to 4 years, but brook trout can live for 14 years in high elevation streams (Kennedy et al. 2003).

Food and Feeding: Brook trout are opportunistic visual feeders that eat a wide variety of aquatic macroinvertebrates during daylight hours. Diptera, trichoptera, and ephemeroptera are an important part of the diet (Allan 1981). They also eat terrestrial insects, crustaceans, and annelids. Larger adults may occasionally eat smaller fish or eggs and are sometimes cannabalistic.

References

Allan JD (1981) Determinants of diet of brook trout (*Salvelinus fontinalis*) in a mountain stream. Can J Fish Aquat Sci 38:184–192. https://doi.org/10.1139/f81-024

Baldwin NS (1956) Food consumption and growth of brook trout at different temperatures. Trans Am Fish Soc 86:323–328. https://doi.org/10.1577/1548-8659(1956)86[323:FCAGOB]2.0.CO;2

Becker GC (1983) Fishes of Wisconsin. University of Wisconsin Press, Madison

Benson NG (1955) Observation on anchor ice in a Michigan trout stream. Ecology 36:529–530. https://doi.org/10.2307/1929599

Bernier-Borugault I, Magnan P (2002) Factors affecting red site selection, hatching, and emergence of brook charr, *Salvelinus fontinalis*, in an artificially enhanced site. Environ Biol Fish 64:333–341. https://doi.org/10.1023/A:1016006303854

Blanchfield PJ, Ridgway MS (1999) The cost of peripheral males in a brook trout mating system. Anim Behav 57:537–544. https://doi.org/10.1006/anbe.1998.1014

Brasch J, McFadden J, Kmiotek S (1958) Brook trout. Life history, ecology, and management. Wis Dep Nat Res Publ 226

Carlson SM, Hendry AP, Letcher BH (2007) Growth rate differences between resident native brook trout and non-native brown trout. J Fish Biol 71:1430–1447. https://doi.org/10.1111/j.1095-8649.2007.01615.x

Coutant CC (1977) Compilation of temperature preference data. J Fish Res Board Can 34:739–745. https://doi.org/10.1139/f77-115

Essington TE, Sorenson PW, Paron DG (1998) High rates of red superimposition by brook trout (*Salvelinus fontinalis*) and brown trout (*Salmo trutta*) in a Minnesota stream cannot be explained by habitat availability alone. Can J Fish Aquat Sci 55:2310–2316. https://doi.org/10.1139/f98-109

Hazzard AS (1932) Some phases of the life history of the eastern brook trout *Salvelinus fontinalus* Mitchill. Trans Am Fish Soc 62:344–350. https://doi.org/10.1577/1548-8659(1932)62[344:SPOTLH]2.0.CO;2

Hoxmeier RJH, Dieterman DJ (2013) Seasonal movement, growth, and survival of brook trout in sympatry with brown trout in Midwestern US streams. Ecol Freshw Fish 22:530–542. https://doi.org/10.1111/eff.12051

Kennedy BM, Peterson DP, Fausch KD (2003) Different life histories of brook trout populations invading mid-elevation and high-elevation cutthroat trout streams in Colorado. West N Am Nat 63:215–223

McCormick JH, Hokanson KEF, Jones BR (1972) Effects of temperature on growth and survival of young brook trout, *Salvelinus fontinalis*. J Fish Res Board Can 29:1107–1112. https://doi.org/10.1139/f72-165

Peterson RH, Sutterlin AM, Metcalfe JL (1979) Temperature preference of several species of *Salmo* and *Salvelinus* and some of their hybrids. J Fish Res Board Can 36:1137–1140. https://doi.org/10.1139/f79-159

Raleigh RF (1982) Habitat suitability index models: brook trout. US Dep Int, Fish Wildl Serv, FWS/OBS-82/10.24

Utz RM, Hartman KJ (2009) Density-dependent individual growth and size dynamics of central Appalachian brook trout (*Salvelinus fontinalis*). Can J Fish Aquat Sci 66:1072–1080. https://doi.org/10.1139/F09-063

Xu CL, Letcher BH, Nislow KH (2010) Size-dependent survival of brook trout *Salvelinus fontinalis* in summer: effects of water temperature and stream flow. J Fish Biol 76:2342–2369. https://doi.org/10.1111/j.1095-8649.2010.02619.x

6.17.10   Lake Trout *Salvelinus namaycush*

(Walbaum, 1792)

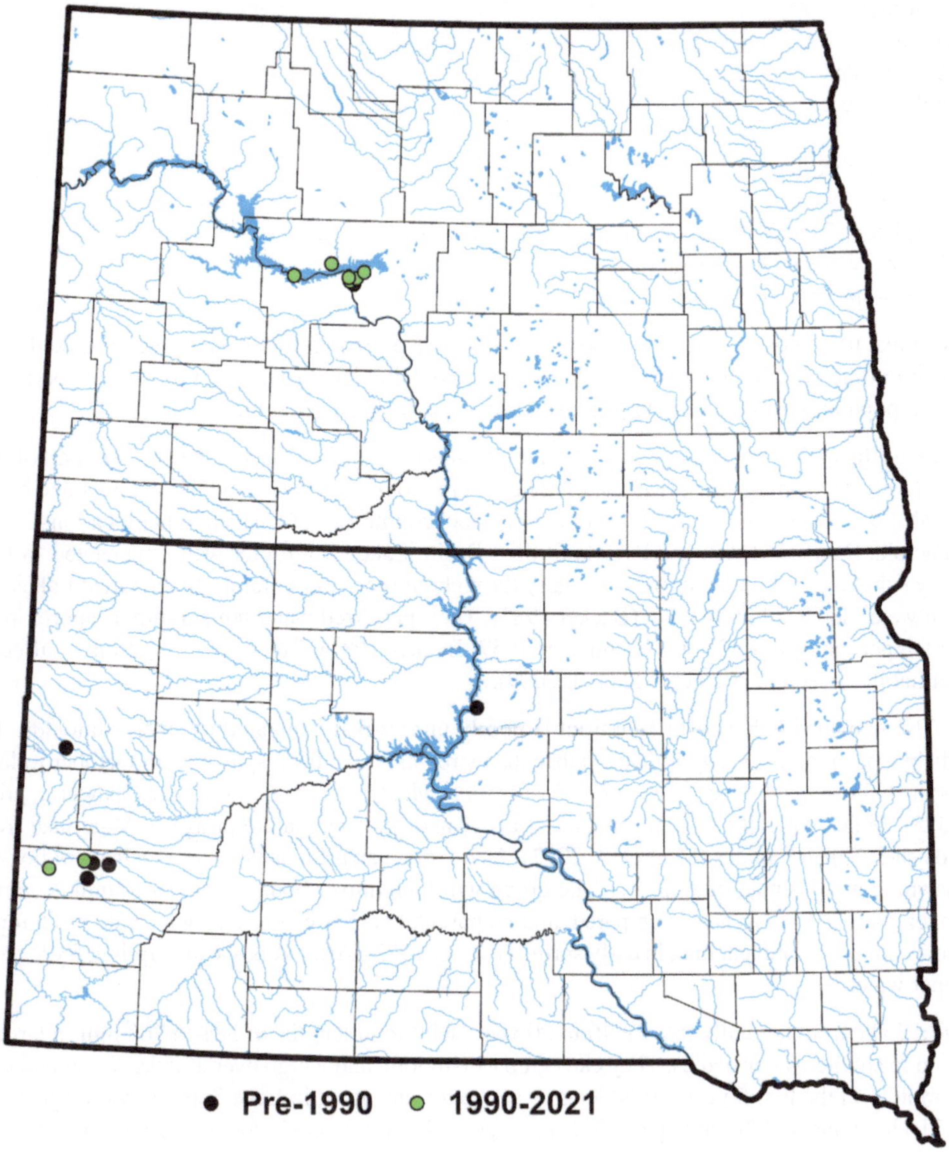

● **Pre-1990**   ○ **1990-2021**

Etymology: *Salvelinus namaycush* – *Salvelinus* = a deviation of the old vernacular name for char; *namaycush* = old Native American name.

Description:

Body – elongated, moderately slender.
Color –

> *Body* – slate-gray to dark emerald-green; creamy white spots present on head, body, and fins, fading to faint white ventral side.
> *Fins* – pectoral, pelvic, anal, caudal – slight orange tint; pectoral, pelvic, anal – white bordering on distal ends.

Head – short.
Opercle – 10 to 13 branchiostegal rays.
Snout – moderately conical, slightly protrudes beyond upper jaw.
Eyes – small.
Mouth – large, terminal; 9 or 10 mandibular pores.

> *Upper jaw* – extends well beyond eye.

Teeth – on upper and lower jaws; point posterior on prevomer.
Gill rakers – 14 to 25.
Dorsal fin – 9 to 14 rays.
Adipose fin – present.
Caudal peduncle – slender.
Caudal fin – deeply forked.
Anal fin – 8 to 11 rays.
Pelvic fins – 8 to 11 rays; axillary process.
Pectoral fins – 13 to 17 rays.
Lateral line – complete; 170 to 220, scales small, cycloid, deeply embedded.
Juveniles – thick, dark, vertical parr marks.
Sexual dimorphism – spawning males have a dark lateral stripe.

Similar Species: Lake trout closely resemble brook trout and splake. Brook trout have a weakly forked caudal fin and lateral sides with blue-outlined red spots. Splake, a hybrid produced using lake trout eggs brook trout sperm, occur sympatrically with lake trout populations in the Dakotas. Splake have a less-forked caudal fin.

Distribution and Habitat: The native range of lake trout includes most of Canada and the northern part of the United States from Montana to Maine, including the Great Lakes. They were stocked in 1973 and 1974 into Lake Sakakawea, North Dakota (Owen et al. 1981). In South Dakota, lake trout were stocked into Lake Oahe and other locations, with current populations in Lake Pactola, Deerfield Lake, and Pactola Stilling Basin. Lake trout prefer well-oxygenated, cold, deeper waters of lakes with rocky substrate. They may move seasonally throughout the water column. In the Great Lakes, lake trout abundance is highest at water depths of 30 to 90 m (Becker 1983). Their preferred water temperature is approximately 10 °C, but they can tolerate temperatures up to 25 °C (Eschmeyer 1957; Becker 1983). Light intensity strongly affects foraging efficiency and winter habitat selection (Blanchfield et al. 2009).

Reproduction: The spawning behavior of lake trout is poorly understood compared to other salmonids. They are unique among salmonids by spawning primarily in lakes rather than streams. Lake trout spawn from October to late November in the late evenings in shallow waters over gravel or rock substrate. Most fish return to the same spawning grounds each year (Binder et al. 2015). Males typically sexually mature at age-4 and females mature at age-5. Male lake trout arrive at the spawning grounds earlier than females (Muir et al. 2012). As a female arrives, one or more males will hover and escort the female to the spawning area (Binder et al. 2014). The group will then quiver bodies together to initiate gamete release and fertilization (Binder et al. 2014). No nest is prepared, and no parental care is given. The demersal, 5 to 6 mm diameter eggs incubate for 15 to 21 weeks before hatching in early spring (Becker 1983). After yolk sac absorption, fry leave spawning beds and move to deeper water.

Age and Growth: Lake trout typically live for 10 to 20 years with total lengths of 381 to 762 mm in larger, deeper lakes, and 304 to 406 mm in smaller water bodies. They can be up to 1-m long and weigh over 22.5 kg (Scott and Crossman 1973). Lake trout grow rapidly in the first 5 years. Total lengths at age for lake trout in Lake Superior were age-1, 102 mm; age-2, 160 mm; age-3, 216 mm; age-4, 279 mm; age-5, 351 mm; age-6, 427 mm; age-7, 500 mm; age-8, 579 mm (Becker 1983).

Food and Feeding: Lake trout are visual feeders. Those in larger lakes eat fish in winter and zooplankton in summer (Blanchfield et al. 2009). In smaller lakes with lower prey fish abundance, they mainly consume zooplankton, crustaceans, and aquatic invertebrates. Lake trout primarily eating zooplankton and invertebrates mature earlier, grow slower, attain a smaller maximum size, and die more quickly than those primarily eating fish (Martin 1966). Juveniles eat crustaceans and aquatic invertebrates.

References

Becker GC (1983) Fishes of Wisconsin. University of Wisconsin Press, Madison

Binder TR, Thompson HT, Muir AM, Riley SC, Marsden JE, Bronte CR, Krueger CC (2014) New insight into the spawning behavior of lake trout, *Salvelinus namaycush,* from a recovering population in the Laurentian Great Lakes. Environ Biol Fish 98:173–181. https://doi.org/10.1007/s10641-014-0247-6

Binder TR, Riley SC, Holbrook CM, Hansen MJ, Bergstedt RA, Bronte CR, He J, Krueger CC (2015) Spawning site fidelity of wild and hatchery lake trout (*Salvelinus namaycush*) in northern Lake Huron. Can J Fish Aquat Sci 73:18–34. https://doi.org/10.1139/cjfas-2015-0175

Blanchfield PJ, Tate LS, Plumb JM (2009) Seasonal habitat selection by lake trout (*Salvelinus namaycush*) in a small Canadian shield lake: constraints imposed by winter conditions. Aqua Ecol 43:777–787. https://doi.org/10.1007/s10452-009-9266-3

Eschmeyer PH (1957) The lake trout (*Salvelinus namaycush*). US Dep Inter, Fish Wildl Serv Fish, Leaflet No. 441

Martin NV (1966) The significance of food habits in the biology, exploitation, and management of Algonquin park, Ontario, lake trout. Trans Am Fish Soc 95:415–422. https://doi.org/10.1577/1548-8659(1966)95[415:TSOFHI]2.0.CO;2

Muir AM, Blackie CT, Marsden JE, Krueger CC (2012) Lake charr *Salvelinus namaycush* spawning behavior: new field observations and a review of current knowledge. Rev Fish Biol Fisher 22:575–593. https://doi.org/10.1007/s11160-012-9258-6

Owen JB, Elsen DS, Russell GW (1981) Distribution of fishes in North and South Dakota Basins affected by the Garrison Diversion Unit. University of North Dakota Press, Grand Forks

Scott WB, Crossman EJ (1973) Freshwater fishes of Canada. Fisheries Research Board of Canada, Bulletin 184

6.18   Osmeridae, Smelt Family

The smelt family consists of 15 extant species in seven genera worldwide. Ten species in six genera occur within North America. Most of the Osmeridae species are entirely marine, originating from the North Pacific and Northern Atlantic oceans near North America and Northern Asia. Some species in the smelt family are anadromous, while other species, such as the rainbow smelt *Osmerus mordax* in the Dakotas, have been introduced in landlocked, freshwater populations.

Smelt are characterized by having a slender and streamlined body, a pointed snout, a small adipose fin anterior to the single dorsal fin, and a deeply forked caudal fin. Most species in the smelt family are small, and only seldom reach total lengths greater than 203 mm. Smelt are distantly related to, and often confused with, species in the salmon family, Salmonidae. However, smelt do not have an axillary process, which is at the anterior base of the pelvic fins in salmonids. Because of the small adipose fin, rainbow smelt may also be misidentified as trout-perch *Percopsis omiscomaycus*, family Percopsidae, in the Dakotas. However, the rainbow smelt lateral line is incomplete, while the trout-perch lateral line is not.

Smelt are pelagic fish and prefer the open, cool waters of oceans, rivers, lakes, and reservoirs. In the Dakotas, rainbow smelt inhabit the deep, dark, cold, stratified waters of the Missouri River reservoirs where they form large schools near dams and intake structures. Rainbow smelt only leave these deep waters during the spawning season, when they migrate upstream into smaller tributaries to spawn during the night in shallow water with gravel substrate. Members of the smelt family are carnivorous, using small teeth on the tongue and both jaws to feed on large zooplankton. Adult smelt are also piscivorous, consuming smaller fishes, larvae, and fry, including other rainbow smelt.

In the Missouri River reservoirs in the Dakotas, rainbow smelt are a dominant pelagic prey item for adult walleye *Sander vitreus*. Rainbow smelt abundance may be the main factor impacting walleye population dynamics. In Lake Oahe, South Dakota, rainbow smelt populations significantly decrease during high discharge periods, increasing entrainment to downstream reservoirs. When this occurs, walleye growth rates increase exponentially. In contrast, when rainbow smelt abundance decreases, walleye condition and growth decreases.

6.18.1 Rainbow Smelt *Osmerus mordax*

(Mitchill, 1814)

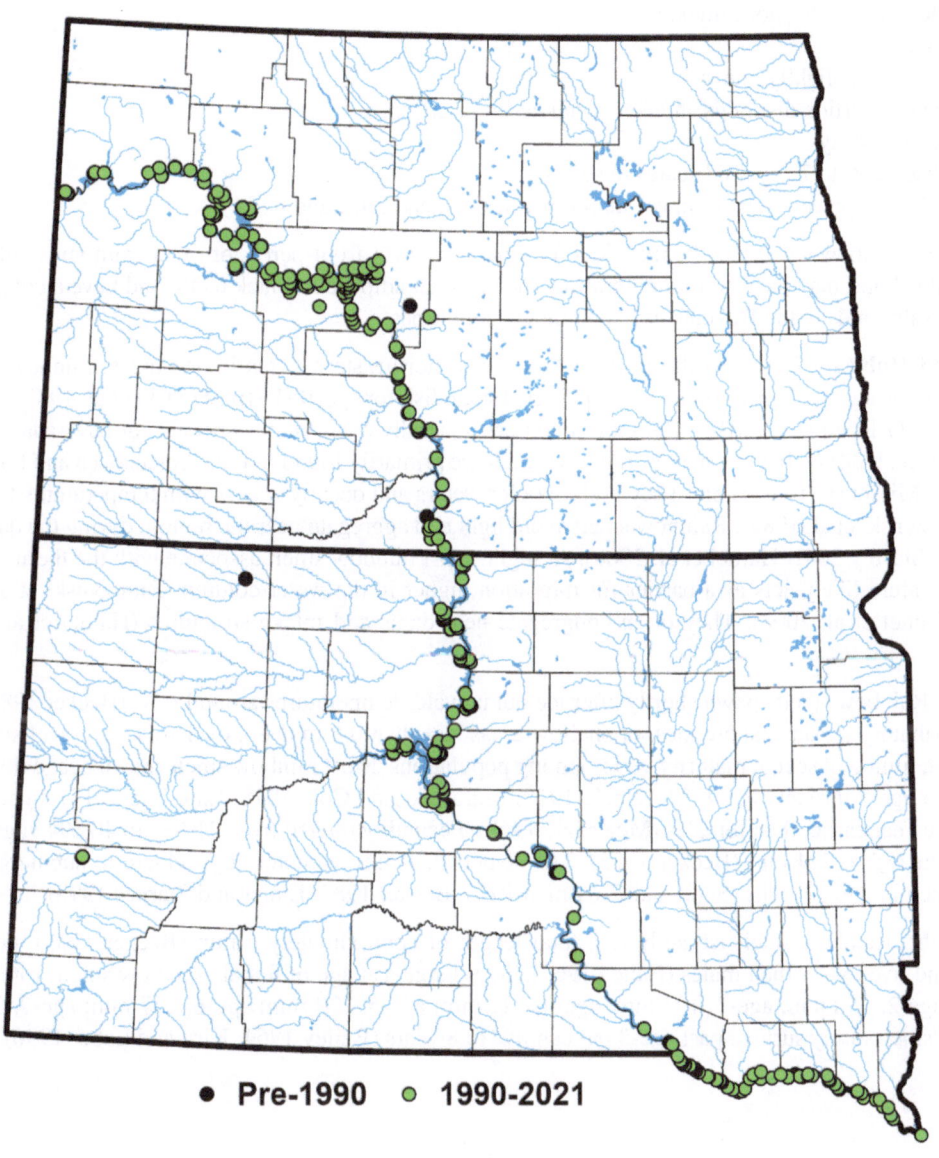

● **Pre-1990**   ○ **1990-2021**

Etymology: *Osmerus mordax* – *Osmerus* = odorous, referencing scent like cucumbers; *mordax* = biting.

Description:

Body – fusiform, slender, laterally compressed.
Color – color fades rather quickly when removed from the water.

>   *Dorsally* – olive-green to dark gray.
>   *Laterally* – silver, hues of purple, blue, and/or light pink iridescence.
>   *Ventrally* – silver to white.
>   *Fins* – clear.

Head – moderate, tapering to a point.
Snout – moderate, tapering to a point.
Eyes – large, laterally on head.
Mouth – terminal, slightly oblique, relatively large.

>   *Upper jaw* – extends to center of eye or just beyond.
>   *Lower jaw* – slightly protrudes past upper jaw.

Teeth – on both jaws and tongue.
Gill rakers – 26 to 34; long, slender.
Dorsal fin – 8 to 11 rays; positioned approximately equidistant between snout and caudal fin.
Adipose fin – small, nearly rounded distal end.
Caudal peduncle – relatively short, thick.
Caudal fin – forked.
Anal fin – 11 to 16 rays; slightly concave.
Pelvic fins – 8 rays; insertion distinctly inferior to dorsal fin insertion.
Pectoral fins – 11 to 14 rays.
Lateral line – absent; 60 to 72 cycloid scales in series.
Sexual dimorphism – spawning males have small tubercles on head, fins, and body.

Similar Species: Rainbow smelt may be mistaken for emerald shiner, trout-perch, or other salmonids with an adipose fin. Although emerald shiner may have similar coloration, they lack an adipose fin, lack teeth, and have a complete lateral line. Trout-perch and salmonids in the Dakotas have complete lateral lines.

Distribution and Habitat: Native to the Pacific, Arctic, and northeastern Atlantic drainages, rainbow smelt have been widely introduced throughout North America as a pelagic forage fish for sportfish (Burczynski et al. 1987; Graeb et al. 2008; Fincel et al. 2014; O'Brien et al. 2014). Rainbow smelt are generally anadromous species but have adapted to landlocked habitats (O'Brien et al. 2014). In the Dakotas, rainbow smelt are primarily found in Lake Sakakawea and Lake Oahe but may entrain into other Missouri River waters. They prefer cooler waters and occupy waters with temperatures from 5 to 14 °C in Lake Oahe (Burczynski et al. 1987). Rainbow smelt avoid light and aggregate near the bottom during the day (Burczynski et al. 1987; Nelson-Stastny 2001; Hamel et al. 2008). At night, most rainbow smelt associate with the thermocline, indicating that water temperature likely acts as a barrier for migration higher in the water column (Burczynski et al. 1987; Nelson-Stastny 2001; Hamel et al. 2008). They often congregate near dams and intake structures (Hamel et al. 2008; Fincel et al. 2016).

Reproduction: Rainbow smelt spawn shortly after ice out in water temperatures around 4 °C (Becker 1983). They migrate up clear, swift tributary streams, shorelines, or shallow areas less than 0.6 m deep, over rock or gravel substrate to spawn (Rupp 1965). The timing of sexual maturity varies among populations. Most rainbow smelt spawn at age-1 but some populations spawn from age-2 through age-4 (McKenzie 1958; Murawski and Cole 1978; Enterline 2013). In some populations, males mature one year earlier than females (McKenzie 1958). Fecundity ranged from 21,534 to 40,894 eggs in females with total lengths from 185 to 224 mm (Becker 1983). The demersal, 1-mm diameter eggs adhere to submerged vegetation or substrate and hatch in 2 to 3 weeks (Scott and Crossman 1973; Becker 1983; Etnier and Starnes 1993).

Age and Growth: Rainbow smelt larvae have a total length of approximately 6 mm (Becker 1983). Females generally become larger and live longer than males (Bailey 1964). Total lengths at age of female rainbow smelt from Wisconsin were age-1, 66 mm; age-2, 150 mm; age-3, 193 mm; age-4, 221 mm; age-5, 239 mm; age-6, 257 mm; age-7, 310 mm (Becker 1983). Rainbow smelt rarely survive past age-3 but can live to 8 years (Bailey 1964; Kirn and Laber 1996).

Food and Feeding: In general, juvenile rainbow smelt are planktivorous, consuming large quantities of copepods, cladocerans, and other zooplankton (Etnier and Starnes 1993). Adults tend to be cannibalistic; they consume the highest number of their own larvae compared to any other fish species (Burbidge 1969). Adults also eat larger crustaceans, immature insects, and fry and larvae of other pelagic forage species (Smith 1979; Lantry and Stewart 2000; Parker Stetter et al. 2007).

References

Bailey MM (1964) Age, growth maturity and sex composition of the American smelt, *Osmerus mordax* (Mitchill), of western Lake Superior. Trans Am Fish Soc 93:382–395. https://doi.org/10.1577/1548-8659(1964)93[382:AGMASC]2.0.CO;2

Becker GC (1983) Fishes of Wisconsin. University of Wisconsin Press, Madison

Burbidge RG (1969) Age, growth, length-weight relationship, sex ratio, and food habits of American smelt, *Osmerus mordax* (Mitchill), from Gull Lake, Michigan. Trans Am Fish Soc 98:631–640. https://doi.org/10.1577/1548-8659(1969)98[631:AGLRSR]2.0.CO;2

Burczynski JJ, Michalets PH, Marrone GM (1987) Hydroacoustic assessment of the abundance and distribution of rainbow smelt in Lake Oahe. N Am J Fish Manag 7:106–116. https://doi.org/10.1577/1548-8659

Enterline CL (2013) Understanding spawning behavior and habitat use by anadromous rainbow smelt (*Osmerus mordax*) using passive integrated transponder systems and telemetry. Dissertation, University of New Hampshire

Etnier DA, Starnes WC (1993) The fishes of Tennessee. University of Tennessee Press, Knoxville

Fincel MJ, Dembkowski DJ, Chipps SR (2014) Influence of variable rainbow smelt and gizzard shad abundance on walleye diets and growth. Lake Reserv Manag 30:258–267. https://doi.org/10.1080/10402381.2014.914989

Fincel MJ, Radigan WJ, Longhenry CM (2016) Entrainment of rainbow smelt through Oahe dam during the 2011 Missouri River flood. N Am J Fish Manag 36:844–851. https://doi.org/10.1080/02755947.2016.1173139

Graeb BD, Chipps SR, Willis DW, Lott JP, Hanten RP, Nelson-Stastny W, Erickson JW (2008) Walleye response to rainbow smelt population decline and liberalized angling regulations in a Missouri River reservoir. In: Allen MS, Sammons S, Maceina MJ (eds) Balancing fisheries management and water uses for impounded river systems. American Fisheries Society, Symposium 62, Bethesda, pp 275–291

Hamel MJ, Brown ML, Chipps SR (2008) Behavioral responses of rainbow smelt to in situ strobe lights. N Am J Fish Manag 28:394–401. https://doi.org/10.1577/M06-254.1

Kirn RA, Laber GW (1996) Growth and survival of rainbow smelt, and their role as prey for stocked salmonids in Lake Champlain. Trans Am Fish Soc 125:87–96. https://doi.org/10.1577/1548-8659

Lantry BF, Stewart DJ (2000) Population dynamics of rainbow smelt (*Osmerus mordax*) in lakes Ontario and Erie: a modeling analysis of cannibalism effects. Can J Fish Aquat Sci 57:1594–1606. https://doi.org/10.1139/f00-092

McKenzie RA (1958) Age and growth of smelt, *Osmerus mordax* (Mitchill), of the Miramichi River, New Brunswick. J Fish Res Board Can 15:1313–1327. https://doi.org/10.1139/f58-071

Murawski SA, Cole CF (1978) Population dynamics of anadromous rainbow smelt *Osmerus mordax*, in a Massachusetts river system. Trans Am Fish Soc 107:535–542. https://doi.org/10.1577/1548-8659

Nelson-Stastny W (2001) Estimates of abundance, biomass, and distribution of rainbow smelt and other pelagic fish in Lake Oahe using hydroacoustic techniques, 1996-1999. S D Game Fish Park Ann Rep 01-06, Pierre

O'Brien TP, Taylor WW, Roseman EF, Madenjian CP, Riley SC (2014) Ecological factors affecting rainbow smelt recruitment in the main basin of Lake Huron, 1976-2010. Trans Am Fish Soc 143:784–795. https://doi.org/10.1080/00028487.2014.880736

Parker Stetter SL, Thomson JLS, Rudstam LG, Parrish DL, Sullivan PJ (2007) Importance and predictability of cannibalism in rainbow smelt. Trans Am Fish Soc 136:227–237. https://doi.org/10.1577/T05-280.1

Rupp RS (1965) Shore-spawning and survival of eggs of the American smelt. Trans Am Fish Soc 94:160–168. https://doi.org/10.1577/1548-8659(1965)94[160:SASOEO]2.0.CO;2

Scott WB, Crossman EJ (1973) Freshwater fishes of Canada. Fisheries Research Board of Canada, Bulletin 184

Smith PW (1979) The fishes of Ohio. University of Illinois Press, Urbana

6.19 Percopsidae, Trout-Perch Family

Trout-perch *Percopsis omiscomaycus* and sand roller *Percopsis transmontana* are the only two fish species of the trout-perch family, Percopsidae. Both species are endemic to North America but only trout-perch occur in the Dakotas. The native range of trout-perch extends from Quebec through Hudson Bay to the Alaskan Yukon, south to Manitoba and the Upper Mississippi River basin to Kentucky, and east to the Hudson River. The sand roller is considered rare and is endemic only in the Columbia River basin in Washington, Oregon, and Idaho.

The trout-perch family belongs to Percopsiformes, a relatively small order of ray-finned fishes. This order has ancestral characteristics such as a small adipose fin and more derived characteristics such as fewer dorsal and anal fin spines. The family name trout-perch comes from a combination of the perch-like features of a large head, teeth, and a spinous first dorsal fin, and the trout-like adipose fin. Both the sand roller and trout-perch have weakly ctenoid scales and rather large pectoral fins that when laid flat against the body overlap the pelvic fin.

Trout-perch and sand roller generally inhabit lotic waters but can also inhabit lentic environments. Trout-perch are often found at shallow and intermediate depths in clear to slightly turbid water with slow to moderate current over sand or rocky substrate. They are often associated with woody debris, roots near undercut banks and other submerged structures. Both species are nocturnal, actively feeding during the night. Adults of both species eat small aquatic insect larvae and occasionally other small fish. Trout-perch spawn in the spring in shallow waters with sand, gravel, or rocky substrate. They are readily consumed by larger predatory fish because of their small size, with lengths typically less than 102 mm.

6.19.1   Trout-Perch *Percopsis omiscomaycus*

(Waldbaum, 1792)

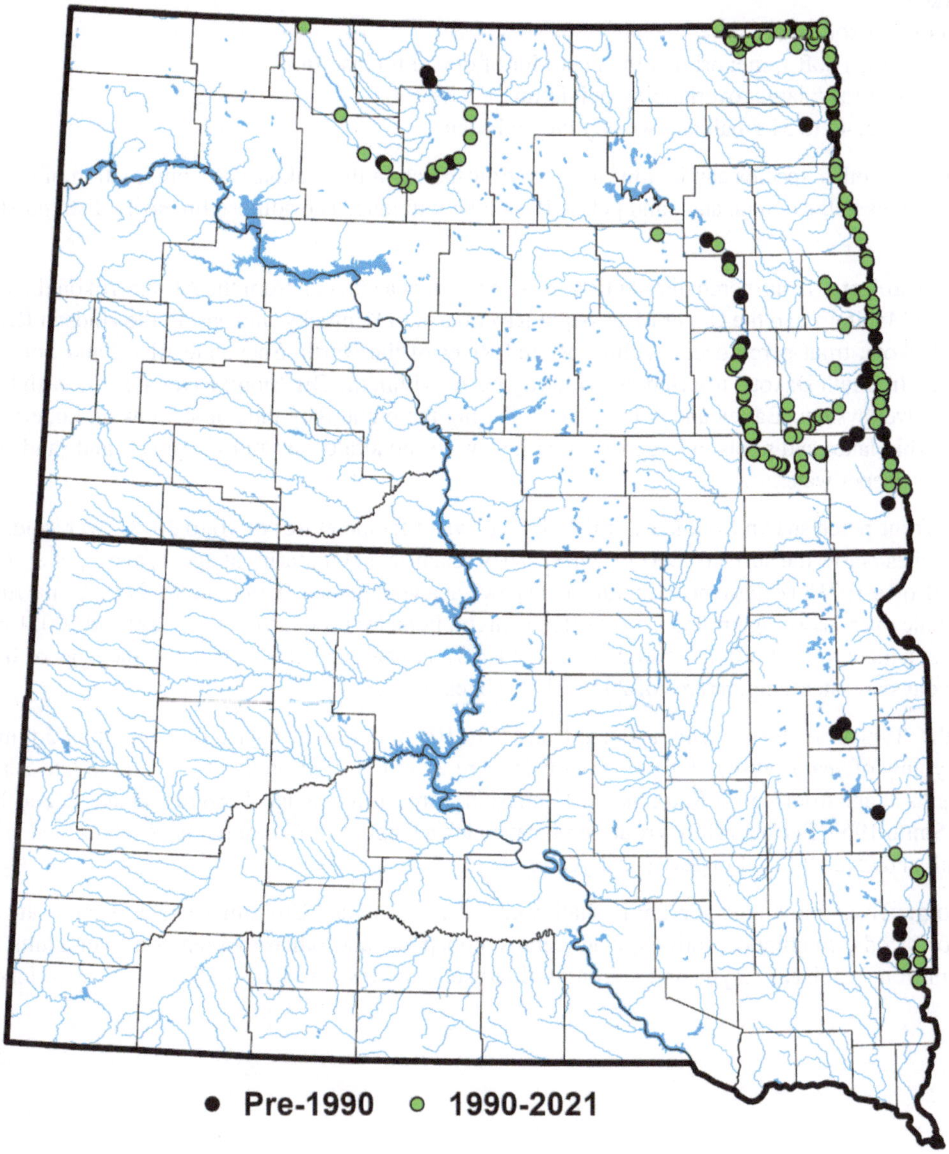

● **Pre-1990**   ○ **1990-2021**

Etymology: *Percopsis omiscomaycus* – *Percopsis* = perch appearance; *omiscomaycus* = most likely an Algonkian Native American name with the root trout.

Description:

Body – elongated, fusiform, cylindrical, slightly compressed.
Color – translucent.

> *Dorsally* – olive-brown.
> *Laterally* – olive-brown; silvery stripe running along lateral line, 9 to 12 dark blotches; several indistinct dark spots above lateral line from head to tail.
> *Ventrally* – fading to silvery-white.
> *Fins* – lightly pigmented.

Head – large, slender, scaleless.
Snout – pointed, long, extending past mouth.
Mouth – large, subterminal; slightly oblique to horizontal; upper lip groove not continuous.
Teeth – minute, villiform; numerous on upper and lower jaws.
Dorsal fin – 9 to 11 rays; 2 soft spines, slightly falcated; nearer tip of snout than caudal fin base.
Adipose fin – small.
Caudal peduncle – long, tapered.
Caudal fin – forked.
Anal fin – 5 to 8 rays; 1 soft spine.
Pelvic fins – 8 or 9 rays; 1 soft spine; originating posterior of dorsal fin insertion.
Pectoral fins – 12 to 15 rays; large, overlapping pelvic fin.
Lateral line – complete; 45 to 51 small and weakly ctenoid scales.

Similar Species: Few other species can be mistaken for trout-perch in the Dakotas. A combination of ctenoid scales, adipose fin, and soft spines in the dorsal, anal, and pelvic fins make it unique. The adipose fin easily distinguishes it from juvenile yellow perch.

Distribution and Habitat: Trout-perch distribution in North America extends from the Alaska Yukon through Hudson Bay to Quebec, south to Manitoba and the Upper Mississippi River basin to Kentucky, and east to the Hudson River (Hrabik et al. 2015). In North Dakota, trout-perch occur within the Red River of the North, Sheyenne River, and Souris River systems (Owen et al. 1981). In South Dakota, trout-perch primarily occur within the Big Sioux River system, with historic reference to the Minnesota River system (Bailey and Allum 1962). During the night, they inhabit shallow and intermediate depths of clear to slightly turbid lakes and pools in large rivers with slow-to-moderate current over gravel and sand substrate. During the day, they prefer deeper water.

Reproduction: Trout-perch spawn from late April to June at water temperatures of 16 to 20 °C over sand, gravel, or rocky substrate a few inches below the surface (Becker 1983). They reach sexual maturity at age-1, age-2, or at total lengths from 5.08 to 6.35 cm (Brown 1971). Males tend to mature and spawn at a younger age than females. During spawning, two to three males court a female over spawning grounds, most often at night. Females produce between 200 and 500, 1.87 mm diameter, demersal eggs that adhere to the bottom substrate (Becker 1983). No parental care is given to the eggs or fry. Eggs hatch in approximately 6 days at water temperatures of 20 to 23 °C (Becker 1983).

Age and Growth: Typical adult total lengths range from 76 to 102 mm. Larvae have a total length of 5 mm (Becker 1983). Total lengths of young of year trout-perch from Wisconsin were 27 mm in June and 43 to 57 mm in late September (Becker 1983). Male lengths at age from Lower Red Lake in Minnesota were 51 mm at age-1, 88 mm at age-2, and 104 mm at age-3 (Magnuson and Smith 1963; Becker 1983). Trout-perch can reach a length of 178 mm but have not exceeded 76 mm in South Dakota (Churchill and Over 1938). Longevity is 4 years.

Food and Feeding: Trout-perch feed at night in shallower waters (Scott and Crossman 1973; Owen et al. 1981). Juveniles consume more ostracods, amphipods, rotifers, cladocerans, copepods, and other zooplankton than adults (Tomlinson and Jude 1977). Adult trout-perch eat mayfly and chironomid larvae, other insects, and small minnows (Carlander 1969; Owen et al. 1981).

References

Bailey RM, Allum MO (1962) Fishes of South Dakota. University of Michigan, Ann Arbor. https://doi.org/10.3998/mpub.9690435

Becker GC (1983) Fishes of Wisconsin. University of Wisconsin Press, Madison

Brown CJD (1971) Fishes of Montana. Montana State University Press, Bozeman

Carlander KD (1969) Handbook of freshwater fishery biology. Iowa State University Press, Ames

Churchill EP, Over WH (1938) Fishes of South Dakota. Brown & Saenger Printers, Sioux Falls

Hrabik RA, Schainost SC, Stasiak RH, Peters EJ (2015) The fishes of Nebraska. University of Nebraska, Lincoln

Magnuson JJ, Smith LL (1963) Some phases of the life history of the troutperch, *Percopsis omiscomaycus*. Ecology 44:83–95. https://doi.org/10.2307/1933183

Owen JB, Elsen DS, Russell GW (1981) Distribution of fishes in North and South Dakota Basins affected by the Garrison Diversion Unit. University of North Dakota Press, Grand Forks

Scott WB, Crossman EJ (1973) Freshwater fishes of Canada. Fisheries Research Board of Canada, Bulletin 184

Tomlinson JC, Jude DJ (1977) Food of the trout-perch, *Percopsis omiscomaycus*, in southeastern Lake Michigan. 20th Conf Great Lakes Res, Int Assoc Great Lakes Res, University of Michigan.

6.20 Gadidae, Cod Family

Worldwide, there are 20 genera and 55 species in the cod family. Approximately 16 species occur in North America, and only one species, burbot *Lota lota*, is found in the Dakotas. The cod family is characterized by two or three spineless dorsal fins, one or two anal fins, and a single barbel on their chin. Burbot have two dorsal fins and one anal fin. Members of the cod family have elongated cylindrical bodies tapering posteriorly to create a slightly laterally compressed appearance, distinguishing them from other fishes, especially in the Dakotas. Cod also have extremely small and embedded cycloid scales, producing a scale-less appearance like catfish.

Most of the cod family live entirely in marine waters, with very few entering freshwater for short periods of time. Burbot is the only cod species to complete its entire life cycle in freshwater. They prefer deep, cold waters of large lakes, reservoirs, and rivers and are found in the Dakotas throughout the Missouri River and Red River of the North. Burbot often hide within dark crevices or under cover during the day, coming out at night to actively feed. Burbot larvae and juveniles eat phytoplankton, zooplankton, and small aquatic insects before undergoing an ontogenetic diet shift to piscivory as adults. They spawn generally at night in the winter and early spring directly under the ice in ball-like groups. Individual fish may not spawn every year.

In northern latitudes, cod are important commercial, highly sought, food fish. In the Dakotas, burbot are considered rough fish and are not popular with recreational anglers. Although its distribution is rather widespread throughout northern North America, Europe, and Asia, many burbot populations are declining because of impoundment construction and pollution.

6.20.1   Burbot *Lota lota*

(Linnaeus, 1758)

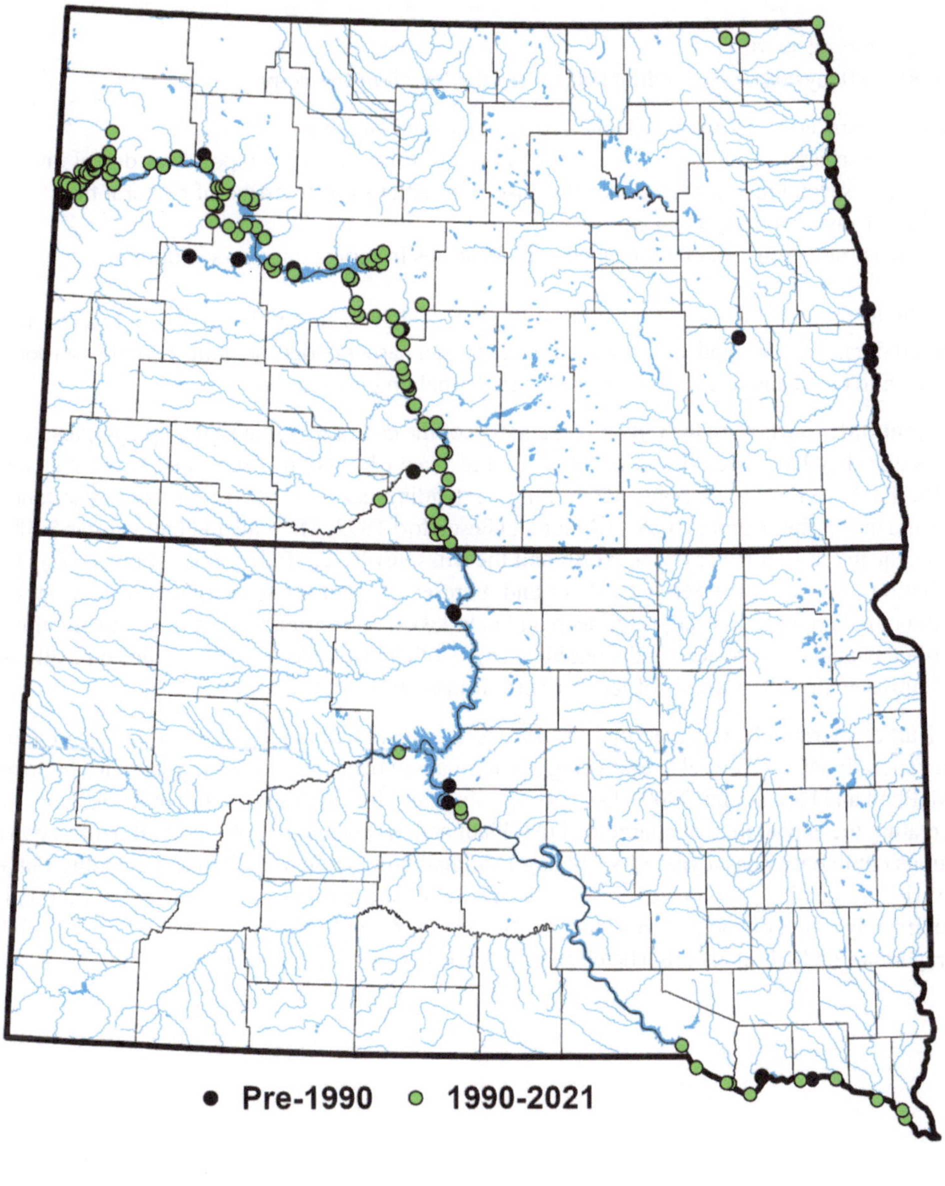

● **Pre-1990**   ● **1990-2021**

Etymology: *Lota lota* – *Lota* = cod or codfish. The common name, burbot stems from the Latin barba, referencing the single barbel or whisker on the chin.

Description:

Body – elongated, anteriorly cylindrical, laterally compressed posteriorly.
Color –

> *Dorsally, laterally* – coloration variable, may be mottled or rather solid mixtures of dark brown, black, tan, or brownish-yellow.
> *Ventrally* – light yellow, cream, or white.

Head – small; moderately dorsoventrally compressed.
Snout – nostrils small, tube like; anteriorly on head.
Eyes – small, laterally on head.
Mouth – large, wide.

> *Jaws* – upper jaw slightly extends past lower jaw.

Teeth – small, villiform, arranged in bands on both jaws.
Barbels – 1 on ventral side of chin.
Dorsal fin – 2.

> *Anterior* – 8 to 16 rays; short.
> *Posterior* – 60 to 80 rays; short, extending nearly half the length of the body.

Caudal fin – rounded distal end; distinctly separated from dorsal and anal fins.
Anal fin – 53 to 65 rays; nearly as long as posterior dorsal fin; insertion posterior to posterior dorsal fin insertion.
Pelvic fins – 5 to 8 rays; second ray elongated; narrow, insertions anterior to insertion of pectoral fins.
Pectoral fins – 17 to 21 rays; rounded distal ends.
Lateral line – complete; numerous small, embedded cycloid scales in series; appears scaleless.

Similar Species: In the Dakotas, American eel are the only fish to slightly resemble burbot, and can easily be distinguished by continuous, fused dorsal, caudal, and anal fins, and the lack of a barbel under the chin. Juvenile burbot can be easily distinguished from juvenile bullheads or catfishes by the single barbel on the chin.

Distribution and Habitat: Burbot are the only representative of the cod family that typically complete their entire lifecycle in freshwater, although they have been found in estuaries and brackish lagoons (Scott and Crossman 1973; McPhail and Paragamian 2000). They have a circumpolar native range extending across northern Europe, Asia, and North America, including Canada and the northern United States (Ryder and Pesendorf 1992; McPhail and Paragamian 2000). In the Dakotas, burbot primarily occur in the Red River of the North and the Missouri River (Bailey and Allum 1962). They likely have a continuous distribution throughout the Missouri River and its reservoirs but are not targeted or routinely sampled, leaving some sections without data points. Burbot prefer deep and cold lakes, reservoirs, large rivers, and streams with a variety of substrates. They often hide in dark crevices under boulders or any other available cover. Juveniles typically inhabit weed beds on rock or gravel substrate in swift currents of larger rivers (Kansas Fishes Committee 2014).

Reproduction: Burbot spawn in lakes, reservoirs, rivers, and streams, often in the winter or early spring under the ice. The spawning season generally lasts 2–3 weeks. Spawning predominately occurs at night but may vary by location (Roach and Evenson 1993; McPhail and Paragamian 2000). During spawning, multiple males releasing milt surround one or two females releasing eggs, creating the appearance of a large ball (McPhail and Paragamian 2000). Not all adults spawn every year and northern populations commonly miss a year of spawning (McPhail and Paragamian 2000). Fecundity ranges from 6,300 to 3,477,699 eggs, varies regionally, and is positively correlated with female age and size (Hesse 1993; McPhail and Paragamian 2000). The approximately 1-mm diameter, semi-buoyant to buoyant, nonadhesive eggs develop faster at higher temperatures (McPhail and Paragamian 2000). Larval burbot hatch near substrate before becoming pelagic (Kansas Fishes Committee 2014).

Age and Growth: Total lengths of larval burbot in North Dakota were 5 mm in April and 81 mm in July (Fisher 2000). Juveniles may reach a total length of 110 to 120 mm by late fall (Scott and Crossman 1973; Sandlund et al. 1985; Fisher 2000; McPhail and Paragamian 2000). Burbot growth in the first year is rapid and varies with growing season duration and food availability. Total length of age-1 burbot from North Dakota was 153 mm (Fisher 2000). Burbot mean total length and weight are 1 m and 8 kg, respectively. Median total length and weight are 300 to 600 mm and 1 to 3 kg (McPhail and Paragamian 2000). Burbot from rivers and reservoirs generally have lower relative weights than lake populations (Fisher et al. 1996; Kansas Fishes Committee 2014). Burbot in Siberia weighed 25 to 30 kg and were 15 to 20 years old (McPhail and Paragamian 2000). Burbot maximum age is 22 years (McPhail and Paragamian 2000).

Food and Feeding: When larvae are big enough to drift through the limnetic zone, they eat larger zooplankton such as rotifers, copepods, and cladocerans (Ghan and Sprules 1993; Fisher 2000). Juveniles continue to eat zooplankton, but will also consume crayfish, macroinvertebrates (odonates, amphipods, plecopterans), and small benthic fishes (Fisher 2000; Jacobs et al. 2010). As burbot grow and reach adulthood, they are increasingly piscivorous and become a top-level predator (Jacobs et al. 2010).

References

Bailey RM, Allum MO (1962) Fishes of South Dakota. University of Michigan, Ann Arbor. https://doi.org/10.3998/mpub.9690435

Fisher SJ (2000) Early life history observation of burbot utilizing two Missouri River backwaters. Burbot biology, ecology, and management. American Fisheries Society, Fish Manage Section Publ 96-0, Bethesda

Fisher SJ, Willis DW, Pope KL (1996) An assessment of burbot (*Lota lota*) weight-length data from North American populations. Can J Zool 74:570–575. https://doi.org/10.1139/z96-063

Ghan D, Sprules WG (1993) Diet, prey selection, and growth of larval and juvenile burbot *Lota lota* (L). J Fish Biol 42:47–64. https://doi.org/10.1111/j.1095-8649.1993.tb00305.x

Hesse LW (1993) The status of Nebraska fishes in the Missouri River, 2. Burbot (Gadidae: *Lota lota*). Trans Nebr Acad Sci Affil Soc 120:67–71

Jacobs GR, Madenjian CP, Bunnell DB, Holuszko JD (2010) Diet of lake trout and burbot in northern Lake Michigan during spring: evidence of ecological interaction. J Great Lakes Res 36:312–317. https://doi.org/10.1016/j.jglr.2010.02.007

Kansas Fishes Committee (2014) Kansas fishes. University Press of Kansas, Lawrence

McPhail JD, Paragamian VL (2000) Burbot biology and life history. In: Paragamian VL, Willis DW (eds) Burbot: biology, ecology, and management. American Fisheries Society, Fish Manage Sect 1, pp 11–23

Roach SM, Evenson MJ (1993) A geometric approach to estimating and predicting fecundity of Tanana River burbot. Alaska Dep Fish Game, Fish Data Series, No. 93-38, Juneau, Alaska

Ryder RA, Pesendorf J (1992) Food, growth, habitat, and community interactions of young-of-the-year burbot, *Lota lota* L., in a Precambrian Shield Lake. Hydrobiologia 243:211–227. https://doi.org/10.1007/978-94-011-2745-5_22

Sandlund OT, Klyve L, Naesje TF (1985) Growth, habitat and food of burbot, *Lota lota*, in Lake Mjoesa. Fauna Blindern 38:37–43

Scott WB, Crossman EJ (1973) Freshwater fishes of Canada. Fisheries Research Board of Canada, Bulletin 184

6.21 Cichlidae, Cichlid Family

Cichlidae is an extremely diverse and large family of fishes. Cichlids are native to the warm, tropical, and temperate waters of Africa, Asia, Central and South America. Body size varies greatly, but body morphology can be used to split the cichlids into two main groups: deeply laterally compressed or cylindrical-and-elongate. Cichlids have a single nostril on each anterior lateral side of the head and fused lower pharyngeal bones. They also have a lateral line divided into an anterior upper section ending below the posterior end of the dorsal fin base and a posterior lower section extending to the center of the caudal peduncle. Because they have highly movable jaws and variations in tooth patterns and shapes, cichlids can efficiently capture and process a wide variety of prey. Most cichlids eat plant matter and small invertebrates.

Cichlids are often aggressive during spawning, with vividly colored males establishing and defending territories. Cichlids can be monogamous or polygamous, with parents protecting the eggs and fry from predators. Some species practice mouth-brooding, whereby one or both parents orally guard eggs and larvae. Bright coloration and unique behaviors make cichlids popular aquarium species, with larger species also highly desirable food and game fish. The fast-growing group of cichlids called tilapiines, are extremely important in commercial aquaculture and consumed worldwide. Peacock bass, genus *Cichia*, are cichlids highly desirable as sport fish. They are native to South America and were introduced to the United States in southern Florida.

Only one species of cichlid occurs in the Dakotas. The nonnative Jack Dempsey *Rocio octofasciata* has established populations in the Fall River and other warm water streams of the southern Black Hills in western South Dakota. They may have been introduced by purposeful or accidental aquarium release.

6.21.1   Jack Dempsey *Rocio octofasciata*

(Regan, 1903)

● **Pre-1990**   ● **1990-2021**

Etymology: *Rocio octofasciata – Rocio* = morning dew, referencing numerous electric blue-cyan spots, named after the author Charles T. Regan's wife; *octofasciata* = eight belts or stripes, referencing 8 to 11 dark, diffuse vertical bars or stripes on lateral sides of young and juvenile individuals. The common name refers to Jack Dempsey, a famous boxer, because of aggressive behavior.

Description:

Body – elongated, moderately deep.
Color – males – dark, horizontal stripes on forehead.

> *Dorsally* – gray olive, dark blue-green to black.
> *Laterally* – olive to brown; 8 to 11 dark, diffuse vertical bars or stripes (most prominent in preserved specimens); two noticeable dark spots, one on middle of body, one on posterior end of the caudal peduncle.
> *Ventrally* – dark brown to olive.
> *Head, body, unpaired fins* – numerous electric cyan spots.
> *Eyes* – gray to bronze.
> *Fins* – caudal, dorsal – bright to deep red.

Head – large.
Snout – conical, short.
Eyes – moderately large, laterally on head.
Mouth – terminal.

> *Upper jaw* – barely extends to anterior edge of eye.
> *Lips* – fleshy.

Dorsal fin – 17 to 19 spines, 9 or 10 rays; elongated.
Caudal peduncle – short, thick.
Caudal fin – slightly rounded distal end.
Anal fin – 8 or 9 spines, 7 or 8 rays; elongated; forming a short, pointed filament on posterior end.
Pelvic fins – thoracic.
Pectoral fins – rounded distal end.
Juveniles – more prominent vertical bars or stripes on lateral sides; fewer electric cyan spots.
Sexual dimorphism – spawning males darker, almost black.

Similar Species: No other species within the Dakotas have similar features to Jack Dempsey.

Distribution and Habitat: Jack Dempsey is the only cichlid in the Dakotas. Fishes in the family Cichlidae are native to the warm, tropical, and temperate waters of Africa, Asia, Central and South America (Cadwallader et al. 1980; Pashkov and Zvorykin 2009). In Central America, Jack Dempsey are native to southern Mexico and Honduras (Page and Burr 1991). Like other cichlids, Jack Dempsey are extremely adaptable to new ecological conditions, allowing it to become established in areas considerably north of its native range. In South Dakota, Jack Dempsey are found in the Fall River near Hot Springs and in other southern Black Hills streams within the Cheyenne River system (Zworykin and Pashkov 2010; Simpson et al. 2014). They were likely introduced by aquarium release. Jack Dempsey do not occur in North Dakota.

Jack Dempsey inhabit warm, clear-to-slightly murky, small spring-fed lakes and shallow rivers with little to no current and large amounts of aquatic vegetation over soft, muddy or sand substrate (Page and Burr 1991). They prefer water temperatures from 22 to 30 °C, lose equilibrium at 9 °C, and have a lower lethal temperature of 8 °C (Shafland and Pestrak 1982; Konings 1989). Jack Dempsey can withstand low oxygen concentrations but cannot tolerate salinities over 8 ppt (Dial and Wainright 1983; Obordo and Chapman 1997).

Reproduction: Little information exists on the natural spawning of Jack Dempsey. They become sexually mature at a total length of 75 mm (Legge 1970; Cadwallader et al. 1980). Fecundity usually does not exceed 800 eggs but can be as high as 1,010 eggs (Sakurai et al. 1993; Pashkov and Zvorykin 2009). Like many other species in the family Cichlidae, both parents protect eggs and fry (Keenleyside 1991). Both parents increase fry food availability and feeding opportunities by fin-digging, stirring up the bottom substrate with their pectoral fins to release detritus, zoobenthos, and phytobenthos (Zworykin 1998; Zworykin and Pashkov 2010). Jack Dempsey are aggressive, with unpretentious behavior (Zworykin and Pashkov 2010).

Age and Growth: Juvenile Jack Dempsey have standard lengths from 38 to 59 mm (Pashkov and Zvorykin 2009). Adult mean total length ranges from 127 to 203 mm. They can reach a total length of 250 mm and have a longevity of 10 years (Legge 1970; Cadwallader et al. 1980; Fuller et al. 1999).

Food and Feeding: As omnivores, adult Jack Dempsey eat a variety of plants and animals. Their main prey is chironomids, trichopterans, odonata larvae, aerial insects, small fish, gastropods, and aquatic invertebrates eggs (Pashkov and Zvorykin 2009). Adults have also eaten sunflower seeds, indicating the ability to adapt to atypical, but abundant and accessible food (Pashkov and Zvorykin 2009). Foraging decreases at 16 °C and ceases at 13 °C (Shafland and Pestrak 1982). There is little information on juvenile diets.

References

Cadwallader PL, Backhouse GN, Fallu R (1980) Occurrence of exotic tropical fish in the cooling pondage of a power station in temperate south-eastern Australia. Australas J Mar Freshw Res 31:541–546. https://doi.org/10.1071/MF9800541

Dial RS, Wainright SC (1983) New distributional records for non-native fishes in Florida. Fla Sci 46:8–15

Fuller PL, Nico LG, Williams JD (1999) Nonindigenous fishes introduced into the inland waters of the United States. American Fisheries Society, Spec Publ 27, Bethesda

Keenleyside MHA (1991) Cichlid fishes: behavior, ecology and evolution. Chapman and Hall, London

Konings A (1989) Cichlids from Central America. TFH Publ, Neptune City

Legge R (1970) The complete aquarists guide to freshwater tropical fishes. Eurobook Ltd, London

Obordo CO, Chapman LJ (1997) Respiratory strategies of a non-native Florida cichlid, *Cichlasoma octofasciatum*. Fla Sci 60:40–52

Page LM, Burr BM (1991) Peterson field guide to freshwater fishes of North America North of Mexico. Houghton Mifflin Harcourt, Boston

Pashkov AN, Zvorykin DD (2009) Some morphological specific features of C*ichlasomine Rocio octofasciata* (Perciformes, Cichlidae) from the population in Lake Staraya Kuban. J Ichthyol 49:383–389. https://doi.org/10.1134/S003294520905004X

Sakurai A, Sakamoto Y, Mori F (1993) Aquarium fish of the world: the comprehensive guide to 650 species. Chron Book, San Francisco

Shafland PL, Pestrak JM (1982) Lower lethal temperatures for fourteen non-native fishes in Florida. Environ Biol Fish 7:149–156. https://doi.org/10.1007/BF00001785

Simpson G, Carreiro J, Galinat G, Davis J, Miller B, Pasbrig C, Fletcher B, Barnes M, Jones D, Bucholz M (2014) Black Hills fisheries management area strategic plan 2014–2018. S D Game Fish Parks, Wildl Div, Pierre

Zworykin DD (1998) Parental fin digging by *Cichlasoma octofasciatum* (Teleostei: Cichlidae), and the effect of parents' satiation state on brood provisioning. Ethology 104:771–779. https://doi.org/10.1111/j.1439-0310.1998.tb00110.x

Zworykin DD, Pashkov AN (2010) Eight-striped *Cichlasoma*-an allochthonous species of cichlid fish (teleostei: cichlidae) from Staraya Kuban Lake. Russ J Biol Inv 1:1–6. https://doi.org/10.1134/S2075111710010017

6.22 Fundulidae, Topminnow and Killifish Family

The topminnow and killifish family has four genera and approximately 40 species. It is distributed in North and Central America from Canada to Cuba and the Bahamas. Banded killifish *Fundulus diaphanus*, northern plains killifish *Fundulus kansae*, and plains topminnow *Fundulus sciadicus* are the three fundulid species occurring in the Dakotas.

Fundulids are in the order Cyprinodontiformes, which means toothed carp in Greek, and was named because of the additional row of many small, unicuspid teeth on both the upper and lower jaws. Fundulids have elongate, slightly laterally compressed bodies dorsoventrally flattened on the head and across the anterior end of the dorsal side. Their small mouths are upturned or superior, with the lower jaw protruding past the upper jaw, and their spineless dorsal fin is far posterior on the body above the anal fin. The rounded distal end of their caudal fin easily differentiates topminnows and killfishes from minnows which have a forked caudal fin.

Within the Dakotas, fundulids occur in the low velocity, relatively shallow areas of small rivers, streams, ponds, sloughs, and lakes. Plains topminnow prefer clear water over silt and sand substrates with nearby dense aquatic vegetation, whereas northern plains killifish and banded killifish are found near sand and gravel substrates. The fundulid body shape and mouth position are adapted to a unique foraging style of skimming just beneath the water surface to feed on a variety of floating small aquatic and terrestrial invertebrates. Male fundulids often display bright nuptial coloration during the spawning season. Spawning habits differ among species, but in general, males become defensive of their spawning territories. No nest or parental care is given to the eggs or larvae.

The historic ranges of some fundulid species within the Great Plains region are decreasing and populations are declining because of habitat loss and alteration, as well as invasive species range expansion. Because of their small size, fundulids are often forage species for piscivorous game fishes such as bass and pike. They also adapt easily to captivity and are an amusing aquarium species. Because they frequently live in stagnant waters and eat larvae, they may serve as a biological control for mosquitos.

6.22.1   Banded Killifish *Fundulus diaphanous*

(Lesueur, 1817)

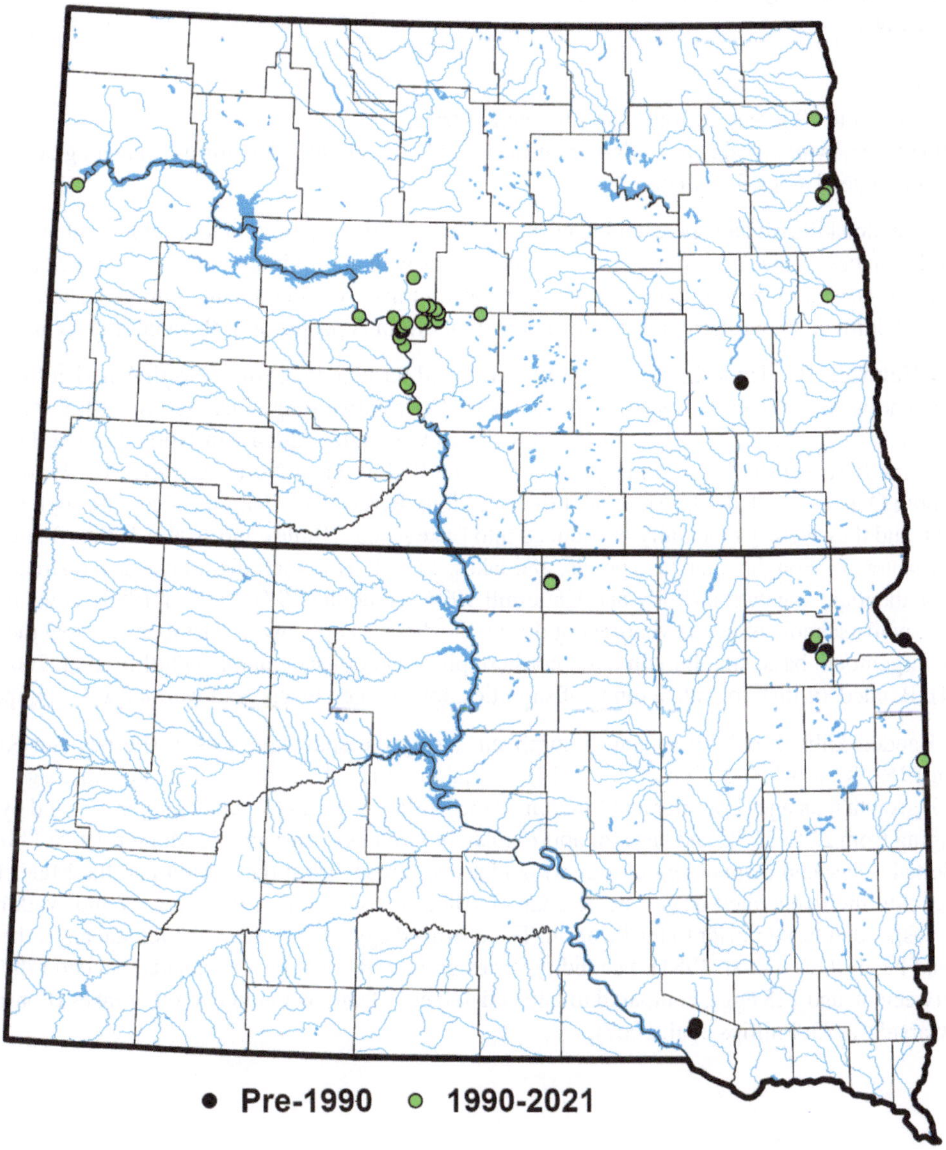

● **Pre-1990**   ○ **1990-2021**

Etymology: *Fundulus diaphanous* – *Fundulus* = bottom, referencing feeding; *diaphanus* = transparent. The common name references distinct, dark vertical bands along lateral sides.

Description:

Body – moderately fusiform, elongated, laterally compressed.
Color –

> *Dorsally* – olive-green.
> *Laterally* – olive-green to silver; 12 to 20 greenish-brown vertical bands; throat greenish-yellow.
> *Ventrally* – silvery-white.
> *Fins* – greenish-yellow to clear.
> *Peritoneum* – silver.

Head – flattened; snout bluntly pointed.
Eyes – larger; medial on the lateral side of head.
Mouth – small, superior; continuous groove between snout and upper lip.

> *Jaws* – lower jaw protruding further than upper jaw.

Teeth – small, sharp; in small rows along both jaws.
Dorsal fin – 10 to 15 rays; base anterior to anal fin base.
Caudal fin – homocercal; slightly rounded to convex.
Anal fin – 10 to 12 rays.
Pelvic fins – 6 rays.
Pectoral fins – rounded.
Lateral line – absent; 39 to 49 larger cycloid scales in lateral series.
Sexual dimorphism – spawning males have darker vertical bands on lateral sides during spawning season; females have darker bands along lateral sides than males.

Similar Species: Banded killifish closely resemble northern plains killifish, central mudminnow, and western mosquitofish. Northern plains killifish have a black peritoneum and smaller scales with approximately 50 to 67 in the lateral series (Kansas Fishes Committee 2014). Central mudminnow and western mosquitofish (not documented in the Dakotas) both have a dark spot below the eye. Additionally, western mosquitofish dorsal fin origin is posterior to the anal fin origin.

Distribution and Habitat: Banded killifish are widely distributed throughout eastern North America from Newfoundland to South Carolina, and west to the eastern side of the Dakotas. Their native range includes Great Lakes, Mississippi River basin, northern Iowa, and northeastern Nebraska (Kansas Fishes Committee 2014). In North Dakota, banded killifish are found in the Sheyenne River, Turtle River, and headwaters of the Red River of North (Copes 1965; Woolman 1896; Owen et al. 1981). In South Dakota, banded killifish inhabit the Big Sioux River, likely gaining access through the Minnesota and Des Moines rivers, and they were accidentally introduced into Lake Andes (Bailey and Allum 1962). Banded killifish prefer quiet and shallow waters of sloughs, marshes, ponds, and lakes, as well as low gradient brooks and streams with gravel or sand substrate and abundant vegetation (Eddy and Underhill 1974; Trautman 1981; Becker 1983). Juveniles school within aquatic vegetation, while schools of adults venture out during the day into open water just inches below the surface, becoming vulnerable to largemouth bass, smallmouth bass, northern pike, and other predators (Phillips et al. 2007; Kansas Fishes Committee 2014). Banded killifish tolerate a range of salinities, low oxygen levels, and warmer water temperatures.

Reproduction: Banded killifish exhibit sexual dimorphism during the spawning season (Forbes and Richardson 1920; Bailey and Allum 1962; Kansas Fishes Committee 2014). Sexual maturity of banded killifish occurs at age-1 or age-2 and is more a function of growth than age (Fournier and Magnin 1975; Phillips et al. 2007). Spawning is dictated by preferred water temperatures ranging from 21 to 23 °C and lasts for approximately 3 weeks (Fournier and Magnin 1975; Phillips et al. 2007). It begins when females suspend a small egg cluster with adhesive threads from the genital papilla. Males then pursue and escort the females into their vegetated or grassy, defended territory (Owen et al. 1981; Stewart and Watkinson 2004). The cluster of eggs is suspended further prior to fertilization. The fertilized eggs detach from the female and adhere to the vegetation to develop and hatch (Owen et al. 1981). Fecundity varies greatly by location. In Canada, banded killifish in two lakes had fecundity between 88 and 128 eggs, compared to 226 and 426 eggs from fish in in two other lakes (Fournier and Magnin 1975; Fritz and Garside 1975; Phillips et al. 2007).

Age and Growth: Adult banded killifish typically have total lengths from 60 to 80 mm but can reach 130 mm (Kansa Fish Committee 2014). Most adults reach age-3 but can live up to 4 years old. Total-lengths-at-age of banded killifish from Lake Erie were 52 to 67 mm at age-1, 63 to 79 mm at age-2, and 93 mm at age-3 (Phillips et al. 2007).

Food and Feeding: All ages of banded killifish feed throughout the water column on cladocerans, benthic macroinvertebrates, and aquatic insect larvae (Owen et al. 1981; Stewart and Watkinson 2004; Phillips et al. 2007). Juveniles primarily eat cladocerans and chironomids. Adults also consume large numbers of cladocerans, but also eat ostracods, copepods, odonatan nymphs, ephemeroptera nymphs, mollusks, and amphipods (Owen et al. 1981; Phillips et al. 2007).

References

Bailey RM, Allum MO (1962) Fishes of South Dakota. University of Michigan, Ann Arbor. https://doi.org/10.3998/mpub.9690435

Becker GC (1983) Fishes of Wisconsin. University of Wisconsin Press, Madison

Copes FA (1965) Fishes of the Red River tributaries of North Dakota. Dissertation, University of North Dakota

Eddy S, Underhill JC (1974) Northern fishes with special reference to the upper Mississippi Valley, 3rd edn. University of Minnesota Press, Minneapolis

Forbes SA, Richardson RE (1920) The fishes of Illinois. Illinois State Legislature, Springfield. https://doi.org/10.5962/bhl.title.5011

Fournier P, Magnin E (1975) Reproduction du petit barré de l'est *Fundulus diaphanus diaphanus* (Le Sueur). Nat Can 102:181–188

Fritz ES, Garside ET (1975) Comparison of age composition, growth, and fecundity between two populations each of *Fundulus heteroclitus* and *F. diaphanus* (Pisces: Cyprinodontidae). Can J Zool 53:300–311. https://doi.org/10.1139/z75-047

Kansas Fishes Committee (2014) Kansas fishes. University Press of Kansas, Lawrence

Owen JB, Elsen DS, Russell GW (1981) Distribution of fishes in North and South Dakota Basins affected by the Garrison Diversion Unit. University of North Dakota Press, Grand Forks

Phillips EC, Ewert Y, Speares PA (2007) Fecundity, age and growth, and diet of *Fundulus diaphanus* (banded killifish) in Presque Isle Bay, Lake Erie. Northeast Nat 14:269–278. https://doi.org/10.1656/1092-6194(2007)14[269:FAAGAD]2.0.CO;2

Stewart KW, Watkinson DA (2004) The freshwater fishes of Manitoba. University of Manitoba Press, Winnipeg

Trautman MB (1981) The fishes of Ohio, revised edn. Ohio State University Press, Columbus

Woolman AJ (1896) Report on ichthyological investigations in western Minnesota and eastern North Dakota. Rep US Fish Comm 1893:343–373

6.22.2   Northern Plains Killifish *Fundulus kansae*

Garman, 1895

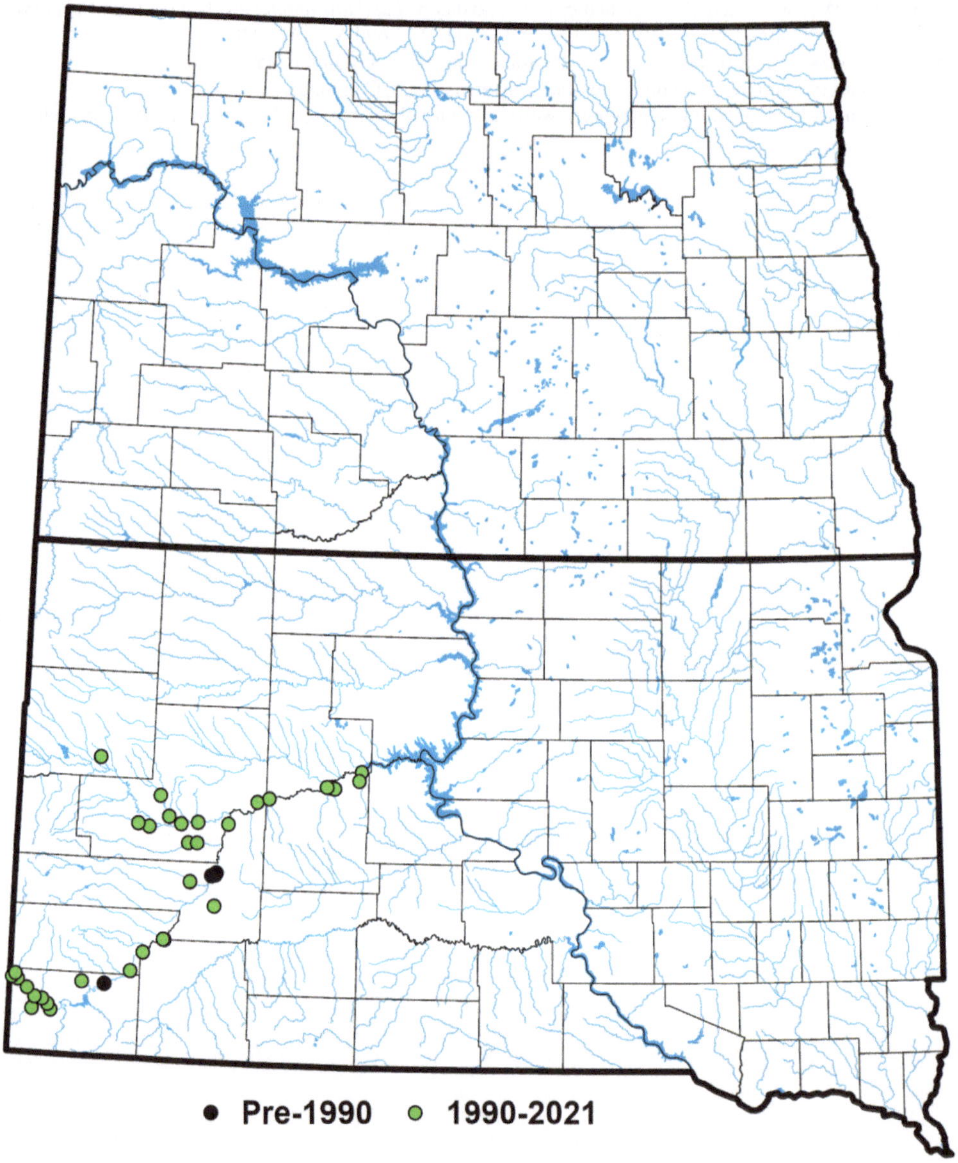

● **Pre-1990**   ● **1990-2021**

Etymology: *Fundulus kansae* – *Fundulus* = bottom, referencing feeding; *kansae* = from Kansas.

Description:

Body – cylindrical, elongated, somewhat laterally compressed, especially posteriorly.
Color –

> *Dorsally* – olive to tan.
> *Laterally* – light olive, silver, or cream; 12 to 28 dark gray vertical bars.
> Ventrally – cream to white.
> *Fins* – clear to light gray.
> *Peritoneum* – black.

Head – flattened dorsally, large scales.
Snout – elongated.
Eyes – large, dorsolaterally on head.
Mouth – small, superior.

> *Jaws* – lower jaw protrudes past upper jaw.
> *Lips* – lower jaw large, fleshy.

Teeth – small, fine, villiform.
Dorsal fin – 13 to 15 rays; rounded distal end; posteriorly on body.
Caudal peduncle – thick, elongated.
Caudal fin – homocercal, nearly straight distal end.
Anal fin – 13 or 14 rays; large, round distal end; insertion directly beneath dorsal fin insertion.
Pelvic fins – small; abdominal.
Pectoral fins – rounded distal end.
Lateral line – absent; 47 to 67 small cycloid scales in lateral line series.
Sexual dimorphism – spawning males have fewer, wider, darker vertical bars on the lateral sides, and bright red-orange caudal, anal, pelvic, pectoral fins.

Similar Species: Northern plains killifish closely resemble banded killifish. Banded killifish have a silver peritoneum and larger cycloid scales with approximately 39 to 49 scales in the lateral series. The two species will rarely, if ever, be in the same locations in the Dakotas. Central mudminnow have an anal fin with seven to nine rays, and 34–37 larger cycloid scales in the lateral series. Plains topminnow lack vertical bars on lateral sides.

Distribution and Habitat: Native to the Great Plains in North America, northern plains killifish range from Montana, east to Missouri, and south to Texas, including the Arkansas River and Missouri River systems. They do not occur in North Dakota. In South Dakota, they are found in the Cheyenne River and its major tributaries (Hoagstrom et al. 2009). Its distribution is patchy because habitat alterations have led to declining populations in parts of its range (Brown 1986). Barriers, like culverts, creating stream fragmentation may be impacting recolonization and dispersal capabilities (Prenosil et al. 2016). Northern plains killifish often school in medium to small rivers, as well as small ponds, shoals, and backwaters with low flow at lower elevations (Quist et al. 2004; Senecal 2009). They prefer shallow water over sand or gravel substrate and avoid areas deeper than 15 cm (Brown 1986; Minckley and Klaassen 1969b). Northern plains killifish tolerate high salinity, alkalinity, and intermittency, but are intolerant of high siltation levels (Baxter and Stone 1995; Pflieger 1997; Gido et al. 2002). They will bury themselves in sand substrate leaving only their eyes and mouth visible to possibly escape predators, ambush prey, or seek cooler temperatures (Minckley and Klaassen 1969a).

Reproduction: Both sexes of northern plains killifish can become sexual mature at age-1, but most mature at age-2 (Minckley and Klaassen 1969b). Females begin developing eggs in February (Minckley and Klaassen 1969b). Spawning occurs from April to August at water temperatures near 26 °C. Northern plains killifish spawn in low current waters less than 10 cm deep with sand or gravel substrate (Minckley and Klaassen 1969b). Males defend small territories where they actively court one female, vibrating their bodies together until eggs are released and fertilized. No nest is constructed, and there is no parental care. Adults retreat to preferred habitat soon after spawning concludes. Eggs are about 2 mm in diameter (Minckley and Klaassen 1969b). Fecundity is from 10 to 100 eggs and increases with female size (Minckley and Klaassen 1969b; Brown 1971). Eggs hatch in water temperatures from 13 to 36 °C (Wilson and Hubbs 1972).

Age and Growth: Newly hatched northern plains killifish larvae have a total length of approximately 10 mm (Brown 1971). They reach total lengths of about 35 mm and 50 mm at the end of the first and second summers, respectively (Minckley and Klaassen 1969b). They can reach a total length of 100 mm and live for 2 years (Eberle 2009).

Food and Feeding: Northern plains killifish are primarily carnivorous, feeding by sight only during the day (Echelle et al. 1971). They forage both throughout the water column and in the benthos, where they dig their snouts into the substrate, take a mouthful of sediment, and retain any organisms (Minckley and Klaassen 1969b). Adults eat adult and larvae aquatic invertebrates, like ephemeropterans and chironomids, and insects (Minckley and Klaassen 1969b; Eberle 2009). They also occasionally consume plant material and diatoms.

References

Baxter GT, Stone MD (1995) Fishes of Wyoming. Wyoming Game and Fish Department, Cheyenne

Brown CJD (1971) Fishes of Montana. Montana State University Press, Bozeman

Brown KL (1986) Population demographic and genetic structure of plains killifish from the Kansas and Arkansas River Basins in Kansas. Trans Am Fish Soc 115:568–576. https://doi.org/10.1577/1548-8659

Eberle ME (2009) Type locality and conservation status of the Northern plains killifish (*Fundulus kansae*: Fundulidae) in Kansas. Trans Kansas Acad Sci 112:87–97. https://doi.org/10.1660/062.112.0211

Echelle AA, Stevenson MM, Echelle AF, Hill LG (1971) Diurnal periodicity of activities in the plains killifish, *Fundulus zebrinus kansae*. Proc Oklak Acad Sci 51:3–7

Gido KB, Guy CS, Strakosh TR, Bernot RJ, Hase KJ, Shaw MA (2002) Long-term changes in the fish assemblages of the Big Blue River Basin 40 years after the construction of Turtle Creek Reservoir. Trans Kansas Acad Sci 105:193–208. https://doi.org/10.1660/0022-8443(2002)105[0193:LTCITF]2.0.CO;2

Hoagstrom CW, Hayer CA, Berry CR (2009) Criteria for determining native distributions of biota: the case of the Northern plains killifish in the Cheyenne River drainage, North America. Aqua Conserv Mari Freshwater Ecosys 19:88–95. https://doi.org/10.1002/aqc.1000

Minckley CO, Klaassen HE (1969a) Burying behavior of the plains killifish, *Fundulus kansae*. Copeia 1969:200–201. https://doi.org/10.2307/1441720

Minckley CO, Klaassen HE (1969b) Life history of the plains killifish, *Fundulus kansae* (Garman) in the Smokey Hill River, Kansas. Trans Am Fish Soc 98:460–465. https://doi.org/10.1577/1548-8659(1969)98[460:LHOTPK]2.0.CO;2

Pflieger WL (1997) The fishes of Missouri, revised edn. Missouri Dep Cons, Jefferson City

Prenosil E, Koupal K, Grauf J, Schoenebeck C, Hoback WW (2016) Swimming and jumping ability of 10 Great Plains fish species. J Freshw Ecol 31:123–130. https://doi.org/10.1080/02705060.2015.1048539

Quist MC, Hubert WA, Rahel FJ (2004) Elevation and stream-size thresholds affect distributions of native and exotic warm-water fish in Wyoming. J Freshw Ecol 19:227–236. https://doi.org/10.1080/02705060.2004.9664536

Senecal AC (2009) Fish assemblage structure and flow regime of the Powder River, Wyoming: an assessment of the potential effects of flow augmentation related to energy development. Dissertation, University of Wyoming

Wilson S, Hubbs C (1972) Developmental rates and tolerances of plains killifish, *Fundulus kansae* and comparison with related fishes. Texas J Sci 23:371–379

6.22.3   Plains Topminnow *Fundulus sciadicus*

Cope, 1865

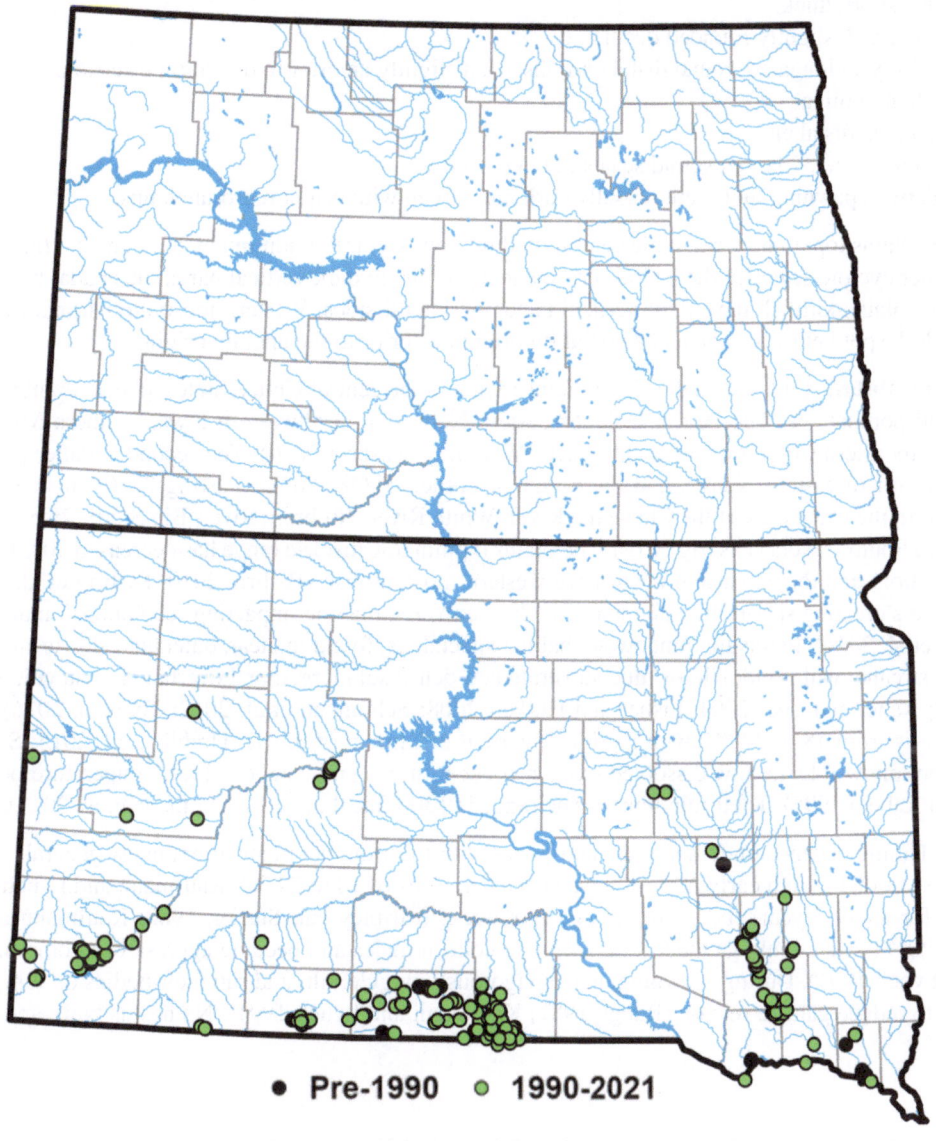

Etymology: *Fundulus sciadicus* – *Fundulus* = bottom, referencing feeding; *sciadicus* = shady, referencing habitat.

Description:

Body – moderately robust, stout anteriorly, laterally compressed posteriorly.
Color –

> *Dorsally* – dark olive-green; prominent gold dorsal stripe anterior of dorsal fin insertion.
> *Laterally* – olive-green to silver.
> *Ventrally* – silver.
> *Head* – small silvery-blue specks.
> *Eyes* – gold iris.
> *Fins* – faint-yellow-orange to clear.

Head – large, broad, dorsoventrally flattened, scaled.
Snout – blunt.
Eyes – moderately large; laterally on head.
Mouth – small, superior.

> *Jaws* – lower jaw slightly protrudes past upper jaw.

Teeth – small, on both jaws.
Dorsal fin – 9 to 11 rays; slightly rounded distal end; posteriorly on body.
Caudal peduncle – short, thick.
Caudal fin – homocercal, slightly rounded distal end.
Anal fin – 12 to 15 rays; elongated, round distal end; insertion slightly anterior to dorsal fin insertion.
Pelvic fins – small, abdominal.
Pectoral fins – rounded distal end.
Lateral line – absent; 33 to 37 large cycloid scales in lateral series.
Sexual dimorphism – spawning males are dorsally dark blue-green with bright red-orange fins.

Similar Species: Plains topminnow resemble central mudminnow. Central mudminnow have an anal fin with seven to nine rays, 34 to 37 larger cycloid scales in the lateral series, and a prominent dark vertical bar at the caudal fin base. Banded killifish and northern plains killifish have dark vertical bands on lateral sides. Western mosquitofish, not documented in the Dakotas, have a dark spot below the eye and a dorsal fin origin posterior to the anal fin origin.

Distribution and Habitat: Endemic to the Great Plains region in the central United States, plains topminnow exist primarily in two isolated populations. The largest includes eastern Nebraska, northwestern Iowa, southern Minnesota, southern South Dakota, northeastern Colorado, and southeastern Wyoming (Pasbrig 2010). The second, smaller population extends from central Missouri to eastern Kansas and the northeast corner of Oklahoma (Pasbrig 2010). It is absent from North Dakota, but found in the tributaries of the Cheyenne River, White River, Niobrara River, Big Sioux River, Vermillion River, and James River in South Dakota (Pasbrig 2012). The plains topminnow historic range has decreased 70% because of habitat loss, habitat alteration, and the expansion of invasive western mosquitofish (Pasbrig 2010; Pasbrig et al. 2012; Schumann et al. 2015). Barriers creating stream fragmentation, such as culverts, may be impacting recolonization and dispersal capabilities (Prenosil et al. 2016). Plains topminnow are habitat specialists, found in clear, quiet, and slower-moving backwaters of small shallow streams with sand, silt, or mud substrate and dense submerged and overhanging aquatic vegetation (Ross and Ultsch 1980; Rahel and Thel 2004; Fisher and Paukert 2008; Schumann et al. 2012; Thiessen 2016). Their optimum water temperature is from 18 to 24 °C, with a critical thermal maximum of 37 °C (Hrabik et al. 2015; Smale and Rabeni 1995). Plains topminnow tolerate low dissolved oxygen concentrations and extended high-temperature periods in isolated pools (Smale and Rabeni 1995). Their minimum home range is 1300 lineal meters (Schumann et al. 2014).

Reproduction: Timing depends on temperature and varies with location, but plains topminnow generally spawn for about 60 days from late March through early August at water temperatures of 18 to 24 °C (Kaufmann and Lynch 1991). They are sexually mature for about 3 years, beginning at age-1 or age-2 (Stribley and Stasiak 1982; Kaufmann and Lynch 1991; Schumann et al. 2012). Males enlarge their gular region to court females and become aggressive to other males by nipping (Kaufmann and Lynch 1991). During spawning, males vigorously wiggle, tilt, and rub their bodies on females near vegetation or substrate to induce gamete release (Baugh 1981; Kaufmann and Lynch 1991). No nest is built, and no parental care is given. The approximately 0.1 to 2-mm diameter, yellow-orange eggs are all broadcast at the same time and adhere to detritus and vegetation (Kaufmann and Lynch 1991; Kinney and Lynch 1991; Pflieger 1997). Fecundity is 50 to 90 eggs (Hrabik et al. 2015). Hatching occurs in 13 or 14 days at 21 to 23 °C (Kaufmann and Lynch 1991; Kinney and Lynch 1991).

Age and Growth: Newly hatched larvae typically have total lengths of 6 to 8 mm (Kaufmann and Lynch 1991). Total-lengths-at-age of plains topminnow from Nebraska were 23 mm at age-0, 47 mm at age-1, 53 mm at age-2, and 62 mm at age-3 (Stribley and Stasiak 1982). Adults have total lengths of 32 to 64 mm but can reach 75 mm (Pflieger 1997; Rahel and Thel 2004). Females are slightly longer and heavier than males of same year class (Stribley and Stasiak 1982). Longevity is 4 years.

Food and Feeding: Plains topminnow primarily feed on the surface in heavily vegetated backwaters (Thiessen et al. 2018). Adults eat adult and larval dipterans, hemipterans, and other small aquatic insects, small crustaceans, and snails (Stribley and Stasiak 1982; Pflieger 1997; Rahel and Thel 2004). Plains topminnow in lentic environments are generalists, while those in lotic systems prefer gastropods over decapods and ephemeropterans (Thiessen et al. 2018). In Missouri, they ate more benthic organisms, indicating that diets may vary seasonally, geographically, and by trophic interactions (Thompson 2014; Thiessen et al. 2018).

References

Baugh TM (1981) In search of plains topminnow. Freshw Mari Aquar 4:39–41

Fisher JR, Paukert CP (2008) Historical and current environmental influences on an endemic Great Plains fish. Am Midl Nat 159:364–377. https://doi.org/10.1674/0003-0031(2008)159[364:HACEIO]2.0.CO;2

Hrabik RA, Schainost SC, Stasiak RH, Peters EJ (2015) The fishes of Nebraska. University of Nebraska, Lincoln

Kaufmann SA, Lynch JD (1991) Courtship, eggs, and development of the plains topminnow in Nebraska (Actinopterygii: Fundulidae). Prairie Nat 23:41–45

Kinney TA, Lynch JD (1991) The fecundity and reproductive season of *Fundulus sciadicus* in Nebraska (Actinopterygii: Fundulidae). Trans Nebr Acad Sci 18:101–104

Pasbrig CA (2010) Reductions in range-wide distribution of plains topminnow, *Fundulus sciaicus*, and production of a broodstock pond. Dissertation, University of Nebraska-Kearney

Pasbrig CA (2012) plains topminnow: the "minnow" that isn't a minnow. S D Conserv Digest 79:2–3

Pasbrig CA, Koupal KD, Schainost S, Hoback WW (2012) Changes in range-wide distribution of plains topminnow *Fundulus sciadicus*. Endangered Spec Res 16:235–247. https://doi.org/10.3354/esr00400

Pflieger WL (1997) The fishes of Missouri, revised edn. Missouri Dep Cons, Jefferson City

Prenosil E, Koupal K, Grauf J, Schoenebeck C, Wyatt Hoback W (2016) Swimming and jumping ability of 10 Great Plains fish species. J Freshw Ecol 31:123–130. https://doi.org/10.1080/02705060.2015.1048539

Rahel FJ, Thel LA (2004) plains topminnow (*Fundulus sciadicus*): a technical conservation assessment. USDA Forest Service, Rocky Mountain Region, golden https://www.fs.usda.gov/Internet/FSE_DOCUMENTS/stelprdb5206793.pdf. Accessed 24 Apr 2023

Ross MJ, Ultsch GR (1980) Temperature and substrate influences on habitat selection in two pleurocerid snails (Goniobasis). Am Midl Nat 103:209–217. https://doi.org/10.2307/2424619

Schumann DA, Pasbrig CA, Koupal KD, Hoback WW (2012) Culture of plains topminnow in a pond constructed for species conservation. N Am J Aquacult 74:360–364. https://doi.org/10.1080/15222055.2012.675989

Schumann DA, Koupal KD, Hoback WW, Schoenebeck CW, Schainost S (2014) Large-scale dispersal patterns and habitat use of plains topminnow *Fundulus sciadicus*: implications for species conservation. J Freshw Ecol 30:311–322. https://doi.org/10.1080/02705060.2014.948083

Schumann DA, Hoback WW, Koupal KD (2015) Complex interactions between native and invasive species: investigating the differential displacement of two topminnows native to Nebraska. Aqua Invas 10:339–346. https://doi.org/10.3391/ai.2015.10.3.09

Smale MA, Rabeni CF (1995) Hypoxia and hyperthermia tolerances of headwater stream fishes. Trans Am Fish Soc 124:698–710. https://doi.org/10.1577/1548-8659

Stribley JA, Stasiak RH (1982) Age, growth and food habits of the plains topminnow, *Fundulus sciadicus* Cope, in Keith County, Nebraska. Proc Nebr Acad Sci Affil Soc 92:17–18

Thiessen JD (2016) Conservation of plains topminnow, *Fundulus sciadicus*, reestablishment success and limiting factors of persistence of reintroduced populations in Nebraska. Dissertation, University of Nebraska-Kearney

Thiessen J, Koupal KD, Schoenebeck CW, Shaffer JJ (2018) Food habits of imperiled plains topminnow and diet overlap with invasive Western mosquitofish in the central Great Plains. Trans Nebr Acad Sci Affil Soc 38:1–9. https://doi.org/10.13014/K2319T31

Thompson GT (2014) Ecology of a declining Great Plains fish, *Fundulus sciadicus*, in the Missouri Ozarks. Dissertation, Missouri University of Science and Technology

6.23   Centrarchidae, Sunfish Family

The sunfish family is native only to North America. It contains eight genera and more than 30 species. Only one centrarchid, the Sacramento perch *Archoplites interruptus*, has a native range west of the continental divide. Because many sunfish family species are highly valued for sport fishing, they have been introduced worldwide and have even been labeled as invasive species in some locations.

Two genera of centrarchids in the Dakotas, *Lepomis* and *Micropterus*, are well defined by body morphology. *Lepomis* spp. have a deep, more round body shape with smaller mouths, whereas *Micropterus* spp. are more elongate with larger mouths. Centrarchids are often confused with percids (perch) and moronids (temperate bass) but can be differentiated by a fused membrane of the spinous and rayed dorsal fins only slightly separated by a shallow notch; percids and moronids have barely connected or separate spinous and rays of the dorsal fins. Centrarchids also have three or more anal spines, whereas percids only have one or two. Spawning male sunfish also develop bright coloration on the cheeks and ventral sides.

Centrarchids are adapted to warm water and occur in a variety of habitats including swamps, ponds, lakes, reservoirs, and minimal flow areas in small streams and rivers. They occupy various water column depths depending on the species. For example, green sunfish *Lepomis cyanellus* frequent the littoral zone, whereas bluegill *Lepomis macrochirus* are more often found within the limnetic zone. Centrarchids are most often found as individuals or small groups rather than large schools. They typically use partially or fully submerged aquatic and overhanging vegetation as cover from predators or to ambush prey. Centrarchids primarily feed by sight during the day, actively foraging or using ambush behavior. Feeding is species specific, with prey captured from the surface, within the water column, along vegetation and submerged structures, and bottom substrate. They typically consume aquatic invertebrates, terrestrial invertebrates, and small fish, but foraging behavior is heavily influenced by interactions with other species.

Spawning behaviors of centrarchids are intricate, with most species exhibiting courtship, nest construction, and parental care. Males of certain species will use their fins to excavate a saucer-shaped nest in the substrate. The nests may be constructed within large colonies or be more dispersed. After the eggs are deposited in the nest by the female, most centrarchid males will provide parental care by fanning and guarding the incubating eggs and hatched fry. Hybridization is common because many species, especially *Lepomis*, have similar spawning behaviors.

Crappies, bass, and other sunfish are all important game fishes in the Dakotas and throughout the United States. In the Dakotas, centrarchid populations are density dependent, with overpopulation potentially causing the stunting of growth.

6.23.1   Rock Bass *Ambloplites rupestris*

(Rafinesque, 1817)

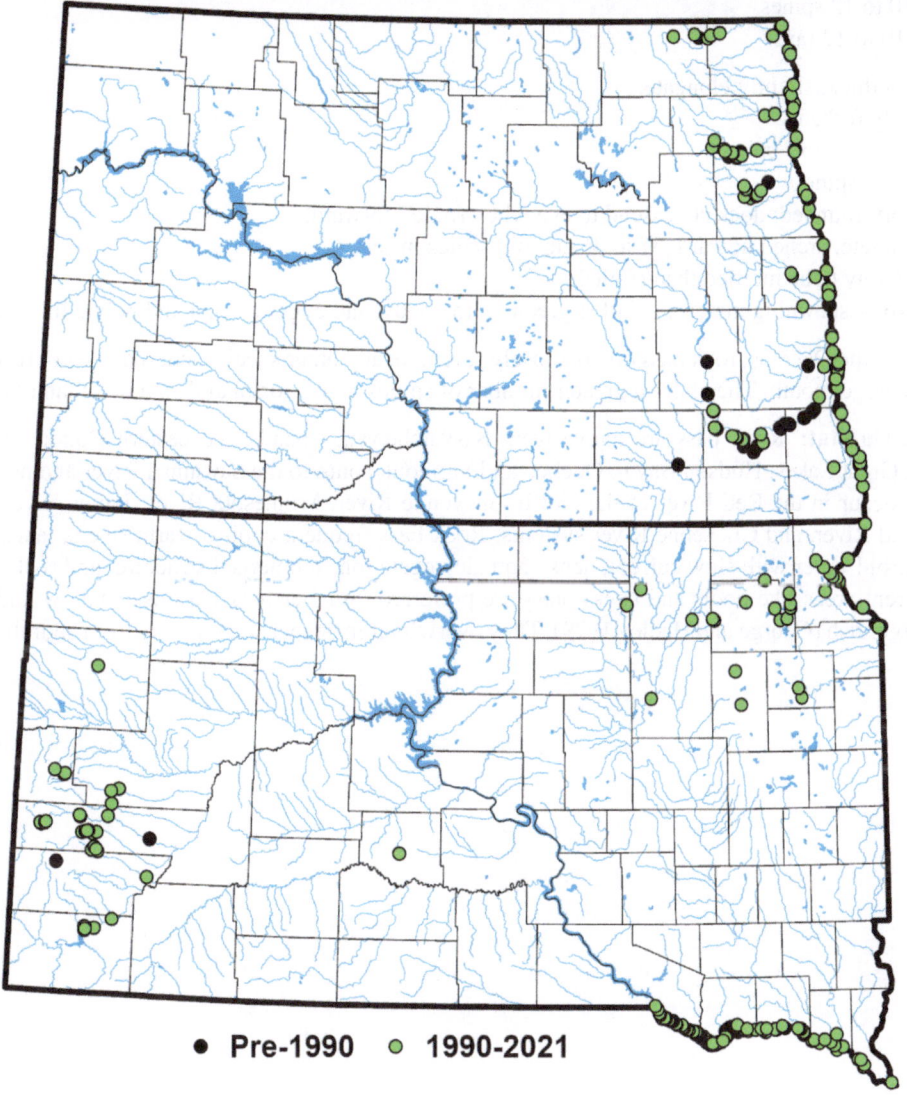

● **Pre-1990**   ○ **1990-2021**

Etymology: *Ambloplites rupestris* – *Ambloplites* = blunt armature; *rupestris* = living among rocks.

Description:

Body – deep, robust, laterally compressed.
Color – capable of changing coloration to camouflage with habitat.

> *Dorsally* – dark brown to olive.
> *Laterally* – solid olive to bronze, may have dark mottling and dark spots on scales below lateral line forming 8 to 10 horizontal rows.
> *Ventrally* – tan to creamy-white
> *Eyes* – partially dark amber-brown.
> *Fins* – dorsal, caudal, anal: dark and light mottling; pelvic, pectoral: dusky.

Head – large, robust.
Snout – conical.
Eyes – large, laterally on head.
Mouth – large, terminal, oblique.

> *Upper jaw* – extends to or beyond middle of pupil.
> *Lower jaw* – protrudes past upper jaw.

Teeth – small, blunt, in pads on upper and lower jaws; tooth patch on tongue.
Gill rakers – 7 to 10; long, thin.
Dorsal fin – 2 lobes, joined appearing as one.

> *Anterior* – 10 to 12 spines.
> *Posterior* – 10 to 12 rays.

Caudal peduncle – thick, slightly elongate.
Caudal fin – slightly forked.
Anal fin – 9 to 11 rays; 5 to 7 spines.
Pelvic fins – 5 rays; 1 spine.
Pectoral fins – short, rounded; does not extend to eye when folded forward.
Lateral line – complete, arched upward, 37 to 46 ctenoid scales in series.
Juveniles – more likely to be mottled than adults.
Sexual dimorphism – spawning males are darker green on the body, and edges of anal and pelvic fins; iris becomes red.

Similar Species: In the Dakotas, rock bass do not closely resemble any other species, but may be confused with green sunfish because of the large mouth. They have a greater number of spines in the dorsal and anal fins than other centrarchids.

Distribution and Habitat: Rock bass are native from Saskatchewan, Canada and eastern Dakotas through the Upper Mississippi basin, Great Lakes, Hudson Bay to Quebec and Vermont, south to the Savannah River and northern Alabama. In the Dakotas, they occur in the Red River of the North, Sheyenne River, Minnesota River, James River, Big Sioux River, Missouri River, Bad River, and Cheyenne River systems. Rock bass frequent cool-to-warm rivers, lakes, and reservoirs in clear-to-slightly turbid water with slow current, dense aquatic vegetation, submerged structure, and rocky substrate. In lotic habitats, slow current areas like pools and backwaters are preferred. Smaller rock bass use shallow and quiet areas more frequently than larger fish (George and Hadley 1979). They overwinter in deeper waters away from shorelines (Becker 1983).

Reproduction: Spawning occurs from May to July. Female rock bass peak spawning behavior occurs from 21 to 23 °C (Gross and Nowell 1980). They sexually mature at age-3 or age-4 in streams and age-5 to age-9 in lakes (Noltie and Keenleyside 1986). Males are territorial, using their anal and pectoral fins to construct 20 to 25 cm saucer-shaped nests in coarse gravel substrate at depths from 0.5 to 1.5 m close to cover (Gross and Nowell 1980; Noltie and Keenleyside 1986; Musch 2007). Nest construction begins at a water temperature of approximately 20 °C, with spawning occurring 2 days later (Scott and Crossman 1973; Gross and Nowell 1980; Becker 1983). Females enter nests without male courting (Gross and Nowell 1980). Spawning is most frequent in the early mornings or late evenings (Gross and Nowell 1980). Males become darker during spawning (Gross and Nowell 1980). Rock bass are fractional spawners. Fecundity ranges from 3,000 to 11,000 eggs. The adhesive, orange eggs are approximately 2 mm in diameter (Becker 1983). Males fan the eggs and defend the nest from predators until larvae leave 9 or 10 days after hatching (Gross and Nowell 1980). Eggs hatch in 5 days at water temperatures of 16 to 22 °C (Gross and Nowell 1980). Spawning ceases when water temperatures reach 26 °C (Scott and Crossman 1973; Becker 1983).

Age and Growth: Newly hatched rock bass larvae have a total length of approximately 6 mm (Buynak and Mohr Jr 1979). Total lengths at age of rock bass from Missouri were age-1, 41 mm; age-2, 86 mm; age-3, 140 mm; age-4, 178 mm; age-5, 203 mm; age-6, 216 mm (Pflieger 1997). They can reach a total length of 432 mm and live for 10 years, although their typical lifespan is from 6 to 8 years (Becker 1983; Pflieger 1997).

Food and Feeding: Rock bass are generalist predators. Young-of-year eat ephemeroptera, cladocerans, copepods, chironomid larvae, amphipods, and isopods (Glessner 1977; Keast 1977). Rock bass with total lengths from 150 to 200 mm mainly consume snails and chironomids (Elrod et al. 1981). Adults primarily eat amphipods, trichopterans, gastropods, decapods, chironomids, and minnows (George and Hadley 1979; Elrod et al. 1981). As mean length increases, consumption of decapods and fish increases and consumption of amphipods and trichopterans decreases (Elrod et al. 1981).

References

Becker GC (1983) Fishes of Wisconsin. University of Wisconsin Press, Madison

Buynak GL, Mohr HW Jr (1979) Larval development of rock bass from the Susquehanna River. Prog Fish Cult 41:39–42. https://doi.org/10.1577/1548-8659(1979)41[39:LDORBF]2.0.CO;2

Elrod JH, Busch WDN, Griswold BL, Schneider CP, Wolfert DR (1981) Food of white perch, rock bass and yellow perch in eastern Lake Ontario. N Y Fish Game J 28:191–201

George EL, Hadley WF (1979) Food and habitat partitioning between rock bass (*Ambloplites rupestris*) and smallmouth bass (*Micropterus dolomieui*) young of the year. Trans Am Fish Soc 108:253–261. https://doi.org/10.1577/1548-8659

Glessner GL (1977) Food habits and growth of the rock bass, *Ambloplites rupestris* (Rafinesque), in Stone Valley Lake, Huntingdon County, Pennsylvania. Dissertation, Pennsylvania State University

Gross MR, Nowell WA (1980) The reproductive biology of rock bass, *Ambloplites rupestris* (Centrarchidae), in Lake Opinicon, Ontario. Copeia 3:482–494. https://doi.org/10.2307/1444526

Keast A (1977) Mechanisms minimizing intraspecific competition in vertebrates, with a quantitative study of the contrasting strategies of two centrarchid fishes, *Ambloplites rupestris* and *Lepomis macrochirus*. Evol Biol 10:333–395. https://doi.org/10.1007/978-1-4615-6953-4_7

Musch AE (2007) Spawning habitat selection of sympatric smallmouth bass (*Micropterus dolomieu*) and rock bass (*Ambloplites rupestris*) in north temperate lakes: habitat separation in space and time. Dissertation, University of Wisconsin-Stevens Point

Noltie DB, Keenleyside MHA (1986) Breeding ecology, nest characteristics, and nest-site selection of stream- and lake-dwelling rock bass, *Ambloplites rupestris* (Rafinesque). Can J Zool 65:379–390. https://doi.org/10.1139/z87-059

Pflieger WL (1997) The fishes of Missouri, revised edn. Missouri Dep Cons, Jefferson City

Scott WB, Crossman EJ (1973) Freshwater fishes of Canada. Fisheries Research Board of Canada, Bulletin 184

6.23.2 Green Sunfish *Lepomis cyanellus*

Rafinesque, 1819

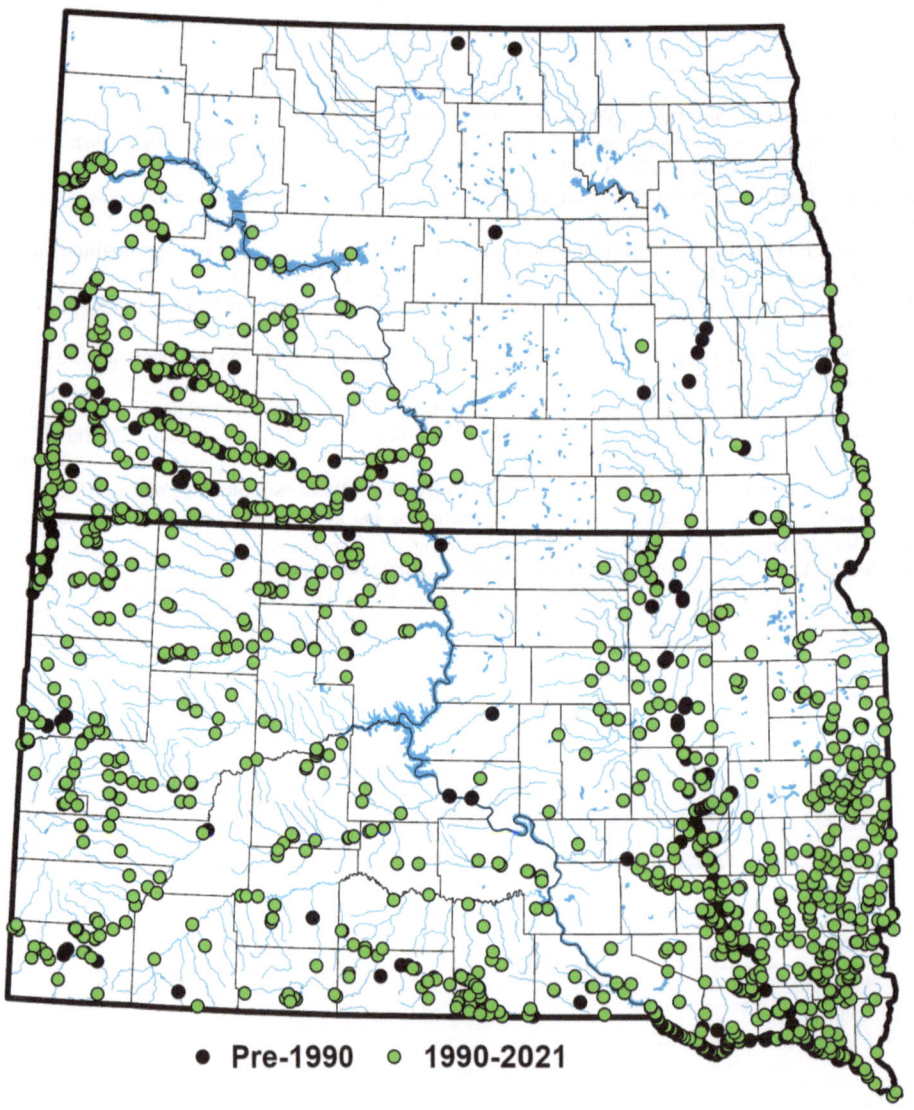

● Pre-1990   ● 1990-2021

Etymology: *Lepomis cyanellus* – *Lepomis* = scaled lid, referencing scaled operculum; *cyanellus* = blue.

Description:

Body – deep, robust, slightly elongated, laterally compressed; arched dorsal profile.
Color –

> *Dorsally* – dark olive to emerald green with emerald reflections.
> *Laterally* – green to light green with blue-green reflections; 7 to 12 extremely faint, dark vertical bars.
> *Ventrally* – yellow to cream.
> *Head* – backward extending wavy light blue-green lines or spots.
> *Opercle* – dark center, light border.
> *Fins* – uniformly pigmented; dark blotch on posterior dorsal fin base.

Head – large, conical.
Opercle – flap elongated, inflexible.
Snout – large, conical.
Eyes – moderately large, laterally on head.
Mouth – terminal, slightly oblique, large.

> *Upper jaw* – extends to middle of eye.
> *Lips* – fleshy, smooth.

Teeth – small, in pads on both jaws; few or none on tongue.
Gill rakers – 11 to 14 on first arch; long, slender.
Dorsal fin – 9 to 11 spines, 10 or 11 rays; elongated.
Caudal peduncle – short, thick.
Caudal fin – slightly forked, rounded lobes.
Anal fin – 8 to 10 rays; 3 spines.
Pelvic fins – 5 rays, 1 spine.
Pectoral fins – 13 to 15 rays; rounded on distal end; when folded forward barely reach eye.
Lateral line – complete, arched upward; 44 to 53 ctenoid scales in series.
Sexual dimorphism – spawning males have more intense coloration and a distinct yellow-white margin on distal end of dorsal, caudal, and anal fins.

Similar Species: Green sunfish closely resemble other *Lepomis* species. Bluegills have a deeper, less elongate body shape, an anal fin with 10 to 12 rays and a short, flexible solid blue opercle. Pumpkinseeds have a single red spot on the opercle posterior edge. Orangespotted sunfish have small, irregular orange spots along the lateral sides, and one pair of large sensory pores between the eyes.

Distribution and Habitat: The native range of green sunfish encompasses the Missouri River system, Mississippi River system, Great Lakes to eastern Pennsylvania, and south to Alabama and eastern New Mexico. Widely introduced across North American, they readily colonize new habitats, particularly after flooding events. In the Dakotas, green sunfish occur in the major tributaries of the Missouri River, as well as the Red River of the North, Sheyenne River, Souris River, and Minnesota River systems. Green sunfish are most abundant in low gradient ponds, lakes, and reservoirs with large littoral areas, abundant submerged vegetation, moderate turbidity, and sand, gravel, or silt substrates (Scott and Crossman 1973). They also inhabit pools in small streams and rivers. Green sunfish avoid water temperatures less than 26 °C and greater than 31 °C and have an optimal water temperature of 28 °C (Beitinger et al. 1975; Stuber et al. 1982). Their mean critical thermal maximum is 37 or 38 °C (Smale and Rabeni 1995). Green sunfish tolerate high turbidity, high alkalinity, low dissolved oxygen, and rapid pH changes, although high turbidity decreases activity (Horkel and Pearson 1976).

Reproduction: Green sunfish spawn from May to early July at water temperatures from 19 to 28 °C. They sexually mature from age-1 to age-3, depending on location. Males construct one or more, approximately 30-cm diameter nests in sand or gravel substrate near vegetation or large rocks in water 4 to 35 cm deep (Hunter 1963; Carlander 1977). Spawning begins 1 or 2 days after nest construction (Hunter 1963). Males court females with a series of grunting sounds or visual cues (Gerald 1971). Males provide parental care for approximately a week until fry flee the nest (Hunter 1963). The demersal, adhesive, 1 mm diameter eggs hatch in 35 to 55 hours in water temperatures of 24 to 27 °C (Taubert 1977).

Age and Growth: Newly hatched green sunfish larvae have a total length of 4 mm (Taubert 1977). Growth rates are highly variable among populations but increase with increasing amounts of woody debris (Quist and Guy 2001). Stunting may occur in overpopulated, small, closed systems where increased competition decreases growth (Werner and Hall 1976). Green sunfish can reach a total length of 305 mm, with a typical life span of 5 or 6 years, and a longevity of 10 years (Carlander 1977).

Food and Feeding: Green sunfish are voracious predators; feeding by sight or using their lateral line system to detect prey in turbid waters (Janssen and Corcoran 1993). At total lengths less than 10 mm, green sunfish feed day and night, but feed more intensely during the day as they age (Barkoh 1984). Fry eat zooplankton and as they grow also consume immature aquatic insects and egg cases (Sadzikowski and Wallace 1976). Prey length increases as fry length increases (Barkoh 1984). In South Dakota, fry primarily selected *Cyclops vernalis* and *Moina brachiate* (Barkoh 1984). Juveniles and adults primarily eat insects, crayfish, and small fish (Pflieger 1997). Reduced competition increases prey size (Werner and Hall 1976). Juveniles can detect chemical alarm signals from other green sunfish whose skin cells are damaged from predators (Brown and Brennan 2000; Golub and Brown 2003). When exposed to these signals, small green sunfish exhibit fin-erect posture, while larger fish may continue to forage (Brown and Brennan 2000; Golub and Brown 2003; Kansas Fishes Committee 2014).

References

Barkoh A (1984) Food selectivity of bluegill and green sunfish fry. Dissertation, South Dakota State University

Beitinger TL, Magnuson JJ, Neill WH, Shaffer WR (1975) Behavioral thermoregulation and activity patterns in the green sunfish, *Lepomis cyanellus*. Anim Behav 23:222–229. https://doi.org/10.1016/0003-3472(75)90067-6

Brown GE, Brennan S (2000) Chemical alarm signals in juvenile green sunfish (*Lepomis cyanellus*, Centrarchidae). Copeia 4:1079–1082. https://doi.org/10.1643/0045-8511(2000)000[1079:CASIJG]2.0.CO;2

Carlander KD (1977) Handbook of freshwater fishery biology, 2nd edn. Iowa State Univ Press, Ames

Gerald JW (1971) Sound production during courtship in six species of sunfish (Centrarchidae). Evolution 25:75–87. https://doi.org/10.2307/2406500

Golub JL, Brown GE (2003) Are all signals the same? Ontogenetic change in the response to conspecific and heterospecific chemical alarm signals by juvenile green sunfish (*Lepomis cyanellus*). Behav Ecol Sociobiol 54:113–118. https://doi.org/10.1007/s00265-003-0629-9

Horkel JD, Pearson WD (1976) Effects of turbidity on ventilation rates and oxygen consumption of green sunfish, *Lepomis cyanellus*. Trans Am Fish Soc 105:107–113. https://doi.org/10.1577/1548-8659

Hunter JR (1963) The reproductive behavior of the green sunfish, *Lepomis cyanellus*. Zoology 48:13–24. https://doi.org/10.5962/p.203306

Janssen J, Corcoran J (1993) Lateral line stimuli can override vision to determine sunfish strike trajectory. J Exp Biol 176:299–305. https://doi.org/10.1242/jeb.176.1.299

Kansas Fishes Committee (2014) Kansas fishes. University Press of Kansas, Lawrence

Pflieger WL (1997) The fishes of Missouri, revised edn. Missouri Dep Cons, Jefferson City

Quist MC, Guy CS (2001) Growth and mortality of prairie stream fishes: relations with fish community and instream habitat characteristics. Ecol Freshw Fish 10:88–96. https://doi.org/10.1034/j.1600-0633.2001.100203.x

Sadzikowski MR, Wallace DC (1976) A comparison of the food habits of size classes of three sunfishes (*Lepomis macrochirus* Rafinesque, *L. gibbosus* (Linnaeus) and *L. cyanellus* Rafinesque). Am Midl Nat 95:220–225. https://doi.org/10.2307/2424252

Scott WB, Crossman EJ (1973) Freshwater fishes of Canada. Fisheries Research Board of Canada, Bulletin 184

Smale MA, Rabeni CF (1995) Hypoxia and hyperthermia tolerances of headwater stream fishes. Trans Am Fish Soc 124:698–710. https://doi.org/10.1577/1548-8659

Stuber RJ, Gebhart G, Maughan OE (1982) Habitat suitability index models: green sunfish. US Dep Inter, Fish Wildl Serv, FWS/OBS-82/10.15

Taubert BD (1977) Early morphological development of the green sunfish, *Lepomis cyanellus*, and its separation from other larval *Lepomis* species. Trans Am Fish Soc 106:445–448. https://doi.org/10.1577/1548-8659

Werner CA, Hall DJ (1976) Niche shifts in sunfishes: experimental evidence and significance. Science 191:404–406. https://doi.org/10.1126/science.1246626

6.23.3 Pumpkinseed *Lepomis gibbosus*

(Linnaeus, 1758)

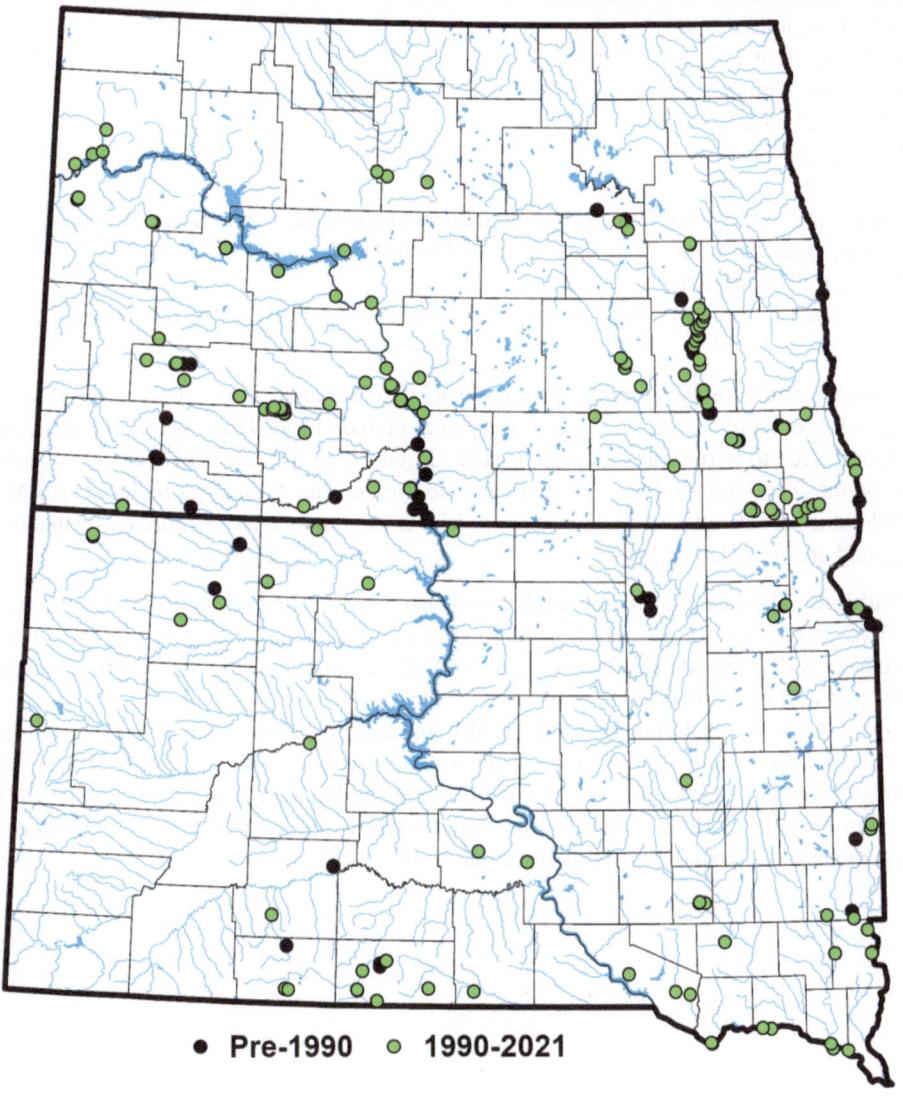

● **Pre-1990** ○ **1990–2021**

Etymology: *Lepomis gibbosus* – *Lepomis* = scaled lid, referencing scaled operculum; *gibbosus* = full moon, referencing body shape.

Description:

Body – deep, laterally compressed, arched dorsal profile.
Color –

> *Dorsally* – golden brown to olive.
> *Laterally* – tan; 7 to 10 faint, dark, diffuse vertical bars (most prominent in females) with orange, emerald, or yellow reflections.
> *Ventrally* – yellow-orange anteriorly fading to light yellow-cream posteriorly;
> *Head* – 3 to 5 wavy blue-green lines extending backward to gill flap.
> *Opercle* – dark black center, white-cream margin separated by single red spot on posterior edge.
> *Fins* – dorsal, caudal, anal – dark with faint to prominent small, brown spots or blotches.

Head – large, conical.
Opercle – opercular flap short, flexible.
Snout – large, conical.
Eyes – moderately large, laterally on head.
Mouth – small, terminal, slightly oblique.

> *Upper jaw* – barely extends to anterior end of eye.
> *Lower jaw* – slightly protrudes past upper jaw
> *Lips* – fleshy.

Teeth – small, in pads on both jaws; absent on tongue.
Gill rakers – 9 to 12; short, thick.
Dorsal fin – 10 to 12 rays; 10 or 11 spines; elongated.
Caudal peduncle – slightly elongate, thick.
Caudal fin – slightly forked with rounded lobes.
Anal fin – 9 or 10 rays; 3 spines.
Pelvic fins – 5 rays; 1 spine.
Pectoral fins – 12 to 14 rays; when folded forward pointed distal end extends to anterior end of eye.
Lateral line – complete, arched upward; 36 to 43 ctenoid scales in series.
Juveniles – less intense, dull coloration.
Sexual dimorphism – spawning males have very bright coloration, especially on throat and red spot on opercle posterior edge.

Similar Species: Pumpkinseed closely resemble other *Lepomis* species but are easily distinguished by a single red spot on posterior opercle edge. Redear sunfish lack blue-green wavy lines on their head. They also have a short and stiff opercle flap with a black center, and a red to burnt orange, crescent-shaped margin on the posterior edge. Orangespotted sunfish have orange spots on the lateral sides, large sensory pores between the eyes, and are typically smaller in size. Young of year pumpkinseed may resemble bluegill but lack a solid dark blue opercle. Green sunfish have a more elongated body, 44 to 53 scales in lateral line series, and are often darker.

Distribution and Habitat: The native range of pumpkinseed encompasses the eastern Dakotas, southeastern Manitoba, the upper Mississippi River basin, Great Lakes, Hudson Bay to New England, and south through the Atlantic Slope to South Carolina. They have been widely introduced elsewhere. In the Dakotas, pumpkinseed occur throughout the main tributaries of the Missouri River, as well as the Red River of the North, Sheyenne River, Souris River, and Minnesota River systems. They inhabit clear, quiet waters of ponds, lakes, and reservoirs, and low flow areas of small streams and rivers. Pumpkinseed are often associated with abundant aquatic vegetation. They actively swim in midwater (Miller 1963). They tolerate a wide variety of environmental conditions, including low dissolved oxygen and moderate turbidity, but are less tolerant of warmer water temperature than bluegill (O'Hara 1968). The preferred water temperature range for pumpkinseed is 24 to 32 °C (Evans 1977). Home ranges vary from 0.2 to 1.1 ha (Fish and Savitz 1983).

Reproduction: Pumpkinseed spawn from May to August at water temperatures near 20 °C (Miller 1963; Reed 1971; Becker 1983). Males sexually mature in 3.5 years or a total length of 104 mm, while females mature at 3.4 years and 100 mm (Fox 1994). Larger pumpkinseed mature earlier than same-age smaller fish in the same population (Fox 1994). Males construct and defend saucer-shaped nests, singularly or in colonies, near submerged structure, in quiet areas under 1 m deep, with partial sunlight, little to no aquatic vegetation, and sand, gravel, or mud substrate (Miller 1963). The nest diameter is approximately twice the length of the male (Miller 1963). Females remain in deeper water until nest construction is completed (Miller 1963). Pumpkinseed are fractional spawners and females spawn in more than one nest. Males vigorously defend and fan the eggs until hatch (Miller 1963). Fecundity may reach 7,000 eggs, depending on female size and age (Becker 1983). The demersal, adhesive, creamy-amber-white eggs are approximately 1 mm in diameter (Miller 1963; Becker 1983). Eggs hatch in 2 or 3 days at a water temperature of 28 °C (Breder 1936).

Age and Growth: Growth varies among pumpkinseed populations depending on density, temperature, and water quality. Stunting occurs frequently in high-density populations. Pumpkinseed total lengths are typically 127 to 203 mm, but they can reach 254 mm (Etnier and Starnes 1993). Longevity is 8 to 10 years.

Food and Feeding: Pumpkinseed primarily eat larval and adult aquatic insects, snails, small crustaceans and other benthic invertebrates, and small fish (Miller 1963; Mittelbach 1984). They shift from littoral to open-water prey at standard lengths from 45 to 70 mm (Mittelbach 1984). Mollusks are crushed with pharyngeal teeth, which change morphologically with snail abundance (Wainwright et al. 1991). Pumpkinseed also will eat large numbers of invasive zebra mussels during spring and summer (Colborne et al. 2015). Juveniles are the most food limited in lakes with bluegill, while adults are most food limited in lakes without bluegill (Osenberg et al. 1992).

References

Becker GC (1983) Fishes of Wisconsin. University of Wisconsin Press, Madison

Breder CM (1936) The reproductive habits of the north American sunfishes (family Centrarchidae). Zoology 21:1–48. https://doi.org/10.5962/p.203693

Colborne SF, Clapp ADM, Longstaffe FJ, Neff BD (2015) Foraging ecology of native pumpkinseed (*Lepomis gibbosus*) following the invasion of zebra mussels (Dreissena polymorpha). Can J Fish Aquat Sci 72:983–990. https://doi.org/10.1139/cjfas-2014-0372

Etnier DA, Starnes WC (1993) The fishes of Tennessee. University of Tennessee Press, Knoxville

Evans DO (1977) Seasonal changes in standard metabolism, upper and lower thermal tolerance and thermoregulatory behavior of the pumpkinseed, *Lepomis gibbosus*, Linnaeus. Dissertation, University of Toronto

Fish PA, Savitz J (1983) Variations in home ranges of largemouth bass, yellow perch, bluegills, and pumpkinseeds in an Illinois lake. Trans Am Fish Soc 112:147–153. https://doi.org/10.1577/1548-8659

Fox MG (1994) Growth, density, and interspecific influences on pumpkinseed sunfish life histories. Ecololgy 75:1157–1171. https://doi.org/10.2307/1939439

Miller HC (1963) The behavior of the pumpkinseed sunfish, *Lepomis gibbosus* (Linneaus), with notes on the behavior of other species of *Lepomis* and the pigmy sunfish, *Elassoma evergladei*. Behavior 22:88–151. https://doi.org/10.1163/156853963X00329

Mittelbach GG (1984) Predation and resource partitioning in two sunfishes (Centrarchidae). Ecology 65:499–513. https://doi.org/10.2307/1941412

O'Hara JJ (1968) Influence of weight and temperature on metabolic rate of sunfish. Ecology 49:159–161. https://doi.org/10.2307/1933575

Osenberg CW, Mittelbach GG, Wainwright PC (1992) Two-stage life histories in fish: the interaction between juvenile competition and adult performance. Ecology 73:255–267. https://doi.org/10.2307/1938737

Reed RJ (1971) Underwater observations of the population density and behavior of pumpkinseed, *Lepomis gibbosus* (Linnaeus) in cranberry pond, Massachusetts. Trans Am Fish Soc 100:350–353. https://doi.org/10.1577/1548-8659

Wainwright PC, Osenberg CW, Mittelbach GG (1991) Trophic polymorphism in the pumpkinseed sunfish (*Lepomis gibbosus* Linnaeus): effects of environment on ontogeny. Funct Ecol 5:40–55. https://doi.org/10.2307/2389554

6.23.4 Orangespotted Sunfish *Lepomis humilis*

(Girard, 1858)

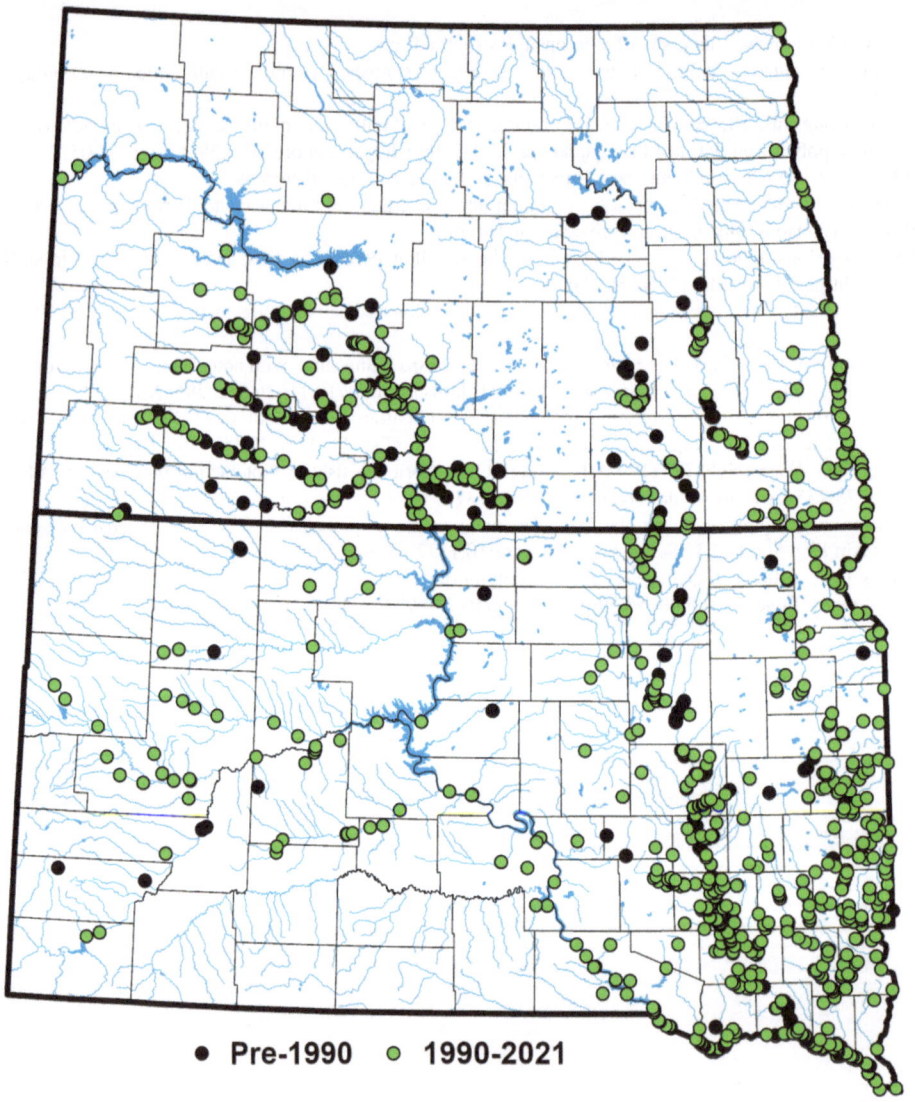

Etymology: *Lepomis humilis – Lepomis* = scaled lid, referencing scaled operculum; *humilis* = humble or insignificant, possibly referencing small size.

Description:

Body – moderately deep, laterally compressed, arched dorsal profile.
Color –

> *Dorsally* – olive to blue-brown.
> *Laterally* – bluish-gray to blue-green; small, irregular orange spots; 5 to 7 extremely faint, dark vertical bands.
> *Ventrally* – white, yellow, or light orange.
> *Opercle* – dark center and light border.
> *Lower jaw* – occasional wavy blue lines extending backward under eye.
> *Fins* – dorsal, caudal, anal: dark, small, faint to dark blotches or spots present, often near base.

Head – large, conical.
Opercle – flap short, flexible.
Snout – large, conical.
Eyes – moderately large, laterally on head; single pair of large sensory pores between eyes above upper lip.
Mouth – terminal, oblique, moderately large.

> *Upper jaw* – extends to or slightly past middle of eye.
> *Lower jaw* – slightly protrudes past upper jaw.
> *Lips* – fleshy.

Teeth – small, in pads on both jaws; absent on tongue.
Gill rakers – 10 to 15; long, thin.
Dorsal fin – 9 or 10 rays; elongated; 10 or 11 spines.
Caudal peduncle – slightly elongated, thick.
Caudal fin – slightly forked, rounded lobes.
Anal fin – 8 or 9 rays; 3 spines.
Pelvic fins – 5 rays; 1 spine.
Pectoral fins – 14 or 15 rays; slightly pointed; when folded forward extends to or barely reaches anterior end of eye.
Lateral line – complete, arched upward; 32 to 41 ctenoid scales in series.
Juveniles – less intense coloration and spots; more prominent, dusky vertical bars on lateral sides.
Sexual dimorphism – spawning males have more intense coloration, especially on fins, lateral, and ventral sides; pelvic and anal fins have a black border on distal end; eye is red-orange.

Similar Species: Orangespotted sunfish closely resemble other *Lepomis* species but are easily distinguished by the large sensory pores between the eyes. Green sunfish lack orange spots on lateral sides, are much darker, more elongate, and have 44 to 53 scales in lateral line series. Pumpkinseeds have a single red spot on opercle posterior edge. Bluegills have a solid dark blue opercle without a light border.

Distribution and Habitat: The native range of orangespotted sunfish encompasses the eastern Dakotas, east to Ohio including the Mississippi River system and Great Lakes, and south from Alabama to Texas. They are widely distributed in the Dakotas but most abundant east of the Missouri River. Orangespotted sunfish prefer pools and backwaters in small to large streams with mud, gravel, or sand substrate. They are less abundant in ponds and lakes than other sunfish species. Orangespotted sunfish tolerate siltation, high water temperatures, and low dissolved oxygen levels. Their critical thermal maximum is 38 °C, and critical dissolved oxygen minimum is 0.6 mg/L (Smale and Rabeni 1995).

Reproduction: Orangespotted sunfish spawn from May to July in water temperatures from 18 to 32 °C (Barney and Anson 1923; Miller 1963; Cross 1967). They become sexually mature at age-2 or age-3. Males excavate 15 to 18 cm diameter, 3 to 4 cm deep, saucer-shaped nests that are much smaller than those of other sunfish species (Barney and Anson 1923; Miller 1963). These nests are often constructed in colonies in shallow, quiet areas with silt, sand, or gravel substrate. Males court females with a series of species-specific grunting sounds (Gerald 1971). Males defend the eggs and keep them silt-free by spawning until hatch (Barney and Anson 1923). While males defend the nest from other predators, darters and minnows eat the eggs (Barney and Anson 1923; Cross 1967). Fecundity increases with female size. A female with a total length of 105 mm produced 4,700 eggs (Barney and Anson 1923). The slightly adhesive, approximately 1-mm diameter, amber eggs hatch in 5 days at water temperatures of 18 to 21 °C. (Barney and Anson 1923).

Age and Growth: In the Big Sioux River in eastern South Dakota, most orangespotted sunfish had total lengths from 50 to 60 mm (Rasmus et al. 2008). In the Cheyenne River in western South Dakota, most orangespotted sunfish had total lengths from 60 to 70 mm (Rasmus et al. 2008). While lengths at age were similar between the Big Sioux and Cheyenne River populations, all age groups in the Big Sioux were consistently larger (Rasmus et al. 2008). Total lengths at age of orangespotted sunfish from the Big Sioux River system were 49 mm at age-1, 54 mm at age-2, 61 mm at age-3, and 73 mm at age-4 (Rasmus et al. 2008). They can reach total lengths over 120 mm and live for 7 years, although their typical lifespan is 3 to 4 years (Barney and Anson 1923).

Food and Feeding: Orangespotted sunfish are generalist predators, consuming a wide variety of prey sizes with their relatively large mouth (Hegrenes 2001). Adults mainly eat midge larvae, small crustaceans, and other aquatic invertebrates. Diet and prey size influence morphological plasticity (Hegrenes 2001). Age-0 orangespotted sunfish eating tiny, planktonic prey developed long, fusiform bodies and sharply angled snouts, while those eating larger prey had deeper bodies and blunt snouts (Hegrenes 2001).

References

Barney RL, Anson BJ (1923) Life history and ecology of the orangespotted sunfish (*Lepomis humilis*). Appendix XV, Rep US Comm Fish 1922:1–16

Cross FB (1967) Handbook of fishes in Kansas. University Press of Kansas, Lawrence

Gerald JW (1971) Sound production during courtship in six species of sunfish (Centrarchidae). Evolution 25:75–87. https://doi.org/10.2307/2406500

Hegrenes S (2001) Diet-induced phenotypic plasticity of feeding morphology in the orangespotted sunfish, *Lepomis humilis*. Ecol Freshw Fish 10:35–42. https://doi.org/10.1034/j.1600-0633.2001.100105.x

Miller HC (1963) The behavior of the pumpkinseed sunfish, *Lepomis gibbosus* (Linneaus), with notes on the behavior of other species of Lepomis and the pigmy sunfish, *Elassoma evergladei*. Behavior 22:88–151. https://doi.org/10.1163/156853963X00329

Rasmus RA, Phelps QE, Duehr JP, Berry CR Jr (2008) Population characteristics of lotic orangespotted sunfish. J Freshw Ecol 23:459–461. https://doi.org/10.1080/02705060.2008.9664224

Smale MA, Rabeni CF (1995) Hypoxia and hyperthermia tolerances of headwater stream fishes. Trans Am Fish Soc 124:698–710. https://doi.org/10.1577/1548-8659

6.23.5   Bluegill *Lepomis macrochirus*

Rafinesque, 1819

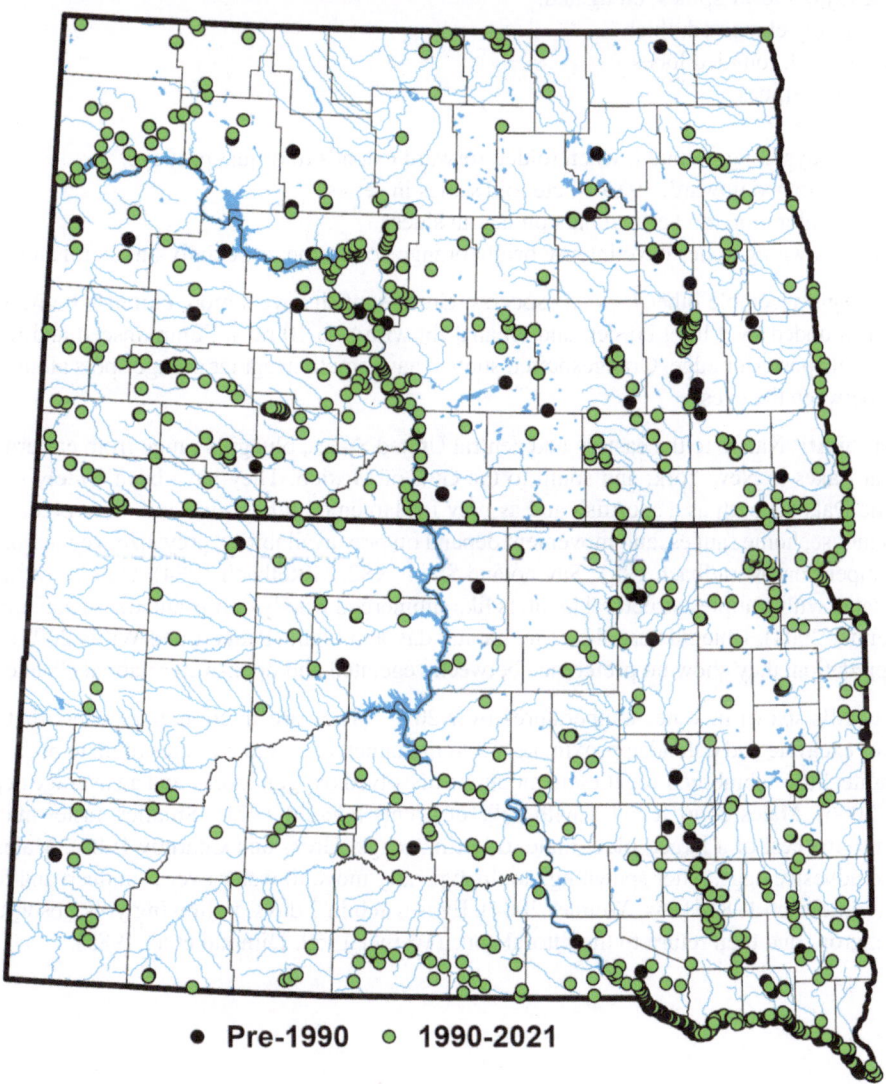

● **Pre-1990**   ○ **1990-2021**

Etymology: *Lepomis macrochirus* – *Lepomis* = scaled lid, referencing scaled operculum; *macrochirus* = long hand, referencing the large pectoral fin.

Description:

Body – deep, robust, strongly laterally compressed, arched dorsal profile.
Color – varies with habitat.

> *Dorsally* – dark olive-brown.
> *Laterally* – olive-brown to dark blue-brown, sometimes small blue-purple reflections; 7 to 10 faint, dark, vertical bars.
> *Ventral/Throat* – white, yellow, or light orange-red.
> *Opercle* – solid dark blue.
> *Lower jaw* – backward extending wavy pale blue lines.
> *Fins* – uniformly pigmented; dorsal, anal – dark blotch on posterior base.

Head – large, conical.
Opercle – flap short, flexible.
Snout – large, conical.
Eyes – moderately large, laterally on head.
Mouth – strongly oblique, small.

> *Upper jaw* – does not extend to anterior end of eye.
> *Lips* – fleshy, smooth.

Teeth – small, in pads on both jaws; typically absent on tongue.
Gill rakers – 13 to 16; long, slender.
Dorsal fin – 10 to 12 rays; 9 to 11 spines; elongated.
Caudal peduncle – slightly elongated, thick.
Caudal fin – slightly forked, rounded lobes.
Anal fin – 10 to 12 rays; 3 spines.
Pelvic fins – 5 rays; 1 spine.
Pectoral fins – 13 or 14 rays; long, pointed; when folded forward extends to or just past eye.
Lateral line – complete, arched upward; 39 to 44 ctenoid scales in series.
Juveniles – less intense coloration, no vertical bars on lateral sides.
Sexual dimorphism – spawning males have intense bright orange coloration, especially on lateral and ventral sides.

Similar Species: Bluegill resemble other *Lepomis* species. Green sunfish have a more elongate body, an inflexible opercle with a dark center surrounded by a light border, and an anal fin with 8 to 10 rays. Pumpkinseed and redear sunfish have a single red spot on opercle posterior edge. Orangespotted sunfish have small, irregular orange spots along the lateral sides and large sensory pores between the eyes.

Distribution and Habitat: Native to the eastern and central United States, bluegills range from eastern South Dakota, east throughout the Great Lakes to New York, and south to the Gulf of Mexico. They have been widely introduced across the United States and the Dakotas both as a sportfish and as prey for largemouth bass fish and other predators (Guy and Willis 1990). Bluegill habitat use, home ranges, and movement depend on prey availability, prey size, vegetation density, predators, and interspecific competition (Mittelbach 1981; Savino and Stein 1982; Mittelbach 1984). Outside of the spawning season, they are often associated with complex structural habitats like submerged woody debris and dense aquatic vegetation (Savino and Stein 1982; Weimer 2004). Bluegills use the same habitat day and night (Paukert and Willis 2002). Because adults are less susceptible to predation, they show no preference between vegetated and open-water habitat (Paukert and Willis 2002).

Reproduction: Bluegills spawn in water temperatures over 20 °C. They are fractional spawners that sexually mature at age-2 or age-3. Males excavate and defend approximately 30-cm diameter nests in colonies of up to 50, along shorelines with gravel or sand substrate, little submerged vegetation, moderate dissolved oxygen levels, and an average depth of 1 m (Weimer 2004; Gosch et al. 2006). Males court females with calls or grunts (Gerald 1971). Smaller, faster-maturing males called sneakers may slip into the nest of a larger, older male to fertilize eggs (Gross and Charnov 1980). Males fan and guard the small, demersal, and adhesive eggs. After spawning, adults typically move to water over 5-m deep and feed on large, open-water zooplankton (Mittelbach 1981, 1984; Weimer 2004). Eggs hatch in 2 days. Following yolk absorption, larvae migrate to the limnetic zone, grow, and then return to the littoral zone (Werner 1969; Dimond et al. 1985).

Age and Growth: Water quality, lake morphometry, and temperature affect bluegill growth (Tomcko and Pierce 1997, 2001; Weimer 2004; Tomcko and Pierce 2005). Many populations become stunted with poor size structure because of high recruitment and lack of predation. Bluegill growth is faster in lakes and impoundments than streams, but deep, large lakes are less likely to produce fast-growing or strong year classes (Tomcko and Pierce 2001). Yellow perch, walleye, black bullhead, brown bullhead, and other predators can decrease bluegill densities and subsequently increase growth (Weimer 2004). A high density of largemouth bass with total lengths less than 300 mm can create a high proportion of bluegills with total lengths of 150 mm (Guy and Willis 1990). Growth is highly variable among populations. Total lengths at age of bluegill from South Dakota were age-1, 55 mm; age-2, 103 mm; age-3, 141 mm; age-4, 166 mm; and age-5, 180 mm (Willis et al. 2001). They can reach a total length of 330 mm and live for 10 years.

Food and Feeding: Bluegills are generalist, primarily diurnal, sight feeders (Sarker 1977). Their optimum feeding temperature ranges from 27 to 31 °C (Kitchell et al. 1974). Young bluegill eat zooplankton. Adults mainly consume daphnia, other zooplankton, chironomids, amphipods, trichopterans, gastropods, and other macroinvertebrates, and small fish (Mittelbach 1981; Engel 1988; Harris et al. 1999; Olson et al. 2003). Bluegill macroinvertebrate consumption is likely dictated by availability and may be higher in lakes with considerable vegetation (Olson et al. 2003). Zooplankton are more likely to be consumed in open water when foraging efficiency is greater and predation risk is limited (Mittelbach 1981). Bluegill may compete for food with yellow perch (Kaemingk and Willis 2012; Kaemingk et al. 2012, 2014).

References

Dimond WF, Storck TW, Kruse KC (1985) A turbidity-related delay in bluegill (*Lepomis macrochirus*) reproduction and some size dependent differences in the spatial distribution of bluegill fry. Trans Ill State Acad Sci 78:49–56

Engel S (1988) The role and interaction of submersed macrophytes in a shallow Wisconsin lake. J Freshw Ecol 4:329–341. https://doi.org/10.1080/02705060.1988.9665182

Gerald JW (1971) Sound production during courtship in six species of sunfish (Centrarchidae). Evolution 25:75–87. https://doi.org/10.2307/2406500

Gosch NJC, Phelps QE, Willis DW (2006) Habitat characteristics at bluegill spawning colonies in a South Dakota glacial lake. Ecol Freshw Fish 15:464–469. https://doi.org/10.1111/j.1600-0633.2006.00178.x

Gross MR, Charnov WA (1980) Alternative male life history in bluegill sunfish. Proc Natl Acad Sci U S A 77:6937–6940. https://doi.org/10.1073/pnas.77.11.6937

Guy CS, Willis DW (1990) Structural relationships of largemouth bass and bluegill populations in South Dakota ponds. N Am J Fish Manag 10:338–343. https://doi.org/10.1577/1548-8675

Harris NJ, Galinat GF, Willis DW (1999) Seasonal food habits of bluegills in Richmond Lake, South Dakota. Proc S D Acad Sci 78:79–85

Kaemingk MA, Willis DW (2012) Mensurative approach to examine potential interactions between age-0 yellow perch (*Perca flavescens*) and bluegill (*Lepomis macrochirus*). Aquat Ecol 46:353–362. https://doi.org/10.1007/s10452-012-9406-z

Kaemingk MA, Jolley JC, Willis DW, Chipps SR (2012) Priority effects among young-of-the-year fish: reduced growth of bluegill sunfish (*Lepomis macrochirus*) caused by yellow perch (*Perca flavescens*)? Freshw Biol 57:654–665. https://doi.org/10.1111/j.1365-2427.2011.02728.x

Kaemingk MA, Stahr KJ, Jolley JC, Holland RS, Willis DW (2014) Evidence for bluegill spawning plasticity obtained by disentangling complex factors related to recruitment. Can J Fish Aquat Sci 71:93–105. https://doi.org/10.1139/cjfas-2013-0282

Kitchell JF, Koonce JF, O'Neill RV, Shugart HH Jr, Magnuson JJ, Booth RS (1974) Model of fish biomass dynamics. Trans Am Fish Soc 103:786–798. https://doi.org/10.1577/1548-8659

Mittelbach GG (1981) Foraging efficiency and body size: a study of optimal diet and habitat use by bluegills. Ecology 62:1370–1386. https://doi.org/10.2307/1937300

Mittelbach GG (1984) Predation and resource partitioning in two sunfishes (Centrarchidae). Ecology 65:499–513. https://doi.org/10.2307/1941412

Olson NW, Paukert CP, Willis DW (2003) Prey selection and diets of bluegill Lepomis macrochirus with differing population characteristics in two Nebraska natural lakes. Fish Manag Ecol 10:31–40. https://doi.org/10.1046/j.1365-2400.2003.00323.x

Paukert CP, Willis DW (2002) Seasonal and diel habitat selection by bluegills in a shallow natural lake. Trans Am Fish Soc 131:1131–1139. https://doi.org/10.1577/1548-8659

Sarker AL (1977) Feeding ecology of the bluegill, *Lepomis macrochirus*, in two heated reservoirs of Texas III. Time of day and patterns of feeding. Trans Am Fish Soc 106:596–601. https://doi.org/10.1577/1548-8659

Savino JF, Stein RA (1982) Predator-prey interaction between largemouth bass and bluegills as influenced by simulated submersed vegetation. Trans Am Fish Soc 111:255–266. https://doi.org/10.1577/1548-8659

Tomcko CM, Pierce RB (1997) Bluegill growth rates in Minnesota. Minn Dep Nat Res, Sec Fish, Invest Rep 458, St. Paul

Tomcko CM, Pierce RB (2001) The relationship of bluegill growth, lake morphometry, and water quality in Minnesota. Trans Am Fish Soc 130:317–321. https://doi.org/10.1577/1548-8659

Tomcko CM, Pierce RB (2005) Bluegill recruitment, growth, population size structure, and associated factors in Minnesota lakes. N Am J Fish Manag 25:171–179. https://doi.org/10.1577/M04-054.1

Weimer EJ (2004) Bluegill seasonal habitat selection, movement, and relationship to angler locations in a South Dakota glacial lake. Dissertation, South Dakota State University

Werner RG (1969) Ecology of limnetic bluegill (*Lepomis macrochirus*) fry in Crane Lake, Indiana. Am Midl Nat 81:164–181. https://doi.org/10.2307/2423658

Willis DW, Isermann DA, Hubers MJ, Johnson BA, Miller WH, St Sauver TR, Sorensen JS, Unkenholz EG, Wickstrom GA (2001) Growth of South Dakota fishes: a statewide summary with means by region and water type. S D Dep Game Fish Park Special Rep 01-05, Pierre

6.23.6 Redear Sunfish *Lepomis microlophus*

(Günther, 1859)

● Pre-1990   ○ 1990-2021

Etymology: *Lepomis microlophus* – *Lepomis* = scaled lid, referencing scaled operculum; *microlophus* = small nape. Also referred to as shellcrackers because of crushing mollusks with specialized pharyngeal teeth.

Description:

Body – deep, laterally compressed, arched dorsal profile.
Color –

    Dorsally – golden brown to olive.
    Laterally – tan to gray; 5 to 10 faint, dark olive, diffuse vertical bars.
    Ventrally – yellow-green to cream.
    Head – dark brown to orange spots on lateral sides.
    Opercle – dark black center; red to burnt-orange crescent-shaped margin on posterior edge.
    Eyes – amber.
    Fins – no prominent spots.

Head – large, conical.
Opercle – flap short, stiff.
Snout – large, conical.
Eyes – moderately large, laterally on head.
Mouth – terminal, small

    Upper jaw – does not extend past anterior edge of eye.
    Lower jaw – slightly protrudes past upper jaw.
    Lips – fleshy.

Teeth – pharyngeal teeth moliform.
Gill rakers – 9 to 11; on first gill arch short, stout.
Dorsal fin – 11 or 12 rays; 9 to 11 spines; elongated.
Caudal peduncle – slightly elongated, thick.
Caudal fin – slightly forked, rounded lobes.
Anal fin – 10 or 11 rays; 3 spines.
Pelvic fins – insertion slightly posterior to pectoral fin insertion.
Pectoral fins – elongated; when folded forward extends past anterior end of eye.
Lateral line – complete, arched upward; 34 to 45 ctenoid scales in series.
Sexual dimorphism – spawning males have bright yellow breast, light orange on the pectoral and pelvic fins, and bright amber eye iris.

Similar Species: Redear sunfish resemble pumpkinseed and bluegill. Pumpkinseeds have three to five blue-green wavy lines extending backwards to the gill flap, a short, flexible opercle flap with a black center and white-cream margin separated by a single red spot on the posterior edge, and a pectoral fin that only extends to the anterior edge of eye when folded forward. Bluegills have a solid dark blue opercle, without any red spots.

Distribution and Habitat: The native range of redear sunfish extends from the Mississippi River basin in Missouri and Indiana, south to North Carolina and Florida, and west into eastern Texas. They have been widely introduced for sportfishing. Redear sunfish do not occur in North Dakota but have been stocked in a limited number of lakes within the Big Sioux River, Missouri River, and Cheyenne River systems in South Dakota. Redear sunfish inhabit clear, quiet, warm waters of ponds, marshes, lakes, and reservoirs, and pools, backwaters, or oxbows of low gradient streams and large rivers (Twomey et al. 1984). They are often associated with aquatic vegetation or submerged structure, which provide cover and feeding benefits. Adults frequent deep waters near the bottom. Although turbidity negatively affects redear sunfish, they can tolerate it better than bass or bluegill (Twomey et al. 1984; Warren Jr 2009; Kansas Fishes Committee 2014).

Reproduction: Reproduction of redear sunfish is similar to other *Lepomis* species. Adults migrate from deeper waters to shallow areas near shorelines to spawn in May through August (Pflieger 1997). Sexual maturity is most likely associated with size rather than age (Twomey et al. 1984). Males defend and excavate saucer-shaped nests with their caudal fins. Nests are often constructed in colonies along edges of aquatic vegetation or submerged structure often exposed to sun in quiet waters at depths ranging from 4 to 83 cm with silt, sand, or mud substrate (Twomey et al. 1984). Males court females into the nesting site with a series of grunting sounds made by the jaw and pharyngeal bones (Gerald 1971). Redear sunfish are fractional spawners with females depositing eggs in more than one nest. Males defend and fan the eggs until hatching 6 to 10 days after fertilization (Twomey et al. 1984). They continue to defend the nest for approximately 1 week after hatch when the fry depart (Lipinot 1961; Twomey et al. 1984). Optimal incubation water temperatures range from 21 to 24 °C, but successful hatching has occurred in temperatures as high as 32 °C (Twomey et al. 1984). Fecundity depends on female size. Females with total lengths from 151 to 213 mm produced from 7,500 to 25,500 eggs (Warren Jr 2009; Kansas Fishes Committee 2014). Eggs are approximately 2 mm in diameter (Meyer 1970).

Age and Growth: One of the larger species of sunfish, redear sunfish can grow faster and become larger than sympatric bluegill (Pflieger 1997; Sammons et al. 2006). Redear sunfish in turbid waters often grow slower than those in clear waters (Boschung Jr and Mayden 2004; Warren Jr 2009; Kansas Fishes Committee 2014). The optimal water temperatures for growth are from 24 to 27 °C, and growth ceases below 10 °C (Hill et al. 1975; Twomey et al. 1984). Adults can reach a total length of 380 mm but are often shorter (Kansas Fishes Committee 2014). Longevity is 6 years in northern populations and 11 years in the south (Pflieger 1997; Sammons et al. 2006; Kansas Fishes Committee 2014).

Food and Feeding: Larvae and young redear sunfish eat zooplankton, green algae, and small aquatic insects like chironomid larvae (Twomey et al. 1984; Etnier and Starnes 1993). Adults are opportunistic, primarily eating snails, clams, and mussels, which they crush with their moliform pharyngeal teeth (Huckins 1997; Kansas Fishes Committee 2014). They also consume aquatic insects like midge larvae (Twomey et al. 1984). Redear sunfish rarely feed along the surface.

References

Boschung HT Jr, Mayden RL (2004) Fishes of Alabama. Smithsonian Books, Washington, DC

Etnier DA, Starnes WC (1993) The fishes of Tennessee. University of Tennessee Press, Knoxville

Gerald JW (1971) Sound production during courtship in six species of sunfish (Centrarchidae). Evolution 25:75–87. https://doi.org/10.2307/2406500

Hill LG, Schnell GD, Pigg J (1975) Thermal acclimation and temperature selection in sunfishes (*Lepomis*, Centrarchidae). Southwest Nat 20:177–184. https://doi.org/10.2307/3670435

Huckins CJF (1997) Functional linkages among morphology, feeding performance, diet, and competitive ability in molluscivorous sunfish. Ecology 78:2401–2414. https://doi.org/10.1890/0012-9658(1997)078[2401:FLAMFP]2.0.CO;2

Kansas Fishes Committee (2014) Kansas fishes. University Press of Kansas, Lawrence

Lipinot A (1961) The red-ear sunfish. Ill Wildl 17:3–4

Meyer FA (1970) Development of some larval centrarchids. Prog Fish Cult 32:130–136. https://doi.org/10.1577/1548-8640(1970)32[130:DOSLC]2.0.CO;2

Pflieger WL (1997) The fishes of Missouri, revised edn. Missouri Dep Cons, Jefferson City

Sammons SM, Partridge DG, Maceina MJ (2006) Differences in population metrics between bluegill and redear sunfish: implications for the effectiveness of harvest restrictions. N Am J Fish Manag 26:777–787. https://doi.org/10.1577/M05-159.1

Twomey KA, Gebhart G, Maughan OE, Nelson PC (1984) Habitat suitability index models and instream flow suitability curves: Redear sunfish. US Dep Inter, Fish Wildl Serv, FWS/OBS-82/10.79

Warren ML Jr (2009) Centrarchid identification and natural history. In: Cooke SJ, Phillip DP (eds) Centrarchid fishes: diversity, biology, and conservation. Wiley-Blackwell, West Sussex, pp 375–533. https://doi.org/10.1002/9781444316032.ch13

6.23.7   Smallmouth Bass *Micropterus dolomieu*

Lacepède, 1802

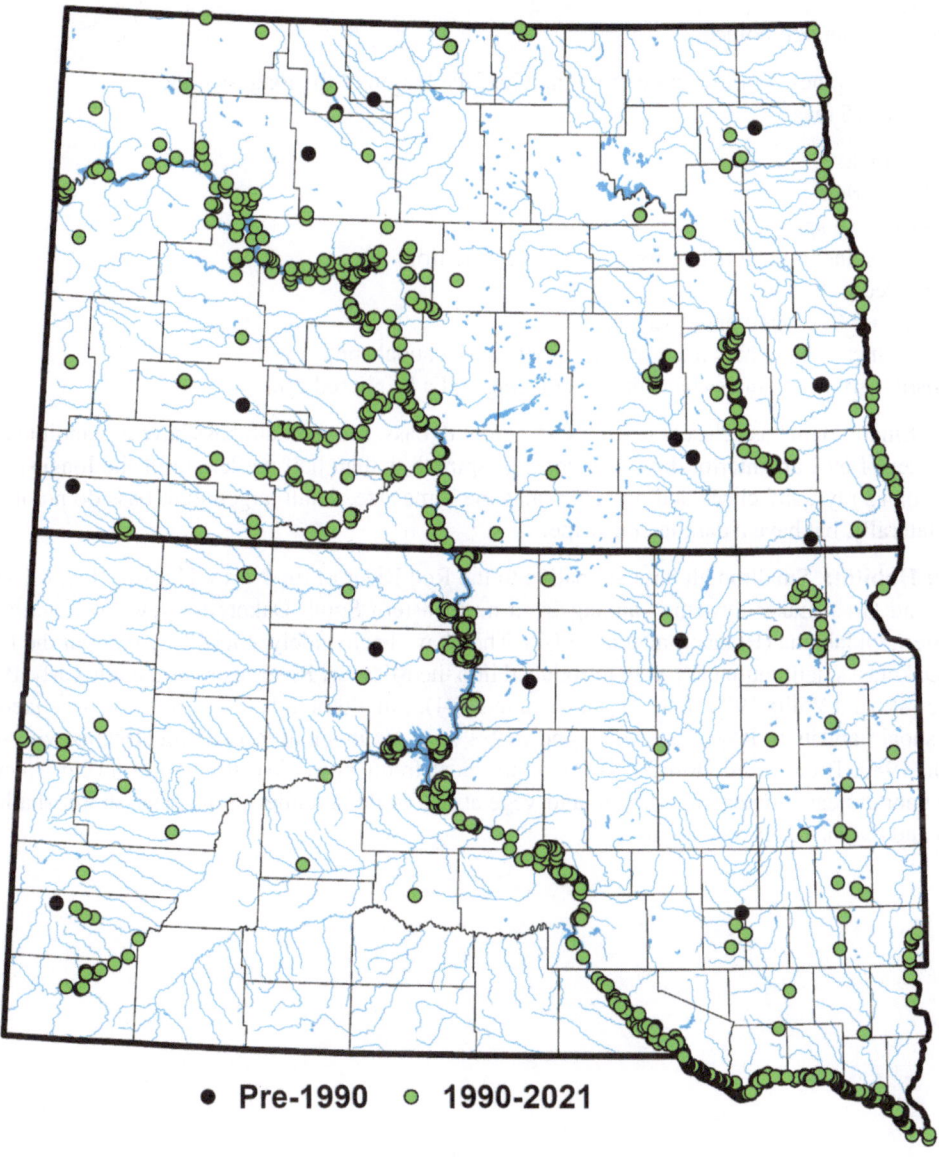

● **Pre-1990**   ○ **1990-2021**

Etymology: *Micropterus dolomieu – Micropterus* = small fin, referencing a torn fin on the type specimen; *dolomieu* = referencing French mineralogist M. Dolomieu.

Description:

Body – fusiform, slender, slightly laterally compressed, moderately deep.
Color –

  Dorsally – brown or olive-green.
  Laterally – bronze; dark mottling or faint vertical bars.
  Ventrally – cream to white.
  Opercle – 3 or 4 dark bars present on cheek and opercle; dark spot roughly size of pupil on posterior edge.
  Eyes – amber.
  Fins – dusky; no distinct markings.

Head – large; scales on cheek smaller than scales on opercle.
Snout – blunt.
Eyes – large.
Mouth – slightly oblique.

  Lower jaw – protrudes past upper jaw.
  Upper jaw – extending to middle or posterior edge of eye, never further.

Teeth – small, in pads on upper and lower jaws; absent on tongue.
Gill rakers – 6 to 8.
Dorsal fin – 2 lobes, moderately joined.

  Anterior – 9 or 10 spines, shortest spine more than half the length of the longest spine.
  Posterior – 13 to 15 rays.

Caudal peduncle – thick.
Caudal fin – slightly forked.
Anal fin – 10 or 11 rays; 3 spines.
Pelvic fins – 5 rays; 1 spine; insertion anterior to dorsal fin insertion.
Pectoral fins – rounded.
Lateral line – complete; 68 to 80 ctenoid scales.
Juveniles – more distinct vertical bars laterally, prominent dark caudal spot.
Sexual dimorphism – spawning males have darker coloring and a bright red eye.

Similar Species: Smallmouth bass closely resemble largemouth bass. Largemouth bass have a mandible extending to posterior edge or further of eye, an anterior dorsal fin shortest spine less than half the length of the longest spine, a posterior dorsal fin with 12 or 13 rays, and cheek and opercle scales the same size. Adult largemouth bass do not have dark mottling nor vertical bars laterally, but have a dark lateral stripe.

Distribution and Habitat: Smallmouth bass are native to the Red River of the North, Hudson Bay, Mississippi River, St. Lawrence River, and Great Lakes systems, ranging from northeastern South Dakota to New York and south to northern Alabama and eastern Oklahoma (Hoagstrom et al. 2007). They have been widely stocked throughout the United States as a sportfish. In the Dakotas, smallmouth bass were introduced into the Missouri River reservoirs in the early 1980s and are now widespread (Milewski and Willis 1990; Berry Jr and Young 2004). Smallmouth bass establish well-defined home ranges in rocky littoral or shoal habitats in lakes, reservoirs, and riffles or pools in small to medium streams with moderate current (Todd and Rabeni 1989; Brewer and Orth 2014). They are most frequent in clear to slightly turbid waters. Adults prefer deeper off-shore habitat near submerged structure and vegetation, while juveniles prefer near-shore, shallow habitats with slower velocities and overhanging vegetation.

Reproduction: In lower Lake Oahe, smallmouth bass over 300 mm long were more likely captured in the spring during onshore–offshore movements, suggesting spawning migrations (Lott 2000). Spawning occurs in late spring and early summer within quiet, shallow, in-shore waters. Males defend and construct nests in rocky or sandy substrate near submerged or overhanging structure. Smallmouth bass are fractional spawners. The adhesive, demersal, pale-yellow eggs are approximately 2 to 4 mm in diameter (Brewer and Orth 2014). Fecundity varies with female size. Eggs hatch in 1 week. Males provide parental care until juveniles disperse from the nest.

Age and Growth: Growth rates of smallmouth bass are highly variable. Length at age varies among populations, depending on habitat and geography. In Lake Poinsett, South Dakota, age-1 smallmouth bass grew the fastest during mid-summer on a macroinvertebrates diet (Gangl et al. 1997). Smallmouth bass growth rates from lakes and reservoirs are often more rapid than those from stream-dwelling fish (Brewer and Orth 2014). Typical total lengths range from 305 to 508 mm. Smallmouth bass longevity is from 6 to 12 years.

Food and Feeding: Smallmouth bass are opportunistic, apex predators eating a wide variety of prey items (Lott 1996; Gangl et al. 1997). They undergo an ontogenetic diet shift from larvae to adult. Larval smallmouth bass primarily consume zooplankton and chironomids. Juveniles shift to larger aquatic macroinvertebrates, crayfish, and small fishes. In eastern South Dakota lakes, adult smallmouth bass consumed invertebrates in the spring, shifted to a fish diet of mainly age-0 yellow perch in early summer, and then consumed black crappie and bluegill in mid and late summer (Bacula 2009; Dembkowski et al. 2015). Yellow perch recruitment in eastern South Dakota lakes does not appear to be impacted by smallmouth bass predation (Dembkowski et al. 2015). In both eastern South Dakota lakes and Lake Sharpe, the minimal smallmouth bass predation on walleye is not likely detrimental to walleye recruitment (Bacula 2009; Wuellner 2009; Wuellner et al. 2010; Galster et al. 2012). Prey fish availability and size likely determine smallmouth bass consumption (Bacula 2009).

References

Bacula TD (2009) Smallmouth bass seasonal dynamics in northeastern South Dakota glacial lakes. Dissertation, South Dakota State University

Berry CR Jr, Young B (2004) Fishes of the Missouri National Recreational River, South Dakota, and Nebraska. Great Plains Res 14:89–114

Brewer SK, Orth DJ (2014) Smallmouth bass *Micropterus dolomieu* Lacepède, 1802. Am Fish Soc Symp 82:9–26

Dembkowski DJ, Willis DW, Blackwell BG, Chipps SR, Bacula TD, Wuellner MR (2015) Influence of smallmouth bass predation on recruitment of age-0 yellow perch in South Dakota glacial lakes. N Am J Fish Manag 35:736–747. https://doi.org/10.1080/02755947.2015.1044629

Galster BJ, Wuellner MR, Graeb BDS (2012) Walleye *Sander vitreus* and smallmouth bass *Micropterus dolomieu* interactions: an historic stable-isotope analysis approach. J Fish Biol 81:135–147. https://doi.org/10.1111/j.1095-8649.2012.03318.x

Gangl RS, Pope KL, Willis DW (1997) Seasonal trends in food habits and growth of smallmouth bass in Lake Poinsett, South Dakota. S D Game Fish Park, Wildl Div Special Rep 97-5, Pierre

Hoagstrom CW, Wall SS, Kral JG, Blackwell BG, Berry CR Jr (2007) Zoogeographic patterns and faunal changes of South Dakota fishes. West N Am Nat 67:161–184. https://doi.org/10.3398/1527-0904(2007)67[161:ZPAFCO]2.0.CO;2

Lott JP (1996) Relationships between smallmouth bass feeding ecology and population structure and dynamics in lower Lake Oahe, South Dakota. S D Game Fish Park, Wildl Div, Compl Rep No. 96-3, Pierre

Lott JP (2000) Smallmouth bass movement, habitat use and electrofishing susceptibility in lower Lake Oahe, South Dakota. S D Game Fish Park, Wildl Div, Ann Rep No 00-15, Pierre

Milewski CL, Willis DW (1990) A statewide summary of smallmouth bass sampling data from South Dakota waters. S D Dep Game Fish Park, Wildl Div, Fish Rep 90-9, Pierre

Todd BL, Rabeni CF (1989) Movement and habitat use by stream-dwelling smallmouth bass. Trans Am Fish Soc 118:229–242. https://doi.org/10.1577/1548-8659

Wuellner MR (2009) Exploring competitive interactions between walleye and smallmouth bass in South Dakota waters. Dissertation, South Dakota State University

Wuellner MR, Chipps SR, Willis DW, Adams WE Jr (2010) Interactions between walleyes and smallmouth bass in a Missouri River reservoir with consideration of the influence of temperature and prey. N Am J Fish Manag 30:445–463. https://doi.org/10.1577/M09-066.1

6.23.8   Largemouth Bass *Micropterus nigricans*

(Cuvier, 1828)

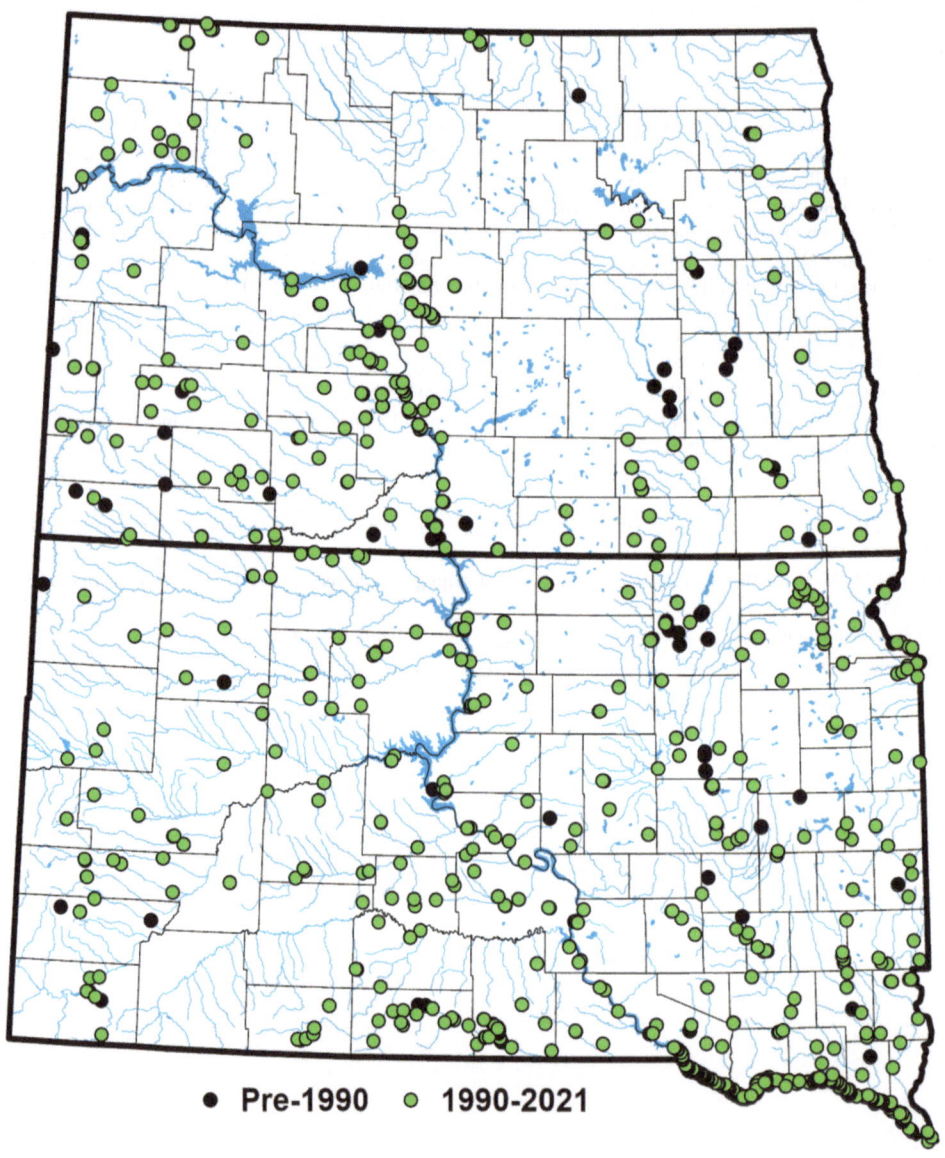

● **Pre-1990**   ○ **1990-2021**

Etymology: *Micropterus nigricans* – *Micropterus* = small fin, referencing a torn fin on the type specimen; *nigricans* = meaning black.

Description:

Body – fusiform, robust, slightly laterally compressed, deep.
Color –

> *Dorsally* – dark olive-green to brown.
> *Laterally* – brown to tan; thick, dark, and irregular (zig-zag) stripe.
> *Ventrally* – cream to white.
> *Opercle* – dark spot roughly size of pupil.
> *Eyes* – golden brown.
> *Fins* – tan with no spots or markings.

Head – large; cheek and opercle scales similar size.
Snout – blunt.
Eyes – large.
Mouth – slightly oblique, large.

> *Lower jaw* – thick, protrudes well past maxillary.
> *Upper jaw* – extends to posterior edge or further of eye.

Teeth – small, conical; in pads on upper and lower jaws; absent on tongue.
Gill rakers – 7 to 9.
Dorsal fin – 2 nearly separate lobes.

> *Anterior* – 9 or 10 spines; shortest spine less than half the length of longest spine.
> *Posterior* – 12 or 13 rays.

Caudal peduncle – thick.
Caudal fin – slightly forked.
Anal fin – 10 to 12 rays; 3 spines.
Pelvic fins – 5 rays; 1 spine; insertion anterior to dorsal fin insertion.
Pectoral fins – rounded.
Lateral line – complete; 60 to 68 ctenoid scales.
Juveniles – prominent vertical blotches compose the lateral stripe.
Sexual dimorphism – spawning males are darker.

Similar Species: Largemouth bass closely resemble smallmouth bass. The mandible of smallmouth bass rarely extends past the posterior edge of the eye, the scales on the cheek are smaller than the scales on the opercle, the shortest anterior dorsal fin spine is more than half the length of the longest spine, and the posterior fin has 13 to 15 rays. Adult smallmouth bass also lack a dark lateral stripe, and instead have mottling and vertical bars laterally. Largemouth bass and Florida bass *Micropterus salmoides* were recently recognized as separate species because of genetic analysis. Previously, these bass were subspecies, commonly referred to as the northern strain (largemouth bass) and the southern strain (Florida bass) (Kim et al. 2022).

Distribution and Habitat: The native distribution of largemouth bass extends from the St. Lawrence River, Great Lakes, Hudson Bay, Red River of the North, and the Mississippi River systems; more specifically, south to the Rio Grande systems in southern Texas and the southern tributaries of the Mississippi River to the Perdido River system in Alabama and Florida (Kim et al. 2022). In the Dakotas, largemouth bass are native to the Red River of North, Minnesota River, and Big Sioux River systems and have been widely introduced across both states (Hoagstrom et al. 2007). Largemouth bass inhabit warmwater ponds, lakes, and reservoirs, as well as low current areas of creeks and backwaters of medium-to-large rivers. They prefer shallow, clear to slightly turbid waters with sand or gravel substrates and aquatic vegetation. Submerged macrophytes are often associated with quality largemouth bass populations (Durocher et al. 1984). They rarely are found more than 6.1 m deep. Largemouth bass are moderately tolerant of turbidity and salinity.

Reproduction: In the Dakotas, reproductive success and recruitment varies among populations and are negatively impacted by severe winters and drought (Beck 1986; Willis and Guy 1991; Kolander 1992; Lucchesi et al. 2004; Spengler 2010). The timing of sexual maturity is also variable, but typically occurs at lengths of 250 to 300 mm (Kansas Fishes Committee 2014).

Spawning occurs in spring and is highly dependent on water temperature. Males move substrate with their fins to make nests in shallow water with submerged or overhanging structure. Males court females by nudging and nipping. Largemouth bass are fractional spawners. The adhesive, demersal, 2 mm diameter, pale-yellow eggs hatch in 7 days (Scott and Crossman 1973). Fecundity varies greatly by female size. Males do not feed while caring for the eggs and hatched fry.

Age and Growth: Largemouth bass growth rates are highly variable and differ among populations. Increased densities, particularly in water bodies under 40 ha, often lead to slower growth, decreased condition, and depressed size structures (Paukert and Willis 2004). In South Dakota, the presence of bluegill and growing season length affect growth rates (Guy and Willis 1990). Largemouth bass increase rapidly in length in first 2 years with increases in weight occurring later (Becker 1983). Females are often larger than males. Largemouth bass typical lengths range from 254 to 508 mm. They have a longevity of 15 to 20 years.

Food and Feeding: Largemouth bass are apex, keystone, opportunistic predators that help regulate panfish populations (Guy and Willis 1990). They eat a wide array of prey, including fish, aquatic insects, crustaceans, small terrestrial mammals, and birds. Largemouth bass feed both day and night and are often found near aquatic vegetation.

References

Beck RD (1986) Growth, survival, and reproductive success of largemouth bass stocked with selected forage fishes in South Dakota ponds. Dissertation, South Dakota State University

Becker GC (1983) Fishes of Wisconsin. University of Wisconsin Press, Madison

Durocher PP, Provine WC, Kraai JE (1984) Relationship between abundance of largemouth bass and submerged vegetation in Texas reservoirs. N Am J Fish Manag 4:84–88. https://doi.org/10.1577/1548-8659

Guy CS, Willis DW (1990) Structural relationships of largemouth bass and bluegill populations in South Dakota ponds. N Am J Fish Manag 10:338–343. https://doi.org/10.1577/1548-8675

Hoagstrom CW, Wall SS, Kral JG, Blackwell BG, Berry CR Jr (2007) Zoogeographic patterns and faunal change of South Dakota fishes. West N Am Nat 67:161–184. https://doi.org/10.3398/1527-0904(2007)67[161:ZPAFCO]2.0.CO;2

Kansas Fishes Committee (2014) Kansas fishes. University Press of Kansas, Lawrence

Kim D, Taylor AT, Near TJ (2022) Phylogenetics and species delimitation of the economically important black basses (*Micropterus*). Sci Rep 12:9113. https://doi.org/10.1038/s41598-022-11743-2

Kolander TD (1992) Factors limiting overwinter survival of young-of-the-year largemouth bass in South Dakota. Dissertation, South Dakota State University

Lucchesi DO, St. Sauver TR, Johnson BA, Hoffman KA, Willis DW (2004) Region III small impoundments strategic plan, 2004-2009. S D Dep Game Fish Park, Wild Div, Sp Rep No 04-13, Pierre

Paukert CP, Willis DW (2004) Environmental influences on largemouth bass *Micropterus salmoides* populations in shallow Nebraska lakes. Fish Manag Ecol 11:345–352. https://doi.org/10.1111/j.1365-2400.2004.00387.x

Scott WB, Crossman EJ (1973) Freshwater fishes of Canada. Fisheries Research Board of Canada, Bulletin 184

Spengler DE (2010) Natural reproductive cycle of northern largemouth bass in the upper Midwest, with applications to off-season spawning. Dissertation, South Dakota State University

Willis DW, Guy CS (1991) Largemouth bass management in South Dakota: comparison with waters further south and east. In: Cooper JL, Hamre RH (eds) Warmwater fisheries symposium I. US Dep Ag, Forest Serv, Gen Tech Rep RM-207, pp 336–342

6.23.9   White Crappie *Pomoxis annularis*

Rafinesque, 1818

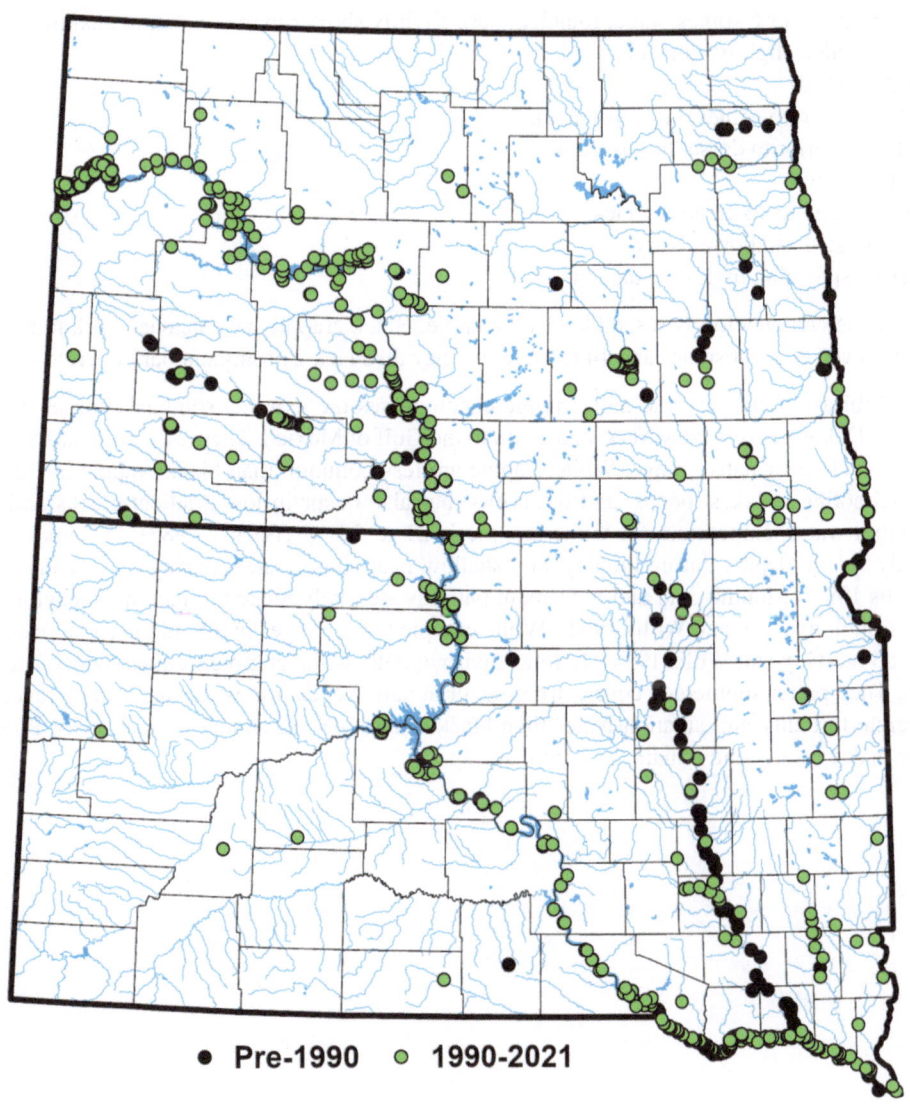

● Pre-1990   ● 1990-2021

Etymology: *Pomoxis annularis* – *Pomoxis* = sharp opercle, referencing the sharp, pointed opercle or gill flap; *annularis* = annular or with rings, referencing the lateral side rings or bars,

Description:

Body – strongly laterally compressed, deep.
Color –

> *Dorsally* – dark olive-gray to bluish-green.
> *Laterally* – silver to light green; 8 to 10 dark, vertical, mottled bars (paler in individuals inhabiting turbid water).
> *Ventrally* – silver to white.
> *Opercle* – dark spot, slightly larger than eye.
> *Eye* – golden yellow to olive-green.
> *Fins* – dorsal, caudal, anal – heavily mottled, nearly round, alternating dark and light spots.

Head – small, short, strongly concave in profile.
Snout – short, pointed.
Eyes – moderately large, laterally on head.
Mouth – large, moderately oblique.

> *Lower jaw* – extending past upper jaw.
> *Upper jaw* – extending at least to middle of eye.

Teeth – on both jaws in pads; small, conical.
Gill rakers – 25 to 32; long, slender.
Dorsal fin – 12 to 16 rays; 5 or 6 spines; large, rounded, base slightly shorter than or equal to anal fin base.
Caudal peduncle – slender, slightly elongate.
Caudal fin – slightly forked.
Anal fin – 16 to 18 rays; 5 or 6 spines; large, rounded.
Pelvic fins – insertion anterior to dorsal fin insertion.
Pectoral fins – 14 to 16 rays.
Lateral line – complete; 36 to 46 ctenoid scales.
Juveniles – more elongated, shallower body.
Sexual dimorphism – spawning males are darker.

Similar Species: White crappie closely resemble black crappie. Black crappie have heavier, dark mottling on lateral sides and a dorsal fin with seven or eight spines and 14 to 16 rays. They also lack a prominent dark opercle spot and vertical bars.

Distribution and Habitat: White crappie native range encompasses the extreme southeast corner of North Dakota and eastern edge of South Dakota, east to New York, and south to the Gulf of Mexico. They are a popular sportfish and have been widely introduced outside their native range. White crappie are less common than black crappie in the Dakotas. They are found most often schooling in lakes, ponds, and reservoirs, but also occur in low gradient creeks and moderate to large streams. The optimum water temperature for white crappie is 27 to 29 °C (Becker 1983). In lentic environments, white crappie are typically at mid-depths during the day, and shallow near-shore waters during the evening and night (Pflieger 1997; Guy and Willis 1994). Monthly and diel movement patterns are likely related to feeding behavior and may vary with habitat and lake morphology (Guy and Willis 1994). White crappie home ranges in a South Dakota glacial lake varied from less than 0.1 ha to 85 ha (Guy and Willis 1994). In lotic environments, white crappie frequent pools, backwaters, and other low-velocity areas. They prefer submerged cover, structure, or aquatic vegetation with sand, mud, gravel, or silt substrates. White crappie tolerate turbidity and siltation better than black crappie (Scott and Crossman 1973; Pflieger 1997). White crappie overwinter in deep water and remain there until spring.

Reproduction: White crappie and black crappie have similar reproductive behavior. Southern populations of white crappie spawn earlier, but in the Dakotas spawning occurs from May to June in water temperatures from 14 to 23 °C in shallow waters with shelter or vegetation (Becker 1983). White crappie sexually mature at age-2 or age-3. Males excavate colonies of nests, small depressions in water 0.2 to 6.1 m deep (Siefert 1968; Pflieger 1997). Males are very territorial and aggressively defend the nests (Becker 1983). They guard and fan the nest until fry leave a few days after hatching. Fecundity ranges from approximately 10,000 to 180,000 eggs and increases with female age and condition (Smith 1979; Bunnell et al. 2005). The approximately 1-mm diameter, demersal eggs adhere to the substrate or aquatic vegetation around the nest (Siefert 1968; Becker 1983). Hatching occurs in 2 to 5 days depending on water temperature (Siefert 1968).

Age and Growth: Growth rates are highly variable among white crappie populations. During the first years, growth rates are density-dependent, and stunting may occur (Pope et al. 2004). Total-lengths-at-age of white crappie in South Dakota were age-1, 93 mm; age-2, 183 mm; age-3, 221 mm; age-4, 252 mm; age-5, 275 mm (Willis et al. 2001). They can reach a total length of 508 mm and live for 12 years, although few survive past age-5.

Food and Feeding: White crappie and black crappie have similar feeding behavior and diets. Young mainly feed during daylight on rotifers, cladocerans, copepod nauplii, and other suspended zooplankton (Becker 1983; O'Brien et al. 1984). Adults mostly feed between dusk and dawn in shallower waters near, or just below, the surface, or in aquatic vegetation. Adults eat larger zooplankton like daphnia, insects like chironomids and corixids, and small fish like age-0 gizzard shad (Unkenholz 1971; Busiahn 1977). After white crappies reach a total length of approximately 150 mm, more fish are eaten (Ellison 1984; O'Brien et al. 1984). Small fish are more important in the diet of white crappie than black crappie.

References

Becker GC (1983) Fishes of Wisconsin. University of Wisconsin Press, Madison

Bunnell DB, Scantland MA, Stein RA (2005) Testing for evidence of maternal effects among individuals and populations of white crappie. Trans Am Fish Soc 134:607–619. https://doi.org/10.1577/T04-094.1

Busiahn TR (1977) Food, growth, and reproduction of white crappies (*Pomoxis annularis*) and black crappies (*P. nigromaculatus*) in Lake Poinsett, South Dakota. Dissertation, South Dakota State University

Ellison DE (1984) Trophic dynamics of a Nebraska black crappie and white crappie population. N Am J Fish Manag 4:355–364. https://doi.org/10.1577/1548-8659

Guy CS, Willis DW (1994) Biotelemetry of white crappies in a South Dakota glacial lake. Trans Am Fish Soc 123:63–70. https://doi.org/10.1577/1548-8659

O'Brien WJ, Loveless B, Wright D (1984) Feeding ecology of young white crappie in a Kansas reservoir. N Am J Fish Manag 4:341–349. https://doi.org/10.1577/1548-8659

Pflieger WL (1997) The fishes of Missouri, revised edn. Missouri Dep Cons, Jefferson City

Pope KL, Wilde GR, Durham BW (2004) Age-specific growth patterns in density-dependent growth of white crappie, *Pomoxis annularis*. Fish Manag Ecol 11:33–38. https://doi.org/10.1111/j.1365-2400.2004.00360.x

Scott WB, Crossman EJ (1973) Freshwater fishes of Canada. Fisheries Research Board of Canada, Bulletin 184

Siefert RE (1968) Reproductive behavior, incubation, and mortality of eggs, and postlarval food selection in the white crappie. Trans Am Fish Soc 97:252–259. https://doi.org/10.1577/1548-8659(1968)97[252:RBIAMO]2.0.CO;2

Smith PW (1979) The fishes of Ohio. University of Illinois Press, Urbana

Unkenholz D (1971) Food habits of black crappies, white crappies, yellow perch, and white suckers in a small impoundment in northeastern South Dakota. Dissertation, South Dakota State University

Willis DW, Isermann DA, Hubers MJ, Johnson BA, Miller WH, St. Sauver TR, Sorensen JS, Unkenholz EG, Wickstrom GA (2001) Growth of South Dakota fishes: a statewide summary with means by region and water type. S D Dep Game Fish Park, Wildl Div, pp 1–5

6.23.10   Black Crappie *Pomoxis nigromaculatus*

(Lesueur, 1829)

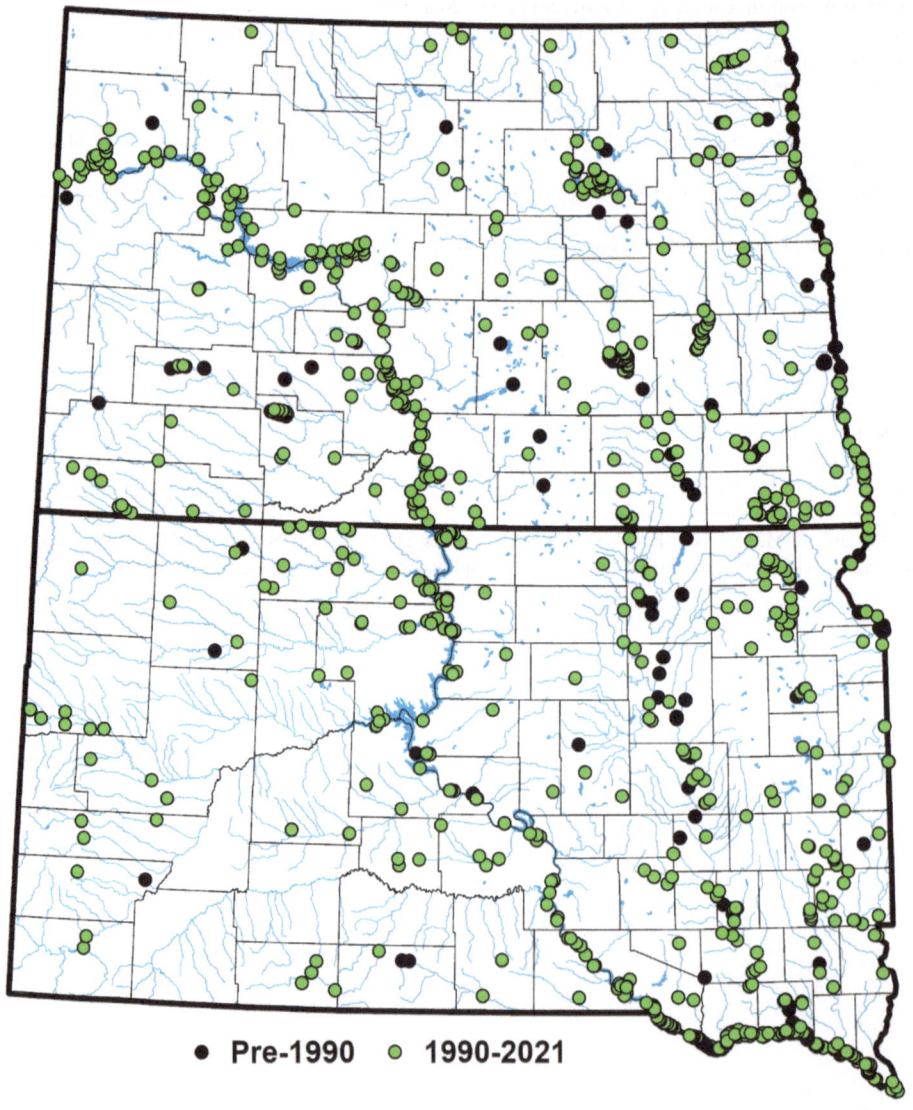

● Pre-1990   ○ 1990-2021

Etymology: *Pomoxis nigromaculatus* – *Pomoxis* = sharp opercle, referencing the sharp, pointed opercle or gill flap; ***nigro-maculatus*** = black spotted.

Description:

Body – strongly laterally compressed, deep.
Color –

> *Dorsally* – dark olive to bluish-green.
> *Laterally* – light olive to silvery-green; heavy, irregular dark green and silver speckled mottling (paler in individuals inhabiting turbid water).
> *Ventrally* – silver to white.
> *Eyes* – yellow-brown.
> *Fins* – dorsal, caudal, anal – heavily mottled, nearly round alternating dark and light spots.

Head – small, short, strongly concave in profile.
Snout – short, pointed.
Eyes – moderately large, laterally on head.
Mouth – large, terminal, oblique.

> *Lower jaw* – extending past upper jaw.
> *Upper jaw* – extending at least to middle of eye.

Teeth – on both jaws in pads; small, conical.
Gill rakers – 27 to 32; long, slender.
Dorsal fin – 14 to 16 rays; 7 or 8 spines; large, rounded, base roughly equal to anal fin base.
Caudal peduncle – slender, slightly elongated.
Caudal fin – slightly forked.
Anal fin – 17 or 18 rays; 6 or 7 spines; large, rounded.
Pelvic fins – insertion well anterior to dorsal fin insertion.
Pectoral fins – 13 to 15 rays.
Lateral line – complete; 38 to 44 ctenoid scales.
Juveniles – more elongated, shallower body.
Sexual dimorphism – spawning males are darker.

Similar Species: Black crappie closely resemble white crappie. White crappie have a dark opercle spot, a dorsal fin with five or six spines and 12 to 16 rays, and lighter lateral sides with eight to ten dark, vertical, mottled bars.

Distribution and Habitat: The native range of black crappie encompasses Manitoba, Canada, south to the Gulf of Mexico, and east to the Atlantic coast. They are native to the eastern Dakotas and have been widely introduced throughout the United States. Black crappie frequently school in lakes, ponds, reservoirs, and low gradient creeks, streams, and rivers. In lentic environments, they frequent mid-depths during the day, and migrate to shallow near-shore waters during the evening and night (Guy et al. 1993). In lotic environments, black crappie occur in pools, backwaters, and other low-velocity areas. They prefer clear water with abundant cover or aquatic vegetation and are less tolerant of turbidity and siltation than white crappie (Scott and Crossman 1973). Black crappie migrate to deeper waters in the late fall and remain there until spring.

Reproduction: In the Dakotas, black crappie spawn from May to June in water temperatures ranging from 14 to 20 °C in sheltered, vegetated bays (Becker 1983). Southern populations spawn earlier. Black crappie become sexually mature at age-2 or age-3. Males excavate colonies of nests, small 200 to 230 mm diameter depressions, in gravel, silt, or sand substrate at depths of 0.4 to 0.8 m, in sheltered areas by woody debris, cattails, or other structure (Becker 1983; Pope and Willis 1997). Males remain in nesting sites for several weeks, guarding and fanning the nest until fry leave a few days after hatching (Pope and Willis 1997). Fecundity increases with female age. Fecundity of age-3 and age-4 black crappie from Lake Poinsett, South Dakota was 46,364 eggs (Busiahn 1977). The approximately 1 mm diameter eggs adhere to the substrate or aquatic vegetation around the nest and hatch in 2 to 4 days (Merriner 1971; Pflieger 1997).

Age and Growth: Growth is highly variable among black crappie populations (Guy and Willis 1993). Turbidity reduces juvenile growth and survival (Pope 1996). Low-density populations grow faster than high density ones (Busiahn 1977; Guy and Willis 1995). In South Dakota, black crappie grow slower in impoundments than natural lakes (Guy and Willis 1993). Total lengths at age of black crappie in South Dakota were age-1, 77 mm; age-2, 151 mm; age-3, 198 mm; age-4, 227 mm;

age-5, 248 mm; age-6, 262 mm; and age-7, 264 mm (Guy and Willis 1993). They can reach a total length of 381 mm and live up to 9 years.

Food and Feeding: Black crappie are sight feeders, eating the largest, slowest-moving organisms that can be captured using the least energy (Busiahn 1977). They mostly feed in the morning and night when migrating from deeper to shallower waters. Black crappie with total lengths under 10 mm eat calanoid copepods, those from 10 to 29 mm also consume dapnia, and those from 29 to 59 mm eat only daphnia (Pope and Willis 1998). Larval black crappie prey selectively and zooplankton availability may determine year class strength (Pope and Willis 1998). Adults forage on zooplankton like daphnia, insects like corixids, and small fish, near or just below the surface, or among aquatic vegetation (Busiahn 1977). Small fish are less important in the diet of black crappie than white crappie.

References

Becker GC (1983) Fishes of Wisconsin. University of Wisconsin Press, Madison

Busiahn TR (1977) Food, growth, and reproduction of white crappies (*Pomoxis annularis*) and black crappies (*P. nigromaculatus*) in Lake Poinsett, South Dakota. Dissertation, South Dakota State University

Guy CS, Willis DW (1993) Statewide summary of sampling data for black and white crappies collected from South Dakota waters. S D Dep Game Fish Park, Comp Rep 93-12, Pierre

Guy CS, Willis DW (1995) Population characteristics of black crappies in South Dakota waters: a case for ecosystem-specific management. N Am J Fish Manag 15:754–765. https://doi.org/10.1577/1548-8675

Guy CS, Neumann RM, Willis DW (1993) Seasonal and diel movements of adult black crappies in a South Dakota natural lake. S D Dep Game Fish Park, Comp Rep 93-14, Pierre

Merriner JV (1971) Egg size as a factor in intergeneric hybrid success of centrarchids. Trans Am Fish Soc 100:29–32. https://doi.org/10.1577/1548-8659

Pflieger WL (1997) The fishes of Missouri, revised edn. Missouri Dep Cons, Jefferson City

Pope KL (1996) Factors affecting recruitment of black crappies in South Dakota waters. Dissertation, South Dakota State University

Pope KL, Willis DW (1997) Environmental characteristics of black crappie (*Pomoxis nigromaculatus*) nesting sites in two South Dakota waters. Ecol Freshw Fish 6:183–189. https://doi.org/10.1111/j.1600-0633.1997.tb00161.x

Pope KL, Willis DW (1998) Early life history and recruitment of black crappie (*Pomoxis nigromaculatus*) in two South Dakota waters. Ecol Freshw Fish 7:56–68. https://doi.org/10.1111/j.1600-0633.1998.tb00172.x

Scott WB, Crossman EJ (1973) Freshwater fishes of Canada. Fisheries Research Board of Canada, Bulletin 184

6.24 Moronidae, Temperate Bass Family

The temperate bass family is a small family of six species within the Perciformes, the largest vertebrate order with approximately 160 families and over 10,000 species. This order includes popular and recognizable sport fish such as walleye *Sander vitreus* in the family Percidae and largemouth bass *Micropterus nigricans* in the family Centrarchidae. Only four temperate bass occur in North America, and only white bass *Morone chrysops*, yellow bass *Morone mississippiensis*, and striped bass *Morone saxatilis* are in the Mississippi River basin. White bass are the only species native to the Dakotas.

Temperate bass have laterally compressed and rather deep bodies with two dorsal fins. The anterior dorsal fin is spinous, while the posterior has soft rays. Morinids also have a small, flat spine on the posterior edge of each gill cover. Temperate bass are very similar in appearance to centrarchids, especially *Micropterus* spp., and percids. They differ from centrarchids by having defined separation or only slight connection between the two dorsal fins, whereas the two dorsal fins are clearly connected in centrarchids. Percids have one or two spines in the anal fin, whereas moronids have three or more.

Temperate bass are either entirely freshwater or anadromous, occurring in North America, Europe, and Africa. White bass and yellow bass are strictly inland freshwater species, while striped bass are anadromous. Temperate bass are often found schooling in the open waters of lakes, reservoirs, and medium-to-large rivers during the summer and winter. In the spring and fall, they are often found closer to shore, likely because of feeding and spawning. White bass are opportunistic sight feeders eating zooplankton and aquatic insects. Upon reaching adulthood, they also become much more piscivorous. Moronids have small, well-developed sharp teeth on both jaws. Moronids do not create a nest or provide parental care after spawning.

White bass are important sportfish in the Dakotas and have been widely stocked across the United States. Striped bass, which were introduced in the Dakotas and are now likely extirpated, will hybridize with white bass to produce hybrids called wipers.

6.24.1 White Bass *Morone chrysops*

(Rafinesque, 1820)

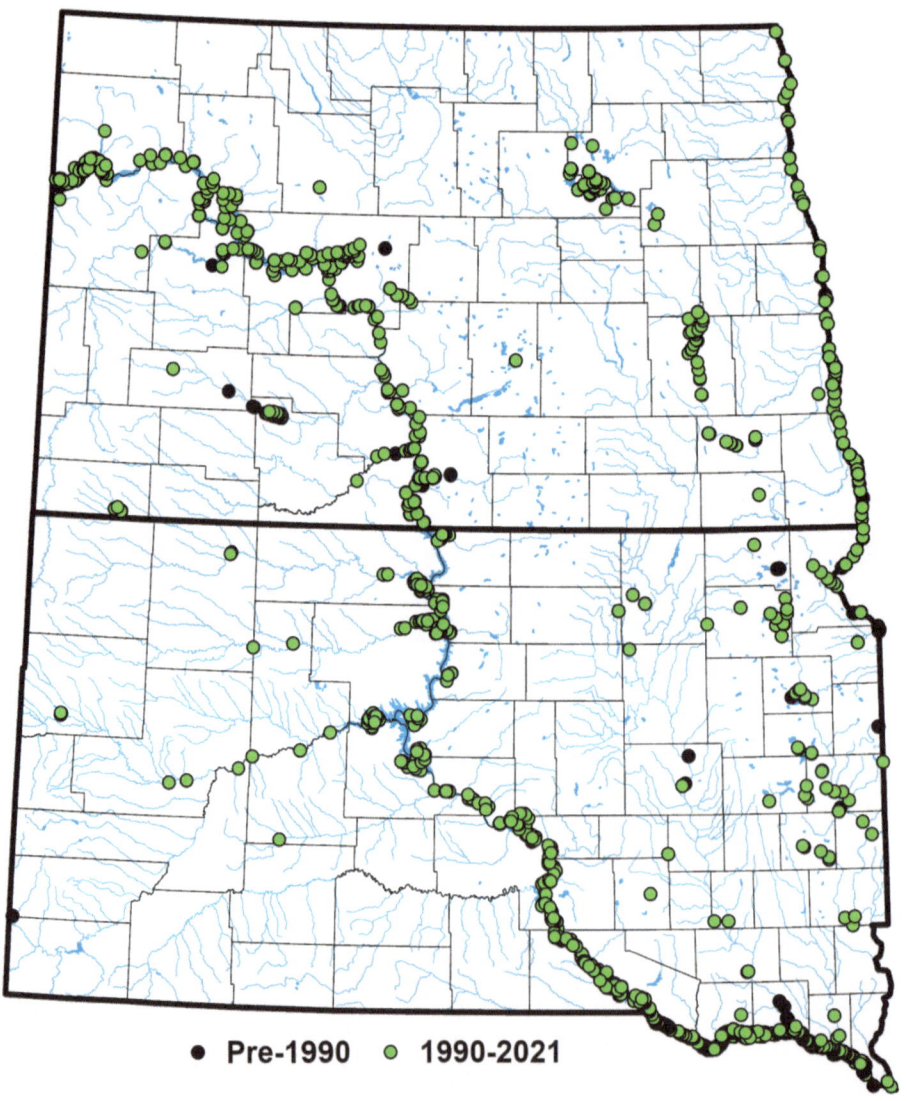

● **Pre-1990**   ○ **1990-2021**

Etymology: *Morone chrysops* – *Morone* = origin unknown; *chrysops* = golden eye.

Description:

Body – deep, laterally compressed, arched anteriorly.
Color –

> *Dorsally* – dark blue-gray to dark silver.
> *Laterally* – silver to pale-green, 6 to 8 dusky to prominent dark horizontal stripes, unbroken above lateral line.
> *Ventrally* – white.
> *Eye* – yellow gold tint.
> *Pelvic fin* – white to light gray.
> *Remaining fins* – dark gray to brown.

Head – small, conical.
Opercle – sharp spine at rear end.
Snout – pointed.
Eyes – moderately large, laterally on head.
Mouth – terminal, oblique.

> *Upper jaw* – extends below eye.
> *Lower jaw* – slightly extends past upper jaw.

Teeth – small, sharp; in patches on both jaws; single patch on tongue posterior end.
Gill rakers – 20 to 25.
Dorsal fin – 2 lobes, close together but not joined.

> *Anterior* – 9 spines.
> *Posterior* – 12 to 14 rays, 1 spine.

Caudal peduncle – short, thick.
Caudal fin – slightly forked.
Anal fin – 11 to 13 rays; 3 spines in ascending order.
Pelvic fins – thoracic.
Pectoral fins – 15 to 17 rays.
Lateral line – complete; 50 to 60 ctenoid scales in series.
Juveniles – shallower body than adults.
Sexual dimorphism – spawning adults may have light blue tint on lower jaw.

Similar Species: Although not common in the Dakotas, yellow bass and striped bass closely resemble white bass. Yellow bass have an anal fin with eight to ten rays and three spines, with the first spine less than one-third the length of the second, and the second and third spines nearly equal in length. Yellow bass also have two dorsal fin lobes that are slightly connected by a membrane and a lower jaw equal in length to the upper jaw. Striped bass have a more elongate body and two elongate and parallel patches of teeth on the tongue.

Distribution and Habitat: The native range of white bass extends from the Minnesota and Big Sioux River systems of eastern South Dakota, through the Mississippi River, Hudson Bay, St. Lawrence River, and Great Lakes systems to New York, and south to Lousiana (Willis et al. 2002). They have been introduced across the Dakotas including the Missouri River reservoirs and eastern glacial lakes (Bailey and Allum 1962). White bass inhabit the open waters of lakes, reservoirs, and medium-to-large rivers of varying turbidity and moderate flow. In eastern South Dakota, white bass prefer offshore habitat during summer and winter, and near-shore shallows in the spring and fall, likely because of feeding and spawning behaviors (Willis et al. 2002).

Reproduction: White bass spawn from May to June at water temperatures of 14 to 22 °C in tributaries or windswept shorelines of lakes and reservoirs with moderate current, often near the surface over rocky substrates (Ruelle 1977; Pflieger 1997; Quist et al. 2002). In Lewis and Clark Lake, South Dakota, females became sexually mature at age-4 or total lengths of 250 to 339 mm (Ruelle 1971). Males mature at age-3 and 220 to 312 mm (Ruelle 1971). No nest is prepared, and no parental care is given. A female with a total length of 390 mm may produce 429,000 to 608,000 eggs (Ruelle 1977). Fecundity increases with female length (Ruelle 1977). The spawning season may last for several weeks (Guy et al. 2002a; Quist et al. 2002). Females release approximately 50% of their eggs, with the smaller eggs reabsorbed later in the summer (Ruelle 1977). The

1 mm diameter, demersal, adhesive eggs hatch in about 2 days (Pflieger 1997; Beck et al. 1999). Recruitment is highly variable, and in eastern South Dakota, it is positively related to spring precipitation and air temperature (Guy et al. 2002b; Willis et al. 2002). Large water discharges and sudden water temperature declines may negatively impact spawning activity (Starnes et al. 1983; Quist et al. 2002).

Age and Growth: The total length of newly hatched white bass larvae is approximately 2 mm (Becker 1983). Growth of age-0 white bass from Lewis and Clark Lake, South Dakota is positively related to water temperature and food abundance (Ruelle 1971). Total lengths at age from South Dakota populations were age-1, 145 mm; age-2, 243 mm; age-3, 298 mm; age-4, 339 mm; age-5, 358 mm; age-6, 384 mm TL; age-7, 394 mm; and age-8, 403 mm (Willis et al. 1997). Females tend to grow faster than males (Guy et al. 2002a). In South Dakota, growth rates are fastest in eastern South Dakota glacial lakes, intermediate in Missouri River reservoirs, and slowest in western reservoirs (Willis et al. 1997). White bass can reach a total length of 533 mm and live 14 years (Willis et al. 2002).

Food and Feeding: White bass are opportunistically feeders. Age-0 white bass primarily eat zooplankton, especially daphnia and calanoid copepods, but may also consume small fish like fathead minnows and Johnny darters (Beck et al. 1999; Blackwell et al. 1999; Willis et al. 2002). Fish are typically not an important prey source until white bass reach a total length around 40 mm (Ruelle 1971; Quist et al. 2002). In Lewis and Clark Lake, South Dakota, white bass with total lengths from 41 to 80 mm commonly consumed emerald shiners and age-0 gizzard shad, while white bass over 90 mm predominantly ate fish (Ruelle 1971). While adults are primarily piscivorous, they will also eat amphipods, corixids, dipterans, and other aquatic macroinvertebrates (Ruelle 1971; Willis et al. 2002).

References

Bailey RM, Allum MO (1962) Fishes of South Dakota. University of Michigan, Ann Arbor. https://doi.org/10.3998/mpub.9690435

Beck HD, Starostka AB, Willis DW (1999) Early life history of white bass in Lake Poinsett. S D Dep Game Fish Park, Comp Rep 99-14, Pierre

Becker GC (1983) Fishes of Wisconsin. University of Wisconsin Press, Madison

Blackwell BG, Soupir CA, Brown ML (1999) Seasonal diets of walleye and diet overlap with other top-level predators in two South Dakota lakes. S D Dep Game Fish Park, Wildl Div, Rep 99-23, Pierre

Guy CS, Schultz RD, Cox CA (2002a) Variation in gonad development, growth, and condition of white bass in Fall River reservoir, Kansas. J Fish Manag 22:643–651. https://doi.org/10.1577/1548-8675

Guy CS, Schultz RD, Colvin MA (2002b) Ecology and management of white bass. N Am J Fish Manag 22:606–608. https://doi.org/10.1577/1548-8675

Pflieger WL (1997) The fishes of Missouri, revised edn. Missouri Dep Cons, Jefferson City

Quist MC, Guy CS, Bernot RJ (2002) Ecology of larval white bass in a large Kansas reservoir. N Am J Fish Manag 22:637–642. https://doi.org/10.1577/1548-8675

Ruelle R (1971) Factors influencing growth of white bass in Lewis and Clark Lake. In: Hall GE (ed) Reservoir fisheries and limnology. American Fisheries Society, Pub 8, Bethesda, pp 411–423

Ruelle R (1977) Reproductive cycle and fecundity of white bass in Lewis and Clark Lake. Trans Am Fish Soc 106:67–76. https://doi.org/10.1577/1548-8659

Starnes LB, Hackney PA, McDonough TA (1983) Larval fish transport: a case study of white bass. Trans Am Fish Soc 112:390–397. https://doi.org/10.1577/1548-8659

Willis DW, Paukert CP, Blackwell BG (2002) Biology of white bass in eastern South Dakota glacial lakes. N Am J Fish Manag 22:627–636. https://doi.org/10.1577/1548-8675

Willis DW, Beck HD, Soupir CA, Johnson BA, Simpson GD, Wickstrom GA (1997) Growth of white bass in South Dakota waters. S D Dep Game Fish Park, Comp Rep 96-16, Pierre

6.24.2   Yellow Bass *Morone mississippiensis*

Jordan & Eigenmann, 1887

© Matt Thomas, KDFWR

● **Pre-1990**   ○ **1990-2021**

Etymology: *Morone mississippiensis* – *Morone* = origin unknown; *mississippiensis* = of the Mississippi River.

Description:

Body – deep, laterally compressed, somewhat arched anteriorly.
Color –

> *Dorsally* – brassy yellow.
> *Laterally* – olive or silvery-white, yellow iridescences, 6 to 8 dark horizontal lateral stripes, 3 above lateral line.
> *Ventrally* – white.
> *Fins* – olive, gray, or brown.

Head – small, conical.
Opercle – sharp spine at upper posterior end.
Snout – pointed.
Eyes – moderately large, laterally on head.
Mouth – terminal, oblique.

> *upper jaw* – extends to anterior edge of eye.

Teeth – small, in irregular rows on both jaws; absent on tongue.
Dorsal fin – 2 lobes, connected by a membrane.

> *Anterior* – 9 spines.
> *Posterior* – 11 to 13 rays; 1 spine.

Caudal peduncle – short, thick.
Caudal fin – moderately forked.
Anal fin – 8 to 10 rays; 3 spines, first is less than one-third the length of the second, second and third nearly equal in length.
Pelvic fins – thoracic; insertions anterior to first dorsal fin lobe insertion.
Lateral line – complete, 51 to 55 ctenoid scales in series.
Juveniles – shallower body.
Sexual dimorphism – spawning adults may be more intensely yellow.

Similar Species: Yellow bass closely resemble white bass and striped bass. White bass have graduated anal fin spines, with the first spine about half the length of the second spine, and the second spine noticeably shorter than the third spine. White bass also have two separate dorsal fin lobes and a lower jaw that slightly extends past the upper jaw. Striped bass have a more elongate body and two elongate, parallel patches of teeth on the tongue.

Distribution and Habitat: Yellow bass are native to thirteen Mississippi River basin states, from Minnesota and Wisconsin in the north, to Louisiana, Alabama, and Texas in the south (Page and Burr 1991). Introduced throughout the United States for sportfishing, they tend to overpopulate and are sometimes considered a nuisance (Tomelleri and Eberle 1990; Page and Burr 1991; Pflieger 1997). In the Dakotas, there has only been a single occurrence in 2010 from a manmade backwater area of the lower Missouri River in South Dakota. Yellow bass generally inhabit clear to turbid pools and backwaters of large rivers, oxbows, reservoirs, and lakes (Helm 1964; Pflieger 1997; Driscoll and Miranda 1999). They prefer open water free of abundant aquatic vegetation, with sand, gravel, silt, or mud substrates (Becker 1983).

Reproduction: Spawning occurs from May through June (Atchison 1967; Becker 1983). Changing water temperature induces yellow bass spawning migrations to tributary streams or areas around 1 m deep (Bulkley 1970). Spawning begins at water temperatures around 15 °C (Bulkley 1970). Males migrate prior to females (Atchison 1967; Bulkley 1970). In southern populations, males become sexually mature as early as age-1, and females mature at age-2 (Stein 2001). Fecundity of yellow bass from Clear Lake, Iowa ranged from 51,300 to 282,400 eggs from females with total lengths and weights of 182 mm and 101 g, and 197 mm and 149 g, respectively (Bulkley 1970). No nest is prepared, and no parental care is given (Atchison 1967). Semibuoyant, 1 mm diameter eggs hatch in 4 to 6 days at a water temperature of 21 °C (Bulkley 1970; Becker 1983).

Age and Growth: Newly hatched yellow bass larvae have a total length of approximately 3 mm (Becker 1983). Total-lengths-at-age of yellow bass from the southern Missouri River were age-1, 88 mm; age-2, 169 mm; age-3, 207 mm; age-4, 237 mm; age-5, 258 mm; age-6, 275 mm; age-7, 284 mm; age-8, 290 mm (Stein 2001). They can reach a total length of 310 mm and live for 8 years (Collier 1963; Stein 2001; Zervas 2010).

Food and Feeding: Yellow bass feeding habits are varied and flexible (Kutkuhn 1955; Collier 1963; Kraus 1963; Driscoll and Miranda 1999; Zervas 2010). Foraging generally occurs in the evening and at night (Kraus 1963). Young-of-year yellow bass from Clear Lake, Iowa primarily consumed entomostracans, chironomids, and *Hyallella* sp. (Ridenhour 1960; Kraus 1963). Amphipods, cladocerans, copepods, and other zooplankton, as well as dipteran larvae and other immature insects, are an important part of adult yellow bass diets (Kraus 1963). Adults feed primarily on forage fish, including young of year gizzard shad and yellow bass, during the spring and summer, and then shift to aquatic insect larvae during fall and winter (Collier 1963; Zervas 2010). Yellow bass also consume large numbers of unidentifiable fish eggs (Driscoll and Miranda 1999).

References

Atchison GJ (1967) Contribution to the life history of the yellow bass, *Roccus mississippiensis* (Jordan and Eigenmann), in Clear Lake, Iowa. Dissertation, Iowa State University

Becker GC (1983) Fishes of Wisconsin. University of Wisconsin Press, Madison

Bulkley RV (1970) Changes in yellow bass reproduction associated with environmental conditions. Iowa State J Sci 45:137–180

Collier JE (1963) Life history studies of yellow bass in North Twin Lake, Iowa. Dissertation. Iowa State University

Driscoll MP, Miranda LE (1999) Diet ecology of yellow bass, *Morone mississippiensis*, in an oxbow of the Mississippi River. J Freshw Ecol 14:477–486. https://doi.org/10.1080/02705060.1999.9663706

Helm WT (1964) Yellow bass in Wisconsin. Trans Wis Acad Sci 53:109–125

Kraus R (1963) Food habits of the yellow bass, *Roccus mississippiensis*, Clear Lake, Iowa, Summer 1962. Proc Iowa Acad Sci 70:209–215

Kutkuhn JH (1955) Food and feeding habits of some fishes in a degraded Iowa lake. Proc Iowa Acad Sci 62:576–588

Page LM, Burr BM (1991) Peterson field guide to freshwater fishes of North America North of Mexico. Houghton Mifflin Harcourt, Boston

Pflieger WL (1997) The fishes of Missouri, revised edn. Missouri Dep Cons, Jefferson City

Ridenhour RL (1960) Abundance, growth and food of young game fishes of Clear Lake, Iowa, 1949 to 1957. Iowa State J Sci 35:1–23

Stein JE (2001) Biology of nonindigenous white perch and yellow bass in an oxbow of the Missouri River. Dissertation, Emporia State University

Tomelleri JR, Eberle ME (1990) Fishes of the Central United States. University Press of Kansas, Lawrence

Zervas PG (2010) Age, reproduction, growth, condition and diet of the introduced yellow bass, *Morone mississippiensis*, in Barren River Lake, Kentucky. Dissertation, Western Kentucky University

6.24.3 Striped Bass *Morone saxatilis*

(Waldbaum, 1792)

© Uland Thomas

● Pre-1990   ○ 1990-2021

Etymology: *Morone saxatilis – Morone* = origin unknown; *saxatilis* = dwelling among rocks.

Description:

Body – fusiform, rather elongated, moderately laterally compressed.
Color –

>   *Dorsally* – dark blue-gray to dark olive-green.
>   *Laterally* – silver to pale green, 6 to 9 continuous, prominent, and thin black stripes horizontally along the body.
>   *Ventrally* – cream to white.
>   *Eye* – faint yellow-gold.

Head – small, conical.
Opercle – 2 small, distinct spines at posterior end.
Snout – acute.
Eyes – moderately large, laterally on head.
Mouth – large, terminal.

>   *Upper jaw* – extends to anterior edge of the eye.
>   *Lower jaw* – slightly extends past upper jaw.

Teeth – 2, elongated, parallel patches of small, sandpaper-like teeth on tongue posterior end.
Gill rakers – 21 to 23.
Dorsal fin – 2 separate lobes.

>   *Anterior* – 8 to 12 spines.
>   *Posterior* – 9 to 13 rays; 1 spine.

Caudal peduncle – moderately elongated, thick.
Caudal fin – weakly forked.
Anal fin – 7 to 13 rays; 3 spines in ascending length, anterior spine being shortest.
Pelvic fins – 5 rays; 1 spine.
Pectoral fins – 13 to 17 rays.
Lateral line – complete, 57 to 68 ctenoid scales in series.
Juveniles – lack distinct horizontal stripes; have faint, dusky vertical bars down lateral sides of the body.

Similar Species: In the Dakotas, striped bass closely resemble white bass and yellow bass. White bass have a deeper body, a single tooth patch on the tongue posterior end, one small spine at the opercle rear end, and six to eight dusky to prominent, dark, horizontal stripes unbroken above the lateral line but fainter and more broken below the lateral line. Yellow bass lack tooth patches on the tongue and have a thin membrane slightly connecting the two dorsal fin lobes.

Distribution and Habitat: Striped bass are native to the North American Atlantic Coast from the St. Lawrence River, Canada, to northern Florida, as well as the major rivers of western Florida, Alabama, Mississippi, and Louisiana that drain into the Gulf of Mexico (Setzler et al. 1980). In 1879, striped bass were introduced to the lower Sacramento River in California and are now found along the Pacific Coast from British Columbia to Mexico (Setzler et al. 1980). Striped bass are anadromous, but it was discovered in 1940 that they can complete their lifecycle entirely in freshwater (Stevens 1958; Setzler et al. 1980). Since then, striped bass have been introduced to large rivers and reservoirs throughout North America for sport fishing, with only a few self-sustaining populations established. Striped bass were stocked into Devils Lake, North Dakota in the late 1970s, but are now likely extirpated because of inadequate natural reproduction. Striped bass have also been reported from the Little Knife River and Little Muddy River in North Dakota but are likely not from established populations. They do not occur in South Dakota. In marine environments, striped bass often school within shallow bays and along shores with large, submerged rocks and boulders. They also frequent grass and mussel beds, or substrates of detritus, sand, or gravel (Hill et al. 1989). In the summer, large striped bass are restricted to a narrow zone right above the thermocline with both cooler temperatures and adequate oxygen (Coutant 1985; Matthews et al. 1985, 1989). In reservoirs, striped bass school in cool, open, free-flowing waters during the summer, but may experience mass die-offs if dissolved oxygen levels decrease (Cheek et al. 1985; Matthews et al. 1985; Moss 1985).

Reproduction: Striped bass spawn in spring at water temperatures from 14 to 21 °C (Hill et al. 1989). Marine populations migrate into freshwater rivers, and freshwater reservoir populations migrate upstream into tributaries with turbulent flow (Hampton 1985). Males and females reach sexual maturity at total lengths of approximately 300 mm and 500 mm, respectively (Westin and Rogers 1978; Hill et al. 1989). Multiple males escort a single female into shallower waters, and splash briefly near the surface while eggs and milt are released (Hill et al. 1989). No nest is prepared, and no parental care is given. Striped bass are broadcast spawners (Hill et al. 1989). They are highly fecund, producing 160,000 eggs/kg of female body weight (Lewis and Bonner Jr 1966). The semi-buoyant, nonadhesive, nearly transparent eggs range from 1 to 5 mm in diameter after water hardening (Albrecht 1964; Murawski 1969; Setzler et al. 1980; Hill et al. 1989). Hatching occurs in approximately 30 hours or 80 hours at water temperatures of 22 °C and 11 °C, respectively (Hill et al. 1989).

Age and Growth: Larval striped bass reach a total length of 30 mm within 20 to 30 days after hatch (Hill et al. 1989). Growth rates depend on location, dissolved oxygen, temperature, and other environmental factors, increasing along a north to south gradient as growing seasons increase (Setzler et al. 1980; Hill et al. 1989). Females tend to grow larger than males (Setzler et al. 1980). Total lengths at age of striped bass from Tennessee were age-1, 216 mm; age-2, 404 mm; age-3, 528 mm; age-4, 625 mm; age 5, 701 mm; and age-6, 731 mm (Weaver 1975; Etnier and Starnes 1993). They can reach 1524 mm long and weigh 34.9 kg. Striped bass longevity in the Gulf Coast is 12 years (Wooley and Crateau 1983).

Food and Feeding: Larvae and juvenile striped bass primarily eat zooplankton and phytoplankton. As they grow, they shift to eating larger aquatic invertebrates and eventually as adults, smaller schooling fish (Setzler et al. 1980; Hill et al. 1989). In reservoirs, striped bass eat gizzard shad (Dorosomatidae) (Matthews et al. 1988). They feed just after dark and right before dawn (Raney 1952; Setzler et al. 1980; Hill et al. 1989).

References

Albrecht AB (1964) Some observations on factors associated with survival of striped bass eggs and larvae. Calif Fish Game 50:100–113

Cheek TE, Van Den Avyle MJ, Coutant CC (1985) Influence of water quality on distribution of striped bass in a Tennessee River impoundment. Trans Am Fish Soc 114:67–76. https://doi.org/10.1577/1548-8659

Coutant CC (1985) Striped bass, temperature, and dissolved oxygen: a speculative hypothesis for environmental risk. Trans Am Fish Soc 114:31–61. https://doi.org/10.1577/1548-8659

Etnier DA, Starnes WC (1993) The fishes of Tennessee. University of Tennessee Press, Knoxville

Hampton K (1985) Movements of striped bass (*Morone saxatilis*) tracked in Wilson Reservoir, Kansas. Dissertation, Fort Hays State University

Hill J, Evans JW, Van Den Avyle MJ (1989) Species profiles: life histories and environmental requirements of coastal fishes and invertebrates (South Atlantic)-Striped bass. US Dep Inter, Fish Wildl Serv, Biol Rep 82 (11.118). https://doi.org/10.21236/ADA226928

Lewis RM, Bonner RR Jr (1966) Fecundity of the striped bass, *Roccus saxatilis* (Waldbaum). Trans Am Fish Soc 95:328–331. https://doi.org/10.1577/1548-8659(1966)95[328:FOTSBR]2.0.CO;2

Matthews WJ, Hill LG, Schellhaass SM (1985) Depth distribution of striped bass and other fish in Lake Texoma (Oklahoma-Texas) during summer stratification. Trans Am Fish Soc 114:84–91. https://doi.org/10.1577/1548-8659

Matthews WJ, Hill LG, Edds DR, Hoover JJ, Heger TG (1988) Trophic ecology of striped bass, *Morone saxatilis*, in a freshwater reservoir (Lake Texoma, USA). J Fish Biol 33:273–288. https://doi.org/10.1111/j.1095-8649.1988.tb05470.x

Matthews WJ, Hill LG, Edds DR, Gelwick FP (1989) Influence of water quality and season on habitat use by striped bass in large southwestern reservoir. Trans Am Fish Soc 118:243–250. https://doi.org/10.1577/1548-8659

Moss JL (1985) Summer selection of thermal refuges by striped bass in Alabama reservoirs and tailwaters. Trans Am Fish Soc 114:77–83. https://doi.org/10.1577/1548-8659

Murawski WS (1969) The distribution of striped bass, *Roccus saxatilis*, eggs and larvae in the lower Delaware River. N J Dep Cons Econ Dev, Div Fish Game, Bureau Fish, Nacote Creek

Raney EC (1952) The life history of the striped bass, *Roccus saxatilis* (Waldbaum). Bingham Oceanographic Laboratory, Yale University 14:5–97

Setzler EM, Boynton WR, Wood KV, Zion HH, Tucker L, Mihursky JA (1980) Synopsis of biological data on striped bass, *Morone saxatilis* (Waldbaum). US Dept Comm, NOAA Tech Rep, NMFS Circ 443, FAO Fish Syn, No 121

Stevens RE (1958) The striped bass of the Santee-Cooper reservoir. In: Proceedings of the annual conference Southeastern Association of Game and Fish Commissioners, vol 11, pp 253–264

Weaver OR (1975) A study of the striped bass, *Morone saxatilis*, in J Percy Priest Reservoir, Tennessee. Dissertation, Tennessee Technical University

Westin DT, Rogers BA (1978) Synopsis of the biological data on the striped bass. Univ R I Marine Tech Rep No. 67

Wooley CM, Crateau EJ (1983) Biology, population estimates and movement of native and introduced striped bass, Apalachicola River, Florida. N Am J Fish Manag 3:383–394. https://doi.org/10.1577/1548-8659

6.25   Percidae, Perch and Darter Family

The perch family consists of ten genera and more than 200 species native to North America and Eurasia. In the Dakotas, this family includes popular sportfish like walleye *Sander vitreus*, yellow perch *Perca flavescens*, and sauger *Sander canadensis*. However, most species in Percidae are much smaller fish collectively called darters, of which five occur in the Dakotas. The family Percidae also includes logperch *Percina caprodes*, a nongame species found in both eastern North Dakota and South Dakota. Zander *Sander lucioperca*, native to western Eurasia, were introduced into Spiritwood Lake, North Dakota in 1989 for sport fishing. Although not plentiful, zander are established in the lake, with yearlings to age-2 individuals recently observed.

Percids may be confused with moronids and centrarchids but are easily differentiated by having only one or two anal spines. Moronids have three and centrarchids have three or more. Furthermore, in contrast to temperate bass, perch lack well-developed opercle spines and their lateral line does not extend onto the caudal fin upper lobe. Percids also have an anterior spinous dorsal fin and a posterior rayed dorsal fin frequently and distinctly separated. Larger percids, such as walleye, sauger, and perch, have easily seen teeth on both sets of jaws and fully developed swim bladders, which distinguishes them from the toothed-darters. Without a swim bladder, darters are bottom-dwelling and often perched upon bottom substrate. Larger percids inhabit the pelagic zone.

In the Dakotas, larger percids primarily inhabit lakes, reservoirs, and rivers, while darters prefer lakes, creeks, and streams. Many larger percids undergo an ontogenetic diet change, primarily consuming invertebrates as juveniles before switching to piscivory as adults. Darters generally eat aquatic invertebrates. Spawning also differs between larger percids and darters. Walleye, yellow perch, and sauger randomly or broadcast spawn over gravel or rocky substrate with no parental care. Darters have more specific spawning locations and also provide some form on parental care to eggs, fry, or both.

6.25.1 Iowa Darter *Etheostoma exile*

(Girard, 1859)

Nuptial Male
Female

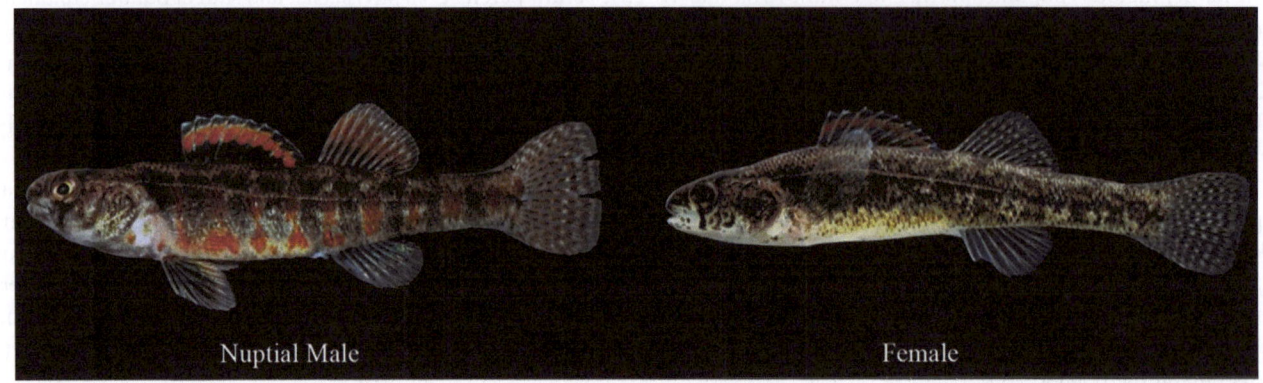

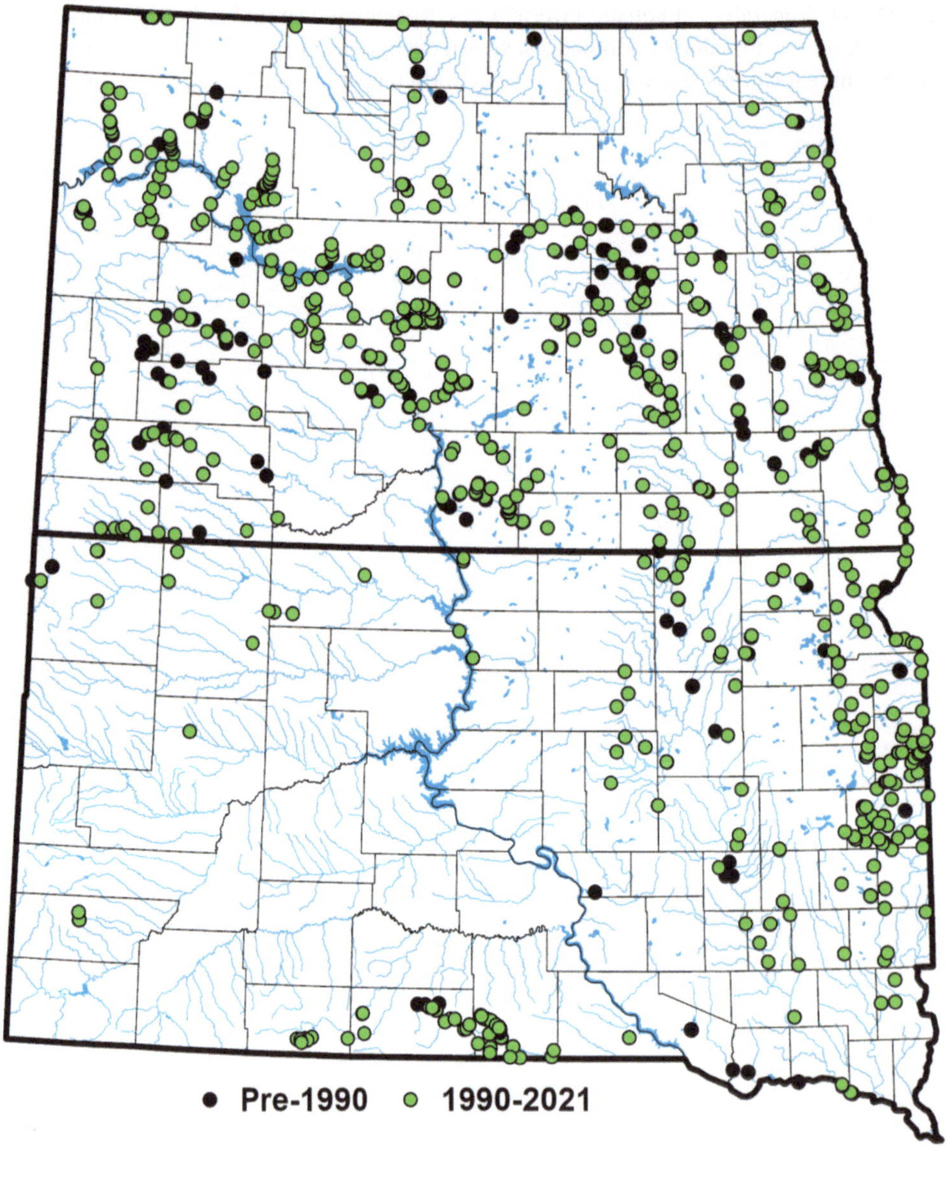
● Pre-1990 ● 1990-2021

Etymology: *Etheostoma exile* – *Etheostoma* = strain mouth; *exile* = slim.

Description:

Body – elongated, slender, slightly laterally compressed.
Color –

> *Males:*
>
>> *Dorsally* – olive-brown to dark brown.
>> *Laterally* – 9 to 12 prominent vertical bars, dark red to burnt orange blotches in interspaces.
>> *Ventrally* – fading to yellowish-white.
>
> *Females:* no bright colors; lateral bars less distinct.
>
>> *Dorsally* – olive to yellow-brown; 8 dark blotches.
>> *Laterally* – 9 to 12 prominent vertical bars.
>
> *Unisex:*
>
>> *Laterally* – distinct black suborbital bar.
>> *Fins* -
>>
>>> *Dorsal* – anterior: 3 color bands: bottom – slate gray, middle – broad and light colored, outer – dusky brown; posterior: light with small dark spots arranged in rows.
>>> *Caudal* – small dark spots giving wavy appearance.
>>> *Anal, pelvic, pectoral* – transparent, dark pigment along rays.

Head – moderate.
Snout – blunt.
Eyes – large.
Mouth – small, slightly oblique to horizontal; premaxillae nonprotractile; upper lip groove non continuous over snout; frenum.
Teeth – tiny, sharp; in narrow bands on upper and lower jaws; infraorbital canal with 8 pores.
Gill rakers – approximately 7; short.
Dorsal fin – 2 lobes.

> *Anterior* – 8 to 12 spines; spiny.
> *Posterior* – 10 to 12 rays; soft.

Caudal peduncle – long.
Caudal fin – squarish; rounded corners.
Anal fin – 7 to 9 rays; 2 spines.
Pelvic fins – base anterior to dorsal fin insertion.
Pectoral fins – fan shaped.
Lateral line – incomplete; 55 to 65 scales in the series; fewer than 35 pored.
Skin – scales ctenoid; opercles, nape and cheeks scaled; belly scaled; breast slightly scaled.
Sexual dimorphism – breeding males have 9 to 12 bright blue lateral bars with red interspaces, spiny dorsal fin layered with bright blue spots along basal half, then bright-orange to red band, with narrow blue stripe on outer edge.

Similar Species: Iowa darter differs from Johnny darter, blackside darter, river darter, and slenderhead darter by lacking a complete lateral line. Johnny darters have numerous, small "x" and "w" markings along the lateral sides. Blackside darter have connected black blotches along the lateral line creating a stripe. River darter lack a frenum, or it is absent when mouth is closed. Slenderhead darter have a pointed snout.

Distribution and Habitat: The native range of Iowa darter extends throughout south-central Canada and the north central United States from Montana east to the Great Lakes and south to the Hudson Bay. In North Dakota, Iowa darter occur in all river systems except Devils Lake (Owen et al. 1981). In South Dakota, they occur primarily east of the Missouri River, and in many of the western tributaries of the Missiouri River (Bailey and Allum 1962). Iowa darters inhabit calm or standing semi-turbid to clear natural lakes, ponds, and slow-moving streams. They prefer abundant submerged aquatic vegetation, filamentous-algae-covered surfaces, and sand, gravel, and silty substrates (Becker 1983).

Reproduction: Colder water temperatures may delay Iowa darter migrations from deeper water to shallower shoreline spawning grounds in late spring. Males migrate before females. Iowa darters spawn over fibrous root material and rooted aquatic vegetation from late April to June (Owen et al. 1981; Becker 1983). Males establish semicircle-shaped spawning territories, with the base along the shoreline, which they only defend from conspecifics (Becker 1983). A single male escorts a single female to the spawning site where he mounts her anteriorly with his posterior end curved downward around her. Their bodies vibrate while gametes are released, with the eggs fertilized and deposited. Fewer than ten eggs are frequently released at a single spawning episode, with approximately 900 eggs being deposited throughout the season (Hrabik et al. 2015). Females spawn with several males and return to deeper water when finished (Becker 1983). Males continue to defend the spawning territory after egg deposition.

Age and Growth: Female Iowa darters are often larger than males. Adults typically have total lengths from 46 to 64 mm, but sometimes reach 70 mm. They can live up to 4 years. Total lengths at age of Iowa darter from northern Wisconsin were 34 to 55 mm at age-1, 48 to 61 mm at age-2, 58 to 68 mm at age-3, and 68 to 69 mm at age-4 (Becker 1983; Lutterbie 1976).

Food and Feeding: Adult Iowa darter eat primarily mayfly, chironomid, dipteran, and other aquatic insect adults and larvae, and occasionally fish eggs. Juveniles eat small insects and crustaceans, including copepods and amphipods.

References

Bailey RM, Allum MO (1962) Fishes of South Dakota. University of Michigan, Ann Arbor. https://doi.org/10.3998/mpub.9690435
Becker GC (1983) Fishes of Wisconsin. University of Wisconsin Press, Madison
Hrabik RA, Schainost SC, Stasiak RH, Peters EJ (2015) The fishes of Nebraska. University of Nebraska, Lincoln
Lutterbie GW (1976) The darters (Pices: Percidae: Etheostomatinae) of Wisconsin. Dissertation, University of Wisconsin-Stevens Point
Owen JB, Elsen DS, Russell GW (1981) Distribution of fishes in North and South Dakota Basins affected by the Garrison Diversion Unit. University of North Dakota Press, Grand Forks

6.25.2   Johnny Darter *Etheostoma nigrum*

Rafinesque, 1820

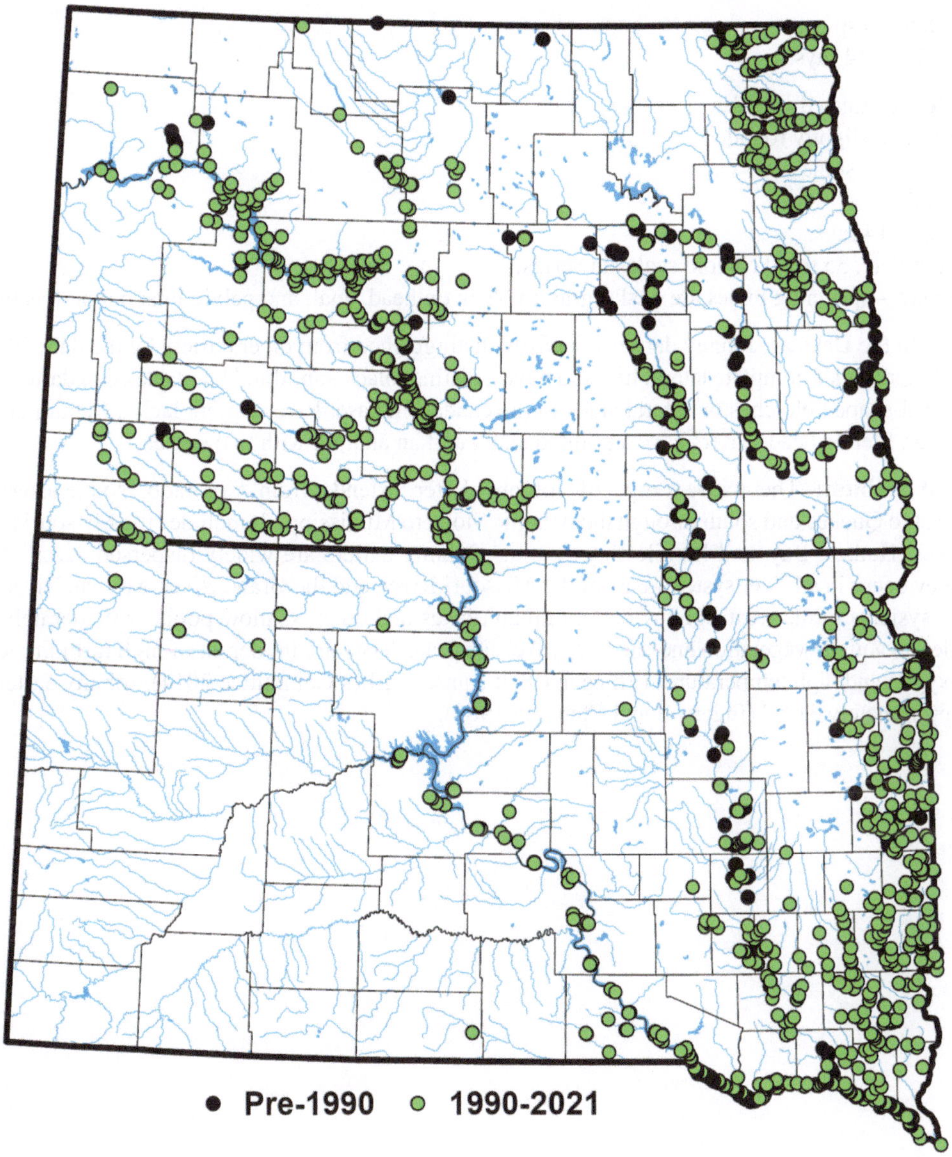

● **Pre-1990**   ○ **1990-2021**

Etymology: *Etheostoma nigrum* – *Etheostoma* = strain mouth; *nigrum* = black, possibly referencing the darkness of spawning males.

Description:

Body – fusiform, elongated, slender, slightly laterally compressed.
Color –

>   *Dorsally* – brown to golden tan, 4 to 7 dark saddles.
>   *Laterally* – light brown to tan, dark "w" or "x" markings.
>   *Ventrally* – creamy yellow to white
>   *Fins* – dorsal/caudal – spotted; others – lightly pigmented to transparent.

Head – moderate.
Opercle – partially to completely scaled.
Snout – blunt.
Eyes – large, dorsolaterally on head.
Mouth – terminal to slightly subterminal; frenum absent.

>   *Upper jaw* – extends to anterior end of eye.

Teeth – small, sharp, in bands on upper and lower jaws.
Gill rakers – 5 to 11; short, stout.
Dorsal fin – 2 lobes scarcely joined.

>   *Anterior* – 8 to 10 spines.
>   *Posterior* – 11 to 14 rays.

Caudal peduncle – elongated, slender.
Caudal fin – square to slightly rounded.
Anal fin – 7 to 9 rays; 1 spine.
Pelvic fins – thoracic.
Pectoral fins – 10 to 14 rays.
Lateral line – complete, 35 to 50 ctenoid scales in series.
Sexual dimorphism – spawning males are dusky-gray to black on head, anal, and pelvic fins; pelvic fins have white tips.

Similar Species: In the Dakotas, Johnny darters are easily distinguished from other darters by the dark "w" and "x" lateral markings. Iowa darters lack a complete lateral line and have a distinct black suborbital bar. Blackside darters have six to eight large, irregular, oval-shaped blotches on lateral sides to the caudal fin base. River darters lack a frenum and have a distinct black suborbital bar. Slenderhead darters have a pointed snout and an anal fin with two spines.

Distribution and Habitat: The native range of Johnny darter extends from Colorado, southeastern Montana, and Saskatchewan, east to Quebec and south through the Atlantic Slope to Mississippi. It includes the Missouri River, Mississippi River, Great Lakes, Hudson Bay, and Mobile Bay systems. Johnny darters are widely dispersed across North Dakota. In South Dakota, they occur in all river systems east of the Missouri River, and the Grand River, Moreau River, Niobrara River, and Ponca River systems in the west. Johnny darters inhabit lakes and quiet, shallow pools or moderately flowing runs of rivers, with moderate aquatic vegetation and sand or gravel substrate. They are intolerant of high turbidity levels but tolerate low dissolved oxygen and high temperatures (Pflieger 1997; Kansas Fishes Committee 2014). Johnny darter critical thermal maximum is approximately 31 °C (Ingersoll and Claussen 1984).

Reproduction: Johnny darters spawn from April to June at water temperatures from 12 to 24 °C (Becker 1983; Garcia et al. 2018). They may spawn earlier at southern latitudes (Parrish et al. 1991). Quick changes in temperature, increased siltation, and rising water levels may postpone or interrupt spawning (Becker 1983). Johnny darters are fractional spawners. After migrating to spawning areas before females, males use their fins to brush sediment from the underside of submerged rocks or other structures (Atz 1940; Winn 1958a, b). These nests are in slow current, often in pools or shallow water (Atz 1940; Winn 1958a). Either sex initiates spawning by inverting themselves to attract the other sex (Atz 1940). Gametes are released when both fish are in an inverted position against a rock or submerged structure and quiver in side by side or head to tail positions (Atz 1940; Winn 1958a). Nests contain from 30 to 1,589 eggs, but females rarely release more than one egg at a time (Winn 1958a, b; Grant and Colgan 1983). Fecundity increases with female size and age, with females depositing 50 to 200 eggs per season (Speare 1965; Becker 1983). The approximately 2 mm diameter eggs are amber with a dark oil globule (Atz 1940; Speare 1965). Males fan, rub, and defend the eggs from predators until hatching (Atz 1940; Grant and Colgan 1984). Males chase away nest predators using darting motions, an open mouth, and quivering (Atz 1940). Fungus-covered eggs are often consumed by the male. Eggs hatch in 6 days at 23 °C, 10 days at 20 °C, and 16 days at 13 °C (Speare 1965).

Age and Growth: Newly hatched Johnny darter larvae have a total length of approximately 5 mm (Becker 1983). Growth is rapid during the first 2 years. Total lengths at age of Johnny darter from Iowa were 29 mm at age-1, 43 mm at age-2, 53 mm at age-3, and 56 mm at age-4 (Karr 1963). Males are typically larger than females. Johnny darters have typical total lengths of 38 to 64 mm, can reach 77 mm, and live for 4 years (Karr 1963; Becker 1983).

Food and Feeding: Johnny darters are visual predators that eat sedentary benthic prey (Roberts and Winn 1962; Paine et al. 1981). They mostly feed during the day (Emery 1973). A protrusible premaxillae allowing the upper jaw to extend well beyond the lower jaw assists in feeding (Paine et al. 1981). Adults primarily consume chironomid larvae and aquatic insects like dipteran, ephemeropterans, trichopterans, and odonates (Karr 1963; Paine et al. 1981). To a lesser extent, Johnny darter will also eat ostracods, cladocerans, copepods, and other microcrustaceans (Karr 1963; Paine et al. 1981).

References

Atz JW (1940) Reproductive behavior in the eastern Johnny darter, *Boleosoma nigrum olmstedi* (Storer). Copeia 2:100–106. https://doi.org/10.2307/1439050

Becker GC (1983) Fishes of Wisconsin. University of Wisconsin Press, Madison

Emery AR (1973) Preliminary comparisons of day and night habits of freshwater fish in Ontario lakes. J Fish Res Board Can 30:761–774. https://doi.org/10.1139/f73-131

Garcia C, Schumann DA, Howell J, Graeb BDS, Bertrand KN, Klumb RA (2018) Seasonality, floods and droughts structure larval fish assemblages in prairie rivers. Ecol Freshw Fish 27:389–397. https://doi.org/10.1111/eff.12354

Grant JWA, Colgan PW (1983) Reproductive success and mate choice in the Johnny darter, *Etheostoma nigrum* (Pisces: Percidae). Can J Zool 61:437–446. https://doi.org/10.1139/z83-058

Grant JWA, Colgan PW (1984) Territorial behavior of the male Johnny darter, *Etheostoma nigrum*. Environ Biol Fish 10:261–269. https://doi.org/10.1007/BF00001479

Ingersoll CG, Claussen DL (1984) Temperature selection and critical thermal maxima of the fantail darter, *Etheostoma flabellare*, and Johnny darter, *E. nigrum*, related to habitat and season. Environ Biol Fish 11:131–138. https://doi.org/10.1007/BF00002262

Kansas Fishes Committee (2014) Kansas fishes. University Press of Kansas, Lawrence

Karr JR (1963) Age, growth, and food habits of Johnny, slenderhead and blacksided darters of Boone County, Iowa. Proc Iowa Acad Sci 70:228–236

Paine MD, Dodson JJ, Power G (1981) Habitat and food resource partitioning among four species of daters (Percidae: *Etheostoma*) in a southern Ontario stream. Can J Zool 60:1635–1641. https://doi.org/10.1139/z82-214

Parrish JD, Heines DC, Baker JA (1991) Reproductive season, clutch parameters, and oocyte size of the Johnny darter *Etheostoma nigrum* from southwestern Mississippi. Am Midl Nat 125:180–186. https://doi.org/10.2307/2426221

Pflieger WL (1997) The fishes of Missouri, revised edn. Missouri Dep Cons, Jefferson City

Roberts NJ, Winn HE (1962) Utilization of the senses in feeding behavior of the Johnny darter, *Etheostoma nigrum*. Copeia 1962:567–570. https://doi.org/10.2307/1441180

Speare EP (1965) Fecundity and egg survival of the central Johnny darter (*Etheostoma nigrum nigrum*) in southern Michigan. Copeia 3:308–314. https://doi.org/10.2307/1440792

Winn HE (1958a) Observation on the reproductive habits of darters (*Pisces*-Percidae). Am Midl Nat 59:190–212. https://doi.org/10.2307/2422384

Winn HE (1958b) Comparative reproductive behavior and ecology of fourteen species of darters (*Pisces*-Percidae). Ecol Monogr 28:155–191. https://doi.org/10.2307/1942207

6.25.3 Yellow Perch *Perca flavescens*

(Mitchill, 1814)

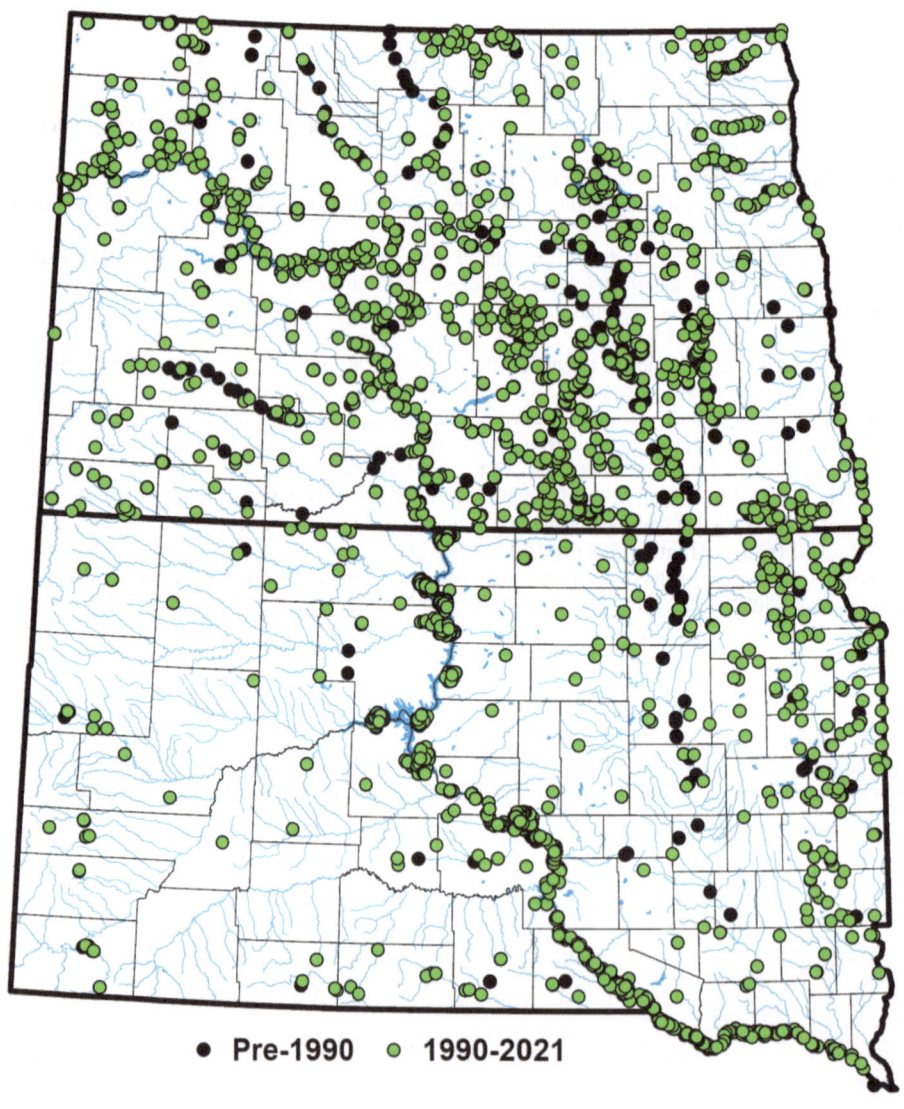

● **Pre-1990**   ○ **1990-2021**

Etymology: *Perca flavescens – Perca* = perch; *flavescens* = yellowish.

Description:

Body – fusiform, laterally compressed, moderately deep, oval in cross section.
Color –

Dorsally – brown to olive, 6 to 9 dark saddles extend vertically down lateral sides.
Laterally – yellow-green to golden bronze.
Ventrally – white.
Eyes – golden.
Fins – dorsal – dusky gray to black; pigment on first 2 and last 3 to 5 membranes; anal, pelvic, pectoral – pale yellow to light orange-red.

Head – small, elongate.
Snout – conical, blunt.
Eyes – laterally on head.
Mouth – moderately large, terminal, slightly oblique; jaws nearly equal in length.

Lower jaw – occasionally slightly longer than upper jaw.
Upper jaw – extends to pupil of eye.

Teeth – small, pointed, in pads on both jaws.
Gill rakers – 18 to 20.
Dorsal fin – 2 distinct lobes.

Anterior – 12 to 14 spines.
Posterior – 12 to 15 rays; 1 or 2 spines.

Caudal peduncle – slender, elongated.
Caudal fin – moderately forked.
Anal fin – 7 or 8 rays; 2 spines.
Pelvic fins – thoracic, insertion posterior to insertion of posterior dorsal fin.
Pectoral fins – 13 to 15 rays.
Lateral line – complete; 52 to 60 ctenoid scales.

Similar Species: No other species within the Dakotas closely resemble yellow perch. Sunfish are much deeper-bodied, with two lobes of the dorsal fin distinctly joined. Bass lack saddles, or vertical bars on lateral sides, and have an anal fin with three spines. Walleye and sauger possess large, distinct canine teeth.

Distribution and Habitat: Yellow perch are native to central Canada and the United States east of the Rocky Mountains from North Dakota and east-central South Dakota, east throughout the Mississippi River, Great Lakes, and Atlantic River systems, and south to South Carolina, Ohio, and Nebraska. They are widely distributed across the Dakotas. Yellow perch adapt to a variety of habitat types and are often found schooling in clear-to-slightly turbid lakes, ponds, sloughs, reservoirs, and streams with sand and gravel substrates and submerged aquatic vegetation, submerged structure, or both vegetation and structure. Eurythermal and temperate, yellow perch have an optimal water temperature ranging from 20 to 27 °C (Brown et al. 2002; Kansas Fishes Committee 2014).

Reproduction: Yellow perch typically spawn in late spring, with peak egg viability in water temperatures from 8 to 11 °C (Hokanson 1977; Kaemingk et al. 2014). In South Dakota, they prefer to spawn in water approximately 0.6 m deep, 3 to 30 m from shore, with rocky and gravel substrate, and periphyton-free submerged woody structure (Fisher et al. 1996). Yellow perch males and females sexually mature at age-2 and age-3, respectively, but will mature earlier if rapidly growing (Hasler 1945; Jansen 1996; Purchase et al. 2005). They mostly spawn at night over a 5 to 11-day period (Breder and Rosen 1966; Isermann and Willis 2008). No nest is prepared, and no parental care is given. Females release a gelatinous, tubular strand of eggs, which becomes entangled within submerged vegetation and structure (Fisher et al. 1996). Fecundity is highly variable and increases with female size and age (Heyer et al. 2001). Females with total lengths over 350 mm can produce from 100,000 to 150,000, 1 to 2 mm diameter eggs (Thorpe 1977; Jansen 1996). Warmer water temperatures decrease incubation duration, reduce larval abnormalities, and increase hatching success (Hokanson and Kleiner 1974). Eggs hatch in 9 days at 13 °C, 23 or 24 days at 9 to 10 °C, and 35 days at 8 °C (Huff et al. 2004). Recruitment is often inconsistent and highly variable among populations, resulting in strong, weak, or missing year classes (Koonce et al. 1977; Newsome and Alto 1987;

Isermann et al. 2007). Recruitment is primarily influenced by temperature and other abiotic factors like lake morphology or water level fluctuations (Pope et al. 1996; Isermann 2003; Dembkowski et al. 2014, 2017; Kaemingk et al. 2014). It increases in years of increased precipitation and less variable temperatures (Pope et al. 1996; Kaemingk et al. 2014). Recruitment is also affected by prey availability, prey size, and other biotic factors (Fisher and Willis 1997; Jolley et al. 2010).

Age and Growth: Newly hatched yellow perch larvae have total lengths from 5 to 6 mm (Thorpe 1977). Optimal growth for age-0 yellow perch occurs at a water temperature of 23 °C (Fisher and Willis 1997). Thermal requirements vary among populations, but in South Dakota, maximum growth occurs at a water temperature of 25 °C (Brown et al. 2002; Brown and Smith 2004). Although growth is frequently correlated with temperature and fluctuation in water levels, it is mostly influenced by prey availability, predation, intra- and interspecific competition, and other biotic factors (Power and Van Den Heuvel 1999; Paukert et al. 2002; Pierce et al. 2006; Schoenebeck and Brown 2010; Kaemingk and Willis 2012; Dembkowski et al. 2014, 2017; Kaemingk et al. 2014). Total-lengths-at-age of yellow perch from South Dakota were age-1, 86 mm; age-2, 145 mm; age-3, 190 mm; age-4, 220 mm; age-5, 242 mm (Willis et al. 2001). They can reach a total length of 406 mm and a weigh up to 0.9 kg, with a longevity of 10 years.

Food and Feeding: Yellow perch are generalist predators. Larvae with total lengths of up to approximately 7 mm, primarily eat rotifers, copepod nauplii, and other zooplankton (Fisher and Willis 1997). Larvae with total lengths from 20 to 40 mm mainly consume adult copepods and cladocerans, while those from 40 to 60 mm select higher energy benthic macroinvertebrates (Fisher and Willis 1997; Graeb et al. 2006). Larval growth and survival may be positively related to copepod densities (Graeb et al. 2004). Diets of adult yellow perch in South Dakota are highly variable among locations and months, but chironomids are a major food source in all populations (Lott 1991). Fish are initially consumed around age-1 at total lengths of 80 to 150 mm, but yellow perch are not primarily piscivorous until age-4 (Keast 1985; Fullhart et al. 2002; Graeb et al. 2004).

References

Breder CM, Rosen DE (1966) Modes of reproduction in fishes. American Museum of Natural History, New York

Brown ML, Smith KA (2004) Temperature-dependent growth models for South Dakota yellow perch, *Perca flavescens*, fingerling production. J Appl Aquacult 16:105–112. https://doi.org/10.1300/J028v16n01_09

Brown PB, Wetzel JE, Mays J, Wilson KA, Kasper CS, Malison J (2002) Growth differences among stocks of yellow perch, *Perca flavescens*, are temperature dependent. J Appl Aquacult 12:43–56. https://doi.org/10.1300/J028v12n01_02

Dembkowski DJ, Weber MJ, Wuellner MR (2017) Factors influencing recruitment and growth of age-0 yellow perch in eastern South Dakota glacial lakes. Fish Manag Ecol 24:372–381. https://doi.org/10.1111/fme.12240

Dembkowski DJ, Chipps SR, Blackwell BG (2014) Response of walleye and yellow perch to water-level fluctuations in glacial lakes. Fish Manag Ecol 21:89–95. https://doi.org/10.1111/fme.12047

Fisher SJ, Willis DW (1997) Early life history of yellow perch in two South Dakota glacial lakes. J Freshw Ecol 12:421–429. https://doi.org/10.1080/02705060.1997.9663552

Fisher SJ, Pope KL, Templeton LJ, Willis DW (1996) Yellow perch spawning habitats in Pickerel Lake, South Dakota. Prairie Nat 28:65–75

Fullhart HG, Parsons BG, Willis DW, Reed JR (2002) Yellow perch piscivory and its possible role in structuring littoral zone fish communities in small Minnesota lakes. J Freshw Ecol 17:37–43. https://doi.org/10.1080/02705060.2002.9663866

Graeb BDS, Dettmers JM, Wahl DH, Cáceres CE (2004) Fish size and prey availability affect growth, survival, prey selection, and foraging behavior of larval yellow perch. Trans Am Fish Soc 133:504–514. https://doi.org/10.1577/T03-050.1

Graeb BDS, Mangan MT, Jolley JC, Wahl DH, Dettmers JM (2006) Ontogenetic changes in prey preference and foraging ability of yellow perch: insights based on relative energetic return of prey. Trans Am Fish Soc 135:1493–1498. https://doi.org/10.1577/T05-063.1

Hasler AD (1945) Observations on the winter perch population of Lake Mendota. Ecology 26:90–94. https://doi.org/10.2307/1931918

Heyer CJ, Miller TJ, Binkowski FP, Caldarone EM, Rice JA (2001) Maternal effects as a recruitment mechanism in Lake Michigan yellow perch (*Perca flavescens*). Can J Fish Aquat Sci 58:1477–1487. https://doi.org/10.1139/f01-090

Hokanson KEF (1977) Temperature requirements of some percids and adaptations to the seasonal temperature cycle. J Fish Res Board Can 34:1524–1550. https://doi.org/10.1139/f77-217

Hokanson KEF, Kleiner CF (1974) Effects of constant and rising temperatures on survival and developmental rates of embryonic and larval yellow perch, *Perca flavescens* (Mitchill). In: Blaxter JHS (ed) The early life history of fishes. Springer, New York, pp 437–448. https://doi.org/10.1007/978-3-642-65852-5_36

Huff DD, Grad G, Williamson CE (2004) Environmental constraints on spawning depth of yellow perch: the roles of low temperature and high solar ultraviolet radiation. Trans Am Fish Soc 133:718–726. https://doi.org/10.1577/T03-048.1

Isermann DA (2003) Population dynamics and management of yellow perch populations in South Dakota glacial lakes. Dissertation, South Dakota State University

Isermann DA, Willis DW (2008) Emergence of larval yellow perch, *Perca flavescens*, in South Dakota lakes: potential implications for recruitment. Fish Manag Ecol 15:259–271. https://doi.org/10.1111/j.1365-2400.2008.00610.x

Isermann DA, Willis DW, Blackwell BG, Lucchesi DO (2007) Yellow perch in South Dakota: population variability and predicted effects of creel limit reductions and minimum length limits. N Am J Fish Manag 27:918–931. https://doi.org/10.1577/M06-222.1

Jansen WA (1996) Plasticity in maturity and fecundity of yellow perch, *Perca flavescens* (Mitchill): comparisons of stunted and normal-growing populations. Ann Zool Fenn 33:403–415

Jolley JC, Willis DW, Holland RS (2010) Match-mismatch regulation for bluegill and yellow perch larvae and their prey in Sandhill lakes. J Fish Wildl Manag 1:73–85. https://doi.org/10.3996/062010-JFWM-018

Kaemingk MA, Willis DW (2012) Mensurative approach to examine potential interactions between age-0 yellow perch (*Perca flavescens*) and bluegill (*Lepomis macrochirus*). Aqua Ecol 46:353–362. https://doi.org/10.1007/s10452-012-9406-z

Kaemingk MA, Graeb BDS, Willis DW (2014) Temperature, hatch date, and prey availability influence age-0 yellow perch growth and survival. Trans Am Fish Soc 143:845–855. https://doi.org/10.1080/00028487.2014.886622

Kansas Fishes Committee (2014) Kansas fishes. University Press of Kansas, Lawrence

Keast A (1985) The piscivore feeding guild in small freshwater ecosystems. Environ Biol Fish 12:119–129. https://doi.org/10.1007/BF00002764

Koonce JF, Bagenal TB, Carline RF, Hokanson KEF, Nagiec M (1977) Factors influencing year-class strength of percids: a summary and a model of temperature effects. J Fish Res Board Can 34:1890–1899. https://doi.org/10.1139/f77-256

Lott JP (1991) Food habits of yellow perch in eastern South Dakota glacial lakes. Dissertation, South Dakota State University

Newsome GE, Alto SK (1987) An egg-mass census method for tracking fluctuations in yellow perch (*Perca flavescens*) populations. Can J Fish Aquat Sci 44:1221–1232. https://doi.org/10.1139/f87-145

Paukert CP, Willis DW, Klammer JA (2002) Effects of predation and environment on quality of yellow perch and bluegill populations in Nebraska Sandhill lakes. N Am J Fish Manag 22:86–95. https://doi.org/10.1577/1548-8675

Pierce RB, Tomcko CM, Negus MT (2006) Interactions between stocked walleyes and native yellow perch in Lake thirteen, Minnesota: a case history of percid community dynamics. N Am J Fish Manag 26:97–107. https://doi.org/10.1577/M05-034.1

Pope KL, Willis DW, Lucchesi DO (1996) Differential relations of age-0 black crappie and yellow perch to climatological variables in a natural lake. J Freshw Ecol 11:345–350. https://doi.org/10.1080/02705060.1996.9664457

Power M, Van Den Heuvel MR (1999) Age-0 yellow perch growth and its relationship to temperature. Trans Am Fish Soc 128:687–700. https://doi.org/10.1577/1548-8659

Purchase CF, Collins NC, Morgan GE, Shuter BJ (2005) Sex-specific covariation among life-history traits of yellow perch (*Perca flavescens*). Evol Ecol Res 7:549–566

Schoenebeck CW, Brown ML (2010) Potential importance of predation, competition, and prey on yellow perch growth from two dissimilar population types. Prairie Nat 42:32–37

Thorpe JE (1977) Morphology, physiology, behavior, and ecology of *Perca fluviatilis* L and *Perca flavescens* Mitchill. J Fish Res Board Can 34:1504–1514. https://doi.org/10.1577/1548-8659

Willis DW, Isermann DA, Hubers MJ, Johnson BA, Miller WH, St. Sauver TR, Sorensen JS, Unkenholz EG, Wickstrom GA (2001) Growth of South Dakota fishes: a statewide summary with means by region and water type. S D Dep Game Fish Park, Spec Rep 01-05, Pierre

6.25.4   Logperch *Percina caprodes*

(Rafinesque, 1818)

● **Pre-1990**   ● **1990-2021**

Etymology: *Percina caprodes* – *Percina* = perch; *caprodes* = pig-like, referencing the snout shape.

Description:

Body – fusiform, elongated, slightly laterally compressed.
Color –

> *Dorsally* – olive to tan.
> *Laterally* – tan to light olive; 14 to 25 dark olive to gray vertical bars of alternating long and short lengths
> *Ventrally* – cream to white; suborbital may be dark to faint.
> *Fins* – dorsal – dark, dusky speckles forming bands; caudal – dark, dusky speckles forming bands; dark spot at base.

Head – moderate.
Opercle – scaled.
Snout – pointed, conical, elongated, overhangs mouth.
Eyes – moderately large, laterally on upper portion of head.
Mouth – subterminal, horizontal; frenum present.

> *Upper jaw* – does not extend to anterior edge of eye.
> *Lips* – fleshy.

Teeth – small, sharp, on both jaws.
Gill rakers – 12 to 15; short.
Dorsal fin – 2 lobes.

> *Anterior* – 13 to 15 spines.
> *Posterior* – 14 to 17 rays.

Caudal peduncle – elongated, thick.
Caudal fin – square to slightly forked.
Anal fin – 9 to 11 rays; 2 spines.
Pelvic fins – thoracic.
Pectoral fins – 13 to 15 rays
Lateral line – complete; 76 to 82 ctenoid scales in series
Sexual dimorphism – spawning males have an enlarged anal fin, single row of enlarged modified scales ventrally along the midline between pelvic fin insertion and vent, and minute tubercles on the posterior ventral side of body, caudal peduncle, caudal, anal, pelvic, and pectoral fins; females typically lack enlarged modified scales ventrally.

Similar Species: Logperch may be confused with blackside darter and slenderhead darter. Blackside darters have a shorter, less dramatic pointed snout, and six to eight somewhat large and irregular oval shaped blotches almost connected on lateral sides. Slenderhead darters have a light yellow-tangerine to orange submarginal band on the spinous dorsal fin.

Distribution and Habitat: The native range of logperch encompasses Saskatchewan and the Mississippi River system, east through the Great Lakes and St. Lawrence River to Quebec and Vermont, and south to Louisiana and the Rio Grande River system in southern Texas (Cooper 1978). They are native to the Red River of North, Minnesota River, and Big Sioux River systems of the Dakotas. Logperch inhabit lakes and medium-to-large rivers with clear to slightly turbid water, in pools, sandy sholes, or riffles, 0.6 to 1.5 m deep, with little to swift current, and sand, gravel, and cobble substrates (Winn 1958; Bailey and Allum 1962; Morris and Page 1981; Becker 1983). They have been found in the Great Lakes at depths of 9 to 40 m (Trautman 1957; Wells 1968; Becker 1983).

Reproduction: Logperch spawn from early morning to early evening from April to early July (Cross 1967; Eddy and Underhill 1974; Lutterbie 1976; Cooper 1978; Becker 1983). They sexually mature at age-1 or age-2 (Becker 1983). In lakes, spawning grounds have sand or gravel substrate and run parallel to the shore for up to 30 m (Winn 1958). Spawning grounds are much smaller in streams (Winn 1958). To enter the spawning site, ripe females swim through schools of males (Cooper 1978). One or more males court a single female and create a moving territory (Winn 1958). With their tails pressed together, the male clasps the female with pelvic fins in an upright horizontal position, and their bodies vibrate to initiate gamete release and subsequent egg fertilization (Winn 1958; Cooper 1978). Their spawning behavior often causes logperch to become almost completely buried in the bottom substrate (Winn 1958). No nest is prepared, and no parental care is given, and exposed eggs are frequently consumed by other male logperch (Winn 1958; Becker 1983). They are fractional spawners, with females spawning with more than one male (Cooper 1978). After spawning ceases, adult logperch migrate back to deeper waters (Winn 1958). Fecundity increases with female size and age and ranges from 1,000 to 3,000 eggs (Winn 1958). The adhesive, demersal, approximately 1-mm diameter, amber eggs contain one large and several small oil globules (Cooper 1978). In an aquarium, hatching occurred in 200 to 205 hours at a water temperature of 17 °C (Cooper 1978).

Age and Growth: Newly hatched logperch larvae have a total length of approximately 5 mm (Simon 1985). Larvae at total lengths of 18 to 24 mm are classified as juveniles (Simon 1985; Cooper 1978). Total-lengths-at-age of logperch from Illinois were 72 mm at age-1, 106 mm at age-2, and 122 mm at age-3 (Thomas 1970). They can reach a total length of 180 mm and live for 4 years (Trautman 1957).

Food and Feeding: Logperch are sedentary, littoral, benthic, generalized omnivores, with little seasonal dietary variation (Turner 1921; Mullan et al. 1968; Phillips and Kilambi 1996). Adults mainly consume aquatic insects (larvae, pupae, and adults), fish eggs, amphipods, small crustaceans, and small mollusks (Winn 1958; Becker 1983; Phillips and Kilambi 1996). Logperch forage using their conical snouts to turn over pebbles and rocks (Pflieger 1997). Larvae are surface feeders, mainly consuming copepods, cladocerans, and other microcrustaceans (Turner 1921).

References

Bailey RM, Allum MO (1962) Fishes of South Dakota. University of Michigan, Ann Arbor. https://doi.org/10.3998/mpub.9690435

Becker GC (1983) Fishes of Wisconsin. University of Wisconsin Press, Madison

Cooper JE (1978) Eggs and larvae of the logperch, *Percina caprodes* (Rafinesque). Am Midl Nat 99:257–269. https://doi.org/10.2307/2424804

Cross FB (1967) Handbook of fishes in Kansas. University Press of Kansas, Lawrence

Eddy S, Underhill JC (1974) Northern fishes with special reference to the upper Mississippi Valley, 3rd edn. University of Minnesota Press, Minneapolis

Lutterbie GW (1976) The darters (Pisces: Percidae: Ethestomatinae) of Wisconsin. Dissertation, University of Wisconsin-Stevens Point

Morris MA, Page LM (1981) Variation in western logperches (Pisces: Percidae), with description of a new subspecies from the Ozarks. Copeia 1981:95–108. https://doi.org/10.2307/1444044

Mullan JW, Applegate RL, Rainwater WC (1968) Food of logperch (*Percina caprodes*), and brook silverside (*Labidesthes sicculus*), in a New and Old Ozark Reservoir. Trans Am Fish Soc 97:300–305. https://doi.org/10.1577/1548-8659(1968)97[300:FOLPCA]2.0.CO;2

Pflieger WL (1997) The fishes of Missouri, revised edn. Missouri Dep Cons, Jefferson City

Phillips EC, Kilambi RV (1996) Food habits of four benthic fish species (*Ethestoma spectabile, Percina caprodes, Noturus exilis, Cottus carolinae*) from northwest Arkansas streams. Southwest Nat 41:69–73

Simon TP (1985) Descriptions of larval Percidae inhabiting the upper Mississippi River Basin (*Osteichthys: Ethestomatini*). Dissertation, University of Wisconsin

Thomas DL (1970) An ecological study of four darter species of the genus *Percina* (Percidae) in the Kaskaskia River, Illinois. Ill Nat Hist Surv, Biol Notes 70:1–18

Trautman MB (1957) The fishes of Ohio. Ohio State University Press, Columbus

Turner CL (1921) Food of the common Ohio darters. Ohio J Sci 22:41–62

Wells L (1968) Seasonal depth distribution of fish in southeastern Lake Michigan. US Dep Inter, Fish and Wildl Serv, Fish Bull 67:1–15

Winn HE (1958) Comparative reproductive behavior and ecology of fourteen species of darters (Pisces-Percidae). Ecol Monogr 28:155–191. https://doi.org/10.2307/1942207

6.25.5   Blackside Darter *Percina maculata*

(Girard, 1859)

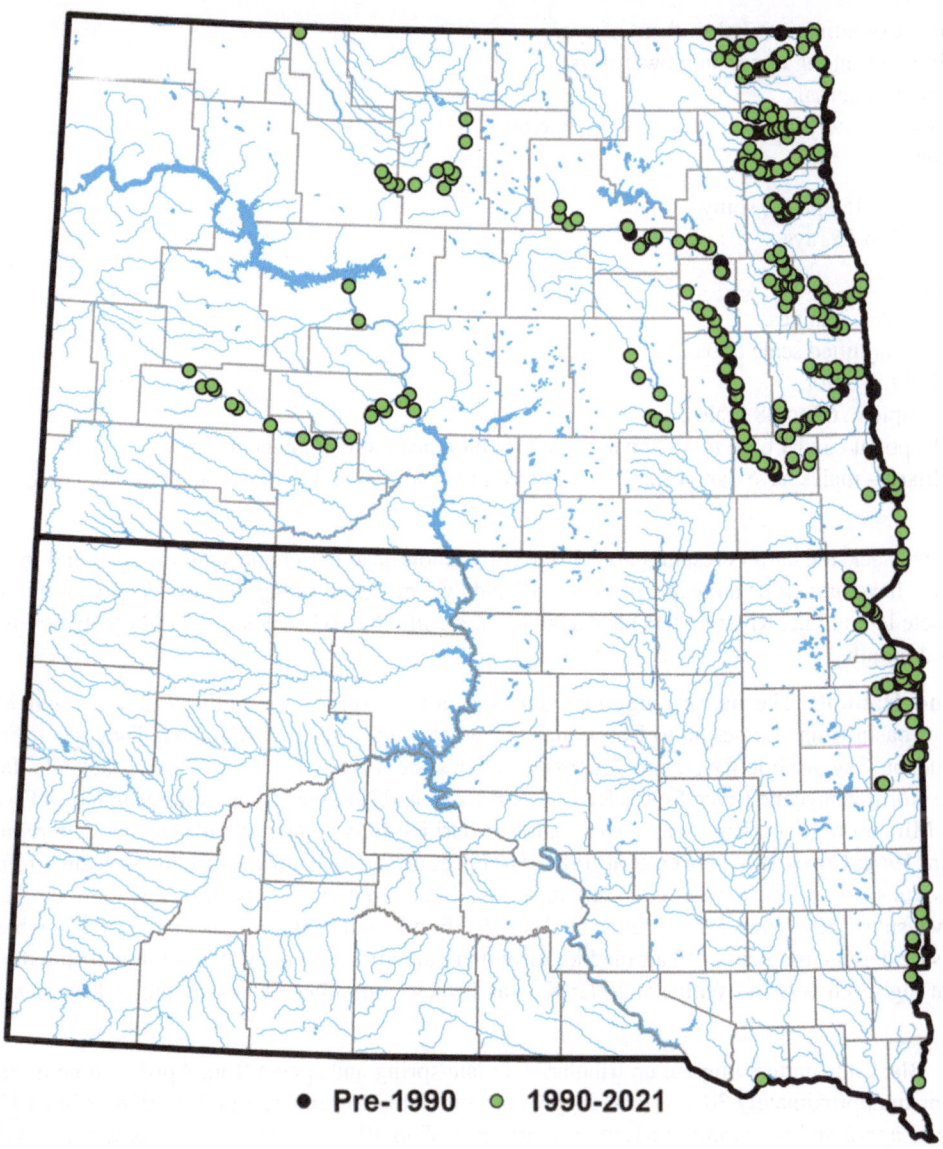

Etymology: *Percina maculata* – *Percina* = perch; *maculata* = spotted, referencing lateral blotches.

Description:

Body – elongated, moderately robust, slightly laterally compressed.
Color –

> *Dorsally* – olive-green to light brown and yellowish; wavy checkerboard design; 6 to 11 dusky crossbars; ctenoid scales.
> *Laterally* – yellow, lateral stripe black to brown running from snout through eye; 6 to 8 somewhat large, irregular oval shaped blotches, almost connected, running longitudinally to the body to the caudal fin base; suborbital bar vertical and dusky.
> *Ventrally* – yellowish white.
> *Fins* – dorsal – dark pigment along the anterior base fading posteriorly, dark dusty spot on first three membranes; caudal – light bar pattern; anal – clear to yellowish-white; pelvic – clear to yellowish.

Head – pointed.
Opercle – cheeks; nape and anterior portion of ventral side few to none embedded scales.
Snout – short, pointed.
Mouth – terminal to subterminal, slightly oblique; upper lip groove noncontinuous over point of snout; frenum; wide on upper lip.

> *Upper jaw* – extending into below front of eye.

Teeth – small, in bands along upper and lower jaws.
Air bladder – small, internal.
Gill rakers – approximately 10; first arched; long, narrow.
Dorsal fin – 2 lobes.

> *Anterior* – 12 to 15 spines; spiny.
> *Posterior* – 12 to 15 rays.

Caudal fin – squarish.
Anal fin – 9 to 11 rays; 2 spines.
Pelvic fins – large, modified scale at base.
Pectoral fins – light in color.
Lateral line – complete; 62 to 68 scales.
Juveniles – dark spot at caudal base more apparent; lateral blotches more connected.
Sexual dimorphism – males have darker fins, a dark brown to gold iris, and a thicker and sturdier spine on anterior edge of pelvic fins.

Similar Species: Blackside darters resemble other darters. Slenderhead darters have an orange band in the anterior dorsal fin. Johnny darters have numerous large and obvious "x" and "w" markings along the lateral sides of the body. Iowa darters have an incomplete lateral line. River darters have a lateral line with 46 to 62 scales, and either lack a frenum or it is not visible with a closed mouth.

Distribution and Habitat: The native range of blackside darter spans from southern Ontario and Manitoba into the Mississippi River basin from Minnesota to Louisiana, and east to Arkansas and Alabama. They are found throughout the Great Lakes, Hudson Bay, Mobile River, and Tennessee River systems (Mettee et al. 1996). Blackside darters occur in the Red River of the North, Sheyenne River, Souris River, Heart River and Missouri River systems in North Dakota, and the Bois de Sioux River, Minnesota River, Big Sioux River, and Missouri River systems in South Dakota (Churchill and Over 1938; Bailey and Allum 1962; Copes 1965; Owen et al. 1981; Hrabik et al. 2015). They are rare in lakes and prefer undercut banks in shallow pools, runs, and riffles in clear to slightly turbid medium-to-large streams with slow to moderate flow, vegetation, and sand or gravel substrates (Winn 1958; Mettee et al. 1996; Hrabik et al. 2015). Unlike most darters which tend to be near substrate, blackside darters use their small air bladder to swim in the middle of the water column during the day and then rest on the bottom at night (Gilbert and Williams 2002; Kansas Fishes Committee 2014). Juveniles hide in vegetation to avoid predators.

Reproduction: Blackside darters migrate up tributaries in late spring and spawn from April to June in gravel or sand substrate depressions in approximately 30 to 60 cm deep water with slow currents (Petravicz 1938; Winn 1958). They become sexually mature at age-2 and very rarely mature as yearlings (Winn 1958). Spawning begins when females swim into the substrate depression. A male creates a moving territory around the female and becomes defensive to other blackside darter males. After the male clasps the female, the pair vibrate together in a horizontal position for several seconds and gametes are released (Winn 1958). The adhesive, approximately 2-mm diameter, fertilized eggs are clear or transparent with a colorless oil globule (Winn 1958). Eggs remain unattended in the depression until hatch (Petravicz 1938; Winn 1958). Fecundity ranges from 1,000 to 2,000 eggs. Females spawn with several males during the season, releasing at least 10 eggs each time (Winn 1958). Larvae hatch in 6 to 8 days, and drift in the water column for up to 3 weeks (Hrabik et al. 2015).

Age and Growth: Newly hatched blackside darters have a total length of approximately 6 mm (Petravicz 1938). Total lengths for age-1 blackside darters from Iowa and Wisconsin were 35 mm and 47 to 51 mm, respectively (Karr 1963; Becker 1983; Petravicz 1938). Age-4 total lengths from the same two states were 77 to 79 mm and 101 to 108 mm (Karr 1963; Becker 1983; Petravicz 1938). They are generally larger than other darters. Males and females are similar in size, with total lengths ranging from 51 to 109 mm (Petravicz 1938; Owen et al. 1981; Becker 1983). Blackside darters can live up to 4 years (Karr 1963; Becker 1983).

Food and Feeding: Blackside darters mainly forage on, or just below, the surface of the water for mayfly larvae, caddisfly larvae, blackfly larvae, chironomids, copepods, and small amounts of plant material (Petravicz 1938; Owen et al. 1981; Kansas Fishes Committee 2014). They also eat other small fish. Juveniles primarily eat copepods and other very small crustaceans.

References

Bailey RM, Allum MO (1962) Fishes of South Dakota. University of Michigan, Ann Arbor. https://doi.org/10.3998/mpub.9690435

Becker GC (1983) Fishes of Wisconsin. University of Wisconsin Press, Madison

Churchill EP, Over WH (1938) Fishes of South Dakota. Brown & Saenger Printers, Sioux Falls

Copes FA (1965) Fishes of the Red River tributaries of North Dakota. Dissertation, University of North Dakota

Gilbert CR, Williams JD (2002) National Audubon Society field guide to fishes: North America, revised edn. Knopf Doubleday Publishing Group, New York

Hrabik RA, Schainost SC, Stasiak RH, Peters EJ (2015) The fishes of Nebraska. University of Nebraska, Lincoln

Kansas Fishes Committee (2014) Kansas fishes. University Press of Kansas, Lawrence

Karr JR (1963) Age, growth, and food habits of johnny, slenderhead, and blackside darters of Boone County, Iowa. Proc Iowa Acad Sci 70:228–236

Mettee MF, O'Neil PE, Pierson JM (1996) Fishes of Alabama and the Mobile Basin. Oxmoor House, Tuscaloosa

Owen JB, Elsen DS, Russell GW (1981) Distribution of fishes in North and South Dakota Basins affected by the Garrison Diversion Unit. University of North Dakota Press, Grand Forks

Petravicz WP (1938) The breeding habits of the Black-sided Darter, *Hadropterus maculatus* Girard. Copeia 1938:40–44. https://doi.org/10.2307/1435522

Winn HE (1958) Comparative reproductive behavior and ecology of fourteen species of darters (Pices-Percidae). Ecol Monogr 28:155–191. https://doi.org/10.2307/1942207

6.25.6 Slenderhead Darter *Percina phoxocephala*

(Nelson, 1876)

● **Pre-1990**   ○ **1990-2021**

Etymology: *Percina phoxocephala* – *Percina* = perch; *phoxocephala* = tapered head.

Description:

Body – elongated, slightly laterally compressed.
Color –

> *Dorsally* – light brown to tan.
> *Laterally* – light olive to tan; 10 to 16 faint, dark, dusky blotches, dark, backward extending lateral stripe on head from tip of snout through eye.
> *Ventrally* – cream to white.
> *Fins* – anterior dorsal – light yellow-tangerine to orange submarginal band; caudal – dark spot at base.

Head – elongated.
Opercle – scaled.
Snout – conical, moderately pointed.
Eyes – large, dorsolaterally on head.
Mouth – terminal to slightly subterminal; frenum.

> *Upper jaw* – extends to anterior end of eye.

Teeth – small, sharp, in bands on upper and lower jaws.
Gill rakers – 9 to 15.
Dorsal fin – 2 lobes.

> *Anterior* – 11 or 12 spines.
> *Posterior* – 11 to 14 rays.

Caudal peduncle – elongated, slender.
Caudal fin – square to slightly forked.
Anal fin – 8 or 9 rays; 2 spines.
Pelvic fins – thoracic.
Pectoral fins – 12 to 15 rays.
Lateral line – complete; 60 to 75 ctenoid scales in series.
Juveniles – may have a larger head and eyes in relation to body size.
Sexual dimorphism – males have scaled breast on posterior end only, enlarged scales from mid-pelvic area to the anus, and bright orange submarginal band on spinous dorsal fin; females have a naked breast except for 1 or 2 large ctenoid scales.

Similar Species: Slenderhead darters are easily distinguished from other darter species in the Dakotas by the light yellow-tangerine to orange submarginal band on the spinous dorsal fin. Iowa darter lack a complete lateral line and have a distinct black suborbital bar. Blackside darter have six to eight large, irregular, oval-shaped blotches on lateral sides to the caudal fin base. River darter lack a frenum and have a distinct black suborbital bar. Johnny darter have dark "w" or "x" shaped markings on lateral sides and a blunter snout.

Distribution and Habitat: The native range of Slenderhead darter extends from the Minnesota River in northeastern South Dakota, throughout the Upper Mississippi River system to western Pennsylvania, and south to northern Alabama and northern Oklahoma (Page and Smith 1971). They are mainly absent from the Great Lakes (Page and Smith 1971). Slenderhead darters are absent from North Dakota and only found in the Minnesota River system in South Dakota. They are most common in shallow, clear to turbid waters near riffles or swift current in small to medium, moderate-gradient streams with sand or gravel substrate (Trautman 1957; Cross 1967; Page and Smith 1971). Slenderhead darters migrate to deeper areas after spawning and remain there during the winter (Page and Smith 1971). They are intolerant of high siltation but more tolerant of turbidity than other darter species (Trautman 1957; Thompson 1980; Miller and Robinson 2004).

Reproduction: Slenderhead darters spawn from April to June when water temperature hovers near 21 °C over 15 to 60 cm deep riffles with gravel substrate (Page and Smith 1971). High water levels and changing water temperatures may postpone or interrupt spawning (Thomas 1970; Page and Smith 1971). Males migrate to spawning locations from late March through early April and likely establish territories (Page and Smith 1971). Females arrive later. Spawning colors develop more quickly in larger males (Page and Smith 1971). Slenderhead darters sexually mature at standard lengths from 40 to 80 mm (Page and Smith 1971). Fecundity ranges from 50 to 1,000 eggs, with larger females producing more eggs (Karr 1963; Page and Smith 1971). The adhesive, approximately 1-mm diameter eggs are transparent with one large oil droplet (Page and

Smith 1971). Eggs hatch in 14 days. Juveniles remain in spawning habitat for about a month after hatch and then migrate to deeper water in the main channel (Page and Smith 1971).

Age and Growth: At 2 weeks post-hatch, Slenderhead darter larvae have a total length of approximately 22 mm (Page and Smith 1971). Total lengths at age of Slenderhead darter from Iowa were 34 mm at age-1, 47 mm at age-2, 49 mm at age-3, and 56 mm at age-4 (Karr 1963). Size and sex are not related (Page and Smith 1971). They can reach a total length of approximately 100 mm and live for 4 years (Trautman 1957).

Food and Feeding: Slenderhead darters primarily feed during the day, with little feeding at night (Thomas 1970). Most of the food eaten for the entire year is consumed in the spring prior to spawning, with minimal feeding after spawning and during the winter (Page and Smith 1971). Juveniles and young primarily eat midge larvae (Thomas 1970). Adults primarily eat midge larvae, black fly larvae, caddisfly larvae, and mayfly larvae, and less frequently eat terrestrial insects, fish eggs, and amphipods (Page and Smith 1971).

References

Cross FB (1967) Handbook of fishes in Kansas. University Press of Kansas, Lawrence

Karr JR (1963) Age, growth, and food habits of Johnny, slenderhead and blacksided darters of Boone County, Iowa. Proc Iowa Acad Sci 70:228–236

Miller RJ, Robinson HW (2004) Fishes of Oklahoma, revised edn. University of Oklahoma Press, Norman

Page LM, Smith PW (1971) The life history of the slenderhead darter, *Percina phoxocephala*, in the Embarras River, Illinois. Illinois Dep Reg Ed, Nat Hist Surv Div. https://doi.org/10.5962/bhl.title.15873

Thomas DL (1970) An ecological study of four darter species of the genus *Percina* (Percidae) in the Kaskaskia River, Illinois. Ill Nat Hist Surv, Biol Notes 70:1–18

Thompson BA (1980) *Percina phoxocephala* (Nelson), slenderhead darter. In: Lee DS, Gilbert CR, Hocutt CH, Jenkins RE, McAllister DE, Stauffer JR Jr (eds) Atlas of North American freshwater fishes. North Carolina Biological Survey Publication 1980-12, p 737

Trautman MB (1957) The fishes of Ohio. Ohio State University Press, Columbus

6.25.7 River Darter *Percina shumardi*

(Girard, 1859)

Nuptial Male   Female

● **Pre-1990**   ○ **1990-2021**

Etymology: *Percina shumardi* – *Percina* = perch; *shumardi* = referencing George C. Shumard, a U.S. Pacific Railroad Survey surgeon and naturalist who discovered the species.

Description:

Body – fusiform, elongated, slightly laterally compressed.
Color – thick, black suborbital bar.

> *Dorsally* – dark olive-brown; 5 to 7 dark diffuse saddles or large irregular shaped spots.
> *Laterally* – light olive to tan; 8 to 13 dark lateral bars anteriorly, forming blotches posteriorly.
> *Ventrally* – cream to white.
> *Fins* – anterior dorsal – distinct, dark gray to black spot on first membrane posterior to first spine and a dark blotch on membranes on the posterior end; posterior dorsal – small speckles forming bars; caudal – small speckles forming bars; faint spot.

Head – elongated.
Opercle – scaled nape of neck, few to no scales.
Snout – short, rounded.
Eyes – moderately large, dorsolaterally on head.
Mouth – terminal to slightly subterminal, slightly oblique; frenum absent or weakly visible when mouth is closed.

> *Upper jaw* – extends to anterior edge of eye.

Teeth – small, sharp, in bands on upper and lower jaws.
Gill rakers – 12 to 17; short.
Dorsal fin – 2 lobes.

> *Anterior* – 10 to 12 spines.
> *Posterior* – 11 to 16 rays.

Caudal peduncle – elongated, slender.
Caudal fin – square to slightly forked.
Anal fin – 10 to 13 rays; 2 spines.
Pelvic fins – thoracic.
Pectoral fins – 13 to 15 rays.
Lateral line – complete; 46 to 60 ctenoid scales in series.
Sexual dimorphism – spawning males are generally darker and have an enlarged anal fin, small tubercles on head, pelvic, anal, and caudal fins, and modified ventral scales between pelvic fins and vent; females lack or have fewer ventral scales.

Similar Species: River darters are easily distinguished from other darter species in the Dakotas by the thick, black suborbital bar and distinct dark gray to black spot and blotch on anterior dorsal fin. Blackside darters have a more pointed head and six to eight large and irregular oval-shaped blotches running longitudinally on the lateral sides. Iowa darters have 9 to 12 prominent vertical bars on lateral sides and dorsal fins with bands of color. Johnny darters have dark "w" or "x" shaped markings on lateral sides. Slenderhead darters have a light yellow-tangerine to orange submarginal band on the spinous dorsal fin and a conical, moderately pointed snout.

Distribution and Habitat: The native range of river darter extends from Manitoba to Ontario, east to Ohio, and south to the Gulf of Mexico. It includes the Mississippi River, Mobile Bay River, Great Lakes, and Hudson Bay systems. In the Dakotas, river darters have only been documented from the Red River of the North in North Dakota. They inhabit large rivers and the rocky shorelines of lakes and reservoirs (Becker 1983; Etnier and Starnes 1993). Smaller streams also provide suitable habitat, especially during winter and the spawning season (Etnier and Starnes 1993). River darters occur most frequently in moderate to swift current, like chutes and riffles, 2 to 5 m deep over sand, gravel, or cobble substrates (Pratt et al. 2016; Scott and Crossman 1973; Thomas 1970). Juveniles prefer locations alongside the main channel with little to moderate current and sand or gravel substrate (Simon 1985). River darters from Manitoba and Ontario were sampled in water temperatures from 9 to 16 °C, but they have been observed in temperatures up to 26 °C (Pratt et al. 2016). They tolerate continuously high turbidity (Pflieger 1997; Pratt et al. 2016).

Reproduction: River darter spawn in April in Kansas, from April to May in Illinois, April to June in North Dakota and Wisconsin, and June to July in Manitoba (Cross 1967; Thomas 1970; Scott and Crossman 1973; Becker 1983). Spawning begins at a water temperature of 10 °C in the upper Mississippi River basin (Simon 1985). They become sexually mature at age-1 (Thomas 1970; Pratt et al. 2016). River darter spawn over gravel or rubble substrate in high velocity at water depths greater than 0.5 m (Simon 1985). They are fractional spawners, depositing more than one clutch of eggs during a season (Simon 1985). The demersal, slightly adhesive, approximately 1 mm diameter, translucent eggs with a single oil globule are typically buried within sand and gravel (Becker 1983; Simon 1985). Eggs hatched in an aquarium in 6 or 7 days at a water temperature of 22 °C (Simon 1985).

Age and Growth: Newly hatched river darter larvae have a total length of approximately 4 mm (Simon 1985). Young-of-year from Illinois were 36 to 43 mm long in June, with age-1 and age-2 river darter from Wisconsin 50 mm and 67 mm long, respectively (Thomas 1970; Lutterbie 1976). River darters from Manitoba and northwestern Ontario grow approximately 10 mm per year (Pratt et al. 2016). Typical total lengths range from 58 to 84 mm, but river darter can reach 89 mm (Pflieger 1997). Longevity is 4 years (Smith 1979; Pratt et al. 2016).

Food and Feeding: River darter feed during daylight, and their diet may change seasonally based on availability (Pratt et al. 2016). Adults mainly eat dipterans, trichopterans, ephemeropterans, crustaceans, fish eggs, and zooplankton (Thomas 1970; Balesic 1971; Pratt et al. 2016). Juveniles primarily eat microcrustaceans. Snails may also been an important dietary component (Balesic 1971; Pratt et al. 2016).

References

Balesic H (1971) Comparative ecology of four species of darters (Etheostominae) in Dauphin and its tributary, the Valley River. Dissertation, University of Manitoba

Becker GC (1983) Fishes of Wisconsin. University of Wisconsin Press, Madison

Cross FB (1967) Handbook of fishes in Kansas. University Press of Kansas, Lawrence

Etnier DA, Starnes WC (1993) The fishes of Tennessee. University of Tennessee Press, Knoxville

Lutterbie GW (1976) The darters (Pisces: Percidae; Ethestomatinae) of Wisconsin. Dissertation, University of Wisconsin

Pflieger WL (1997) The fishes of Missouri, revised edn. Missouri Dep Cons, Jefferson City

Pratt TC, Gardner WM, Watkinson DA, Bouvier LD (2016) Ecology of the river darter in Canadian waters: distribution, relative abundance, life-history traits, diet, and habitat characteristics. Diversity 8:22. https://doi.org/10.3390/d8040022

Scott WB, Crossman EJ (1973) Freshwater fishes of Canada. Fisheries Research Board of Canada, Bulletin 184

Simon TP (1985) Descriptions of the larval Percidae inhabiting the upper Mississippi River basin (Osteichthyes: Etheostomatini). Dissertation, University of Wisconsin

Smith PW (1979) The fishes of Ohio. University of Illinois Press, Urbana

Thomas DL (1970) An ecological study of four darter species of the genus *Percina* (Percidae) in the Kaskaskia River, Illinois. Ill Nat Hist Survey, Biol Notes 070

6.25.8   Sauger *Sander canadensis*

(Griffith & Smith, 1834)

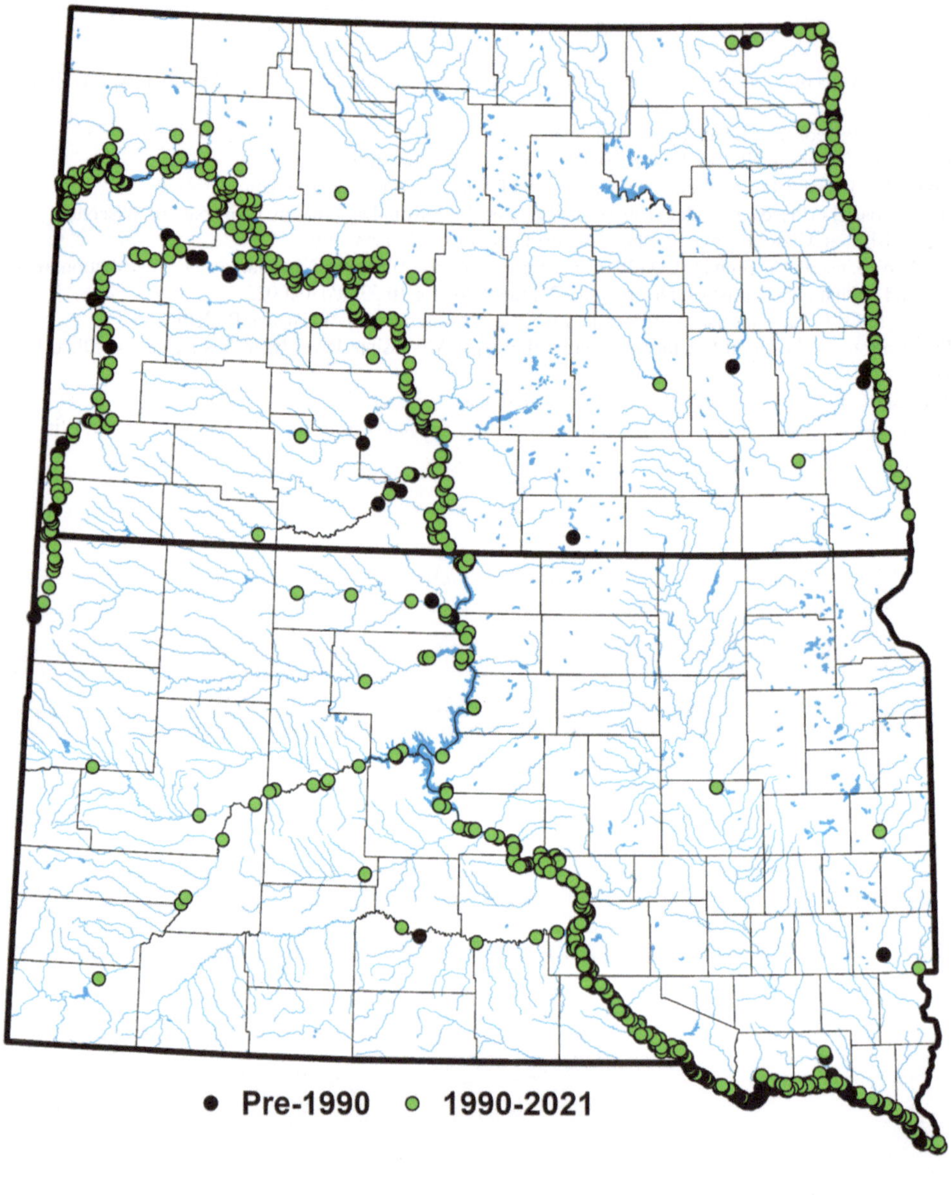

● **Pre-1990**   ○ **1990-2021**

Etymology: *Sander canadensis* – *Sander* = pike-perch, referencing resemblance to the nonrelated pike family (Esocidae); *canadensis* = of Canada.

Description:

Body – fusiform, elongated, cylindrical in cross section, slightly laterally compressed.
Color –

> *Dorsally* – dark gray to brown; dark, diffused saddles extending down lateral sides.
> *Laterally* – dark gray to golden brown; small, speckled dark spots.
> *Ventrally* – cream to white.
> *Fins* – anterior dorsal – distinct, black spots arranged in horizontal rows;
> posterior dorsal – rows of small, dark spots.

Head – large, elongated.
Opercle – cheek scaled.
Snout – conical, blunt.
Eyes – large, glassy; laterally on upper portion of head; silver reflecting layer (*tapetum lucidum*).
Mouth – large, terminal, oblique; jaws nearly equal in length.

> *upper jaw* – extends to posterior edge of eye.

Teeth – large, sharp, canine, on both jaws.
Gill rakers – 7 to 11.
Dorsal fin – 2 lobes.

> *Anterior* – 12 or 13 spines.
> *Posterior* – 17 to 19 rays; 1 or 2 spines.

Caudal peduncle – elongated, moderately thick.
Caudal fin – slightly forked.
Anal fin – 11 to 13 rays; 2 spines; insertion posterior to posterior dorsal fin insertion.
Pelvic fins – thoracic.
Pectoral fins – 14 to 16 rays.
Lateral line – complete; 77 to 87 ctenoid scales in series.
Juveniles – more elongated.

Similar Species: Sauger closely resemble and occur sympatrically with walleye. Walleye have a slightly deeper body, cheeks with few to no scales, white tip on the caudal fin lower lobe, and lack a spotted anterior dorsal fin. Zander lack an opercular spine and have few to no cheek scales.

Distribution and Habitat: The native range of sauger encompasses southeastern Alberta, Montana, and Wyoming, east to southwestern Quebec and Vermont, and south to the southern Mississippi River in Louisiana, and the Tennessee River in northern Alabama (Scott and Crossman 1973; Billington et al. 2011). Sauger are continuously distributed throughout the mainstem of the Missouri River, including most of its major tributaries in the Dakotas. Sauger are also in the Red River of the North and Sheyenne Rivers in eastern North Dakota. Sauger populations in the Dakotas are stable, but have declined elsewhere because of hybridization with walleye, exploitation, and dam construction blocking access to historical spawning sites and altering thermal and hydrologic regimes (Pegg et al. 1996; Nelson and Walburg 1977; Graeb 2006; Bellgraph et al. 2008; Bozek et al. 2011b). Sauger prefer the deep, highly turbid, main channels of large rivers with sand and silt substrates but also inhabit large, shallow lakes (Gangl et al. 2000; Bozek et al. 2011b). Young of year select water depths of less than 3.7 m (Nelson 1968a). Sauger have an optimal water temperature of 20 to 22 °C (Bozek et al. 2011b).

Reproduction: In Missouri River reservoirs, sauger spawn in deltas, secondary channels, and tributary streams under 1.5 m deep, with warmer water temperatures, flowing water, and high turbidity (Graeb 2006). Upstream spawning migrations can be quite long, with movements of up to 260 km observed in the middle Missouri River (Bozek et al. 2011b). Adults return to their original location after the spawning. Sauger sexually mature at age-3 or age-4, with females maturing later than males (Nelson 1969). They spawn at temperatures similar to, or warmer than, walleyes (Bozek et al. 2011b). Sauger spawn in Lewis and Clark Lake, South Dakota in the evenings from late April to early May, beginning at a water temperature of 6 °C (Nelson 1968a). Peak spawning activity lasts for 5 to 7 days, but spawning may occur for 14 days (Nelson 1968a). Sauger broadcast spawn and do not construct nests nor provide parental care. While fecundity increases with fish length, egg quality decreases (Graeb et al. 2007). Sauger from Lewis and Clark Lake averaged 65,250 eggs/kg of body weight (Nelson 1969). The demersal, adhesive, approximately 1 to 2 mm diameter eggs hatch in 21 days at water temperatures of 8 to 9 °C (Nelson 1968a, b; Walburg 1972; Bozek et al. 2011a). In Missouri River reservoirs, annual recruitment may be higher in years with reduced flow and warmer spring and early summer water temperatures (Graeb 2006). Year-class strength depends on fluctuating water levels (Nelson 1968a).

Age and Growth: Newly hatched sauger larvae have a total length of approximately 5 mm (Nelson et al. 1965). Females grow faster than males and are often larger as adults. Total lengths at age-1 range from 74 to 244 mm (Bozek et al. 2011a). Optimal growth occurs at a temperature of 22 °C (Bozek et al. 2011a). Sauger can reach a total length of 711 mm and weight of 5.4 kg. In the northern part of their range, they can live for 13 years (Scott and Crossman 1973; Etnier and Starnes 1993).

Food and Feeding: Sauger are top-level predators. Larvae with total lengths from 17 to 80 mm primarily eat copepods (Nelson 1968b). Larger larvae and juveniles consume benthic invertebrates, copepods, and cladocerans (Nelson 1968b). Fish become the primary food when sauger reach a total length of 70 to 110 mm (Nelson 1968a). Adult diets are similar to walleye and consist primarily of fish-like gizzard shad, yellow perch, freshwater drum, and rainbow smelt (Scott and Crossman 1973; Swenson 1977; Mero 1992; Graeb 2006; Bellgraph et al. 2008). Adults will also continue to consume chironomids and other invertebrates (Scott and Crossman 1973; Swenson 1977; Mero 1992; Graeb 2006; Bellgraph et al. 2008). Sauger diets change seasonally with changes in prey abundance and availability (Mero 1992).

References

Bellgraph BJ, Guy CS, Gardner WM, Leathe SA (2008) Competition potential between saugers and walleye in nonnative sympatry. Trans Am Fish Soc 137:790–800. https://doi.org/10.1577/T07-102.1

Billington N, Wilson CC, Sloss BL (2011) Distribution and population genetics of walleye and sauger. In: Barton BA (ed) Biology, management, and culture of walleye and sauger. American Fisheries Society, Bethesda, pp 105–132. https://doi.org/10.47886/9781934874226.ch4

Bozek MA, Baccante DA, Lester NP (2011a) Walleye and sauger life history. In: Barton BA (ed) Biology, management, and culture of walleye and sauger. American Fisheries Society, Bethesda, pp 233–286. https://doi.org/10.47886/9781934874226.ch7

Bozek MA, Haxton TJ, Raabe JK (2011b) Walleye and sauger habitat. In: Barton BA (ed) Biology, management, and culture of walleye and sauger. American Fisheries Society, Bethesda, pp 133–179. https://doi.org/10.47886/9781934874226.ch5

Etnier DA, Starnes WC (1993) The fishes of Tennessee. University of Tennessee Press, Knoxville

Gangl RS, Pereira DL, Walsh RJ (2000) Seasonal movements, habitat use, and spawning areas of walleye *Stizostedion vitreum* and sauger *S. canadense* in pool 2 of the upper Mississippi River. Minnesota Dep Nat Res, Inv Rep 482, St. Paul

Graeb BDS (2006) Sauger population ecology in three Missouri River mainstem reservoirs. Dissertation, South Dakota State University

Graeb BDS, Kaemingk MA, Willis DW (2007) Early life history of sauger in Missouri River reservoirs. SD Dep Game Fish Parks, Compl Rep 07-08, Pierre

Mero SW (1992) Food habits of walleye and sauger in Lake Sakakawea, North Dakota. Dissertation, South Dakota State University

Nelson WR (1968a) Reproduction and early life history of sauger, *Stizostedion canadense*, in Lewis and Clark Lake. Trans Am Fish Soc 97:159–166. https://doi.org/10.1577/1548-8659(1968)97[159:RAELHO]2.0.CO;2

Nelson WR (1968b) Embryo and larval characteristics of sauger, walleye, and their reciprocal hybrids. Trans Am Fish Soc 97:167–174. https://doi.org/10.1577/1548-8659(1968)97[167:EALCOS]2.0.CO;2

Nelson WR (1969) Biological characteristics of the sauger population in Lewis and Clark Lake, No. 21. US Dep Inter, Fish Wildl Serv

Nelson WR, Walburg CH (1977) Population dynamics of yellow perch (*Perca flavescens*), sauger (*Stizostedion canadense*), and walleye (*S. vitreum*) in four main-stem Missouri River reservoirs. J Fish Res Board Can 34:1748–1763. https://doi.org/10.1139/f77-240

Nelson WR, Hines NR, Beckman LG (1965) Artificial propagation of saugers and hybridization with walleyes. Prog Fish Cult 27:216–218. https://doi.org/10.1577/1548-8640(1965)27[216:APOSAH]2.0.CO;2

Pegg MA, Layzer JB, Bettoli PW (1996) Angler exploitation and movements of anchor-tagged sauger in the lower Tennessee River. N Am J Fish Manag 16:218–223. https://doi.org/10.1577/1548-8675

Scott WB, Crossman EJ (1973) Freshwater fishes of Canada. Fisheries Research Board of Canada, Bulletin 184

Swenson WA (1977) Food consumption of walleye (*Stizostedion vitreum*) and sauger (*S. canadense*) in relation to food availability and physical conditions in Lake of the Woods, Minnesota, Shagwa Lake, and western Lake Superior. J Fish Res Board Can 34:1643–1654. https://doi.org/10.1139/f77-229

Walburg CH (1972) Some factors associated with fluctuation in year-class strength of sauger, Lewis and Clark Lake, South Dakota. Trans Am Fish Soc 101:311–316. https://doi.org/10.1577/1548-8659

6.25.9   Zander *Sander lucioperca*

(Linnaeus, 1758)

● **Pre-1990**   ○ **1990-2021**

Etymology: *Sander lucioperca* – *Sander* = pike perch, referencing resemblance to the nonrelated pike family (Esocidae); *lucioperca* = pike perch likely suggesting the species resemblance to both the pike family (Esocidae) and perch family (Percidae). Commonly referred to as the pikeperch.

Description:

Body – fusiform, elongated, laterally compressed.
Color –

> *Dorsally* – dark olive-green to gray.
> *Laterally* – silvery-green to gray; 8 to 12 dark transverse bars running vertically along the body.
> *Ventrally* – white.
> *Fins* – translucent olive to gray; dorsal, caudal – rows of dark spots.

Head – moderate, conical.
Opercle – strongly serrated preopercle.
Snout – elongate, blunt tip.
Eyes – large; laterally on upper portion of head; silver reflecting layer (*tapetum lucidum*).
Mouth – large, terminal.
Teeth – large; pronounced canine teeth arranged in narrow rows on both jaws.
Dorsal fin – 2 lobes.

> *Anterior* – 13 to 20 spines.
> *Posterior* – 18 to 24 soft rays; 1 or 2 spines.

Caudal peduncle – thick, short.
Caudal fin – 17 soft rays; deeply forked, rounded distal ends.
Anal fin – 10 to 14 soft rays; 2 or 3 spines; insertion directly under or slightly posterior to posterior dorsal fin insertion.
Pelvic fins – insertions anterior to anterior dorsal fin insertion.
Lateral line – complete; 80 to 97 cycloid scales in series.

Similar Species: Zander closely resemble walleye and sauger in the Dakotas. Walleye have a prominent opercular spine and a white patch on the caudal fin lower lobe. They lack rows of dark spots on the dorsal fin membranes. Sauger have a prominent opercular spine, and the rows of spots on the dorsal fin membranes are more of a sandy, dull brown color with three to four, dark, oblong saddle band marks.

Distribution and Habitat: Native to coastal brackish waters, large rivers, and lakes of continental Europe to western Siberia, zanders have the most widespread distribution of the three species of Eurasian *Sander* species (Fuller 2011; Haponski and Stepien 2013; Larsen and Berg 2014). They have been widely introduced throughout Eurasia for both recreational and commercial fishing (Haponski and Stepien 2013; Larsen and Berg 2014). In 1989, a one-time stocking of 180,000 fry and 1,050 juveniles occurred in Spiritwood Lake, Stutsman County, North Dakota to create a recreational fishery (Haponski and Stepien 2013). Spiritwood Lake is a completely enclosed water body. Although angler catches of zander are rare in Spiritwood Lake, adults and young-of-the-year are occasionally captured (Fuller 2011). Zander thrive in moderately eutrophic, turbid waters with high dissolved oxygen levels (Larsen and Berg 2014).

Reproduction: Zander spawn at dawn or night during April and May at water temperatures of 10 to 14 °C in lakes and rivers with sand, gravel, or clay substrates (Kottelat and Freyhof 2007; Larsen and Berg 2014). Spawning migrations are typically 10 to 30 km, but can be up to 250 km, and begin a month prior to spawning (Lappalainen et al. 2003; Larsen and Berg 2014). In Europe, zander in brackish waters move inland to freshwater for spawning (Larsen and Berg 2014). Males construct approximately 0.5-m diameter, 5 to 10 cm deep nests in sand or stone substrates (Lappalainen et al. 2003). Males also guard and fan the eggs and fry (Lappalainen et al. 2003). Spawning is monogamous, with a single female releasing all of her eggs at once with a single male (Haponski and Stepien 2013). Fecundity varies greatly depending on female size and food supply (Lappalainen et al. 2003). The approximately 1-mm diameter egg adhere to exposed plant roots in the nest (Lappalainen et al. 2003). Zander sexually mature from age-2 to age-5, depending on growth rates (Lappalainen et al. 2003; Larsen and Berg 2014). Optimal egg incubation water temperatures are from 12 to 20 °C (Lappalainen et al. 2003; Larsen and Berg 2014). Hatching occurs at approximately 110-degree days (Larsen and Berg 2014).

Age and Growth: Newly hatched zander larvae have a total length of approximately 5 mm (Larsen and Berg 2014). A large zander from Spiritwood Lake was 889 mm long and weighed 7.2 kg. They can reach a length of 1,300 mm and weight of 20 kg (Larsen and Berg 2014). In Europe, northern populations generally grow slower and live longer than southern populations, reaching ages of 20 to 24 years compared to only 8 or 9 years (Lappalainen et al. 2003; Larsen and Berg 2014).

Food and Feeding: Zander larvae eat small zooplankton but start piscivory behavior between the lengths of 10 to 25 mm. After reaching a length of 100 mm, they eat almost entirely fish (Larsen and Berg 2014). Adults are top predator, pelagic piscivores, feeding heavily on smaller fish like pelagic minnows and smelt. Because of this piscicory, zander are considered a pest in some waters and have also been used to reduce the number of unwanted fish (Lappalainen et al. 2003).

References

Fuller P (2011) *Sander lucioperca* (Linnaeus, 1758). US Geological Survey, Nonindigenous Aquatic Species Database, Gainesville, Florida. https://nas.er.usgs.gov/queries/FactSheet.aspx?SpeciesID=830. Accessed 24 Apr 2023

Haponski AE, Stepien CA (2013) Phylogenetic and biogeographical relationships of the sander pikeperches (Percidae: Perciformes): patterns across North America and Eurasia. Biol J Linn S 110:156–179. https://doi.org/10.1111/bij.12114

Kottelat M, Freyhof J (2007) Handbook of European freshwater fishes. Kottelate, Cornol

Lappalainen J, Dörner H, Wysujack K (2003) Reproduction biology of pikeperch (*Sander lucioperca* (L)) – a review. Ecol Freshw Fish 12:95–106. https://doi.org/10.1034/j.1600-0633.2003.00005.x

Larsen LK, Berg S (2014) *Stizostedion lucioperca* - NOBANIS – Invasive alien species fact sheet. https://www.nobanis.org/globalassets/speciesinfo/s/sander-lucioperca/sander_lucioperca_2014.pdf. Accessed 30 Oct 2019

6.25.10   Walleye *Sander vitreus*

(Mitchill, 1818)

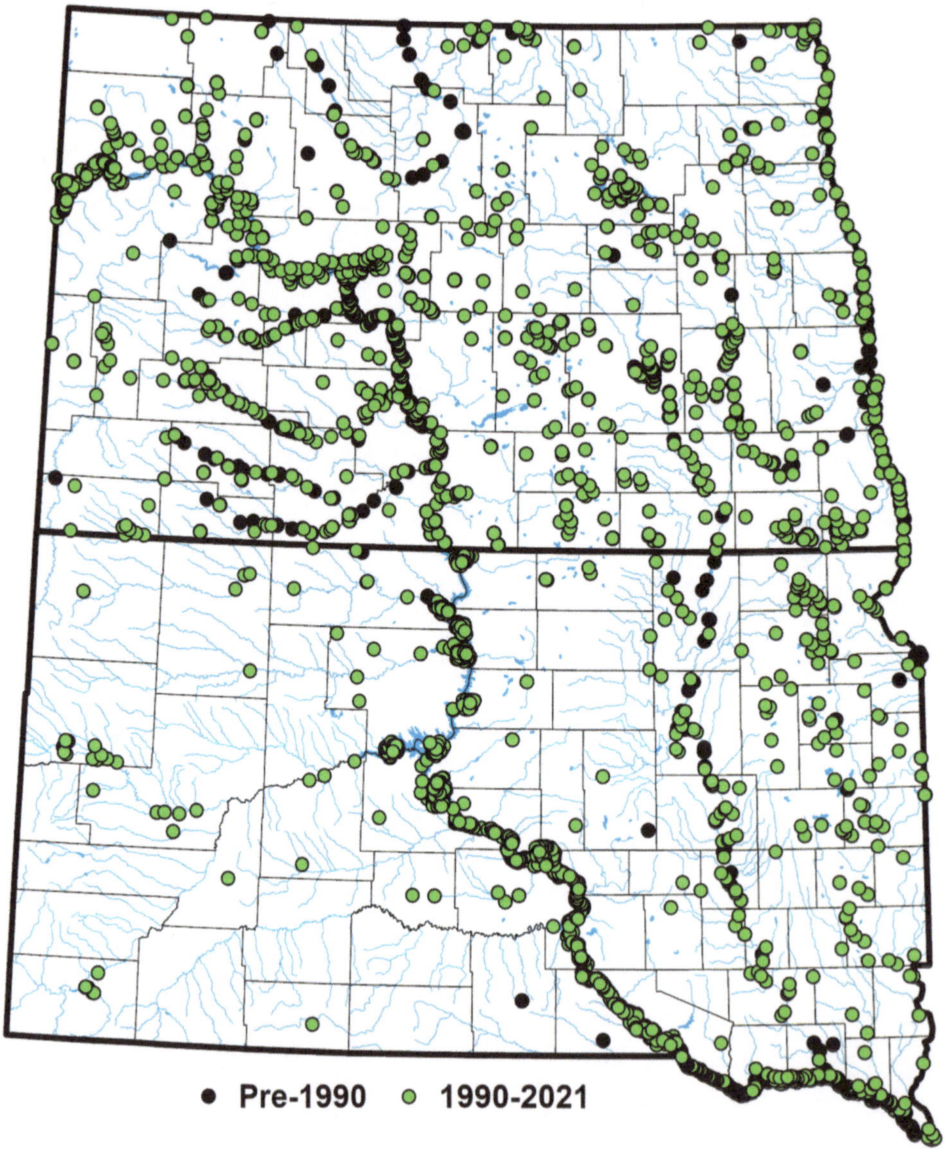

● Pre-1990   ○ 1990-2021

Etymology: *Sander vitreus – Sander* = pike-perch, referencing resemblance to the nonrelated pike family (Esocidae); *vitreum* = glassy, referencing the reflective, glassy eyes.

Description:

Body – fusiform, elongated, slightly laterally compressed, cylindrical in cross section.

Color –

> *Dorsally* – golden brown to olive.
> *Laterally* – golden brown to olive; small, speckled dark spots.
> *Ventrally* – cream to white.
> *Fins* – anterior dorsal – no distinct horizontal spots, dark posterior membranes; posterior dorsal – rows of small, dark spots; caudal – white tip on lower lobe.

Head – large, elongated.

Opercle – cheek with few to no scales.

Snout – conical, blunt.

Eyes – large, glassy; laterally on upper portion of head; silver reflecting layer (*tapetum lucidum*).

Mouth – large, terminal, oblique; upper jaw extends past pupil of eye; jaws nearly equal in length; lower jaw occasionally slightly extending past upper jaw.

Teeth – large, sharp, canine, on both jaws.

Gill rakers – 8 to 11; long, thin.

Dorsal fin – 2 lobes.

> *Anterior* – 13 or 14 spines.
> *Posterior* – 19 to 22 soft rays; 1 or 2 spines.

Caudal peduncle – elongated, thick.

Caudal fin – forked.

Anal fin – 12 to 14 rays; 2 spines; insertion posterior to posterior dorsal fin insertion.

Pelvic fins – thoracic.

Pectoral fins – 13 to 16 rays.

Lateral line – complete; 77 to 90 ctenoid scales in series.

Juveniles – more elongated with dark, dusky bands on dorsal and lateral sides.

Similar Species: Walleye closely resemble and occur sympatrically with sauger. Sauger have a distinctly scaled check and a spotted anterior dorsal fin. They lack dark pigment on anterior dorsal fin posterior membranes and do not have a white tip on the caudal fin lower lobe. Zander do not have an opercular spine.

Distribution and Habitat: The native range of walleye encompasses most of Canada and the northern United States south to Arkansas and Alabama, including the Missouri River, Mississippi River, St. Lawrence River, Artic River, and Great Lakes systems. In the Dakotas, walleye occur in all river systems and are routinely and widely stocked. They have optimal water temperatures from 20 to 24 °C (Bozek et al. 2011a). Walleye are pelagic until reaching a total length of 25 to 30 mm, when they reside deeper near the bottom (Becker 1983). They tolerate a wide range of environmental conditions (Bozek et al. 2011a, b). In late spring, summer, and early autumn, walleye prefer cool, clear, deep, and dark open-waters of large lakes, reservoirs, and rivers with sand, gravel, cobble, or rocky substrates (Paragamian 1989; Bozek et al. 2011a, b). During the warmest summer days, walleye move into the cooler temperatures and higher dissolved oxygen concentrations of dark, deep pools (Paragamian 1989). Mortality occurs at water temperatures above 32 °C (Carlander 1997). Walleye activity decreases while overwintering in deep pools or sheltered areas with little current at water temperatures around 5 °C (Paragamian 1989). They prefer dissolved oxygen levels above 5 mg/L (Bozek et al. 2011a, b).

Reproduction: Walleye often undergo extensive migration days before spawning (Paragamian 1989). They spawn at night soon after ice out in water temperatures from 7 to 10 °C in shallow, wave-washed lake shorelines or flowing rivers with rocky substrate (Bozek et al. 2011a, b; Colby et al. 1979). Walleye broadcast spawn, and neither construct nests nor provide parental care. Fecundity ranges from 20,000 to 80,000 eggs/kg of body weight (Bozek et al. 2011a). The adhesive, 2-mm diameter eggs hatch in 7 days at a water temperature of 14 °C or up to 14 days at colder temperatures (Cross 1967; Colby et al. 1979; Becker 1983). Hatching success depends on spawning date but increases with maternal age (Johnston et al. 2007). Walleye recruitment is highly variable, as evidenced by the routine stockings in the Dakotas (Ellison and Franzin 1992; Lucchesi 1997; Grote et al. 2015). Annual recruitment depends on spring water level fluctuations, discharge, wind, maximum winter temperature, and other factors (Colby et al. 1979; Carlander 1997; DeBour et al. 2013; Quist et al. 2003). Walleye require water temperatures below 10 °C for gonadal maturation (Colby and Nepszy 1981). If winter temperatures are not cold enough for a long-enough duration, walleye may skip spawning that year (Colby and Nepszy 1981).

Age and Growth: Newly hatched walleye larvae have total lengths from approximately 6 to 9 mm. In northern populations, most of the annual growth occurs from mid-June to August (Kelso and Ward 1972; Carlander 1997). Relative weights increase during high-water periods, and females are often larger than males (Dembkowski et al. 2014). Total lengths at age of walleye from South Dakota were 169 mm at age-1, 279 mm at age-2, 360 mm at age-3, 425 mm at age-4, and 490 mm at age-5 (Willis et al. 2001). Walleye can reach a total length of 1,067 mm and a weight of 8.2 kg. They typically live for 5 to 7 years but can live as long as 18 years (Becker 1983).

Food and Feeding: Walleyes are opportunistic feeders. They eat mainly at night, with feeding influenced by water levels and temperature (Bryan et al. 1995). Larvae with total lengths of 8 or 9 mm primarily consume cladocerans, copepods, and other small zooplankton (Chipps and Graeb 2011). Piscivory increases with increased walleye length (Slipke and Duffy 1997; Starostka 1999). Age-0 walleyes eat fish up to 33% of their length but can consume spinous fish that are 40% of their length (Einfalt and Wahl 1997; Knight et al. 1984). Walleyes at total lengths of 20 to 40 mm consume fish (Walker and Applegate 1976; Chipps and Graeb 2011). In an eastern South Dakota lake, fathead minnows were found in the diet of walleyes with total lengths over 62 mm and were the primary food for walleyes over 106 mm (Walker and Applegate 1976; Chipps and Graeb 2011). Adult walleyes are mainly piscivorous, consuming yellow perch, white bass, fathead minnows, gizzard shad, rainbow smelt, and other smaller fish (Bryan et al. 1995; Slipke and Duffy 1997; Starostka 1999; Graeb et al. 2008; Wuellner 2009). They also eat crayfish, frogs, dipteran larvae, and other macroinvertebrates (Bryan et al. 1995; Slipke and Duffy 1997; Starostka 1999; Graeb et al. 2008; Wuellner 2009).

References

Becker GC (1983) Fishes of Wisconsin. University of Wisconsin Press, Madison

Bozek MA, Baccante DA, Lester NP (2011a) Walleye and sauger life history. In: Barton BA (ed) Biology, management, and culture of walleye and sauger. American Fisheries Society, Bethesda, pp 233–301. https://doi.org/10.47886/9781934874226.ch7

Bozek MA, Haxton TJ, Raabe JK (2011b) Walleye and sauger habitat. In: Barton BA (ed) Biology, management, and culture of walleye and sauger. American Fisheries Society, Bethesda, pp 133–179. https://doi.org/10.47886/9781934874226.ch5

Bryan SD, Hill TD, Lynott ST, Duffy WG (1995) The influence of changing water levels and temperatures of the food habits of walleye in Lake Oahe, South Dakota. J Freshw Ecol 10:1–10. https://doi.org/10.1080/02705060.1995.9663411

Carlander KD (1997) Handbook of freshwater fishery biology, 3rd edn. Iowa State University Press, Ames

Chipps SR, Graeb BDS (2011) Feeding ecology and energetics. In: Barton BA (ed) Biology management, and culture of walleye and sauger. American Fisheries Society, Bethesda, pp 303–319. https://doi.org/10.47886/9781934874226.ch8

Colby PJ, Nepszy SJ (1981) Variation among stocks of walleye (*Stizostedion vitreum vitreum*): management implications. Can J Fish Aquat Sci 38:1814–1831. https://doi.org/10.1139/f81-228

Colby PJ, McNichol RE, Ryder RA (1979) Synopsis of biological data on the walleye *Stizostedion v. vitreum*. FAO Fish Synop 119, Rome, Italy

Cross FB (1967) Handbook of fishes in Kansas. University Press of Kansas, Lawrence

DeBoer JA, Pope KL, Koupal KD (2013) Environmental factors regulating the recruitment of walleye *Sander vitreus* and white bass *Morone chrysops* in irrigation reservoirs. Ecol Freshw Fish 22:43–54. https://doi.org/10.1111/eff.12000

Dembkowski DJ, Chipps SR, Blackwell BG (2014) Response of walleye and yellow perch to water-level fluctuations in glacial lakes. Fish Manag Ecol 21:89–95. https://doi.org/10.1111/fme.12047

Einfalt LM, Wahl DH (1997) Prey selection by juvenile walleye as influenced by prey morphology and behavior. Can J Fish Aquat Sci 54:2618–2626. https://doi.org/10.1139/f97-172

Ellison DG, Franzin WG (1992) Overview of the symposium on walleye stocks and stocking. N Am J Fish Manag 12:271–275. https://doi.org/10.1577/1548-8675

Graeb BDS, Chipps SR, Willis DW, Lott JP, Hanten RP, Nelson-Stastny W, Erickson JW (2008) Walleye response to rainbow smelt population decline and liberalized angling regulations in a Missouri River reservoir. In: Allen MS, Sammons S, Maceina MJ (eds) Balancing fisheries management and water uses for impounded river systems. American Fish Society, Bethesda, pp 275–292

Grote JD, Wuellner MR, Dembkowski DJ, Blackwell BG, Lucchesi DO (2015) Evaluating factors that affect recruitment of young-of-year walleye in eastern South Dakota natural lakes. S D Dep Game Fish Parks, Comp Rep 16-02, Pierre

Johnston TA, Wiegand MD, Leggett WC, Pronyk RJ, Dyal SD, Watchorn KE, Kollar S, Casselman JM (2007) Hatching success of walleye embryos in relation to maternal and ova characteristics. Ecol Freshw Fish 16:295–306. https://doi.org/10.1111/j.1600-0633.2006.00219.x

Kelso JRM, Ward FJ (1972) Vital statistics, biomass, and seasonal production of an unexploited walleye (*Stizostedion vitreum vitreum*) population in West Blue Lake, Manitoba. J Fish Res Board Can 29:1043–1052. https://doi.org/10.1139/f72-150

Knight RL, Margraf FJ, Carline RF (1984) Piscivory by walleyes and yellow perch in western Lake Erie. Trans Am Fish Soc 113:677–693. https://doi.org/10.1577/1548-8659

Lucchesi DO (1997) Evaluation of large walleye fingerling stocking in eastern South Dakota lakes. S D Dep Game Fish Parks, Comp Rep 97-18, Pierre

Paragamian VL (1989) Seasonal habitat use by walleye in a warmwater river system, as determined by radio telemetry. N Am J Fish Manag 9:392–401. https://doi.org/10.1577/1548-8675

Quist MC, Guy CS, Stephen JL (2003) Recruitment dynamics of walleyes (*Stizostedion vitreum*) in Kansas reservoirs: generalities with natural systems and effects of a centrarchid predator. Can J Fish Aquat Sci 60:830–839. https://doi.org/10.1139/f03-067

Slipke JW, Duffy WG (1997) Food habits of walleye in Shadehill reservoir, South Dakota. J Freshw Ecol 12:11–17. https://doi.org/10.1080/02705060.1997.9663504

Starostka AB (1999) Food habits and diet overlap of age-1 and older walleye and white bass in Lake Poinsett, South Dakota. Dissertation, South Dakota State University

Walker RE, Applegate RL (1976) Growth, food, and possible ecological effects of young-of-the-year walleyes in a South Dakota prairie pothole. Prog Fish Cult 38:217–220. https://doi.org/10.1577/1548-8659(1976)38[217:GFAPEE]2.0.CO;2

Willis DW, Isermann DA, Hubers MJ, Johnson BA, Miller WH, Sauver TR St., Sorensen JS, Unkenholz EG, Wickstrom GA (2001) Growth of South Dakota fishes: a statewide summary with means by region and water type. S D Dep Game Fish Parks, Spec Rep 01-05, Pierre

Wuellner MR (2009) Exploring competitive interactions between walleye and smallmouth bass in South Dakota waters. Dissertation, South Dakota State University

6.26 Sciaenidae, Drum and Croaker Family

The drum and croaker family contains over 280 species worldwide, of which approximately 80 occur in North America. This large family is almost entirely marine species, with many in the tropical and coastal waters of the Atlantic, Indian, and Pacific oceans. While a few freshwater species occur in South America, only one, the freshwater drum *Aplodinotus grunniens*, is found in North America. Freshwater drum originated from the Gulf of Mexico before adapting entirely to freshwater. Over time, they moved up the Mississippi River and its major tributaries, eventually reaching North Dakota and South Dakota.

Drums and croakers have a complete lateral line extending to the end of the caudal fin, one or two anal fin spines, and a dorsal fin usually separated by a deep notch into two parts. The family has large ear bones, or otoliths. The family common names of drums and croakers comes from the loud, drumming, grunting, and croaking noises emitted during the spawning season, or when the fish are removed from the water. Contracting muscles vibrating against the swim bladder create the noises. Sciaenids also have unusually large and robust pharyngeal jaws and teeth to crush mollusks, which are an important dietary component for adults. They also eat crayfish and fish.

In the Dakotas, freshwater drum occur primarily in low current areas of large turbid rivers. They are also found in streams, lakes, and impoundments over mud and silty substrates. Freshwater drum are pelagic, broadcast spawners with extremely high fecundity. Although freshwater drum are not desired by anglers nor a sport fish in the Dakotas, the large sizes of certain marine sciaenids makes them important sportfish and commercially valuable.

6.26.1   Freshwater Drum *Aplodinotus grunniens*

Rafinesque, 1819

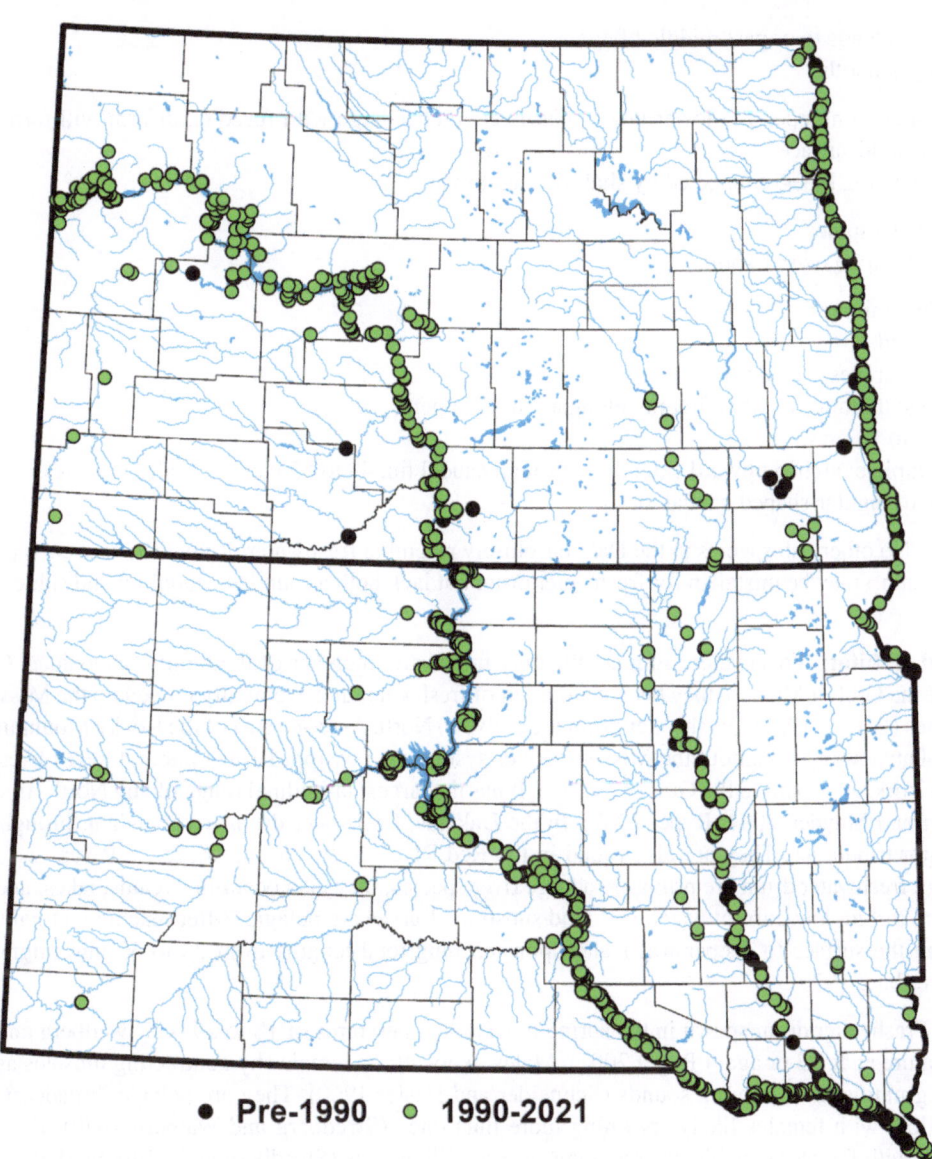

● **Pre-1990**   ● **1990-2021**

Etymology: *Aplodinotus grunniens – Aplodinotus* = single back, referencing the elongated dorsal fin; ***grunniens*** = grunting, referencing the sound of muscles contracting against the swim bladder. Other common names include sheephead, grunter, and gray bass.

Description:

Body – deep, strongly dorsolaterally compressed, concave, arched, steep profile anterior to dorsal fin insertion.
Color –

> *Dorsally* – bronze to olive-brown.
> *Laterally* – light brown to silver; iridescent sheen.
> *Ventrally* – silver to white.
> *Fins* – pectoral, pelvic – light gray to white; remaining – dusky gray-brown.

Head – small, conical, fully scaled.
Opercle – scaled.
Snout – rounded, slightly extending past subterminal mouth.
Eyes – large, laterally on top half of head.
Mouth –

> *Upper jaw* – extends to or past middle of eye.
> *Lips* – fleshy, smooth.

Teeth – conical, small, on upper and lower jaws; pharyngeal arches large with fused cardiform, villiform, and molariform teeth forming a solid surface.
Dorsal fin – elongated; 2 lobes, connected by shallow membrane.

> *Anterior* – 8–10 spines.
> *Posterior* – 24 to 32 rays; 1 spine.

Caudal peduncle – thin, short.
Caudal fin – rounded.
Anal fin – 7 rays; 2 spines, second spine elongate.
Pelvic fins – 5 rays; thoratic; 1 spine; first ray elongate into filament.
Pectoral fins – 17 to 18 rays.
Lateral line – complete; arched upward extending through caudal fin, 48 to 53 ctenoid scales in series.
Juveniles – more triangular-shaped caudal fin.

Similar Species: No other fish species in the Dakotas closely resemble freshwater drum. Carpsuckers (*carpiodes* spp.) and buffalos (*Ictiobus* spp.) may be mistaken for freshwater drum but lack both a rounded caudal fin and spines in the dorsal and anal fins.

Distribution and Habitat: Freshwater drum are the only freshwater member of the family Sciaenidae. Originating in the brackish water of the Gulf of Mexico, they became tolerant of freshwater and eventually moved up the Mississippi River and its major tributaries (Barney 1926). Freshwater drum are native to North America east of the Rocky Mountains from Montana and southern Saskatchewan throughout the Missouri River system, Mississippi River system, Great Lakes, Hudson Bay to Quebec and New York, and south to the Gulf of Mexico. It has the largest latitudinal range of any North American freshwater fish (Boschung Jr and Mayden 2004; Rypel 2007). In the Dakotas, freshwater drum are abundant throughout the Missouri River and its major tributaries. They are also found in the Red River of the North, Sheyenne River, and Minnesota River systems. Although freshwater drum are primarily a large river species, they also occur in streams, lakes, and impoundments. They prefer warm, turbid, low-current water, and mud substrate. Larvae are pelagic. After reaching a total length of 10 mm, they migrate from the surface to deeper water and upon reaching total lengths from 20 to 30 mm migrate to floodplains (Swedberg and Walburg 1970).

Reproduction: Freshwater drum spawn in the spring at water temperatures of 18 to 25 °C (Swedberg and Walburg 1970). They sexually mature at age-3 or age-4 Rypel 2007). Males likely attract females by contracting muscles against their swim bladder to make grunting or drumming sounds (Schneider and Hasler 1960). They are pelagic, broadcast spawners over a 6- to 7-week period, with females likely spawning more than once (Swedberg and Walburg 1970). Freshwater drum are extremely fecund, with females capable of producing over a million eggs (Swedberg and Walburg 1970; Rypel 2007). The semi-buoyant, transparent, approximately 1.5 mm diameter eggs have a single oil droplet and are often seen floating near the surface (Swedberg and Walburg 1970). Hatching occurs 1 or 2 days after fertilization.

Age and Growth: At hatching, freshwater drum larvae have a total length of approximately 3 mm (Walburg 1976). Juvenile growth is rapid. Total lengths at age from Lewis and Clark Lake, South Dakota were age-1, 91 mm; age-2, 166 mm; age-3, 216 mm; age-4, 252 mm; age-5, 285 mm; age-6, 308 mm; and age-7, 325 mm (Swedberg 1968). They typically reach total lengths from 305 to 508 mm but can reach a total length of 711 mm and a weight over 27 kg (Witt Jr 1960). Freshwater drum are long lived, with an average lifespan of 8 to 10 years and a longevity of 72 years (Pereira et al. 1992). Females are generally larger and have faster growth rates than males (Rypel 2007).

Food and Feeding: Freshwater drum are opportunistic omnivores that feed day and night. Their diet shifts from soft to hard prey as they grow larger (Griswold and Tubb 1977; Wahl et al. 1988; Essner Jr et al. 2014). Larvae are highly dependent on copepods, cladocerans, and other zooplankton, and they will select larger prey as body and gape sizes increase (Swedberg and Walburg 1970; Schael et al. 1991; Sullivan et al. 2012). Juveniles eat chironomids, mayfly larvae, and caddisfly larvae. Adults primarily suction feed along the bottom, consuming fish, mollusks, and crayfish (Diaber 1950; Essner Jr et al. 2014). They feed diurnally, consuming zooplankton during the day and benthic organisms during the night (Swedberg 1968). Freshwater drum diets change during the year, with a greater number of fish eaten from August through November (Griswold and Tubb 1977). Robust pharyngeal jaws and molariform teeth help adults crush mollusks.

References

Barney RL (1926) The distribution of the freshwater sheephead, *Aplodinotus grunniens*, in respect to the glacial history of North America. Ecology 7:351–364. https://doi.org/10.2307/1929317

Boschung HT Jr, Mayden RL (2004) Fishes of Alabama. Smithson Book, Washington, DC

Diaber FC (1950) The life history and ecology of the sheephead, *Aplodinotus grunniens* Rafinesque, in western Lake Erie. Dissertation, Ohio State University

Essner RL Jr, Patel R, Reilly SM (2014) Ontogeny of body shape and diet of freshwater drum (*Aplodinotus grunniens*). Trans Ill Acad Sci 107:27–30

Griswold BL, Tubb RA (1977) Food of yellow perch, white bass, freshwater drum, and channel catfish in Sandusky Bay, Lake Erie. Ohio J Sci 77:43–47

Pereira DL, Cohen Y, Spangler GR (1992) Dynamics and species interactions in the commercial fishery of the Red Lakes, Minnesota. Can J Fish Aquat Sci 49:293–302. https://doi.org/10.1139/f92-033

Rypel AL (2007) Sexual dimorphism in growth of freshwater drum. Southeast Nat 6:333–342. https://doi.org/10.1656/1528-7092(2007)6[333:SDIGOF]2.0.CO;2

Schael DM, Rudstam LG, Post JR (1991) Gape limitation and prey selection in larval yellow perch (*Perca flavescens*), freshwater drum (*Aplodinotus grunniens*), and black crappie (*Pomoxis nigromaculatus*). Can J Fish Aquat Sci 48:1919–1925. https://doi.org/10.1139/f91-228

Schneider H, Hasler AD (1960) Laute und Lauterzeugung beim Süsswassertrommler *Aplodinotus grunniens* Rafinesque (Sciaenidae, Pisces). Z Vergl Physiol 43:499–517. https://doi.org/10.1007/BF00298073

Sullivan, C.L, Koupal KD, Hoback WW, Peterson BC, Schoenebeck CW (2012) Food habits and abundance of larval freshwater drum in a South Central Nebraska irrigation reservoir. J Freshw Ecol 27:111–121. https://doi.org/10.1080/02705060.2011.628515

Swedberg DV (1968) Food and growth of the freshwater drum in Lewis and Clark Lake, SD. Trans Am Fish Soc 97:442–447. https://doi.org/10.1577/1548-8659(1968)97[442:FAGOTF]2.0.CO;2

Swedberg D, Walburg C (1970) Spawning and early life history of the freshwater drum in Lewis and Clark Lake, Missouri River. Trans Am Fish Soc 99:560–570. https://doi.org/10.1577/1548-8659

Wahl DH, Bruner KA, Nielsen LA (1988) Trophic ecology of freshwater drum in large rivers. J Freshw Ecol 4:483–491. https://doi.org/10.1080/02705060.1988.9665198

Walburg CH (1976) Changes in the fish populations of Lewis and Clark Lake, Missouri River, 1956-1962. US Dep Inter, Fish Wildl Serv, Spec Rep, Fish No. 482, p 27

Witt A Jr (1960) Length and weight of ancient freshwater drum, *Aplodinotus grunniens*, calculated from otoliths found in Indian Middens. Copeia 3:181–185. https://doi.org/10.2307/1439653

6.27   Gasterosteidae, Stickleback Family

The stickleback family has five genera and approximately 16 species, of which six occur in the United States. With a holarctic distribution, the gasterosteidae are spread across North America, Asia, and Europe. Most stickleback species are marine and prefer coastal and brackish waters. Brook stickleback *Culaea inconstans* are the only strictly freshwater gasterosteidid species, and the only species of sticklebacks present in the Dakotas.

Sticklebacks are small, scaleless fishes with an ovate and laterally compressed body shape. They are distinguished from other fish families by dorsal spines with small, trailing membranes. Sticklebacks also have slightly elongate and narrow caudal peduncles, a fan-shaped caudal fin, and a rayed dorsal fin far posterior on the body behind the dorsal spines. Although sticklebacks are scaleless, brook stickleback, and other species have weakly developed dermal or body plates.

The life histories of stickleback species are intricate and species-specific. For example, the three-spined stickleback, *Gasterosteus aculeatus*, which lives in the Great Lakes and Pacific and Atlantic coasts, has captivating social, spawning, and antipredator behaviors. All male sticklebacks use kidney secretions to construct and secure a nest to submerged aquatic vegetation. Nests are hollow on the inside, with a small hole that males use to lure females. After spawning inside the nest, females of most stickleback species create a hole in the back to exit, which the male later patches. Males guard and tend to the incubating eggs and post-hatch larvae.

Sticklebacks are often prey for sunfish, pike, salmon, and other piscivorous sport fishes. They are also eaten by numerous avian predators. The unique appearance and intricate life history traits of sticklebacks make them an amusing group of aquarium fishes.

6.27.1   Brook Stickleback *Culaea inconstans*

(Kirtland, 1840)

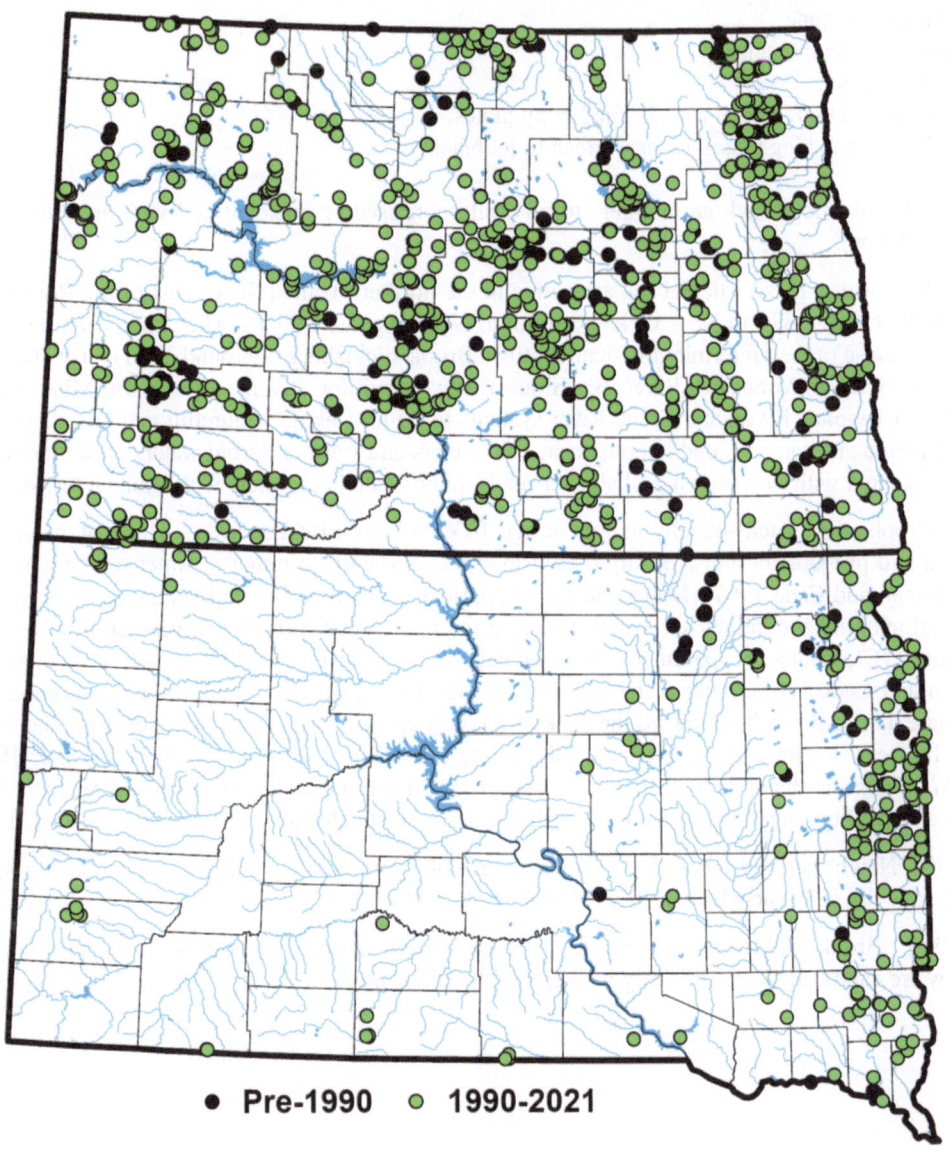

● Pre-1990   ○ 1990-2021

Etymology: *Culaea inconstans* – *Culaea* = *Eucalia* – good nest, referencing intricate nest structure; *inconstans* = variable.

Description:

Body – ovate, laterally compressed.

Color – can rapidly change color and camouflage to blend in with surroundings.
    Dorsally, Laterally – dark olive-green; dark, irregular mottled pattern.
    Ventrally – cream to white.

Head – elongated, conical.
Snout – short, conical.
Eyes – moderately large, laterally on head.
Mouth – small, oblique.

    Jaws – do not extend past anterior end of eye.
Teeth – small, sharp; on both jaws.
Dorsal fin – 14 to 16 rays; 4 to 6 spines spaced equally apart with trailing membranes (finlets).
Caudal peduncle – narrow.
Caudal fin – truncated, rounded edges.
Anal fin – 9 to 11 rays; 1 spine.
Pelvic fins – 1 ray; 1 spine; thoracic.
Pectoral fins – 9 to 11 rays.
Lateral line – no scales, single lateral row of small bony plates.
Sexual dimorphism – spawning males are dark green to jet black.

Similar Species: Brook stickleback are unlikely to be easily confused or misidentified with any other species in North Dakota and South Dakota.

Distribution and Habitat: Brook stickleback are native to the northern United States east of the Rocky Mountains from Montana through the Great Lakes to New York and south to Nebraska and Indiana. They have been introduced outside of their native range because of their intermingling with baitfish. Brook stickleback are found in all North Dakota river systems, and the Bois de Sioux River, Minnesota River, Big Sioux River, Vermillion River, James River, Missouri River, Niobrara River, White River, Cheyenne River, and Grand River systems in South Dakota (Churchill and Over 1938). Brook stickleback are very adaptable and can be found in both clear, cool waters and warmer, turbid waters with little dissolved oxygen. They are often associated with slow-moving, shallower pools, ponds, and backwaters with abundant aquatic vegetation.

Reproduction: Brook stickleback sexually mature at age 1 (Winn 1960; Becker 1983). The prolonged spawning season varies by location and photoperiod but typically occurs in spring and early summer at water temperatures of 15 to 19 °C (Winn 1960; Reisman and Cade 1967). Males construct spherical, walnut-sized, single-opening nests in shallower waters using aquatic plant material and kidney secretions, securing the nest to vertically rooted vegetation or a stick (Winn 1960; Reisman and Cade 1967). Males are extremely territorial and are aggressive toward other male sticklebacks and other fish species (Reisman and Cade 1967). After males escort a female into the nest, the male prods the female to stimulate egg release. The female then makes a hole toward the back of the nest and exits while the male fertilizes the eggs (Winn 1960; Becker 1983). The clear to light yellow, demersal, adhesive eggs hatch in 8 or 9 days at 18 °C (Barker 1918; Winn 1960; McKenzie 1974). Males fan the eggs with their pectoral fins and keep the larvae in the nest after hatching (Winn 1960; McKenzie 1974). Fecundity ranged from 104 to 451 eggs from brook stickleback in Manitoba, Canada (Moodie 1986).

Age and Growth: Newly hatched larvae have a total length of approximately 5 mm, and adults rarely exceed 89 mm (Barker 1918). Longevity is 3 years (Brown 1971; Hrabik et al. 2015).

Food and Feeding: Brook stickleback are opportunistic carnivores, eating primarily small aquatic insects, larvae, and crustaceans. They are cannibalistic, particularly with eggs (Moodie 1986).

References

Barker EE (1918) The brook stickleback. Sci Mon 6:526–529

Becker GC (1983) Fishes of Wisconsin. University of Wisconsin Press, Madison

Brown CJD (1971) Fishes of Montana. Montana State University Press, Bozeman

Churchill EP, Over WH (1938) Fishes of South Dakota. Brown & Saenger Printers, Sioux Falls

Hrabik RA, Schainost SC, Stasiak RH, Peters EJ (2015) The fishes of Nebraska. University of Nebraska, Lincoln

McKenzie JA (1974) The parental behavior of the male brook stickleback *Culaea inconstans* (Kirtland). Can J Zool 52:649–652. https://doi.org/10.1139/z74-083

Moodie GEE (1986) The population biology of *Culaea inconstans*, the brook stickleback, in a small prairie lake. Can J Zool 64:1709–1717. https://doi.org/10.1139/z86-258

Reisman HM, Cade TJ (1967) Physiological and behavioral aspects of reproduction in the brook stickleback, *Culaea inconstans*. Am Midl Nat 77:257–295. https://doi.org/10.2307/2423344

Winn HE (1960) Biology of the brook stickleback *Eucalia inconstans* (Kirtland). Am Midl Nat 63:424–438. https://doi.org/10.2307/2422804

Correction to: Species Accounts

Kathryn E. Schlafke, Matthew D. Wagner, and Chelsey A. Pasbrig

Correction to:
Chapter 6 in: M. Barnes (ed.), *Fishes of the Dakotas*,
https://doi.org/10.1007/978-3-031-38040-2_6

In Chap. 6, the maps listed in the page numbers below were incorrectly published in the original version of the book. The correct maps are now updated in this revised version.

The updated version of this chapter can be found at
https://doi.org/10.1007/978-3-031-38040-2_6

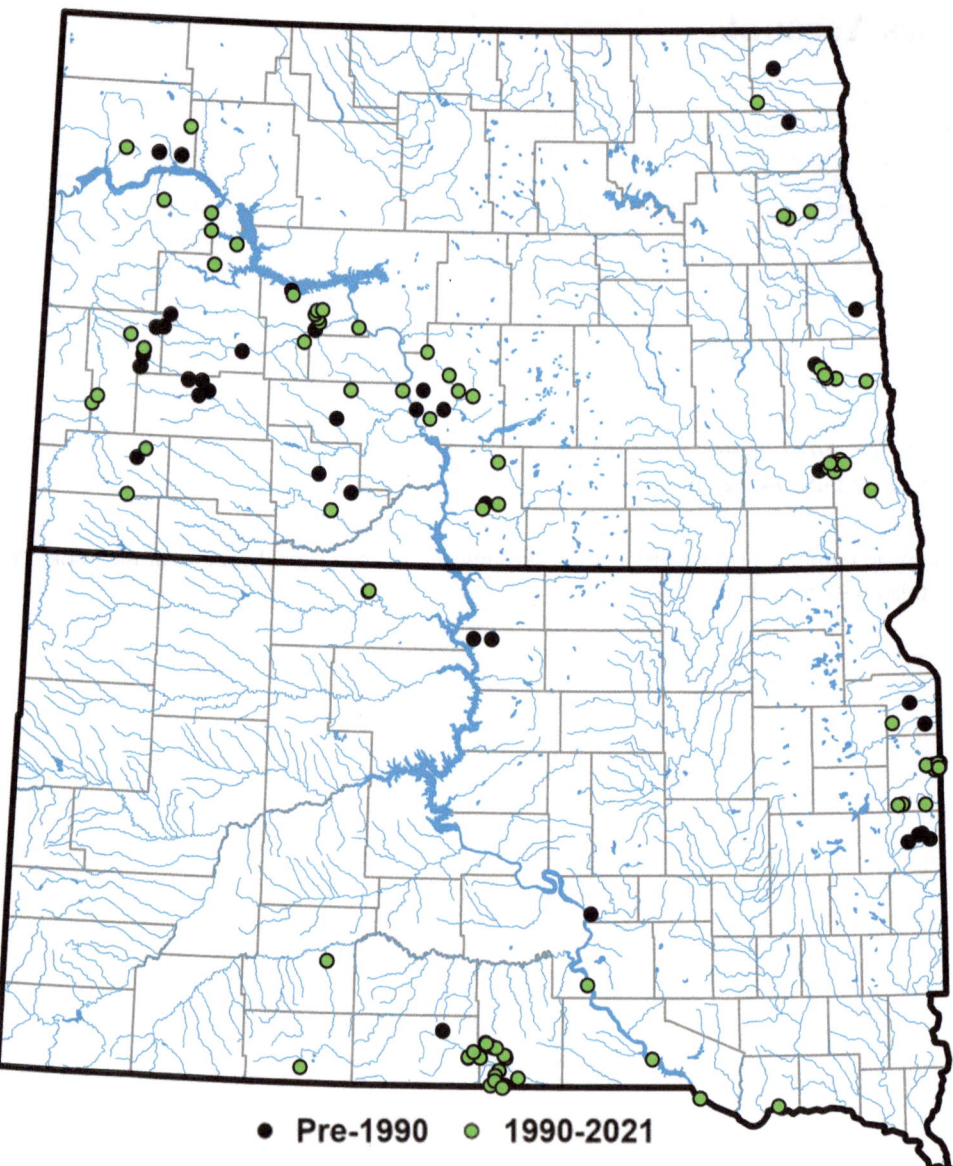

● **Pre-1990**   ● **1990-2021**

Page: 217 Northern Redbelly Dace

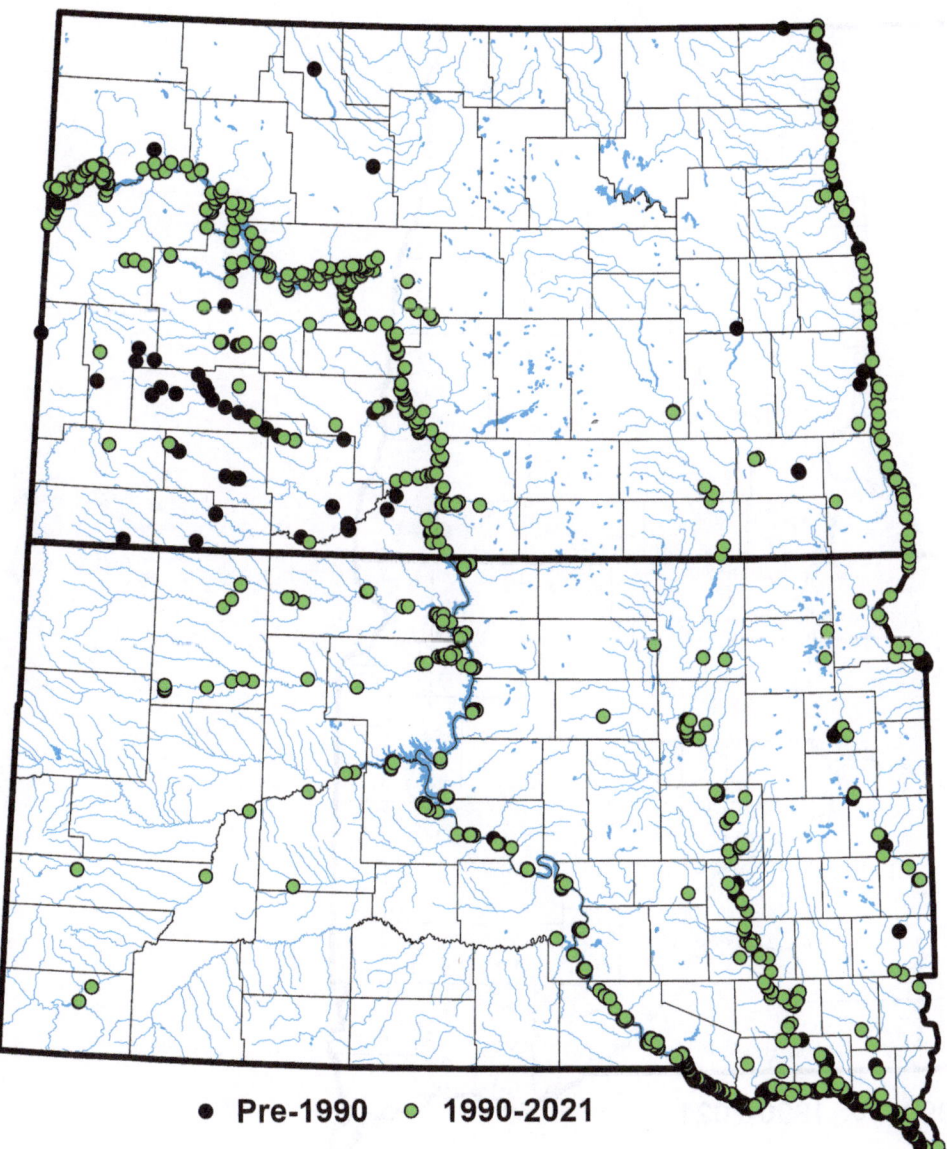

● **Pre-1990**   ○ **1990-2021**

Page: 283 Emerald Shiner

● **Pre-1990** ○ **1990-2021**

Page: 348 Redfin Pickerel

Glossary

Abdominal Pertaining to the ventral surface between the pelvic fins and anal fin. Pelvic fins are referred to as abdominal when they are inserted distinctly posterior to the insertion of the pectoral fins, nearer to the origin of the anal fin. Compare to thoracic.

Adipose eyelid Translucent tissue that partially covers the eye, as on herrings and shads

Adipose fin Fleshy fin on the back between the dorsal and caudal fins that lacks spines or rays

Anal fin Median, ventral, unpaired fin usually just posterior to the anus

Anterior Nearer to the front, especially situated in the front of the body or nearer to the head

Anus The terminal opening of the digestive tract or opening of the cloaca, also called a vent

Axillary process Thin, flap of membrane (modified scale) at the inner base of the pectoral or pelvic fin

Barbel Flexible, flattened, cylindrical, or conical process near the mouth that functions in touch or taste, as on catfishes, sturgeons, some minnows, and other fish

Basal Base or bottom

Basicaudal On the posterior end of the caudal peduncle or at the base of the caudal fin

Basioccipital process Hindmost bone on the underside of the skull. The process of this bone and the point of attachment of the associated muscles are sometimes used to distinguish species of minnows in the genus *Hybognathus*.

Belly Ventral surface between the pelvic fins and anal fin. Compare to breast.

Bicuspid Having two cusps or points, usually referring to teeth in lampreys

Blotch Irregular patch or unsightly mark on a surface, typically the skin

Body depth Measurement from the dorsal side to the ventral side (back to the belly)

Branchiostegal ray One of the slender bones supporting the branchiostegal membranes on the underside of the head

Breast Ventral surface anterior to the base of the pelvic fins. Compare to belly.

Buccal funnel Funnel/disc-shaped sucking mouth that is part of a lamprey usually containing keratinized teeth for gripping

Bulbous Fat, round, or bulging

Cartilaginous ridge Referring to a hard structure made of cartilage inside the mouth of *Campostoma* spp. and *Catostomus platyrhynchus* that is used to scrape algae off rocks. Sometimes appearing as a non-fleshy second inner lower lip

Caudal fin Tail fin; also see heterocercal tail and homocercal tail

Caudal peduncle Narrow region of the body anterior to the caudal fin, from the posterior base of the anal fin to the base of the caudal fin

Chevron Line or stripe in the shape of a V. Usually sideways on the caudal fin of some *Notropis*, especially *Notropis Topeka*

Circumoral Teeth in the buccal funnel of a lamprey that surround the mouth opening (oral opening) in a circular pattern

Circumpenduncular Around the caudal peduncle, typically describing a scale count

Cloaca A common chamber into which the digestive tract and urogenital ducts discharge in some fishes and other vertebrates

Concave Curved inward; also see falcate. Compared to convex

Conical Having the shape of a cone

Convex Curved (arched) outward. Compare to concave

Crescent-shaped Curved sickle shape of the waxing or waning moon

Crosshatch Intersecting sets of parallel lines

Ctenoid scales Scales with serrate posterior margin. Compare to cycloid scale and ganoid scale

Cycloid scales Scales with entire (smooth) posterior margin. Compare to ctenoid scale and ganoid scale

Cylindrical Solid geometric figure with straight parallel sides and a circular or oval section

Disc Shape or surface that is round and flat in appearance

Distal tip Farthest from the point of attachment, such as the margin of a fin

Dorsal/dorsad Pertaining to the back

Elongated Long and narrow, lengthened, or stretched

Emarginate Having a shallow notch or indentation, as on the margin of the caudal fin

Equidistant Equal distances

Eye to fork length Length of the fish measured from the anterior eye to the caudal fin fork. This measurement is used for paddlefish instead of the total or standard length because rostrum or caudal fins are frequently damaged. Abbreviated as EFL

Falcate Sickle-shaped, concave margin, usually in reference to a fin margin

Filament Slender threadlike object or fiber

Finlet Small rayless nonretractable fins between the last dorsal or anal fin and the caudal fin of some fishes

Fleshy Resembling flesh in appearance or texture

Fork length Measurement from the tip of the jaw with a closed mouth to the center of the fork in the tail

Frenum Fleshy connection between the upper jaw and snout that interrupts the otherwise continuous groove separating the two

Ganoid scales Scales covered with hard enamel (as on gars). Compare to ctenoid scale and cycloid scale

Gill arch Bone or cartilage structure to which gill filaments and gill rakers are attached

Gill opening/gill cleft Opening of the gill chamber, usually used in reference to lampreys, sharks, and rays

Gill rakers Projection from the concave anterior surface of the gill arch, often modified to aid in the capture and digestion of food

Herringbone Arrangement or design consisting of columns of short parallel lines, with all the lines in one column sloping one way and all the lines in the next column sloping the other way so as to resemble the bones in a fish

Heterocercal Caudal fin in which the vertebral column curves upward into the dorsal lobe of the fin. Compare to homocercal

Homocercal Caudal fin in which neither lobe of the fin is invaded by the vertebral column. Compare to the heterocercal tail

Hyoid teeth Teeth on the tongue of fishes

Inferior mouth Mouth on the ventral surface of the head, commonly opening downward. Compare to the oblique mouth, subterminal mouth, terminal mouth, and superior mouth

Insertion Point of attachment of a fin

Isthmus Contracted part of the breast that projects anteriorly between (and separates) the gill chambers

Keel Fleshy or bony ridge, usually along the center of the ventral surface but rarely on the scales on the sides

Lateral line Part of the lateralis system of sensory canals communicating to the body surface by pores. Typically refers to the longitudinal row of scales along the side that has pores. A complete lateral line has scales with pores to the base of the caudal fin (or onto the caudal fin of the freshwater drum (*Aplodinotus grunniens*)). An incomplete lateral line has pores only on the anterior scales, and an interrupted lateral line has gaps between the exposed pores. The lateral line might be straight, arched toward the dorsal surface, or decurved toward the ventral surface.

Lateral stripes Stripes of dark pigment that run parallel to the longitudinal axis of the fish (on its sides)

Laterally compressed Anatomical description of something compacted from the sides

Lobe Referring to a part of the body bisected into two parts, usually the lips or tail

Longitudinal ridge Referring to a keel or crease running from the anterior end to the posterior end of a scale in *Macrhybopsis* spp.

Marbled Having a streaked and patterned appearance like that of variegated marble

Margin (fin margin) Referring to either the basal or distal area of a structure (commonly fins)

Maxilla Bone on each side of the upper jaw that is immediately above (or behind) and parallel to the premaxilla

Median Situated in the middle

Melanophore Cell (chromatophore) producing black pigment; when contracted, cells appear as small spots, and, when expanded, a larger area is dark

Membrane Referring to the pliable sheet-like structure connecting spines or rays in the fins of fishes

Mid-dorsal stripe Stripe along the midline of the back, which usually starts at the base of the skull and terminates at either the dorsal fin or the beginning of the caudal fin

Modally Most common number in a sequence of numbers. Typically used in counts of fin rays when most of the time the count is a certain number, but some variation is occasionally observed

Molariform Having the form of a molar tooth

Mottled Referring to patterns with spots or smears of color

Myomeres Blocks of skeletal muscle tissue. They are commonly zigzag or "W-" or "V"-shaped muscle fibers

Nape Dorsal part of the body from the occiput (base of the skull) to the origin of the dorsal fin

Nare/Naris Nostril, usually with an anterior and posterior opening. Used to detect smell/taste, not to breathe

Nuptial Referring to fishes in reproductive condition (commonly associated with tubercle development, skin swelling, color changes, etc.)

Oblique Referring to a slanting mouth, in between a superior mouth and a terminal mouth. Compare to the inferior mouth, subterminal mouth, terminal mouth, and superior mouth

Opercle Large posterior bone of the gill cover

Operculum Gill cover

Orbit Eye socket. Orbital diameter is measured from the anterior to the posterior bony rim of the orbit, whereas eye diameter is measured across the cornea (slightly less than the orbital diameter)

Origin Anterior end of the base of a fin

Papillose/papillae Covered with small bumps (papillae). Compare to plicate

Pectoral fin Paired fin on the side or breast, behind the head

Pelvic fin Paired fin below the pectoral fin or between the pectoral fin and anal fin

Peritoneum Membrane lining the interior of the body cavity

Pharyngeal arch Fifth gill arch, which does not function in respiration and is embedded in tissues behind the gill-bearing arches

Pharyngeal teeth Variously shaped bony projections from the fifth gill arch, which does not function in respiration and is embedded in tissues behind the gill-bearing arches. Sometimes referred to as throat teeth

Plicate Having parallel folds or soft ridges and grooves. Compare to papillose

Pore Minute opening on the skin through which gases and liquids can pass

Posterior Toward the tail (rear)

Predorsal Anterior to/in front of the dorsal fin

Predorsal scales Scales on the back between the head and the origin of the dorsal fin

Premaxillae Paired bones at the front of the upper jaw. Right and left premaxillae join anteriorly to form all or part of the margin of the upper jaw

Principle rays Fin rays that extend to the distal margin of the dorsal, anal, and caudal fins, especially if those fins have a straight anterior edge. Only one unbranched anterior ray and subsequent branched rays are counted along the base of the fin. Compare to rudimentary rays and spine

Pyloric caeca Finger-shaped pouches attached to the stomach, which secrete digestive enzymes and absorb nutrients

Radii Referring to scales; straight lines that extend (or radiate) from the center of the scale to the perimeter a circle or sphere

Rays Sometimes referred to as soft rays, usually segmented (cross-striated) support for the fin membrane comprising two lateral parts. Often branched and often flexible. Also see principal rays and rudimentary rays. Compare to spine

Relative weight Ratio of the actual weight of a fish to what a rapidly growing healthy fish of the same length should weigh. Abbreviated as Wr

Robust Sturdy in construction

Rudimentary rays Small, continuous rays at the anterior bases of the dorsal, anal, and caudal fins of some fishes that are excluded from counts of the principal rays. Also called procurrent rays. Compare to principal rays and spine

Serrae Sharp-pointed tooth-like structures, like the teeth of a saw

Serrate Toothed margin, similar to the margin of a saw blade

Slender Having little width in proportion to the height or length; long and thin

Snout Part of the head anterior to the eye but not including the lower jaw

Soft dorsal fin Posterior dorsal fin with segmented rays instead of spines; fused to the spinous dorsal fin in the genera *Rocio*, *Lepomis*, and *Pomoxis*

Spine Unsegmented (not cross-segmented), unbranched, usually stiff support for the fin membrane not composed of two lateral parts. Compare to rays

Spinous dorsal fin Anterior dorsal fin with hard spines instead of rays; fused to the soft dorsal fin in *Rocio*, *Lepomis*, and *Pomoxis*

Spot Round area consisting of dark pigment

Standard length Length of the fish measured from the anterior tip of the head to the last vertebra at the base of the caudal fin (where the central rays originate). Typically not used for fish with a heterocercal tail. Abbreviated as SL

Submandibular pores Pores found in some species on the underside of the lower jaw, used in identification of *Esox* spp.

Suboperculum Bony plate immediately below the opercle on the gill cover

Subterminal Mouth that opens slightly ventrally rather than at the tip of the snout (snout projects farther forward than the mouth). The lower jaw closes within the upper jaw rather being equal in its anterior extent. Compare to the inferior mouth, oblique mouth, subterminal mouth, and superior mouth

Superior Mouth in which the lower jaw extends upward and the mouth opens dorsally. Compare to the inferior mouth, subterminal mouth, and terminal mouth

Terminal Horizontal mouth (as viewed from the side) that opens at the tip of the snout. Upper and lower jaws usually equal at their medial tips. Compare to the inferior mouth, oblique mouth, subterminal mouth, and superior mouth

Thoracic Pertaining to the thorax (chest). Pelvic fins are referred to as thoracic when they are inserted below the pectoral fins, much nearer to the insertion of the pectoral fins than the origin of the anal fin. Compare to abdominal

Total length Length of the fish measured from the anterior tip of the head to the tip of the folded lobes of the caudal fin. Abbreviated as TL

Transitory Not permanent; referring to barbels in the genera *Pimephales* and *Chrosomus*

Truncate Ending abruptly as if cut off across the base or tip

Tubercles Hardened, often conical (sometimes ridge-like) projection on scales or fin rays, typically on breeding males of some species. Sometimes referred to as breeding tubercles or nuptial tubercles. Referred to as pearl organs in older references

Unicuspid Having one cusp or point; referring to the teeth of *Ichthyomyzon* spp.

Ventral Relating to the underside of a fish

Ventrolateral Situated toward the junction of the ventral and lateral sides

Vermiculation Color or pattern describing wormlike marks

Vomer bone Median, usually unpaired bone at the roof of the mouth, in front of the parasphenoid; commonly used in the Salmonid identification

Wet weight Weight of a fish has not been dried, frozen, or preserved. Abbreviated as WW